Mathematics for Chemists

Mathematics for Chemists

CHARLES L. PERRIN

University of California, San Diego
La Jolla, California

WILEY-INTERSCIENCE
a Division of John Wiley & Sons, Inc.
New York · London · Sydney · Toronto

Library of Congress Catalogue Card Number: 79-112850

ISBN 0 471 68069 9

Printed in the United States of America

10 9 8 7 6 5 4 3 2

to Marilyn

PREFACE

Most of the material in this book has developed out of a one-quarter course given at the University of California, San Diego, to seniors and first-year graduate students in chemistry and related fields. Nearly all of it can be completed in a one-semester course that requires a great deal of work on the part of the student. I hope that this text will stimulate the spread of such mathematics courses for chemists, preferably taught by chemists. I have also tried to make it suitable for self-study.

This textbook differs from most others on applied mathematics in that it is oriented exclusively toward chemists rather than physicists. The coverage is somewhat more elementary, with less emphasis on such topics as special functions, boundary-value problems, and complex-variable theory. The examples and problems are drawn chiefly from physical chemistry, and therefore the student should have had some exposure to that subject. The text, however, is not intended solely for physical chemists but for all chemistry majors, even those who lack an adequate comprehension of the usual two years of college mathematics.

Included is a thorough review of the *useful* portions of elementary calculus, with some new material admixed, followed by a discussion of various topics in applied mathematics. This should be sufficient to enable the student to go on to advanced courses in kinetics, thermodynamics, statistical mechanics, and quantum mechanics and to understand most of the mathematical techniques used in the current chemical literature. Also included are brief exposures to several more advanced methods, so that a chemist may become aware of their existence when he encounters the need for one of them in his research. Although chemical physicists must go beyond what is covered here, this material provides a bridge between elementary calculus and the advanced material found in specialized texts.

This is not primarily a reference book but rather a text to be studied. I am interested only in the application of mathematical methods to chemistry, so I make no attempt at the luxury of rigor. In many cases the theorems are merely stated, with neither proof nor even any indication of the conditions under which they are valid. The pathological situations that so delight mathematicians do not arise in problems of chemical interest, and therefore the theorems are applicable as stated. A short bibliography is provided to guide the student toward proofs, further discussion, and standard references.

The subject matter is organized according to the mathematics and should be studied in sequence. Since the mathematics applicable to various chemical subjects is developed throughout the book, the detailed index should be consulted to guide the student to those sections that deal with a topic of particular interest to him.

The burden of the pedagogy rests with the exercises, since mathematics cannot be learned merely by reading textual material. Although many examples have been included to illustrate new concepts when they are presented, few worked-out problems are provided on the assumption that students will learn more by struggling with a problem than by a cursory reading of its solution. Indeed, much expository material that would ordinarily be in the body of a book has been transferred to the exercises. Although I have tried to avoid repeating the same problem in different guises merely to reinforce a point, some repetition is included for practice in developing facility at mathematical manipulation. If the student feels that he cannot proceed because he does not yet understand some part of the material, he is encouraged to devise similar problems and solve them for additional practice. Exercises marked with an asterisk may be treated as optional. Some cover details, special cases, or proofs. Others are of interest only for their relevance to particular topics not of general interest. Yet many are instructive and are optional only because their elaborateness precludes their assignment in a one-semester course. Also, many exercises are phrased "Convince yourself." I mean exactly that—provide a "proof" that satisfies you or perhaps just construct a few examples. These are generally useful results, which you will find easier to understand and to invoke if you think about why they are reasonable.

I am grateful to my wife, my colleagues, and my students for advice and encouragement. Special thanks are due my grader, Michael Scott. I am pleased to acknowledge the help of Mrs. Dorothy Prior and Mrs. Gweneth Robbins, who so graciously and so splendidly typed the manuscript.

CHARLES L. PERRIN

La Jolla, California
April 1970

CONTENTS

Mathematics for Chemists

1

FUNCTIONS OF A SINGLE VARIABLE

This chapter reviews the basic concepts of elementary calculus: function, derivative, integral, and infinite series. We begin with a careful explanation of the nature of a function and we describe to composite and inverse functions. We then review the familiar notions of derivative and differentiation and immediately introduce operators and operator notation, with differentiation serving as a familiar example. Although this may be new material to some students, it is not so difficult conceptually; therefore we introduce it at this very early point and continue to use it throughout what follows. We digress to consider the problem of solving for the roots of an equation; opportunities for practice on a desk calculator are included. We then turn to a discussion of integration, which includes two important results concerning integrals of even and odd functions. Leibniz' rule is introduced at this point; it may be new, but it is quite easy to apply. Two important techniques of integration are presented briefly. Next we turn to the expansion of functions as infinite series and formally introduce the elementary transcendental functions: exponential, logarithm, trigonometric functions, and hyperbolic functions. New ideas presented in exercises include the noncommutation of operators, the Laplace and Fourier transforms, linear dependence, orthogonality of functions, convolutions, and generating functions. Since the material of this chapter is basic to all of physical science, it is aimed at no particular area of chemistry. However, there are exercises that demonstrate applications to quantum mechanics, molecular structure, spectroscopy, equilibrium-constant problems (solution thermodynamics), and statistical mechanics.

From the beginning we shall not restrict ourselves to the real numbers. The student who is unfamiliar with complex numbers is advised to read Appendix 4 before starting this chapter. The student who has difficulty in grasping the precise meaning of mathematical statements may find Appendix 1 helpful.

1.1 Definitions and Notation

A SET[1] is a collection of entities, usually specified by some property of its members, or ELEMENTS.

A FUNCTION is a rule that relates elements of two sets, such that for any element of the first set (called the DOMAIN of the function), the function specifies a unique element of the second set (called the RANGE of the function). An element of the domain is called an ARGUMENT, and the corresponding element of the range is called the VALUE of the function at that argument (see Fig. 1.1). A variable that ranges over all possible arguments is called an INDEPENDENT VARIABLE, and a variable that ranges over all possible values is a DEPENDENT VARIABLE.

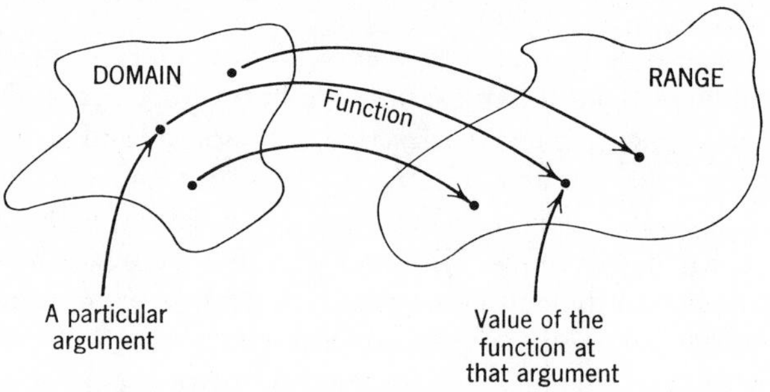

Figure 1.1

It is essential to distinguish carefully between the function and the value of the function. For example, the function "square" is that rule which assigns 1 to 1, 4 to 2, 9 to 3, etc., and the possible values of the function are all numbers. Thus we may write $9 = f(3) = 3^2$, which states that the value of the function "square," at the argument 3, is 9. Similarly we may specify the function "square" by the equation $y = f(x) = x^2$, written in terms of dependent and independent variables. It is important to recognize that in both these equations the function is symbolized by f, and not by $f(x)$ or x^2. The symbols $y, f(x)$, and x^2 represent the value of the function at the arbitrary argument x, just as $9, f(3)$, and 3^2 represent the value of the function at the particular argument 3.

It is clear from this notation that the symbols used to represent the variables are completely arbitrary and that f represents the same function even when it is specified in terms of other variables, such as $\sigma = f(t) = t^2$

[1] Whenever a term is defined, it is set in capital letters.

You should be familiar with representing a function of a real variable by a graph, and it is obvious that the graph of a function looks the same no matter what names the axes are given.

A common notation is to represent the function and the dependent variable by the same symbol, as in $y = y(x) = x^2$. But this should not confuse you if you keep in mind that a function is a rule and not a set of values. It is to avoid such confusion that this discussion is so elaborate.

Examples of Functions Whose Range and Domain Are the Complex Numbers

1. The function "identity": $f(z) = z$.
2. The functions "constant": $f(z) = z_0$. There are an infinity of such functions and the range of each contains only a single element.
3. The function "twice square": $f(z) = 2z^2$.
4. The functions "Nth-order polynomial": $f(z) = P_N(z) = a_0 + a_1 z + \cdots + a_N z^N$.
5. The function "reciprocal": $f(z) = 1/z$. Zero is excluded from the domain and range.
6. The function "complex conjugate": $f(z) = z^*$.
7. The function "absolute value": $f(z) = |z| = +(zz^*)^{1/2}$.
8. The function "real part": $f(z) = \text{Re}\,(z) = (z + z^*)/2$.
9. The function "imaginary part": $f(z) = \text{Im}\,(z) = (z - z^*)/2i$.

Notice that the range of the last three functions is the real numbers.

Examples of Functions Involving Nonnumerical Sets

10. The function "temperature," which assigns to each point in a room a real number.
11. The function "number of sides," which assigns to each polygon a positive integer.
12. The function "height," which assigns to each person a positive real number.
13. The function "mother," which assigns to each person some female.
14. The function "parity," which assigns to each integer either the word "even" or the word "odd."

These last five examples are included in order to emphasize the notion of function as a rule. Functions involving nonnumerical sets are less frequently encountered, but it is necessary not to restrict the definition of a function. We shall have occasion to use functions whose range or domain may be vectors or matrices or elements of an abstract group or the set of possible outcomes of an "experiment."

Notice the requirement of uniqueness in the definition of a function. There must be no ambiguity as to the value of a function at any argument. For example, square root is not a function if there is no way of deciding whether $\sqrt{4}$ is $+2$ or -2. But square root may be made a function if we explicitly restrict the range to positive numbers. Likewise, although the values of the function "mother" are determinable unambiguously, "child" does not qualify as a function whose domain is the set of mothers. If a mother has more than one child, there is no way of deciding which child is the value of this "function." On the other hand, it *is* possible to define the function "eldest child," which involves no such difficulty.

The requirement of uniqueness is not inflexible. There often arise so-called

MULTIPLE-VALUED FUNCTIONS, which must be handled with special care. For example, we may accept square root as such a function but must establish its actual range from the context of the particular problem at hand.

Next we consider COMPOSITE FUNCTIONS. What can we say about $z = f(y)$, the value of a function of the dependent variable y in the equation $y = g(x)$? Clearly, once x is chosen the value of y is determined by the function g. But once y is determined the value of z is determined by the function f. Therefore the value of z depends ultimately on the choice of x; to each x there corresponds a z. But this means that z must be the value of some function of x. This function is neither f nor g, but is in general a different function F.

Examples of Composite Functions

15. With $f(x) = P_N(x)$ and $g(x) = 1/x$, $F(x) = P_N(1/x) = a_0 + a_1x^{-1} + \cdots + a_Nx^{-N}$.
16. With $f(x) = 1/x$ and $g(x) = P_N(x)$, $F(x) = 1/P_N(x) = 1/(a_0 + a_1x + \cdots + a_Nx^N)$.
17. With f and g both the function "mother," the composite function is the function "maternal grandmother."

We may summarize this discussion of composite functions by the equations

$$z = f[g(x)] = F(x) \tag{1.1}$$

but here it is especially important to use different symbols for functions and values. Although f and F are different functions, $F(x)$ and $f(y)$ have the same value, z, when $y = g(x)$. It could be quite confusing to use the same symbol z to represent two different functions.

Exercises

1. Learn the Greek alphabet.
2. (a) How would you define the sum, $(f + g)$, of two functions, f and g, which have the same domain?
   (b) Is this sum also a function?
   (c) How would you define the complex conjugate, f^*, of a function f?
3. Express each of the following functions.
   (a) The energy of a photon as a function of its frequency.
   (b) The rate constant of a reaction as a function of temperature.
   (c) The potential of a silver electrode as a function of the concentration of silver ion.
   *(d) The fraction of solute remaining in V_W ml of water after extraction with V_E ml of ether, as a function of n, the total number of extractions, each with V_E/n ml of ether. (Let partition coefficient = K = moles solute per liter ether/moles solute per liter water.)

*4. If you feel you need the practice or the review, sketch a graph of each of the following functions:
   (a) $f(x) = x^3 - 3x^2 - x + 3$.
   (b) $f(x) = \sin^2 x$.
   (c) $f(x) = (1 - x)/(1 + x)$.
   (d) $f(x) = (a^2 + b^2x^2)^{1/2}$.
   Remember that it helps to find $f(0)$, $f(\pm\infty)$, and the zeros, maxima, and minima of $f(x)$.

Next we investigate how we may go from a value $f(x)$ back to that argument x which led to it. The function f is a "one-way street" that takes a member of the domain and produces a member of the range but does not work in the opposite direction. The rule that does produce x unambiguously from $f(x)$ is likewise a function since it assigns a member of one set to a member of another set. This function is called the INVERSE FUNCTION of f and is denoted by f^{-1} (read "f inverse"). In symbols, if $y = f(x)$, then $x = f^{-1}(y)$ since f^{-1} is the rule that must be applied to y to give x back again. Substituting, we obtain

$$x = f^{-1}[f(x)] \qquad \text{and} \qquad y = f[f^{-1}(y)] \tag{1.2}$$

which shows that f^{-1} is what undoes the action of f and vice versa. Although f and f^{-1} are in general different functions, with different graphs for $y = f(x)$ and $y = f^{-1}(x)$, the equations $y = f(x)$ and $x = f^{-1}(y)$ have exactly the same graph since if the point (x, y) satisfies one equation, it also satisfies the other.

Examples of Inverse Functions

18. $y = f(x) = x^2 \leftrightarrow x = f^{-1}(y) = y^{1/2}$. The function inverse to "square" is "square root," although we should choose the positive square root in order to guarantee that f^{-1} will be single-valued. Alternatively, we could accept square root as a double-valued function in order to allow for negative values of x.

19. $y = f(x) = \sin x \leftrightarrow x = f^{-1}(y) = \sin^{-1} y$, but again we must be careful about the range of f^{-1}. Do not be confused by this notation; the -1 means inverse and not reciprocal.

20. $y = f(x) = 1/x \leftrightarrow x = f^{-1}(y) = 1/y$. The function "reciprocal" is its own inverse.

21. $y = f(x) = 1 - x^3 \leftrightarrow x = f^{-1}(y) = (1 - y)^{1/3}$. The function inverse to "one minus cube of" is "cube root of one minus."

Notice how cumbersome verbal description of a function can be. Yet it does emphasize the notion of function as a rule and shows that f is the same function whether it is specified by $f(x) = 1 - x^3$ or $f(y) = 1 - y^3$, and that f^{-1} is the same function whether it is specified by $f^{-1}(y) = (1 - y)^{1/3}$ or $f^{-1}(x) = (1 - x)^{1/3}$. What is important is the relationship of f^{-1} to f.

Exercises

5. Convince yourself that the domain of f^{-1} is the range of f, but that the range of f^{-1} is not necessarily the domain of f.
6. If f is such that whenever $x_1 \neq x_2$, $f(x_1) \neq f(x_2)$, then f is said to be ONE-TO-ONE, otherwise MANY-TO-ONE. Convince yourself that if f is many-to-one, there will be uniqueness problems in specifying f^{-1}.
7. What is the inverse function corresponding to each of the following:
   (a) $f(x) = e^x - 1$.
   (b) $f(x) = \tan x$.
   (c) $f(x) = \cos(\ln x)$.
   (d) $f(x) = ax^2 + bx + c$.
   *(e) $f(x) = (e^x - e^{-x})/(e^x + e^{-x})$. Hint: Solve for e^x by the quadratic formula.
   (f) $f(z) = z^*$.
   Be sure to indicate which ones lead to problems of uniqueness.

A function f is said to be PERIODIC of period p if and only if, for all x, $f(x + p) = f(x)$.

A function f is said to be an EVEN FUNCTION if and only if, for all x, $f(-x) = f(x)$. A function f is said to be an ODD FUNCTION if and only if, for all x, $f(-x) = -f(x)$.

Exercises

8. Convince yourself that the graph of an even function is symmetric with respect to reflection in the y-axis, and that the graph of an odd function is symmetric with respect to inversion through the origin.
9. Convince yourself that
   (a) The product of two even functions is an even function.
   (b) The product of two odd functions is an even function.
   (c) The product of an even function by an odd function is odd.

*10. Show that any function may be expressed as the sum of an even function plus an odd one. [*Hint:* For any function f, the sum $f(x) + f(-x)$ is even.]

The LIMIT of $f(x)$ as x approaches x_0 is said to have the value v if we can make $f(x)$ as close to v as we wish, merely by making x sufficiently close to x_0. The function f is said to be CONTINUOUS on an interval if it is "smooth" in that interval. You have encountered more rigorous definitions of these two terms, but it is sufficient for our purposes for you to have a qualitative feel for their meaning.

1.2 Differentiation

The DERIVATIVE of the function f, at $x = x_0$, is a number denoted by $f'(x_0)$ or $df/dx|_{x=x_0}$ and defined[1] by

$$f'(x_0) \equiv \lim_{x \to x_0} \frac{f(x) - f(x_0)}{x - x_0}. \tag{1.3}$$

An alternative notation is dy/dx, but this is based on the confusing use of the same symbol to represent the function and the dependent variable.

Since we can calculate the number $f'(x_0)$ at every x_0, we have a rule for associating numbers to numbers. Therefore, we have another function f' called the DERIVATIVE of f, such that the value of this new function at the argument x is $f'(x)$. Similarly, we can take the derivative of f', to produce yet another function, the SECOND DERIVATIVE, denoted by f''. Further applications of this process lead to higher derivatives.

We have shown that there exists a process called DIFFERENTIATION which associates to the arbitrary function f the function f'. Thus we have created a new kind of function whose domain and range are sets of functions.

[1] The symbol $\equiv$ means "is defined to be equal to."

Something which assigns functions to functions is not called a function, though, but an OPERATOR. We shall represent operators by capital letters. The application of an operator to a function is indicated by juxtaposition, with the operator acting from the left: Ωf. Repeated application of operators is indicated as $\Omega_2\Omega_1 f$, which means $\Omega_2(\Omega_1 f)$: first apply Ω_1 to f and then apply Ω_2 to the resultant function $\Omega_1 f$. Since $\Omega_2\Omega_1 f$ is still a function, we may conclude that the "product" $\Omega_2\Omega_1$ is also an operator.

Let us write the familiar rules of differentiation in operator notation with the differentiation operator represented by D. The first two rules merely serve to define D and D^n:

1. $Df = f'$.
2. $D^2 f = D(Df) = Df' = f''$; D^n similarly.
3. $D(af + bg) = aDf + bDg$.
4. $Dc = 0$.
5. $D(fg) = fDg + gDf$.
6. $D(f/g) = (gDf - fDg)/g^2$.

Notice that these are abbreviated forms of theorems and that the general symbols may be replaced by any suitable particular examples of interest. For example, the complete form of rule 3 is "If f and g are functions and a and b are numbers, then $D(af + bg) = aDf + bDg$." The complete form of rule 4 is "If c is a constant function, which assigns to every x the same value c, then $Dc = 0$, the zero function, which assigns to every x the same value 0."

Any operator Ω such that $\Omega(af + bg) = a\Omega f + b\Omega g$ is said to be a LINEAR OPERATOR. Rule 3 states that differentiation is a linear operator.

The converse of rule 4 is an important theorem: The only functions whose derivative is the zero function are the constant functions.

Do not be confused by the notation $Df(x_0)$. This cannot mean the derivative of $f(x_0)$ since the operator D cannot act on the *number* $f(x_0)$. The operator D acts on the *function* f to produce the function $Df = f'$, which may be evaluated at x_0 to obtain the value $Df(x_0) = f'(x_0)$.

The familiar CHAIN RULE states that the derivative of the composite function $F(x) = f[g(x)]$, where $y = g(x)$, is

$$F'(x) = f'[g(x)]g'(x)$$

or

$$\frac{dF}{dx} = \frac{df}{dy}\frac{dg}{dx}. \tag{1.4}$$

If $z = F(x) = f(y)$, this may also be written in the mnemonic forms

$$\frac{dz}{dx} = \frac{dz}{dy}\frac{dy}{dx} \quad \text{or} \quad \frac{d}{dx} = \frac{dy}{dx}\frac{d}{dy}, \tag{1.4'}$$

but it should now be clear that using the same symbol for two different functions and their common value can be a dangerous practice.

We may use the chain rule to express the derivative of the inverse function. Differentiating both sides of the equation $f^{-1}[f(x)] = f^{-1}(y) = x$ with respect to x gives

$$\frac{df^{-1}}{dy} \cdot \frac{df}{dx} = 1$$

or

$$\frac{df^{-1}}{dy} = \left(\frac{df}{dx}\right)^{-1}. \tag{1.5}$$

The derivative of the inverse function f^{-1}, with respect to its variable, is the reciprocal of the derivative of f with respect to its variable. However, this equation gives df^{-1}/dy as a function of x, and this derivative must be converted to a function of y. (See Example 29, below.)

Examples of Derivatives

22. $Dx^n = nx^{n-1}$.
23. $De^x = e^x$.
24. $D \ln x = 1/x$.
25. $D \sin x = \cos x$.
26. $D \cos x = -\sin x$.
27. $D(\sin^2 x + \cos^2 x) = 2 \sin x \cos x - 2 \sin x \cos x = 0$. The linearity of D and the chain rule, with $f(y) = y^2$ and $g(x) = \sin x$ or $\cos x$, have been invoked. Also, by the converse of rule 4, $\sin^2 x + \cos^2 x$ is a constant, which may be evaluated as 1.
28. $D \tan x = D(\sin x/\cos x) = (\sin^2 x + \cos^2 x)/\cos^2 x = \sec^2 x$. Rule 6 and Example 1.27 have been invoked.
29. $D \tan^{-1} x = 1/(1 + x^2)$, since if $y = f^{-1}(x) = \tan^{-1} x$, then by Eq. 1.5, $D \tan^{-1} x = df^{-1}/dx = 1/(df/dy) = 1/\sec^2 y = \cos^2 y$. But by trigonometry, if $x = \tan y$, then $\cos^2 y = 1/(1 + x^2)$; see Figure 1.2.

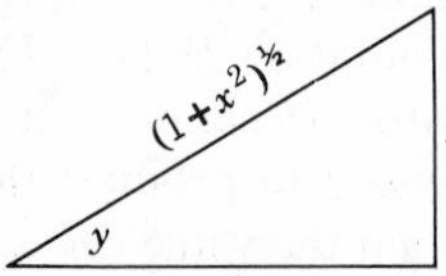

Figure 1.2

30.
$$D \ln \left[\sin \frac{ax}{(1 + x^2)^{1/2}}\right] = \frac{a}{(1 + x^2)^{1/2}} \cot \frac{ax}{(1 + x^2)^{1/2}}$$

by repeated application of the chain rule, and simplification.

Exercises

11. Find the derivatives of the following functions.
    (a) $\sin^{-1} x$.
    (b) $[\cos (e^{1/2 x^2} - 1)]^{1/2}$.
    (c) $e^{2x} \sin x \cos x(1 + \ln x)$.

12. If $f(\theta) = (3 + 2\cos\theta)/[\theta(1 + \tan 2\theta)]$, find $f'(\pi/2), f'(\pi)$.
13. (a) How would you define the sum of two operators, Ω_1 and Ω_2?
    (b) Is this sum also an operator?
    (c) How would you define $\Omega_1 - \Omega_2$?
14. Convince yourself that the operator "twice," defined by $Tf(x) = 2f(x)$, is a linear operator, but that the operator "square," defined by $Sf(x) = [f(x)]^2$, is not.
15. The *operator X* means "multiply by x"; it is an operator because the result of multiplying the function f by x is a new function Xf such that $Xf(x) = xf(x)$.
    (a) Is it a linear operator?
    (b) Convince yourself that $DX \neq XD$. Thus these operators do not COMMUTE; when they are applied to an arbitrary function, the result depends upon the order of operation.
    (c) Show that the difference, $DX - XD$, known as the COMMUTATOR of D and X and written $[D, X]$, is equal to the IDENTITY OPERATOR, 1, which operates on any f to produce $f: 1f = f$. (This result is of considerable significance in connection with the Heisenberg uncertainty principle.)
16. Equations involving operators may be manipulated just as ordinary equations, except that due account must be taken of the noncommutation of operators. Thus, if $\Omega_1 = \Omega_2$, we may multiply by Ω_3 from the left to obtain $\Omega_3\Omega_1 = \Omega_3\Omega_2$, or from the right to obtain $\Omega_1\Omega_3 = \Omega_2\Omega_3$, but we must not derive $\Omega_1\Omega_3 = \Omega_3\Omega_2$ or $\Omega_3\Omega_1 = \Omega_2\Omega_3$. Now multiply the result of Exercise 1.15 by D, both from the left and from the right, and then add the two equations to evaluate $[D^2, X]$.
17. To represent the two possible orientations of the spin of a single electron or proton, only two functions need be considered. These two functions are written as α and β (spin up and spin down, respectively) without ever specifying the independent variable. The effect of SPIN OPERATORS is as follows:

$$s_x\alpha = \tfrac{1}{2}\beta, \qquad s_y\alpha = \frac{i}{2}\beta, \qquad s_z\alpha = \tfrac{1}{2}\alpha,$$

$$s_x\beta = \tfrac{1}{2}\alpha, \qquad s_y\beta = -\frac{i}{2}\alpha \qquad s_z\beta = -\tfrac{1}{2}\beta.$$

Further operators are defined by $s_+ = s_x + is_y$, $s_- = s_x - is_y$, $s^2 = s_x^2 + s_y^2 + s_z^2$.
    (a) Find $s_+\alpha$, $s_+\beta$, $s_-\alpha$, and $s_-\beta$. Thus s_+ and s_- are called raising and lowering operators, respectively.
    (b) Find $s^2\alpha$ and $s^2\beta$.
    (c) Find $[s_x, s_y]$, $[s_y, s_z]$, and $[s_z, s_x]$. (*Hint:* Apply the first commutator to both α and β and determine what operator has the same effect. Then guess the other commutators "by symmetry.")
    (d) Relate s_+s_- to $s_x^2 + s_y^2$ and a commutator. Use the above result to simplify your answer.
    (e) Do the same as (d) for s_-s_+.
    (f) Use the above results to express s^2 in *two* alternative forms, both involving s_z and either s_+s_- or s_-s_+.
    (g) Use the above results to find $[s_+, s_-]$.
    *(h) Show that $[s^2, s_x] = [s^2, s_y] = [s^2, s_z] = 0$. Then show that $[s^2, s_\pm] = 0$.

The geometric interpretation of $f'(x_0)$ is that it equals the slope of the tangent to the curve $y = f(x)$ at the point $(x_0, f(x_0))$, as in Figure 1.3. A necessary condition that $f(x)$ be an EXTREMUM (maximum or minimum) at $x = x_0$

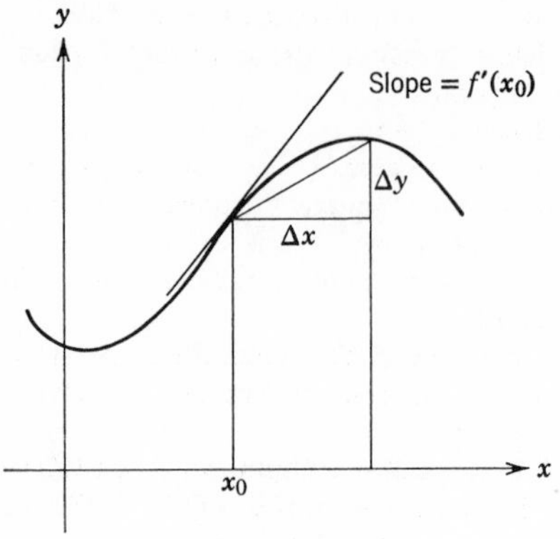

Figure 1.3

is that $f'(x_0) = 0$. There are further requirements on $f''(x_0)$ that ensure that $f(x)$ is indeed maximal or minimal at x_0. Also, whenever $f''(x_0) = 0$, $f(x)$ has an INFLECTION POINT at x_0.

When x changes by the amount Δx, the value $y = f(x)$ changes by an amount Δy, (See Fig. 1.3) given approximately by

$$\Delta y \sim f'(x)\,\Delta x. \tag{1.6}$$

In the limit that Δx becomes "infinitesimally small,"[1] this equation may be written in terms of DIFFERENTIALS, "infinitesimally small" changes of x and y:

$$dy = f'(x)\,dx \tag{1.7}$$

Notice that dy depends upon both x and dx. And we may interpret the derivative $f'(x)$ as the ratio of dy, the change in the value of f, to dx, the infinitesimal change in x.

Exercises

*18. What can you conclude about the derivative of an even function? Of an odd function?

19. The Morse function for the potential energy of vibration of a diatomic molecule is $V(r) = D_e[1 - e^{-a(r-r_e)}]^2$. Evaluate d^2V/dr^2 at that value of r which minimizes $V(r)$.

20. A Gaussian line shape has an intensity distribution

$$I(\nu) = \frac{1}{\sigma\sqrt{2\pi}}\, e^{-(\nu-\nu_0)^2/2\sigma^2}$$

whereas a Lorentzian line shape has an intensity distribution

$$I(\nu) = \frac{1}{\pi}\,\frac{\sigma\sqrt{3}}{(\nu - \nu_0)^2 + 3\sigma^2}\,.$$

[1] Whenever a term is vague or undefined, it is set in quotes. (But there are other uses for quotes.)

For both shapes, find
(a) The frequency at which $I(\nu)$ is maximum, and I_{max}.
(b) The frequencies at which $I(\nu)$ has an inflection.
(c) The width of the band at half-height, that is, $\Delta\nu$ from $I = I_{max}/2$ to $I_{max}/2$.

*21. The Beer-Lambert law states that

$$\log_{10} (I^\circ/I) = \varepsilon lc,$$

where $\varepsilon = \varepsilon(\lambda)$, and $\log_{10} x \equiv \ln x/\ln 10$. Assume we have a substance whose solutions obey the Beer-Lambert law for monochromatic light of wavelength λ_1 (extinction coefficient ε_1) and also for monochromatic light of wavelength λ_2 (extinction coefficient ε_2). Show that for a beam of nonmonochromatic light, composed of wavelengths λ_1 and λ_2 (incident intensities I_1° and I_2°, respectively), the Beer-Lambert law is not obeyed, that is, if I° and I are *total* incident and transmitted intensities, respectively, $\log (I^\circ/I)$ is not linear in (lc). [*Hint:* If $\log (I^\circ/I)$ were linear in (lc), its derivative with respect to (lc) would be a constant and its second derivative would be zero.]

*22. Use the chain rule to find the second derivative of the inverse function: If $y = f(x)$, what is

$$\frac{d^2f^{-1}}{dy^2} = \frac{d}{dy}\left(\frac{1}{df/dx}\right)$$

as a function of x?

*23. The titration curve of a weak base (acid dissociation constant of its conjugate acid = K) with a strong acid is given by

$$[\mathrm{H}^+] + M_0 \frac{[\mathrm{H}^+]}{[\mathrm{H}^+] + K} - \frac{K_w}{[\mathrm{H}^+]} = M$$

where M_0 is the initial concentration of base and M is the concentration of added acid. (The volume change on addition has been neglected.) This equation is a cubic in $[\mathrm{H}^+]$ and cannot be solved for either $[\mathrm{H}^+]$ or $\mathrm{pH} \equiv \log_{10} [\mathrm{H}^+]$ as a function of M. Use the above result to find an equation to be solved to determine those values of $[\mathrm{H}^+]$ which maximize or minimize $d(\mathrm{pH})/dM$.

*24. The coefficient of thermal expansion of a substance is defined by $\alpha = (1/V)(dV/dT)$, where V is the volume and T the temperature. Use the table in the *Handbook of Chemistry and Physics* to estimate α for water at 70°C. (*Hint:* V is tabulated only at special values of T, and the derivative must be approximated by $\Delta V/\Delta T$. Be careful that you aren't estimating α at 70.5°C.)

*25. When n moles of NaCl is dissolved in 1000.00 g H_2O at 25°C and 1 atm, the volume of the resulting solution and the total heat absorbed are given by the empirical equations

$$V = 1002.93 + 16.61n + 2.153n^{3/2} \text{ ml.}$$
$$\Delta H = 923n + 476n^{3/2} - 726n^2 + 243n^{5/2} \text{ cal.}$$

Let us imagine that we have a very large volume of this n-molal NaCl solution, and let us dissolve one more mole of NaCl in the solution. Let the volume be so large ("infinitely large") that this one extra mole does not change the molality of the solution.
(a) Convince yourself that the volume of the solution increases by dV/dn, and evaluate this quantity, known as the PARTIAL MOLAL VOLUME, V_{NaCl}, as a function of n.
(b) What is V_{NaCl} in a 1.000-molal aqueous solution?

(c) How much heat is absorbed on dissolving 1 mole of NaCl in an infinite amount of an n-molal NaCl solution? This quantity is known as the DIFFERENTIAL HEAT OF SOLUTION, ΔH_{NaCl}.
(d) What is ΔH_{NaCl} in a 1.000-molal aqueous solution?
(e) How much heat is absorbed when one *molecule* of NaCl is dissolved in a 1.000-molal aqueous NaCl solution?

*26. Use the *Handbook of Chemistry and Physics* tables (46th ed or later) of the densities of aqueous NaOH solutions to estimate the partial molal volume of NaOH in a 1% solution at 20°C. (*Hints:* The table lists the concentrations C_{NaOH} and C_{water}; the sum of these is the density in g/l. First determine the volume, in ml, that would be occupied by a 1% solution containing exactly 1000 g water. Do the same for 0.5% and 1.5% solutions. Then convert these percentages to molalities, and approximate dV/dn by $\Delta V/\Delta n$.) Notice that V_{NaOH} is negative.

1.3 Roots of Equations

Consider the situation in which we are given $v = f(x)$, where v is some experimental value and the corresponding value of x is to be determined. In special cases f^{-1} is also a well-known function, so that the desired value is simply $x = f^{-1}(v)$. For example, the solution of $x^2 + 3x + 2 = v$ is given by the quadratic formula as $x = [-3 \pm (1 + 4v)^{1/2}]/2$. The solution of $C = C_0 \exp(-kt)$ is $k = \ln(C_0/C)/t$.

But in general it is not possible to specify f^{-1} directly. For example, $x + e^x = v$ cannot be solved for x since the function inverse to the function defined by $f(x) = x + e^x$ is not a well-known, tabulated function. Therefore it is necessary to use a numerical method.

Let $F(x) = f(x) - v$ for the given value of v. Then our problem is to find the ROOTS of the equation $F(x) = 0$, that is, those values of x for which $F(x) = 0$.

The method of solution is by successive approximations. From a guess, x_i, what is the next approximation, x_{i+1}, which will be closer to the true root? There are two slightly different formulas for obtaining x_{i+1} from x_i, both of which may be derived by setting $y + \Delta y$ equal to zero, with Δy given by Eq. 1.6.

$$F(x_i) + (x_{i+1} - x_i)F'(x_i) \simeq F(x_{i+1}) = 0.$$

Solving for x_{i+1} in the extremities of this "equation" leads to the NEWTON-RAPHSON FORMULA

$$x_{i+1} = x_i - \frac{F(x_i)}{F'(x_i)}. \qquad (1.8)$$

The REGULA FALSA (approximate roots) formula arises by approximating $F'(x_i)$ by $\Delta F/\Delta x$. To do so requires two approximate roots, x_{i1} and x_{i2}, so that $F'(x_i)$ in Eq. 1.8 is replaced by $[F(x_{i2}) - F(x_{i1})]/(x_{i2} - x_{i1})$. The resulting formula is

$$x_{i+1} = x_{i1} - F(x_{i1}) \frac{x_{i2} - x_{i1}}{F(x_{i2}) - F(x_{i1})}. \tag{1.9}$$

Repeated application of either formula will give closer and closer approximation to the root until the desired accuracy is obtained.

The Newton-Raphson method is usually to be preferred since the derivative is evaluated directly without the need for two additional subtractions and an extra multiplication. Furthermore, for many functions it is easier to evaluate F' than to evaluate F; for example, in finding the roots of an Nth-order polynomial the Newton-Raphson method requires evaluation of an Nth-order polynomial and an $(N - 1)$st-order polynomial.

1.4 Some Comments on the Use of a Desk Calculator

Since there is such variety in the range of versatility of commercial desk calculators, it is not possible to provide detailed instructions on their use. Instead, you must familiarize yourself with the machine at hand. First read the instruction manual to learn the techniques for computation and the capabilities of the machine. Then try a few simple operations whose results you know—$1 + 2$, 2×3, $2 \times 3 \times 4$, $(3 \times 4)/2$, $(1 \times 2) + (3 \times 4)$, $(\frac{12}{4}) - (\frac{6}{3})$—to make sure that you know how all the buttons work. These exercises also serve to set the decimal point empirically on each register if you do not wish to pore over the manual to learn where the decimal point goes.

The two major considerations in using a desk calculator are, of course, accuracy and speed. In all computational work it is advisable to keep in mind the simple observation that the majority of inaccuracies, and the greatest loss of time, arise in entering numbers on the keyboard and in copying intermediate results. Therefore you will save a considerable amount of time and error if you plan your calculation in advance so as to *minimize the number of keyboard entries and transcriptions*.

For example, some desk calculators are equipped with a memory register, which should be used for a recurrent number. Thus in the evaluation of the polynomial $P_4(x) = a + bx + cx^2 + dx^3 + ex^4$, it would be possible to enter x into the memory, evaluate and transcribe x^2, x^3, and x^4, and add the terms. A better method is to rewrite in the form $P_N(x) = a + x\{b + x[c + x(d + ex)]\}$ and evaluate starting at the innermost parentheses. On a versatile machine with provision for multiplication of a sum by a keyboard entry or a number in the memory register, it is possible to evaluate a polynomial without transcription of any intermediate results.

In statistical analysis of data it is often necessary to evaluate sums of the form $\sum x_i^2$ and $\sum x_i y_i$. One technique would be to evaluate each such product, transcribe the result, and finally re-enter the products and add them. But most machines can perform accumulative multiplication, automatically adding the product to the sum of products already accumulated. Also, it is usually possible to accumulate $\sum x_i$ or $\sum y_i$ simultaneously since the sum of the multipliers may be produced in another register.

27. Find $\sqrt{2}$ to 8 significant figures.
28. Solve $x + e^x = 2$ to three decimal points. Use $x_0 = 1$ as an initial guess and look up values for e^x in the *Handbook of Chemistry and Physics*.
*29. According to the Planck radiation formula, the energy density of black-body radiation, as a function of frequency and temperature, is given by

$$\rho(\nu) = (8\pi h/c^3)\nu^3/[\exp(h\nu/kT) - 1].$$

Find the constant of proportionality between the frequency at which ρ is maximum and the temperature.
*30. What is the solubility (to 3 "significant" figures) of $CaCO_3$ in pure water?

$$K_{sp} = 8.7 \times 10^{-9}. \qquad K_2(H_2CO_3) = 4.4 \times 10^{-11}.$$

*31. What is $[H^+]$ (2 significant figures) in an $0.1M$ solution of $KHSO_4$?

$$K_{HSO_4^-} = a_{H^+}a_{SO_4^{2-}}/a_{HSO_4^-} = [H^+]\gamma_{H^+}[SO_4^{2-}]\gamma_{SO_4^{2-}}/[HSO_4^-]\gamma_{HSO_4^-} = 2 \times 10^{-2}.$$

(a) Approximating activities by molarities ($\gamma = 1$).
(b) Approximating activity coefficients by the Debye-Hückel limiting law:

$$\tfrac{1}{2}(\log_{10}\gamma_{K^+} + \log_{10}\gamma_{HSO_4^-}) = -\tfrac{1}{2}\sqrt{I}$$

$$\tfrac{1}{3}(\log_{10}\gamma_{K^+} + \log_{10}\gamma_{H^+} + \log_{10}\gamma_{SO_4^{2-}}) = -\sqrt{I}$$

where $I = \frac{1}{2}([K^+] + [H^+] + [HSO_4^-] + 4[SO_4^{2-}])$.
*32. An alternative to the Newton-Raphson method is the METHOD OF SUCCESSIVE SUBSTITUTIONS, which is applicable to finding the solutions of $x = f(x)$ by $x_{i+1} = f(x_i)$. Show that a necessary condition for this method to succeed is $|f'(x_i)| \ll 1$. Then solve Exercise 1.29 by this method.

1.5 Integration

The DEFINITE INTEGRAL of a function f, over the interval from a to b, is a number defined by

$$\int_a^b f(x)\,dx \equiv \lim_{\Delta x_i \to 0} \sum_i f(x_i)\,\Delta x_i, \qquad (a \leqslant x_i \leqslant b). \tag{1.10}$$

What this means is that we choose a set of points between a and b, evaluate f at each point, multiply that value by the spacing between that point and the next point, add up all these products, and take the limit as the points become infinitely dense. See Figure 1.4.

Exercise

33. The translational partition function for a gas in a one-dimensional box of length L is $Q_{trans}^{[1]} = \sum_{n=1}^{\infty} \exp(-n^2h^2/8mL^2kT)$, which cannot be summed explicitly. However, it

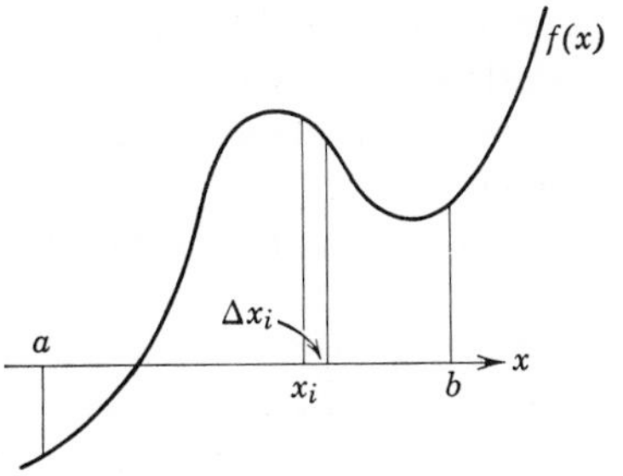

Figure 1.4

is possible to find $\lim_{T\to\infty} Q^{[1]}_{\text{trans}}/\sqrt{T}$ by the following procedure: Let $x = n/\sqrt{T}$. Then as n goes from 1 to ∞ in steps of unity, x goes from $1/\sqrt{T}$ to ∞ in steps of $\Delta x = 1/\sqrt{T}$. Thus as $T \to \infty$, $\Delta x \to 0$ and the limit becomes an integral with respect to the continuous variable x. Find the integral; we shall evaluate it later. Then do the same for $\lim_{T\to\infty} Q_{\text{rot}}/T$ where $Q_{\text{rot}} = \sum_{J=0}^{\infty} (2J + 1) \exp[-J(J + 1)h^2/2IkT]$, which is the rotational partition function for a heteronuclear diatomic molecule. (*Hint:* Let $x = J/T$.)

Geometrically, the definite integral may be interpreted as the area under the curve $y = f(x)$ and above the x-axis, and bounded by the straight lines $x = a$ and $x = b$; regions where the curve $y = f(x)$ lies below the x-axis contribute negatively to the integral.

Exercises

*34. On graph paper draw a large and accurate graph of $f(x) = 1/x$ on the interval $1 < x < 2$. Carefully cut out the piece of paper below this graph and above the x-axis, and weigh it. Also cut out and weigh a rectangle of known dimensions. By comparing the weights you should be able to estimate $\int_1^2 dx/x$ by GRAPHICAL INTEGRATION. Compare your estimate with the exact value.

35. Convince yourself that if f is a periodic function of period p

$$\int_a^{a+np} f(t)\,dt = n\int_0^p f(t)\,dt.$$

Some of the rules for dealing with definite integrals are

1. $\displaystyle\int_a^b f(x)\,dx + \int_b^c f(x)\,dx = \int_a^c f(x)\,dx$

2. $\displaystyle\int_b^a f(x)\,dx = -\int_a^b f(x)\,dx$

3. $\displaystyle\int_a^b [\alpha f(x) + \beta g(x)]\,dx = \alpha\int_a^b f(x)\,dx + \beta\int_a^b g(x)\,dx$

4. $\displaystyle\int_a^b f(x)\,dx = \int_a^b f(y)\,dy = \int_a^b f(t)\,dt$

From rule 2 it follows immediately that $\int_a^a f(x)\,dx = 0$. Rule 4 states that the variable of integration is a DUMMY VARIABLE, one whose name may be changed at will (except, of course, that it may be confusing to give it the same symbol as that of some other variable in the problem). This is reasonable since the definite integral depends upon the function f and the limits a and b, but not upon the name of the axis.

We may use some of these rules to derive a useful result: The integral of an odd function over a symmetric interval is zero. Let o be an odd function, and let $I = \int_{-a}^{a} o(x)\,dx$. Since x is a dummy variable it may be replaced by $-y$ to give $I = \int_{-a}^{a} o(-y)\,d(-y)$. But $d(-y)$ is $-dy$ and o is odd, so that $I = \int_a^{-a} o(y)\,dy$. Notice that when the variable $-y$ runs from $-a$ to a, the variable y runs from a to $-a$; the limits of integration must be consistent with the variable of integration. But from rule 2, $I = -\int_{-a}^{a} o(y)\,dy$, and by rule 4 again, $I = -\int_{-a}^{a} o(x)\,dx$. Therefore $I = -I$, which implies that

$$\int_{-a}^{a} o(x)\,dx = 0.$$

The geometric interpretation of this result is that for every $o(x_i)\,\Delta x_i$ which is positive, there is a product $o(-x_i)\,\Delta x_i$, which is equal in magnitude but negative, so that all products cancel in pairs. Similar reasoning leads to the conclusion that the integral of an even function e over a symmetric interval satisfies

$$\int_{-a}^{a} e(x)\,dx = 2\int_0^a e(x)\,dx.$$

Exercise

*36. If o is an odd function, what can you conclude about $\int_{-\infty}^{x} o(x)\,dx$?

Notice that the definite integral is a number! However, this number depends upon the upper (and lower) limit of integration, which can be variable. Therefore $\int_a^x f(t)\,dt$ is the value of some function F, evaluated at x. Do not be confused if this value is written as $\int_a^x f(x)\,dx$. The variable of integration is still a dummy variable, and the x-dependence arises from the variability of the limit of integration.

The fundamental theorem of calculus states

$$\frac{d}{dx}\int_a^x f(t)\,dt = f(x) \tag{1.11}$$

or

$$DF = f.$$

LEIBNIZ' RULE is the application of the chain rule to the evaluation of the derivative of an integral whose limits of integration are dependent variables. To evaluate

$$\frac{d}{dx}\int_{f(x)}^{g(x)} h(t)\,dt = \frac{d}{dx}\left[\int_a^{g(x)} h(t)\,dt - \int_a^{f(x)} h(t)\,dt\right]$$

where a is arbitrary, it is sufficient to show how the first term is evaluated. Let $y = g(x)$, so that

$$\frac{d}{dx}\int_a^{g(x)} h(t)\,dt = \frac{dy}{dx}\frac{d}{dy}\int_a^{y} h(t)\,dt = \frac{dy}{dx}\,h(y) = h[g(x)]g'(x)$$

by Eqs. 1.4′ and 1.11. Therefore

$$\frac{d}{dx}\int_{f(x)}^{g(x)} h(t)\,dt = h[g(x)]g'(x) - h[f(x)]f'(x). \tag{1.12}$$

Exercise

37. What is $d/dT \int_0^{\Theta/T} (x^3\,dx)/(e^x - 1)$? This integral is known as the Debye function and is important in the theory of heat capacities of crystals.

Next let us try to define an INTEGRATION OPERATOR, I, which converts f back to F. We seek an operator inverse to differentiation, such that $DI = 1 = ID$. The fundamental theorem states that $DIf = f$, so that the first equality is satisfied. But if $DF = f$, so also does $D(F + c) = f$, where c is any of the constant functions. Applying I to both sides of this last equation gives $ID(F + c) = F$, which shows that $ID \neq 1$ and that the integration operator must be arbitrary to the extent of addition of a constant function: $If = F + c$.

Nevertheless, we may use the integration operator to express the definite integral as

$$\int_a^b f(x)\,dx = [If(x)]_a^b = If(b) - If(a) = F(b) - F(a) \tag{1.13}$$

since the arbitrary constant function in If cancels in the subtraction.

Exercise

38. The heat capacity of silver metal is given by the empirical formula

$$C_P = 5.09 + 1.02 \times 10^{-3}T + 0.36 \times 10^5 T^{-2} \text{ kcal/mole-deg.}$$

The enthalpy and entropy are given by

$$H_T^\circ - H_{298}^\circ = \int_{298}^{T} C_P\,dT \text{ kcal/mole,}$$

$$S_T^\circ - 10.20 = \int_{298}^{T} \frac{1}{T} C_P\,dT \text{ e.u.}$$

Find $H_{373}^\circ - H_{298}^\circ$ and S_{373}°.

All that is necessary for the evaluation of definite integrals is a table of pairs of functions related by $If = F$. But it is not possible to include in a table of integrals all the functions whose integrals may be required. Indeed, there are many functions f such that F is not a well-known function. However, by suitable tricks, such as integration by substitution and integration by parts, some integrals not included in tables may be reduced to integrals which are included.

Integration by parts and integration by substitution are two techniques that are applicable to those integrals of the form $\int f(x)g(x)\,dx$ for which the integral of one factor, say $g(x)$, is simple. Integration by parts is to be preferred if f' is simpler than f. Integration by substitution is to be preferred if Ig is "closely related" to f.

The formula for integration by parts may be derived from the formula for the derivative of a product, $D(fG) = fDG + GDf = fg + Gf'$, by rearranging and integrating to obtain

$$\int f(x)g(x)\,dx = f(x)G(x) - \int G(x)f'(x)\,dx. \tag{1.14}$$

Thus, if the derivative of f is simpler than f itself, and if G, the integral of g, is no more complicated than g, then integration by parts may convert the problem to a simpler integration.

Examples

31. To evaluate $\int xe^x\,dx$, set $f(x) = x$, $f'(x) = 1$, $g(x) = e^x$, and $G(x) = Ig(x) = e^x$. Then

$$\int xe^x\,dx = xe^x - \int e^x\,dx = xe^x - e^x.$$

32. To evaluate $\int x^n \cos \omega x\,dx$, set $f(x) = x^n$, $f'(x) = nx^{n-1}$, $g(x) = \cos \omega x$, and $G(x) = (1/\omega) \sin \omega x$. Then

$$\int x^n \cos \omega x\,dx = \frac{1}{\omega} x^n \sin \omega x - \frac{n}{\omega}\int x^{n-1} \sin \omega x\,dx.$$

The latter integral may be further reduced by another integration by parts. Each application of this procedure decreases the exponent by 1, until eventually $\int \sin \omega x\,dx$ or $\int \cos \omega x\,dx$ is reached.

For simplifying definite integrals by integration by parts, the general formula is

$$\int_a^b f(x)g(x)\,dx = f(b)G(b) - f(a)G(a) - \int_a^b G(x)f'(x)\,dx. \tag{1.15}$$

Example

33. To evaluate $\int_0^1 x^2 \ln x\,dx$, set $f(x) = \ln x$, $f'(x) = 1/x$, $g(x) = x^2$, and $G(x) = \frac{1}{3}x^3$. Then

$$\int_0^1 x^2 \ln x\,dx = [\tfrac{1}{3}x^3 \ln x]_0^1 - \tfrac{1}{3}\int_0^1 x^2\,dx = -\tfrac{1}{9}$$

since $\ln 1 = 0$ and $\lim_{x\to 0} x^3 \ln x = 0$.

The general formula for simplifying $\int f(x)g(x)\,dx$ by the substitution $y = G(x) = Ig(x)$ is

$$\int f(x)g(x)\,dx = \int f(x)\,dG(x) = \int f\{G^{-1}[G(x)]\}\,dG(x) = \int f[G^{-1}(y)]dy \tag{1.16}$$

where we have used the identity $x = G^{-1}[G(x)]$. This formula is more complicated in its generality than it is in specific cases, so do not let it frighten you; in those cases for which integration by substitution is applicable, $f[G^{-1}(y)]$ is quite simple and usually obvious.

Examples

34. $\int \sin^2 x \cos x\,dx = \int \sin^2 x\,d(\sin x)$. Here $f(x) = \sin^2 x, g(x) = \cos x, G(x) = Ig(x) = \sin x = y$, and $x = G^{-1}(y) = \sin^{-1} y$, so $f(x)$ becomes $f[G^{-1}(y)] = \sin^2(\sin^{-1} y) = y^2$. But it is more obvious *without* the symbolism that the two functions, sin and $\sin^2$, in the integral $\int \sin^2 x\,d(\sin x)$ are "closely related," and that the substitution $y = \sin x$ leads to the simple $\int y^2\,dy = \frac{1}{3}y^3 = \frac{1}{3}\sin^3 x$.

35. Likewise, $\int \cos^3 x\,dx = \int \cos^2 x \cos x\,dx = \int \cos^2 x\,d(\sin x) = \int (1 - \sin^2 x)\,d(\sin x) = \int (1 - y^2)\,dy = y - \frac{1}{3}y^3 = \sin x - \frac{1}{3}\sin^3 x$. Notice that the trick here is to separate the integrand into two appropriate factors and to recognize how "closely related" $\sin x$ is to $\cos^2 x$.

36. $$\int \frac{\sin(\ln x)}{x}\,dx = \int \sin(\ln x)\,d(\ln x) = \int \sin y\,dy = -\cos y = -\cos(\ln x).$$

37.

$$\int \frac{\cos x\,dx}{a + b\sin x} = \int \frac{d(\sin x)}{a + b\sin x} = \frac{1}{b}\int \frac{d(a + b\sin x)}{a + b\sin x} = \frac{1}{b}\int \frac{dy}{y} = \frac{1}{b}\ln y = \frac{1}{b}\ln(a + b\sin x).$$

Notice here the trick of incorporating the constants into the new variable.

38. The first step in simplifying $\int \sec^2 \sqrt{x}\,dx$ is to eliminate the square root in the argument by setting $\theta = \sqrt{x}$, $x = \theta^2$, and $dx = 2\theta\,d\theta$, giving $2\int \theta \sec^2\theta\,d\theta$. Next we integrate by parts, with $f(\theta) = \theta, f'(\theta) = 1, g(\theta) = \sec^2\theta$, and $G(\theta) = \tan\theta$, so that we obtain for the integral $\theta\tan\theta - \int \tan\theta\,d\theta$. Finally we write

$$\int \tan\theta\,d\theta = \int \frac{1}{\cos\theta}\sin\theta\,d\theta = -\int \frac{d(\cos\theta)}{\cos\theta} = -\ln(\cos\theta).$$

In summary, $\int \sec^2\sqrt{x}\,dx = 2\sqrt{x}\tan\sqrt{x} + 2\ln(\cos\sqrt{x})$.

It is clear that there are many tricks in this game. Of course, it is necessary to remember the simple integrals and these rules of manipulation. But it is also necessary to develop a feeling for simplicity so that you can recognize whether a manipulation has made the integral easier or harder, more likely or less likely to be included in a table of integrals.

In evaluating definite integrals by substitution, it is necessary to change the limits of integration. With $y = G(x) = Ig(x)$, the general formula is

$$\int_a^b f(x)g(x)\,dx = \int_{G(a)}^{G(b)} f[G^{-1}(y)]\,dy \tag{1.17}$$

since as x runs from a to b the new variable of integration, y, runs from $G(a)$ to $G(b)$.

Example

39. $$\int_e^{e^2} \frac{dx}{x \ln x} = \int_{\ln e}^{\ln e^2} \frac{d(\ln x)}{\ln x} = \int_1^2 \frac{dy}{y} = \ln 2 - \ln 1 = \ln 2.$$

Exercises

39. The total radiant energy emitted by a black body at temperature T is given by the integral

$$E(T) = \frac{8\pi h}{c^3} \int_0^\infty \frac{\nu^3\,d\nu}{\exp\,(h\nu/kT) - 1}.$$

Express $E(T)$ in the form $E(T) = aT^n$. Leave a in terms of a definite integral which we shall learn to evaluate later.

40. Evaluate
   (a) $\int x(\ln x)^2\,dx$.
   (b) $\int \cos x \ln(\sin x)\,dx$.
   (c) $\displaystyle\int \frac{dx}{(x - \lambda_1)(x - \lambda_2)}$ (expand integrand in partial fractions).
   (d) $\displaystyle\int_0^1 \frac{x\,dx}{a + bx^2}$.
   (e) $\displaystyle\int_{\pi/6}^{\pi/3} \frac{d\theta}{\sin\theta\cos\theta} = \int_{\pi/6}^{\pi/3} \frac{\cos\theta}{\sin\theta} \frac{d\theta}{\cos^2\theta}$.
   (f) $\int_0^1 x^2 e^{ax}\,dx$ (cf. part a).
   (g) $I = \int e^x \cos x\,dx$ (integrate twice by parts).
41. In Exercise 1.25 you found the partial molal volume V_{NaCl} as a function of n_{NaCl}. It can be shown that for aqueous NaCl solutions at constant temperature and pressure

$$n_{H_2O}\,dV_{H_2O} + n_{NaCl}\,dV_{NaCl} = 0.$$

Solve for dV_{H_2O} and integrate to find the partial molal volume V_{H_2O} as a function of n_{NaCl}. (*Hints:* Convert the integral with respect to V_{NaCl} to one with respect to n_{NaCl}. Start with 1000 g of H_2O so that n_{H_2O} is a constant, 1000/18.016. At 25°C, V_{H_2O} in pure water is 18.069 ml.)

*42. To evaluate $\int (1 + x^2)^{-1/2}\,dx$:
   (a) Substitute $x = \tan\theta$.
   (b) Multiply both top and bottom of the resulting integrand by $\sec\theta + \tan\theta$ to obtain $\int (\sec^2\theta + \sec\theta\tan\theta)\,d\theta/(\sec\theta + \tan\theta)$.
   (c) For help in solving this integral, find $D(\sec\theta + \tan\theta)$.

(d) Substitute x back into the result. We shall find that hyperbolic functions offer a less cumbersome form for this integral.

*43. Evaluate

$$\int_0^{\pi} \frac{k^2 + k\cos\phi}{(1 + k^2) + 2k\cos\phi}\,d\phi.$$

Without loss of generality, k may be taken as positive, but distinguish the cases $k < 1$, $k = 1$, $k > 1$. (*Hints:* $k^2 + k\cos\phi = (\frac{1}{2} + \frac{1}{2}k^2 + k\cos\phi) + (\frac{1}{2}k^2 - \frac{1}{2})$. Then substitute $x = \tan\frac{1}{2}\phi$.)

44. Define an operator L by $L[f(t)] = F(s) = \int_0^\infty e^{-st}f(t)\,dt$. (The operator L takes the function f into the function F. The independent variables are given different symbols only for clarity.)

(a) Convince yourself that L is a linear operator (known as the LAPLACE TRANSFORM).

(b) Show $L[c] = \dfrac{c}{s}$ $\qquad L[e^{at}] = \dfrac{1}{s - a}\ (s > a)$

$$L[t] = \frac{1}{s^2} \qquad L[f(at)] = \frac{1}{a}F\left(\frac{s}{a}\right)$$

$$L\left[\frac{df}{dt}\right] = sF(s) - f(0) = sL[f(t)] - f(0).$$

1.6 Infinite Series

A SEQUENCE is a collection of objects in order, as distinguished from a set, in which ordering is irrelevant. We shall use the brace notation for sequences: $\{u_n\}_{n=1}^N$ represents the sequence whose elements are $u_1, u_2, \ldots, u_N$. In cases in which the indices are obvious they may be omitted. It is possible for N to be infinite and it is possible to refer to a sequence of functions, $\{f_n\}$, and the sequence of their values $\{f_n(x)\}$.

The SUM NOTATION, $\sum_{n=1}^N u_n$, is an abbreviation for the sum of all the elements of the sequence $\{u_n\}_{n=1}^N$. Such a sum is usually called a SERIES, and again N may be infinite. We shall be primarily concerned with infinite series which are sums of the elements of a sequence of functions $\{f_n\}_{n=0}^\infty$. Just as $f_n(x)$ is the value of the function f_n at the argument x, the sum $\sum_{n=0}^\infty f_n(x)$ is a number which depends upon x and therefore represents the value, $F(x)$, of some function $F = \sum_{n=0}^\infty f_n$ at the argument x.

Of course, there is always the problem of convergence of such infinite series since there is no automatic guarantee that the sum of an infinite number of numbers is finite. Tests for convergence are discussed in mathematics texts. We shall merely point out that a necessary condition for the convergence of $\lim_{N\to\infty} \sum_{n=0}^N f_n(x)$ is that $f_n(x)$ approach zero as n approaches infinity. And we note that it is possible for this condition to be met for some values of x but not for others.

PRODUCT NOTATION, $\prod_{n=0}^{N} u_n$ or $\prod_{n=0}^{N} f_n(x)$, is entirely analogous to sum notation except that the elements are to be multiplied together. For example, the FACTORIAL function of the integer N is defined by $N! \equiv \prod_{n=1}^{N} n$.

An important feature of all these three notations is that the index n is a DUMMY INDEX, whose purpose is to range over a set of integers, but whose name may be changed at will.

Exercises

45. Convince yourself that $0! = 1$. (*Hint:* What is $(N-1)!/N!$?)
46. Multiply numerator and denominator of $1 \cdot 3 \cdot 5 \cdots (2n-1)/2 \cdot 4 \cdot 6 \cdots (2n)$ by appropriate quantities and simplify this fraction by the use of factorial notation.
*47. Use the $\sum$ and $\prod$ notation to express in compact form
   (a) The arithmetic mean of a set of numbers.
   (b) The geometric mean of a set of numbers.
   (c) $D^n(fg)$. (*Hint:* Remember the binomial formula.)
48. Use $\sum$ and factorial notation to express in compact form
   (a) $(1+x)^{-2} = 1 - 2x + 3x^2 - 4x^3 + \cdots$.
   (b) $\cos 2x = 1 - \dfrac{2 \cdot 2}{1 \cdot 2} x^2 + \dfrac{2 \cdot 2 \cdot 2 \cdot 2}{1 \cdot 2 \cdot 3 \cdot 4} x^4 - \dfrac{2 \cdot 2 \cdot 2 \cdot 2 \cdot 2 \cdot 2}{1 \cdot 2 \cdot 3 \cdot 4 \cdot 5 \cdot 6} x^6 + \cdots$.
   (c) $\sinh^{-1} x = x - \dfrac{1}{2}\dfrac{x^3}{3} + \dfrac{1 \cdot 3}{2 \cdot 4}\dfrac{x^5}{5} - \dfrac{1 \cdot 3 \cdot 5}{2 \cdot 4 \cdot 6}\dfrac{x^7}{7} + \cdots$. (*Hint:* Use the result of Exercise 1.46.)
49. Sum notation may be extended to indicate double sums:
   (a) Convince yourself that $(\sum_i x_i)(\sum_j y_j) = \sum_{i,j} x_i y_j \neq \sum_i x_i y_i$.
   (b) Convince yourself that $(\sum_i x_i)^2 = \sum_i x_i^2 + \sum_{i \neq j} x_i x_j = \sum_i x_i^2 + 2\sum_{i>j} x_i x_j$. (*Note:* $\sum_{i \neq j}$ is a short form of the double summation $\sum_j \sum_{i(i \neq j)}$, etc. Such abbreviated notation should be avoided if it is not clear from context whether j is to be summed over or not.)
*50. (a) Convince yourself that $\sum_{n=0}^{\infty} (-1)^n n!/x^n$ cannot converge for any finite x.
   (b) Let us write such a series as $\sum_{u=0}^{N} (-1)^n c_n/x^n$, and investigate its behavior, not as N goes to infinity, but as x goes to infinity. If

$$\lim_{x \to \infty} x^N \Big[F(x) - \sum_{n=0}^{N} (-1)^n c_n/x^n\Big] = 0$$

   then $\sum (-1)^n c_n/x^n$ is said to be ASYMPTOTICALLY CONVERGENT to $F(x)$. In particular, $\sum (-1)^n n!/x^n \sim xe^x \int_x^{\infty} t^{-1}e^{-t}\,dt$. And if each $c_n > 0$, then it can be shown that the error associated with terminating the series after N terms is less than the magnitude of the $(N+1)$st term. Thus, as $N \to \infty$, this error goes to infinity. Now convince yourself that for large x, $F(x)$ can be calculated to high accuracy by terminating the series just before the smallest term.
   (c) Use these results to estimate $\int_{10}^{\infty} t^{-1}e^{-t}\,dt$ and the percent error in your estimate.
51. A LINEAR COMBINATION of the set of functions, $\{f_i\}$, is the sum $\sum c_i f_i$, also a function. A set of functions, $\{f_i\}$, is said to be LINEARLY DEPENDENT if there exists a nontrivial linear combination of them, $\sum c_i f_i$, equal to the zero function; of course there always exists the trivial linear combination with every $c_i = 0$. Convince

yourself that if a set of functions is linearly dependent, then at least one of them is a linear combination of the others.

We next turn to the problem of representing an arbitrary function F as a linear combination of the sequence $\{f_n\}$: $F(x) = \sum c_n f_n(x)$. For now we shall restrict the functions f_n to power functions, such that $F(x)$ is expanded in powers of $(x - a)$:

$$F(x) = \sum_{n=0}^{\infty} c_n(x - a)^n. \tag{1.18}$$

We know that the coefficients c_n are given by TAYLOR'S FORMULA:

$$c_n = \frac{1}{n!} D^n F(a). \tag{1.19}$$

Therefore the TAYLOR SERIES expansion of F about the point $x = a$ is

$$F(x) = \sum_{n=0}^{\infty} \frac{1}{n!} D^n F(a)(x - a)^n. \tag{1.20}$$

The MacLAURIN SERIES is the special case with $a = 0$.

Examples

40. $1/(1 + x) = 1 - x + x^2 + \cdots = \sum_{n=0}^{\infty} (-1)^n x^n$, as you know from the sum of an infinite geometrical progression. Alternatively, this could be evaluated by repeated differentiation: $D[(1+x)^{-1}] = -(1 + x)^{-2}$, $D^2[(1 + x)^{-1}] = 2(1 + x)^{-3}$, $D^n[(1 + x)^{-1}] = (-1)^n n!(1 + x)^{-1-n}$, $D^n F(0) = (-1)^n n!$. Notice that only if $|x| < 1$ does $f_n(x)$ approach zero, so that this series can converge only for $|x| < 1$. Memorize this series.

41. Since $D \tan^{-1} x = 1/(1 + x^2)$, and $\tan^{-1} 0 = 0$, $\tan^{-1} x = \int_0^x dt/(1 + t^2)$. Expanding the integrand according to the above result (with x replaced by t^2), reversing the order of integration and summation, and integrating term-by-term leads to

$$\tan^{-1} x = \int_0^x \frac{dt}{1 + t^2} = \int_0^x \sum_{n=0}^{\infty} (-1)^n t^{2n}\, dt = \sum_{n=0}^{\infty} (-1)^n \int_0^x t^{2n}\, dt = \sum_{n=0}^{\infty} (-1)^n \frac{x^{2n+1}}{2n + 1}$$

42. Let us define a function f by the following:

$$Df = f \qquad f(0) = 1 \tag{1.21}$$

Repeated differentiation leads to $D^2 f = Df = f$, $D^n f = f$, $D^n f(0) = 1$. Therefore the Maclaurin series expansion of this function is

$$f(x) = \sum_{n=0}^{\infty} \frac{x^n}{n!} \tag{1.22}$$

This function is called the EXPONENTIAL FUNCTION and its value is written exp (x) or e^x. For all real x, $e^x > 0$ (See Figure 1.5). Further properties include

$$\exp(1) \simeq 2.718\ldots, \text{ called } e$$

$$\exp(x + y) = \exp(x)\exp(y) \qquad \text{or} \qquad e^{x+y} = e^x e^y$$

$$\exp(xy) = [\exp(x)]^y \qquad\qquad e^{xy} = (e^x)^y$$

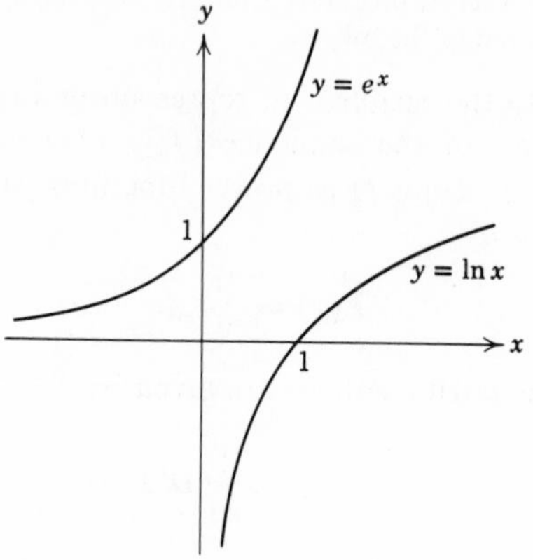

Figure 1.5

Let us define another function, ln, called LOGARITHM, as the function inverse to exp:

$$y = f(x) = \exp x \qquad x = f^{-1}(y) = \ln y \qquad e^{\ln y} = y = \ln (e^y) \tag{1.23}$$

But since $df/dx = y$, by Eq. 1.5 $df^{-1}/dy = 1/y = d(\ln y)/dy$. Therefore an alternative form is $\ln y = I(1/y)$, and we may set the constant of integration by noting that $\exp (0) = 1$ implies $\ln (1) = 0$, or

$$\ln y = \int_1^y \frac{dt}{t} \tag{1.24}$$

Memorize Figure 1.5. Further properties include

$$\ln (e) = 1$$
$$\ln (xy) = \ln x + \ln y$$
$$\ln (x^y) = y \ln x$$

43. The Maclaurin series for $\ln (1 + x)$ may be obtained by replacing y in Eq. 1.24 by $(1 + x)$, substituting s for $t - 1$, expanding the integrand, and integrating term-by-term:

$$\ln (1 + x) = \int_1^{1+x} \frac{dt}{t} = \int_0^x \frac{ds}{1 + s} = \int_0^x \sum_{n=0}^{\infty} (-1)^n s^n \, ds = \sum_{n=0}^{\infty} (-1)^n \frac{x^{n+1}}{n + 1}$$

Exercises

52. Sometimes it is necessary to find $\lim_{x \to a} f(x)/g(x)$ when $f(a) = 0 = g(a)$. Assume that f and g may be expanded in Taylor series about $x = a$, and show that this limit equals $f'(a)/g'(a)$ [this is l'Hospital's rule and may be extended: (*1*) If $f'(a) = 0 = g'(a)$, etc., the process may be repeated to ratios of higher derivatives. (*2*) If $a = \pm\infty$, the substitution $y = 1/x$ shows that the conclusion is still valid. (*3*) If $|f(a)| = \infty = |g(a)|$, the substitutions $F(x) = 1/f(x)$, $G(x) = 1/g(x)$ show that the conclusion is still valid.]

53. (a) Use the previous result to find $\lim_{x \to 0} \dfrac{x}{e^x - 1}$.

*(b) Find $\lim_{x \to 0} x^x$. (*Hint:* Take the logarithm.)

*(c) Find $\lim_{T\to 0} \dfrac{T^{-4}}{1 - \exp(\Theta/T)}$.

54. Use the expansion of $(1 - x)^{-1}$ to simplify the partition function for a harmonic oscillator

$$\sum_{n=0}^{\infty} e^{-(n+\frac{1}{2})h\nu/kT} = e^{-\frac{1}{2}h\nu/kT} \sum_{n=0}^{\infty} (e^{-h\nu/kT})^n$$

*55. What is the series representation for $(1 + x)^n$, where n is a positive integer? Use $\sum$ and factorial notation.

56. Find the first three terms of the Taylor's series expansion of $f(t) = -t + x \ln t$ about its maximum. (x is a positive constant.)

*57. According to the Debye model, the energy of a crystal containing N atoms is given by

$$E(T) = E_0 + \frac{9Nk}{\Theta^3} T^4 \int_0^{\Theta/T} \frac{x^3\,dx}{e^x - 1}.$$

The heat capacity of the crystal is given by $C_V = dE/dT$. Show that the Maclaurin series of C_V begins

$$C_V(T) = \left(\frac{36Nk}{\Theta^3} \int_0^{\infty} \frac{x^3\,dx}{e^x - 1}\right) T^3.$$

(*Hint:* Use the results of Exercise 1.37 and 1.53c.) We shall evaluate the definite integral later. What is important is the T^3-dependence of C_V at low temperatures.

58. Multiply the series expansion of $1/(1 - x)$ by x^N to find the series expansion of $x^N/(1 - x)$. Then subtract these equations to find the sum of the finite geometric progression $\sum_{n=0}^{N-1} x^n$.

*59. Find by direct evaluation the Maclaurin series expansion for $(1 + x)^{-\frac{1}{2}}$. Write the coefficient of x^n as a product of n ratios, then use $\sum$ notation and the result of Exercise 1.46 to express this series in compact form. Next find the series expansion of $(1 - x)^{-\frac{1}{2}}$.

*60. Use the results of Exercises 1.11a and 1.59 to find a series expansion for $\sin^{-1} x$.

61. What are the series representations for

(a) $\displaystyle\int_0^x e^{-t^2}\,dt.$

(b) $\displaystyle f(x) = \int_0^1 \frac{1 - e^{-tx}}{t}\,dt.$

Notice that the integrals cannot be solved in terms of "ordinary" functions, but the series expansions provide a means for evaluation.

*62. If $b^2 \gg ac$, one root of the quadratic $ax^2 + bx + c = 0$ is nearly zero. Find a two-term approximation to this root by each of the following methods:

(a) Neglect ax^2 and solve for x_0, a one-term approximation; then substitute x_0 into the ax^2 term and solve. (Cf. Exercise 1.32)

(b) Substitute $x = x_0(1 + \delta)$ and find the first term in the expansion of δ as a series in ac/b^2.

(c) Expand the square root in the quadratic formula as a power series in ac/b^2.

*63. The Bernoulli numbers, $\{B_n\}$, are defined by

$$\frac{x}{e^x - 1} = \sum_{n=0}^{\infty} \frac{B_n}{n!} x^n.$$

Expand the denominator as a truncated Maclaurin series and use polynomial long division to find the first five Bernoulli numbers. These numbers arise as coefficients in the series expansions of many other functions.

64. Define the displacement operator $e^{hD} = \sum_{n=0}^{\infty} (h^n/n!)D^n$. What is $e^{hD}f(x)$?

1.7 Complex Exponential and Related Functions

The exponential function may be extended to complex argument: $e^z \equiv \sum_{n=0}^{\infty} z^n/n!$, and all the above rules for exponentials still hold. If z is separated into real and imaginary parts, $e^z = e^{x+iy} = e^x e^{iy}$, so e^z may be separated into a real, positive factor and the exponential of a pure imaginary.

Let us evaluate the square of the absolute value of this second factor:

$$|e^{iy}|^2 = e^{iy}(e^{iy})^* = e^{iy}e^{-iy} = e^0 = 1.$$

Therefore e^{iy} lies on the unit circle in the complex plane of Figure 1.6.

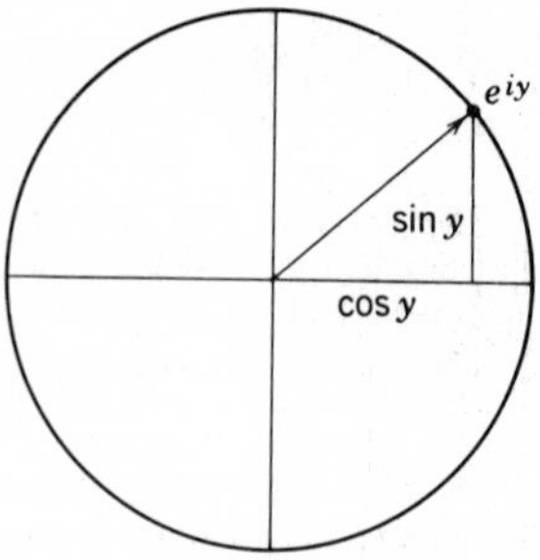

Figure 1.6

Define the TRIGONOMETRIC FUNCTIONS, SINE and COSINE, by

$$\cos y \equiv \frac{e^{iy} + e^{-iy}}{2}, \tag{1.25}$$

$$\sin y \equiv \frac{e^{iy} - e^{-iy}}{2i}. \tag{1.26}$$

Since y is real, it is clear that $\cos y = \mathrm{Re}(e^{iy})$ and $\sin y = \mathrm{Im}(e^{iy})$, as indicated in Figure 1.6. Thus these definitions are consistent with the usual ones in terms of ratios of sides of right triangles. Memorize Figure 1.7.

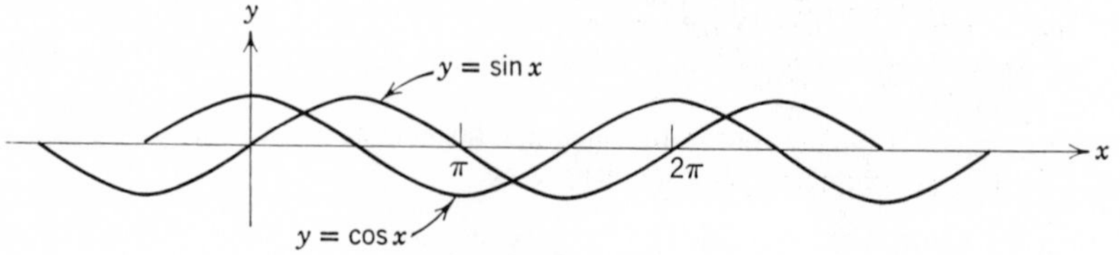

Figure 1.7

Reconstituting e^{iy} from its real and imaginary parts leads to Euler's formula

$$e^{iy} = \cos y + i \sin y. \tag{1.27}$$

Some special values of the complex exponential are $e^{\pi i/2} = i$, $e^{\pi i} = -1$, and $e^{2\pi i} = 1$. Since $e^{\theta+2\pi i} = e^{\theta}e^{2\pi i} = e^{\theta}$, we may conclude that exp is a periodic function of period $2\pi i$. This statement should serve as a warning about the uniqueness of the inverse function, logarithm, when complex numbers are involved.

The complex exponential provides a convenient notation for the n nth roots of unity:

$$1^{1/n} = e^{2\pi ik/n} \qquad k = 0, 1, 2, \ldots, (n-1), \tag{1.28}$$

since $(e^{2\pi ik/n})^n = e^{2\pi ik} = 1$. These numbers are the n roots of the equation $z^n - 1 = 0$, and by a theorem of algebra, the sum of the roots equals minus the coefficient of z^{n-1}, which is zero. Therefore, as in Exercise A4.4,

$$\sum_{k=0}^{n-1} e^{2\pi ik/n} = 0. \tag{1.29}$$

We may obtain the series expansions of sine and cosine from the series expansion of exponential:

$$e^{iy} = \sum_{n=0}^{\infty} \frac{(iy)^n}{n!} = \sum_{n=0}^{\infty} \frac{i^n y^n}{n!},$$

$$e^{-iy} = \sum_{n=0}^{\infty} \frac{(-iy)^n}{n!} = \sum_{n=0}^{\infty} (-1)^n \frac{i^n y^n}{n!},$$

$$\cos y = \sum_{\substack{n=0 \\ n=2m}}^{\infty} \frac{i^n y^n}{n!} = \sum_{m=0}^{\infty} \frac{i^{2m} y^{2m}}{(2m)!} = \sum_{m=0}^{\infty} (-1)^m \frac{y^{2m}}{(2m)!}, \tag{1.30}$$

$$\sin y = \frac{1}{i} \sum_{\substack{n=0 \\ n=2m+1}}^{\infty} \frac{i^n y^n}{n!} = \frac{1}{i} \sum_{m=0}^{\infty} \frac{i^{2m+1} y^{2m+1}}{(2m+1)!} = \sum_{m=0}^{\infty} (-1)^m \frac{y^{2m+1}}{(2m+1)!}. \tag{1.31}$$

Here we have used the facts that on addition of e^{iy} and $(e^{iy})^*$ only even powers of i remain and that on subtraction of $(e^{iy})^*$ from e^{iy} only odd powers of i remain. From Eqs. 1.25 and 1.26 or from Eqs. 1.30 and 1.31, it follows that $D \sin = \cos$ and $D \cos = -\sin$.

The further trigonometric functions—$\tan = \sin/\cos$, $\cot = 1/\tan$, $\sec = 1/\cos$, and $\csc = 1/\sin$—are defined in terms of these, along with the six inverse trigonometric functions. It is also possible to extend any of these functions to complex argument, and all the familiar theorems on trigonometric functions still hold.

Exercises

*65. (a) What is a^{x+iy} where a is real and positive?
(b) What is i^i?

66. Find cos 5θ and sin 5θ in terms of products of powers of sin θ and cos θ. [*Hint:* $e^{5i\theta} = \cos 5\theta + i \sin 5\theta = (\cos\theta + i\sin\theta)^5$.]

67. (a) Consider the geometric series $\sum_{j=1}^{n} e^{ij\theta}$, in which each term may be obtained from the previous one by multiplying by $e^{i\theta}$. Use the result of Exercise 1.58 to sum the series and then evaluate $\sum_{j=1}^{n} \sin j\theta$ and $\sum_{j=1}^{n} \cos j\theta$. (*Hint:* After you sum the series, convert the denominator to sin $(\theta/2)$ before proceeding.)
*(b) Hückel molecular orbital (HMO) calculations on cyclic C_{2+4n} hydrocarbons give orbital energies $\varepsilon_j = \alpha + \beta\cos[j\pi/(2n+1)]$. Since $\beta < 0$, the lowest-energy orbitals, which are doubly occupied in the ground state, correspond to $j = 0$, $\pm 1, \pm 2, \ldots, \pm n$. Find the total π energy of the hydrocarbon: $E_\pi = 2\sum_{j=-n}^{n} \varepsilon_j$.

68. (a) What is the period of cos 2θ? Of $\sin^2\theta$?
(b) What is sin $n\pi$? Cos $n\pi$? $\mathrm{Sin}^{-1}\,0$? $\mathrm{Sin}^{-1}\,1$?

69. Show that for all integral m and n, $\int_{-\pi}^{\pi} e^{-imx} e^{inx}\,dx$ equals zero if $m \neq n$ and equals 2π if $m = n$. Memorize this result. (*Note:* Two functions, f and g, are said to be ORTHOGONAL on the interval from a to b if $\int_a^b f*(x)g(x)\,dx = 0$.)

*70. Use the previous result to help show that for all integral m and n

$$\int_{-\pi}^{\pi} \sin mx \cos nx\,dx = 0; \qquad \int_{-\pi}^{\pi} dx = 2\pi.$$

$$\int_{-\pi}^{\pi} \sin mx \sin nx\,dx = 0, \qquad n \neq m; \qquad \int_{-\pi}^{\pi} \sin^2 nx = \pi \quad (n \neq 0).$$

$$\int_{-\pi}^{\pi} \cos mx \cos nx\,dx = 0, \qquad n \neq m; \qquad \int_{-\pi}^{\pi} \cos^2 nx = \pi \quad (n \neq 0).$$

71. Define an operator F:

$$F\{f(x)\} = g(y) = \frac{1}{\sqrt{2\pi}} \int_{-\infty}^{\infty} f(x) e^{-ixy} dx.$$

(a) Convince yourself that F is a linear operator (known as the FOURIER TRANSFORM).
(b) If $f(\pm\infty) = 0$, relate $F\{f'(x)\}$ to $F\{f(x)\}$.
(c) Find the Fourier transform of $f(x) = e^{-a|x|}$, $a > 0$. (Be careful.)
(d) Find the Fourier transform of the function $f(x) = 1$ for $|x| < \pi$, and $f(x) = 0$ for $|x| > \pi$.

*72. If f is real and even, show that $g(y) = F\{f(x)\} = \sqrt{2/\pi}\int_0^\infty f(x)\cos xy\,dx$. What can you conclude if f is real and odd?

73. The CONVOLUTION of two functions, f_1 and f_2, is a function $f_1 * f_2$, defined by

$$f_1 * f_2(x) = \frac{1}{\sqrt{2\pi}} \int_{-\infty}^{\infty} f_1(t) f_2(x - t)\,dt$$

(a) Convince yourself that $(c_1 f_1 + c_2 f_2) * f_3 = c_1 f_1 * f_3 + c_2 f_2 * f_3$.

(b) Show that $f_1 * f_2 = f_2 * f_1$.

*(c) Show that $(f_1 * f_2) * f_3 = f_1 * (f_2 * f_3)$.

(d) Relate $F\{f_1 * f_2\}$ to $g_1 = Ff_1$ and $g_2 = Ff_2$. (*Hint:* Eliminate x by substituting $u = x - t$.)

*74. A set of functions $\{f_n\}$ is often defined in terms of a GENERATING FUNCTION $G(x, t)$ by the relation $G(x, t) = \sum_{n=0}^{\infty} f_n(x)t^n$. Different generating functions lead to different sets $\{f_n\}$. What set arises from the generating function $G(x, t) = \sin x/(1 - 2t \cos x + t^2)$? (*Hint:* Expand in partial fractions, then expand each partial fraction as a power series in t.)

75. Use the result of Exercise 1.52 to find

(a) $\lim_{x \to 0} \dfrac{\sin nx}{\sin x}$.

(b) $\lim_{x \to 0} \dfrac{1 - \cos x + x \sin x}{x^2}$.

76. Find the series representation of $\int_0^x (\sin t\, dt)/t$.

*77. Convince yourself that for a convergent infinite series whose terms alternate in sign, the error introduced by terminating the series after N terms is less than the absolute value of the $(N + 1)$st term. Then use this result to compute $\sin(\frac{1}{2})$ to seven decimal places. How many terms are required to calculate ln 2 to three decimal places from the series expansion of $\ln(1 + x)$?

The HYPERBOLIC FUNCTIONS may be defined and expanded in power series as

$$\cosh x \equiv \frac{e^x + e^{-x}}{2} = \sum_{n=0}^{\infty} \frac{x^{2n}}{(2n)!}, \tag{1.32}$$

$$\sinh x \equiv \frac{e^x - e^{-x}}{2} = \sum_{n=0}^{\infty} \frac{x^{2n+1}}{(2n + 1)!}, \tag{1.33}$$

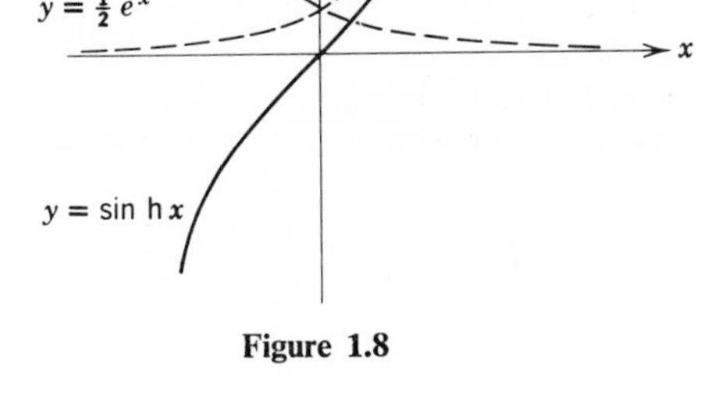

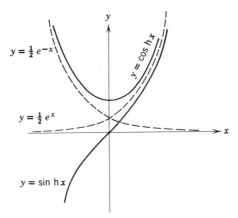

Figure 1.8

and it is clear that $D \sinh = \cosh$ and $D \cosh = \sinh$. These functions are graphed in Figure 1.8. The further functions—tanh, coth, sech, csch—and the inverse functions are defined just as the trigonometric functions.

Comparison of Eqs. 1.32 and 1.33 with Eqs. 1.25 and 1.26 shows that

$$\cosh (ix) = \cos x,$$
$$\sinh (ix) = i \sin x.$$

Substitution of $-iy$ for x leads to

$$\cosh y = \cos (-iy) = \cos (iy),$$
$$i \sinh y = -\sin (-iy) = \sin (iy).$$

These four equations provide simplifying formulas for dealing with trigonometric and hyperbolic functions of imaginary argument.

Exercises

78. Show that $\cos nx$ and $\cosh x$ are even functions, and that $\sin nx$ and $\sinh x$ are odd functions. What is e^x?

*79. Show that for all x:
   (a) $\sin^2 x + \cos^2 x = 1$.
   (b) $\cosh^2 x - \sinh^2 x = 1$.
   (c) $\sin 2x = 2 \sin x \cos x$.
   (d) $\cos 2x = \cos^2 x - \sin^2 x$.
   (e) What are the corresponding formulas for $\sinh 2x$ and $\cosh 2x$?

*80. Solve $\cos \theta = i$. (*Hint:* Obviously θ cannot be real, so let $\theta = x + iy$.)

*81. (a) Differentiate both sides of $x = \sinh y$ with respect to x.
   (b) Also solve $x = \sinh y$ for e^y by the quadratic formula; since $e^y > 0$, only the positive square root is meaningful.
   (c) Substitute and simplify to find $D \sinh^{-1} x$. (Cf. Exercise 1.42.)

82. From $L[e^{at}] = 1/(s - a)$ and the fact that L is a linear operator, find
   (a) $L[\cos \omega t]$.
   (b) $L[\sin \omega t]$.
   (c) $L[\cosh at]$.
   (d) $L[\sinh at]$.
   Simplify your answers.

*83. Starting from $\tan^{-1} x = \int_0^x dt/(1 + t^2)$ and the relationship between trigonometric and hyperbolic functions of imaginary argument, show that $\tanh^{-1} x = \int_0^x dt/(1 - t^2)$ and show that this may be expanded as $\sum_{n=0}^{\infty} x^{2n+1}/(2n + 1)$. Then show that $\frac{1}{2} \ln [(1 + x)/(1 - x)]$ has the same series expansion.

*84. Show that $\mathrm{Re}\,[e^{-3i\theta} \sin (e^{i\theta})] = \cos 3\theta \sin (\cos \theta) \cosh (\sin \theta) + \sin 3\theta \cos (\cos \theta) \sinh (\sin \theta)$. Can you integrate this horror from 0 to 2π? In Appendix 5 we show how to evaluate many such definite integrals.

85. Express the partition function of Exercise 1.54 in terms of $\mathrm{csch}\,(h\nu/2kT)$.

2

FUNCTIONS OF SEVERAL VARIABLES

This chapter extends our discussion of functions to more than one variable. After describing such functions, we proceed to consider various aspects of partial differentiation—the equality of the mixed second partial derivatives, the necessity of subscript notation, and chain rules. Then we introduce several new concepts: We show how Lagrange's method of undetermined multipliers may be applied to extremum problems with constraints. We introduce the notion of homogeneous functions, and Euler's theorem concerning them, and we indicate their relevance to thermodynamics. Indeed, although the material of this chapter is basic to all of physical science, it is particularly important for thermodynamics. Therefore we provide a short discussion to help bridge the gap between familiar mathematics and familiar thermodynamics. Then we return to describe multiple integrals and show how they are evaluated. Also included are Leibniz' Rule (Part II) and transformation of variables. Polar coordinates are introduced. Exercises demonstrate applications to quantum mechanics, thermodynamics, spectroscopy, and statistical mechanics.

2.1 Introduction

There is no difficulty in extending the notion of a function to several independent variables. The domain of such a function is a set of ordered pairs, triplets, etc. Alternatively, a function of N variables takes an element of set one, an element of set two, . . . , and an element of set N, and unambiguously produces an element of set $(N + 1)$. If x, y, . . . represent independent variables that range over the elements of set one, set two, . . . , then the value of the function F is $F(x, y, \ldots)$; for example, the value of the function "twice square of first, minus thrice second" at the particular argument (2, 1) is 5 and at the arbitrary argument (x, y) is $2x^2 - 3y$.

There are several ways of illustrating a function of two variables: (*1*) Graph the surface $z = F(x, y)$. (*2*) Indicate the value $F(x, y)$ at certain points spaced at regular intervals. (*3*) Draw CONTOUR LINES (LEVEL LINES, isowhatevers), along each of which the value $F(x, y)$ is constant. (*4*) Indicate the magnitude of $F(x, y)$ by the density of dots or shading in the neighborhood of (x, y). (See Fig. 2.1.)

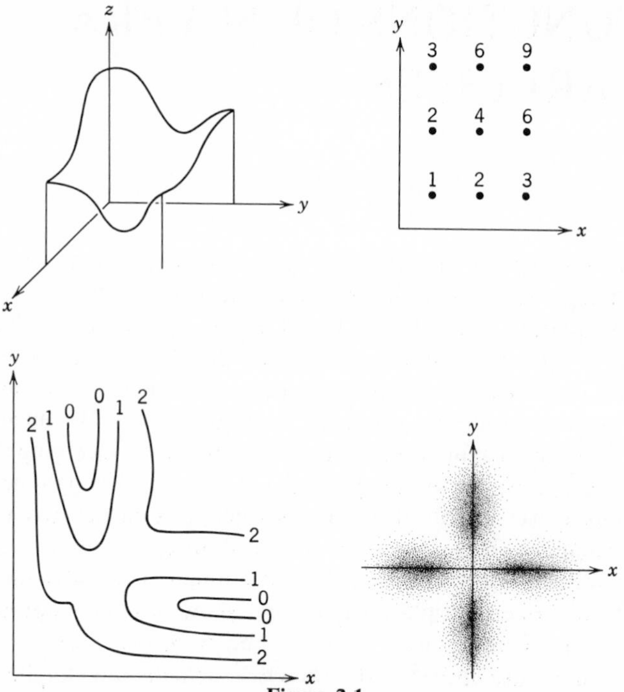

Figure 2.1

For three variables any of these methods may be used, except the first. (*2*) At regularly spaced points in three-dimensional space affix tabs indicating the value $F(x, y, z)$; for example, suspend thermometers throughout a room to illustrate the function "temperature." (*3*) Draw contour surfaces satisfying the equation $F(x, y, z) = c$, for regularly spaced constants, c. (*4*) You should be familiar with the illustration of the probability density, $\psi^2(x, y, z)$, of an electron in the various hydrogenic orbitals. However, the simplest and most familiar illustration of an orbital is as in (*3*), but with only one contour of constant ψ^2; the constant value of ψ^2 is chosen so that the probability of finding an electron inside that contour is, say, 0.90.

Exercises

1. Sketch $F(x, y) = (x^2 - y^2)/(x^2 + y^2)$ and $V(T, P) = RT/P$ in whatever art form you find congenial.
*2. Sketch $z = F(x, y) = \tan^{-1}(y/x)$: $-\pi/2 < z < \pi/2$.
*3. Sketch $\psi = \exp[-(x^2 + y^2 + z^2)^{1/2}]$ and $\psi = z \exp[-(x^2 + y^2 + z^2)^{1/2}]$.

The notion of a function of more than one variable enables us to extend our notation of functions of a single variable. Consider the equation $F(x, y) = 0$. If a value for x is chosen, only a specific value of y will satisfy the equation. Therefore this equation specifies y as the value of some IMPLICIT FUNCTION of x even though it may not be possible to rewrite the equation in the form $y = f(x)$. For example, in the quantum-mechanical problem of a particle in a potential well, there arises the equation

$$\varepsilon^{1/2} \cot \varepsilon^{1/2} - (v - \varepsilon)^{1/2} \tanh (v - \varepsilon)^{1/2} = 0,$$

which is not soluble, except numerically, for ε as a function of v.

2.2 Partial Differentiation

Define the PARTIAL DERIVATIVE of the function of two variables, F, with respect to its first variable, at the point (x_0, y_0) by

$$\left.\frac{\partial F}{\partial x}\right|_{x_0, y_0} \equiv \lim_{x \to x_0} \frac{F(x, y_0) - F(x_0, y_0)}{x - x_0}. \tag{2.1}$$

We may define $\partial F/\partial y|_{x_0, y_0}$ and partial derivatives of functions of more variables analogously. By comparison of this definition with the definition of the derivative of a function of a single variable (Eq.1.3), it is clear that partial differentiation is just ordinary differentiation with the other variable(s) held constant.

Since we can calculate these partial derivatives at each argument, we have constructed two new functions $\partial F/\partial x$ and $\partial F/\partial y$ (sometimes written F_x and F_y, or F_1 and F_2, or, if $z = F(x, y)$, $\partial z/\partial x$ and $\partial z/\partial y$). Furthermore, we may define the PARTIAL DIFFERENTIATION OPERATORS, D_x and D_y, which act on F to produce $\partial F/\partial x$ and $\partial F/\partial y$, respectively. And we may apply these operators in succession, to obtain higher partial derivatives, $D_x^2 F \equiv D_x(D_x F)$, $D_x D_y F \equiv D_x(D_y F)$, etc.

There is an important theorem which guarantees that the order of application of the partial differentiation operators is immaterial, so that the mixed partials are equal:

$$D_x D_y F \equiv \frac{\partial^2 F}{\partial x\, \partial y} = \frac{\partial^2 F}{\partial y\, \partial x} \equiv D_y D_x F. \tag{2.2}$$

Thus we see that the operators D_x and D_y commute.

Exercises

4. (a) Convince yourself that the operator X ("multiply by x") commutes with D_y and D_z but not with D_x, etc.
   (b) Let $L_x = YD_z - ZD_y$, $L_y = ZD_x - XD_z$, $L_z = XD_y - YD_x$. Use the result of Exercise 1.15c to help find the commutator $[L_x, L_y]$. (*Note:* These are closely related to angular momentum operators in quantum mechanics.)
   *(c) Find $[L_y, L_z]$ and $[L_z, L_x]$ or guess them by symmetry.
5. To describe the spin state of an N-electron system, a function of N variables must be specified. Such a function may always be expressed as a linear combination of products of one-electron spin functions. Again, the independent variables are not mentioned except to indicate which electrons have α spin and which have β. An example of such a function, with $N = 3$, is $\alpha(1)\beta(2)\alpha(3) - \alpha(1)\alpha(2)\beta(3)$, which is often abbreviated to $\alpha\beta\alpha - \alpha\alpha\beta$. The many-electron spin operators are defined as sums of one-electron spin operators, s_{xi}, etc., where s_{xi} acts only on the spin function of the ith electron:

$$S_x = \sum s_{xi} \qquad S_y = \sum s_{yi} \qquad S_z = \sum s_{zi}$$
$$S_\pm = S_x \pm iS_y = \sum s_{\pm i} \qquad S^2 = S_x{}^2 + S_y{}^2 + S_z{}^2 \neq \sum s_i{}^2.$$

   (a) Find $[S_x, S_y]$. (*Note:* One-electron spin operators for one electron commute with all one-electron spin operators of all other electrons.)
   *(b) Show that the other results of Exercises 1.17c–h are also extendable to the many-electron spin operators.
   (c) Find $S_z\alpha\alpha$, $S_z\alpha\beta$, $S_z\beta\alpha$, and $S_z\beta\beta$.
   (d) Find $S_+\alpha\alpha$, $S_+\alpha\beta$, $S_+\beta\alpha$, and $S_+\beta\beta$. Find $S_-\alpha\alpha$, $S_-\alpha\beta$, $S_-\beta\alpha$, $S_-\beta\beta$.
   (e) Use the extension of either result of Exercise 1.17f to find $S^2\alpha\alpha$, $S^2\alpha\beta$, $S^2\beta\alpha$, and $S^2\beta\beta$.
   (f) Find $S^2\alpha\alpha\alpha$, $S^2\alpha\alpha\beta$, $S^2\alpha\beta\beta$, and $S^2\beta\beta\beta$.
6. Find

$$\frac{\partial}{\partial\phi}\left[\left(\frac{n}{x^2} + \frac{\cos\phi}{x}\right)\sin\,(x\sin\phi - n\phi)\right].$$

*7. Express $\partial^2F/\partial x\,\partial y$ and $\partial^2F/\partial y\,\partial x$ as double limits. Convince yourself that if $z = F(x, y)$ is a "reasonable" surface, then these mixed partial derivatives are equal.

The geometrical significance of the partial derivative $\partial F/\partial x|_{x_0,y_0}$ is that it equals the slope of the tangent to the curve formed by the intersection of the surface $z = F(x, y)$ and the vertical plane $y = y_0$ (See Fig. 2.2.). A necessary

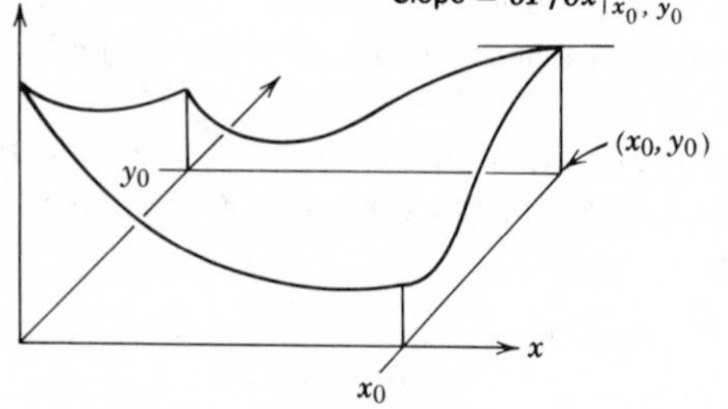

Figure 2.2

condition for $F(x_0, y_0)$ to be an extremum is that $\partial F/\partial x|_{x_0,y_0}$ and $\partial F/\partial y|_{x_0,y_0}$ both be zero; there are further conditions on the second derivatives (Exercise 2.8).

The TOTAL DIFFERENTIAL of the function F represents the "infinitesimally small" change in the value of F when x and y are changed by the "infinitesimally small" amounts, dx and dy, respectively:

$$dF = \frac{\partial F}{\partial x}\,dx + \frac{\partial F}{\partial y}\,dy. \tag{2.3}$$

Taylor's series in two variables is most conveniently expressed as

$$F(x_0 + h, y_0 + k) = \sum_{n=0}^{\infty} \frac{1}{n!}\,[hD_x + kD_y]^n F(x_0, y_0). \tag{2.4}$$

Exercises

8. (a) If $z = x^2 - y^2$, find $\partial z/\partial x$, $\partial z/\partial y$, $\partial^2 z/\partial x^2$, $\partial^2 z/\partial y^2$, and $\partial^2 z/\partial x\,\partial y$.
   (b) Do the same for $z = xy$.
   (c) Recall that the further condition that guarantees $F(x, y)$ to be maximum (minimum) at a point where $D_xF = 0 = D_yF$ is that the surface $z = F(x, y)$ be concave downward (upward) there. To be concave downward, it is sufficient that $D_x{}^2F < 0$, $D_y{}^2F < 0$, and $(D_x{}^2F)(D_y{}^2F) > (D_xD_yF)^2$. To be concave upward, it is sufficient that $D_x{}^2F > 0$, $D_y{}^2F > 0$, and $(D_x{}^2F)(D_y{}^2F) > (D_xD_yF)^2$.
   (d) Use these conditions to convince yourself that both of these surfaces have a "saddle point," rather than an extremum, at (0, 0).

*9. (a). Write out the first six terms of the Taylor series expansion of a function of two variables.
   (b) The potential energy V of an N-atomic molecule is a function of the $3N$ coordinates $\{x_i, y_i, z_i\}_{i=1}^{N} = \{X_j\}_{j=1}^{3N}$. Convince yourself that the Taylor series expansion of $V(\{X_j\})$ about its minimum at $\{X_j{}^\circ\}$ begins

$$V(\{X_j\}) = V(\{X_j{}^\circ\}) + \frac{1}{2}\sum_{i=1}^{3N}\sum_{j=1}^{3N} \left.\frac{\partial^2 V}{\partial X_i\,\partial X_j}\right|_{\{X_j{}^\circ\}} (X_i - X_i{}^\circ)(X_j - X_j{}^\circ) + \cdots.$$

   Why are there no first-order terms? What are the third-order terms? We shall return to this expansion at the end of Chapter 8 when we consider molecular vibrations. For small vibrations this expansion may be truncated before the third-order terms.

*10. Generalize the Newton-Raphson formula to find an iterative method of solving the simultaneous equations $F(x, y) = 0$ and $G(x, y) = 0$, neither of which can be solved for one variable in terms of the other.

A further element of notation is to indicate by subscripts the variables which are held constant in the partial differentiation. Thus $\partial F/\partial x$ is usually written as $(\partial F/\partial x)_y$. The necessity of this notation is demonstrated in the following three examples:

1. If $v = F(x, y, z)$ and $z = f(x, y)$, then $v = F[x, y, f(x, y)] = G(x, y)$. Since F and G are different functions, there is no ambiguity in denoting their partial derivatives by $\partial F/\partial x$

and $\partial G/\partial x$, which are, of course, also different. But if we dispense with the symbols F and G and focus only on their common value v, then $\partial v/\partial x$ is ambiguous. To avoid the confusion that results from using the same symbol for two functions and their common value, it is necessary to distinguish the partial derivatives by subscripts, $\partial F/\partial x$ by $(\partial v/\partial x)_{y,z}$ and $\partial G/\partial x$ by $(\partial v/\partial x)_y$.

2. The Sackur-Tetrode equation for the translational entropy of a perfect gas as a function of temperature and pressure is

$$S = nC_P \ln T - nR \ln P + \text{const.} \tag{2.5}$$

Also, for a perfect gas the equation of state is $PV = nRT$. Solving for P and substituting gives

$$\begin{aligned} S &= nC_P \ln T - nR \ln T + nR \ln V - nR \ln nR + \text{const.} \\ &= n(C_P - R) \ln T + nR \ln V + \text{new const.} \end{aligned} \tag{2.6}$$

expressing S as a function of T and V. Notice that Eqs. 2.5 and 2.6 are of the form $S = F(T, P)$ and $S = G(T, V)$. However, there are no symbols introduced to represent the functions, since the common value S is the quantity of interest. Therefore it is essential to use subscripts. From Eq. 2.5, $(\partial S/\partial T)_P = nC_P/T$, and from Eq. 2.6, $(\partial S/\partial T)_V = n(C_P - R)/T$, a different quantity because it is the partial derivative of a different *function*.

3. POLAR COORDINATES are defined by

$$x = r \cos \phi \qquad y = r \sin \phi \qquad (0 \leqslant \phi \leqslant 2\pi). \tag{2.7}$$

Solving for r gives $r = x \sec \phi$ or $r = (x^2 + y^2)^{1/2}$. But what is $\partial r/\partial x$? From the former equation, $(\partial r/\partial x)_\phi = \sec \phi$. From the last equation, $(\partial r/\partial x)_y = x(x^2 + y^2)^{-1/2} = \cos \phi$. That these partial derivatives are different should also be clear from Figure 2.3.

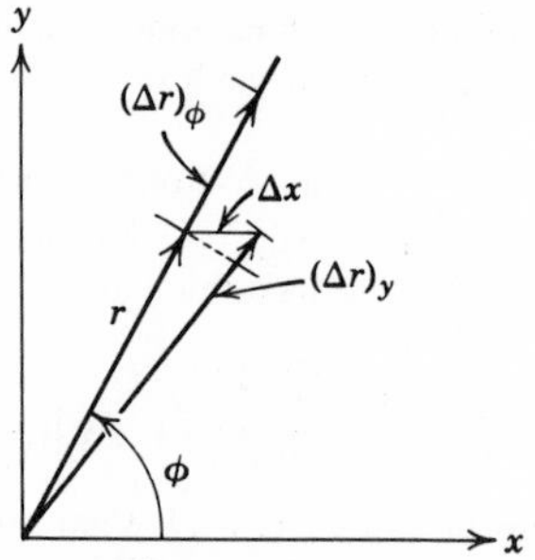

Figure 2.3

Exercises

11. Find $(\partial V/\partial P)_T$ and $(\partial V/\partial T)_P$ as functions of P and V for a gas satisfying each of the following equations of state:

(a) $PV = RT$.

(b) $P(V - b) = RT$.

(c) $\left(P + \dfrac{a}{V^2}\right)(V - b) = RT$.

Notice that it is not easy to solve for $V = V(T, P)$ in the last case, but that the partial derivatives may (in all cases) be evaluated by partial differentiation of both sides of the equation.

*12. At the critical point of a nonideal gas, $(\partial P/\partial V)_T = 0 = (\partial^2 P/\partial V^2)_T$. Show that at the critical point of a van der Waals gas, whose equation of state is that in Exercise 2.11c, T, V, and P are given by

$$T = 8a/27bR \qquad V = 3b \qquad P = a/27b^2.$$

With several variables, there are three chain rules for derivatives of composite functions:

1. If $F(x, y) = f[g(x, y)]$, then

$$\left(\frac{\partial F}{\partial x}\right)_y = f'[g(x, y)]\left(\frac{\partial g}{\partial x}\right)_y. \tag{2.8}$$

This is a trivial rule since if y is held constant, both F and f become functions of only one variable and our previous chain rule (Eq.1.4) is still valid.

2. If $x = f(t)$, $y = g(t)$, and $z = h(t)$, then $F(x, y, z) = \phi(t)$ and

$$\frac{d\phi}{dt} = \left(\frac{\partial F}{\partial x}\right)_{y,z}\frac{df}{dt} + \left(\frac{\partial F}{\partial y}\right)_{x,z}\frac{dg}{dt} + \left(\frac{\partial F}{\partial z}\right)_{x,y}\frac{dh}{dt}. \tag{2.9}$$

For mnemonic purposes, let u be the common value and divide the total differential du by dt:

$$\frac{du}{dt} = \left(\frac{\partial u}{\partial x}\right)_{y,z}\frac{dx}{dt} + \left(\frac{\partial u}{\partial y}\right)_{x,z}\frac{dy}{dt} + \left(\frac{\partial u}{\partial z}\right)_{x,y}\frac{dz}{dt}.$$

3. If $x = f(u, v)$ and $y = g(u, v)$, then $F(x, y) = G(u, v)$ and

$$\left(\frac{\partial G}{\partial u}\right)_v = \left(\frac{\partial F}{\partial x}\right)_y\left(\frac{\partial f}{\partial u}\right)_v + \left(\frac{\partial F}{\partial y}\right)_x\left(\frac{\partial g}{\partial u}\right)_v \tag{2.10}$$

and similarly for $(\partial G/\partial v)_u$. For mnemonic purposes, let z be the common value and divide the total differential dz by ∂u (or ∂v):

$$\left(\frac{\partial z}{\partial u}\right)_v = \left(\frac{\partial z}{\partial x}\right)_y\left(\frac{\partial x}{\partial u}\right)_v + \left(\frac{\partial z}{\partial y}\right)_x\left(\frac{\partial y}{\partial u}\right)_v.$$

Exercises

13. If $z = F(x, y)$, then there is a type of inverse function G such that $x = G(y, z)$. Convince yourself that $(\partial F/\partial x)_y = [(\partial G/\partial z)_y]^{-1}$, or in mnemonic form

$$\left(\frac{\partial z}{\partial x}\right)_y \cdot \left(\frac{\partial x}{\partial z}\right)_y = 1.$$

*14. It can be shown that if a substance is stable to phase separation, then its entropy S must be maximum in the sense that the surface $S = S(E, V)$ is concave downward.
(a) If $(\partial S/\partial E)_V = 1/T$ and $(\partial S/\partial V)_E = P/T$, use the conditions of Exercise 2.8c to derive three inequalities.

(b) Then use the result of Exercise 2.13 to conclude that for any phase

$$C_V \equiv \left(\frac{\partial E}{\partial T}\right)_V > 0,$$

$$\frac{T\left(\dfrac{\partial P}{\partial V}\right)_E - P\left(\dfrac{\partial T}{\partial V}\right)_E}{\left(\dfrac{\partial E}{\partial T}\right)_V} + \left(\frac{\partial T}{\partial V}\right)_E^2 < 0.$$

(*Hints:* By definition $T > 0$. Recall the rules for working with inequalities: If $a > b$, then $a + c > b + c$. If $a > b$ and $c > 0$, then $ac > bc$. If $a > b$ and $c < 0$, then $ac < bc$.)

15. Given $u(r, \phi) = v(x, y)$, with $x = r\cos\phi$, $y = r\sin\phi$, find $(\partial v/\partial x)_y$ and $(\partial v/\partial y)_x$ in terms of $(\partial u/\partial r)_\phi$ and $(\partial u/\partial \phi)_r$. Express the result in operator form, showing D_x and D_y in terms of D_r and D_ϕ, and functions of r and ϕ.

*16. Solve the previous problem "backwards": Find $(\partial u/\partial r)_\phi$ and $(\partial u/\partial \phi)_r$ in terms of $(\partial v/\partial x)_y$ and $(\partial v/\partial y)_x$, and functions of r and ϕ. Then solve the simultaneous equations for $(\partial v/\partial x)_y$ and $(\partial v/\partial y)_x$. You may find this method easier because it is simple to find $(\partial x/\partial r)_\phi$, etc., instead of $(\partial \phi/\partial x)_y$, etc.

*17. Use the previous result, in operator form, to relate

$$\frac{\partial^2 v}{\partial x^2} + \frac{\partial^2 v}{\partial y^2} = \left[\frac{\partial}{\partial x}\left(\frac{\partial v}{\partial x}\right)_y\right]_y + \left[\frac{\partial}{\partial y}\left(\frac{\partial v}{\partial y}\right)_x\right]_x$$

to partials of $u(r, \phi)$. (*Warning:* Remember that $(\partial u/\partial r)_\phi$ is still a function of ϕ, so that the product rule must be used to evaluate such terms as $(\partial/\partial\phi)[\cos\phi(\partial u/\partial r)_\phi]$.) It is clear that this method for transforming partial derivatives is quite tedious, and there does exist a quicker method which involves "metric coefficients."

*18. In the "calculus of variations" there arises the equation $(\partial F/\partial y) - (d/dx)(\partial F/\partial y') = 0$, where $y' = dy/dx$ and $F = F(x, y, y')$. For the special case $\partial F/\partial x = 0$, multiply this equation by y' and use the expression for the total derivative dF/dx to rewrite this equation in the form $(d/dx)(\text{something}) = 0$.

19. If $z = f(x, y)$, then $G(x, z) = F(x, y)$.
    (a) Use the third chain rule to express $(\partial F/\partial x)_y$ in terms of partial derivatives of G and f.
    (b) Use this result to convince yourself that

$$\left(\frac{\partial E}{\partial T}\right)_P = \left(\frac{\partial E}{\partial T}\right)_V + \left(\frac{\partial E}{\partial V}\right)_T\left(\frac{\partial V}{\partial T}\right)_P$$

    (c) Express $(\partial P/\partial V)_E$ in terms of $(\partial P/\partial V)_T$ and $(\partial P/\partial T)_V$.

20. If $z = f(x, y)$, then $F(x, y, z) = G(x, y)$. Express $(\partial G/\partial x)_y$ in terms of partial derivatives of F.

Given the equation $F(x, y) = 0$, which implicitly defines y as a function f of x, we may use the second chain rule (with $t = x$ and z absent) to find $f'(x)$:

$$\frac{dF}{dx} = \left(\frac{\partial F}{\partial x}\right)_y\frac{dx}{dx} + \left(\frac{\partial F}{\partial y}\right)_x\frac{df}{dx}$$

But no matter how x varies, y adjusts so that $F(x, y)$ remains zero. Therefore $dF/dx = 0$, and we may solve for

$$\frac{df}{dx} = -\frac{(\partial F/\partial x)_y}{(\partial F/\partial y)_x}. \tag{2.11}$$

If $z = F(x, y)$, this is often written as

$$\left(\frac{\partial y}{\partial x}\right)_z = -\frac{(\partial z/\partial x)_y}{(\partial z/\partial y)_x}. \tag{2.11'}$$

Notice that z goes into the "numerator" of both numerator and denominator of the fraction, x "remains in the denominator," and y goes into the "denominator" of the denominator. We may use the result of Exercise 2.13 to express this equation in the more symmetrical form

$$\left(\frac{\partial y}{\partial x}\right)_z \left(\frac{\partial x}{\partial z}\right)_y \left(\frac{\partial z}{\partial y}\right)_x = -1 \tag{2.12}$$

but be sure not to forget the minus sign in these equations.

Exercises

21. Use Eq. 2.11 to find the slope of the ellipse $9x^2 + 16y^2 = 1$ at the point $(1/5, -1/5)$.
22. Relate $(\partial P/\partial T)_V$ to $\alpha = (1/V)(\partial V/\partial T)_P$, the coefficient of thermal expansion, and $\beta = -(1/V)(\partial V/\partial P)_T$, the coefficient of compressibility.
23. Use Eq. 2.11′ and then Exercise 2.13 to do Exercise 2.11c.
24. (a) If $\gamma \equiv (\partial S/\partial T)_P/(\partial S/\partial T)_V$, simplify $(\partial P/\partial V)_S$.
    (b) Then, if $PV = nRT$, evaluate $(1/P)(\partial P/\partial V)_S$ as a function of V.
    (c) Find the ratio of final pressure P_1 to initial pressure P_0 on expanding n moles of an ideal gas adiabatically (at constant S) from initial volume V_0 to final volume V_1. (Assume γ is constant. *Hint:* What is $\int dP/P$?)
    (d) Then find the work done: $W = \int_{V_0}^{V_1} P(V)\, dV$.

2.3 Extremum Problems with Constraints

Previously we implied that a necessary condition for an extremum of $F(x, y, z)$ is $\partial F/\partial x = \partial F/\partial y = \partial F/\partial z = 0$. However, it may be that we are not interested in the most general extremum, but that the only extrema of interest are those which also satisfy the constraint $G(x, y, z) = 0$.

One method is to solve $G(x, y, z) = 0$ for z as a function of x and y, substitute this expression into $F(x, y, z)$ to produce $H(x, y)$, and solve $\partial H/\partial x = 0 = \partial H/\partial y$.

An alternative method (LAGRANGE'S METHOD), which preserves "symmetry" among the variables, is to form the function

$$u(x, y, z, \lambda) \equiv F(x, y, z) + \lambda G(x, y, z) \tag{2.13}$$

where λ is an as-yet UNDETERMINED MULTIPLIER (LAGRANGIAN MULTIPLIER), and solve the three equations

$$\left(\frac{\partial u}{\partial x}\right)_{y,z,\lambda} = \left(\frac{\partial u}{\partial y}\right)_{x,z,\lambda} = \left(\frac{\partial u}{\partial z}\right)_{x,y,\lambda} = 0 \tag{2.14}$$

along with the equation of constraint. We shall justify this procedure later.

Example

4. The energy of a quantum-mechanical particle in a three-dimensional rectangular box of size $a \times b \times c$ is

$$E = \frac{h^2}{8m}\left(\frac{n_x{}^2}{a^2} + \frac{n_y{}^2}{b^2} + \frac{n_z{}^2}{c^2}\right),$$

where n_x, n_y, and n_z are quantum numbers. If the sum of the quantum numbers is constrained to be a constant N, what is the minimum energy of the particle? We shall assume that N, and also n_x, n_y, and n_z, are so large that they may be considered as continuous variables, with respect to which we may differentiate. Setting $u = E + \lambda(n_x + n_y + n_z - N)$ and setting the partial derivatives with respect to n_x, n_y, and n_z equal to zero gives

$$\frac{\partial u}{\partial n_x} = \frac{h^2}{4m}\frac{n_x}{a^2} + \lambda = 0,$$

$$\frac{\partial u}{\partial n_y} = \frac{h^2}{4m}\frac{n_y}{b^2} + \lambda = 0,$$

$$\frac{\partial u}{\partial n_z} = \frac{h^2}{4m}\frac{n_z}{c^2} + \lambda = 0.$$

Solving the first equation for n_x, the second for n_y, and the third for n_z, and adding gives

$$n_x + n_y + n_z = N = -\frac{4m}{h^2}(a^2 + b^2 + c^2)\lambda$$

or

$$\lambda = -\frac{h^2}{4m}\frac{N}{a^2 + b^2 + c^2}.$$

Multiplying the first equation by n_x, the second by n_y, and the third by n_z, and adding gives

$$\frac{h^2}{4m}\left(\frac{n_x{}^2}{a^2} + \frac{n_y{}^2}{b^2} + \frac{n_z{}^2}{c^2}\right) + \lambda(n_x + n_y + n_z) = 0$$

or

$$2E + \lambda N = 0.$$

Substituting λ from the previous equation into this last equation and solving for E then gives

$$E = \frac{h^2}{8m}\frac{N^2}{a^2 + b^2 + c^2}$$

Notice that this method requires a little ingenuity to find the proper manipulations of the equations, but the "symmetry" is often helpful. Also, the method is capable of determining the extremum value directly without the necessity of solving for those values of the variables that produce it.

Exercises

25. Find the minimum distance from the point (x_0, y_0, z_0) to the plane $ax + by + cz = d$. [*Hint:* Minimize $D^2 = (x - x_0)^2 + (y - y_0)^2 + (z - z_0)^2$.]
26. A light beam is to travel from A to B across the phase boundary of Fig. 2.4. In the A phase, velocity $v = c/n_A$ (index of refraction n), and in the B phase, $v = c/n_B$. Use Lagrange's method to show that the time the light beam must take is minimum for

$$n_A \sin \theta_A = n_B \sin \theta_B.$$

This is Snell's law of refraction.

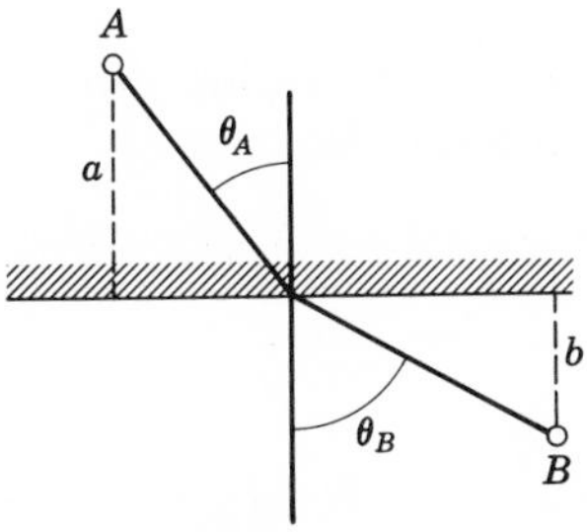

Figure 2.4

*27. The entropy of mixing of N ideal gases is $-R \sum X_i \ln X_i$, where X_i is the mole fraction of the ith gas. Use Lagrange's method of undetermined multipliers to find the maximum entropy of mixing.
*28. Maximize c_1c_2 subject to the constraint $c_1{}^2 + c_2{}^2 = 1$. (This is an alternative statement of the HMO problem of ethylene; the multiplier is simply related to the orbital energy.)
*29. An ammonia reactor cannot operate above a maximum pressure P_{max}. Therefore it is not possible to push the reaction to completion, and at equilibrium there is always some H_2 and N_2 left unreacted. Let r equal the ratio of the partial pressure of unreacted H_2 to the partial pressure of unreacted N_2.
    (a) Use two Lagrangian multipliers to find that value of r which maximizes the equilibrium partial pressure of NH_3. You may approximate fugacities for partial pressures.
    (b) Use two Lagrangian multipliers to find that value of r which minimizes the cost per mole of NH_3 produced. (*Warning:* Remember that you mustn't divide by zero.)

2.4 Homogeneous Functions

A function F is said to be HOMOGENEOUS of degree n if and only if, for all x, y, and t,

$$F(tx, ty) = t^n F(x, y). \tag{2.15}$$

For example:

$$\frac{xy}{x+y} \quad (n = 1) \qquad x^2y \ln (y/x) \quad (n = 3) \qquad x^2 + y^2 + z^2 \quad (n = 2)$$

$$\frac{x}{x^2+y^2} \quad (n = -1) \qquad \frac{x^2-y^2}{x^2+y^2} \quad (n = 0) \qquad x^{1/3} + xy^{-2/3} \quad (n = \tfrac{1}{3})$$

EULER'S THEOREM states that if F is homogeneous of degree n, then

$$x\left(\frac{\partial F}{\partial x}\right)_y + y\left(\frac{\partial F}{\partial y}\right)_x = nF. \tag{2.16}$$

The extension to more than two variables is obvious.

In thermodynamics a homogeneous function of degree one is known as an EXTENSIVE VARIABLE. The following two examples should provide a feeling for what such a function "looks like."

5. The volume of a solution containing n_A moles of species A and n_B moles of species B is some function V of n_A and n_B. A solution containing tn_A moles of A and tn_B moles of B would have a volume t times as great; that is, $V(tn_A, tn_B) = tV(n_A, n_B)$. And since V is homogeneous of degree one,

$$n_A\left(\frac{\partial V}{\partial n_A}\right)_{n_B} + n_B\left(\frac{\partial V}{\partial n_B}\right)_{n_A} = V.$$

This result is quite reasonable. The partial derivative $\partial V/\partial n_A$ is the partial molal volume of A in a solution whose A/B ratio is n_A/n_B. If an infinitesimal quantity of A were added to this solution, the volume would increase by $(\partial V/\partial n_A)\, dn_A$. Furthermore, if we were to prepare such a solution by alternately mixing infinitesimal quantities of A and B such that the A/B ratio never varies from n_A/n_B, then the total volume would be $V = (\partial V/\partial n_A) \int dn_A + (\partial V/\partial n_B) \int dn_B = (\partial V/\partial n_A)n_A + (\partial V/\partial n_B)n_B$, as also given by Euler's theorem.

6. The Gibbs free energy G is a function of temperature, pressure, and n_i, the number of moles of each chemical species. At constant T and P, doubling each n_i doubles the free energy content. Therefore G is a homogeneous function of degree one in $\{n_i\}$, and by Euler's theorem

$$G = \sum_i n_i\left(\frac{\partial G}{\partial n_i}\right)_{T,P,\{n_j\}_{j\neq i}}$$

The partial derivatives are usually symbolized by μ_i, the chemical potential. We may derive a further bonus: The total differential of G is

$$dG = d\sum n_i\mu_i = \sum n_i\, d\mu_i + \sum \mu_i\, dn_i.$$

But the total differential of $G(T, P, \{n_i\})$ is also given by

$$dG = -S\, dT + V\, dP + \sum \mu_i\, dn_i$$

with $(\partial G/\partial T) = -S$ and $(\partial G/\partial P) = V$. Subtracting the first equation for dG from the second gives the Gibbs-Duhem equation

$$-S\, dT + V\, dP - \sum n_i\, d\mu_i = 0.$$

Exercises

*30. Prove Euler's theorem. [*Hint:* Differentiate $F(tx, ty) = t^n F(x, y)$ with respect to t.]

31. Convince yourself that (at constant pressure) temperature, considered as a function of the number of moles of each chemical species, is a homogeneous function of order zero. In thermodynamics such a function is known as an INTENSIVE VARIABLE.

32. Convince yourself that if V is a homogeneous function of order one in the variables $\{n_i\}$, then $V/\sum n_i$ is a homogeneous function of order zero.

*33. Let us divide the Gibbs-Duhem equation, with $dT = 0 = dP$, by $n_1 + n_2$. If we define the mole fractions $x_1 = n_1/(n_1 + n_2)$ and $x_2 = n_2/(n_1 + n_2)$, then on rearrangement we obtain

$$d\mu_2 = -\frac{x_1}{x_2}\, d\mu_1.$$

Thus if we know μ_1 as a function of x_1, we can integrate to find μ_2 as a function of $x_1 = 1 - x_2$. The chemical potential of Hg in thallium amalgams at $T = 325°\text{C}$ is given by the empirical equation

$$\mu_{\text{Hg}} = \mu_{\text{Hg}}^{\circ} + RT \ln x_{\text{Hg}} - RT\alpha\left(\beta + \frac{x_{\text{Hg}}}{1 - x_{\text{Hg}}}\right)^{-2}$$

where μ_{Hg}° is a constant, $\alpha = 3.2$, and $\beta = 3.8$. Find μ_{Tl}, expressed as a function of x_{Tl}. Choose the constant of integration so that at $x_{\text{Tl}} = 1$, $\mu_{\text{Tl}} = \mu_{\text{Tl}}^{\circ}$. (*Hints:* First convert to an integral with respect to x_{Hg}. There are two terms in this integral; one is easy. To evaluate the other, substitute $r = x_{\text{Hg}}/(1 - x_{\text{Hg}})$ and look up the resulting integral in a table.)

*34. Derive the equation of Exercise 1.41.

*35. If $u = H(x, y)$, with H homogeneous of order one, what can you conclude about the contour lines of constant u in the xy-plane.

2.5 Exact Differentials

A DIFFERENTIAL FORM (PFAFFIAN FORM) in two variables

$$M(x, y)\, dx + N(x, y)\, dy$$

is said to be an EXACT DIFFERENTIAL if there exists a function u whose total differential du is this form, that is, such that

$$\left(\frac{\partial u}{\partial x}\right)_y = M \qquad \left(\frac{\partial u}{\partial y}\right)_x = N. \tag{2.17}$$

A necessary and sufficient condition that $M\,dx + N\,dy$ be exact is

$$\left(\frac{\partial M}{\partial y}\right)_x = \left(\frac{\partial N}{\partial x}\right)_y \tag{2.18}$$

since these partial derivatives are $\partial^2 u/\partial y\, \partial x$ and $\partial^2 u/\partial x\, \partial y$, respectively, and we have said that the mixed second partial derivatives are equal.

Examples

7. The differential form $(y + 2x^2)\, dx - x\, dy$ is not exact since $\partial M/\partial y = 1$ and $\partial N/\partial x = -1$. Inexact differentials are often distinguished from total differentials by a lower case

delta. The notation δu enables us to consider a differential form without specifying which variables are involved; for example, if the variables are transformed by $x = f(s, t)$, $y = g(s, t)$, then $M\,dx + N\,dy$ becomes $P\,ds + Q\,dt$, and these equivalent forms may be represented by their common "value" δu.

8. From thermodynamics, the differentials of heat and work, $\delta q = C_V\,dT + T(\partial P/\partial T)_V\,dV = C_P\,dT - T(\partial V/\partial T)_P\,dP$ and $\delta W = P\,dV = P(\partial V/\partial T)_P\,dT + P(\partial V/\partial P)_T\,dP$, respectively, are inexact differentials. One form of the first law of thermodynamics is a statement that their difference, $\delta q - \delta W$, is an exact differential dE. One form of the second law of thermodynamics is a statement that the differential of entropy, dS, is exact. Other exact differentials are those of enthalpy, dH, and free energy, dA and dG.

9. The statement that $-S\,dT + V\,dP$ is an exact differential, dG, means that $(\partial G/\partial T)_P = -S$ and $(\partial G/\partial P)_T = V$. Furthermore, from Eq. 2.18, or from the identity of the mixed second partials, $\partial^2 G/\partial T\,\partial P$ and $\partial^2 G/\partial P\,\partial T$, we may conclude that $(\partial S/\partial P)_T = -(\partial V/\partial T)_P$.

Exercises

36. (a) From $dE = T\,dS - P\,dV$ and $dA = -S\,dT - P\,dV$, show that $(\partial E/\partial V)_T = T(\partial P/\partial T)_V - P$.
(b) Next, show that

$$\left(\frac{\partial T}{\partial V}\right)_E = \frac{[P - T(\partial P/\partial T)_V]}{(\partial E/\partial T)_V}.$$

*37. Apply the methods of the previous exercise to simplify $(\partial T/\partial P)_H$.

38. If $H \equiv E + PV$, use the results of Exercise 2.19b and 2.36a to show that

$$C_P - C_V \equiv \left(\frac{\partial H}{\partial T}\right)_P - \left(\frac{\partial E}{\partial T}\right)_V = T\left(\frac{\partial P}{\partial T}\right)_V \left(\frac{\partial V}{\partial T}\right)_P$$

*39. Use the result of Exercise 2.19c and then the result of Exercise 2.36b to simplify the latter inequality of Exercise 2.14 to $T(\partial P/\partial V)_T/(\partial E/\partial T)_V < 0$. Notice that in Exercise 2.14 we have already shown that $(\partial E/\partial T)_V > 0$. Therefore we may also conclude that for any stable phase the compressibility $-(1/V)(\partial V/\partial P)_T > 0$.

There is a theorem which states that for any differential form in two variables, $M\,dx + N\,dy$, there exists an INTEGRATING FACTOR $\mu(x, y)$ such that $\mu M\,dx + \mu N\,dy$ is an exact differential.

Examples

10. The inexact differential $\delta u = (y + 2x^2)\,dx - x\,dy$ of Example 2.7 has an integrating factor $1/x^2$ since

$$\frac{\delta u}{x^2} = \left(\frac{y}{x^2} + 2\right) dx - \frac{x}{x^2}\,dy = d\left(-\frac{y}{x} + 2x\right).$$

11. Integrating factors for the inexact differential $\delta u = -y\,dx + x\,dy$ include $1/x^2$, $1/y^2$, $1/xy$, and $1/(x^2 + y^2)$ since

$$\frac{\delta u}{x^2} = \frac{-y\,dx + x\,dy}{x^2} = d\left(\frac{y}{x}\right) \qquad \frac{\delta u}{y^2} = \frac{-y\,dx + x\,dy}{y^2} = d\left(-\frac{x}{y}\right)$$

$$\frac{\delta u}{xy} = -\frac{dx}{x} + \frac{dy}{y} = d\left(\ln\frac{y}{x}\right) \qquad \frac{\delta u}{x^2 + y^2} = d\left(\tan^{-1}\frac{y}{x}\right).$$

Indeed, there always exist an infinity of integrating factors (Exercise 2.41). However, the theorem only guarantees that there exists an integrating factor, but it does not provide a recipe for finding one.

Next we turn to differential forms in three variables. Again, the differential form

$$M(x, y, z)\, dx + N(x, y, z)\, dy + P(x, y, z)\, dz$$

is said to be exact if this expression is the total differential of some function u, that is, if there exists a function u such that

$$\left(\frac{\partial u}{\partial x}\right)_{y,z} = M \qquad \left(\frac{\partial u}{\partial y}\right)_{x,z} = N \qquad \left(\frac{\partial u}{\partial z}\right)_{x,y} = P.$$

A necessary and sufficient condition for this differential form to be exact may be obtained from the requirement that the six mixed second partials of u be equal in pairs:

$$\frac{\partial M}{\partial y} = \frac{\partial N}{\partial x} \qquad \frac{\partial N}{\partial z} = \frac{\partial P}{\partial y} \qquad \text{and} \qquad \frac{\partial P}{\partial x} = \frac{\partial M}{\partial z}.$$

It is not necessary that there exist an integrating factor for a differential form in three or more variables. Those differential forms for which an integrating factor *does* exist are said to be INTEGRABLE. One form of the second law of thermodynamics states that the inexact differential, δq, in any number of variables, is integrable, and that in particular the integrating factor $1/T$ converts it to the exact differential $dS = \delta q/T$.

Exercises

*40. Test for exactness, and find an integrating factor if inexact:
(a) $y[x \sinh(x + y) + \cosh(x + y)]\, dx + x[y \sinh(x + y) + \cosh(x + y)]\, dy$.
(b) $V\, dP + P\, dV - \dfrac{PV}{T}\, dT$.

*41. If μ is an integrating factor which converts $M\, dx + N\, dy$ into the total differential $d\phi$, show that $\mu(x, y) f[\phi(x, y)]$, with f an arbitrary function of one variable, is also an integrating factor.

2.6 The Dimensionality of Thermodynamics

Students often have considerable difficulty in applying their knowledge of mathematics to thermodynamics. Much of this difficulty arises because it is not obvious how theorems about functions of x, y, and z are extended to functions of P, V, T, E, S, H, A, and G. For example, it may not be obvious whether the total differential of S is $dS = (\partial S/\partial P)\, dP + (\partial S/\partial V)\, dV$ or $dS = (\partial S/\partial P)\, dP + (\partial S/\partial V)\, dV + (\partial S/\partial T)\, dT$ or a sum of still more terms. It is understood why $(\partial P/\partial T)_V$ for an ideal gas is nR/V, but what is the significance of the symbolism $dP/dT = \Delta H/T\Delta V$ in the derivation of the Clapeyron-Clausius equation for phase equilibrium? What are those triangular coordinate systems, used for phase diagrams of ternary mixtures?

In the most elementary situations (one phase, one component), it is convenient to consider thermodynamic quantities in terms of the three variables, T, P, and V. Then the point (T_0, P_0, V_0) in T-P-V space corresponds to a particular condition of a thermodynamic system. In fact, of course, there is another variable, n, the number of moles of material. However, it is convenient to eliminate this variable by restricting consideration to only 1 mole of material. This arbitrary restriction may then be removed in practice because T and P are intensive variables and all the rest—V, E, S, H, A, and G—extensive. Therefore the volume, or the energy, or the entropy, etc., corresponding to n moles is just n times the value corresponding to 1 mole, which is what is tabulated.

For this 1 mole of material there is always an experimentally determined equation of state relating T, P, and V:

$$F(T, P, V) = 0.$$

For example, for an ideal gas, $F(T, P, V) = PV - RT$; for a van der Waals gas $F(T, P, V) = (P + a/V^2)(V - b) - RT$. Therefore if the amount of material is fixed, not all points in T-P-V space are accessible. The locus of points that do satisfy $F(T, P, V) = 0$ is some surface in this space, as illustrated in Figure 2.5. And in many cases it is possible to solve this equation for one variable in terms of the other two: $T = T(P, V)$ or $P = P(T, V)$ or $V = V(T, P)$, where we have acquiesced to popular usage and used the same symbol for a function and its value.

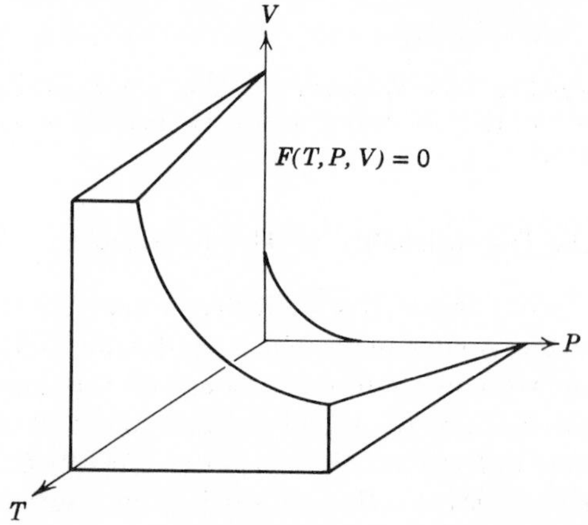

Figure 2.5

At each point of the surface $F(T, P, V) = 0$ each of the remaining thermodynamic quantities, E, S, H, A, and G, may be evaluated:

$$\begin{aligned} E &= E(T, P, V) \\ &= E(T, P, V(T, P)) = E(T, P) \\ &= E(T, P(T, V), V) = E(T, V) \\ &= E(T(P, V), P, V) = E(P, V) \end{aligned}$$

etc. Again, we have acquiesced to popular usage and used the same symbol (*1*) for a value; (*2*) for a function of T, P, and V; (*3*) for a function of T and P; (*4*) for a function of T and V; and (*5*) for a function of P and V.

What we now have is a set of eight variables, related by six constraints—the equation of state and the functional relationships giving E, S, H, A, and G. There remain two independent variables, which we have chosen from T, P, and V. But we may choose *any* two variables from among the eight to be the independent variables. These are then the two variables whose specification is necessary and sufficient to describe the state of the system. Each of the other six variables may then be specified as a function of the two independent variables. Thus we see why differential forms in elementary thermodynamics are written in two variables, and why we must always indicate for every partial derivative the other independent variable, which is held constant.

The choice of independent variables is generally made for convenience. For example, it is natural to express E as a function of S and V because in the expression $dE = (\partial E/\partial S)_V\, dS + (\partial E/\partial V)_S\, dV$ for the total differential of E, the two partial derivatives, $(\partial E/\partial S)_V$ and $(\partial E/\partial V)_S$, have the simple forms T and $-P$, respectively. Or we might be considering an isothermal process, whereupon T would be a natural choice for one independent variable. It is then convenient to express E as a function of T and V since $(\partial E/\partial T)_V$ and $(\partial E/\partial V)_T$ have the simple forms C_V and $T(\partial P/\partial T)_V - P$, respectively. On the other hand, we would not ordinarily express E as a function of T and G because $(\partial E/\partial T)_G$ and $(\partial E/\partial G)_T$ are not "simple."

Exercise

*42. Show that if the function $S(E, V)$ is known, then all the remaining thermodynamic variables, including the equation of state, may be determined from this function alone. This is the approach of statistical mechanics.

Next we consider the geometric interpretation of these considerations. Since there are a total of eight variables and two degrees of freedom, the graph of the system of relationships is some surface in an eight-dimensional space. Such a graph is beyond our capacity to draw or imagine, but it is usually possible to deal with only four variables in the following fashion: Let us call the two independent variables x and y and the two "relevant" dependent variables u and v. A frequent problem in thermodynamics is to determine the change in the value of u on going from the point (x_0, y_0) to a point (x_1, y), where y is determined by the requirement that the value of v remains unchanged. (Of course, it would be most convenient to know $u(x, v)$, but that may not be the function given. For example, we might wish to find the temperature change accompanying adiabatic expansion; then we are given $T(V, P)$ when it is $T(V, S)$ that is desired.) The intersection of the surface $v = v(x, y)$ in Figure 2.6 with the horizontal surface $v = v_0$ is a curve which defines a relationship between x and y. It is along this curve that the system changes from the state specified by (x_0, y_0) to the state specified by (x_1, y).

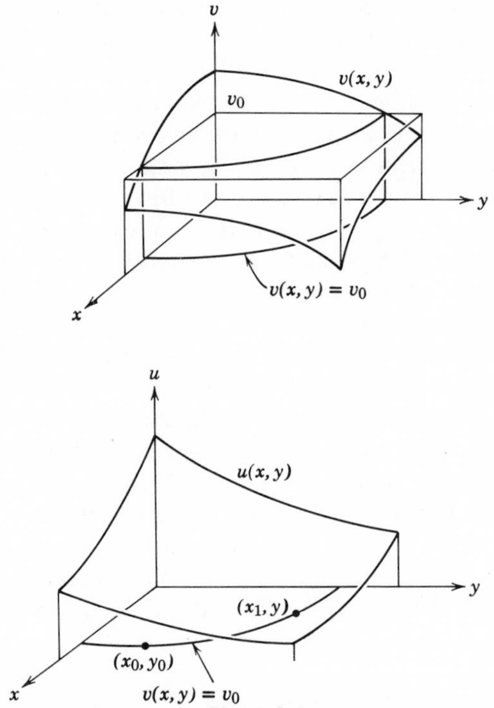

Figure 2.6

Next let us consider the situation with two phases present with half a mole of material in each. The requirement of equilibrium between the two phases imposes an additional constraint, say $\phi(T, P, V) = 0$, on the system, so that the variables that were previously independent can no longer be specified independently. In T-P-V space the surface of points satisfying the equation of state $V = V(T, P)$ is not the same as the surface defined by the additional constraint: Only the curve of intersection of these two surfaces corresponds to states of the system for which two phases are in equilibrium. In regions of Figure 2.7 marked "Vapor," both T and P may be varied independently, and V adjusts accordingly. But vapor is in equilibrium with liquid only at those points on the boundary curve between "Vapor" and "Liquid" regions. Specifying T, and requiring that the two phases be in

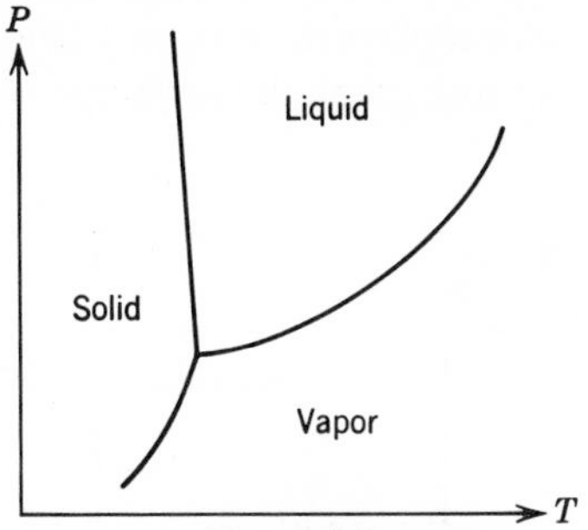

Figure 2.7

equilibrium, then serves to determine both P and V. Furthermore, a derivative dP/dT corresponds to the slope of one of these boundary curves; for consistency it ought to be written as $(\partial P/\partial T)_\phi$, but since there is only one degree of freedom, it is often written as a total derivative.

Exercise

*43. Sketch (or imagine) the graph of the surface $V = V(T, P)$ for an ideal gas (cf. Fig. 2.5) that can condense to a liquid and to a solid. For simplicity, let the condensed phases be incompressible, $(\partial V/\partial P)_T = 0$, and non-expanding, $(\partial V/\partial T)_P = 0$, but take account of the discontinuities at the phase boundaries: $V_{\text{gas}} > V_{\text{solid}} > V_{\text{liq}}$ (for H_2O). Then sketch (or imagine) the graph of the surface $G = G(T, P)$ for the same substance. There are no discontinuities at the phase boundaries, but remember that $(\partial G/\partial T)_P = -S$ and $(\partial G/\partial P)_T = V$, with $S_{\text{gas}} > S_{\text{liq}} > S_{\text{solid}}$.

Finally, let us allow for more than one component but only one phase. Now we must include N additional independent variables $\{n_i\}_{i=1}^N$, where n_i is the number of moles of the ith component, no longer held constant. To describe the state of a system we must specify each n_i, plus two additional quantities. Each of the other six quantities is then determined as a function of $N + 2$ variables. And a total differential is now a differential form in $N + 2$ variables. For example

$$dG = \left(\frac{\partial G}{\partial T}\right)_{P,\{n_i\}} dT + \left(\frac{\partial G}{\partial P}\right)_{T,\{n_i\}} dP + \sum_i \left(\frac{\partial G}{\partial n_i}\right)_{T,P,\{n_j\}_{j\neq i}} dn_i.$$

It is not possible to graph any of these functions of $N + 2$ variables. Therefore let us first reduce the dimensionality of the problem by holding the two variables T and P fixed while we allow $\{n_i\}$ to vary. Let us also take $\sum n_i$ and the mole fractions $X_i \equiv n_i/\sum n_i$ as N new variables and then discard the variable $\sum n_i$. We have thus introduced the additional constraint $\sum X_i = 1$ and thereby further reduced the number of independent variables to $N - 1$. Again, this arbitrary constraint may be removed in practice because all the remaining variables—V, E, S, H, A, and G—are extensive. For $N = 2$, the surface $u = u(n_1, n_2)$ is replaced by $u = u(X_2)$, the curve of the intersection of this surface with the vertical plane $n_1 + n_2 = 1$ (Fig. 2.8). It is then possible to reintroduce another axis and graph the surface $u = u(X_2, T)$ or the surface

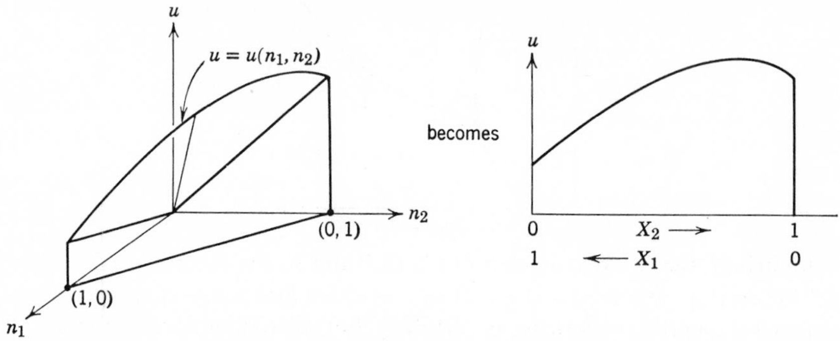

Figure 2.8

$u = u(X_2, P)$. For $N = 3$, it is not possible to draw the three-dimensional "surface" $u = u(n_1, n_2, n_3)$, and we must graph this function by drawing contours of constant u in n_1-n_2-n_3 space. But it is possible to graph as a surface a function defined at every point on the plane $n_1 + n_2 + n_3 = 1$ (Fig. 2.9).

Exercises

*44. Mentally draw a straight line from the X_1 vertex to the opposite side of the equilateral triangle that serves as a coordinate system for a ternary phase diagram. Convince yourself that this line is the line of constant ratio $X_2:X_3$.

*45. How might you construct a coordinate system for a quaternary phase diagram?

2.7 Multiple Integrals

We define the DOUBLE INTEGRAL of the function F, of two variables, over the region A of the xy-plane (Figure 2.10) by

$$\iint_A F(x, y)\, dx\, dy \equiv \lim_{\substack{\Delta x_i \to 0 \\ \Delta y_j \to 0}} \sum_{i,j} F(x_i, y_j)\, \Delta x_i\, \Delta y_j. \tag{2.19}$$

In words, we (*1*) divide A into many little rectangles, of dimension $\Delta x_i \times \Delta y_j$; (*2*) evaluate F at a point (x_i, y_j) within each rectangle and multiply the

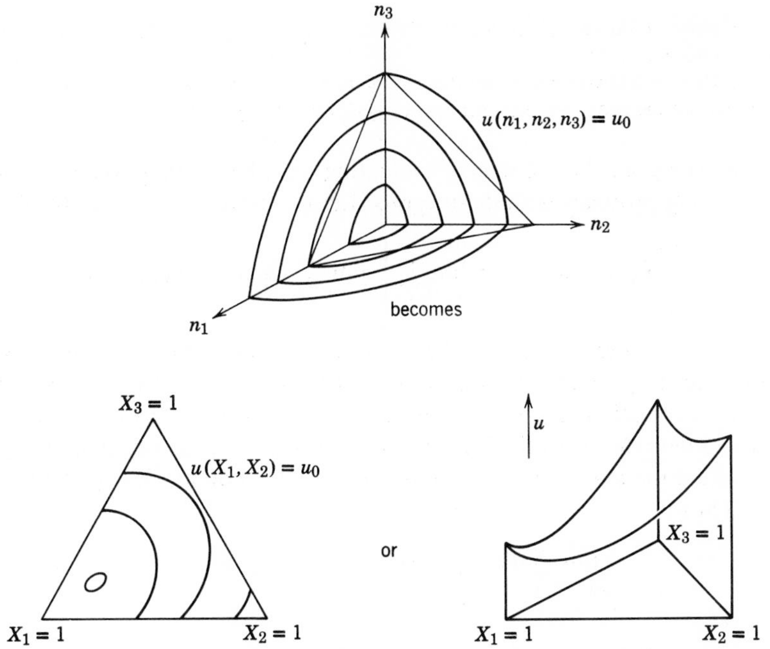

Figure 2.9

value by the area of that rectangle; (*3*) add all the products; and (*4*) take the limit as every rectangle becomes infinitely small. Geometrically, this double integral equals the volume of the solid above A and between the surface $z = F(x, y)$ and the xy-plane. Volumes below the xy-plane contribute negatively to the integral. The extension to more than two variables is obvious, even if the geometric interpretation is not.

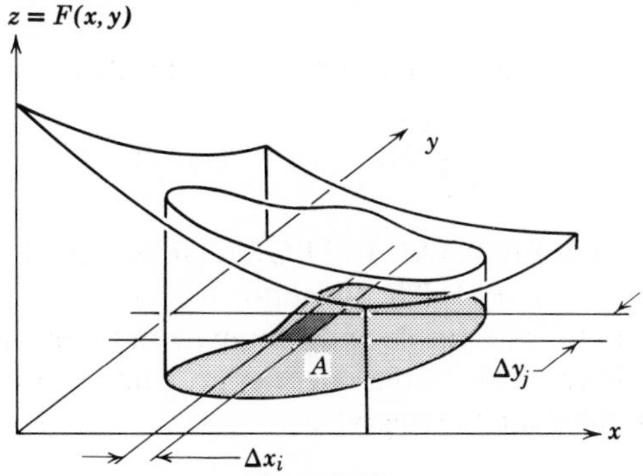

Figure 2.10

Exercises

46. (a) How would you generalize the Fourier transform operator to convert $f(x, y, z)$ into $g(\xi, \eta, \zeta)$?
  *(b) How would you generalize the convolution of $f_1(x, y, z)$ and $f_2(x, y, z)$?
47. What are the geometric significances of $\iint_A dx\,dy$ and $\iiint_V dx\,dy\,dz$?

We must reduce the problem of evaluating a double integral to the problem of evaluating ordinary definite integrals. Let us rearrange Eq. 2.19 to obtain

$$\iint_A F(x, y)\,dx\,dy = \lim_{\Delta x_i \to 0} \sum_i \left[\lim_{\Delta y_j \to 0} \sum_j F(x_i, y_j)\,\Delta y_j \right] \Delta x_i. \tag{2.20}$$

The quantity in brackets is just the definition of an ordinary definite integral with respect to y, with x held fixed at x_i. But we must be careful about the limits of integration. Later we shall allow x to run from a to b, so that in order to integrate over all of A, and over only A, y must run from $f(x_i)$ to $g(x_i)$, where the curves $y = f(x)$ and $y = g(x)$ are boundaries of A (Figure 2.11). Thus

$$\lim_{\Delta y_j \to 0} \sum_j F(x_i, y_j)\,\Delta y_j = \int_{f(x_i)}^{g(x_i)} F(x_i, y)\,dy.$$

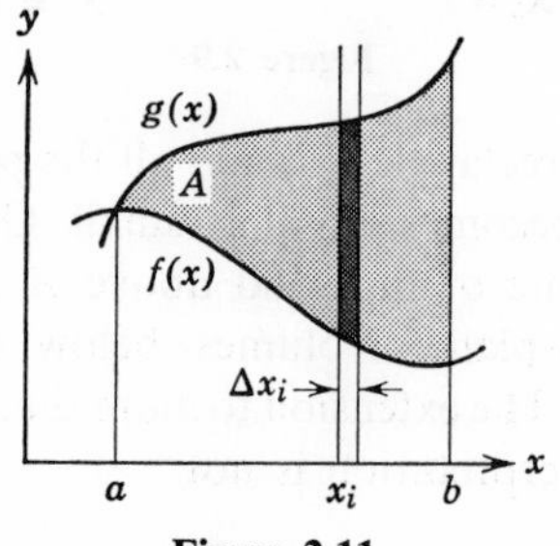

Figure 2.11

If we now extend the fundamental theorem of calculus to conclude that

$$\frac{\partial}{\partial y} \int_a^y \Phi(x, t)\,dt = \Phi(x, y) \tag{2.21}$$

we may then define a PARTIAL INTEGRATION OPERATOR, I_y, as the operator such that $D_y I_y = 1$. This operator is analogous to the integration operator I, except that whereas I is indeterminate to the extent of an arbitrary constant, I_y is indeterminate to the extent of an arbitrary function of x. Then we may evaluate the definite integral as

$$\int_{f(x_i)}^{g(x_i)} F(x_i, y)\,dy = I_y F[x_i, g(x_i)] - I_y F[x_i, f(x_i)]$$

since the arbitrary function of x cancels in the subtraction. Finally, substituting this result into Eq. 2.20 gives

$$\iint_A F(x, y)\, dx\, dy = \lim_{\Delta x_i \to 0} \sum_i \{I_y F[x_i, g(x_i)] - I_y F[x_i, f(x_i)]\} \Delta x_i$$

$$= \int_a^b \{I_y F[x, g(x)] - I_y F[x, f(x)]\}\, dx,$$

again by the definition of an ordinary definite integral. And we may also write the double integral in the form of two ordinary definite integrals if we are careful to indicate which variable is integrated over first:

$$\iint_A F(x, y)\, dx\, dy = \int_a^b \left[\int_{f(x)}^{g(x)} F(x, y)\, dy \right] dx$$

$$= \int_a^b dx \int_{f(x)}^{g(x)} dy F(x, y). \tag{2.22}$$

In both these expressions the notation is intended to indicate that the integration is first performed with respect to y; the brackets are often omitted. Analogous forms are obtained if the integration is first performed with respect to x.

The extension to more than two variables should be obvious:

$$\iiint_V \Phi(x, y, z)\, dV = \int_a^b dx \int_{f(x)}^{g(x)} dy \int_{F(x,y)}^{G(x,y)} dz \Phi(x, y, z).$$

In practice, the only new feature of double integrals is the problem of determining the limits of integration when the functions f and g are not given explicitly. The only remedy is to determine the boundaries of A as a function of one of the variables (the one that is integrated over *last*). For example, it is readily apparent that the double integral of F over the triangular area of Figure 2.12 is

$$\int_0^1 dx \int_0^x dy F(x, y) \qquad \text{or} \qquad \int_0^1 dy \int_y^1 dx F(x, y).$$

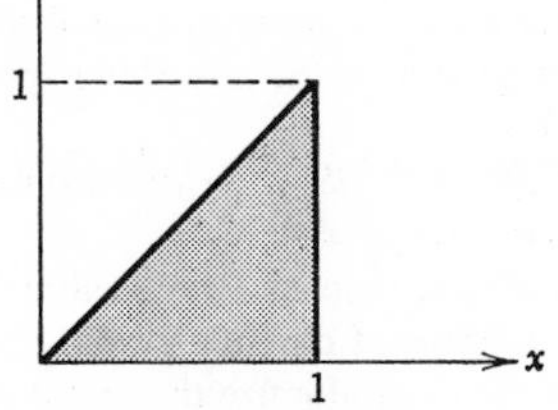

Figure 2.12

Exercises

48. What are the limits of integration for each of the regions in Figure 2.13:
    (a) If the first integration is with respect to y?
    (b) If the first integration is with respect to x?

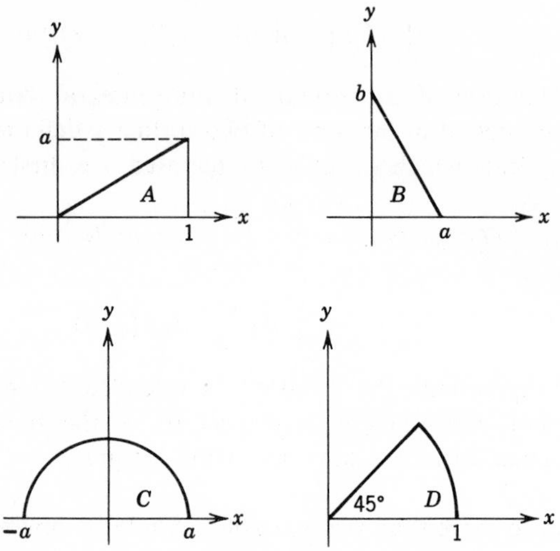

Figure 2.13

49. (a) What is the volume V of the tetrahedron cut from the first octant by the plane $x/a + y/b + z/c = 1$?
    (b) The coordinates of its center of mass are given by

$$x_{\text{com}} = \frac{1}{V}\iiint x\,dV \qquad y_{\text{com}} = \frac{1}{V}\iiint y\,dV \qquad z_{\text{com}} = \frac{1}{V}\iiint z\,dV.$$

What are these coordinates? (*Hint:* You should be able to guess y_{com} and z_{com} "by symmetry.")

We state without proof a theorem that is another part of Leibniz' rule:

$$\frac{d}{dx}\int_a^b F(x, y)\,dy = \int_a^b \frac{\partial F}{\partial x}\,dy. \tag{2.23}$$

Exercises

*50. Extend the result of Exercise 1.21 to a continuous distribution of incident intensity, $I^\circ(\lambda)$; let the extinction coefficient of the absorbing species be $\varepsilon(\lambda)$. (Show again that $\log(I^\circ/I)$ is not linear in lc.)

*51. Differentiate both sides of $\int_{-\infty}^{\infty} e^{-ax^2}\,dx = (\pi/a)^{1/2}$ n times with respect to a to evaluate $\int_{-\infty}^{\infty} x^{2n}e^{-ax^2}\,dx$. Do the same to $\int_0^{\infty} xe^{-ax^2}\,dx$.

52. (a) Evaluate $(d/dx)\ln\int_{-1}^{1} e^{xy}\,dy$; then see what Leibniz' rule gives for this expression.
    (b) In an electric field E, a dipole of dipole moment μ tends to orient with the field, but thermal motion tends to randomize the orientation. The average value of the

component of dipole moment among the direction of the field is given by

$$\bar{\mu}_z = \frac{\displaystyle\int_0^{\pi} \mu e^{x \cos \theta} \cos \theta \sin \theta \, d\theta}{\displaystyle\int_0^{\pi} e^{x \cos \theta} \sin \theta \, d\theta}$$

where $x = \mu E/kT$. Substitute $y = \cos \theta$ and use the result of part (a) to evaluate $\bar{\mu}_z$.

(c) What is the first nonzero term in the expansion of $\bar{\mu}_z$ about $x = 0$? (*Caution:* Expand $e^{\pm x}$ to four terms.)

Next we consider transformation of variables in multiple integrals. The double integral $\iint F(x, y)\, dx\, dy$ equals $\lim \sum F(x_i, y_j)\, \Delta A$, where ΔA is the tiny rectangular area bounded by the four straight lines $x = x_i$, $x = x_i + \Delta x_i$, $y = y_j$, and $y = y_j + \Delta y_j$. If we transform to polar coordinates (Eq. 2.7), $F(x, y)$ becomes $G(r, \phi)$, and we may evaluate the definite integral as $\lim \sum G(r_i, \phi_j)\, \Delta A$, where ΔA is the tiny area bounded by the circles $r = r_i$ and $r = r_i + \Delta r_i$ and the rays $\phi = \phi_j$ and $\phi = \phi_j + \Delta\phi_j$ in Figure 2.14. However, simple considerations of geometry show that ΔA is not simply

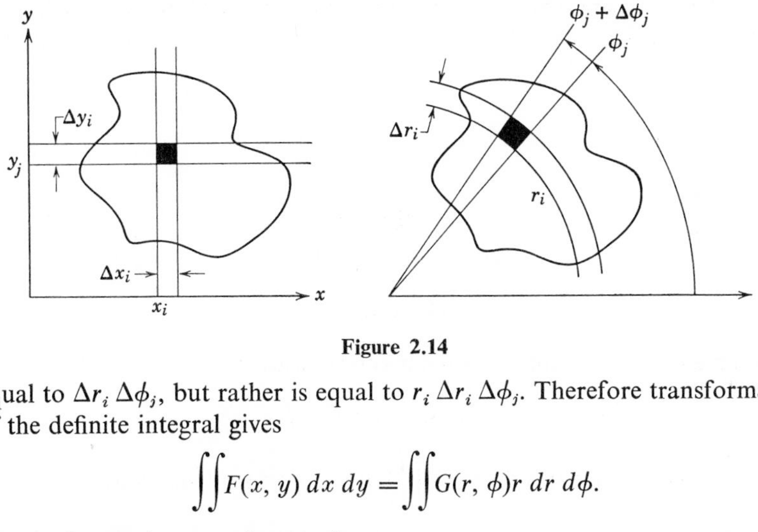

Figure 2.14

equal to $\Delta r_i\, \Delta\phi_j$, but rather is equal to $r_i\, \Delta r_i\, \Delta\phi_j$. Therefore transformation of the definite integral gives

$$\iint F(x, y)\, dx\, dy = \iint G(r, \phi) r\, dr\, d\phi.$$

Do not forget the extra factor of r.

In three dimensions there are two sets of polar coordinates. The first is quite similar to polar coordinates in the plane. CYLINDRICAL POLAR COORDINATES (Fig. 2.15) are defined by

$$x = r \cos \phi \qquad y = r \sin \phi \qquad z = z. \tag{2.24}$$

The formula for the transformation of the volume element to cylindrical

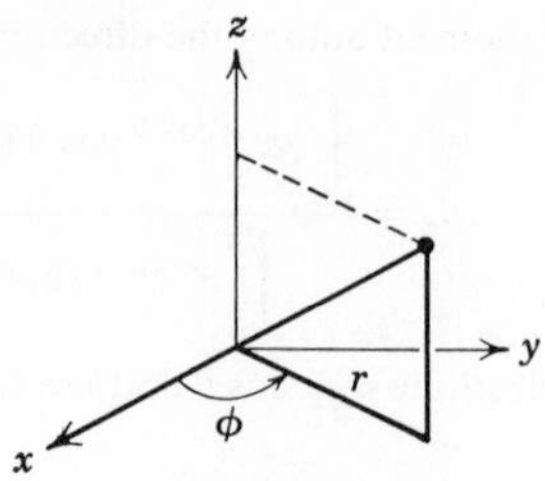

Figure 2.15

polar coordinates is

$$dV = r\,dr\,d\phi\,dz$$

$$\iiint F(x, y, z)\,dx\,dy\,dz = \iiint G(r, \phi, z)r\,dr\,d\phi\,dz.$$

SPHERICAL POLAR COORDINATES are defined by

$$x = r \sin\theta \cos\phi \qquad y = r \sin\theta \sin\phi \qquad z = r\cos\theta \tag{2.25}$$

where θ is measured down from the z-axis and ϕ is measured from the x-axis around toward the y-axis (Fig. 2.16). The formula for the transformation of

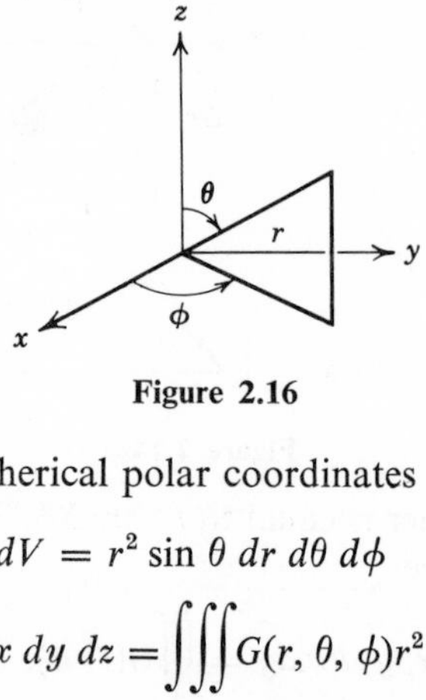

Figure 2.16

the volume element to spherical polar coordinates is

$$dV = r^2 \sin\theta\,dr\,d\theta\,d\phi$$

$$\iiint F(x, y, z)\,dx\,dy\,dz = \iiint G(r, \theta, \phi)r^2 \sin\theta\,dr\,d\theta\,d\phi.$$

Above we have merely stated the formulas for the volume element in the new coordinates. All of them may be calculated from the general formula

$$dV = dx\,dy\,dz = \left|\frac{\partial(x, y, z)}{\partial(u, v, w)}\right| du\,dv\,dw \tag{2.26}$$

where we have substituted $x = x(u, v, w)$, $y = y(u, v, w)$, $z = z(u, v, w)$. Here the vertical lines mean absolute value, and the JACOBIAN of the transformation is a function of u, v, and w, defined by the 3×3 determinant

$$\frac{\partial(x, y, z)}{\partial(u, v, w)} = \begin{vmatrix} \left(\frac{\partial x}{\partial u}\right)_{v,w} & \left(\frac{\partial x}{\partial v}\right)_{u,w} & \left(\frac{\partial x}{\partial w}\right)_{u,v} \\ \left(\frac{\partial y}{\partial u}\right)_{v,w} & \left(\frac{\partial y}{\partial v}\right)_{u,w} & \left(\frac{\partial y}{\partial w}\right)_{u,v} \\ \left(\frac{\partial z}{\partial u}\right)_{v,w} & \left(\frac{\partial z}{\partial v}\right)_{u,w} & \left(\frac{\partial z}{\partial w}\right)_{u,v} \end{vmatrix}. \tag{2.27}$$

We defer the justification of this formula until we consider vectors.

Of course, in all transformations of variables we must adjust the limits of integration so that we integrate over the same region of space. For example, the integral of $F(x, y) = G(r, \phi)$ over the wedge-shaped region of Figure 2.17 is given by

$$\int_0^a dx \int_0^{\sqrt{a^2-x^2}} dyF(x, y) = \int_0^a dr \int_0^{\frac{1}{2}\pi} d\phi\; rG(r, \phi) = \int_0^{\frac{1}{2}\pi} d\phi \int_0^a dr\; rG(r, \phi).$$

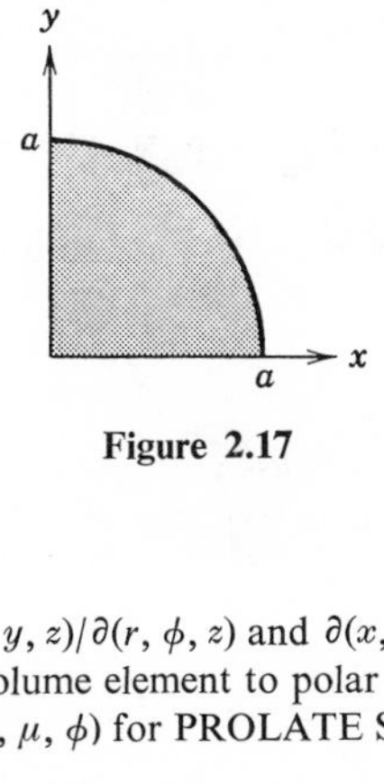

Figure 2.17

Exercises

*53. Evaluate the Jacobians $\partial(x, y, z)/\partial(r, \phi, z)$ and $\partial(x, y, z)/\partial(r, \theta, \phi)$ and thereby check the transformation of the volume element to polar coordinates.

*54. Calculate $J = \partial(x, y, z)/\partial(\lambda, \mu, \phi)$ for PROLATE SPHEROIDAL COORDINATES (a is a constant):

$$z = a\lambda\mu \qquad x = a\sqrt{(\lambda^2 - 1)(1 - \mu^2)}\cos\phi \qquad y = a\sqrt{(\lambda^2 - 1)(1 - \mu^2)}\sin\phi$$

*55. Expand the 2×2 determinants and multiply to show that the product of Jacobians, $\partial(x, y)/\partial(u, v) \cdot \partial(u, v)/\partial(x, y) = 1$. This result may be extended to Jacobians in three variables.

*56. What are the limits of integration, in polar coordinates, for the regions of Figure 2.13?

57. Evaluate $I = \int_0^\infty e^{-x^2}\, dx$ by writing $I^2 = (\int_0^\infty e^{-x^2}\, dx)(\int_0^\infty e^{-y^2}\, dy)$, transforming to polar coordinates, and evaluating. This integral arises frequently; memorize it.

58. Evaluate the integrals of Exercise 1.33 and use the results to provide approximate forms for $Q_{\text{trans}}^{[1]}$ and Q_{rot}.

*59. The CORRELATION FUNCTION G, of a function f, is defined by $G(\tau) = \int f(t)f(t + \tau)\, dt$, where the integration is over the domain of f. See how G is a measure of the correlation ("resemblance") of the value of f from argument to nearby argument to distant argument by calculating G for each of the following:

(a) $f(t) = e^{-kt}$ $\quad 0 \leqslant t < \infty$.
(b) $f(t) = \sin nt$ $\quad -\pi \leqslant t \leqslant \pi$ (n integral).
(c) $f(t) = \exp(-t^2/2\sigma^2)$ $\quad -\infty < t < \infty$.
(d) $f(t) = a,\ -t_0 < t < t_0;\ f(t) = 0$ elsewhere. [*Hint:* Sketch $f(t)$ and $f(t + \tau)$.]

60. If V is a spherical volume and G a function only of r, reduce $\iiint_V G(r)\, dV$ to an ordinary definite integral.
61. Find the volume of a sphere of radius r.
*62. Find the average distance between the proton and the electron in the 1s state of the hydrogen atom, given by the triple integral over all space

$$\frac{1}{\pi a_0^3}\iiint (x^2 + y^2 + z^2)^{1/2} \exp\left[\frac{-2(x^2 + y^2 + z^2)^{1/2}}{a_0}\right] dx\, dy\, dz.$$

*63. The overlap integral S between two 1s orbitals on hydrogen atoms separated by a distance $2a$ is given by the integral over all space

$$S = \frac{1}{\pi a_0^3}\iiint \exp\left\{\frac{-[x^2 + y^2 + (z - a)^2]^{1/2}}{a_0}\right\} \exp\left\{\frac{-[x^2 + y^2 + (z + a)^2]^{1/2}}{a_0}\right\} dx\, dy\, dz$$

Use the result of Exercise 2.54 to transform this integral to prolate spheroidal coordinates and thereby show that

$$S = \left(1 + 2\frac{a}{a_0} + \frac{4}{3}\frac{a^2}{a_0^2}\right) \exp(-2a/a_0)$$

(*Hint*: The limits are $1 < \lambda < \infty$, $-1 < \mu < 1$, $0 < \phi < 2\pi$.)

3

SOME SPECIAL FUNCTIONS

This short chapter introduces the student to a few simple special functions, especially the error function, elliptic integrals, the gamma function, and the Dirac delta function. Also included are brief discussions of Stirling's approximation to the factorial and of interpolation. Some of the unfamiliar topics introduced in the previous chapters are developed further. The exercises are designed to familiarize the student with these topics, but will not develop proficiency at them. The impatient student may wish to skip most of this chapter. However, he is advised to learn about both Stirling's formula and the delta functions, and to note what else is included, since subsequent chapters continually draw upon material in this one.

3.1 Functions Defined as Integrals

We have already encountered a few integrals that we didn't recognize as integrals of any of the familiar functions. For example, we had to resort to a trick to evaluate the definite integral $\int_0^\infty \exp(-t^2)\,dt$, and we still haven't evaluated $\int_0^\infty x^3\,dx/(e^x - 1)$. Also, the integral $\int dx/x$ is the exception to the rule $\int x^n\,dx = x^{n+1}/(n + 1)$ and is unknown in terms of algebraic functions. We chose to define logarithm as the function inverse to exponential, but we could just as well have defined $\ln x = \int_1^x dx/x$. From either definition it is possible to derive a series expansion whereby the values of the function may be calculated. Of course, the values are tabulated and very readily available, so that any problem that can be reduced to $\int dx/x$ is considered solved.

Whenever we encounter an integral that is not reducible to any familiar function (powers, roots, all functions related to exp), we may *define* that integral to be a new function. Then, to solve the problem, all we need do is determine a convenient power series and calculate the integral as a function of its upper limit. Often the integral has been encountered previously, and someone has already calculated and tabulated the values, so that we need merely find the table. Alternatively, the integral is *reducible* to one that has been encountered previously, and must be reduced to the standard form in which it has been tabulated.

There is a problem here if we don't know what integrals are tabulated. For example, only $\int dx/x$ is tabulated, and not $\int dx/(1 + x)$ or $\int x\,dx/(1 + x^2)$ or other integrals related to $\ln x$, but we know we should aim for $\int dx/x$ when we simplify related integrals. However, if we don't know what integrals are tabulated, we don't know what standard forms to aim for in simplifying an unknown integral. Indeed, we might even be so ignorant of the tabulated integrals that we would try in vain to reduce a tabulated integral to familiar functions. Therefore it is desirable to gain an introduction to the most commonly encountered tabulated integrals. We merely mention these functions so that they may be available for reference when you encounter a baffling integral. However, just as there are many helpful relations among the trigonometric and hyperbolic functions, there are many helpful relations involving tabulated integrals, and you should investigate these relations whenever you must use a tabulated integal.

A very important function, especially in statistics, is the ERROR FUNCTION

$$\operatorname{erf}(x) \equiv \frac{2}{\sqrt{\pi}} \int_0^x \exp(-t^2)\,dt$$

the area under the "bell-shaped curve." In Exercise 2.57 you showed that $\operatorname{erf}(\infty) = 1$, and in Exercise 1.61a you found a series expansion that could be used to evaluate the integral. For purposes in statistics this function is often tabulated as the related integral

$$\Phi(x) = \frac{1}{\sqrt{2\pi}} \int_0^x \exp(-\tfrac{1}{2}t^2)\,dt$$

A function closely related to the error function is its complement

$$\operatorname{erfc}(x) \equiv 1 - \operatorname{erf}(x) = \frac{2}{\sqrt{\pi}} \int_x^\infty \exp(-t^2)\,dt$$

Other related integrals are the FRESNEL INTEGRALS

$$C(x) \equiv \int_0^x \cos(\pi t^2/2)\,dt \qquad S(x) \equiv \int_0^x \sin(\pi t^2/2)\,dt$$

which arise in diffraction problems and which can be considered as error functions of complex argument. Alternative forms may be tabulated.

Exercises

1. Simplify $\int_0^x t^{-1/2}e^{-t}\,dt$.
*2. Relate Φ to erf, and evaluate $\Phi(\infty)$.
*3. Integrate $\operatorname{erfc} x = (2/\sqrt{\pi})\int_x^\infty e^{-t^2}\,dt = (2/\sqrt{\pi})\int_x^\infty te^{-t^2}\,dt/t$ thrice by parts to obtain the first three terms of a series expansion involving inverse powers of x. (This series may be used to approximate erfc x for large x, and you will note that the three terms decrease. However, after some number of terms, they will begin to increase again and, for fixed x, the series actually diverges. This series is therefore an asymptotic expansion and, for purposes of calculation, must be terminated after a finite number of terms.)
*4. Find the Laplace transform of erf. (*Hints:* Reverse the order of integration—watch the limits—and then complete the square in the exponent.)
5. Find the Laplace transform of erfc $(t^{-1/2})$ by reversing the order of integration—watch the limits—and looking up the resulting definite integral in a table. Then use the results of Exercise 1.44 to find $L[c \operatorname{erfc}(1/\sqrt{at})]$, where c and a are constants.

In LCAO MO and VB calculations of the H_2 molecule, and in many other contexts, there arises the EXPONENTIAL INTEGRAL

$$\operatorname{Ei}(x) \equiv \int_{-\infty}^{x} \frac{e^t\,dt}{t}$$

Related functions include the SINE INTEGRAL and the COSINE INTEGRAL

$$si(x) \equiv -\int_x^\infty \frac{\sin t\,dt}{t} \qquad Ci(x) \equiv -\int_x^\infty \frac{\cos t\,dt}{t}$$

Exercises

*6. Show that $-\operatorname{Ei}(-x) = \int_x^\infty (e^{-t}\,dt)/t$ and that $\operatorname{Ei}(ix) = \operatorname{Ci}(x) + i\operatorname{si}(x)$, analogous to Eq. 1.27. (*Note*: You may set $\int_{i\infty}^{\infty}(e^{iu}\,du/u) = 0$; see Exercise A5.20.)
*7. Integrate by parts to show that $-\operatorname{Ei}(-x) \sim e^{-x}\sum_{n=0}^{N}(-1)^n(n!/x^{n+1})$. (Cf. Exercise 1.50.)
*8. (a) Show that $-\operatorname{Ei}(-x) = \int_1^\infty (e^{-xu}\,du/u)$.
(b) Use this result to find the Laplace transform of Ei $(-x)$.
9. Simplify the LOGARITHMIC INTEGRAL, $\operatorname{li}(x) \equiv \int_0^x (du/\ln u)$.

Often there arise integrals of the general forms

$$\int (a\cos x + b\sin x + c)^{\pm 1/2}\,dx \qquad \text{and} \qquad \int R(x, P_4(x))\,dx$$

where P_4 is a (third- or) fourth-order polynomial and R is a function involving no worse than ratios of polynomials. Such integrals can be reduced to elliptic integrals, about which there exists a vast body of information which should be consulted as necessary. We define ELLIPTIC INTEGRALS of the first and second kind, respectively, by

$$F(k, \phi) \equiv \int_0^\phi \frac{d\phi}{\sqrt{1 - k^2 \sin^2 \phi}} \qquad |k| < 1$$

$$E(k, \phi) \equiv \int_0^\phi \sqrt{1 - k^2 \sin^2 \phi}\, d\phi \qquad |k| < 1$$

There are also COMPLETE ELLIPTIC INTEGRALS

$$K(k) \equiv F(k, \pi/2) \qquad E(k) \equiv E(k, \pi/2)$$

Again, there are alternative forms resulting from substitutions $k = \sin \theta$ or $\phi = \sin^{-1} x$.

Exercise

*10. Reduce $\int_0^x [(1 - k^2x^2)(1 - x^2)]^{-1/2}\, dx$ and $\int_0^x [(1 - k^2x^2)/(1 - x^2)]^{1/2}\, dx$ to elliptic integrals.

One of the most frequently encountered integrals is the GAMMA FUNCTION

$$\Gamma(x) \equiv \int_0^\infty e^{-t} t^{x-1}\, dt.$$

Clearly $\Gamma(1) = \int_0^\infty e^{-t}\, dt = 1$. Let us investigate $\Gamma(x + 1)$ by integrating by parts, with $f(t) = t^x$, $f'(t) = xt^{x-1}$, $g(t) = e^{-t}$, $G(t) = -e^{-t}$.

$$\Gamma(x + 1) = \int_0^\infty e^{-t} t^x\, dt = [-t^x e^{-t}]_0^\infty + \int_0^\infty x t^{x-1} e^{-t}\, dt = x \int_0^\infty t^{x-1} e^{-t}\, dt$$

or

$$\Gamma(x + 1) = x\Gamma(x) \tag{3.1}$$

since $e^{-\infty} = 0 = 0^x$. Application of this formula serves to evaluate the gamma function for all positive integers: $\Gamma(2) = 1\Gamma(1) = 1$, $\Gamma(3) = 2\Gamma(2) = 2$, $\Gamma(4) = 3\Gamma(3) = 6$, or, in general, if x is an integer

$$\Gamma(x) = (x - 1)!$$

Also, since $\Gamma(1) = 1$, we may set $0! = 1$ to keep the factorial function consistent with the gamma function. Furthermore, we may extend the gamma function to negative numbers by the equation $\Gamma(x) = \Gamma(x + 1)/x$, except that the gamma function is not defined for nonpositive integers.

Notice that although the factorial function is defined only for nonnegative integers, the gamma function is a continuous function defined almost everywhere. For non-integral arguments, the definite integral may be evaluated numerically. Values are tabulated for $1 \leqslant x \leqslant 2$; other values may be obtained by application of Eq. 3.1.

Exercises

11. Evaluate $\Gamma(\frac{1}{2})$, $\Gamma(\frac{3}{2})$. What is $\Gamma(n + \frac{1}{2})$, $n \geqslant 0$? (Use Exercise 1.46 for simplification.)
*12. What is the Laplace transform of t^n? What if n is non-integral?
13. Simplify $F(x) = \int_0^\infty \exp(-t^{1/x})\, dt$ $(x > 0)$.
*14. Evaluate $\int_0^1 (\ln x)^n\, dx$ for n integral.
15. We have twice encountered (Exercises 1.39 and 1.57) the definite integral

$$\int_0^\infty \frac{x^3\, dx}{e^x - 1} = \int_0^\infty \frac{x^3 e^{-x}\, dx}{(1 - e^{-x})}.$$

Expand the denominator as an infinite series (Watch convergence!) and integrate term-by-term to express the integral as an infinite sum.
*16. In statistical mechanics there arises the problem of finding the "volume," V_n, of an n-dimensional sphere of radius r. Let $V_n = c_n r^n$, and evaluate c_n by the following procedure: Let $I_n = \int\int \cdots \int_0^\infty \exp(-\sum x_i^2) \prod dx_i = \prod_{i=1}^n \int_0^\infty \exp(-x_i^2)\, dx_i$ and evaluate I_n. Next transform to polar coordinates and integrate over all angular coordinates: $I_n = \int_0^\infty e^{-r^2}(dV_n/dr)\, dr = nc_n \int_0^\infty r^{n-1}e^{-r^2}\, dr$. Finally, evaluate this integral and compare the two expressions for I_n. Check V_2 and V_3.
17. Show that

$$I_n = \int_0^{\pi/2} \sin^n \theta\, d\theta = \tfrac{1}{2}\sqrt{\pi}\,\Gamma\left(\frac{n+1}{2}\right)\Big/\Gamma\left(\frac{n}{2} + 1\right) = \int_0^{\pi/2} \cos^n \theta\, d\theta = I_n'$$

by the following procedure:
(a) Multiply both I_n and I_n' by $\int_0^\infty r^{n+1}e^{-r^2}\, dr$.
(b) Convert the double integrals to Cartesian coordinates.
(c) One integral is easy; evaluate it.
(d) The remaining integrals may be converted to gamma functions by an appropriate substitution.
*18. Use the results of Exercises 1.59 and 4.17 to find a series expansion for $K(k)$.
*19. (a) Rewrite $\Gamma(x)$ in an alternative form by substituting $t = u^2$.
(b) Next express $\Gamma(x)\Gamma(y)$ as a double integral and transform to polar coordinates.
(c) Finally, substitute $t = \cos^2 \phi$, and thereby show that the BETA FUNCTION

$$\mathrm{B}(x, y) \equiv \int_0^1 t^{x-1}(1 - t)^{y-1}\, dt = \Gamma(x)\Gamma(y)/\Gamma(x + y).$$

*20. Use the substitution $u = t/(1 - t)$ and the above result to express $\int_0^\infty u^{a-1}(1 + u)^{-b}\, du$ in terms of gamma functions.
*21. Relate $-\mathrm{Ei}(-x)$ and $\mathrm{erfc}(x)$ to the incomplete gamma function, whatever you suppose *that* to be.

We may use the gamma function to find an approximation (STIRLING'S FORMULA) to $x!$ when x is large. First write the integrand as a single exponential

$$x! = \Gamma(x + 1) = \int_0^\infty e^{-t}t^x\, dt = \int_0^\infty e^{-t + x \ln t}\, dt.$$

Since the exponent $-t + x \ln t$ approaches $-\infty$ as t approaches either zero or infinity, the integrand approaches zero near the limits of integration.

When x is large, the integrand is large only near its maximum at $t = x$, and rapidly approaches zero elsewhere. Therefore the principal contribution to the integral comes from $t \simeq x$. In Exercise 1.56 you found that the first terms in the expansion of $f(t) = -t + x \ln t$ about $t = x$ are $(-x + x \ln x) - (t - x)^2/2x$. Therefore we may approximate the integral by

$$x! \sim \int_0^\infty e^{-x+x \ln x-(t-x)^2/2x}\, dt = e^{-x+x \ln x}\sqrt{2x}\int_{-(x/2)^{1/2}}^{\infty} e^{-u^2}\, du$$

where we have used the substitution $u = (t - x)/(2x)^{1/2}$. Also, if x is large, very little error is introduced by replacing $-(x/2)^{1/2}$ in the lower limit of integration by $-\infty$. (Cf. erfc x when x is large.) Thus

$$x! \sim e^{-x+x \ln x}\sqrt{2x}\int_{-\infty}^{\infty} e^{-u^2}\, du = e^{-x+x \ln x}\sqrt{2\pi x}$$

or

$$x! \sim x^{x+1/2}e^{-x}\sqrt{2\pi}. \tag{3.2}$$

Further analysis would show that this result represents the first term of the asymptotic series

$$x! \sim x^{x+1/2}e^{-x}\sqrt{2\pi}\left(1 + \frac{1}{12x} + \frac{1}{288x^2}\cdots\right)$$

Another useful form is

$$\ln x! \sim (x + \tfrac{1}{2}) \ln x - x + \tfrac{1}{2} \ln 2\pi. \tag{3.3}$$

For x especially large, $x \gg \frac{1}{2} \ln 2\pi > \frac{1}{2}$, so that such small terms may be dropped to obtain a cruder approximation, which is quite simple

$$\ln x! = x \ln x - x$$

Exercises

22. Find $\lim_{n\to\infty} (n!)^{1/n}/n$.

*23. Find $\lim_{n\to\infty} \dfrac{n^{1/2}(2n)!}{2^{2n}n!^2} = \lim_{n\to\infty} \dfrac{1 \cdot 3 \cdot 5 \cdots (2n-3)(2n-1)}{2 \cdot 4 \cdot 6 \cdots (2n-2)2n}\sqrt{n}.$

*24. Show that

$$P(r) = \lim_{\substack{n\to\infty \\ p\to 0}} \frac{n!}{r!\,(n-r)!}p^r(1-p)^{n-r} = \frac{(np)^r e^{-np}}{r!}$$

where the limiting process is carried out so as to keep (np) fixed. You will need to show

$$\lim_{\substack{n\to\infty \\ p\to 0}} (1-p)^n = e^{-np}$$

in the course of this problem. This formula is known as the Poisson approximation to the binomial distribution and is important in probability and statistics.

*25. Let

$$P(r) = \frac{n!}{r!\,(n-r)!} p^r q^{n-r} \qquad 0 < p < 1,\, 0 < q < 1,\, n \text{ large}$$

Let $\delta = r - np$ and use Stirling's approximation (better form) and Taylor's expansion of logarithm (to second order in the small quantities δ/np and δ/nq), to show that $-\ln P(r) \sim \delta^2/2npq + \frac{1}{2}\ln(2\pi npq)$. This formula is known as the normal approximation to the binomial distribution and is important in probability and statistics.

*26. What, approximately, is 1000!? Estimate the percent error of this approximation.

3.2 Interpolation

We have mentioned several functions whose values must be looked up in tables. I'm sure you've had occasion to look up values of $\log_{10} x$, or $\sin\theta$, or e^x. However, tables do not tabulate $f(x)$ for the infinity of possible values of x, but only for certain values of x at regular intervals, that is, the table lists values of $f(x)$ for $x_n = x_0 + n\,\Delta x$; $n = 0, 1, 2, \ldots$, with Δx the interval. And it is often necessary to find $f(x)$ for an x which is not in the table.

To find the value $f(x)$ for a given x, there are two simple formulas.

1. Linear interpolation: f is approximated by a straight line over a short

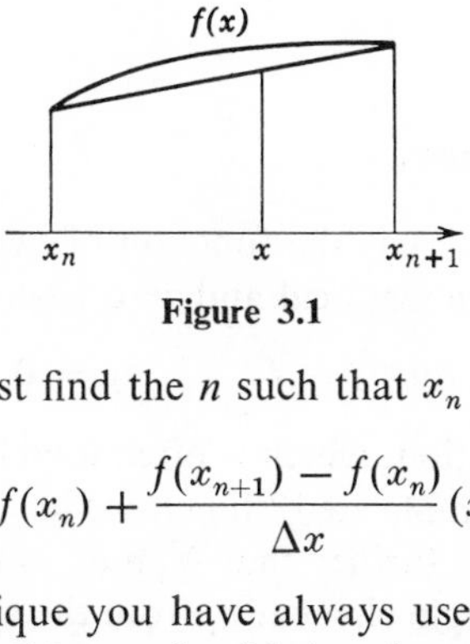

Figure 3.1

interval (Figure 3.1). First find the n such that $x_n < x < x_{n+1}$. Then

$$f(x) \simeq f(x_n) + \frac{f(x_{n+1}) - f(x_n)}{\Delta x}(x - x_n).$$

This is the simple technique you have always used automatically. For any table with sufficiently small intervals, this is a decent approximation. Higher-order interpolation formulae may be derived by noticing the similarity of the above equation to the Taylor series expansion of f about $x = x_n$:

2. Second-order (quadratic, parabolic) interpolation:

$$f(x) \simeq f(x_n) + \left.\frac{df}{dx}\right|_{x_n}(x - x_n) + \frac{1}{2}\left.\frac{d^2f}{dx^2}\right|_{x_n}(x - x_n)^2$$

where

$$\left.\frac{df}{dx}\right|_{x_n} \simeq \left.\frac{\Delta f}{\Delta x}\right|_{x_n} = \frac{f(x_{n+1}) - f(x_n)}{\Delta x}$$

and

$$\left.\frac{d^2f}{dx^2}\right|_{x_n} \simeq \frac{d}{dx}\frac{\Delta f}{\Delta x} \simeq \frac{\Delta\Delta f}{(\Delta x)^2} = \frac{\dfrac{f(x_{n+1}) - f(x_n)}{\Delta x} - \dfrac{f(x_n) - f(x_{n-1})}{\Delta x}}{\Delta x}$$

$$= \frac{f(x_{n+1}) + f(x_{n-1}) - 2f(x_n)}{(\Delta x)^2}.$$

This formula is necessary only when $f(x)$ varies rapidly with x. If this is still inadequate, you are advised to locate a finer table.

Exercise

*27. Use the tables in the *Handbook of Chemistry and Physics* to find
(a) $\log_{10} 10.84505$ (8 s.f.).
(b) exp (0.4731) (5 s.f.).
(c) sinh (0.1014), cosh (0.1014) (5 s.f.).
(d) erf (1.000) (4 s.f.).
(e) $\log_{10} \sin (18' 24'')$ (6 s.f.).
(f) $K(\frac{1}{2})$.
(g) $K(0.2)$.
(h) $E(0.2, 45°)$.
(i) $E(0.2, 31.7°)$.
(j) $\Gamma(3.125)$.

3.3 Delta Functions

The KRONECKER DELTA is the function of two integers whose value is unity when the integers are identical and zero when they are different

$$\delta_{ij} = 1 \quad \text{for} \quad i = j \qquad \delta_{ij} = 0 \quad \text{for} \quad i \neq j.$$

(Notice the subscript notation, which is often used for functions of integers.) This is certainly an "uncomplicated" function; its chief utility is for simplification of notation. Notice further that $\sum_j a_j\delta_{ij} = a_i$; this summing device serves to pluck out the single element a_i from the sequence $\{a_j\}$.

Exercise

28. Use the Kronecker delta function to abbreviate the notation of Exercise 1.70 to only three equations. Memorise these results.

The DIRAC DELTA FUNCTION, $\delta(x)$, is a much more remarkable beast. It is a function whose value is zero everywhere but at the origin. At the origin the value of the function is so large an infinity that $\int_{-\infty}^{\infty} \delta(x)\, dx = 1$.

Such a definition is not immediately believable, since it asserts that the area of a rectangle of zero width is nonzero. This delta function may be made more visualizable by presenting it as the limit of a sequence of continuous functions. One example of this approach is as follows: Since $f(x) = e^{-x^2/a^2}$ has points of inflection at $x = \pm a/\sqrt{2}$, we may say that the width of this curve is $a\sqrt{2}$. Also

$$\int_{-\infty}^{\infty} e^{-x^2/a^2}\,dx = a\int_{-\infty}^{\infty} e^{-y^2}\,dy = a\sqrt{\pi}.$$

Next define $D_a(x) = e^{-x^2/a^2}/a\sqrt{\pi}$, still of width $a\sqrt{2}$, but such that $\int_{-\infty}^{\infty} D_a(x)\,dx = 1$. Now we may set

$$\delta(x) = \lim_{a\to 0} D_a(x) = \lim_{a\to 0} \frac{1}{a\sqrt{\pi}} e^{-x^2/a^2},$$

which does have the properties specified above.

Some of the rules for manipulating the delta function are

1. $\delta(-x) = \delta(x)$.
2. $\int_{-\infty}^{\infty}\delta(x - a)\,dx = 1$.
3. $\int_{-\infty}^{\infty} f(x)\delta(x - a)\,dx = f(a)$.
4. $f(x)\delta(x - a) = f(a)\delta(x - a)$.

Rule 1 states that the delta function is an even function. Rule 2 is readily proved by a change of variable; it states that the area under the delta function is unity no matter where the infinity is.

The proof of rule 3 depends upon the fact that there can be no contribution to the integral from regions of integration where the value of the delta function is zero. The only contribution comes from the immediate vicinity of $x = a$, where $f(x)$ takes on the constant value $f(a)$ and may be removed from the integral

$$\int_{-\infty}^{\infty} f(x)\delta(x - a)\,dx = \int_{-\infty}^{\infty} f(a)\delta(x - a)\,dx = f(a)\int_{-\infty}^{\infty} \delta(x - a)\,dx = f(a).$$

This is the most important property of the delta function; multiplying a function by a delta function, then integrating, serves to pluck out a particular value of the function of interest. (Cf. the Kronecker delta.)

Although the delta function is an unusual function, in practice it arises in such fashion that it will eventually be integrated and removed from further consideration. It is with this eventuality in mind that we may consider rule 4 valid; integration of both sides leads to rule 3.

Exercises

29. Sketch a graph of the UNIT-STEP FUNCTION, $\int_{-\infty}^{x} \delta(x)\,dx$.

*30. Show that $\delta(ax) = \delta(x)/a$ and that $x\delta(x) = 0$ in the sense that integration of both sides of either equation leads to an identity.

31. (a) What is $\int_{-\infty}^{\infty} \delta(x - a)\delta(x - b)\,dx$?
    (b) Show that although $\int_{-\infty}^{\infty} \delta(x)\,dx$ is finite, $\int_{-\infty}^{\infty} [\delta(x)]^2\,dx$ is not.

32. Express $f(x)$ as a linear combination of $\{\delta(x - x')\}_{x'=-\infty}^{\infty}$. (*Hint:* A sum over a continuous variable must become an integral with respect to x'.)

*33. Integrate by parts to show that

$$\int_{-\infty}^{\infty} f(x) D^n \delta(x - a)\, dx = (-1)^n D^n f(a)$$

*34. (a) What is the convolution $\delta(x - a) * f(x)$?
(b) Sketch a graph of $f_1 * f_2$ if $f_1(x) = \delta(x + 1) + \delta(x - 1)$ and $f_2(x) = e^{-x^2}$.
(c) Sketch $f_1 * f_2$ if $f_1(x) = \frac{1}{2}\delta(x + 2) + \delta(x + 1) + 2\delta(x) + \delta(x - 1) + \frac{1}{2}\delta(x - 2)$ and $f_2(x) = e^{-x^2}$.
(d) What is the convolution $[\sum c_i \delta(x - a_i)] * f(x)$?
(e) Notice further that, according to Exercise 4.32, *any* function may be expressed as a linear combination of delta functions. Now try sketching $e^{-x^2} * e^{-x^2}$ and compare your sketch with the convolution evaluated analytically.

*35. If $f_1(x, y, z) = \sum_i \sum_j \sum_k \delta(x - x_i)\delta(y - y_j)\delta(z - z_k)$, where $\{x_i, y_j, z_k\}$ are the points of a regular "crystal lattice" and $f_2(x, y, z)$ is the electron density within a single unit cell whose corners are the points of the lattice, what is the physical significance of $f_1 * f_2$?

*36. Find the Laplace transform of $\delta(t - a)$. (*Caution:* What if $a = 0$? If $a < 0$?)

4

DIFFERENTIAL EQUATIONS

This chapter opens with a discussion of differential equations and how they may be classified. A presentation of some of the elementary ways of solving them follows, with special attention to differential equations of interest to chemists. Section 4.5 concludes our review of basic calculus. We proceed to material that may be largely unfamiliar—series solution of differential equations, and the special functions of Legendre and Bessel. We conclude the chapter with the topic of eigenvalue-eigenfunction problems, which are rarely discussed in elementary mathematics courses but which are so important in physical science. The exercises in this chapter demonstrate applications to spectroscopy, kinetics and rate processes, classical mechanics, atomic and molecular structure, and quantum mechanics (including the basis for a more sophisticated approach to the Pauli exclusion principle).

4.1 Definitions

A DIFFERENTIAL EQUATION is an equation involving derivatives of functions. The solution of a differential equation is a relationship between the variables. For example, we have already encountered the equation $dy/dx = y$, whose solution is $y = Ce^x$, where C is an arbitrary constant. An alternative notation is $Df = f$, whose solution is $f(x) = Ce^x$.

Notice that the relationship we seek is functional. The problem here is to find one variable as a function of the other: $y = f(x)$, or at least to reduce the differential equation to the form $F(x, y) = 0$, which implicitly specifies one variable as a function of the other. [Contrast this situation with the problem of finding a number which solves the numerical equation $f(x) = 0$.] But just as a numerical equation may be checked by substituting the answer back into the equation, so may a differential equation be checked. Since differential equations are often solved by lengthy manipulations, this practice is advisable.

Not all differential equations may be solved in terms of familiar functions, and often numerical methods must be used. However, if a differential equation has frequently arisen, someone has evaluated the function. For example, the above equation, $Df = f$, led us to define a "new" function, exp, whose power series expansion we obtained and whose values are tabulated.

The ORDER of a differential equation is the order of the highest derivative which occurs. For example, $Df = f$ is a first-order differential equation because only the first derivative of the unknown function occurs. An example of a second-order differential equation is $(D^2f)^2 + p(x)(Df)^3 + q(x)f^4 = r(x)$.

Exercises

1. Find a simple second-order differential equation which both $\sin \omega x$ and $\cos \omega x$ satisfy. Convince yourself that any linear combination of these two functions also satisfies the differential equation. Convince yourself that the particular linear combinations $e^{i\omega x}$ and $e^{-i\omega x}$ are obvious solutions.
2. Find a simple second-order differential equation which is satisfied by any linear combination of $\sinh x$ and $\cosh x$.

A simple geometrical picture is possible for first-order differential equations. The most general form of such an equation is $F(x, y, y') = 0$. At each point of the xy-plane we may solve for y', the slope of the unknown function. Connecting adjacent points in appropriate fashion leads to a graph of the function. The example of $y' = -xy$ is indicated in Figure 4.1.

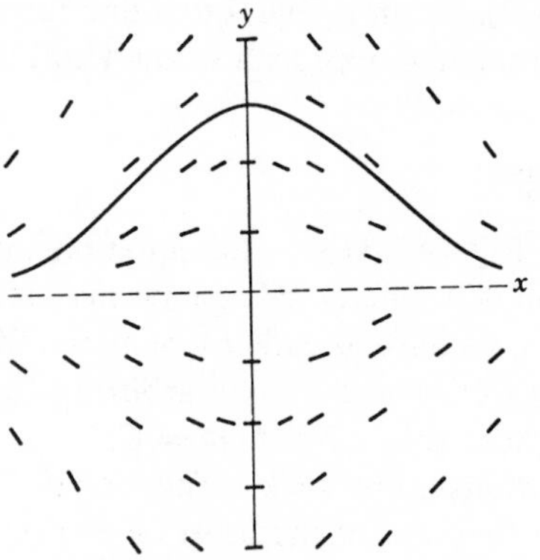

Figure 4.1

The DEGREE of a differential equation is the power to which the highest derivative is taken. For example, the second-order differential equation given above is of the second degree.

A differential equation is said to be LINEAR if the dependent variable (unknown function) and all its derivatives occur to only the first power, with no products of such quantities. Therefore a linear differential equation is necessarily of the first degree. The most general form of the Nth-order linear differential equation is $\sum_{n=0}^{N} M_n(x)D^n y = N(x)$. We shall write this equation as $L_N y = N(x)$, where the linear operator $L_N \equiv \sum_{n=0}^{N} M_n(x)D^n$. For a standard form, we may take $M_N(x) = 1$ without loss of generality.

Exercises

3. Convince yourself that the operator $\sum_{n=0}^{N} M_n(x)D^n$ is a linear operator.
*4. Find a linear second-order differential equation which is satisfied by $e^{-kt} \cos \omega t$. Try for the "simplest" possible.
*5. Find a linear second-order differential equation which is satisfied by $\exp(\sin^{-1} x)$. Try for one of the form $L_2 y = 0$.
*6. Convince yourself that the differential equations of chemical kinetics are all first order, but that first-order kinetics leads to linear differential equations.

Since solving a differential equation involves "eliminating" the differentiation, finding the solution must somehow involve integration. Therefore there results a constant of integration for each differentiation which must be eliminated, so that the GENERAL SOLUTION of an Nth-order differential equation contains N arbitrary constants.

Often in physical problems only a PARTICULAR SOLUTION is required and the arbitrary constants are fixed by specifying N independent BOUNDARY CONDITIONS. For example, the general solution of $Df = f$ is $f(x) = Ce^x$, but when we first encountered this differential equation we specified the particular solution $f(x) = e^x$ by the boundary condition $f(0) = 1$.

There is no general method applicable to solving all differential equations. Different types yield to different approaches, and we next describe several fruitful approaches.

4.2 First-Order Equations of the First Degree

A first-order equation of the first degree may be written in the form $y' = F(x, y)$. An alternative form, which preserves the equivalence of x and y, is

$$M(x, y)\,dx + N(x, y)\,dy = 0. \tag{4.1}$$

If $D_y M = D_x N$, then the differential form $M\,dx + N\,dy$ is exact and equal to a total differential du. Then the solution to the differential equation $du = 0$ is simply $u = C$, where C is an arbitrary constant. Furthermore, u may be determined by integration since $u = I_x M + f(y) = I_y N + g(x)$; the arbitrary functions of partial integration are chosen "by inspection" to satisfy the last equality.

Even if the differential form $M\,dx + N\,dy$ is inexact, there is a theorem which guarantees the existence of an integrating factor, $\mu(x, y)$, such that $\mu M\,dx + \mu N\,dy$ is an exact differential. In Example 2.10 we showed that the integrating factor $\mu = 1/x^2$ converts $\delta u = (y + 2x^2)\,dx - x\,dy$ to the exact differential $d[-y/x + 2x]$, so the solution of $\delta u = 0$ is $-y/x + 2x = C$ or $y = 2x^2 - Cx$. The difficulty is that the theorem guarantees the existence of an integrating factor, but does not present any method of finding it. There is no recipe for finding an integrating factor for a general first-order differential equation of the first degree. Often considerable ingenuity must be devoted to finding one, and a thorough knowledge of derivatives is required.

Exercises

*7. Solve for the general solutions of the following first-order differential equations. First test the differential form for exactness, then find an integrating factor if necessary.
(a) $(3x^2 - y)\,dx + (3y^2 - x)\,dy = 0$.
(b) $y[x + \coth(x + y)]\,dx + x[y + \coth(x + y)]\,dy = 0$. (Cf. Exercise 2.40.)

*8. Occasionally there arise TOTAL DIFFERENTIAL EQUATIONS in three (or more) variables

$$P(x, y, z)\,dx + Q(x, y, z)\,dy + R(x, y, z)\,dz = 0.$$

(Do not confuse these total differential equations with partial differential equations, which also involve more than two variables and which are discussed in the next chapter.) Clearly, if the differential form $P\,dx + Q\,dy + R\,dz$ is exact and equal to a total derivative du, then the general solution to this equation is $u(x, y, z) = C$. Even if $P\,dx + Q\,dy + R\,dz$ is inexact, it may be integrable such that

$$\mu(P\,dx + Q\,dy + R\,dz)$$

is exact and equal to dv; then the general solution to the differential equation is $v(x, y, z) = C$. Notice that in both these cases the general solution is a set of surfaces in three-dimensional space, each characterized by the parameter C. Solve the following differential equations "by inspection":

(a) $-S\,dT + V\,dP + \sum \mu_i\,dn_i = 0$.

(b) $V\,dP + P\,dV - \dfrac{PV}{T}\,dT = 0$.

In contrast to these cases, there do exist differential forms that are not even integrable, so that no one-parameter family of curves is a solution to the differential equation. In Section 7.14 we shall provide a test for integrability. For now we just indicate the significance of these considerations by stating Carathéodory's formulation of the second law of thermodynamics: The general solution to the differential equation $\delta q = 0$, in *any* number of variables, *is* a family of surfaces, each characterized by a parameter S which may be identified as entropy. Alternatively, in the "neighborhood" of a state of a thermodynamic system, there exist states which cannot be reached by any adiabatic path (one satisfying $\delta q = 0$), because those states are on different surfaces, characterized by different values of S.

The special case in which $M(x, y) = m(x)u(y)$ and $N(x, y) = n(x)v(y)$ occurs quite frequently in physical problems. For this case $1/n(x)u(y)$ is an integrating factor, and the differential equation becomes

$$\frac{m(x)}{n(x)}\,dx + \frac{v(y)}{u(y)}\,dy = 0. \tag{4.2}$$

For obvious reasons this special case is called the case of SEPARABLE VARIABLES. Integrating both sides of this equation reduces the differential equation to the problem of evaluating the two integrals $\int m(x)\,dx/n(x)$ and $\int v(y)\,dy/u(y)$. Although these integrals may be difficult or impossible to evaluate in terms of familiar functions, nevertheless the problem is no longer one of solving a differential equation, but only one of integration.

Exercises

9. A light beam of incident intensity I_0 passes through a thickness L of an absorbing solution. The loss of intensity per unit distance traveled, $-dI/dx$, is proportional to the concentration of the absorbing species and to the intensity of light remaining. Let the proportionality constant be $\varepsilon \ln 10$, and solve the resulting differential equation to obtain the Beer-Lambert law.
10. Find the general solutions of the differential equations:
    (a) $y' = e^{-y}$.
    (b) $y' = (y \ln y)/x$.
11. (a) Solve the first-order kinetics $A \xrightarrow{k_1}$ product ($dA/dt = -k_1 A$), with $A(0) = A_0$.
    (b) Express the answer in an alternative form in terms of the RELAXATION TIME τ, the time required for A to decrease to $1/e$ of its initial value.
    (c) Solve the second-order kinetics $2A \xrightarrow{k_2}$ product ($dA/dt = -2k_2 A^2$), with $A(0) = A_0$.
    (d) For both cases, what is the HALF-LIFE, the time required for A to decrease to half its initial value?

*12. Find the particular solution of the differential equation implied by the second-order kinetics $A + B \xrightarrow{k}$ product, subject to the initial conditions $A(0) = A_0 \neq B(0) = B_0$. (Cf. Exercise 1.40c.)

*13. Solve the differential equation

$$\frac{dy}{dx} = \sqrt{1 - y^2}\,\sqrt{1 - k^2 y^2} \qquad y(0) = 0$$

Notice that it is easy to solve for x as a function of y (incomplete elliptic integral), but the inverse function is not familiar. (Compare $y' = \sqrt{1 - y^2}$, which by separation of variables immediately gives $x = \sin^{-1} y$.) The inverse function is known as an ELLIPTIC FUNCTION, written $y = \operatorname{sn} x$, and there are many similarities to trigonometric functions.

14. The Fourier transform of $f(x) = \exp(-x^2/2\sigma^2)$ may be evaluated by direct integration, but due account of complex variables must be taken. Find $g(y) = F\{f(x)\}$ by a different procedure: Differentiate $g(y)$ with respect to y, then integrate by parts to relate the integral on the right to $g(y)$. Solve the resulting differential equation and determine $g(0)$ in order to set the constant of integration.

*15. Many problems involve finding the path $y = y(x)$ which minimizes the line integral (defined later) $I = \int_{x_0,y_0}^{x_1,y_1} F(x, y, dy/dx)\, dx$. Euler's equation of the calculus of variations states that the path which minimizes (or maximizes) I is the solution to the differential equation, $(\partial F/\partial y) - (d/dx)(\partial F/\partial y') = 0$. For example, Fermat's principle states that a light beam travels along the path which takes the shortest time, that is, along that path which minimizes $t = (1/c) \int_{x_0,y_0}^{x_1,y_1} n(x, y) \sqrt{1 + (dy/dx)^2}\, dx$, where c = speed of light in vacuum, and $n(x, y)$ = refractive index of medium. The refractive index of a solution in an ultracentrifuge is not constant but varies with its position along the centrifuge tube. Let $n(x, y) = n_0 e^{-y/a}$. Then a beam of light bends on traveling through the solution, and the light source appears to be somewhere higher along the tube than it really is (Figure 4.2). Use the result of Exercise 2.18 to find the path the light beam takes, and find θ, the indicated angle of refraction. (*Hint:* By symmetry, $(dy/dx)|_{x=0} = 0$.)

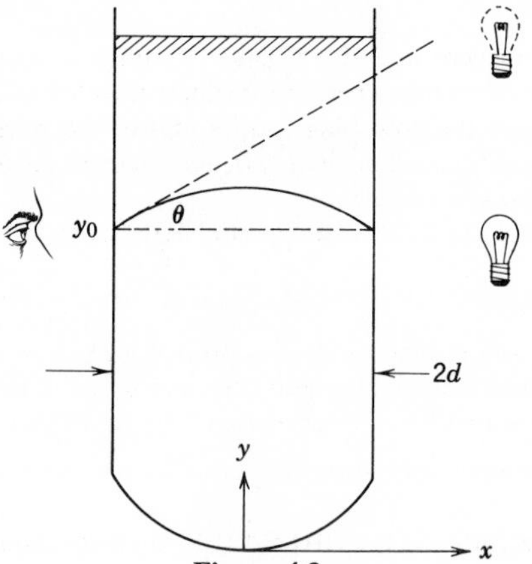

Figure 4.2

*16. Find what seems to be the general solution to the nonlinear differential equation $y' = -(a^2 - y^2)^{1/2}/y$. Then show that $y = a$ and $y = -a$ are further SINGULAR SOLUTIONS which are not obtained by the usual methods.

*17. The first-order differential equation $y' = H(x, y)$ may be solved in the case that H is a homogeneous function of order zero. Show that the substitution $u = y/x$ leads to a differential equation in which the variables are separable, and reduce the equation to an integral. Then solve $y' = xy/(x^2 + y^2)$.

4.3 Linear First-Order Equations

For the general linear first-order differential equation, $y' + M(x)y = N(x)$, there *is* a rule for determining the integrating factor: Multiplying by $\mu = e^{\int M(x)dx}$ leads to the general solution

$$y = e^{-\int M(x)dx} \int N(x)e^{\int M(x)dx}\, dx + Ce^{-\int M(x)dx} \tag{4.3}$$

Example

1. Consecutive first-order chemical reactions $A \xrightarrow{k_1} B \xrightarrow{k_2} C$ with initial conditions $A(0) = A_0$, $B(0) = 0 = C(0)$: The differential equations implied by these kinetics are

$$\frac{dA}{dt} = -k_1 A$$

$$\frac{dB}{dt} = k_1 A - k_2 B$$

$$\frac{dC}{dt} = k_2 B$$

In the first equation the variables are separable:

$$-\frac{dA}{A} = k_1\, dt \quad \rightarrow \quad \int_{A_0}^{A} \frac{dA}{A} = -k_1 \int_0^t dt \quad \rightarrow \quad \ln A - \ln A_0 = -k_1 t$$

or

$$A = A_0 e^{-k_1 t}$$

Substituting into the second equation and rearranging to standard form gives

$$\frac{dB}{dt} + k_2 B = k_1 A_0 e^{-k_1 t}$$

This equation has the integrating factor $\exp\left[\int k_2\, dt\right] = e^{k_2 t}$. Multiplication by the integrating factor gives

$$e^{k_2 t} \frac{dB}{dt} + k_2 e^{k_2 t} B = k_1 A_0 e^{(k_2 - k_1)t}$$

$$d(Be^{k_2 t}) = k_1 A_0 e^{(k_2 - k_1)t}\, dt$$

$$Be^{k_2 t} - B(0) = k_1 A_0 \int_0^t e^{(k_2 - k_1)t}\, dt$$

$$Be^{k_2 t} = \frac{k_1 A_0}{k_2 - k_1} [e^{(k_2 - k_1)t}]_0^t = \frac{k_1 A_0}{k_2 - k_1} [e^{(k_2 - k_1)t} - 1]$$

or

$$B = \frac{k_1 A_0}{k_2 - k_1} [e^{-k_1 t} - e^{-k_2 t}].$$

It is now possible to determine $C(t)$ from $dC/dt = k_2 B$. However, this integration is unnecessary, since $C(t)$ is given by the stoichiometry. Adding the three initial differential equations gives

$$\frac{d}{dt}(A + B + C) = 0 \qquad \text{or} \qquad A + B + C = \text{const.} = A_0.$$

Exercises

*18. The integrating factor for the linear first-order differential equation

$$dy + \{M(x)y - N(x)\}\, dx = 0$$

is given as $\exp\{\int M(x)\, dx\}$. Show that this is correct and derive Eq. 4.3.

19. Find the general solution of
 (a) $y' + y \cos x = \sin 2x = 2 \sin x \cos x$.
 (b) $(x + 1)y' + 2y = (x + 1)^4$.
 (c) $y' - y = \cos x - \sin x$.

*20. The differential equation for a particle thrown upward and subject to a gravitational force and also a frictional deceleration proportional to its velocity is

$$\frac{d^2z}{dt^2} + k\frac{dz}{dt} = -g \qquad z(0) = 0,\ z'(0) = v_0$$

Solve the differential equation for z as a function of t. (If the first integration baffles you, let $p = dz/dt$. Check that your answer approaches the expected limiting behavior as k approaches zero.) Also, what is the limiting velocity of the particle subject to friction as t approaches infinity? (Notice that the limiting velocity may be determined very simply by setting $d^2z/dt^2 = 0$ and solving the simplified differential equation.) This equation is a basis for Milliken's oildrop experiment, except that the gravitational force must be corrected for buoyancy and the frictional force must be related to air viscosity.

*21. What function is given by the series expansion

$$y = f(x) = x + \frac{2}{3}x^3 + \frac{2 \cdot 4}{3 \cdot 5}x^5 + \frac{2 \cdot 4 \cdot 6}{3 \cdot 5 \cdot 7}x^7 + \cdots = \sum_{n=0}^{\infty} \frac{2^{2n}(n!)^2}{(2n + 1)!}x^{2n+1}.$$

[*Hint:* Differentiate term by term, then find $y' - x^2y'$ as a function of x and y. Solve this differential equation and evaluate the constant of integration by noting that $f'(0) = 1$.]

*22. Solve the kinetics $A + B \xrightarrow{k_1} C \xrightarrow{k_2} D + E$; $A(0) = A_0 = B(0)$, $C(0) = D(0) = E(0) = 0$ in terms of the exponential integral, Ei.

*23. The kinetics $A \rightarrow B \rightarrow C$ may be solved by Laplace transforms also: Apply the Laplace transform to the first two differential equations, in dA/dt and dB/dt, in order to replace differentiation by simple multiplication. This yields two simultaneous linear equations in LA and LB; solve them. In Exercise A5.17 we present a general method for inverting Laplace transforms, and there is a table of functions and their Laplace transforms given in the *Handbook of Chemistry and Physics*. But for now, use the results cf Exercise 1.44 as a partial table of Laplace transforms to find A and B. (*Hint:* Expand LB in partial fractions.)

4.4 Linear Homogeneous Equations

A linear differential equation is said to be HOMOGENEOUS if $N(x) = 0$, that is, if the differential equation is $L_N y = 0$. Most of the differential equations of physics which lead to special functions are linear homogeneous second-order equations, whose general form is usually written

$$y'' + p(x)y' + q(x)y = 0. \tag{4.4}$$

For example, Bessel functions, Legendre polynomials, and Hermite polynomials arise as solutions to such equations. However, there is *no* general method for reducing such equations to integrals.

An important theorem is the following: If y_1 and y_2 are solutions of $L_N y = 0$, then so is any linear combination of y_1 and y_2. This follows very simply from the fact that L_N is a linear operator:

$$L_N y_1 = 0 \quad \text{and} \quad L_N y_2 = 0 \qquad \rightarrow \qquad L_N(c_1 y_1 + c_2 y_2) = 0 \tag{4.5}$$

And, in particular, if y_1 is a solution of $L_N y = 0$, then so is any multiple of y_1. Furthermore, it can be shown that the Nth-order equation, $L_N y = 0$, has N linearly independent solutions $\{y_i\}$. Then the general solution is a linear combination $\sum c_i y_i$ of these solutions, and the coefficients $\{c_i\}$ are the N arbitrary constants that must arise in the general solution of an Nth-order differential equation. For example, in Exercise 4.1 we have encountered the differential equation $y'' + y = 0$, whose solutions are $\sin x$ and $\cos x$; any linear combination of $\sin x$ and $\cos x$ is also a solution. Likewise, the differential equation $y'' - y = 0$ has as its general solution any linear combination of $\sinh x$ and $\cosh x$ (Exercise 4.2).

Exercises

24. Convince yourself that the zero function is always a solution to any linear homogeneous differential equation. This is a trivial solution which we shall always ignore.

*25. There is no general method of reducing a second-order linear equation to integrals. However, if $y = y_0(x)$ is one solution of the homogeneous equation $y'' + p(x)y' + q(x)y = 0$, the substitution $y = y_0(x)u(x)$ will convert this second-order equation to two integrations. Show that this is so, and then apply this method to solve for the other solution of $x^2y''/(x \tan x - 2) + xy' - y = 0$, whose obvious solution is $y = x$.

A special case of the linear equation arises if the coefficients $M_n(x)$ are not functions of x, but real constants m_n. The resulting differential equation

$$L_N y = \sum_{n=0}^{N} m_n D^n y = 0 \tag{4.6}$$

is readily solvable as follows.

Consider L_N as an Nth-order polynomial $\sum_{n=0}^{N} m_n D^n$ in the symbol D. This Nth-order polynomial has N roots, r_i, satisfying the equation $\sum_{n=0}^{N} m_n r_i^{\,n} = 0$. (These roots need not be distinct or real.) Therefore the polynomial may be factored:

$$\sum_{n=0}^{N} m_n D^n = \prod_{i=1}^{N} (D - r_i)$$

or

$$L_N y = \prod_{i=1}^{N} (D - r_i) y = 0.$$

Notice that the product notation means successive application of the operators. This last form suggests that the solution to $(D - r_i)y = 0$ may also be a solution of $L_N y = 0$. The suggested solution is $y = e^{r_i x}$, and substitution into $L_N y = 0$ shows that it is indeed a correct solution (Exercise 4.26).

Furthermore, we have N solutions in all, one to each root r_i. But since L_N is a linear operator, any linear combination of these solutions is also a solution. Therefore we may conclude that the general solution to the differential equation is

$$y(x) = \sum_{i=1}^{N} c_i e^{r_i x}. \tag{4.7}$$

Two further possibilities must be considered, multiple roots and complex roots.

If two roots are equal, then the polynomial contains the factor $(D - r_i)^2$. The above general solution would imply that there are two identical terms in the sum, so there must be a term missing (in order to insure the existence of N constants of integration). However, it is readily shown that if there are two identical roots, r_i, this root contributes two terms, $c_i e^{r_i x} + c_{i+1} x e^{r_i x}$ (cf. Exercise 4.30).

If the coefficients m_n are real, complex roots occur in pairs, since if $\sum m_n r_i{}^n = 0$, $\sum m_n^*(r_i^*)^n = \sum m_n(r_i^*)^n = 0^* = 0$, that is, if $a + ib$ is a solution, so is $(a + ib)^* = a - ib$. The two terms contributed by these roots are

$$c_1 e^{(a+ib)x} + c_2 e^{(a-ib)x} = e^{ax}[c_1 e^{ibx} + c_2 e^{-ibx}].$$

But if any such linear combination of e^{ibx} and e^{-ibx} is a solution, so is any such linear combination of $(e^{ibx} + e^{-ibx})$ and $(e^{ibx} - e^{-ibx})$. Therefore other forms for these terms are

$$e^{ax}[c_1' \cos bx + c_2' \sin bx]$$
$$ce^{ax} \cos (bx + \delta).$$

Example

2. Find the general solution of the differential equation $y''' + 5y'' + 11y' + 15y = 0$. Setting the operator polynomial equal to zero gives

$$D^3 + 5D^2 + 11D + 15 = 0$$

which has an obvious root $r_1 = -3$ and the factor $(D + 3)$. Polynomial long division leads to the factorization

$$(D + 3)(D^2 + 2D + 5) = 0$$

with the remaining roots at $-1 \pm 2i$. Therefore the general solution is

$$y = c_1 e^{-3x} + c_2 e^{-x} \cos 2x + c_3 e^{-x} \sin 2x.$$

Exercises

*26. Convince yourself that the factorization $\sum m_n D^n = \prod (D - r_i)$ is indeed legitimate for operators. Then convince yourself that if $(D - r_j)y_j = 0$, then $\prod_i (D - r_i)y_j = 0$.

27. Find the general solutions of
    (a) $y'' + 4y = 0$.
    (b) $y'' - 4y' + 5y = 0$.
    (c) $y'' + y' - 2y = 0$.
    (d) $y'' + 2y' + y = 0$.
28. Find the particular solutions of
    (a) $y'' + y' - y = 0;\ y(0) = 1,\ y(-\infty) = 0$.
    (b) $y'' - 2y' + 2y = 0;\ y(0) = 1,\ y(\pi/2) = 1$.
29. A particle of mass m at a distance x from the origin is attracted to the origin by a spring which exerts a force $F = -kx$.
    (a) Newton's second law states that $F = m(d^2x/dt^2)$. Find $x(t)$, subject to the initial conditions $x(0) = x_0$, $(dx/dt)|_0 = 0$.
    (b) The kinetic energy T is given by $T = \frac{1}{2}m(dx/dt)^2$. Find $T(t)$ and $T_{\max}$, the maximum value of T.
    (c) The potential energy V is defined by $V = T_{\max} - T$. Find V as a function of time.
    (d) Find V as a function of x.

*30. Generalize the comment about multiple roots of the Nth-order polynomial arising in the solution of an Nth-order linear homogeneous differential equation: Solve the third-order differential equation $(D - r)^3 y = 0$, whose only obvious solution is $y = ce^{rx}$. (*Hint:* Try for a solution of the form $y = c(x)e^{rx}$.)

4.5 Linear Inhomogeneous Equations

We return to the more general problem $L_N y = N(x)$. To obtain the general solution to this equation, all we need is one particular solution to it and the general solution to the homogeneous equation $L_N y = 0$. Let y_0 be a particular solution satisfying $L_N y_0 = N(x)$, and let $(c_1 y_1 + c_2 y_2)$ be the general solution to $L_N y = 0$.

Then the general solution to $L_N y = N(x)$ is $y_0 + c_1 y_1 + c_2 y_2$ since $L_N(y_0 + c_1 y_1 + c_2 y_2) = L_N y_0 + L_N(c_1 y_1 + c_2 y_2) = L_N y_0 = N(x)$. This theorem simplifies the solution of such inhomogeneous equations since only one particular solution need be found to it, and the problem of finding the general solution is a problem involving a homogeneous equation, which is simpler. Furthermore, if the general solution to the homogeneous equation is known, there is a general method for finding a particular solution to the inhomogeneous equation. The method involves the technique of "variation of parameters" and is presented in Exercise 4.34.

Example

3. Find the general solution of $y''' + 5y'' + 11y' + 15y = 3x$. A trial solution of the form $y = \alpha x + \beta$ leads to the simple particular solution $y = \frac{1}{5}x - \frac{11}{75}$. In Example 4.2 we derived the general solution to the homogeneous equation. Therefore the general solution to the inhomogeneous equation is

$$y(x) = \tfrac{1}{5}x + c_1 e^{-3x} + c_2 e^{-x}\cos 2x + c_3 e^{-x}\sin 2x - \tfrac{11}{75}.$$

Exercises

31. (a) Solve Exercise 4.19c by considering it in the form $L_1y = N(x)$.
   *(b) Find the general solution of $(D^4 + 1)y = 2\cos x$.

*32. A light beam of frequency ω exerts a periodic force on a diatomic molecule whose natural frequency of harmonic oscillation is ω_0. If x is the displacement of the internuclear distance from its equilibrium value, the differential equation to be solved for the classical forced vibration of the molecule is $x'' + \omega_0{}^2x = F_0 \sin \omega t$. Solve for the motion of the molecule if it is initially at rest in its equilibrium configuration. (*Hint:* Try for a particular solution oscillating with the forcing frequency ω.) What forcing frequency leads to maximum vibrational amplitude? This is an example of resonance.

*33. A harmonic oscillator initially at rest is smacked at $t = 0$. The resulting motion is described by the differential equation, $x'' + \omega^2 x = F_0\delta(t)$. (Notice how the delta function may be used to represent a force of infinitesimal duration but finite effect.) Take Laplace transforms of both sides, simplify the transform of a second derivative, refer to Exercise 3.36 for $L\{\delta(t)\}$, solve for $L\{x(t)\}$, and use the result of Exercise 1.82 to invert the Laplace transform and find $x(t)$.

*34. Let the general solution of $L_2y = y'' + p(x)y' + q(x)y = 0$ be $c_1y_1 + c_2y_2$ such that $W(x) \equiv y_1y_2' - y_2y_1' \neq 0$. To find a particular solution of the *inhomogeneous* equation $L_2y = r(x)$, try for a solution of the form $y = c_1(x)y_1(x) + c_2(x)y_2(x)$ such that $c_1'y_1 + c_2'y_2 = 0$. Find a second equation which c_1' and c_2' must satisfy (*Hint:* To help eliminate c_1'' and c_2'', differentiate the equation of constraint.), solve the two simultaneous equations, and reduce the problem to integrals.

*35. One of the most important nonlinear differential equations is the equation of motion of a pendulum $d^2\theta/dt^2 = -(g/l)\sin\theta$, subject to the initial condition $\theta(0) = \theta_0$, $(d\theta/dt)|_0 = 0$. The usual approximation is to linearize the equation by replacing $\sin\theta$ by θ. However, this equation may be reduced to an elliptic integral: Let $p(t) = d\theta/dt$. Then $p(t)$ is an integrating factor for the resulting first-order equation. After this is solved, the resulting first-order equation is separable. Then convert the integral to a standard form. (*Hint:* $\cos\phi = 1 - \sin^2(\phi/2)$.)

4.6 Solution by Power-Series Expansion

A general technique for solving differential equations involves expanding the solution in a power series and determining the coefficients. This technique is quite general, but is most suited to linear differential equations in which $N(x)$ and each $M_n(x)$ are such as to facilitate collection of terms. It is quite cumbersome to present the technique in full generality, but the method and its limitations should be clear from the following examples.

Example

4. Solution of $y'' + y = 0$ by power series expansion: This is a very simple linear second-order differential equation whose solutions are linear combinations of $\sin x$ and $\cos x$. But what would we do if we weren't already familiar with these functions or didn't recognize them as solutions? The approach we shall try is to assume that $y(x)$ may be expanded as a

power series:

$$y = \sum_{n=0}^{\infty} a_n x^n.$$

$$y' = \sum_{n=0}^{\infty} n a_n x^{n-1} = \sum_{n=1}^{\infty} n a_n x^{n-1}.$$

$$y'' = \sum_{n=1}^{\infty} n(n-1) a_n x^{n-2} = \sum_{n=2}^{\infty} n(n-1) a_n x^{n-2}.$$

Next we change the dummy summation index by letting $m = n - 2$ and then let $n = m$.

$$y'' = \sum_{m=0}^{\infty} (m+2)(m+1) a_{m+2} x^m = \sum_{n=0}^{\infty} (n+2)(n+1) a_{n+2} x^n.$$

(Enumeration of the terms in these expansions of y'' shows that the series are identical.) Substitution into the differential equation and collection of coefficients of x^n gives

$$\sum_{n=0}^{\infty} [(n+2)(n+1)a_{n+2} + a_n] x^n = 0.$$

This is a remarkable equation—the left-hand side depends upon x but the right-hand side is zero, independent of x. The only way in which the left-hand side can also be independent of x is if *every* coefficient $(n+1)(n+2)a_{n+2} + a_n$ is equal to zero. (Another way of demonstrating this requirement is by noting that the left-hand side is the Maclaurin series expansion of the zero function, all of whose derivatives, evaluated at zero, are zero.) Therefore

$$a_{n+2} = \frac{-a_n}{(n+1)(n+2)}$$

is a recursion formula, which gives the $(n+2)$nd coefficient in terms of the nth. Choosing a value for a_0 serves to specify $a_2 = -a_0/1 \cdot 2$, $a_4 = -a_2/3 \cdot 4 = a_0/4!$, $a_6 = \cdots$, the coefficients of all even powers of x. Choosing a value for a_1 serves to specify $a_3, a_5, \ldots$, the coefficients of all odd powers of x. These first two coefficients may be chosen independently, as is reasonable, since a second-order differential equation has two constants of integration. Then an investigation of "what the recursion formula does" leads to the general formulas $a_n = (-1)^{n/2} a_0/n!$ for n even and $a_n = (-1)^{(n-1)/2} a_1/n!$ for n odd. Thus

$$y = a_0 \sum_{\substack{n=0 \\ n \text{ even}}}^{\infty} (-1)^{n/2} \frac{x^n}{n!} + a_1 \sum_{\substack{n=1 \\ n \text{ odd}}}^{\infty} (-1)^{(n-1)/2} \frac{x^n}{n!}$$

$$= a_0 \sum_{n=0}^{\infty} (-1)^n \frac{x^{2n}}{(2n)!} + a_1 \sum_{n=0}^{\infty} (-1)^n \frac{x^{2n+1}}{(2n+1)!}.$$

Exercises

36. If $a_1 = 1$ and $a_n = [(1-n)/n]a_{n-1}$, what is the general expression for a_n?
37. Find the series expansion of the even function which satisfies $xy'' + y' + xy = 0$.

4.7 Legendre Polynomials

The linear homogeneous second-order differential equation

$$(1 - x^2)y'' - 2xy' + l(l + 1)y = 0$$

is not solvable in terms of familiar functions. Let us try for a series solution of the form $y = \sum_{n=0}^{\infty} a_n x^n/n!$. (Notice the variant.) Then

$$y' = \sum_{n=0}^{\infty} na_n x^{n-1}/n! = \sum_{n=1}^{\infty} a_n x^{n-1}/(n - 1)!$$

$$y'' = \sum_{n=2}^{\infty} a_n x^{n-2}/(n - 2)!.$$

Substituting into the differential equation gives

$$\sum_{n=2}^{\infty} \frac{a_n x^{n-2}}{(n-2)!} - \sum_{n=2}^{\infty} \frac{a_n x^n}{(n-2)!} - 2\sum_{n=1}^{\infty} \frac{a_n x^n}{(n-1)!} + l(l+1)\sum_{n=0}^{\infty} \frac{a_n x^n}{n!} = 0.$$

Setting $n' = n - 2$ in the first term, then setting $n = n'$, converts this term to $\sum_{n=0}^{\infty} a_{n+2}x^n/n!$. Collecting the coefficients of x^0, x^1, and x^n $(n \geqslant 2)$ leads to

$$[a_2 + l(l + 1)a_0] + [a_3 - 2a_1 + l(l + 1)a_1]x + \sum_{n=2}^{\infty}\left[\frac{a_{n+2}}{n!} - \frac{a_n}{(n-2)!} - \frac{2a_n}{(n-1)!} + \frac{l(l+1)a_n}{n!}\right]x^n = 0$$

By the same reasoning as before, the coefficient of every power of x must be zero. Therefore

$$\begin{aligned} a_2 &= -l(l + 1)a_0. \\ a_3 &= -[l(l + 1) - 2]a_1 = -(l - 1)(l + 2)a_1. \\ a_{n+2} &= -[l(l + 1) - 2n - n(n - 1)]a_n = -(l - n)(l + n + 1)a_n. \end{aligned}$$

Inspection shows that this recursion formula holds for $n = 0$ or 1 as well. If a_0 is specified, every even coefficient may be evaluated by application of the recursion formula. Likewise, once a_1 is specified, every odd coefficient is determined. *Two* coefficients may be specified independently since two independent solutions are to be expected for a linear second-order differential equation.

In the case that l is a nonnegative integer, it follows that $a_{l+2} = 0$, as are all further coefficients which are proportional to a_{l+2}. This means that one series solution ends with the term x^l while the other continues as an infinite series. For most practical purposes, it is the lth-order polynomial solution which is desired. This polynomial is known as the lth Legendre polynomial $P_l(x)$. If l is an even integer, a_1 is set equal to zero and the polynomial contains only even powers of x. If l is an odd integer, a_0 is set equal to zero and the polynomial contains only odd powers of x.

Instead of specifying the first nonzero coefficient, it is conventional to arrange that $P_l(1) = 1$, which requires that the coefficient of x^l be $(2l)!/2^l l!^2$. Coefficients of lower terms are then obtained by application of the recursion formula. Then

$$P_l(x) = \frac{(2l)!}{2^l l!^2}\left[x^l - \frac{l(l-1)}{2(2l-1)}x^{l-2} + \frac{l(l-1)(l-2)(l-3)}{2\cdot 4\cdot(2l-1)(2l-3)}x^{l-4}\cdots\right].$$

The first few Legendre polynomials are

$$P_0(x) = 1 \qquad P_2(x) = \tfrac{1}{2}(3x^2 - 1)$$

$$P_1(x) = x \qquad P_3(x) = \tfrac{1}{2}(5x^3 - 3x).$$

Further polynomials may be obtained from the recursion formula $(l+1)P_{l+1}(x) - (2l+1)xP_l(x) + lP_{l-1}(x) = 0$. The generating function for the Legendre polynomials is

$$(1 - 2xt + t^2)^{-1/2} = \sum_{l=0}^{\infty} P_l(x)t^l.$$

It can also be shown that $P_l(x) = D^l[(x^2-1)^l]/2^l l!$.

The Legendre polynomials are not unfamiliar to you. The angular dependence of the hydrogenic orbital wave function with quantum numbers n, l, m is given by $P_l^{|m|}(\cos\theta)e^{im\phi}$, where $P_l^{|m|}$ is an ASSOCIATED LEGENDRE FUNCTION defined by $P_l^{|m|}(x) = (1-x^2)^{|m|/2}D^{|m|}P_l(x)$. In particular, with $m = 0$, the wave function for an s, p_z, d_{z^2}, or f_{z^3} orbital is an appropriate function of r times 1, $\cos\theta$, $\frac{1}{2}(3\cos^2\theta - 1)$, or $\frac{1}{2}(5\cos^3\theta - 3\cos\theta)$, respectively. Indeed, the functions $P_l^{|m|}(\cos\theta)e^{im\phi} \propto Y_{lm}(\theta, \phi)$ arise in many problems showing spherical symmetry; these functions are called SPHERICAL HARMONICS or SURFACE HARMONICS.

Exercises

*38. (a) Expand $1 - x^2$ as a sum of terms, $P_l(x)$.
(b) Expand $\sin^2\theta$ as a sum of terms, $P_l(\cos\theta)$.

39. Use the generating function of the Legendre polynomials to show that $P_l(1) = 1$.

*40. Use the generating function to show that the Legendre polynomial P_l is even or odd according as l is even or odd.

41. Use the recursion formula relating the various Legendre polynomials to find $P_4(x)$.

*42. Express $1/r_{12} = [r_1^2 + r_2^2 - 2r_1r_2\cos\theta]^{-1/2}$ as a power series involving Legendre polynomials. (*Caution:* For definiteness let $r_1 > r_2$ and watch convergence.)

43. Starting from the differential equation for the Legendre polynomials, show that the functions P_l and P_m, $l \neq m$, are orthogonal on the interval -1 to 1. Then let $x = \cos\theta$ to obtain a generalized orthogonality condition on $\{P_l(\cos\theta)\}$.

44. What are the spherical harmonics $Y_{11}(\theta, \phi)$, $Y_{1-1}(\theta, \phi)$, $Y_{21}(\theta, \phi)$, $Y_{2-1}(\theta, \phi)$, $Y_{22}(\theta, \phi)$, $Y_{2-2}(\theta, \phi)$, and $Y_{30}(\theta, \phi)$? Neglect proportionality constants. Convince yourself that (except perhaps for a factor of i) $Y_{11} \pm Y_{1-1}$ "look like" p_x and p_y orbitals, $Y_{21} \pm Y_{2-1}$ like d_{xz} and d_{yz} orbitals, and $Y_{22} \pm Y_{2-2}$ like $d_{x^2-y^2}$ and d_{xy}

orbitals. Use the form of Y_{30} to sketch an f_{z^3} orbital. (*Hint:* What are the surfaces on which $Y_{30}(\theta, \phi) = 0$?) Sketch $f_{x^2-y^2,z}$ and f_{xyz} orbitals, which look like $Y_{32} \pm Y_{3-2}$.

4.8 Bessel Functions

Bessel's differential equation is $x^2y'' + xy' + (x^2 - n^2)y = 0$, which may be written in standard form as $y'' + (1/x)y' + (1 - n^2/x^2)y = 0$. We encounter further difficulty with this equation because the coefficient functions, $p(x) = 1/x$ and $q(x) = (1 - n^2/x^2)$, do not have power series expansions about $x = 0$. The following generalization is applicable to such a problem in the case where $xp(x)$ and $x^2q(x)$ do have power series expansions about $x = 0$.

Let us try for a solution of the form $y = \sum_{k=0}^{\infty} a_k x^{r+k}$, where r is a number (not necessarily integral) yet to be determined, such that the first nonzero term in the expansion is $a_0 x^r$. Then

$$y' = \sum_{k=0}^{\infty}(r + k)a_k x^{r+k-1} \qquad \text{and} \qquad y'' = \sum_{k=0}^{\infty}(r + k)(r + k - 1)a_k x^{r+k-2}.$$

Substituting into the first form of the differential equation and collecting terms, we obtain

$$\sum_{k=0}^{\infty}[(r + k)(r + k - 1) + (r + k) - n^2]a_k x^{r+k} + \sum_{k=0}^{\infty} a_k x^{r+k+2} = 0.$$

Setting $k' = k + 2$ in the last term, then setting $k = k'$ and simplifying the first term gives

$$\sum_{k=0}^{\infty}[(r + k)^2 - n^2]a_k x^{r+k} + \sum_{k=2}^{\infty} a_{k-2} x^{r+k} = 0.$$

Again, the coefficient of every power of x must be zero. The coefficient of x^r is $(r^2 - n^2)a_0$. Since we have postulated that $a_0 \neq 0$, we may conclude that

$$r^2 - n^2 = 0 \qquad \text{or} \qquad r = \pm n.$$

This equation is called the INDICIAL EQUATION. It arises from setting equal to zero the coefficient of $a_0 x^r$ in the expansion of the zero function. It is always a quadratic in r and serves to determine two values of r, one for each independent solution. (If the two roots happen to be identical, only one solution can be found by this technique.) We shall consider only the solution for which r is nonnegative, to ensure that y be continuous at $x = 0$.

The coefficient of x^{r+1} is $[(r + 1)^2 - n^2]a_1 = [(r^2 - n^2) + 2r + 1]a_1$, which equals $(2r + 1)a_1$ since $r^2 - n^2 = 0$. From the fact that this coefficient must also equal zero, we may conclude that $a_1 = 0$ except for the special case that $r = -\frac{1}{2}$. (This special case arises only if $|n| = \frac{1}{2}$ and is considered in Exercise 4.46.)

Setting further coefficients equal to zero leads to the recursion formula

$$a_k = -\frac{a_{k-2}}{r^2 - n^2 + 2rk + k^2} = -\frac{a_{k-2}}{k(2r + k)}.$$

Unless $|n| = \frac{1}{2}$, a_1 and all odd coefficients are necessarily zero. It is conventional to set $a_0 = 1/2^n n!$.

Therefore the solution of interest, denoted by $J_n(x)$, the nth-order Bessel function of the first kind, is

$$J_n(x) = a_0 x^n\left[1 - \frac{x^2}{2(2n + 2)} + \frac{x^4}{2(2n + 2)4(2n + 4)} - \frac{x^6}{2(2n + 2)4(2n + 4)6(2n + 6)} \pm \cdots\right]$$

$$= \frac{x^n}{2^n}\sum_{k=0}^{\infty}(-1)^k \frac{x^{2k}}{2^{2k}k!(n + k)!}$$

which is readily generalized to nonintegral n by replacing $(n + k)!$ by $\Gamma(n + k + 1)$. Since this was a second-order differential equation, there is another solution, $N_n(x)$, the nth-order Bessel function of the second kind, but this solution is discontinuous at $x = 0$ and is usually disregarded.

I have introduced the special functions of Legendre and Bessel because they occur as solutions to commonly encountered second-order differential equations which cannot be solved in terms of familiar functions. Similarly, the differential equation $y'' + y = 0$ leads to the functions sin and cos, which arise so frequently in elementary contexts. Therefore their values are tabulated, and a problem may be considered solved if it can be reduced to sin and cos. Many less familiar functions which are encountered in physical problems are also tabulated, and a problem may be considered solved if it can be reduced to such a function. You should not be discouraged when you encounter an unfamiliar function; to the extent that the values are tabulated, it is no more terrifying than sin or cos.

Exercises

*45. Use the result of Exercise 2.6 to show that when n is integral,

$$J_n(x) = \frac{1}{\pi}\int_0^\pi \cos(x \sin\phi - n\phi)\, d\phi$$

satisfies Bessel's differential equation.

*46. Solve Bessel's equation for the special case $n = \frac{1}{2}$ by setting $y = x^{-1/2}u$ and solving the resulting differential equation for u. Which solution is finite at $x = 0$?

*47. Solve Laguerre's differential equation

$$L_n''(x) + \frac{1 - x}{x} L_n'(x) + \frac{n}{x} L_n(x) = 0$$

by expansion in power series. Find only that solution which is continuous at $x = 0$.

*48. An operator L is said to be SELF-ADJOINT if it is of the form

$$L[u(x)] = [p(x)u'(x)]' - q(x)u(x).$$

The STURM-LIOUVILLE DIFFERENTIAL EQUATION (of what order?) is $L[u] + \lambda w(x)u = 0$, where the solution $u(x)$ is defined on the interval, $a \leqslant x \leqslant b$. Only certain solutions are "allowable": for u and v to be allowable it is required that $v(a)p(a)u'(a) = v(b)p(b)u'(b)$ and $v'(a)p(a)u(a) = v'(b)p(b)u(b)$. (This requirement restricts the possible values of λ.)

(a) Show that $\int_a^b vL[u]\,dx = \int_a^b uL[v]\,dx$.

(b) Show that $(\lambda_i - \lambda_j)\int_a^b u_i u_j w(x)\,dx = 0$. This means that acceptable solutions belonging to different values of λ are said to be ORTHOGONAL WITH RESPECT TO THE WEIGHT FUNCTION $w(x)$ on the interval $a \leqslant x \leqslant b$.

(c) Write Legendre's differential equation in Sturm-Liouville form.

(d) Write Bessel's differential equation in Sturm-Liouville form.

4.9 Eigenvalue-Eigenfunction Problems

Many differential equations, especially in quantum mechanics, arise in the form of an EIGENVALUE-EIGENFUNCTION EQUATION

$$\Omega\psi = \lambda\psi \tag{4.8}$$

where Ω is a linear operator. An operator is something which assigns functions to functions; what this equation asks is, "What functions are assigned themselves, except for a constant multiplier?"

The constant multiplier λ is known as the EIGENVALUE. The corresponding function ψ is known as the EIGENFUNCTION. If ψ is known, λ may be determined by applying Ω to ψ; in this sense the eigenvalue may be said to belong to the eigenfunction. Nevertheless, it is customary to say that it is the eigenfunction that belongs to the eigenvalue. It is possible that more than one independent function belong to a single eigenvalue; such an eigenvalue is said to be DEGENERATE of a MULTIPLICITY equal to the number of linearly independent functions belonging to it.

Any eigenfunction is arbitrary to the extent of a constant factor: If $\Omega\psi = \lambda\psi$, then since Ω is linear, $\Omega(C\psi) = \lambda(C\psi)$; if ψ is an eigenfunction belonging to eigenvalue λ, then so is $C\psi$. Any linear combination of eigenfunctions belonging to a degenerate eigenvalue is also an eigenfunction belonging to that eigenvalue: if $\Omega\psi_1 = \lambda\psi_1$ and $\Omega\psi_2 = \lambda\psi_2$, then $\Omega(C_1\psi_1 + C_2\psi_2) = \lambda(C_1\psi_1 + C_2\psi_2)$, again because Ω is linear.

Not all numbers and functions are allowable as eigenvalues and eigenfunctions, respectively. In most situations there are boundary conditions which an eigenfunction must satisfy, and these restrictions eliminate many otherwise legitimate eigenvalues.

Examples

5. $\Omega = D$. The solution of $D\psi = \lambda\psi$ is $\psi(x) = Ce^{\lambda x}$, with λ arbitrary. However, if it is required that $\psi(x)$ approach zero as x approaches infinity, then λ is restricted to negative values. Alternatively, if it is required that $|\psi(x)|$ be bounded (less than some maximum value for all x), then λ is restricted to imaginary values.

6. $\Omega = xD$. The solution of $xD\psi = \lambda\psi$ is $\psi(x) = Cx^{\lambda}$. If it is required that ψ be continuous everywhere, then λ cannot be negative. If it is required that ψ and all its derivatives be continuous everywhere, then λ must be a nonnegative integer.

7. $\Omega = x$. The only possible solution to the equation $x\psi = \lambda\psi$ is $\psi(x) = C\delta(x - \lambda)$ since $\int x\delta(x - \lambda)\, dx = \lambda \int \delta(x - \lambda)\, dx$. All values of λ are possible; the usual boundary conditions cannot be applied.

In quantum mechanics the usual boundary conditions are (*1*) ψ must be a genuine (single-valued) function, (*2*) ψ must be continuous (and therefore finite-valued), and (*3*) $\int |\psi(x)|^2\, dx = 1$. Although it would seem that these boundary conditions are not very demanding, they are actually quite restrictive and permit only certain eigenvalues. Condition (*3*) states that ψ must be NORMALIZED TO UNITY; this requirement also fixes the arbitrary factor in the eigenfunction.

Exercises

49. Write Legendre's equation and Bessel's equation in the form of eigenvalue-eigenfunction problems.
50. Use the results of Exercise 1.17 to find the eigenvalues and eigenfunctions of the spin operators s^2 and s_z. Notice that the eigenvalue of s^2 is doubly degenerate.
51. Convince yourself that if the eigenfunction ψ is not normalized to unity, division by $(\int |\psi(x)|^2\, dx)^{1/2}$ converts ψ to one that is.
52. Find the eigenvalues and corresponding eigenfunctions of the operator $-D^2$ if it is required that for a solution to be acceptable, $\psi(0) = 0 = \psi(L)$ and $\int_0^L \psi^*(x)\psi(x)\, dx = 1$. (*Note:* This problem and the next one are quite closely related to particle-in-a-box models for the electronic structure of linear and cyclic polyenes.
53. Find the eigenvalues and corresponding eigenfunctions of the operator $-D^2$ if it is required that in order for a solution to be acceptable, $\psi(\theta + 2\pi) = \psi(\theta)$, and $\int_{-\pi}^{\pi} \psi^*(\theta)\psi(\theta)\, d\theta = 1$. Notice that all eigenvalues except one are doubly degenerate.
54. The quantum-mechanical treatment of the harmonic oscillator leads to the eigenvalue-eigenfunction problem $(-D^2 + x^2)\psi = \lambda\psi$, where $\lambda = 2E/h\nu$. Substitute $\psi(x) = \exp(-x^2/2)H(x)$, and solve the resulting differential equation by expansion in power series. It can be shown that in order to satisfy the boundary condition that $\int \psi^2\, dx$ be finite, it is necessary to terminate the power series after a finite number of terms, so that $H(x)$ is an nth-order polynomial. What form of λ does this condition require? What is $H_n(x)$? (*Note:* Instead of choosing a value for the coefficient a_0 or a_1, it is customary to set $a_n = 2^n$. Then the recursion formula may be used to find coefficients of all lower powers of x.)
55. (a) Show that if ψ_i is an eigenfunction of the linear operator Ω with eigenvalue λ_i, then the operator $\Omega - \lambda_i 1$ acts on ψ_i to produce the zero function, or to ANNIHILATE ψ_i.
    (b) Define the operator $P_i \equiv \prod_{j \neq i} (\Omega - \lambda_j 1)/(\lambda_i - \lambda_j)$. For an arbitrary function ϕ, which is a linear combination of normalized eigenfunctions of Ω, what is $P_i\phi$?

(c) What is $P_iP_j\phi$ if $i \neq j$?

(d) What is $P_i^2\phi$? (*Note:* Since ϕ is arbitrary, this result must hold for every function. Therefore we may conclude that $P_i^2 = P_i$. Such an operator is said to be IDEMPOTENT, or a PROJECTION OPERATOR.)

56. (a) Find all eigenvalues and describe all eigenfunctions $\psi(x)$ of the INVERSION OPERATOR I, defined by $If(x) = f(-x)$. (*Hint:* If ψ is an eigenfunction of I with eigenvalue λ, what is $I^2\psi$?)
    (b) For each eigenvalue, construct an operator that will act on an arbitrary function ϕ to produce an eigenfunction of I with that eigenvalue. (*Hint:* Use the result of Exercise 1.10 or 4.55b.)
57. Use the results of Exercise 2.5 to find all eigenvalues and eigenfunctions of the two-electron spin operators S^2 and S_z. (*Note:* The eigenvalues of the general S^2 are always of the form $S(S + 1)$, where S is integral or half-integral. The corresponding eigenfunctions are called SINGLET, DOUBLET, TRIPLET, etc., according as $(2S + 1)$, which is the multiplicity of the eigenvalue, is 1, 2, 3, etc.)
58. The function $\alpha\alpha\beta$ is a linear combination of doublet and quartet eigenfunctions of S^2. Construct projection operators P_2 and P_4 which will act on $\alpha\alpha\beta$ to produce eigenfunctions of S^2 with eigenvalues $\frac{3}{4}$ and $\frac{15}{4}$, respectively. Find these eigenfunctions.

*59. The function $\alpha\beta\alpha\beta$ is a linear combination of singlet, triplet, and quintet eigenfunctions of S^2. Construct the projection operator P_1 and find $P_1\alpha\beta\alpha\beta$.

60. Define INTERCHANGE OPERATORS acting on a function of n variables by

$$P_{ij}\Phi(x_1, x_2, \ldots, x_i, \ldots, x_j, \ldots, x_n) = \Phi(x_1, x_2, \ldots, x_j, \ldots, x_i, \ldots, x_n)$$

where the indices on P indicate which *variables* are to be interchanged, rather than which *slots* exchange variables.
(a) Show that in general, $[P_{ij}, P_{jk}] \neq 0$.
(b) What operator is $P_{ij}^2 = P_{ij}P_{ij}$?
(c) Give the eigenvalue(s) of P_{ij}^2.
(d) Give the eigenvalue(s) of P_{ij}.
(e) For $n = 2$, construct two operators which will act on $\Phi(x_1, x_2)$ to produce two independent eigenfunctions of P_{12}. What are the resulting functions? What is the eigenvalue corresponding to each?
These results are quite important in connection with the Pauli exclusion principle.

*61. For $n = 3$, construct a projection operator that will act on $\Phi(x_1, x_2, x_3)$ to produce a function that is an eigenfunction of *each* P_{ij}, with eigenvalue -1. (*Hint:* The operator is a product of three operators.)

*62. An electronic wave function for the ground state of the helium atom is a function of the coordinates of electron 1 and the coordinates of electron 2; we shall abbreviate such a function to $\Phi(x_1, x_2)$. The trial function $\phi_1\alpha\phi_2\beta = \phi_1(x_1)\alpha(x_1)\phi_2(x_2)\beta(x_2)$ is not an acceptable wave function because (*1*) it must be an eigenfunction of P_{12} with eigenvalue -1, and (*2*) it must be an eigenfunction of S^2 with eigenvalue zero. Apply projection operators to construct an acceptable function.

*63. Apply the above reasoning to the ground state of the lithium atom, for which an acceptable function must satisfy $P_{ij}\psi = -\psi$, $S^2\psi = \frac{3}{4}\psi$, and $S_z\psi = \frac{1}{2}\psi$.

*64. (a) Let Ω be the operator that converts $f(x)$ into $-f''(x) + x^2f(x)$. What is $\Omega f(y)$?
    (b) Show that Ω commutes with the Fourier transform operator F. (*Hints:* Extend the result of Exercise 1.71b to $F[D^2f]$. Use Leibniz' Rule to find $D^2(Ff)$.)
    (c) We have shown in Exercise 4.54 that there exists an eigenfunction ψ_n of Ω belonging to the nondegenerate eigenvalue $\lambda_n = 2n + 1$. Use merely the result of Part (b) to show that $F\psi_n$ must also be eigenfunction of Ω belonging to λ_n.

(d) Use the fact that λ_n is nondegenerate to convince yourself that $F\psi_n$ must be a multiple of ψ_n. (*Hint:* What does it mean to say that ψ_n and $F\psi_n$ are linearly dependent?)

(e) Use the result of Part (d) to convince yourself that ψ_n must also be an eigenfunction of F.

In Sections 8.14 and 9.3 we shall pursue the phenomenon that commuting operators have the same eigenfunctions.

5

PARTIAL DIFFERENTIAL EQUATIONS

This chapter concludes our discussion of functions. It opens with a discussion of partial differential equations and some elementary ways of solving them, especially the method of separation of variables. We then use the topic of partial differential equations as an entry to discussing several other topics. We review the topic of wave motion in some detail since it is basic to both spectroscopy and quantum mechanics. The discussion of Fourier series and Fourier transforms does not adequately demonstrate the power of these methods, for it is intended only to introduce them to chemists and to indicate a few chemical applications. Finally, we describe expansion in complete sets of orthonormal functions but we do not investigate the detailed properties of these sets. We indicate the similarity of the expansion of a function to the expansion of a vector, and in Chapter 9 we shall pursue this similarity in considerable detail. The material of this chapter is aimed primarily at quantum chemistry, but there are also exercises indicating applications to diffusion processes and polarography, to X-ray diffraction, and to spectroscopy.

5.1 Definitions

A PARTIAL DIFFERENTIAL EQUATION is a differential equation involving partial derivatives. The solution of a partial differential equation is a functional relationship among several variables. We shall be concerned exclusively with linear homogeneous partial differential equations, whose form is $Lu = 0$, where L is a linear operator and u is a function of several variables. Often partial differential equations arise as eigenvalue-eigenfunction problems, $\Omega\psi = \lambda\psi$, where Ω is a linear partial differential operator and ψ is a function of several variables. Again, the order of a partial differential equation is the order of the highest derivative.

In solving an Nth-order differential equation, N arbitrary constants of integration are introduced. However, with a partial differential equation these constants are not numbers but arbitrary functions. As a result, the general solution of a partial differential equation is much too general for any practical purpose. It is necessary to specify some boundary conditions to obtain a less general solution. To obtain a particular solution of an Nth-order partial differential equation, N functions must be specified as boundary conditions.

Examples

1. The general solution of Laplace's equation in two dimensions,

$$\frac{\partial^2 u}{\partial x^2} + \frac{\partial^2 u}{\partial y^2} = 0,$$

is $u(x, y) = f(x + iy) + g(x - iy)$, with f and g completely arbitrary, as is clear on substitution. This solution is too general for any practical use.

2. The general solution of the second-order partial differential equation,

$$\frac{\partial^2 u}{\partial t^2} = c^2 \frac{\partial^2 u}{\partial x^2},$$

is $u(x, t) = f(x - ct) + g(x + ct)$, with f and g arbitrary. It is instructive to sketch a graph of $f(x - ct)$, plotted vs. x, with t identified as time (Figure 5.1). At $t = 0$, $f(x)$ has some shape. After a time interval t has elapsed, $f(x - ct)$ has the same shape but is displaced a distance ct in the $+x$ direction. Therefore we may identify the solution $f(x - ct)$ as the

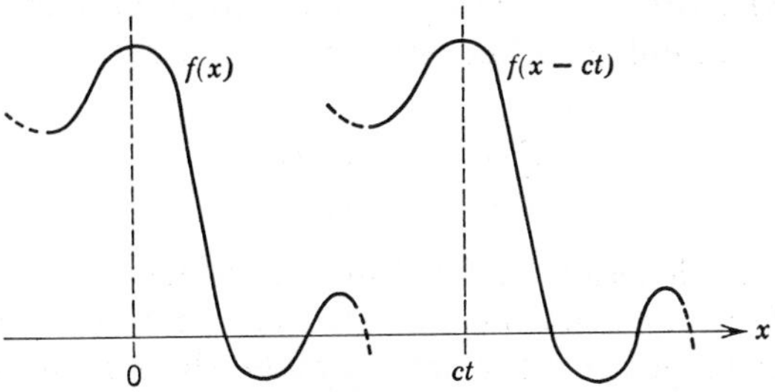

Figure 5.1

curve $f(x)$, traveling in the $+x$ direction with speed c. Likewise, the solution $g(x + ct)$ is a curve traveling in the $-x$ direction with speed c. The general solution represents a wave traveling with speed c; for this reason the differential equation is known as the WAVE EQUATION in one dimension. In the next sections we consider less general solutions of this equation.

Exercises

*1. What is the general solution to $\partial u/\partial x = \partial u/\partial y$?

2. Convince yourself that the graph of $g(x + ct)$ is indeed displaced a distance ct in the *negative* x-direction.

3. Solve the DIFFUSION EQUATION, $\partial c/\partial t = D(\partial^2 c/\partial x^2)$, where $c(x, t)$ is the concentration of some substance at time t in a tube stretching from $x = 0$ to $x = \infty$, subject to the boundary conditions $c(x, 0) = C_0 = c(\infty, t)$ and $c(0, t) = 0$. (These boundary conditions are appropriate for the diffusion of, say, Ag^+ to a cathode located at $x = 0$. Before electrolysis begins, the concentration of Ag^+ is everywhere C_0, but after electrolysis begins the concentration of Ag^+ at the electrode is reduced to zero. Therefore Ag^+ diffuses toward the electrode, but there is no depletion of Ag^+ an infinite distance away.) For convenience, let $u(x, t) = C_0 - c(x, t)$. The resulting partial differential equation may be converted to an ordinary one by taking the Laplace transform with respect to t: $U(x, s) = L[u(x, t)] = \int_0^\infty e^{-st}u(x, t)\,dt$. Use the result of Exercise 1.44 to simplify $L[\partial u/\partial t]$. Solve the resulting second-order equation for $U(x, s)$. Set the constants of integration by determining $U(0, s)$ and $U(\infty, s)$ from the boundary conditions. Finally, use the result of Exercise 3.5 to find $u(x, t)$ and then $c(x, t)$.

5.2 The Vibrating String

A string stretched along the x-axis can vibrate in a direction perpendicular to this axis. Let $u(x, t)$ represent the displacement of the string from its equilibrium position, $u = 0$. The equation of motion of this string is just the wave equation, $(\partial^2 u/\partial t^2) = c^2(\partial^2 u/\partial x^2)$. Here $c^2 = T/\rho$, where T is the tension in the string, ρ is the linear density of the string in g/cm, and c is the speed at which a disturbance is propagated along the string. Let us further assume that the string is of length L and is held fixed at the ends so that $u(0, t) = 0 = u(L, t)$.

We shall solve this partial differential equation by the method of SEPARATION OF VARIABLES. Let us try to find a solution of the form $u(x, t) = X(x)T(t)$. Then the differential equation becomes $XT'' = c^2X''T$. Dividing by $u = XT$ gives

$$\frac{T''(t)}{T(t)} = c^2\frac{X''(x)}{X(x)}.$$

This is a remarkable result; the left-hand side is not a function of x and the right-hand side is not a function of t, yet the two sides are equal. Next we make the crucial argument. The only way whereby a quantity that does not depend upon x can equal a quantity that does not depend upon t is if neither quantity depends upon either x or t. If the left-hand side were to depend on t, it could not be equal to the right-hand side, which is independent of t; if the right-hand side were to depend on x, it could not be equal to the left-hand side, which is independent of x. Therefore we may set both sides equal to a constant, which we shall call $-\omega^2$. This constant is as yet arbitrary, but it must be negative for reasons which will become apparent. Later we shall identify ω as an angular frequency.

We have now reduced our partial differential equation to the two ordinary differential equations, $T''(t)/T(t) = -\omega^2$ and $c^2X''(x)/X(x) = -\omega^2$. The solution to the first equation is any linear combination of $\cos \omega t$ and $\sin \omega t$; for convenience we choose to write this function in the form $T(t) = \cos(\omega t + \delta)$. It is now clear that the constant, $-\omega^2$, must be negative in order to ensure oscillatory motion of the string.

The general solution to the second ordinary differential equation is the linear combination $X(x) = A \sin \omega x/c + B \cos \omega x/c$. The boundary condition $u(0, t) = X(0)T(t) = 0$ requires that $B = 0$. The boundary condition $u(L, t) = X(L)T(t) = 0$ requires that $\sin \omega L/c = 0$ or $\omega L/c = n\pi$ (Exercise 1.68b). Thus we have found that the arbitrary constant ω is restricted to the values $n\pi c/L$. A solution to the partial differential equation, subject to the specified boundary conditions, is

$$u(x, t) = X(x)T(t) = A \sin \frac{n\pi x}{L} \cos\left(\frac{n\pi c}{L} t + \delta\right). \tag{5.1}$$

Since the partial differential equation is linear and homogeneous, any linear combination of solutions is also a solution. Therefore the general solution of the equation, subject to the boundary conditions, is

$$u(x, t) = \sum_{n=1}^{\infty} A_n \sin \frac{n\pi x}{L} \cos\left(\frac{n\pi c}{L} t + \delta_n\right). \tag{5.2}$$

(We have omitted the terms with n zero and negative since these are not linearly independent of the terms included.)

Exercises

4. Find a particular solution to the linear partial differential equation $r(\partial u/\partial r) + (\partial^2 u/\partial \phi^2) = 0$ by the method of separation of variables, subject to the single-valuedness condition, $u(r, \phi + 2\pi) = u(r, \phi)$, for all r and ϕ. What is the general solution to this equation (again subject to the condition)?

*5. Convince yourself that a solution of Laplace's equation $(\partial^2\phi/\partial x^2) + (\partial^2\phi/\partial y^2) + (\partial^2\phi/\partial z^2) = 0$ can have no extremum (except perhaps at the boundary). (*Hint:* What are the requirements on the second derivatives that insure an extremum?)

*6. Find the *general* solution of Laplace's equation. (*Hint:* Try for a particular solution in which the variables are separable. Also, try to treat all variables in a symmetric fashion by the device of an equation of constraint on the constants of integration.)

*7. The Schrödinger equation for the electronic wave function of a hydrogenic atom is

$$\frac{1}{r^2 \sin^2\theta}\frac{\partial^2\psi}{\partial\phi^2} + \frac{1}{r^2\sin\theta}\frac{\partial}{\partial\theta}\left(\sin\theta\frac{\partial\psi}{\partial\theta}\right) + \frac{1}{r^2}\frac{\partial}{\partial r}\left(r^2\frac{\partial\psi}{\partial r}\right) + \frac{8\pi^2\mu}{h^2}\left(E + \frac{Ze^2}{r}\right)\psi = 0.$$

Try for a separable solution of the form $\psi(r, \theta, \phi) = R_{nlm}(r)\Theta_{lm}(\theta)\Phi_m(\phi)$, and reduce this partial differential equation to three ordinary differential equations.

8. Find the eigenvalues and eigenfunctions of the partial differential operator $-D_x^{\,2} - D_y^{\,2}$, subject to the requirements that for $\Psi(x, y)$ to be acceptable, $\Psi(0, y) = 0 = \Psi(L, y)$, $\Psi(x, 0) = 0 = \Psi(x, M)$, and $\int_0^L \int_0^M \Psi^*\Psi\, dy\, dx = 1$. (*Hint:* Try for a solution in which the variables are separable.)

*9. Find the eigenvalues and eigenfunctions of the partial differential operator

$$-\frac{1}{r}\frac{\partial}{\partial r}r\frac{\partial}{\partial r}-\frac{1}{r^2}\frac{\partial^2}{\partial\phi^2},$$

subject to the requirements that for $\Psi(r, \phi)$ to be acceptable, Ψ must be continuous, $\Psi(r, \phi + 2\pi) = \Psi(r, \phi)$, and $\Psi(r_0, \phi) = 0$. [*Hint:* Try for a solution of the form $\Psi(r, \phi) = R(x_{nm}r/r_0)\Phi(\phi)$, where x_{nm} is the nth root of the equation $J_m(x) = 0$.]

10. The total Hamiltonian operator for a molecule acts on (a function of) all the coordinates, but it may be approximated by a sum of translational, rotational, vibrational, and electronic Hamiltonian operators, each of which acts on only a certain set of coordinates:

$$H_{\text{total}} = H_{\text{trans}}(Q_{\text{trans}}) + H_{\text{rot}}(Q_{\text{rot}}) + H_{\text{vib}}(Q_{\text{vib}}) + H_{\text{el}}(Q_{\text{el}})$$

(For example, H_{trans} acts only on the coordinates of the center of mass, and H_{el} acts only on electronic coordinates.) Convince yourself that the eigenvalue-eigenfunction problem, $H_{\text{total}}\Psi(Q) = E_{\text{total}}\Psi(Q)$, is separable, by trying for a solution of the form

$$\Psi(Q) = \psi_{\text{trans}}(Q_{\text{trans}})\psi_{\text{rot}}(Q_{\text{rot}})\psi_{\text{vib}}(Q_{\text{vib}})\psi_{\text{el}}(Q_{\text{el}})$$

Notice that E_{total} is the sum of the eigenvalues of the separate Hamiltonian operators.

*11. The partial differential operator which represents the Hamiltonian for a many-electron atom is a sum of one-electron terms, plus a sum over electron pairs:

$$H = -\frac{h^2}{8\pi^2 m}\sum_i \nabla_i^2 - Ze^2\sum_i \frac{1}{r_i} + e^2\sum_{i>j}\frac{1}{r_{ij}}$$

where $\nabla_i^2 = (\partial^2/\partial x_i^2) + (\partial^2/\partial y_i^2) + (\partial^2/\partial z_i^2)$, r_i is the distance between electron i and the nucleus, and r_{ij} is the distance between electrons i and j. The eigenfunction we seek is a function of the coordinates of all the electrons. Convince yourself that this eigenfunction is not separable into a product of one-electron functions. Then convince yourself that the eigenvalue-eigenfunction problem is separable into a set of equations, one for each electron, if the double sum is approximated by a sum of one-electron terms:

$$e^2\sum_{i>j}\frac{1}{r_{ij}} \sim \sum_i V_i^{\text{effective}}(x_i, y_i, z_i).$$

This approximation is known as the ORBITAL APPROXIMATION.

5.3 Standing Waves

Let us consider the behavior of the single term

$$u(x, t) = A_n \sin\frac{n\pi x}{L}\cos\left(\frac{n\pi c}{L}t + \delta_n\right). \tag{5.1'}$$

The quantity $|A_n|$ is called the AMPLITUDE of the vibration; it represents the maximum displacement of the string from its equilibrium position. The quantity δ_n is called the PHASE of the vibration.

The cosine factor acts only to change $u(x, t)$ with time, while the "shape" of the curve remains the same. All the points on the string oscillate in unison, and at any instant the string describes a sine curve. Since there is no motion of the curve along the x-axis, such a curve is called a STANDING WAVE (Figure 5.2).

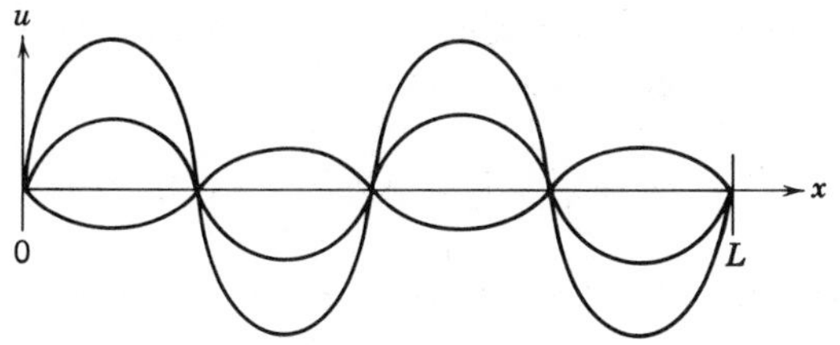

Figure 5.2

Since $\cos [(n\pi c/L)(t + 2L/nc) + \delta_n] = \cos [(n\pi c/L)t + \delta_n]$, the string returns to a given position after the time interval $2L/nc$. This time interval is known as the PERIOD, T, measured in seconds. The reciprocal of the period is the FREQUENCY, $\nu = nc/2L$, measured in cycles/second, or hertz. The ANGULAR FREQUENCY, ω, equals $2\pi\nu$.

There are certain points on the string which remain stationary. These points satisfy the equation $\sin n\pi x/L = 0$, corresponding to $x = kL/n$, with $k = 0, 1, \ldots, n$. Such points are called NODES. If the endpoints are included, there are $(n + 1)$ nodes in all, evenly spaced and separated by L/n. Double the separation between the nodes is known as the WAVELENGTH, $\lambda = 2L/n$, measured in centimeters. One complete cycle of the sine wave requires a length of string λ. Comparing the formulas for ν and λ, we obtain the important equation $\nu\lambda = c$.

In terms of these quantities, an alternative form for the sine factor is $\sin 2\pi x/\lambda$, and alternative forms for the cosine factor are $\cos (\omega t + \delta_n)$, $\cos (2\pi\nu t + \delta_n)$, and $\cos (2\pi t/T + \delta_n)$.

Exercises

*12. Show that the standing wave $A_n \sin (\omega x/c) \cos \omega t$ is of the form $f(x - ct) + f(x + ct)$, that is, that a standing wave is a superposition of two identical traveling waves, traveling in opposite directions.

*13. Discuss the dependence of the frequency of vibration of a violin string on tension, density, length, and number of nodes.

5.4 Vibrating String Subject to Initial Conditions

In general, when a stretched string is set vibrating, its motion is not simply that of a standing wave but a more complicated motion which must be described as a superposition of standing waves. We must therefore return to the general solution of the wave equation and determine the particular solution which results from a given set of initial conditions.

The initial conditions we shall take to be $u(x, 0) = f(x)$ and $(\partial u/\partial t)(x, 0) = 0$. At time zero the string is released from some initial shape, $f(x)$, with zero velocity. (These conditions are appropriate to a harpsichord or banjo string, which is plucked, rather than a piano string, which is struck.)

Applying the second boundary condition gives

$$\frac{\partial u}{\partial t}(x, 0) = -\omega \sum_{n=1}^{\infty} A_n \sin \frac{n\pi x}{L} \sin \delta_n = 0,$$

which can be satisfied for all x if every $\delta_n = 0$. Therefore

$$u(x, t) = \sum_{n=1}^{\infty} A_n \sin \frac{n\pi x}{L} \cos \omega t.$$

According to the first boundary condition

$$u(x, 0) = f(x) = \sum_{n=1}^{\infty} A_n \sin \frac{n\pi x}{L}. \tag{5.3}$$

The problem remains to find the appropriate numerical values of the coefficients A_n. To evaluate the coefficients, multiply Eq. 5.3 by $\sin (m\pi x/L)$ (m a positive integer) and integrate from 0 to L:

$$\int_0^L f(x) \sin \frac{m\pi x}{L}\, dx = \sum_{n=1}^{\infty} A_n \int_0^L \sin \frac{m\pi x}{L} \sin \frac{n\pi x}{L}\, dx.$$

Each integral on the right is the integral of an even function, and therefore equals half the integral of the same integrand over the interval $-L$ to L. To evaluate each integral, substitute $\theta = \pi x/L$ and use a result of Exercise 3.28. Then simplify the sum over the Kronecker delta, to obtain

$$\int_0^L f(x) \sin \frac{m\pi x}{L}\, dx = \frac{L}{2\pi} \sum_{n=1}^{\infty} A_n \int_{-\pi}^{\pi} \sin m\theta \sin n\theta\, d\theta = \frac{L}{2\pi} \sum_{n=1}^{\infty} A_n \pi \delta_{mn} = \frac{L}{2} A_m$$

or

$$A_n = \frac{2}{L} \int_0^L f(x) \sin \frac{n\pi x}{L}\, dx. \tag{5.4}$$

Thus we can evaluate each coefficient as a definite integral. The amplitude of a given frequency in the overall motion of the string depends upon the extent to which the initial displacement "looks like" the standing wave of that frequency.

Exercises

*14. Sketch the position, at time intervals $L/4c$ apart, of a string of length L, released with zero velocity from the initial position $u(x, 0) = \sin (\pi x/L) + \sin (2\pi x/L)$.

15. (a) Solve the diffusion equation $\partial c/\partial t = D(\partial^2 c/\partial x^2)$, where $c(x, t)$ is the concentration of some substance at time t in a tube stretching from $x = 0$ to L, subject to the boundary condition $(\partial c/\partial x)(0, t) = 0 = (\partial c/\partial x)(L, t)$.
(b) Find the particular solution to this equation, given the initial condition that all the material is in the center of the tube, that is, $c(x, 0) = N\delta(x - \frac{1}{2}L)$.

5.5 Fourier Series

What we have done above is a specific example of a general technique, the expansion of an arbitrary function, not as a power series, but as a trigonometric series. In the above example the function was defined on the interval from 0 to L. It is more common to treat a function as being periodic, with period 2π, and expand it on the interval from $-\pi$ to π. The expansion is usually written in the form

$$f(x) = \frac{a_0}{2} + \sum_{n=1}^{\infty}(a_n \cos nx + b_n \sin nx). \tag{5.5}$$

Again there remains the problem of finding the coefficients $\{a_n\}$ and $\{b_n\}$. First let us multiply both sides of the expansion equation by $\sin mx$ $(m \neq 0)$ and integrate from $-\pi$ to π.

$$\int_{-\pi}^{\pi} f(x) \sin mx\, dx$$
$$= \frac{a_0}{2}\int_{-\pi}^{\pi} \sin mx\, dx + \sum_{n=1}^{\infty}\left(a_n \int_{-\pi}^{\pi} \sin mx \cos nx\, dx + b_n \int_{-\pi}^{\pi} \sin mx \sin nx\, dx\right).$$

But from Exercise 3.28

$$\int_{-\pi}^{\pi} \sin mx\, dx = 0 = \int_{-\pi}^{\pi} \sin mx \cos nx\, dx$$

and

$$\int_{-\pi}^{\pi} \sin mx \sin nx\, dx = \pi\delta_{mn}.$$

Therefore

$$\int_{-\pi}^{\pi} f(x) \sin mx\, dx = \sum_{n=1}^{\infty} b_n \pi \delta_{mn} = \pi b_m$$

or

$$b_n = \frac{1}{\pi}\int_{-\pi}^{\pi} f(x) \sin nx\, dx. \tag{5.6}$$

Likewise, multiplying the expansion equation by $\cos mx$ $(m \neq 0)$ and integrating from $-\pi$ to π gives

$$\int_{-\pi}^{\pi} f(x) \cos mx\, dx$$
$$= \frac{a_0}{2}\int_{-\pi}^{\pi} \cos mx\, dx + \sum_{n=1}^{\infty}\left(a_n \int_{-\pi}^{\pi} \cos mx \cos nx\, dx + b_n \int_{-\pi}^{\pi} \cos mx \sin nx\, dx\right).$$

But

$$\int_{-\pi}^{\pi} \cos mx \, dx = 0 = \int_{-\pi}^{\pi} \cos mx \sin nx \, dx$$

and

$$\int_{-\pi}^{\pi} \cos mx \cos nx \, dx = \pi \delta_{mn}.$$

Therefore

$$\int_{-\pi}^{\pi} f(x) \cos mx \, dx = \sum_{n=1}^{\infty} a_n \pi \delta_{mn} = \pi a_m$$

or

$$a_n = \frac{1}{\pi} \int_{-\pi}^{\pi} f(x) \cos nx \, dx. \tag{5.7}$$

Finally, on integrating the expansion (Eq. 5.5) from $-\pi$ to π, only the first term on the right is nonzero. Since $\int_{-\pi}^{\pi} dx = 2\pi$, the value of this term is πa_0, so that we find that Eq. 5.7 is applicable even if $n = 0$. It is to guarantee the applicability of this formula to a_0 that the first term in the expansion is written as $a_0/2$, rather than a_0.

An expansion of an arbitrary function as a trigonometric series, for which the coefficients are determined by Eqs. 5.6 and 5.7, is known as a FOURIER EXPANSION.

Example

3. To expand $f(x) = x$ on the interval $-\pi < x < \pi$: First, every $a_n = (1/\pi) \int_{-\pi}^{\pi} x \cos nx \, dx$ must be zero since x is odd and $\cos nx$ is even. To evaluate b_n, integrate by parts:

$$b_n = \frac{1}{\pi} \int_{-\pi}^{\pi} x \sin nx \, dx = \frac{1}{\pi} \left[-\frac{1}{n} x \cos nx \right]_{-\pi}^{\pi} + \frac{1}{n\pi} \int_{-\pi}^{\pi} \cos nx \, dx$$

$$= -\frac{1}{n\pi} [\pi \cos n\pi - (-\pi) \cos n\pi] = -\frac{2}{n} \cos n\pi = -\frac{2}{n} (-1)^n$$

or

$$x = 2 \sum_{n=1}^{\infty} (-1)^{n+1} \frac{\sin nx}{n}.$$

Notice that the right-hand side is a periodic function of period 2π. The function $f(x) = x$ is represented by this trigonometric series only on the interval $-\pi$ to π; elsewhere the function is expanded as though it were periodic of period 2π.

It is also possible to generalize the Fourier expansion to the finite interval from $-L$ to L

$$f(x) = \frac{a_0}{2} + \sum_{n=1}^{\infty} \left(a_n \cos \frac{n\pi x}{L} + b_n \sin \frac{n\pi x}{L} \right) \tag{5.8}$$

with

$$a_n = \frac{1}{L}\int_{-L}^{L} f(x)\cos\frac{n\pi x}{L}\,dx \tag{5.9}$$

and

$$b_n = \frac{1}{L}\int_{-L}^{L} f(x)\sin\frac{n\pi x}{L}\,dx. \tag{5.10}$$

A more "symmetrical" formulation of the Fourier expansion results if we expand a function, not as a series of sine and cosine terms, but as a linear combination of the related complex exponentials

$$f(x) = \frac{1}{\sqrt{2\pi}}\sum_{n=-\infty}^{\infty} c_n e^{inx}. \tag{5.11}$$

The factor of $1/\sqrt{2\pi}$ has been included for "symmetry." Then the coefficients are given by

$$c_n = \frac{1}{\sqrt{2\pi}}\int_{-\pi}^{\pi} e^{-inx} f(x)\,dx. \tag{5.12}$$

Exercises

16. What can you conclude about the Fourier expansion of an even function? Of an odd function?
17. Find the Fourier series expansions of the functions
    (a) $f(x) = \delta(x);\ -\pi < x < \pi$.
    (b) $f(x) = x^2;\ -\pi < x < \pi$. Use this result to evaluate $\sum_{n=1}^{\infty} (-1)^n(1/n^2)$.
    (c) $f(x) = -1,\quad -L < x < 0;\ f(x) = 1,\quad 0 < x < L;$ that is, $f(x) = -1 + 2\int_{-L}^{x} \delta(t)\,dt$; can this function be expanded in power series?
    *(d) $f(x) = (\pi/2 - |x|)\cos x + (1 + x^2)|\sin x| - \pi x \sin x;\ -\pi < x < \pi$. Use this result to evaluate
    $$\sum_{n=1}^{\infty} \frac{n^2}{(n^2 - \frac{1}{4})^3}.$$
*18. Find the Fourier series expansion of x^4 $(-\pi \leqslant x \leqslant \pi)$. Then set $x = \pi$ and use the result of Exercise 5.17b to evaluate the definite integral and the sum of Exercise 3.15.
*19. Derive Eq. 5.12 for the coefficients c_n in the complex form of the Fourier expansion. Then generalize this expansion to the interval from $-L$ to L.
*20. (a) Find the complex form of the Fourier expansion of $\delta(y)$ on the interval from $-\pi$ to π. Note that since the expansion is periodic of period 2π, it is in fact equal to $\sum_{m=-\infty}^{\infty} \delta(y - 2\pi m)$.
    (b) Use this result, and that of Exercise 3.30, to show that the Fourier *transform* of the one-dimensional LATTICE FUNCTION, $\sum_{n=-\infty}^{\infty} \delta(x - nd)$, is
    $$\frac{\sqrt{2\pi}}{d}\sum_{m=-\infty}^{\infty} \delta\left[y - \frac{2\pi}{d}m\right].$$
    This result, that the Fourier transform of a lattice function is another lattice function, is the basis of X-ray crystallography. Notice that the periods d and $2\pi/d$ of the two lattice functions are inversely related.

5.6 Fourier Transforms

We next consider the expansion of an arbitrary function on the interval $-L$ to L and investigate what happens as we let L become infinite. For purposes of maximum "symmetry" we choose the appropriate complex form of the Fourier expansion, which from Exercise 5.19 is

$$f(x) = \frac{1}{\sqrt{2L}} \sum_{n=-\infty}^{\infty} c_n e^{in\pi x/L}, \tag{5.13}$$

where

$$c_n = \frac{1}{\sqrt{2L}} \int_{-L}^{L} f(x')e^{-in\pi x'/L}\, dx'. \tag{5.14}$$

We have written the dummy variable of integration as x' since we next substitute this formula for c_n into Eq. 5.13:

$$f(x) = \sum_{n=-\infty}^{\infty} \int_{-L}^{L} f(x')e^{-in\pi x'/L}\, dx' e^{in\pi x/L} \frac{1}{2L}.$$

Let $y = n\pi/L$. Then as the integer n proceeds from $-\infty$ to ∞ in steps of unity, y proceeds from $-\infty$ to ∞ in steps $\Delta y = \pi/L$. When we allow L to become infinite, Δy becomes so small that y may be considered as a continuous variable. Thus the sum over n becomes an integral with respect to y, and

$$f(x) = \frac{1}{2\pi} \int_{-\infty}^{\infty} \int_{-\infty}^{\infty} f(x')e^{-ix'y}\, dx' e^{ixy}\, dy, \tag{5.15}$$

the FOURIER INTEGRAL THEOREM.

But we have defined the Fourier transform of f by

$$F\{f(x')\} = g(y) = \frac{1}{\sqrt{2\pi}} \int_{-\infty}^{\infty} f(x')e^{-ix'y}\, dx', \tag{5.16}$$

so that on substituting $g(y)$ for this integral in the Fourier integral theorem we obtain

$$f(x) = \frac{1}{\sqrt{2\pi}} \int_{-\infty}^{\infty} g(y)e^{ixy}\, dy. \tag{5.17}$$

This result tells us that if we wish to invert (undo) the Fourier transform and recover the original function f, we should multiply by $e^{+ixy}/\sqrt{2\pi}$ and integrate. We shall symbolize the operator which inverts the Fourier transform by F^{-1}, such that $F^{-1}F = 1 = FF^{-1}$. Then

$$F^{-1}[F\{f(x)\}] = F^{-1}\{g(y)\} = \frac{1}{\sqrt{2\pi}} \int_{-\infty}^{\infty} g(y)e^{ixy}\, dy = f(x). \tag{5.18}$$

Fourier transforms arise quite frequently in physical situations. In quantum mechanics we usually solve the Schrödinger equation, $H\psi = E\psi$, for eigenfunctions, $\psi(x, y, z)$, which are functions of position. The physical meaning of such an eigenfunction is that $|\psi(x, y, z)|^2\, dx\, dy\, dz$ represents the probability of finding a particle within the volume element $dx\, dy\, dz$ about the point (x, y, z). It can be shown that the probability of finding the particle with momentum between (p_x, p_y, p_z) and $(p_x + dp_x, p_y + dp_y, p_z + dp_z)$ is $|\chi(p_x, p_y, p_z)|^2\, dp_x\, dp_y\, dp_z$, where

$$\chi(p_x, p_y, p_z) = F\psi(x, y, z) = (2\pi)^{-3/2} \iiint \psi(x, y, z) e^{-i(xp_x + yp_y + zp_z)}\, dx\, dy\, dz.$$

In X-ray crystallography the diffracted wave is equal to a Fourier transform of the electron density within the crystal. If it were possible to observe this wave, the structure of the crystal could be determined merely by inverting the Fourier transform. Unfortunately, only the absolute value of this transform can be determined, while the complex phase factors of magnitude unity are lost, so that more elaborate techniques are required to solve for the structure.

Exercises

*21. Apply the Fourier integral theorem to $f_1(x)$ in order to show that the Fourier transform of the product $f_1 f_2$ is the convolution of the transforms $g_1 = Ff_1$ and $g_2 = Ff_2$.

*22. Relate $\int_{-\infty}^{\infty} g_1^*(y) g_2(y)\, dy$ to $\int_{-\infty}^{\infty} f_1^*(x) f_2(x)\, dx$. (*Caution:* Although $g_1(y) = F[f_1(x)]$, $g_1^*(y) \neq F[f_1^*(x)]$.)

23. Rewrite the Fourier integral formula as

$$f(x) = \int_{-\infty}^{\infty} f(x') \left[\frac{1}{2\pi} \int_{-\infty}^{\infty} e^{i(x-x')y}\, dy \right] dx'.$$

(a) Which familiar function is equal to the expression in brackets?

(b) Use this result to find $F(e^{\pm i\omega x})$, $F(\cos \omega x)$, and $F(\sin \omega x)$.

(c) What is $F(\sum a_n e^{in\omega x})$?

*24. The oscillatory behavior of a monochromatic light beam of angular frequency ω, passing by a given point, is conveniently represented by $ae^{i\omega t}$. The most general polychromatic light beam is a sum of such terms, one for each frequency. But, of course, such a sum over a continuous variable becomes the integral $f(t) = \int a(\omega) e^{i\omega t}\, d\omega$. Find the Fourier transform of this light beam. (*Note:* A spectroscope is a Fourier transformer which determines the extent to which each frequency is represented in a polychromatic light beam. The intensity of light of a given frequency is the square of the absolute value of the Fourier transform of the light beam.)

*25. (a) Find the absolute value of the Fourier transform of the FINITE WAVE TRAIN (light pulse), $f(t) = ae^{i\omega_0 t}$, $0 < t < t_0$, $f(t) = 0$ at other times. Notice that this light is not monochromatic.

(b) What is the principal angular frequency represented?

(c) Convince yourself that only an infinite wave train can be truly monochromatic, and that no measurement of finite duration can determine a frequency with complete accuracy.

(d) What frequencies are not represented in the light pulse?
(e) As a measure of the uncertainty in frequency $\Delta\omega$, find the difference between the principal frequency and the closest frequency which is absent.
(f) What is $t_0\, \Delta\omega$? This is an example of the uncertainty principle.

26. Solve the diffusion equation for the concentration $c(x, t)$ of a substance in an infinite tube, subject to the initial condition $c(x, 0) = f(x)$:
(a) Try for a solution in which the variables are separable, and call the arbitrary constant $-y^2$; keep D in the t equation and use only e^{-iyx} for the solution to the x equation. Since the partial differential equation is linear, the general solution is a linear combination of particular solutions. However, since there are no boundary conditions, there are no restrictions on y, and the general linear combination is not a sum but an integral, which we shall write

$$\frac{1}{\sqrt{2\pi}} \int_{-\infty}^{\infty} g(y) X(x) T(t)\, dy.$$

(b) Express this solution as a Fourier transform.
(c) Apply the initial condition and use the rule for inverting Fourier transforms to solve for $g(y)$.
(d) Express this particular solution in symbolic notation using Fourier transform operators.

*27. Use the result of Exercise 4.14 to express e^{-Dy^2t} in the form $\gamma(y) = F^{-1}\{\phi(x)\}$. Then use the result of Exercise 5.21 to express the solution of Exercise 5.26 as a convolution.

*28. Use the above result to solve the diffusion equation subject to the initial condition $c(x, 0) = N\delta(x)$, that is, all the material at the center of the infinite tube. Sketch the way the substance spreads out as t progresses. Notice that the width of the distribution varies as $t^{1/2}$.

5.7 Generalized Fourier Series

We have seen that an arbitrary function may be expanded as an infinite series of sine and cosine terms, and that each coefficient may be determined as a definite integral. The essential feature of sin nx and cos nx which enabled us to derive a formula for the coefficients was the orthogonality properties of these functions. However, the trigonometric functions are not unique in this respect. Let us define an ORTHONORMAL SET OF FUNCTIONS, $\{\phi_n(x)\}$, by the requirement

$$\int \phi_m^*(x)\phi_n(x)w(x)\, dx = \delta_{mn} \tag{5.19}$$

where the integrals are to be taken over some appropriate interval. The function w is known as the WEIGHT FUNCTION; it can, if convenient, be absorbed into the orthonormal function by redefining $\Phi_n(x) = \phi_n(x)[w(x)]^{1/2}$. We require further that this set be COMPLETE, such that any "reasonable" function f can be expanded as an infinite series of these functions:

$$f(x) = \sum_n^{\infty} c_n \phi_n(x). \tag{5.20}$$

The set of functions used in such an expansion is called a BASIS SET or SET OF BASIS FUNCTIONS. (More advanced methods are required to prove that a set is complete.) To find the coefficients, multiply both sides of this expansion by $\phi_m^*(x)w(x)$ and integrate

$$\int \phi_m^*(x)f(x)w(x)\,dx = \sum c_n \int \phi_m^*(x)\phi_n(x)w(x)\,dx = \sum c_n\delta_{mn} = c_m$$

or

$$c_n = \int \phi_n^*(x)f(x)w(x)\,dx. \tag{5.21}$$

There are many such complete orthonormal sets. We list the most important:

Functions	**Symbol**	**Interval**	w
Trigonometric	$\cos nx$, $\sin nx$	$-\pi$ to π	1
Complex exponential	$e^{in\pi x/L}$	$-L$ to L	1
Legendre polynomials	$P_l(x)$	-1 to 1	1
Legendre functions	$P_l(\cos\theta)$	0 to π	$\sin\theta$
Hermite polynomials	$H_n(x)$	$-\infty$ to ∞	e^{-x^2}
Hermite functions	$H_n(x)e^{-x^2/2}$	$-\infty$ to ∞	1
Laguerre polynomials	$L_n(x)$	0 to ∞	e^{-x}
Chebyshev polynomials	$U_n(x)$	-1 to 1	$(1-x^2)^{-1}$
Spherical harmonics	$Y_{lm}(\theta, \phi)$	$\begin{Bmatrix} 0 < \theta < \pi \\ 0 < \phi < 2\pi \end{Bmatrix}$	$\sin\theta$
Bessel functions	$J_n(x)$	Orthogonality more complicated	

Actually, these functions, as conventionally defined, are orthogonal but not normalized, so that it is always necessary to check the value of the normalization integral $\int \phi_n^*(x)\phi_n(x)w(x)\,dx$. Just as there are many helpful relations among the trigonometric functions, so there are many properties of these orthogonal functions. We have omitted the discussion of these properties but you should investigate them in an advanced text whenever you use such functions.

Notice the similarity of the expansion $f(x) = \sum c_n\phi_n(x)$ to the expansion of a vector $\mathbf{V}$ in terms of its coordinates. In N dimensions this expansion becomes $\mathbf{V} = \sum_{i=1}^N v_i\hat{\mathbf{i}}$, where $\hat{\mathbf{i}}$ is a unit vector in the direction of the ith coordinate axis. The coefficient c_n represents the component or projection of f in the direction of ϕ_n, just as the coefficient v_i represents the component or projection of $\mathbf{V}$ in the direction of $\hat{\mathbf{i}}$. Indeed, the use of the term "orthogonal" is taken from the context of vectors. Later we shall pursue this analogy in more detail.

In practice there is always the question of which basis set to choose. However, this is not a fundamental question, but one of suitability and convenience. We have already seen examples of finding power series solutions to differential equations. Although $\{x^n\}$ is not an orthonormal set, sometimes there are advantages to such an expansion, especially when it is easier to find coefficients by differentiation than by integration. In our investigation of the vibrating string (Section 5.4), a Fourier sine expansion was a natural expansion for expressing the initial displacement of the string. Expansion in some other basis set would be entirely possible but less convenient. Similarly, the initial displacement of a two-dimensional analog, a vibrating circular membrane, may be specified as a series expansion in a complete orthonormal set of functions of r and θ, but the natural basis functions, which arise automatically in the solution (Exercise 5.9), are Bessel functions and complex exponentials. In both these vibration problems the natural functions for the expansion have some direct connection with the solution, in that a string or membrane initially displaced in the pattern of such a basis function continues to oscillate with a single frequency rather than in some more complicated fashion. In more difficult problems the "best" choice is not so obvious, but a helpful, if vague, rule is to use a basis set which "has something to do with the problem." For example, the solutions of the Schrödinger equation for the helium atom are more conveniently expanded as a linear combination of hydrogenic wave functions rather than functions appropriate for, say, a three-dimensional oscillator. The solutions of the Schrödinger equation for the electronic wave functions of the hydrogen molecule are conveniently expressed as linear combinations of the atomic-orbital wave functions, but there are many other possibilities, each with its own merits. Indeed, much of the art of quantum chemistry consists in the choice of a convenient set of basis functions.

Exercises

29. Show that the set $\phi_x(\xi) = \delta(\xi - x)$ is an orthogonal (not normalized, though) set on the interval $-\infty < \xi < \infty$. (Notice that the different functions of the set are indexed by the continuous variable x, rather than an integer n, so you will have to extend Eq. 5.19 to cover this case.)
30. The generating function for the Laguerre polynomials is

$$\sum_{n=0}^{\infty} L_n(r)x^n = \frac{1}{1-x} \exp\left(\frac{-rx}{1-x}\right).$$

Apply the Laplace transform operator to both sides of this equation, evaluate the integral on the right, expand the result as a power series in x [*Hint:* Convert to an expansion of the form of $(1+t)^{-1}$.], and compare coefficients of x^n on both sides to find the Laplace transform of L_n.
31. Obtain a formula for determining the coefficients in the expansion of an arbitrary function as a linear combination of Laguerre polynomials, for which the normalization

integral $\int_0^\infty [L_n(r)]^2 e^{-r}\,dr = 1$. Then use the integrals evaluated in the previous exercise to expand $f(r) = e^{-ar}$ as such a linear combination.

*32. The generating function for the Hermite polynomials is

$$e^{-t^2+2xt} = \sum_{n=0}^{\infty} H_n(x)t^n/n!$$

Find the recursion formula relating the polynomials $H_{n+1}(x)$, $H_n(x)$, and $H_{n-1}(x)$ by the following procedure: Differentiate both sides of this equation with respect to t, express the left-hand side as an infinite series, and change dummy indices of summation so that coefficients of $t^n/n!$ may be compared.

*33. Use the generating function for the Hermite polynomials to evaluate

$$\int_{-\infty}^{\infty} H_m(x)H_n(x)e^{-x^2}\,dx:$$

Rewrite the formula, with t replaced by s and n replaced by m, then multiply the two equations, and integrate both sides. (*Hint:* Complete the square in the exponent.) Then compare coefficients of $s^m t^n$.

*34. Use the previous result to obtain a formula for determining the coefficients in the expansion of an arbitrary function as a linear combination of Hermite functions:

$$f(x) = \sum_{n=0}^{\infty} c_n H_n(x)e^{-x^2/2}.$$

*35. Use the above result to expand $f(x) = e^{-3/2 x^2}$ as a linear combination of Hermite functions. (*Hint:* The generating function for $\{H_n\}$ is an aid in evaluating the integrals.)

6

PROBABILITY AND STATISTICS

There are two main goals of this chapter. The first is to introduce the student to basic notions of probability in order to facilitate the study of statistical mechanics. The second is to introduce him to some simple aspects of statistics for use in handling experimental data. We begin with the concepts and methods of probability, with many examples and exercises designed to familiarize the student with these notions and to develop a facility at applying these methods. Unfortunately, most scientific problems that involve non-trivial aspects of probability require elaborate presentation of the science involved. Therefore most examples and exercises are drawn from gambling situations—coin throwing, card games, and dice—which are familiar to everyone. Nevertheless, there are also some instructive examples and exercises demonstrating applications to magnetic resonance, genetics, statistical mechanics, diffusion, radioactivity, chromatography, polymerization, and quantum mechanics. Section 6.6, on averages and related quantities, and Section 6.7, on continuous distributions, form the bridge from probability theory (which involves deducing or deriving probabilities of events in terms of known or assumed probabilities of other events) to statistics (which involves inducing or inferring unknown probabilities on the basis of experimental observations). Further applications to diffusion, chromatography, radioactivity, polymerization, quantum mechanics, kinetic theory of gases, and polarography are indicated. Section 6.8, on statistical inference, is aimed primarily at chemists in the "less exact" areas of chemistry. This short section is not sufficient to develop a facility at all the complicated statistical methods, but it is intended to illustrate some of the simplest methods and to indicate what kinds of problems may yield to a statistical approach. The remaining sections are devoted to the handling of experimental data, although in a rather abstract fashion, in order to concentrate on principles.

There is little material in this chapter that is required for subsequent chapters. Therefore a student who is interested in quantum mechanics has the option to defer this chapter, except that he is advised to review the notion of average before proceeding.

6.1 The Laws of Probability

An outcome of an "experiment" is called a SAMPLE POINT. The set of all sample points, or the set of all possible outcomes of an experiment, is called the SAMPLE SPACE of the experiment. A RANDOM VARIABLE is a function whose domain is a sample space. Often such a function is just a means for labeling each outcome of an experiment with a number or a set of numbers. It should be explained now that according to the nomenclature that we have been using, this name is a misnomer. Also, it should be noted that there are many functions that can be defined on a given sample space, although some are more "meaningful" than others. Some examples are:

Experiment	Random Variable	Values
Throw one die	Number of points	1, 2, 3, 4, 5, 6
Throw two dice	Number of points on each die	{1, 1}, {1, 2}, {2, 1}, . . .
Throw two dice	Sum of the points	2, 3, . . . , 12
Choose a digit	Digit chosen	0, 1, 2, . . . , 9
Choose a digit	Parity	Even, odd
Pick a card, any card	Suit	S, H, D, C
Pick a card, any card	Face	A, 2, 3, . . . , 10, J, Q, K
Measure s_z of an electron	$s_z/\hbar$	$+\frac{1}{2}, -\frac{1}{2}$
Measure S_z of N electrons	$\sum s_z/\hbar$	$\frac{1}{2}N, \frac{1}{2}N - 1, \ldots, -\frac{1}{2}N$
Deal 5 cards from a deck of 52	Number of spades	0, 1, 2, 3, 4, 5
Deal 5 cards from a deck of 52	Number of aces	0, 1, 2, 3, 4
Determine C^{14} content of 44 g CO_2	Disintegrations per minute	$0, 1, 2, \ldots, 6 \times 10^{23}$
Measure energy of a quantum oscillator	Energy	$\frac{1}{2}h\nu, \frac{3}{2}h\nu, \ldots$
Titrate 0.1 N acid	Titer of H^+	Continuum clustered about 0.1000 · · ·
Locate a particle	Position	Coordinates of any point in space

We shall return to random variables in Section 6.3. Meanwhile, we consider a special one.

We may define PROBABILITY as that particular random variable which assigns to each outcome the frequency with which that outcome occurs. What this means is that if an experiment is performed N times and the number of times the outcome i occurs is n_i, then the probability of that outcome is

$$P_i \equiv \lim_{N\to\infty} \frac{n_i}{N}. \tag{6.1}$$

It follows that for all outcomes $0 \leqslant P_i \leqslant 1$; an outcome for which $P_i = 0$ is said to be IMPOSSIBLE, and an outcome for which $P_i = 1$ is said to be CERTAIN. It also follows from the definition that the sum over all sample space

$$\sum_i P_i = 1. \tag{6.2}$$

In words, whenever the experiment is performed, *some* outcome must result. Furthermore, we point out that whenever the experiment is performed, a *unique* outcome results. For example, when a coin is flipped it lands either heads or tails, not half of each. But because we do not know how to predict the outcome of the experiment, we must content ourselves with statements about probabilities.

Next we investigate how probabilities of "complex" outcomes may be calculated in terms of known or assumed probabilities of "simple" outcomes. We shall assume the sample space has been specified in such fashion that every sample point is SIMPLE (INDECOMPOSABLE), such that it is not composed of two or more distinguishable outcomes. For example, on throwing a die the outcome "obtain less than three" may be decomposed into the two simple outcomes "obtain an ace" and "obtain a two." This way of specifying sample points often leads to the following simplification.

A UNIFORM sample space is one whose every sample point has the same probability. Then if there are N points in the sample space, the probability of any one of them is

$$P_i = \frac{1}{N}. \tag{6.3}$$

This situation occurs frequently when the sample points are all chosen to be simple. In many cases it is then "obvious" that the sample space is uniform since no result seems special or favored. For example, there is no reason to expect a perfect cube to prefer to land on one face more than another, so we accept the value 1/6 for the probability of obtaining an ace (or any other simple outcome) on throwing one die. Likewise, if we flip two coins in succession, we would expect each of the four simple outcomes—HH, HT, TH, and TT—to be equally likely. Notice how choosing simple sample points facilitates the analysis of probabilities; it becomes clear that the nonsimple outcome "obtain one head and one tail" is twice as probable as the outcome "obtain two heads." Nothing would be changed if we were dropping two stones at random into two boxes, marked H and T, and the four simple outcomes above are still equally likely. On the other hand, if we were randomly dropping electrons, which are indistinguishable, into the two boxes, it is no longer clear whether both HT and TH should be separate sample points since these two are not distinguishable. We might have assumed anyway that finding the two electrons in different boxes is twice as probable as finding both in box H, but experimentally these two outcomes are found to be equally probable. We shall consider probabilities involving indistinguishable objects in Section 6.5. Meanwhile, we shall consider only experiments for which the simple sample points are obviously equally likely.

Usually we are not interested in the probability of a simple outcome, but in the probability of an EVENT, some subset of sample space, usually specified by some common property.

Examples of Events

1. Obtaining a total of seven on throwing two dice corresponds to the six points {1, 6}, {2, 5}, {3, 4}, {4, 3}, {5, 2}, and {6, 1} in a 36-point sample space.
2. Four points out of a 52-point sample space correspond to the event "deal an ace from a deck of cards."
3. Only one point in an eight-point sample space corresponds to the event "throw three heads with three coins."
4. To the event "deal a spade from a deck of cards and also throw either a five or a six with one die" there correspond a total of 26 sample points in a 312-point sample space, as is clear on enumeration of all possibilities.
5. Many points correspond to the event "deal a 13-card hand containing four spades."

It follows from our definition of probability that the probability of an event is just the sum of the probabilities of the various outcomes that make up the event. And whenever the sample space is uniform and finite, the probability of the event E is given simply by

$$P(E) = \frac{n(E)}{N}, \tag{6.4}$$

where $n(E)$ is the number of points corresponding to event E and N is the total number of points in the sample space. This result is often stated as follows: If there are N equally likely outcomes, of which n are considered to be "favorable," then the probability of a favorable outcome is n/N. Thus the probabilities of the events of Examples 6.1–6.4 are readily seen to be 6/36, 4/52, 1/8, and 26/312, respectively. It also follows that the probability of an "unfavorable" outcome is $1 - n/N$. Thus the probability of dealing something other than an ace is 48/52, and the probability of throwing fewer than three heads (at least one tail) with three coins is 7/8. Notice that all these probabilities may be determined simply by counting. Later we shall learn how to evaluate the probability of the event of Example 6.5.

Exercises

*1. Dominoes are dice with only two faces. On each face there is a number from 0 to 9. What is the probability of drawing the double-nine from a complete set of dominoes that has one of every piece? (*Note:* The set does not contain both a 0-1 and a 1-0, etc.)

2. The roulette wheel at Monte Carlo has 37 slots, one for each integer from 0 to 36, inclusive. What is the probability that the ball will fall into:
   (a) Slot #36?
   (b) An odd-numbered slot?

3. Four coins are tossed. What is the probability of getting:
   (a) All heads?
   (b) Exactly one head?
   (c) At least one head?
   (d) At least three heads?

*4. Three dice are thrown. What is the probability of getting:
   (a) All aces?
   (b) An ace on the first die, a two on the second, and a three on the third?
   (c) One ace, one two, and one three?
   (d) A total of six ($= 3 + 2 + 1 = 2 + 2 + 2 = 4 + 1 + 1$) on the three dice?

*5. The natural abundance of C^{13} is 1/90.
   (a) What is the probability that both carbon atoms of an ethane molecule are C^{13}?
   (b) What is the probability that exactly one carbon atom of an ethane molecule is C^{13}? (*Hint*: Choose a sample space of 8100 equally likely outcomes.) These probabilities are of interest in connection with nmr spin-spin coupling constants.

*6. Which is the more probable hand to be dealt, all 13 spades or S973 H7543 D64 C8542?

7. Let $\sim E$, read "not E," be the COMPLEMENT of event E, all those points of sample space not corresponding to E. What is $P(\sim E)$?

The probabilities of still more "complex" events are often not readily determined by counting. It is therefore necessary to develop more rules of probability. First we consider the probability of a JOINT EVENT, one characterized by the joint occurrence of two or more "simpler" events. The event of Example 6.4 is clearly such an event. We shall indicate a joint event by juxtaposition (symbolic multiplication), so that E_1E_2 is the event "both E_1 and E_2." The JOINT PROBABILITY of event E_1E_2 is still given by $P(E_1E_2) = n(E_1E_2)/N$, where $n(E_1E_2)$ is the number of points corresponding to both E_1 and E_2. Figure 6.1 illustrates the relation among these events; the square symbolizes the entire sample space, the shaded circles contain those sample points corresponding to events E_1 and E_2, and the crosshatched area common to both circles contains those sample points corresponding to the joint event E_1E_2. To obtain another form of the joint probability, let us multiply and divide by $n(E_1)$:

$$P(E_1E_2) = \frac{n(E_1)}{N}\frac{n(E_1E_2)}{n(E_1)}. \tag{6.5}$$

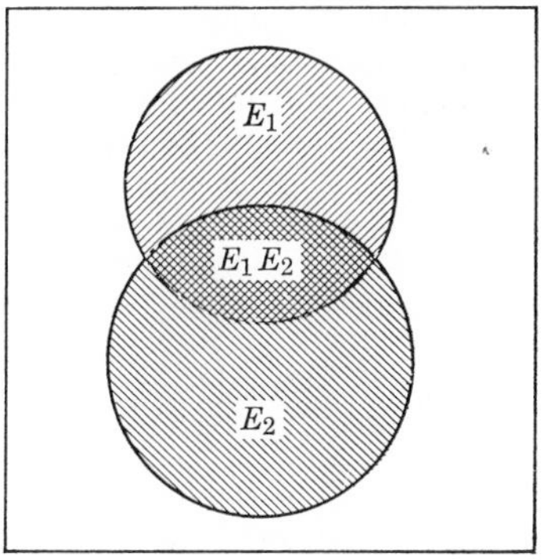

Figure 6.1

The first factor, $n(E_1)/N$, is simply $P(E_1)$, the probability of event E_1. Next we consider the significance of the second factor, $n(E_1E_2)/n(E_1)$. Let us imagine that we know that event E_1 has occurred. Now what is the probability of event E_2? It is no longer $n(E_2)/N = P(E_2)$, but it may be determined according to Eq. 6.4 if we choose a new sample space restricted to those points corresponding to the event E_1. From Figure 6.1 it is clear that there are $n(E_1)$ points in this sample space, of which $n(E_1E_2)$ correspond to the event E_2. Thus we see that the quantity $n(E_1E_2)/n(E_1)$ is the probability of E_2 in the case in which it is known that E_1 has occurred. This probability is called the CONDITIONAL PROBABILITY of E_2 on E_1. It is usually written as $P(E_2 \mid E_1)$ and read "the probability of E_2 given E_1." Thus we may rewrite Eq. 6.5 as

$$P(E_1E_2) = P(E_1)P(E_2 \mid E_1). \tag{6.6a}$$

And by symmetry we may also write

$$P(E_1E_2) = P(E_2)P(E_1 \mid E_2). \tag{6.6b}$$

Clearly we could solve Eq. 6.6 for the conditional probability in terms of the joint probability, but in practice the latter is more often the desired probability and the former the more readily calculated. Finally, we may generalize to the joint probability of n events, or the probability that all n will occur:

$$P\left(\prod_{i=1}^{n} E_i\right) = P(E_1)\prod_{j=1}^{n-1} P\left(E_{j+1} \,\middle|\, \prod_{i=1}^{j} E_i\right) = P(E_1)P(E_2 \mid E_1)P(E_3 \mid E_1E_2)\cdots \tag{6.7}$$

Examples

6. What is the probability that the top two cards of a shuffled deck are both spades? Let S_1 = "top card is a spade" and S_2 = "next card is a spade." We shall calculate $P(S_1S_2)$ by calculating $P(S_1)$ and $P(S_2 \mid S_1)$. Clearly, $P(S_1) = 13/52$. To calculate $P(S_2 \mid S_1)$,

we note that if the top card is a spade, then only 12 cards of the remaining 51 are spades, so $P(S_2 \mid S_1) = 12/51$. Thus $P(S_1S_2) = 13 \cdot 12/52 \cdot 51$. Notice that this is still the answer even if (simultaneously or in succession) two cards are dealt from the deck or picked at random; neither the location nor the timing is relevant. The important feature is that to calculate the conditional probability, $P(S_2 \mid S_1)$, we must consider the problem as though a spade had been removed from the deck.

7. What is the probability of dealing a bridge hand of all 13 spades? By Eq. 6.7 and evaluation of each successive conditional probability, we obtain

$$P(S_1S_2 \cdots S_{13}) = \frac{13}{52}\frac{12}{51}\cdots\frac{1}{40}.$$

8. A card is drawn from a deck of cards, then replaced. The deck is shuffled and another card drawn. What is the probability that both cards were spades? As in Example 6.6, $P(S_1) = 13/52$. But now $P(S_2 \mid S_1) = 13/52$, since the probabilities on the second draw are unaffected by the result of the first. Next we consider the generality of this phenomenon.

Exercises

*8. What is the probability of obtaining exactly 4 heads with 6 coins, given that the first 3 coins all landed heads?

9. What is the probability of being dealt 4 aces in a hand of 4 cards?

10. What is the probability of being dealt:
   (a) A poker hand (5 cards) of 5 spades?
   (b) A bridge hand (13 cards) with no spades?

*11. What is the probability of being dealt a poker hand whose lowest card is a:
   (a) King (i.e., a hand of only aces and kings)?
   (b) Queen (i.e., a hand of aces, kings, and queens)?
   (c) Jack?
   (d) Etc.?

12. What is the probability of being dealt no pair at poker (i.e., all 5 cards different faces)?

13. What is the probability of dealing 4 cards, one of each suit?

14. What is the probability of getting one of each face:
   (a) On throwing 6 dice?
   (b) In a bridge hand?

15. A box contains 15 black balls and 12 white ones. Two balls are withdrawn. What is the probability that:
   (a) Both are black?
   (b) Both are white?
   (c) One is black and the other white? (*Hint:* What is the sum of these three probabilities?)

16. A bridge player holds 2 aces. What is the probability that his partner holds:
   (a) No aces?
   (b) Both the other aces? (*Hint:* If his partner holds the other two aces, how many can their opponents hold?)
   (c) One ace?

17. The incidence of gene a in the population is p, and that of gene A is $(1 - p)$.
   (a) Convince yourself that the incidence of the three genotypes aa, AA, and Aa are p^2, $(1 - p)^2$, and $1 - p^2 - (1 - p)^2 = 2p(1 - p)$, respectively.
   (b) Gene a is recessive, so only an aa genotype shows the trait associated with this gene. Use Eq. 6.6 to find the probability that an individual without this trait has one a gene.

*18. Use Eq. 6.6 to answer the following classical problem of three drawers, one containing two gold coins, one containing two silver coins, and one containing one gold coin and one silver coin: A drawer is picked at random and a coin withdrawn. If the coin is gold, what is the probability that the other coin in the drawer is gold? (*Hint:* What is the probability of choosing a gold coin from among the six coins?)

*19. (a) Show that

$$\frac{P(E_1 \mid E_2)}{P(E_2 \mid E_1)} = \frac{P(E_1)}{P(E_2)}.$$

(b) Show that

$$\frac{P(E_1 \mid E_3)}{P(E_2 \mid E_3)} = \frac{P(E_1)\,P(E_3 \mid E_1)}{P(E_2)\,P(E_3 \mid E_2)}.$$

Two events E_1 and E_2 are said to be INDEPENDENT if $P(E_1E_2) = (PE_1)(PE_2)$. Substitution into Eq. 6.6 justifies this name; if $P(E_2 \mid E_1) = P(E_2)$ and $P(E_1 \mid E_2) = P(E_1)$, then the probability of one event does not depend upon whether the other has occurred.

Examples

9. What is the probability of getting no sixes on throwing 3 dice? Let $\sim S_1$ = "first die is not a six," etc. Clearly, the way one die lands does not affect the probabilities for the other dice, so

$$P(\sim S_1 \sim S_2 \sim S_3) = P(\sim S_1)P(\sim S_2)P(\sim S_3) = (5/6)^3.$$

10. What is the probability of getting at least one six on throwing 3 dice? Since we must get either no six or at least one six, this probability plus that of Example 6.9 must equal unity. Therefore we may conclude that the desired probability is $1 - (5/6)^3 = 91/216$. It is instructive to consider a naive approach to this problem: "The chance of getting a six on the first die is 1/6, and the chance of getting a six on the second die is 1/6, and likewise for the third die. Then the total probability of getting a six on some die is just thrice the probability of getting a six on a single die, since there are three chances of success. Therefore the desired probability is 3/6." That this reasoning cannot be correct may be demonstrated: The same reasoning would lead to the conclusion that the probability of obtaining at least one six on throwing *seven* dice is 7/6; this value is greater than unity, and therefore contrary to the notion of probability. (Although it does seem as though getting at least one six with seven dice is pretty likely, it can't be more than certain.) Next we investigate how we might add probabilities without being trapped by this sort of fallacious reasoning.

Exercises

*20. Show that if E_1 and E_2 are independent, so are E_1 and $\sim E_2$.

21. What is the probability of drawing 4 aces from a deck of cards, WITH REPLACEMENT, that is, if each card is replaced after it is drawn and the deck shuffled? (Cf. Exercise 6.9).

22. A box is composed of two compartments, A and B, of volume V_A and V_B, respectively.
   (a) What is the probability that a given molecule will be found in A?
   (b) What is the probability that all of N molecules will be found in A, if it is assumed that the molecules neither attract nor repel each other?
   (c) If entropy S is related to probability P by $S = k \ln P$, what is the entropy change on expanding N molecules of a gas isothermally from volume V_A to volume $V_A + V_B$?

We wish to determine the INCLUSIVE PROBABILITY of events E_1 and E_2, the probability that E_1 or E_2 or both occur. If there are $n(E_1)$ sample points corresponding to E_1 and $n(E_2)$ corresponding to E_2, let us determine how many sample points correspond to this combined event, which we shall symbolize by $E_1 + E_2$, read "E_1 or E_2." In Figure 6.1 these points are in the lined regions, and it is apparent that

$$n(E_1) + n(E_2) = n(E_1 + E_2) + n(E_1E_2) \tag{6.8}$$

since adding the points within the two lined circles counts the points in the common crosshatched region twice. Dividing by N and rearranging gives

$$P(E_1 + E_2) = P(E_1) + P(E_2) - P(E_1E_2). \tag{6.9}$$

Thus we see that in general we are not justified in merely adding probabilities to determine an inclusive probability. However, $P(E_1 + E_2)$ does simplify in the case that the joint event E_1E_2 is impossible, so that $P(E_1E_2) = 0$. Such events are said to be MUTUALLY EXCLUSIVE. Thus we conclude that if two events are mutually exclusive, then

$$P(E_1 + E_2) = P(E_1) + P(E_2). \tag{6.9'}$$

Equation 6.9 is applicable only to two events, and the general formula for n events is somewhat cumbersome. However, if the events $\{E_i\}$ are all (pairwise) mutually exclusive, then the probability of achieving some E_i becomes a simple sum:

$$P(\sum E_i) = \sum P(E_i). \tag{6.9''}$$

Now we can see why Eq. 6.9″ does not apply to the problem of Example 6.10. The events "get a six on the first die" and "get a six on the second die" are not mutually exclusive, since it is possible for both to happen. Therefore we are not justified in adding probabilities in this case.

Exercises

*23. Use a diagram like Figure 6.1 to relate $P(A + B + C)$ to $P(A)$, $P(B)$, $P(BC)$, $P(ABC)$, etc.

*24. Use diagrams like Figure 6.1 to determine whether $(E_1 + E_2)E_3 = E_1E_3 + E_2E_3$ and whether $E_1 + E_2E_3 = (E_1 + E_2)(E_1 + E_3)$. Convince yourself that $E + E = E$ and $EE = E$. Convince yourself that $\sim(E_1 + E_2) = \sim E_1 \sim E_2$ and $\sim(E_1E_2) = \sim E_1 + \sim E_2$. This sort of algebra of sets is known as BOOLEAN ALGEBRA and has applications in logic and electronic circuitry.

25. Solve Exercise 6.15c by noting that drawing a white ball and then a black ball is mutually exclusive to drawing a black ball and then a white one.

26. What is the probability of obtaining exactly one head on throwing N coins? One ace on throwing N dice?

*27. A box contains 4 black balls, 3 white balls, and 2 red balls. Another box contains 1 black ball, 4 white balls, and 3 red balls. One ball is chosen from each box. What is the probability that both are the same color?

*28. A factory produces widgets, of which 1% are defective. In order to reduce the number of complaints, the company packages 101 widgets in a box labeled as containing 100 perfect widgets.
(a) What fraction of the boxes contain no defective widgets?
(b) What fraction of the boxes contain one defective widget? (*Hint:* What is the probability that the first widget is defective and all the rest good?)
(c) What fraction of the boxes contain fewer than 100 perfect widgets? Obtain a decimal approximation for this.

*29. (a) What is the probability of being dealt AKQJ10 at poker without regard to suit?
(b) What is the probability of being dealt a straight (5 cards in a row, with ace next to 2, as in 5432A, or next to K, as in AKQJ10)?

30. (a) What is the probability of being dealt a flush at poker (5 cards, all of the same suit)?
(b) What is the total probability of getting a flush in draw poker, if it is agreed that if we are not dealt a flush, but only 4 of a suit, we will trade in the fifth card and draw in the hope of completing the flush? (*Hint:* What is the probability that the first four cards dealt are spades and the fifth a nonspade?)

*31. (a) What is the probability of being dealt a bridge hand with no honors (no AKQJ10)?
(b) With no AKQJ109?
(c) Use these results and Eq. 6.9′ to find the probability of being dealt a bridge hand whose highest card is a 9 (one or more).

*32. Let a sample space be composed of the mutually exclusive events $\{E_i\}$, and let A be any arbitrary event. Convince yourself that $P(A) = \sum P(A \mid E_i)P(E_i)$.

*33. One box contains 4 white balls and 3 black balls, and another contains 2 white balls and 6 black balls.
(a) A box is chosen at random and one ball withdrawn from it. Use the result of Exercise 6.32 to find the probability that the ball is white?
(b) A box is chosen at random and two balls withdrawn from it. What is the probability that both balls are white? Black? One of each?
(c) A box is chosen at random and a ball withdrawn from it and not replaced, then a box is again chosen at random and a ball withdrawn from it. What is the probability that both balls are white? Black? One of each?
(d) What are these probabilities if the 15 balls are in the same box and 2 are withdrawn?

*34. The incidence of recessive gene a is p. An individual inherits one gene at random from each parent. Thus we may write the conditional probability of genotype aa on the parental genotypes: $P(aa \mid aa, aa) = 1$, $P(aa \mid aa, Aa) = 1/2$, $P(aa \mid Aa, Aa) = 1/4$, $P(aa \mid aa, AA) = P(aa \mid Aa, AA) = P(aa \mid AA, AA) = 0$. Use the results of Exercises 6.17a and 6.19a to find the probability that the parents of an aa individual are:
(a) Both aa.
(b) Both Aa.
(c) One aa and one Aa.
Use the result of Exercise 6.17b to find the probability that an individual will be aa:
(d) If one parent is aa but the other is not.
(e) If neither parent is aa.
(f) Check that these answers satisfy the result of Exercise 6.32, with event A = "individual is aa genotype" and $\{E_i\}$ = set of all possible parental genotypes.

*35. A pinochle deck contains 8 each of AKQJ109. A pinochle deck and a bridge deck are placed side by side and one of them chosen at random. The top card of that deck is found to be an ace.

(a) Use the result of Exercise 6.19b to find the probability that the pinochle deck was chosen.
(b) The ace is replaced and that deck shuffled. A card is again drawn from it. Use the result of Exercise 6.32 to find the probability that this card is also an ace.

In practice, if you intend to add probabilities, you must always ascertain that the events are indeed mutually exclusive. If they are not, then it may be possible to apply Eq. 6.9 or one of its generalizations. More frequently, an inclusive probability may be determined by calculating the probability of the complementary event, which can often be found by Eq. 6.7. (Cf. Exercise 6.36).

Example

11. Find by two different methods the probability that two cards dealt at random are of the same suit? From Example 6.6, the probability that these two cards are both spades is $13 \cdot 12/52 \cdot 51$. By the same reasoning, the probability that they are both hearts is also $13 \cdot 12/52 \cdot 51$, etc. The four events "both cards are spades," "both cards are hearts," etc., are mutually exclusive. Therefore we may use Eq. 6.9″ to conclude that the desired probability is the sum of four identical terms

$$P(\text{both cards the same suit}) = 4\,\frac{13 \cdot 12}{52 \cdot 51}.$$

Alternatively, let us determine the probability that they are of different suits. It doesn't matter what the first card is, and the probability that the second card is of a different suit from that of the first is 39/51, since 39 of the remaining 51 cards are of a different suit. Therefore

$$P(\text{both cards the same suit}) = 1 - \frac{39}{51}$$

which is in fact the same as the previous result.

Exercises

36. Use a diagram like Figure 6.1 to convince yourself that $P(\sum E_i) = 1 - P(\prod \sim E_i)$.
37. What is the probability of flipping a coin n times and getting at least one head?
38. What is the probability of being dealt at least one ace in a bridge hand?
39. A total of r balls are distributed at random among n boxes.
    (a) What is the probability that any particular box is occupied?
    (b) What is the probability that at least one box has more than one ball in it?

*40. What is the probability that in a group of 24 people at least two will celebrate their birthday on the same day. (Assume all of 365 birthdays are equally likely.) Use Stirling's formula to obtain an approximate numerical answer.

*41. Convince yourself that in general $P(\sum E_i) \leqslant \sum P(E_i)$.

6.2 Combinatorial Quantities

For large sample spaces it is inconvenient to count sample points. Therefore we require compact formulas for various combinatorial quantities. We shall describe these quantities in terms of "objects" and "slots" whose significance depends upon context. We may in practice be assigning balls to boxes or cards to a bridge hand or molecules to energy levels. In this section we shall assume that the objects (and slots) are DISTINGUISHABLE, capable of being labeled so that they may be told apart.

A DISTRIBUTION of r objects to n slots is an assignment of the objects to the slots. There are a total of n^r distinguishable distributions, since the first object may be assigned to any of n slots, the second may be assigned to any of n slots, etc. And if the distributions are all equally likely, then the probability of any one of them is $1/n^r$.

Examples

12. Each outcome of the experiment "flip N coins" is a distribution of N objects to 2 slots, labeled H and T. The probability of any given outcome (e.g., all heads) is $1/2^N$.

13. There are 6^N equally likely outcomes to throwing N dice, and the probability of any given outcome is $1/6^N$.

Exercises

42. A total of r cards are drawn in order, with replacement, from a deck of cards. How many distinguishable sequences of draws are possible? (*Hint:* What are objects and what are slots?) What is the probability that all r cards are the ace of spades?

43. (a) The genetic code is based upon a four-letter alphabet. How many three-letter words are possible?
    (b) A protein may be considered a word written in the twenty-letter alphabet of the amino acids. How many 124-letter proteins are possible?

*44. How many three-letter "words" are possible if a word must contain one or two vowels, a, e, i, o, u?

Next we consider distributions of r objects to n slots, but with multiple occupancy prohibited, that is, subject to the requirement that no more than one object can be in any slot. Of course, we must then have $n \geqslant r$. There are a total of $n!/(n - r)!$ distinguishable distributions of this kind since the first object may be assigned to any of n slots, the second to any of $(n - 1)$, etc., and the rth to any of $(n - r + 1)$. This total is usually symbolized by P_r^n, read "*Pnr*." And if each distribution is equally likely, then the probability of any one is $1/P_r^n = (n - r)!/n!$. For historical reasons a distribution without multiple occupancy is unfortunately called a PERMUTATION of n objects taken r at a time. (Note the reversal!)

Examples

14. There are P_r^n distinguishable ways of selecting r objects from a set of n objects $(n \geqslant r)$ if the order of selection is important. This is the historical sense of the number of permutations of n objects taken r at a time.

15. There are a total of 12 ways of distributing 2 objects, A and B, to 4 slots, without multiple occupancy, namely, *AB*--, *A*-*B*-, *A*--*B*, *BA*--, -*AB*-, -*A*-*B*, *B*-*A*-, -*BA*-, --*AB*,

B--A, *-B-A*, and *--BA*. Alternatively, there are 12 ways of selecting 2 objects in order from the 4 objects *W*, *X*, *Y*, *Z*; namely, *WX*, *WY*, *WZ*, *XW*, *XY*, *XZ*, *YW*, *YX*, *YZ*, *ZW*, *ZX*, and *ZY*.

16. In blackjack, an initial deal consists of dealing each player one card face down and one card face up. If there are N players using a 104-card deck, there are $P_{2N}^{104} = 104!/(104 - 2N)!$ distinguishable initial deals.

17. Let us take the 13 spades from a deck of cards and set them side by side in a row. Each arrangement is a distribution of 13 objects (faces or cards) to 13 slots (positions), without multiple occupancy. If the cards have been set out at random, then the probability that they are arranged in descending order, AKQJ10 · · · 32, is $1/P_{13}^{13} = 1/13!$. A permutation of n objects taken n at a time is generally just called a PERMUTATION of n objects and is merely an arrangement of those objects in order. There are $n!$ such permutations.

Exercises

*45. (a) In five-card stud poker, each player is dealt one card face down and four cards face up. The order of the cards face up is important, since there is betting after each round is dealt. If there are N players and no one drops out, how many deals are possible?
(b) In baseball, a batting order is a list of 9 players in order. How many batting orders are possible if there are 25 players on the team?

46. If baseball teams won at random, what would be the probability that the final standings listed the six teams of a division in alphabetical order?

The most important combinatorial quantity is the number of ways we may choose or select r objects out of n, without regard for the order in which they are chosen. First let us note that if the order of choosing the objects were important, then according to Example 6.14 there would be P_r^n ways of choosing r objects from n. However, since the order is unimportant, no new choice results if we rearrange the order in which the same r objects are chosen. There are a total of $r!$ such rearrangements since there are $r!$ permutations of r objects. To claim that there are P_r^n choices is then an overestimate since each of these $r!$ permutations is not a distinguishable choice. Since they are all the same choice, we must correct this overestimate by dividing by $r!$. Thus we conclude that the number of ways of choosing r objects out of n is $P_r^n/r! = n!/(n - r)!r!$. For historical reasons this quantity is called the number of COMBINATIONS of n objects taken r at a time. It is usually symbolized by C_r^n or $\binom{n}{r}$, but you are advised to remember C as standing for choosings rather than combinations. Finally, if each choice is equally likely, the probability of any one of them is $1/C_r^n = r!(n - r)!/n!$

Examples

18. In Example 6.15, if the order of choice is not important, *WX* and *XW* no longer represent different choices. Indeed, there are only six ways of choosing 2 objects from the 4 objects *W*, *X*, *Y*, *Z*; namely, *WX*, *WY*, *WZ*, *XY*, *XZ*, and *YZ*.

19. How many bridge hands are possible? This number is just the number of ways of choosing 13 cards out of 52, or C_{13}^{52}. The probability of making the particular choice of the

hand with all 13 spades is then $1/C_{13}^{52}$, which is in fact the same answer as we obtained previously in Example 6.7.

20. There are C_5^{13} ways of choosing 5 spades from 13 and therefore there are C_5^{13} distinguishable 5-card hands containing all spades.

21. According to Example 6.12 the number of ways of distributing n objects to 2 slots is 2^n. In how many ways may the n objects be distributed so that r objects are assigned to slot 1 and the remaining $(n - r)$ objects assigned to slot 2? This is just the number of ways that r objects may be chosen from the n objects, or C_r^n. We shall generalize this result in Section 6.4.

Exercises

47. Convince yourself that:
    (a) $C_n^n = C_0^n = 1$.
    (b) $C_1^n = C_{n-1}^n = n$.
    (c) $C_{n-r}^n = C_r^n$.

*48. Show that $C_r^n + C_{r+1}^n = C_{r+1}^{n+1}$ $(r < n)$.

49. (a) In a round-robin chess tournament, every player plays one game against every other player. How many games must be played if there are 10 entrants?
    (b) A qualitative analysis scheme will analyze for 24 cations. How many different unknowns containing 4 cations can be compounded?

50. How many 5-card poker hands are possible? What is the probability of being dealt the AKQJ10 of spades?

Finally, we state explicitly that if events E_1 and E_2 are independent and can be achieved in n_1 and n_2 ways, respectively, then the event E_1E_2 may be achieved in a total of n_1n_2 ways. For example, with 7 cations and 5 anions it is possible to construct 35 salts; also, there are 4 ways to choose a suit and C_5^{13} ways to choose 5 cards of that suit, so there are $4C_5^{13}$ 5-card hands all of the same suit. Furthermore, we state that if the events $\{E_i\}$ are all mutually exclusive and if E_i can be achieved in n_i ways, then the event $\sum E_i$ may be achieved in a total of $\sum n_i$ ways. For example, there are 6 mutually exclusive ways of throwing N dice so that they all show the same face; also, there are C_5^{13} ways of choosing 5 spades and C_5^{13} ways of choosing 5 hearts, etc., so we see again that there are $4C_5^{13}$ 5-card hands all of the same suit.

Exercise

51. (a) How many 6-place license plates are possible if the first 3 places are letters and the last 3 numbers?
    (b) A chemistry department consists of 17 faculty members and 72 graduate students. How many different student–faculty committees may be appointed if each committee consists of 4 faculty members and 3 students?

Next we demonstrate how we may use combinatorial quantities to determine probabilities. All that is necessary is to determine the number of equally likely sample points and the number of those points that correspond to the event of interest.

Examples

22. There are 5^N distinguishable ways of throwing N dice so that no six shows. Since there are a total of 6^N outcomes on throwing N dice, the probability that no six shows on throwing N dice is $5^N/6^N$. (Cf. Example 6.9.)

23. There are n^r ways of distributing r objects to n slots if there are no restrictions, but there are only P^n_r ways of distributing r objects among n slots without multiple occupancy. Therefore if r objects are distributed randomly among n slots, the probability that no slot is multiply occupied is P^n_r/n^r. Alternatively, there are n^r ways of drawing r objects in order from n objects if each object is replaced after being chosen, but there are only P^n_r ways of *withdrawing* r objects in order from n objects. Therefore if from n objects r objects are chosen with replacement, the probability that no object will be chosen again is P^n_r/n^r. This result concerning nonrepetition of choices arises in many guises. For examples, the probability of getting one of each face on throwing 6 dice (Exercise 6.14a) is $P^6_6/6^6 \sim 0.015$. And the probability that 24 people will celebrate their birthdays on 24 different days (the complement of the event of Exercise 6.40) is $P^{365}_{24}/365^{24}$.

24. Since there are $4C^{13}_5$ 5-card hands containing cards all of the same suit and a total of C^{52}_5 5-card hands, the probability of being dealt a flush at poker is $4C^{13}_5/C^{52}_5$. (Cf. Exercise 6.30a).

25. A basket contains 20 apples of which 4 are rotten. What is the probability of picking 5 good apples? We could solve this problem by Eq. 6.7, but let us solve it in terms of numbers of choosings. There are C^{16}_5 ways of choosing 5 good apples from among the 16 good apples. The sample space is composed of C^{20}_5 equally likely sample points, one for each choice of 5 apples from the basket. Therefore the desired probability is C^{16}_5/C^{20}_5.

26. What is the probability of being dealt a bridge hand with at least 7 spades? There are C^{13}_7 ways of choosing 7 spades, and there are C^{39}_6 ways of choosing 6 nonspades. Therefore there are $C^{13}_7C^{39}_6$ distinguishable bridge hands that contain exactly 7 spades. Since there are a total of C^{52}_{13} bridge hands, the probability of being dealt 7 spades at bridge is $C^{13}_7C^{39}_6/C^{52}_{13}$. Furthermore, since being dealt 7 spades is mutually exclusive to being dealt 8 spades, etc., the probability of being dealt at least 7 spades is

$$(C^{13}_7C^{39}_6 + C^{13}_8C^{39}_5 + C^{13}_9C^{39}_4 + C^{13}_{10}C^{39}_3 + C^{13}_{11}C^{39}_2 + C^{13}_{12}C^{39}_1 + C^{13}_{13}C^{39}_0)/C^{52}_{13}.$$

27. The probability of being dealt no pair at poker (Exercise 6.12) is $4^5C^{13}_5/C^{52}_5$, since there are C^{13}_5 ways of choosing 5 different faces from the thirteen possible faces, 4^5 distributions of 5 cards among the four slots S, H, D, C, and C^{52}_5 5-card hands.

Exercises

*52. (a) A football team scored four touchdowns in its latest game. If scoring occurred at random, what would be the probability that one touchdown was scored in each quarter?
(b) What is the probability that 6 random digits are all different?
Surprise yourself by obtaining decimal approximations.

53. (a) What is the probability of being dealt a bridge hand with exactly one ace?
(b) Use Eq. 6.7 to find the probability that each of the four bridge players holds one ace?

54. Use combinatorial quantities to solve Exercises 6.10, 6.13, 6.14b, 6.16, 6.29a, and 6.39b.

*55. (a) What is the probability of being dealt a poker hand with two spades and one of each of the other suits?
(b) With all four suits represented ("a 2-1-1-1 distribution")?

(c) What is the probability of being dealt a bridge hand with 4 spades and 3 of each of the other suits?

(d) With 4 of one suit and 3 of each of the other suits ("a 4-3-3-3 distribution")?

*56. A deck of cards is turned over card-by-card until an ace appears. What is the probability that this ace is the rth card to be turned over?

*57. My partner and I hold a total of 9 spades at bridge. What is the probability that the player on my left holds:

(a) 2 spades?

(b) 1 spade?

(c) 3 spades?

(d) 0 spades?

(e) 4 spades?

(f) What are the probabilities of a 2-2 split (2 spades in each opponent's hand), a 3-1 split (3 spades in one opponent's hand and 1 in the other's), and a 4-0 split? Obtain decimal approximations and note that the more "uniform" 2-2 split is less likely than a 3-1 split.

*58. Play cards tonight. Use the opportunity to devise more probability problems to practice on.

6.3 Frequency Functions

Before we proceed, we wish to simplify our notation. So far we have defined probability as a function on a sample space. It would be more convenient if we could deal with probability as the usual sort of function, whose domain and range are numerical. Therefore let us return to the idea of a random variable as a function that labels points in sample space. In particular, we shall use a random variable to assign the same value to each sample point corresponding to an event, and different values to different events. Now we need no longer talk about the probability of an event. Instead we may take the value of the random variable as the argument for a new probability function. Thus we also define PROBABILITY as a function of a random variable such that its value at a given argument is equal to the probability of the event associated with that argument. Finally, let us neglect to distinguish functions from dependent variables. Then we may see that the random variable is both the dependent variable for a function defined on sample space and also the independent variable for the function probability. It is this duality of role that leads to the name "random variable." Next let i be a variable that ranges over sample space, let x be a random variable, and let P be a variable that ranges over probabilities. Since our original probability was a numerical-valued function on sample space, we may recognize it as a composite function (cf. Eq. 1.1)

$$P[x(i)] = P(i).$$

The probability $P(x)$, as a function of a random variable, is also called the FREQUENCY FUNCTION or the DISTRIBUTION of that random variable. It is unfortunate that "distribution" has two meanings in probability. To add to the confusion, it is common to use the same symbol, P, for *all* these functions, even when several different functions appear in the same problem.

Examples

28. For the experiment "throw one die" and the random variable "number of points," $P(1) = P(2) = P(3) = P(4) = P(5) = P(6) = 1/6$.

29. For the experiment "throw two dice" and the random variable "sum of the points," $P(2) = 1/36 = P(12)$, $P(3) = 2/36 = P(11)$, etc.

30. For the experiment "flip two coins" and the random variable "number of heads," $P(0) = 1/4 = P(2)$ and $P(1) = 1/2$.

31. For the experiment "deal a bridge hand" and the random variable "number of aces," $P(0) = C^{48}_{13}/C^{52}_{13}$, $P(1) = C^4_1 C^{48}_{12}/C^{52}_{13}$, etc.

This approach allows us to draw a graph of probability as a function of the random variable. Figure 6.2 shows graphs of each of the probability

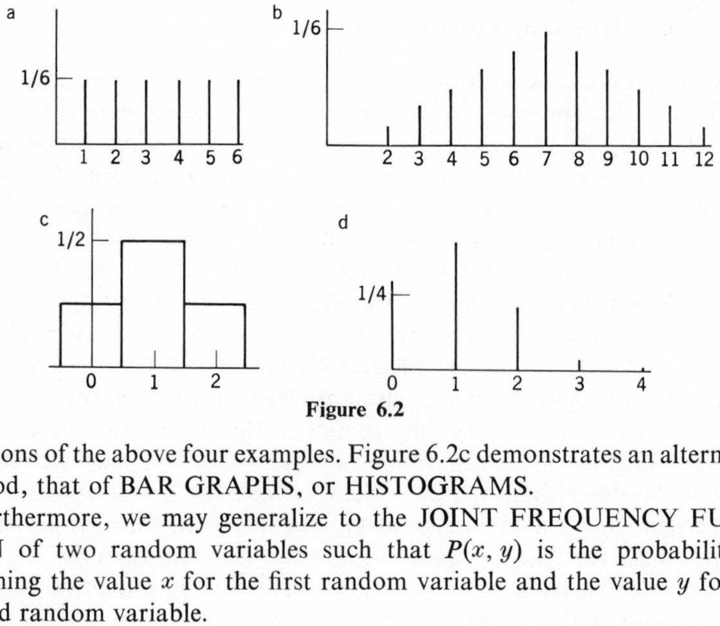

Figure 6.2

functions of the above four examples. Figure 6.2c demonstrates an alternative method, that of BAR GRAPHS, or HISTOGRAMS.

Furthermore, we may generalize to the JOINT FREQUENCY FUNCTION of two random variables such that $P(x, y)$ is the probability of obtaining the value x for the first random variable and the value y for the second random variable.

Examples

32. For the experiment "throw two dice" and the two random variables "number of points on the first die" and "number of points on the second die," $P(1, 1) = P(1, 2) = \cdots = 1/36$.

33. For the experiment "flip two coins" and the two random variables "number of heads" and "number of tails," $P(2, 0) = 1/4 = P(0, 2)$ and $P(1, 1) = 1/2$, with all other $P(x, y) = 0$.

We may transfer the notions of conditional probability and independence to frequency functions of random variables. The CONDITIONAL FREQUENCY FUNCTION $P(y \mid x)$ is the probability of obtaining the value y for one random variable, given that the value x has been obtained for another random variable. It follows that $P(x, y) = P(x)P(y \mid x) = P(y)P(x \mid y)$. Furthermore, two random variables x and y, for which $P(x, y) = P(x)P(y)$, are said to be INDEPENDENT. Those of Example 6.32 are independent, but those of Example 6.33 are not, since obtaining 2 heads with 2 coins precludes obtaining any tails.

Exercises

59. (a) Convince yourself that $\sum P(r) = 1$, where the sum is over all possible values of r.
    *(b) Convince yourself that the area under the histogram of the frequency function of a random variable restricted to integers equals unity.

*60. A random variable n is restricted to nonnegative integers. The probability of obtaining the value n is given by the frequency function $P(n)$.
    (a) Convince yourself that the probability of obtaining a value n_0 or less is given by the CUMULATIVE DISTRIBUTION FUNCTION $\Pi(n_0) = \sum_{n=0}^{n_0} P(n)$.
    (b) What is $\Pi(\infty)$?

61. For the experiment "deal a poker hand," tabulate the frequency function of the random variable "number of aces."

*62. For the experiment "throw two dice," tabulate the values of the joint distribution function of the two random variables "number of aces" and "number of sixes."

*63. Convince yourself that if x and y are independent random variables, then $P(y \mid x) = P(y)$ and $P(x \mid y) = P(x)$.

*64. For the experiment "deal a bridge hand," what is the joint frequency function of the two random variables "number of aces" and "number of kings"? Are these independent? Are the two random variables "number of aces" and "number of spades" independent?

6.4 Some Special Distributions

Next we consider some frequency functions associated with some special distributions of objects to slots. First we consider distributions of n distinguishable objects to 2 slots. Let us agree to call the distribution of n objects the experiment and the distribution of a single object a TRIAL; if an object is assigned to the first slot, we shall call that trial "successful," otherwise "unsuccessful." Let us assume that for each trial the probability of success is p, independent of the results of all other trials. What then is the frequency function of the random variable "number of successes," that is, what is the probability that the experiment had exactly r successful trials or that exactly r objects are in the first slot? Let us first determine the probability of the particular outcome with the first r trials successful and the last $(n - r)$ unsuccessful. Since the trials are independent, this probability is just a product of the probabilities on each trial, or $p^r(1-p)^{n-r}$. But getting all the successes first is not the only way to order r successful trials and $(n-r)$ unsuccessful ones. According to Example 6.21, there are a total of C_r^n ways of assigning r trials to the "successful" slot and $(n-r)$ trials to the "unsuccessful" slot; there are C_r^n ways of choosing r successes from n trials. The probability of each of these is $p^r(1-p)^{n-r}$ and they are all mutually exclusive. Therefore if the probability of success on each trial is p, then the probability of exactly r successes in an experiment consisting of n trials is

$$P(r) = C_r^n p^r(1-p)^{n-r}. \tag{6.10}$$

This frequency function is known as the BINOMIAL DISTRIBUTION since the BINOMIAL COEFFICIENTS $\{C_r^n\}$ are the coefficients in the BINOMIAL EXPANSION (cf. Exercise 1.55)

$$(p+q)^n = \sum_{r=0}^{n} C_r^n p^r q^{n-r} \tag{6.11}$$

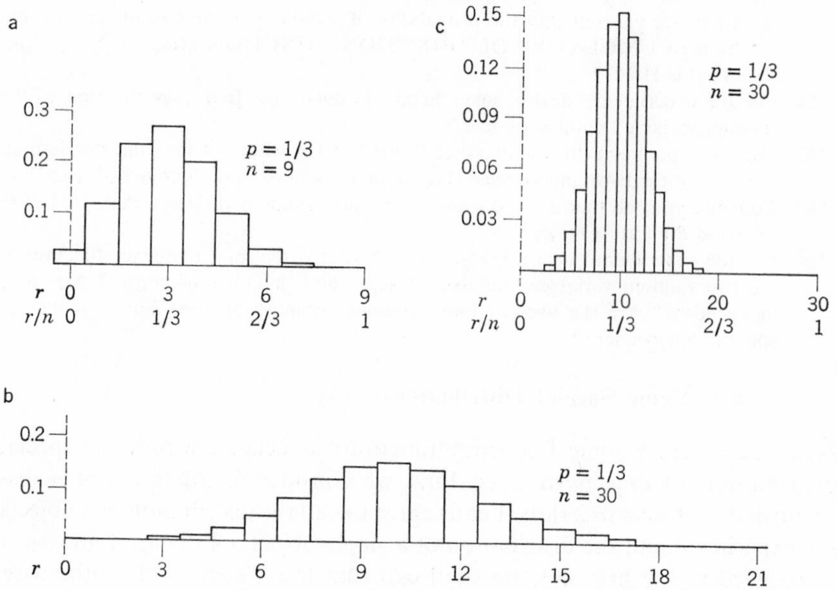

Figure 6.3

Figure 6.3 shows histograms of the binomial distribution with $p = 1/3$ and $n = 9$ or 30. Figures 6.3b and 6.3c are the same but with different scales, as indicated. Figure 6.3b shows that the distribution with $n = 30$ is the "broader" one. Figure 6.3c shows that increasing n from 9 to 30 does not lead to as much broadening as might have been expected. Indeed, we shall find that the "breadth" of the binomial distribution does not increase as n, but only as $n^{1/2}$.

Examples

34. Sixteen coins are tossed. The probability of obtaining exactly 8 heads is $P(8) = C_8^{16}(1/2)^8(1/2)^8 \sim 0.20$. Notice that although we might have expected 8 heads, this number is really quite unlikely.

35. A thousand random digits are sampled. The probability of obtaining exactly r zeros is $P(r) = C_r^{1000}(1/10)^r(9/10)^{1000-r}$. A thousand random digits do not always contain 100 zeros.

36. A total of n objects are distributed at random to m slots. The probability that any given slot contains exactly r objects is $P(r) = C_r^n(1/m)^r[1 - (1/m)]^{n-r}$. Thus if there are 50 red blood cells to be distributed among 25 squares of a hemacytometer, it is unlikely that every square contains 2 cells. Instead, the probability that a square contains r cells is $P(r) = C_r^{50}(1/25)^r(24/25)^{50-r}$.

37. A molecule is constantly being buffeted by surrounding molecules. Let us restrict the molecule to one dimension, and let us assume that upon each encounter with a neighbor the molecule is knocked a distance d (Angstroms), but that it is just as likely to be knocked to the left as to the right. Then, after a total of $2n$ encounters, what is $P(r)$, the probability that the molecule has traveled a distance $2rd$ to the right $(-n \leqslant r \leqslant n)$? To have traveled exactly $2rd$ after n encounters, the molecule must have been knocked $n + r$ times to the right and $n - r$ times to the left. The probability of such an occurrence is

$$P(r) = C_{n+r}^{2n}(1/2)^{n+r}(1/2)^{n-r}$$

This is an example of a "random-walk" problem and can serve as a model for diffusion processes.

Exercises

65. Set $q = 1 - p$ in Eq. 6.11 to convince yourself that for the binomial distribution $\sum_{r=0}^{n} P(r) = 1$, as required for a frequency function.

*66. What is $\sum_{r=0}^{n} C_r^n$? $\sum_{r=0}^{n}(-1)^r C_r^n$?

67. The nmr spectrum of isopropyl chloride shows two "bands." The intense one, due to methyl absorption, is a doublet (two peaks of equal intensity), because in half the molecules the ClC*H* nuclear spin is up and in the other half it is down. Assume that each of the six methyl proton spins may be up or down at random, and determine the relative intensities of the seven peaks in the ClC*H* band.

68. What is the chance of throwing exactly 3 aces with 10 dice?

69. What is the probability of drawing 13 spades in 52 draws, with replacement, from a deck of 52 cards?

70. A 25-square hemacytometer contains 50 red blood cells. Use Eq. 6.7 to find the probability that every square contains exactly 2 cells. Simplify your answer.

*71. Objects are distributed at random among m slots until exactly r objects have accumulated in the first slot. For example, a die is tossed repeatedly until r aces have come up. Show that $P(n)$, the probability that this occurrence requires exactly n objects $(n \geqslant r)$ to have been distributed, is given by $P(n) = C_{r-1}^{n-1}p^r(1 - p)^{n-r}$, where $p = 1/m$. Notice that this is not a binomial distribution.

*72. Ten coins have been flipped into a hat. Four are withdrawn at random and all are found to be heads. Use the result of Exercise 6.19a to show that the probability that r of the 10 coins in the hat are heads $(r \geqslant 4)$ is still a binomial distribution, but now with $n = 6$.

For large n the terms of the binomial distribution become quite inconvenient to evaluate. However, there are two approximations to the binomial distribution, both applicable for large n. The NORMAL DISTRIBUTION, or GAUSSIAN DISTRIBUTION, is applicable when $p \sim 1/2$. According to Exercise 3.25

$$P(r) = \frac{1}{[2\pi np(1-p)]^{1/2}} \exp\left[-\frac{(r-np)^2}{2np(1-p)}\right] \tag{6.12}$$

the familiar, symmetrical bell-shaped curve, centered at $r = np$. This formula is especially convenient for summing terms of the binomial distribution. The POISSON DISTRIBUTION is applicable when p is small. According to Exercise 3.24

$$P(r) = \frac{(np)^r e^{-np}}{r!}. \tag{6.13}$$

This distribution is not symmetrical about its maximum at $r = np$.

Examples

38. On an exam with 100 true-false questions, 6 members of a class of 60 scored at least 90% correct. Can this result be ascribed to chance or did these students demonstrate a knowledge of the material? Let us first consider what may be expected for a single student. The probability p that a student guessing at random would score 90% or better is

$$C_{90}^{100}(1/2)^{90}(1/2)^{10} + C_{91}^{100}(1/2)^{91}(1/2)^{9} + \cdots + C_{100}^{100}(1/2)^{100}.$$

Let us not only approximate the binomial distribution by the normal distribution but also replace the sum by an integral (yet see Exercise 6.76):

$$p \sim \frac{1}{\sqrt{50\pi}} \int_{90}^{100} e^{-(r-50)^2/50}\,dr = \frac{1}{\sqrt{\pi}} \int_{4\sqrt{2}}^{5\sqrt{2}} e^{-t^2}\,dt \sim \frac{1}{\sqrt{\pi}} \int_{4\sqrt{2}}^{\infty} e^{-t^2}\,dt$$
$$= \tfrac{1}{2}\,\mathrm{erfc}\,(4\sqrt{2}) \sim 1.6 \times 10^{-15}.$$

It is now obvious that the performance of the 6 students cannot be attributed to mere chance since the probability that 6 students out of 60 would score at least 90% is only about $C_6^{60} \times (1.6 \times 10^{-15})^6$.

39. We may use the Poisson approximation for the distribution of 50 red blood cells among 25 squares of a hemacytometer; the probability that any given square contains exactly r cells is $P(r) \sim 2^r e^{-2}/r!$. The exact value is given in Example 6.36.

40. Let a sample contain N radioactive atoms and let p be the probability for each to disintegrate during the next minute. Then since N is very large and p generally quite small, we may use the Poisson approximation to determine the probability of exactly r disintegrations in the next minute: $P(r) = (Np)^r e^{-Np}/r!$. Notice that since disintegrations occur at random, they do not occur at regular time intervals.

41. Chromatographic separation is so familiar to chemists that a simplified model of it should make quite clear what the binomial and Poisson distributions "look like." Let us imagine a long column packed with V_s ml of a stationary phase (e.g., alumina, paper, ion-exchange resin, silicone oil). A constant stream of a mobile phase (eluting solvent or helium gas) is flowing through the column. At any time let there be V_m ml of mobile phase

in the column. Let us assume that the column may be divided into N theoretical plates, each of volume $(V_s + V_m)/N$ ml. Within each theoretical plate an added solute equilibrates between stationary and mobile phases, but equilibrium is *not* maintained between theoretical plates. The equilibrium within a theoretical plate is governed by a partition coefficient

$$K \equiv \frac{\text{moles solute/volume of stationary phase}}{\text{moles solute/volume of mobile phase}}.$$

Now let us determine what fraction of solute within a theoretical plate may be found in the mobile phase of that plate. Call this fraction p so that the fraction of solute in the stationary phase is $(1-p)$. Since the volumes of these two phases are V_s/N and V_m/N, the *relative* molarities of solute in these phases are $(1-p)/(V_s/N)$ and $p/(V_m/N)$. And to satisfy the partitioning equilibrium, we must have

$$K = \frac{(1-p)}{V_s/N}\,\frac{V_m/N}{p}.$$

Solving for p gives

$$p = \left(1 + K\frac{V_s}{V_m}\right)^{-1}.$$

Thus we may solve for the fraction in the mobile phase in terms of the characteristics of the column.

Let M_0 moles of solute be applied at plate 0. When equilibrium is established, there are M_0p moles of solute in the mobile phase of plate 0 and $M_0(1-p)$ moles in the stationary phase (Fig. 6.4a). Next let us transfer the V_m/N ml of mobile phase from plate 0 to plate 1, along with the M_0p moles of solute it contains. When equilibrium is again established in plate 1, its mobile phase contains M_0p^2 moles of solute, and its stationary phase contains $M_0p(1-p)$ moles (Fig. 6.4b). The $M_0(1-p)$ moles of solute that remained in plate 0 equilibrate so that $M_0p(1-p)$ moles are in the mobile phase and $M_0(1-p)^2$ in the

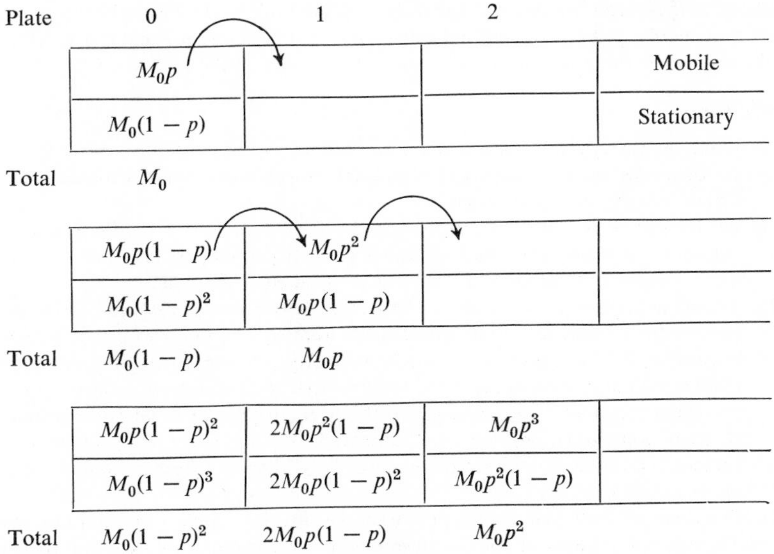

Figure 6.4

stationary phase. Next let us transfer the V_m/N ml of mobile phase, with its solute, from plate 1 to plate 2, and likewise from plate 0 to plate 1. Then the total contents of plates 0, 1, and 2 are $M_0(1-p)^2$, $2M_0p(1-p)$, and M_0p^2, respectively (Fig. 6.4c). Repeated application of this reasoning leads to the conclusion that after nV_m/N ml of mobile phase has passed through the column, the total contents of the rth plate is

$$M(r) = C_r^n M_0 p^r (1-p)^{n-r}.$$

That the binomial distribution should arise in this context is quite reasonable, especially if we interpret p as "the probability that a molecule will succeed in its attempt to move on to the next plate" and $M(r)/M_0$ as "the probability that a molecule will have achieved exactly r successes in n tries." And we may note that the molecule is most likely to be in the (np)th plate.

In practice, K is usually much greater than 1, corresponding to strong adsorption. Then p is very small, and the Poisson distribution becomes quite a good approximation for the distribution of solute among the theoretical plates:

$$M(r) \sim M_0(np)^r e^{-np}/r!.$$

We may use this result to determine the shape of the elution curve—moles of solute eluted as a function of the volume of mobile phase that has passed through the column. After nV_m/N ml of mobile phase has passed through the column, there are $M_0(np)^{N-1}e^{-np}/(N-1)!$ moles of solute in the last theoretical plate [the $(N-1)$st]. Of this material, the fraction p, or $M_0p(np)^{N-1}e^{-np}/(N-1)!$ moles, is in the mobile phase. For mathematical simplicity, we shall insert the detector into the mobile phase of the last theoretical plate. We may then consider these $M_0p(np)^{N-1}e^{-np}/(N-1)!$ moles of solute to be eluted. Thus we may conclude that the probability that the nth (V_m/N)-ml portion of mobile phase will elute a molecule is

$$P(n) = p(np)^{N-1}e^{-np}/(N-1)!.$$

This is neither a binomial nor a Poisson distribution. However, differentiation with respect to n leads to the conclusion that a molecule is most probable to elute with the $[(N-1)/p]$th portion of mobile phase. We shall return to this elution curve in Section 6.6, where we shall consider the sharpness of the peak and the efficiency of the separation.

Exercises

73. What, approximately (Poisson approximation), is the probability of drawing the ace of spades exactly twice in 52 tries with replacement? What is the exact value? Obtain decimal approximations.

*74. Let 10^8 people be distributed at random among 10^4 cities. Use the Poisson approximation to estimate the probability that a given city has a population of 10^6. Do you think that the U.S. urban population is distributed at random?

*75. The amino-acid sequences of two 40-amino-acid polypeptides are determined. If each of the 20 most common amino acids were equally likely to occur, how many "coincidences" of amino acids would you expect if these two polypeptides were chosen at random? (A coincidence requires that the same amino acid is found at a particular position in the sequences of both polypeptides.) What is the probability (Poisson approximation) that two "unrelated" 40-amino-acid polypeptides show 31 coincidences in their amino-acid sequences? Use Stirling's formula to find a decimal approximation.

*76. Convince yourself that in Example 6.38 the integral should run from $89\frac{1}{2}$ to $99\frac{1}{2}$. This sort of adjustment should always be made for more accurate approximation of sums by integrals, but we shall not bother.

77. Use the normal approximation to the binomial to estimate the probability of obtaining:
    (a) Five heads on throwing ten coins.
    (b) Fifty heads on throwing one hundred coins.
    (c) Five hundred heads on throwing one thousand coins.
    Obtain decimal approximations (2 significant figures).

*78. Use the normal approximation to the binomial, and replace sums by integrals, to estimate the probability of obtaining:
    (a) 5 ± 1 heads on throwing 10 coins.
    (b) 50 ± 10 heads on throwing 100 coins.
    (c) 500 ± 100 heads on throwing 1000 coins.
    Obtain decimal approximations and use the result of Exercise 3.3 when necessary.

*79. Let us consider the polymerization of N_m molecules of monomer (e.g., ethylene oxide or styrene) initiated by N_i molecules of nucleophile (e.g., alkoxide or carbanion). After all the monomer has been consumed, what is $P(r)$, the probability that a polymer molecule contains exactly r molecules of monomer? Assume that each polymer chain, no matter what size, has the same reactivity toward monomer. (*Hint:* What are the objects and what are the slots?) Approximate this distribution by the Poisson or the normal distribution, whichever is appropriate.

We have been considering distributions of n objects to 2 slots; next we consider distributions of r objects to n slots. Again we shall call the distribution of r objects the experiment and the distribution of a single object a trial. For generality, we shall assume that the probability of assigning an object to the ith slot is p_i, independent of the results of all other trials; often all slots are equally likely, so we would then have $p_i = 1/n$. Let us determine $P(\{r_i\}_{i=1}^n)$, the probability that there are exactly r_1 objects in the first slot, r_2 objects in the second, . . . , r_i in the ith, . . . , and r_n in the nth; the numbers $\{r_i\}$ are known as OCCUPATION NUMBERS and, of course, we require $\sum r_i = r$. First we note the probability that the first r_1 objects are distributed to the first slot, then the next r_2 objects to the second, . . . , and the last r_n to the nth slot, is $p_1^{r_1}p_2^{r_2}\cdots p_n^{r_n} = \prod p_i^{r_i}$. But this is not the only way to choose r_1 objects for the first slot, r_2 for the second, etc. Let the number of ways of choosing this sequence of occupation numbers be $M^r_{\{r_i\}}$, and let us evaluate this number. For the first slot there are $C^r_{r_1}$ ways of choosing r_1 objects out of r. For the second slot there are then $C^{r-r_1}_{r_2}$ ways of choosing r_2 objects out of the remaining $r - r_1$, etc. Therefore the number of ways of choosing the sequence $\{r_i\}$ is

$$\begin{aligned} M^r_{\{r_i\}} &= C^r_{r_1} C^{r-r_1}_{r_2} C^{r-r_1-r_2}_{r_3} \cdots \\ &= \frac{r!}{r_1!(r-r_1)!}\,\frac{(r-r_1)!}{r_2!(r-r_1-r_2)!}\,\frac{(r-r_1-r_2)!}{r_3!(r-r_1-r_2-r_3)!}\cdots \\ &= \frac{r!}{\prod r_i!} \end{aligned} \tag{6.14}$$

since factors $(r - r_1)!$, etc., cancel. The probability of each of these ways of achieving a distribution with occupation numbers $\{r_i\}$ is still $\prod p_i^{r_i}$, and they are all mutually exclusive. Therefore we conclude that if p_i is the probability of assigning an object to the ith slot, then the probability of obtaining the occupation numbers $\{r_i\}$ on distributing r objects to n slots is

$$P(\{r_i\}_{i=1}^n) = M_{\{r_i\}}^r \prod p_i^{r_i} = \frac{r!}{\prod r_i!} \prod p_i^{r_i} \tag{6.15}$$

where the sequences and the products are over all i from 1 to n, but subject to the restriction $\sum r_i = r$. This frequency function is known as the MULTINOMIAL DISTRIBUTION since the coefficients $M_{\{r_i\}}^r$ are the MULTINOMIAL COEFFICIENTS, which arise in the MULTINOMIAL EXPANSION

$$\left(\sum_{i=1}^n x_i\right)^r = \sum_{\{r_i\}} M_{\{r_i\}}^r \prod_{i=1}^n x_i^{r_i} \tag{6.16}$$

where the sum on the right is over all sequences $\{r_i\}$ of occupation numbers satisfying $\sum r_i = r$.

Examples

42. A bridge deal is a distribution of 52 objects to 4 slots. Of course, not all such distributions are permissible, but only those with occupation numbers {13, 13, 13, 13}. The number of distributions with these occupation numbers is $M_{\{13,13,13,13\}}^{52} = 52!/13!^4$. This is the total number of bridge deals possible.

43. Likewise, if there are 6 players at poker, each deal is a distribution of 52 objects with occupation numbers {5, 5, 5, 5, 5, 5, 22}. Notice that the remainder of the deck must be considered as a slot since the sum of the occupation numbers must always equal the number of objects to be distributed. The number of ways of achieving a distribution of 52 objects to 7 slots, with these occupation numbers, is $52!/5!^6 22!$.

44. What is the probability of throwing exactly 3 aces and also 2 twos with 10 dice? Here the three slots are "ace," "two," and "other" with probabilities 1/6, 1/6, and 4/6, respectively. And the probability of obtaining a distribution with the occupation numbers {3, 2, 5} is $M_{\{3,2,5\}}^{10}(1/6)^3(1/6)^2(4/6)^5 = 10!\,4^5/3!\,2!\,5!\,6^{10}$.

45. A total of N objects are distributed at random among m equally likely slots. What is the most probable set of occupation numbers? What we are seeking is that $\{n_i\}$ which maximizes $P(\{n_i\})$ or, more conveniently, $\ln P(\{n_i\})$. Then

$$P(\{n_i\}) = \frac{N!}{\prod n_i!}\left(\frac{1}{m}\right)^N$$

$$\ln P(\{n_i\}) = \ln N! - \sum \ln n_i! - N \ln m.$$

But we must remember that the product and sum are subject to the constraint $\sum n_i = N$. Therefore we shall use a Lagrangian multiplier λ and construct

$$\begin{aligned} u &= \ln P(\{n_i\}) + \lambda(\sum n_i - N) \\ &= \ln N! - \sum (\ln n_i! - \lambda n_i) - N \ln m - \lambda N. \end{aligned}$$

We shall further assume each n_i to be so large that $n_i!$ may be approximated by the simplest

Stirling's formula and also that n_i may be considered a continuous variable with respect to which we may differentiate. Then, for every n_i

$$0 = \frac{\partial u}{\partial n_i} = \frac{\partial}{\partial n_i}(n_i \ln n_i - n_i - \lambda n_i) = \ln n_i - \lambda$$

or

$$n_i = e^{\lambda}.$$

This equation shows the most probable $\{n_i\}$ is that with all the n_is equal. It can be shown that this is still the answer even if N is small. Indeed, this is the answer we would expect as the most "even." For example, on throwing 60 dice, the most probable outcome is 10 of each face, although this is a rare occurrence.

46. A total of N molecules are distributed among M energy levels; the energy of the ith level is ε_i. Every distribution which conserves energy, that is, for which $\sum n_i\varepsilon_i = E$, is equally likely. What is the most probable set of occupation numbers? Again, we maximize

$$\ln P(\{n_i\}) = \ln N! - \sum \ln n_i! - N \ln M$$

but now subject to the two constraints $\sum n_i = N$ and $\sum n_i\varepsilon_i = E$. Therefore we shall use two Lagrangian multipliers, α and β, and construct

$$\begin{aligned} u &= \ln P(\{n_i\}) + \alpha(\textstyle\sum n_i - N) + \beta(\textstyle\sum n_i\varepsilon_i - E) \\ &= \ln N! - \textstyle\sum (\ln n_i! - \alpha n_i - \beta n_i\varepsilon_i) - N \ln M - \alpha N - \beta E. \end{aligned}$$

Then, with the same assumptions as in the previous example

$$0 = \frac{\partial u}{\partial n_i} = \frac{\partial}{\partial n_i}(n_i \ln n_i - n_i - \alpha n_i - \beta n_i\varepsilon_i) = \ln n_i - \alpha - \beta\varepsilon_i$$

or

$$n_i = e^{\alpha}e^{\beta\varepsilon_i}$$

Finally, we may eliminate α by summing this result over all energy levels: $N = e^{\alpha}\sum e^{\beta\varepsilon_i}$. Substituting and rearranging gives

$$\frac{n_i}{N} = \frac{e^{\beta\varepsilon_i}}{\sum e^{\beta\varepsilon_i}}$$

A distribution with these occupation numbers is known as a MAXWELL-BOLTZMANN DISTRIBUTION. The sum $\sum e^{\beta\varepsilon_i}$ is known as the PARTITION FUNCTION; β may be identified as $-1/kT$.

Exercises

80. Convince yourself that with 2 slots, the formulas for the multinomial distribution reduce to those for the binomial distribution.
81. Expand $(x + y + z)^4$ in order to see what some multinomial coefficients look like.
*82. (a) In a golf tournament with 32 entrants, how many ways are there to arrange 8 foursomes?
    (b) In gin rummy, the dealer deals himself 10 cards and his opponent 11. How many gin rummy deals are possible?
83. (a) In how many ways may six objects be distributed among six slots, one per slot?
    (b) Use this result to solve Exercise 6.14a.
    (c) In how many ways may twelve objects be distributed among six slots, two per slot?
    (d) Use this result to determine the probability of obtaining two of each face with twelve dice.
    (e) Solve Exercise 6.70 by this method.

84. Solve Exercise 6.53b by determining the number of ways 4 aces may be distributed to 4 hands, one per hand, and the number of ways 48 non-aces may be distributed to 4 hands, 12 per hand; then multiply, and divide by the total number of bridge deals.

*85. What is the probability of being dealt a bridge hand with a 4-4-3-2 distribution? A 5-4-3-1 distribution? (Cf. Exercise 6.55)

6.5 Indistinguishable Objects

We have been considering combinatorial quantities as representing the number of ways of achieving various types of arrangements of objects. However, these quantities are appropriate only if the objects are distinguishable. For example, the ace of spades is clearly distinguishable from the ace of hearts, and although two dice look the same, one could be marked to distinguish it from the other. On the other hand, such objects as molecules, atoms, alpha particles, and electrons are not distinguishable. We next wish to consider combinatorial quantities appropriate for such objects.

We have said that there are $r!$ permutations or arrangements of r distinguishable objects. However, let r_1 of the objects be indistinguishable from each other, another r_2 of the objects be indistinguishable from each other, etc., so that there are r_i indistinguishable objects of type i. It would be an overestimate to claim that there are still $r!$ distinguishable arrangements of these r objects, since rearranging indistinguishable objects does not lead to a truly different ordering of the objects. (In particular, if there were only one type of object, then there would be only one distinct way of ordering the r indistinguishable objects, and any other order would "look the same.") Since the $r_1!$ arrangements of r_1 distinguishable objects represent only one distinguishable arrangement of r_1 indistinguishable objects, we have overestimated the number of arrangements of the r objects by a factor of $r_1!$. Also, by the same reasoning, we have overestimated the number of arrangements of the r objects by a factor of $r_2!$, etc. Therefore the true number of distinguishable arrangements of the r objects is $r!/\prod r_i! = M^r_{\{r_i\}}$. Notice that the multinomial coefficients arise in connection with both distributions of r distinguishable objects (r_i to slot i) and arrangements of indistinguishable objects (r_i of type i).

Next we consider distributions among n slots of r indistinguishable objects, all of the same type. There are no longer $M^r_{\{r_i\}}$ ways of achieving a distribution with occupation numbers $\{r_i\}$, but only one distinguishable way since rearranging the objects does not lead to a truly different distribution. Therefore we may specify a distribution of indistinguishable objects merely by specifying $\{n_i\}$. Furthermore, instead of a total of n^r distributions, there are only C_r^{n+r-1} distinguishable distributions, by the following reasoning: Specifying a distribution is equivalent to assigning r objects and $(n-1)$

partitions to $(n + r - 1)$ locations. For example, with $r = 10$ and $n = 3$, the distribution $\{5, 2, 3\}$ is specified by placing a partition in location 6 and

x	x	x	x	x	■	x	x	■	x	x	x

another in location 9, and placing objects in the other 10 locations. But there are exactly $C_{n-1}^{n+r-1} = C_r^{n+r-1}$ ways of choosing the locations of the $(n - 1)$ partitions or of the r objects, and therefore there are C_r^{n+r-1} distinguishable distributions. We may postulate that each of these distinguishable distributions is equally likely, so that the probability of any one of them is $1/C_r^{n+r-1}$. Experimentally, it is found that some particles are distributed among energy levels in just this way. Such particles are called BOSONS and are said to obey BOSE-EINSTEIN STATISTICS. Examples are deuterons, alpha particles, and photons.

There are other particles whose distributions are subject to the further restriction that multiple occupancy of energy levels is forbidden, that is, $r_i = 0$ or 1. Again, rearranging r indistinguishable objects does not lead to a truly different distribution, so there is only one way of achieving each distribution. Furthermore, there are only C_r^n distinguishable distributions, since there are exactly C_r^n ways of choosing the r singly occupied slots from n slots. And we may postulate that each of these distinguishable distributions is equally likely, so that the probability of any one of them is $1/C_r^n$. Experimentally, it is found that some particles are distributed among energy levels in just this way. Such particles are called FERMIONS and are said to obey FERMI-DIRAC STATISTICS. Examples are electrons, protons, neutrons, and particles such as an N^{14} atom, which are composed of an odd number of fermions. However, particles such as an H_2 molecule, which are composed of an even number of fermions, are bosons.

Examples

47. From Example 6.36, if r distinguishable objects are distributed at random to n slots, the probability that a given slot contains k objects is $P(k) = C_k^r(1/n)^k[1 - (1/n)]^{r-k}$. But if the objects are bosons, this probability is $P(k) = C_{r-k}^{n+r-k-2}/C_r^{n+r-1}$ by the following reasoning: There are a total of C_r^{n+r-1} distinguishable distributions of r indistinguishable objects to n slots. If we place k objects in the slot of interest, there remain $r - k$ indistinguishable objects to be distributed among the other $n - 1$ slots. Each distinguishable distribution of these represents one distinguishable distribution of r objects to n slots, subject to the requirement that the given slot contain k objects. There are $C_{r-k}^{(n-1)+(r-k)-1}$ such distributions. It should be noted that if the objects are distinguishable, $P(k)$ is a binomial distribution and therefore is maximum at $k = r/n$. But if the objects are bosons, $P(k)$ may be seen to be maximum for $k = 0$. This result reflects an apparent "attraction" of bosons for each other, such that they accumulate in the same slots and leave other slots unoccupied.

48. How many electronic states arise from a d^2 atomic configuration? This is just the number of distinguishable distributions of 2 electrons to 10 spin orbitals ($m_l = 0, \pm 1, \pm 2$;

$m_s = \pm\frac{1}{2}$) without multiple occupancy, or $C_2^{10} = 45$. Notice that because the electrons are indistinguishable, the answer is not P_2^{10}.

Exercises

86. How many distinguishable 6-letter "words" may be constructed:
    (a) With one each of 6 different letters?
    (b) With 4A's and 2B's?
    (c) With 3A's, 2B's, and a C?
87. Show all distinguishable distributions of 2 objects to 3 slots:
    (a) If the objects are distinguishable.
    (b) If the objects are bosons.
    (c) If the objects are fermions.
    For each case check that the number of distinguishable distributions agrees with the formula.
88. How many distinguishable states result from distributing 4 quanta to a triply degenerate vibration?
89. In the nitrogen atom there are 3 $2p$ electrons to distribute to 6 $2p$ spin orbitals (p_x, p_y, p_z; α or β spin). How many states arise from this p^3 configuration?
*90. If r distinguishable objects are distributed among n slots, what is the probability that all are in the first slot? What is this probability if the objects are bosons? Which probability is larger?
91. According to Maxwell-Boltzmann statistics, the probability of the occupation numbers $\{n_i\}$ is $P(\{n_i\}) = N!/M^N \prod n_i!$. When $M \gg N$, it is very unlikely that there is more than one object (molecule) in a given slot (energy level), so that it is necessary to consider only those distributions for which each n_i is either zero or one. Then $\ln P(\{n_i\}) \to \ln N! - N \ln M$. Use the crudest Stirling's approximation to show that the limiting value of $\ln P(\{n_i\})$ as given by either Fermi-Dirac or Bose-Einstein statistics differs from this value by a quantity only of the order N^2/M. (This condition is valid for all chemically important situations. Exceptions are those for which quantum effects are significant, such as helium at low temperature and conduction electrons in a metal.)

6.6 Parameters of a Distribution

Let $P(x)$ be the frequency function of the random variable x. We define m_k, the kth MOMENT of x about the origin, by

$$m_k \equiv \sum x^k P(x) \tag{6.17}$$

where the sum is over all accessible values of x. Alternatively, let there be $n(x_i)$ sample points labeled by x_i and $\sum n(x_i)$ sample points in all. Then, by an extension of Eq. 6.4 to frequency functions of a random variable, $P(x_i) = n(x_i)/\sum n(x_i)$ and

$$m_k = \frac{\sum n(x_i) x_i^{\,k}}{\sum n(x_i)}\,. \tag{6.18}$$

The most important moment is the first, which is known as the MEAN; it may be symbolized by μ, μ_x, $\bar{x}$, or $\langle x \rangle$:

$$\bar{x} \equiv m_1 = \sum xP(x) = \frac{\sum n(x_i)x_i}{\sum n(x_i)}. \tag{6.19}$$

In some contexts the mean is known as the EXPECTATION VALUE. In common parlance, it is often called the AVERAGE, but there are other means which are also called averages. Furthermore, if f is any function of one variable, we may define the mean (average value) of f by

$$\bar{f} = \langle f \rangle \equiv \sum f(x)P(x) = \frac{\sum n(x_i)f(x_i)}{\sum n(x_i)}. \tag{6.20}$$

In particular, we define v_x, the VARIANCE of the random variable x, as the mean of $f(x) = (x - \bar{x})^2$

$$v_x \equiv \overline{(x - \bar{x})^2} = \sum (x - \bar{x})^2 P(x) = \frac{\sum n(x_i)(x_i - \bar{x})^2}{\sum n(x_i)}. \tag{6.21}$$

For purposes of calculation, we may simplify this expression by expanding:

$$v_x = \sum x^2P(x) - 2\bar{x} \sum xP(x) + \bar{x}^2 \sum P(x) = \overline{x^2} - \bar{x}^2 \tag{6.22}$$

since $\sum xP(x) = \bar{x}$, $\sum P(x) = 1$, and the last term cancels half the middle term. It is more convenient to deal with a quantity whose dimensions are the same as those of x. Therefore we define σ_x, the STANDARD DEVIATION of the random variable x, by

$$\sigma_x \equiv v_x^{1/2} = (\overline{x^2} - \bar{x}^2)^{1/2}. \tag{6.23}$$

Henceforth we shall abandon the symbol v_x and use σ_x^2 for the variance.

We may identify μ_x and σ_x as indicators of the location and width, respectively, of the frequency function, in the sense that when we perform the experiment, we will usually obtain a value of x between $\mu_x - \sigma_x$ and $\mu_x + \sigma_x$. It is readily apparent that the mean is a quantity representative of what x value might be expected; sometimes a value greater than μ_x is obtained, and sometimes a value less than μ_x, but a "typical" experiment should lead to a value "near" μ_x. Also, since $|x - \mu_x|$ is the distance of a representative argument from the mean, σ_x is the "root-mean-square average" of this distance, that is, an estimate of how near μ_x this "typical" outcome may be expected to be. For σ_x to be zero, all the probability must be concentrated into a single x; when σ_x is large, there is a large "spread" to $P(x)$, and a "typical" outcome may be quite different from μ_x.

Examples

49. For the experiment "throw a die" and the random variable x = "number of points"

$$\bar{x} = 1 \cdot \tfrac{1}{6} + 2 \cdot \tfrac{1}{6} + 3 \cdot \tfrac{1}{6} + 4 \cdot \tfrac{1}{6} + 5 \cdot \tfrac{1}{6} + 6 \cdot \tfrac{1}{6} = \tfrac{7}{2}.$$

$$\overline{x^2} = 1 \cdot \tfrac{1}{6} + 4 \cdot \tfrac{1}{6} + 9 \cdot \tfrac{1}{6} + 16 \cdot \tfrac{1}{6} + 25 \cdot \tfrac{1}{6} + 36 \cdot \tfrac{1}{6} = \tfrac{91}{6}.$$

$$\sigma_x{}^2 = \overline{x^2} - \bar{x}^2 = \tfrac{91}{6} - (\tfrac{7}{2})^2 = \tfrac{35}{12}.$$

$$\sigma_x = (\tfrac{35}{12})^{1/2} \sim 1.7.$$

Thus we conclude that one die usually shows 3.5 ± 1.7 points, or between 1.8 and 5.2 points. Indeed, the outcomes 2, 3, 4, and 5 do represent $\frac{4}{6}$ of the total sample space.

50. For the experiment "throw two dice" and the random variable x = "sum of the points"

$$\bar{x} = 2 \cdot \tfrac{1}{36} + 3 \cdot \tfrac{2}{36} + \cdots + 12 \cdot \tfrac{1}{36} = 7.$$

$$\overline{x^2} = 4 \cdot \tfrac{1}{36} + 9 \cdot \tfrac{2}{36} + \cdots + 144 \cdot \tfrac{1}{36} = \tfrac{1974}{36}.$$

$$\sigma_x{}^2 = \overline{x^2} - \bar{x}^2 = \tfrac{1974}{36} - 7^2 = \tfrac{35}{6}.$$

$$\sigma_x = (\tfrac{35}{6})^{1/2} \sim 2.4.$$

Thus we conclude that two dice usually show 7 ± 2.4 points, or between 4.6 and 9.4 points. Indeed, the outcomes 5, 6, 7, 8, and 9 do represent $\frac{24}{36}$ of the total sample space. Notice that with 2 dice, $\bar{x}$ doubles, but σ_x increases by only a factor of $2^{1/2}$. Compare Figure 6.2a with Figure 6.2b to see how the distribution with 2 dice is broader, but not twice as broad. With 2 dice the distribution is more "peaked," so that the eight most probable outcomes represent much more than just $\frac{2}{3}$ of the total sample space.

51. For the binomial distribution $P(r) = C_r^n p^r(1 - p)^{n-r}$, Exercise 6.102 shows that the mean and standard deviation are $\bar{r} = np$ and $\sigma_r = [np(1 - p)]^{1/2}$. For example, for the experiment "toss n coins," we would expect to obtain $\frac{1}{2}n$ heads. Any one experiment might give more or fewer than $\frac{1}{2}n$ heads, but if we perform the experiment many times, those experiments giving more than $\frac{1}{2}n$ heads should "cancel" those giving less, and the average number of heads for all the experiments would approach $\frac{1}{2}n$. Furthermore, since the standard deviation is a measure of the "width" or "spread" of the distribution, we may say that a "typical" experiment results in $\frac{1}{2}n$ heads, "plus-or-minus" $\frac{1}{2}n^{1/2}$, or that most experiments result in between $\frac{1}{2}(n - n^{1/2})$ and $\frac{1}{2}(n + n^{1/2})$ heads. Notice that this standard deviation increases as $n^{1/2}$, not n. Thus, on throwing 100 coins, we would generally expect between 45 and 55 heads, and on throwing 10 000 coins, we would generally expect between 4950 and 5050 heads. Although the spread has increased from 10 to 100, the percent spread has decreased from 10% to 1% of the total number of coins thrown.

52. From the random-walk problem of Example 6.37, substitute $m = n + r$ so that $P(m) = C_m^{2n}(\frac{1}{2})^{2n}$ is a binomial distribution with mean $\bar{m} = n$ and standard deviation $\sigma_m = \frac{1}{2}n^{1/2}$. Then, from Exercise 6.92b, $\bar{r} = \bar{m} - n = 0$, and from Exercise 6.94, $\sigma_r = \sigma_m = \frac{1}{2}n^{1/2}$. Thus we see that after n encounters, a "typical" molecule has not diffused beyond a distance $n^{1/2}d$ from its starting point. This result is consistent with that of Exercise 5.28, where it is found that the width of the concentration distribution increases as $t^{1/2}$, not t.

53. In Example 6.41 we derived the probability that exactly nV_m/N ml of mobile phase has eluted a molecule of solute:

$$P(n) = p(np)^{N-1}e^{-np}/(N - 1)!.$$

Let us determine the average retention volume, that is, the average volume of mobile phase required to elute the solute:

$$\bar{V} = \frac{V_m}{N}\bar{n} = \frac{V_m}{N}\sum nP(n) = \frac{V_m}{N!}\sum_{n=0}^{\infty} np(np)^{N-1}e^{-np} \sim \frac{V_m}{N!}\int_0^{\infty}(np)^N e^{-np}\,dn$$

$$= \frac{V_m}{N!\,p}\int_0^{\infty} x^N e^{-x}\,dx = \frac{V_m}{N!\,p}\Gamma(N+1) = \frac{V_m}{p}$$

where we have invoked the result of Exercise 6.92a and also approximated the sum by an integral. Next let us determine the width of the chromatographic peak:

$$\overline{V^2} = \frac{V_m^{\ 2}}{N^2}\overline{n^2} = \frac{V_m^{\ 2}}{N^2}\sum n^2P(n) \sim \frac{V_m^{\ 2}}{N!N}\int_0^{\infty} n(np)^N e^{-np}\,dn = \frac{V_m^{\ 2}}{N!Np^2}\int_0^{\infty} x^{N+1}e^{-x}\,dx$$

$$= \frac{V_m^{\ 2}}{N!Np^2}\Gamma(N+2) = \frac{V_m^{\ 2}}{p^2}\frac{N+1}{N}.$$

$$\sigma_V^{\ 2} = \overline{V^2} - \bar{V}^2 = \frac{V_m^{\ 2}}{p^2}\frac{N+1}{N} - \frac{V_m^{\ 2}}{p^2} = \frac{V_m^{\ 2}}{Np^2}.$$

$$\sigma_V = \frac{V_m}{N^{1/2}p} = \frac{\bar{V}}{N^{1/2}}.$$

In terms of quantities characteristic of the chromatography column

$$\bar{V} = V_m + KV_s \qquad \sigma_V = \frac{V_m + KV_s}{N^{1/2}}$$

Thus we see that the retention volume is indeed large when the solute is strongly held on the column, and that the peaks are indeed narrower when there are many theoretical plates. But it may not be obvious why the efficiency of a separation is increased when we increase the length of the column. Let us double the length. Then both V_m and V_s are doubled, $\bar{V}$ is doubled, and so is the separation between peaks of two different solutes. But doubling the column length also doubles N, so σ_V is increased by only a factor of $2^{1/2}$. Therefore, although the two peaks are both broader, lengthening the column separates them to a greater extent than it broadens them, and thus leads to a better separation.

Other measures of location and width exist, but they are less often used than the mean and standard deviation. We have already referred to the MOST PROBABLE value; now we define it as that x for which $P(x)$ is maximum. The MEDIAN, x_{med} is that value of x such that

$$\sum_{x<x_{\text{med}}} P(x) = \tfrac{1}{2} = \sum_{x>x_{\text{med}}} P(x). \tag{6.24}$$

In words, we are just as likely to obtain a value less than the median as greater than the median. The median is occasionally suitable as a measure of location for unsymmetrical distributions. Another measure of the width of a distribution is the MEAN DEVIATION:

$$\overline{|x-\bar{x}|} = \sum |x-\bar{x}|\,P(x) = \frac{\sum n(x_i)\,|x_i-\bar{x}|}{\sum n(x_i)} \tag{6.25}$$

This looks as though it is easier to calculate than the standard deviation, but in fact Eq. 6.22 makes it easier to calculate σ, especially with a desk calculator. Furthermore, it is easier to prove theorems concerning the standard deviation. Therefore the mean deviation is rarely used.

Exercises

92. Convince yourself that if c is any constant:
    (a) $\overline{cx} = c\bar{x}$.
    (b) $\overline{x + c} = \bar{x} + c$.

*93. Show that if x and y are random variables with joint frequency function $P(x, y)$, then $\overline{x + y} = \bar{x} + \bar{y}$. [*Hint:* What is $\sum_j P(x_i, y_j)$?]

*94. Show that $\sigma^2_{x+c} = \sigma^2_x$, where $\sigma^2_{x+c} = \overline{[(x + c) - (\overline{x + c})]^2}$. This result should be obvious; it states that the "width" of a frequency function is not changed by shifting its graph horizontally.

95. Let $P(x)$ be the frequency function of the random variable x, with mean μ and standard deviation σ. Let $\xi = (x - \mu)/\sigma$. Use the results of Exercise 6.92 to find $\bar{\xi}$ and $\overline{\xi^2}$.

*96. Convince yourself that μ_x is at the center of mass of the graph of the frequency function $P(x)$, that is, that a figure in the shape of $P(x)$ would balance on a knife-edge at μ_x.

*97. The COVARIANCE of the random variables x and y is defined by $\sigma_{xy} \equiv \overline{xy} - \bar{x}\bar{y}$. Show that if x and y are independent random variables, $\sigma_{xy} = 0$.

*98. The CORRELATION COEFFICIENT of the random variables x and y is defined by $\rho_{xy} \equiv \sigma_{xy}/\sigma_x\sigma_y$. From the previous exercise, $\rho_{xy} = 0$ if x and y are independent. At the other extreme, let $P(y \mid x) = \delta_{y,ax+b}$; the random variable x occurs with probability $P(x)$, but y depends linearly on x. For example, for the experiment "throw N coins" and the random variables $x =$ "number of heads" and $y =$ "number of tails," $y = N - x$. Show that if $P(y \mid x) = \delta_{y,ax+b}$, then $\rho_{xy} = a/|a|$, that is, $\rho_{xy} = \pm 1$, the sign depending upon the sign of a.

99. For the experiment "choose a random digit" and the random variable $x =$ "digit chosen," what is:
    (a) $\bar{x}$?
    (b) $\overline{x^2}$?
    (c) σ_x?

100. For the experiment "choose a pair of random digits" and the random variable $x =$ "half the sum of the digits," what is:
    (a) $\bar{x}$?
    (b) $\overline{x^2}$? (*Suggestion:* Use a desk calculator.)
    (c) σ_x?

*101. (a) A dice player throws a die and wins a dollar for each point that shows. What is the expectation value of his winnings per throw?
    (b) A card player deals a card and wins \$2 if the card is an ace, \$1 if the card is a king, queen, or jack, and otherwise nothing. What is the expectation value of his winnings per deal? Notice that the expectation value is not necessarily a value that can be expected as a possible outcome, but it does represent a fair price to pay for the privilege of playing.

*102. (a) Show that the mean $\sum rP(r)$ for the binomial distribution is np. (*Hint:* Let $q = 1 - p$, take the partial derivative of $\sum C^n_r p^r q^{n-r} = (p + q)^n$ with respect to p, then multiply by p.)

(b) Show that the standard deviation is $[np(1-p)]^{1/2}$. (*Hint:* Differentiate twice with respect to p.)

*103. Convince yourself that $\bar{r}_i$, the expectation value of r_i for the multinomial distribution of r objects to n slots, is given by $\bar{r}_i = rp_i$. For example, on throwing 60 dice, each $\bar{r}_i = 10$. (*Hint:* This result can be obtained by taking the partial derivative of $1 = \sum M^r_{\{r_i\}} \prod p_i^{r_i}$ with respect to p_i, as in Exercise 6.102. However, it is easier to recognize that this same value of $\bar{r}_i$ would be obtained for a binomial distribution with probability p_i of success and probability $1 - p_i = \sum_{j \neq i} p_j$ of failure.) Similarly convince yourself that the variance $\overline{r_i^2} - \bar{r}_i^2 = rp_i(1 - p_i)$, which is approximately rp_i, when n is large and each p_i small.

*104. For the experiment "deal a poker hand" and the random variable n_A = "number of aces," convince yourself that $\bar{n}_A = 5/13$. (*Hints:* What is $\overline{n_A + n_K + n_Q + \cdots + n_2}$? Then extend the result of Exercise 6.93 to 13 random variables and invoke a "symmetry principle.") Notice that $P(n_A)$ is not a binomial distribution, since 5 cards are dealt without replacement and the probability at each trial (card) is not independent of the results of previous trials (cards already dealt). Nevertheless, $\bar{n}_A$ is the same as would be obtained for a binomial distribution with $n = 5$ and $p = 4/52$.

*105. (a) Show that the mean for the Poisson distribution is np. (*Hints:* Remember that the index of summation is a dummy index, and remember the series expansion of e^{np}.)

(b) Show that the standard deviation is $(np)^{1/2}$.

106. For the experiment "throw 600 dice" and the random variable r = "number of aces":

(a) What is $\bar{r}$, the average number of aces per experiment?

(b) What is σ_r, the standard deviation of the number of aces?

(c) What are these quantities if the distribution is approximated by the Poisson distribution? (See Exercise 6.105.)

(d) What are these four numbers if 6000 dice are thrown?

107. (a) Use the results of Example 6.40 and of Exercise 6.105 to find $\bar{r}$, the average number of disintegrations per minute for a sample of N radioactive atoms, if p is the probability that any given atom will disintegrate during the next minute.

(b) If $\bar{r} = 100$, convince yourself that during a "typical" minute, between 90 and 110 disintegrations occur.

108. The probability of finding a one-dimensional harmonic oscillator with n quanta of excitation is given by $P(n) = e^{-nh\nu/kT}(1 - e^{-h\nu/kT})$.

(a) What is $\bar{n}$? (*Hint:* Since $P(n)$ is a frequency function, $\sum_{n=0}^{\infty} P(n) = 1$. Differentiate this equation with respect to ν and solve for $\sum nP(n)$.)

(b) If $E = (n + \frac{1}{2})h\nu$, what is $\bar{E}$?

*109. Use the result of Exercise 6.71 to show that the average number of objects that must be distributed among m slots in order to have r objects assigned to the first slot is given by $\bar{n} = r/p = rm$. [*Hint:* Let $q = 1 - p$, take the partial derivative of $\sum P(n) = 1$ with respect to p, then multiply by p.] Show that $\sigma_n^2 = rm(m-1)$. (*Hint:* Let $q = 1 - p$ in the previous result and differentiate with respect to p, then multiply by p.)

*110. Let us consider the polymerization of a difunctional molecule X—Y (e.g., $H_2N(CH_2)_6COCl$). After the reaction has proceeded, most of the X and Y groups have reacted, but the reaction has not gone to completion. Let p = probability that a Y group has reacted; we shall assume the same p for all Y groups, whether on a monomer or at the end of a growing polymer. Then $1 - p$ = probability that a Y remains unreacted, either on a monomer or at the end of a polymer.

(a) Convince yourself that the probability that a molecule contains r monomers is given by $P(r) = p^{r-1}(1 - p)$.

(b) Show that $\sum_{r=1}^{\infty} P(r) = 1$. [*Hint:* Remember the series expansion of $1/(1 - p)$.]

(c) Show that $\bar{r} = 1/(1 - p)$ and $\sigma_r = p^{1/2}/(1 - p)$. [*Hints:* What are $(d/dp) \sum_{r=0}^{\infty} p^r$ and $(d/dp)p \sum_{r=1}^{\infty} rp^{r-1}$?] Notice that this distribution is much broader than that of Exercise 6.79. For example, if polymerization is carried to 99% completion, a molecule chosen at random contains "100 ± 99.5" monomer units.

*111. In some contexts the HARMONIC MEAN, $h_x \equiv 1/\sum (1/x)P(x)$, is useful. On a 100-mile trip, we traveled 25 miles at 25 mph, 25 miles at 50 mph, and 50 miles at 75 mph. How many hours did the trip take? What value would you give as a meaningful "average speed"? Notice that this is the harmonic mean of the individual speeds.

*112. In some contexts the GEOMETRIC MEAN, $g_x \equiv \prod x^{P(x)}$, is useful.

(a) Convince yourself that $\ln g_x = \overline{\ln x}$.

(b) A four-step synthesis gives yields of 40, 75, 80, and 54% in the successive steps. What value would you quote as an average yield per step?

*113. Expand $M(s) \equiv \sum e^{rs}P(r)$ as a power series in s to see why M is known as the MOMENT-GENERATING FUNCTION for the frequency function P.

*114. Find the moment-generating function for the Poisson distribution. Check the results of Exercise 6.105.

*115. Find the moment-generating function for the binomial distribution $P(r) = C_r^n p^r(1 - p)^{n-r}$. (*Hint:* M is a binomial expansion.)

*116. We may generalize the convolution to functions of discrete variables:

$$f_1 * f_2(\rho) \equiv \sum f_1(r)f_2(\rho - r)$$

where the sum is over all possible values of r.

(a) Convince yourself that if P is the frequency function of the random variable "number of points on a die," then $P * P$ is the frequency function of the random variable "sum of the points on two dice." (*Hints:* Compare the expression for $P * P(8)$ with the five mutually exclusive ways of achieving a sum of eight points with two dice. Or compare Figures 6.2a and 6.2b in the light of Exercise 3.34.) This result is readily generalized to the result that if $P_1, P_2, \ldots, P_N$ are frequency functions of the N random variables $r_1, r_2, \ldots, r_N$, then the frequency function of the sum $\sum r_i$ is the N-fold convolution $P_1 * P_2 * \cdots * P_N$.

(b) Show that the moment-generating function of $P_1 * P_2$ is the product of the moment-generating functions of P_1 and P_2. Generalize to the moment-generating function of a sum of N random variables.

(c) Find the moment-generating function of the random variable "number of heads on throwing a single coin." Then use the previous result to find the moment-generating function of the random variable "number of heads on throwing n coins." Check the result of Exercise 6.115.

(d) Find the moment-generating function of the random variable "sum of the points on throwing N dice." Use the result of Exercise 1.58 to simplify your answer. Notice that for $N > 2$ the frequency function of this random variable is not particularly simple, but its moment-generating function is.

6.7 Continuous Distributions

We wish to generalize to a sample space containing an infinite number of sample points and to random variables that are not restricted to DISCRETE (integer or fractional) values. For example, for the experiments "measure the distance between two points" or "weigh this sample of ore" or "titrate the acid in this solution," *any* real number can in principle be obtained (if we agree not to round off to a finite number of significant figures). We might try to define the frequency function P of the random variable x so that $P(x)$ is the probability of obtaining the value x on performing the experiment. However, the probability of obtaining exactly x must be zero because each time we perform the experiment the value x is terribly unlikely to occur, relative to the infinity of other values. For example, if we were to choose a real number between 0 and 1 by generating an infinite sequence of random digits, it would be extraordinarily unlikely if we obtained the value $\frac{1}{3} = 0.33333 \cdots$. But it is not impossible to obtain an x value within some interval because there are an infinity of points within that interval. For example, if we were to choose a real number between 0 and 1, it would not be so unlikely to obtain a value between 0.3 and 0.4; indeed, if the number is chosen at random, the probability of obtaining a value within this interval is 1/10. Thus we cannot interpret $P(x)$ as the probability of obtaining the value x, but we can use P to indicate the probability of obtaining a value within some interval. We shall thus define a CONTINUOUS FREQUENCY FUNCTION (DENSITY FUNCTION) P by

$$\int_a^b P(x)\,dx \equiv \text{probability of obtaining a value between } a \text{ and } b. \tag{6.26}$$

Then, although $P(x)$ is not the probability of obtaining the value x, we may identify $P(x)\,dx$ as the probability of obtaining a value between x and $x + dx$. It also follows that

$$P(x) \geqslant 0.$$

$$\int_{-\infty}^{\infty} P(x)\,dx = 1. \tag{6.27}$$

Examples

54. The barometric pressure was rounded to the nearest 0.1 mm Hg and recorded as 753.6 mm Hg. In the absence of any other information, we might expect that all barometric pressures between 753.55 and 753.65 mm Hg are equally likely. Then we may express the frequency function for the true barometric pressure as $P(p)\,dp = 10\,dp$, $753.55 < p < 753.65$, $P(p) = 0$ elsewhere.

55. One of a sample of N radioactive atoms has just disintegrated at time zero. What is the probability that the next atom to disintegrate will disintegrate between time t (minutes) and time $t + dt$? For such an occurrence, there must be no disintegrations from time zero to time t, and then one disintegration between time t and time $t + dt$. If p is the probability that any given atom will disintegrate during the next minute, then the probability that that atom will disintegrate during the next half minute is $\frac{1}{2}p$, and likewise for any time period, so that the probability that an atom will disintegrate during the next dt minutes is $p\,dt$. Furthermore, if dt is infinitesimally small, the disintegrations of two different atoms during that time interval are mutually exclusive events. Therefore we may apply Eq. 6.9″

to conclude that the probability that some atom will disintegrate during a time interval dt minutes long is $Np\,dt$. Furthermore, according to Example 6.40 we know that the probability that there are no disintegrations during the first minute is $P(0) = e^{-Np}$. Likewise, the probability that there are no disintegrations during each successive minute is e^{-Np}. And since these events are independent, the probability that there are no disintegrations for the period from time zero to time t minutes is the t-fold product

$$e^{-Np}e^{-Np}\cdots e^{-Np} = e^{-Npt}.$$

Therefore we conclude that the probability that we must wait between t and $t + dt$ minutes for the next disintegration is

$$P(t)\,dt = Npe^{-Npt}\,dt.$$

Exercises

117. What is the frequency function of the random variable $\theta =$ "angle between the minute hand of a clock and the vertical," $0 < \theta < \pi$? What is the probability that the minute hand is *exactly* horizontal?

*118. (a) What is the significance of the CUMULATIVE DISTRIBUTION FUNCTION $\int_{-\infty}^{x} P(x)\,dx$?
(b) How would you define the median of a continuous random variable?

119. According to the Maxwell-Boltzmann distribution, the probability that a particle has energy ε_i is $e^{-\varepsilon_i/kT}/\sum e^{-\varepsilon_i/kT}$. In the earth's gravity, the potential energy of a molecule of mass m at a height h above sea level is mgh. What is the probability of finding the molecule between h and $h + dh$?

We may also extend continuous frequency functions to joint frequency functions, $P(x, y)$, such that $P(x, y)\,dx\,dy$ is the probability of obtaining a value between x and $x + dx$ for the first random variable and also a value between y and $y + dy$ for the second random variable. And again we say that the variables are independent if $P(x, y) = P(x)P(y)$; in terms of conditional probabilities, x and y are independent random variables if $P(x \mid y) = P(x)$ and $P(y \mid x) = P(y)$. For example, for the experiment "measure the height and weight of an individual," the random variables $x =$ "height" and $y =$ "weight" are not independent since a tall person is likely to be heavier than average: $P(y \mid x) \neq P(y)$.

We may define the parameters of a continuous frequency function P in a fashion analogous to that for discrete frequency functions, but all sums must be replaced by integrals:

$$\begin{aligned}
m_k &\equiv \int x^k P(x)\,dx. \\
\bar{x} &\equiv m_1 = \int xP(x)\,dx. \\
\bar{f} &\equiv \int f(x)P(x)\,dx. \\
\sigma_x^2 &\equiv \overline{(x - \bar{x})^2} = \int (x - \bar{x})^2 P(x)\,dx = \overline{x^2} - \bar{x}^2.
\end{aligned} \qquad (6.28)$$

Exercises

120. Find the mean and standard deviation of the UNIFORM, or RECTANGULAR, DISTRIBUTION, $P(x)\,dx = dx/(b-a)$, $a < x < b$; $P(x)\,dx = 0$ elsewhere.

*121. What is the mean of the Lorentzian distribution

$$P(x)\,dx = \frac{1}{\pi}\frac{\sigma\sqrt{3}}{(x-\mu)^2 + 3\sigma^2}\,dx?$$

Justify your answer. What is the standard deviation?

*122. A frequency function is subject to broadening by two different mechanisms. For example, a spectral line is subject to broadening by two different mechanisms or an experimental value is subject to two types of error. If the second mechanism were inoperative, the frequency function would be $P_1(x)$, with mean $\mu_1 = \int xP_1(x)\,dx$ and variance $\sigma_1{}^2 = \int (x-\mu_1)^2 P_1(x)\,dx$. If the first were inoperative, the frequency function would be $P_2(x)$, with mean μ_2 and variance $\sigma_2{}^2$. When both mechanisms are operative, the resultant frequency function is the convolution $P_1 * P_2(x) \equiv \int P_1(t)P_2(x-t)\,dt$. (The truth of this statement is readily apparent if P_1 is a linear combination of delta functions. Cf. Exercise 3.34.) Show that the mean and variance of $P_1 * P_2$ are given by $\mu_{1*2} = \mu_1 + \mu_2$ and $\sigma_{1*2}^2 = \sigma_1{}^2 + \sigma_2{}^2$.

123. The position of a quantum-mechanical particle may be described by the wave function $\psi(x)$, whose significance is that $\psi^*(x)\psi(x)\,dx$ is the probability of finding the particle between x and $x + dx$. The wave function for a particle in a box of length L, with n quanta of energy, is $\psi_n(x) = N_n \sin(n\pi x/L)$, $0 < x < L$; $\psi_n(x) = 0$ elsewhere.
(a) Find the value of N_n which guarantees that $\int P(x)\,dx = 1$.
(b) Find $\bar{x}$.
(c) Find $\overline{x^2}$.
(d) Find σ_x.
Notice that σ_x is a measure of our uncertainty as to the exact position of the particle.

*124. The wave function for a harmonic oscillator with n quanta of energy is $\psi_n(x) = N_n e^{-\frac{1}{2}\alpha^2 x^2} H_n(\alpha x)$, where $\alpha = 2\pi(\nu/h)^{1/2}$, $N_n = (\alpha/2^n n!\pi^{1/2})^{1/2}$, and H_n is the nth Hermite polynomial.
(a) Find $\bar{x}$. (*Hint:* H_n is even or odd according as n is even or odd.)
(b) Use the generating function given in Exercise 5.32 to help find $\overline{x^2}$.

*125. The wave function for a hydrogenic atom in its 1*s* state is

$$\psi(r, \theta, \phi) = (Z^3/\pi a_0{}^3)^{1/2} e^{-Zr/a_0}.$$

Show that $\overline{(1/r)} = Z/a_0$.

126. A wave function has the further significance that if O is any observable, then $\bar{O} = \int \psi^*(x)\Omega\psi(x)\,dx$, where Ω is the operator associated with O. The operator associated with momentum p is $(\hbar/i)(d/dx)$. For the particle in a box of Exercise 6.123 find:
(a) $\bar{p}$.
(b) $\overline{p^2}$. (*Note:* The operator associated with O^2 is $\Omega^2 = \Omega\Omega$.)
(c) σ_p. Notice that σ_p is a measure of our uncertainty as to the exact momentum of the particle.
(d) What is $\sigma_x\sigma_p$? This is an example of the uncertainty principle.

*127. A particle executes harmonic motion, with $x(t) = A\cos\omega t$. The probability $P(x)\,dx$ of finding the particle between x and $x + dx$ is proportional to the time it takes for the particle to go from x to $x + dx$ or from $x + dx$ to x. (Imagine the individual frames of a motion picture of a swinging pendulum.)

(a) Find $P(x)\,dx$. Notice that at the turning points $x = \pm A$, $P(x)$ is infinite.
(b) What is $\bar{x}$?
(c) What is σ_x^2?

*128. The kinetic energy of a molecule of mass m and x-component of velocity v_x is $\frac{1}{2}mv_x^2$, so that according to the Maxwell-Boltzmann distribution, the probability that a molecule has an x-component of velocity between v_x and $v_x + dv_x$ is $P(v_x)\,dv_x = e^{-\frac{1}{2}mv_x^2/kT}\,dv_x/I$, where $I = \int_{-\infty}^{\infty} e^{-\frac{1}{2}mv_x^2/kT}\,dv_x$. Furthermore, the probability that a molecule has a y-component of velocity between v_y and $v_y + dv_y$ is $P(v_y)\,dv_y = e^{-\frac{1}{2}mv_y^2/kT}\,dv_y/I$, independent of its x-component of velocity; the z-component behaves likewise.
(a) Evaluate I.
(b) What is the significance of $P(v_x)P(v_y)P(v_z)\,dv_x\,dv_y\,dv_z/I^3$?
(c) Transform to spherical polar coordinates, $\{0 < v < \infty, 0 < \theta_v < \pi, 0 < \phi_v < 2\pi\}$. What is the significance of $v^2 e^{-\frac{1}{2}mv^2/kT} \sin\theta_v\,dv\,d\theta_v\,d\phi_v/I^3$?
(d) What is the significance of $P(v)\,dv = 4\pi v^2 e^{-\frac{1}{2}mv^2/kT}\,dv/I^3$?
(e) Show that $\bar{v} = (8kT/\pi m)^{\frac{1}{2}}$.
(f) Show that $\overline{v^2} = 3kT/m$.

*129. Sometimes we are given $P(x)$, the frequency function of a random variable, but we want $P[f(x)]$, the frequency function of a function of a random variable. The previous two exercises involve such changes of variable. Find the general formula:
(a) What is the probability of finding a value of the random variable less than x?
(b) What is the probability of finding a value of the function f less than y?
(c) If $f'(x)$ is always positive, so that $f(x)$ always increases as x increases, convince yourself that these two probabilities are equal when $y = f(x)$.
(d) Differentiate both sides of the resulting equation with respect to y, and thus show that $P[f(x)] = P(x)(dx/dy) = P[f^{-1}(y)](df^{-1}/dy)$.
(e) Convince yourself that if $f(x)$ always decreases as x increases, then $P[f(x)] = -P(x)(dx/dy)$, so that we may write the general formula as

$$P[f(x)] = P(x)\,|dx/dy| = P[f^{-1}(y)]\,|df^{-1}/dy|.$$

For a mnemonic, multiply by $|dy|$ to obtain $P(y)\,|dy| = P(x)\,|dx|$. Furthermore, we may generalize this result to $P[u(x, y, z), v(x, y, z), w(x, y, z)]\,du\,dv\,dw = P(x, y, z)\,|\partial(u, v, w)/\partial(x, y, z)|\,dx\,dy\,dz$. (Cf. Eq. 2.26.)

*130. The total kinetic energy of a molecule is given by $E = \frac{1}{2}mv^2 = \frac{1}{2}m(v_x^2 + v_y^2 + v_z^2)$. Use the results of the two previous exercises to show that the probability of finding a molecule with total kinetic energy between E and $E + dE$ is $P(E)\,dE = (2/\pi^{1/2}k^{3/2}T^{3/2})E^{\frac{1}{2}}e^{-E/kT}\,dE$.

131. (a) A TIME AVERAGE of $f(t)$ over the interval from t to $t + \Delta t$ is the average value of f, with all times on this interval considered as equally likely. Convince yourself that

$$\langle f(t)\rangle = \frac{1}{\Delta t}\int_t^{t+\Delta t} f(t)\,dt.$$

(b) In polarography the instantaneous current $i(t)$ is proportional to $t^{\frac{1}{6}}$, where t is the time since the mercury drop began to grow. After a time $t_{\max}$ the drop falls and a new one begins to form. Relate $\langle i(t)\rangle$, the average current during the life of the drop, to $i(t_{\max})$.

132. What is the average value of $\cos^2\theta$:
(a) Averaged over θ from 0 to 2π (i.e., if all θ values are equally likely)?
(b) Averaged over all three-dimensional space (i.e., with $P(\theta)$ proportional to $\sin\theta$)?

*133. Expand the Fourier transform $M = FP$ of a frequency function P as a power series in y to see why M is known as the moment-generating function for the frequency function. (*Note:* If $x > 0$, the Laplace transform may be preferable.)

One of the most important continuous frequency functions is the continuous NORMAL DISTRIBUTION

$$P(x)\,dx = \frac{1}{\sigma\sqrt{2\pi}}\,e^{-(x-\mu)^2/2\sigma^2}, \tag{6.29}$$

the familiar bell-shaped curve, centered at $x = \mu$ and with inflection points at $x = \mu \pm \sigma$. It may readily be shown that μ and σ are the mean and standard deviation, respectively, of this frequency function. Let us determine the probability of obtaining a value within one standard deviation from the mean

$$\int_{\mu-\sigma}^{\mu+\sigma} P(x)\,dx = \frac{1}{\sigma\sqrt{2\pi}}\int_{\mu-\sigma}^{\mu+\sigma} e^{-(x-\mu)^2/2\sigma^2}\,dx = \frac{1}{\sqrt{\pi}}\int_{-1/\sqrt{2}}^{1/\sqrt{2}} e^{-t^2}\,dt$$

$$= \frac{2}{\sqrt{\pi}}\int_0^{1/\sqrt{2}} e^{-t^2}\,dt = \operatorname{erf}(1/\sqrt{2}) \sim 0.6827, \tag{6.30}$$

where we have substituted $t = (x - \mu)/\sigma\sqrt{2}$ and converted the integral over a symmetric interval to one that we recognize as an error function. Thus we conclude that whenever x is distributed NORMALLY (according to the normal distribution), about 68% ($\sim$2/3) of the values are within one standard deviation from the mean. Similarly

$$\int_{\mu-2\sigma}^{\mu+2\sigma} P(x)\,dx \sim 0.9545, \qquad \int_{\mu-3\sigma}^{\mu+3\sigma} P(x)\,dx \sim 0.9973. \tag{6.31}$$

Less than 5% of the values are beyond two standard deviations from the mean and only 0.27% are beyond three standard deviations from the mean. It is sometimes convenient to substitute $\xi = (x - \mu)/\sigma$ and obtain the STANDARD NORMAL DISTRIBUTION

$$P(\xi)\,d\xi = \frac{1}{\sqrt{2\pi}}\,e^{-\xi^2/2}\,d\xi \tag{6.32}$$

with zero mean and unit standard deviation. Also, you should be warned again that many tables do not list erf but

$$\Phi(\xi) = \frac{1}{\sqrt{2\pi}}\int_0^{\xi} e^{-t^2/2}\,dt. \tag{6.33}$$

This function is more convenient in statistical applications since its argument is expressed in standard deviations. For comparison with the above, we note that $\Phi(1) = 0.3413$, $\Phi(2) = 0.4772$, and $\Phi(\infty) = \frac{1}{2}$. See Figure 6.5.

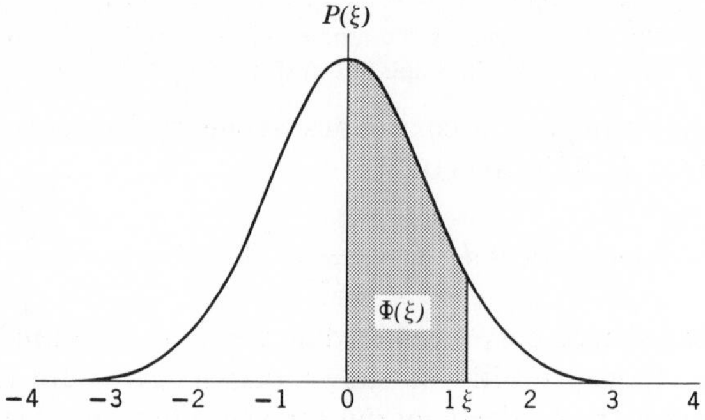

Figure 6.5

Exercises

*134. Show by direct integration that the mean and variance of the continuous normal distribution are indeed μ and σ^2, respectively.

*135. Use the results of Exercises 4.14 and 6.133 to find all even moments of the standard normal distribution.

*136. A SKEWED NORMAL DISTRIBUTION may be expressed by $P(x)\,dx = (1/\sigma\sqrt{2\pi})e^{-\xi^2/2}[1 - (\gamma/2)(\xi - \frac{1}{3}\xi^3)]\,dx$, where $\xi = (x - \mu)/\sigma$. Find $\bar{x}$, $\overline{(x - \bar{x})^2}$, and $\overline{(x - \bar{x})^3}$. (*Hint:* Remember the gamma function or use the result of Exercise 2.51.) Thus $\overline{(x - \bar{x})^3}$ is a measure of the asymmetry of a distribution.

*137. The most general standard normal distribution in two variables is $P(\xi, \eta)\,d\xi\,d\eta = [2\pi(1 - \rho^2)^{1/2}]^{-1} \exp[-(\xi^2 - 2\rho\xi\eta + \eta^2)/2(1 - \rho^2)]\,d\xi\,d\eta$.
(a) Show that $\bar{\xi} = 0 = \bar{\eta}$ and $\sigma_\xi^{\,2} = 1 = \sigma_\eta^{\,2}$.
(b) What is $\overline{\xi\eta}$?
(c) Under what circumstances are ξ and η independent?

Another important continuous frequency function is the chi-square distribution. We have seen that when n and np are large, the binomial distribution $P(r) = C_r^n p^r(1 - p)^{n-r}$ may be approximated by the continuous normal distribution $P(x)\,dx = (2\pi\sigma^2)^{-1/2} \exp(-x^2/2\sigma^2)\,dx$, where $x = r - np$ and $\sigma^2 = np(1 - p)$. Likewise, when r and each r_i are large, the multinomial distribution $P(\{r_i\}) = M_{\{r_i\}}^r \prod p_i^{\,r_i}$ may be approximated by a distribution of the form

$$P(\{x_i\}) \prod_{i=1}^{n} dx_i = C \exp\left(-\tfrac{1}{2}\sum_{i=1}^{n} x_i^{\,2}/\sigma_i^{\,2}\right) \prod_{i=1}^{n} dx_i, \tag{6.34}$$

where $x_i = r_i - rp_i$, $\sigma_i^{\,2} = rp_i$ (cf. Exercise 6.103), and C is a constant which is chosen so that the integral over all allowable $\{x_i\}$ remains unity. In particular, we note that since $\sum r_i = r$ and $\sum p_i = 1$, $\sum x_i = \sum r_i + r\sum p_i = 0$ is a restriction on $\{x_i\}$.

It is helpful to try to imagine a graph of the function P, even though we cannot visualize a graph of a function of more than 3 variables. The value of the function P is maximum at $x_1 = x_2 = \cdots = x_n = 0$ and falls off in every direction as a bell-shaped curve. Indeed, we would expect the most probable outcome to be the one with each $x_i = 0$ or each $r_i = rp_i$. And we would expect large deviations from this most probable outcome to be quite improbable. For example, the most probable outcome on throwing 60 dice is 10 of each face, and an outcome "far" from this one would be unlikely. However, we must remember that not every point in this n-dimensional space corresponds to a possible outcome of the experiment, but only those points that satisfy the restriction $\sum x_i = 0$. For $n = 2$ these points are on the line $x_1 + x_2 = 0$; for $n = 3$ they are on the plane $x_1 + x_2 + x_3 = 0$, etc. For $n = 2$, the section of the surface $z = P(x_1, x_2)$ above the line $x_1 + x_2 = 0$ is a bell-shaped curve, and C is adjusted so that the area under this curve is equal to unity. Notice that this curve is just a normal distribution since when $n = 2$ the multinomial distribution is just the binomial distribution and the chi-square distribution must reduce to a normal distribution. When $n = 3$, the section of $P(x_1, x_2, x_3)$ "above" the plane $x_1 + x_2 + x_3 = 0$ is a bell-shaped surface and C is adjusted so that the volume under this surface is equal to unity. And even for $n > 3$ it is clear that the points satisfying $\sum x_i = 0$ lie in an $(n - 1)$-dimensional space.

Thus we may conclude that $P(\{x_i\}) \prod dx_i$ is the probability of finding a value for the first random variable between x_1 and $x_1 + dx_1$, a value for the second random variable between x_2 and $x_2 + dx_2, \ldots,$ and a value for the nth random variable between x_n and $x_n + dx_n$, but subject to the condition that $\sum x_i = 0$. Alternatively, $P(\{x_i\}) \prod dx_i$ is the probability of finding a set of values within the volume element $dV = \prod dx_i$ and within the $(n - 1)$-dimensional "surface" $\sum x_i = 0$ about the point $(x_1, x_2, \ldots, x_n)$.

In practice, we are usually not interested in the direction of the deviation from the most probable set of values, but only in the magnitude of the deviation. For example, on throwing 60 dice, the outcomes $\{7, 9, 10, 11, 11, 12\}$ and $\{12, 7, 10, 11, 11, 9\}$ are quite different, but both deviate to the same extent from the most probable outcome $\{10, 10, 10, 10, 10, 10\}$. Therefore we define a quantity χ to be a measure of the deviation, or the distance, in an n-dimensional space, of the point $(x_1, x_2, \ldots, x_n)$ from the most probable point:

$$\chi^2 \equiv \sum_{i=1}^{n} x_i^2/\sigma_i^2 = \sum_{i=1}^{n} (r_i - rp_i)^2/rp_i. \tag{6.35}$$

Thus the most probable outcome has $\chi^2 = 0$, but the outcome $\{7, 9, 10, 11, 11, 12\}$ and any other with the same numbers, but rearranged, has

$$\chi^2 = \frac{(7-10)^2}{10} + \frac{(9-10)^2}{10} + \frac{(10-10)^2}{10} + \frac{(11-10)^2}{10} + \frac{(11-10)^2}{10} + \frac{(12-10)^2}{10} = \frac{16}{10}.$$

Finally, we must transform $P(\{x_i\})\,dV$ to $P(\chi^2)\,d\chi^2$, the probability of finding a squared deviation between χ^2 and $\chi^2 + d\chi^2$. Notice that the variable is χ^2, not χ. It can be shown that χ^2 is distributed according to the CHI-SQUARE DISTRIBUTION

$$P(\chi^2)\,d\chi^2 = (\chi^2)^{\frac{1}{2}\nu-1}e^{-\frac{1}{2}\chi^2}\,d\chi^2/2^{\frac{1}{2}\nu}\Gamma(\tfrac{1}{2}\nu), \tag{6.36}$$

where ν is known as the number of DEGREES OF FREEDOM. Here it equals $n - 1$, the dimensionality of the space of allowable points. More generally, ν is the number of independent variables. And just as the normal distribution and its integrals are tabulated, so also are the chi-square distribution and its integrals. Clearly, for each ν, it is possible to calculate the integral $\int_0^{\chi^2} P(\chi^2)\,d\chi^2$ as a function of χ^2. However, the integrals of the chi-square distribution are tabulated in an inverse form, with χ^2 tabulated as a function of ν and $1 - \int_0^{\chi^2} P(\chi^2)\,d\chi^2$. See Figure 6.6. For example,

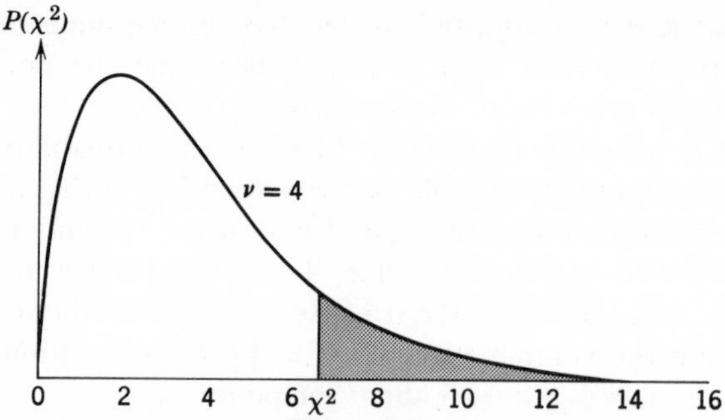

Figure 6.6

inspection of a χ^2 table (included in the *Handbook of Chemistry and Physics*) shows that for $\nu = 5$, $\int_{1.610}^{\infty} P(\chi^2)\,d\chi^2 = 0.90$ and $\int_{4.351}^{\infty} P(\chi^2)\,d\chi^2 = 0.50$. We found that on throwing 60 dice the outcome $\{7, 9, 10, 11, 11, 12\}$ has $\chi^2 = 1.6$. Therefore we conclude that about 90% of the time we may expect a more "uneven" outcome than this one. And 50% of the time we may expect $\chi^2 > 4.351$; for example, the outcome $\{6, 7, 11, 11, 12, 13\}$ has $\chi^2 = 4.0$ and represents an outcome slightly more "even" than a "typical" one.

Exercises

*138. Adapt the method of Exercise 3.25 to approximate the multinomial distribution by Eq. 6.34. (*Warning:* This is quite tedious.)

*139. Show that transformation of variables takes Eq. 6.34 into Eq. 6.36: First we note that $\exp(-\frac{1}{2}\sum x_i^2/\sigma_i^2) = \exp(-\frac{1}{2}\chi^2)$ depends only upon χ, the measure of the distance of the point $(x_1, x_2, \ldots, x_n)$ from the origin, and does not depend upon angular coordinates. Therefore we may transform to ν-dimensional spherical polar coordinates and integrate over all angular coordinates.

(a) Convince yourself that $C\exp(-\frac{1}{2}\sum x_i^2/\sigma_i^2)\,dV = C'\chi^{\nu-1}e^{-\frac{1}{2}\chi^2}\,d\chi$, where C' is another constant. (*Hint:* Use dimensionality arguments or generalize from the cases $\nu = 2$, where circular symmetry reduces dV to $2\pi r\, dr$, and $\nu = 3$, where spherical symmetry reduces dV to $4\pi r^2\, dr$. Cf. Exercises 2.60 and 3.16.)

(b) Next use the result of Exercise 6.129 to transform from χ to χ^2 and show that $P(\chi^2)\, d\chi^2 = \frac{1}{2}C'(\chi^2)^{\frac{1}{2}\nu-1}e^{-\frac{1}{2}\chi^2}\, d\chi^2$.

(c) Finally use the requirement that $\int_0^\infty P(\chi^2)\, d\chi^2 = 1$ to evaluate the constant.

*140. (a) Show that for ν degrees of freedom $\overline{\chi^2} = \nu + 2$ and $\sigma_{\chi^2}{}^2 = 2\nu + 4$. Therefore a "typical" χ^2 value is $\nu + 2$, "plus-or-minus" $(2\nu + 4)^{1/2}$.

(b) Find the column of a chi-square table containing the median values of χ^2. Notice that for all ν, $\overline{\chi^2} > \chi^2_{\text{med}}$.

6.8 An Introduction to Statistical Inference

Frequently when we perform an experiment to test a hypothesis concerning a frequency function we find that the result of the experiment is not exactly what the hypothesis predicts. Then we must decide whether the discrepancy between prediction and observation may reasonably be attributed to chance or whether the discrepancy is so great that we must reject our hypothesis as false. The set of results that are so discrepant that they call for rejection of the hypothesis is known as the CRITICAL REGION. The critical region must always be chosen in advance of the test. If the experimental result is found to lie within the critical region, then the hypothesis must be rejected. If the experimental result is found to lie outside the critical region, then the hypothesis is said to have withstood the test or, loosely speaking, to be accepted. However, it is not possible to deduce that a hypothesis is definitely right or definitely wrong, and we will sometimes reject true hypotheses and accept false ones. A few examples of statistical inference should make the procedure, and its limitations, clear.

Examples

56. An experiment was designed to test the hypothesis that the yellow color of some peas is governed by a single dominant gene. According to the hypothesis, the probability that a given pea is yellow is $p = 3/4$. When the experiment was performed, only 873 of 1200 peas were found to be yellow. Should we infer that the single-gene hypothesis is inadequate to account for these results? Clearly, we can always answer this question No, since we can claim that the single-gene hypothesis is correct and that the discrepancy of 27 peaks from the expected result of 900 yellow peas is just due to chance. Indeed, no matter what number might be found, we could still make the same claim. After all, *any* number of yellow peas is *possible*, since the number of yellow peas corresponds to the number of successes in 1200 trials, with a probability $p = 3/4$ for success on each trial. Thus we see that the distribution of successes in fact must follow a binomial distribution, which we shall approximate by a normal distribution with mean $\bar{r} = np = 900$ and standard deviation $\sigma_r = [np(1 - p)]^{1/2} = 15$. For example, it is possible to obtain fewer than 200 yellow peas, but the probability of such an outcome is only $\sim 10^{-476}$. So if we had obtained fewer than 200 peas, we could claim that even this result is consistent with the hypothesis, but we would be forced to maintain that an extremely improbable event had occurred. Instead, we would prefer to reject the hypothesis on the basis of obtaining fewer than 200 yellow peas, and only once

in every 10^{476} times would we be mistaken in rejecting such a hypothesis on such grounds. On the other hand, if we were to reject the hypothesis whenever we obtained a discrepancy of more than $\sigma_r = 15$ from the expected number of 900, we would be mistaken about once in every three times. This mistake of rejecting a hypothesis when it is true is known as a TYPE I MISTAKE, and the probability of making a Type I mistake with a given test is known as the LEVEL OF SIGNIFICANCE, α, of the test. Choosing $\alpha = 1/3$ is generally not advisable; rarely can we afford to make a Type I mistake as often as once every three times, so we must refuse to reject the hypothesis even when we get discrepancies greater than $\sigma_r = 15$. We can of course reduce α to 10^{-476} by agreeing to reject the hypothesis if and only if we observe fewer than 200 yellow peas. But then we would too often make a TYPE II MISTAKE, that of not rejecting the hypothesis when it is false. We must find some compromise which keeps α low but which also preserves sufficient power to avoid many Type II mistakes. It is convenient to choose $\alpha = 0.05$, and we shall usually do so, but the value to choose does depend upon circumstances and upon the risk the experimenter is prepared to accept. Then the critical region is chosen so that its aggregate probability is α. So whenever we obtain a result in the critical region and reject the hypothesis, we will be wrong with a probability α and right with a probability $(1 - \alpha)$. This probability of being right in rejecting the hypothesis is known as the CONFIDENCE LEVEL; with $\alpha = 0.05$ we may be confident that rejected hypotheses will be wrong 95% of the time. Now we are prepared to answer the original question: With $\alpha = 0.05$, the critical region for the experiment is the two tails of the distribution which contain the values differing from the mean by more than two standard deviations, or 30 (more precisely, 1.96 standard deviations, since $\Phi(1.96) = 0.475$). The probability of obtaining a value in this critical region is 0.05, and we would have inferred that the single-gene hypothesis is inadequate had we obtained a value of less than 870 or more than 930. But we cannot reject the single-gene hypothesis on the basis of the observed value of 873. Notice that we do not know whether the hypothesis is right or wrong. It has withstood our test, but it may nevertheless be wrong.

57. A factory produces test tubes of which 1% are defective. Of a week's output of 10 000 test tubes, 128 were found to be defective. Can we be 99% confident that this result is SIGNIFICANT and cannot be attributed to chance? We shall assume that the production of defective test tubes follows a binomial distribution with $n = 10\,000$ and $p = 1/100$, and we shall approximate this distribution by a normal distribution with mean 100 and standard deviation $\sqrt{99} \sim 10$. According to tables, $\Phi(\xi) = 0.49$ is satisfied by $\xi = 2.33$, so we conclude that the critical region is the one tail of values greater than $np + 2.33\sigma = 123$. Since the observed value lies in this critical region, we must reject the hypothesis that this result is a chance occurrence. Furthermore, we note that the probability of obtaining a value 128 or greater is only about $1/2 - \Phi(2.8) = 0.0026$. Thus we have concluded that the experimental result is significant, but we must remember that we cannot be certain that we are right; indeed, about one time in 400 we will conclude that such results are significant when they are in fact due to chance.

58. A box contains both white and black balls, in the unknown ratio $p:(1 - p)$. A total of $n = 100$ balls are drawn with replacement, and 10 are found to be white. Let us use this observation to estimate p. We might expect p to be $\sim$0.10, but we do not know how much higher or lower than 0.10 it may be. Let us gain an appreciation of how much higher than 0.10 p may be by testing the hypothesis that $p = 0.20$ at the 5% significance level. We shall approximate this binomial distribution by a normal distribution with $\mu = np = 20$ and $\sigma^2 = np(1 - p) = 16$. Since 10, the observed number, is 2.5 standard deviations from 20, the number expected on the basis of $p = 0.20$, and since $\Phi(2.5) = 0.4938 > 0.475$, we reject the hypothesis that $p = 0.20$. Similarly, we would reject the hypothesis that $p = 0.05$. Let us find two values, p_1 and $p_2 > p_1$, such that the hypotheses

$p = p_1$ and $p = p_2$ are both "just barely" rejected at the 5% significance level. Then we require that 10, the observed number, differ from the expected number by exactly 1.96 standard deviations, or

$$(10 - np)^2 = 1.96^2 np(1 - p).$$

The solutions to this quadratic are $p_1 = 0.055$ and $p_2 = 0.174$. Thus, loosely speaking, the probability that p is between 0.055 and 0.174 is 0.95. Of course, the probability that p is between 0.055 and 0.174 must be either 0 or 1; p either is within this interval or it is not! What this result does mean is that in the long run, if we use this experimental observation to infer that p is between 0.055 and 0.174, we will be correct 95% of the time. Only 2.5% of the time will p be less than 0.055 and only 2.5% of the time will p be greater than 0.174. The quantities p_1 and p_2 are known as the 95% CONFIDENCE LIMITS for p, and the interval $p_1 < p < p_2$ is known as the 95% CONFIDENCE INTERVAL for p. This sort of problem arises in many guises; Example 6.56 is another. For precise work, especially when n or p is small, the normal approximation should not be used, and there exist tables and graphs of confidence limits for binomial and Poisson samples. The exact answer for this problem is $p_1 = 0.049$ and $p_2 = 0.176$.

59. To 40 significant figures, $\pi = 3.1415\ 92653\ 58979\ 32384\ 62643\ 38327\ 95028\ 84197$. If these 40 digits are random, then we would expect there to be about 4 of each. We shall test the goodness-of-fit with this expectation at a level of significance $\alpha = 0.05$. However, the approximations used to derive the chi-square distribution require that each rp_i be large. Experience shows that the chi-square distribution may be used when most slots contain at least 5 objects. Therefore we shall combine the digits in pairs:

Digits	0, 1	2, 3	4, 5	6, 7	8, 9
Number of occurrences	4	12	8	6	10
Expected number	8	8	8	8	8

Then $\chi^2 = (4^2 + 4^2 + 0^2 + 2^2 + 2^2)/8 = 5.00$. The critical region is the set of large χ^2 values with aggregate probability 0.05. Inspection of a χ^2 table shows that with $\nu = 4$, the probability of obtaining a value of χ^2 greater than 9.488 is 0.05. Since our observed χ^2 is less than 9.488, we have no basis for supposing that these 40 digits are not random. Indeed, almost 30% of all sets of 40 random numbers will have a value of χ^2 greater than our observed χ^2.

Exercises

141. A coin is tossed 400 times. Would you have cause to suspect that the coin is lopsided if it lands 203 heads? 208 tails? 222 heads?

*142. For the experiment "deal 169 poker hands" and the random variable N = "total number of aces in the 169 hands," we may approximate $P(N)$ by a normal distribution with $n = 169$ and $p = 5/13$. (Cf. Exercise 6.104.)
- (a) What is $\bar{N}$?
- (b) What is σ_N?
- (c) Use a table of $\Phi(x)$ to convince yourself that the probability of obtaining fewer than $\bar{N} + \sigma_N$ aces is 0.84.
- (d) What is the probability of obtaining more than $\bar{N} + 2\sigma_N$ aces?
- (e) When the experiment was performed, one of the players received a total of 88 aces. Can we be 99% confident that this result cannot be attributed to chance?

*143. A 10.72-mg sample of a tritiated ketone is known to have 2.880×10^4 disintegrations per minute (dpm). It was subjected to conditions that might exchange out the tritium. After reisolation and purification, a 0.536-mg sample was counted for 1 min and found to have 14173 dpm. Can we infer with 95% confidence that some tritium has been lost from the sample? [*Hints:* Approximate the Poisson distribution by a normal distribution. Choose the critical region carefully; you should need the value $\Phi(1.645) = 0.45$.]

144. According to hypothesis, a genetics experiment should lead to four different types of peas with relative frequencies 9:3:3:1. When the experiment was performed with 1600 peas, the number of peas of each type was 876, 312, 324, and 88, respectively. Should we reject the hypothesis at a 5% significance level?

*145. A baseball team plays 162 games per season.
   (a) If a total of 972 wins were distributed at random among the 12 teams of the National League, how many wins, on the average, would you expect each team to have?
   (b) Can we be 95% confident that last year's final standings of the teams cannot be attributed to a random distribution of wins? (*Caution:* Some teams may not have played 162 games. Bring each total to 162 by assigning additional wins to the *better* teams.)
   (c) Check the American League standings for randomness also.

*146. A gambler assures me that his dice are not loaded. He claims to have thrown the pair a total of 150 times and obtained a total of 50 aces, 48 twos, 52 threes, 51 fours, 52 fives, and 47 sixes. Do these results seem "too good?" What, approximately, is the probability of a result more "even" than this?

6.9 Estimation of Parameters and Errors

So far we have been considering only THEORETICAL frequency functions of a random variable since we have always known the probability of all possible outcomes or could calculate them in terms of simpler probabilities. But in the laboratory we do not know the frequency function of a random variable because we cannot know the probability of every possible outcome of our experiment. For example, we obtain different values each time we perform the experiment "count the radioactivity in this sample," because we are dealing with random phenomena. Or we obtain different values each time we perform the experiment "determine the rate constant of this reaction," because our measurements are subject to error. And we cannot specify the frequency at which we obtain these various values. Were we able to perform the experiment an infinite number of times, we would be able to determine exactly the probability of each outcome, according to the definition of probability given in Eq. 6.1. But we cannot perform an infinite number of experiments to obtain this TRUE frequency function. All we have is a random SAMPLING from this frequency function, a finite number of experimental results, each drawn with a probability given by the true frequency function. We shall assume that these results have been determined INDEPENDENTLY, such that the result of one experiment does not affect the results of others. Also, we may consider the set of experimental results as an EMPIRICAL FREQUENCY FUNCTION of a random variable x, and we may graph $n(x)/N$ against x, just as we graph the true frequency function $P(x)$ against x.

Examples

60. Let us consider the random variable "random digit." If we could choose an infinite number of random digits, 10% of them would be zeros, 10% would be ones, etc., since the true frequency function is $P(0) = P(1) = \cdots = 1/10$. But a finite sample of, say, 100 random digits generally does not contain exactly 10 of each digit, but is just a random sampling from the theoretical frequency function. Figure 6.7 is a histogram of one empirical frequency function obtained by sampling 100 random digits. If we did not know the frequency function from which these digits have been sampled, we would not be able to unequivocally determine from this sample what that frequency function is. Indeed, it is not even apparent from the figure that the frequency function is a uniform one.

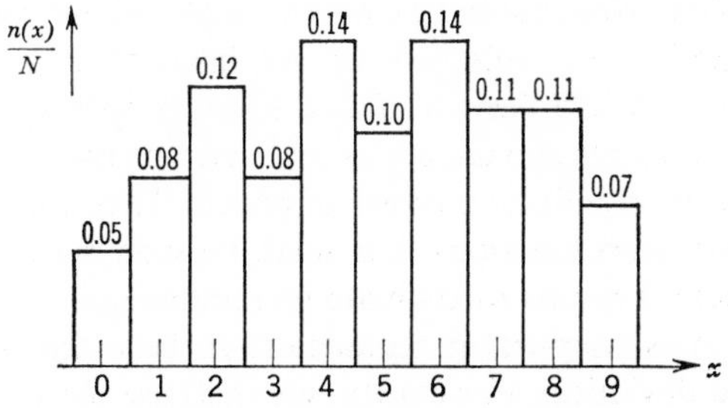

Figure 6.7

61. Independent repetitions of a measurement (e.g., of concentration, rate constant, radioactivity, melting point) produce an empirical frequency function of the quantity being measured. This empirical frequency function represents a sampling of the true frequency function, which would result if we could perform the experiment an infinite number of times. Furthermore, it can be shown that if the experiment is subject to only "random errors," then the true frequency function is a normal distribution whose mean is equal to the true value of the quantity being measured and whose standard deviation is a measure of the error inherent in a "typical" experiment. Although we do not define random error, we may distinguish it from SYSTEMATIC ERROR, an error of technique which biases the results to be generally higher (or lower) than the true value. Systematic error is too fundamental an imperfection of the experimental method to be treated by a statistical approach, so we shall ignore it here. But even if there is only random error, we cannot unequivocally determine the true value of the quantity on the basis of only a finite sample. Thus our task is to use a finite sample to estimate that true value and to provide a measure of how much error may be associated with our estimate.

Exercises

147. Use the following short table of random integers to construct a bar graph of a "typical" empirical frequency function obtained by sampling 24 points from the rectangular distribution from −0.5 to 999.5. So that you can see whether certain intervals predominate, CLASSIFY all numbers from −0.5 to 99.5 into a single

bar of width 100 centered at 49.5, etc.

028	614	887	440	653	260	624	529
618	359	791	287	041	840	201	150
501	653	006	108	492	932	220	565

148. Construct bar graphs of the following "typical" empirical frequency function obtained by sampling 24 points from the standard normal distribution:
(a) Classify the numbers from 0 to 0.25 into a single bar, etc.
(b) Classify the numbers from −0.125 to 0.125 into a single bar. (*Note:* Just as there exist tables of random digits, so also are there tables of random standard normal deviates.)

−1.90	0.29	1.21	−0.15	0.39	−0.64	0.32	0.07
0.30	−0.36	0.81	−0.56	−1.73	1.56	−0.84	−1.03
+0.00	0.39	−2.44	−1.23	−0.02	1.49	−0.77	0.16

Now we consider what estimates we can make about the true frequency function of a random variable, not on the basis of an infinite number of measurements, but on the basis of only a finite sample. Of course, no finite number of measurements can enable us to determine the value of a frequency function at each of an infinite number of points. Therefore we must assume a form for the frequency function—binomial, Poisson, normal, or other—and use our experimental results to estimate its parameters.

Certainly the most important parameters to be estimated are the mean and the standard deviation. For clarity, we shall use the Greek letters μ and σ for the mean and standard deviation, respectively, of the true frequency function. We shall use Roman letters for the parameters of the empirical frequency function

$$m_x \equiv \frac{1}{N} \sum n(x_i)x_i = \bar{x} \tag{6.37}$$

$$s_x^2 = \frac{1}{N} \sum n(x_i)(x_i - m_x)^2 = \overline{(x - \bar{x})^2}, \tag{6.38}$$

where $n(x_i)$ is the number of times the value x_i was obtained experimentally, and $N = \sum n(x_i)$ is the total number of experiments. The quantities m_x, s_x^2, and s_x are called the SAMPLE MEAN, SAMPLE VARIANCE, and SAMPLE STANDARD DEVIATION, respectively. Notice that even if the values are sampled from a continuous distribution, there are no integrals since our sample is always finite. And we may still calculate the sample variance by the simpler formula

$$s_x^2 = \overline{x^2} - \bar{x}^2 = \frac{1}{N} \sum n(x_i)x_i^2 - \left(\frac{1}{N} \sum n(x_i)x_i\right)^2 \tag{6.39}$$

which is especially convenient on a desk calculator equipped for accumulative multiplication. Notice though that this formula gives the variance as the difference of two large numbers, so be careful not to drop what seem to be insignificant figures.

Exercise

149. Evaluate the sample mean and sample variance of the empirical frequency function of:
(a) Exercise 6.147.
(b) Exercise 6.148.

It can be shown that the best estimate of μ_x is $m_x = \bar{x}$, the sample mean, as your intuition tells you. We shall indicate a best estimate by a caret

$$\hat{\mu}_x = \bar{x}. \tag{6.40}$$

But the question remains, "How confident can we be of this estimate?" The answer to this question is embodied in the CENTRAL-LIMIT THEOREM: Let x be a random variable distributed as $P(x)$, with mean μ_x and standard deviation σ_x, and let $\bar{x}$ be the mean of n values sampled from $P(x)$. Then, for large n, no matter what the form of $P(x)$ is, $P(\bar{x})$ approaches a normal distribution with mean $\mu_{\bar{x}} = \mu_x$ and standard deviation $\sigma_{\bar{x}} = \sigma_x/\sqrt{n}$. Alternatively, for large n, the distribution of the variable $(\bar{x} - \mu_x)\sqrt{n}/\sigma_x$ approaches the standard normal distribution. What this theorem means is that although a "typical" x is equal to μ_x "plus-or-minus" σ_x, a "typical" $\bar{x}$ is equal to μ_x "plus-or-minus" $\sigma_x/\sqrt{n}$. Furthermore, since $\bar{x}$ is distributed normally, we may be more precise and state that the probability that $\bar{x}$ is within $\sigma_x/\sqrt{n}$ of μ_x is 0.6827. Of course, we have sampled only a single $\bar{x}$, and we would like to know instead how confident we may be that μ_x is within $\sigma_x/\sqrt{n}$ of that $\bar{x}$. Again, the probability that μ_x is between $\bar{x} - \sigma_x/\sqrt{n}$ and $\bar{x} + \sigma_x/\sqrt{n}$ is not 0.68, but must be either 0 or 1. However, if we infer that μ_x is within this interval, in the long run we will be correct 68% of the time. And if we claim that μ_x is between $\bar{x} - 2\sigma_x/\sqrt{n}$ and $\bar{x} + 2\sigma_x/\sqrt{n}$, we will be correct 95% of the time. Therefore we conclude that the interval

$$\bar{x} - \frac{2\sigma_x}{\sqrt{n}} < \mu_x < \bar{x} + \frac{2\sigma_x}{\sqrt{n}} \tag{6.41}$$

is a 95% confidence interval for μ_x.

Notice the factor $1/\sqrt{n}$ in $\sigma_{\bar{x}}$. As you already know, the error in the average is less than the error associated with a single measurement since in taking the mean of several x values high values cancel low ones, and a "typical" $\bar{x}$ is closer to μ_x than is a "typical" x. However, the accuracy resulting from repetition does not increase as n, but only as $\sqrt{n}$. Thus, 100 determinations give a value only 10 times as accurate as a single determination, and another 9900 determinations must be made to achieve another factor of 10 in accuracy.

Example

62. A gravimetric analysis of silver can be performed with an accuracy (standard deviation) of ±0.06%. Four determinations of the silver content of an alloy gave values of 62.03, 62.16, 62.21, and 62.12%. The mean of these four values is $\bar{x} = 62.13\%$, and $\sigma_{\bar{x}}$, the standard deviation of this mean, is $0.06\%/\sqrt{4}$. Therefore we conclude that the silver content of the alloy is 62.13 ± 0.03%, by which we mean that we can be 68% confident that the silver content is between 62.10% and 62.16%, and 95% confident that it is between 62.07% and 62.19%.

Exercises

*150. Find a larger table of random integers and plot $P(\bar{x})$, each $\bar{x}$ being the average of 12 numbers. You should plot enough values to see that although $P(x)$ is not a normal distribution, $P(\bar{x})$ does look approximately like a Gaussian.

151. Convince yourself that 50% confidence limits for μ_x are $\bar{x} \pm 0.6745\sigma_x/\sqrt{n}$. The quantity $0.6745\sigma_x/\sqrt{n}$ is known as the PROBABLE ERROR, a misnomer since it is in fact the median error. Frequently the probable error is cited instead of the standard deviation; always check what is meant.

Often we do not know the true standard deviation σ_x, so we cannot immediately specify confidence intervals for μ_x. It can be shown that the true variance may be estimated on the basis of the sample variance

$$\hat{\sigma}_x^{\,2} = \frac{n}{n-1}\, s_x^{\,2}. \tag{6.42}$$

It then follows that the standard deviation of the mean may be estimated by

$$\hat{\sigma}_{\bar{x}} = \frac{1}{\sqrt{n}}\,\hat{\sigma}_x = \frac{1}{\sqrt{n-1}}\, s_x. \tag{6.43}$$

If we knew $\sigma_{\bar{x}}$ exactly, we would conclude that the 95% confidence limits for μ_x are $\bar{x} \pm 2\sigma_{\bar{x}}$. But all we have is the estimate $\hat{\sigma}_{\bar{x}}$. For "large" n we are justified in substituting $\hat{\sigma}_{\bar{x}}$ for $\sigma_{\bar{x}}$, and in concluding that the 95% confidence interval for μ_x is

$$\bar{x} - \frac{2s_x}{\sqrt{n-1}} < \mu_x < \bar{x} + \frac{2s_x}{\sqrt{n-1}}. \tag{6.44}$$

For small n (less than ~20), we must take account of the fact that although the variable $(\bar{x} - \mu_x)/\sigma_{\bar{x}}$ is distributed as the standard normal distribution, it can be shown that the variable

$$t \equiv (\bar{x} - \mu_x)/\hat{\sigma}_{\bar{x}} = (\bar{x} - \mu_x)\sqrt{n-1}/s_x$$

with $\sigma_{\bar{x}}$ replaced by $\hat{\sigma}_{\bar{x}}$, is not distributed normally, but as STUDENT's t DISTRIBUTION

$$P(t)\, dt \propto \left(1 + \frac{t^2}{\nu}\right)^{-(\nu+1)/2} dt, \tag{6.45}$$

where $\nu = n - 1$ is called the number of degrees of freedom. Do not be misled by this name; Student was the pseudonym of the statistician who first published this result. This distribution does not fall off as rapidly in the tails as does the normal distribution, as can be seen comparing Figure 6.5 with Figure 6.8. As a result, there is a greater than 5% probability that we are

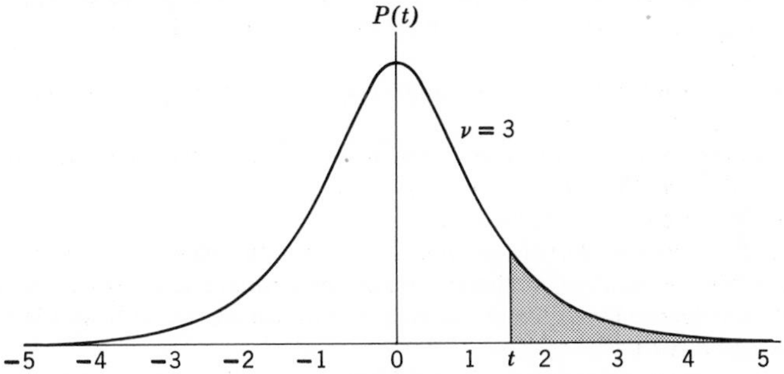

Figure 6.8

wrong in inferring that μ_x is within the interval from $\bar{x} - 2s_x/\sqrt{n-1}$ to $\bar{x} + 2s_x/\sqrt{n-1}$. And to take account of the deviation from normality, we must enlarge confidence intervals by looking up the coefficient of $\hat{\sigma}_{\bar{x}} = s_x/\sqrt{n-1}$ to be used. Integrals of Student's t distribution are tabulated in an inverse form, with t given as a function of $\int_t^\infty P(t)\,dt$ [but sometimes of $2\int_t^\infty P(t)\,dt$ or $\int_{-\infty}^t P(t)\,dt$] and ν. See Figure 6.8. To orient yourself to such a table, notice the entries at the bottom for $\nu = \infty$. For example, the value 1.96 should appear in a column headed 0.025 (or 0.05 or 0.975) since with an infinite sample only 5% of all sample means are beyond $1.96\hat{\sigma}_{\bar{x}}$ of the true mean μ_x. But with a seven-point sample ($\nu = 6$), 10% of all sample means are beyond $1.943\hat{\sigma}_{\bar{x}}$ of μ_x, and 5% of all sample means are beyond $2.447\hat{\sigma}_{\bar{x}}$ of μ_x.

Example

63. If we did not know the accuracy of the gravimetric analysis of Example 6.62, we could estimate it in terms of the sample variance of the four determinations:

$$s_x^2 = [(.03 - .13)^2 + (.16 - .13)^2 + (.21 - .13)^2 + (.12 - .13)^2]/4,$$

and $\hat{\sigma}_x = (4/3)^{1/2} s_x = 0.076\%$, which is to be compared with the true value $\sigma_x = 0.06\%$. And from tables, with $\nu = 3$, the probability is 0.05 of obtaining a value of $|t|$ 3.182 or greater. Therefore the 95% confidence limits for the silver content of the alloy are $\bar{x} \pm 3.182\hat{\sigma}_{\bar{x}} = 62.13 \pm 0.12\%$.

Exercises

*152. Convince yourself that the appearance of the factor $n/(n - 1)$ in Eq. 6.42 is reasonable. (*Hint:* If $n = 1$, what can you legitimately conclude about σ_x?)

153. (a) Use the results of Exercise 6.149b to evaluate the estimates $\hat{\mu}_x$ and $\hat{\sigma}_x$ of the frequency function from which the numbers of Exercise 6.148 were sampled.
(b) Compare $\hat{\sigma}_x$ with $\sigma_x = 1$.
(c) What are the 68 and 95% confidence intervals for μ_x? (*Hint:* This is a "large" sample.)
(d) Remember that the true mean $\mu_x = 0$. Is this value within the 68% confidence interval?

154. (a) Use the results of Exercise 6.149a to evaluate the estimates $\hat{\mu}_x$ and $\hat{\sigma}_x$ of the frequency function from which the numbers of Exercise 6.147 were sampled.
(b) What are the 68 and 95% confidence intervals for μ_x? (*Hint:* This is a "large" sample.)
(c) Compare your answer with the true μ_x, calculated in Exercise 6.120. Is μ_x within the 68% confidence interval?
(d) Compare $\hat{\sigma}_x$ with the true σ_x.

155. A radioactive sample was counted for 10 min and gave an average of 3610 dpm. Approximate the Poisson distribution by the normal and convince yourself that the 68% confidence limits for the activity of this sample are 3591 and 3629 dpm. (Cf. Exercise 6.107.)

*156. Use the result of Exercise 3.20 or A5.24c to show that the proportionality constant of Student's t distribution is $\Gamma[(\nu + 1)/2]/\Gamma(\nu/2)\sqrt{\pi\nu}$.

*157. Use Student's t distribution to find the 50% confidence interval for the mean of the distribution from which the last *eight* numbers of Exercise 6.147 were sampled. Is the true mean within this interval?

*158. It is possible to use the chi-square distribution to determine confidence intervals for a variance: Let x be normally distributed with variance σ_x^2 and let s_x^2 be the sample variance of an n-point sample. Then it can be shown that the variable ns_x^2/σ_x^2 is distributed as $P(\chi^2)$ with $n - 1$ degrees of freedom. Use this result to work the following problem: A new analytical technique for osmium has been developed. Six analyses of standardized samples gave a sample standard deviation of $\pm 0.15\%$. Show that we cannot be 95% confident that the true error (standard deviation) is less than $\pm 0.25\%$. (*Hint:* From chi-square tables, the probability that $6s_x^2/\sigma_x^2$ is less than 1.145 is 0.05.)

*159. The chi-square test of goodness-of-fit is more widely applicable: After an empirical frequency function has been fitted to a theoretical one by estimating some parameters, we would like to know how good the fit is. It can be shown that the chi-square test may still be applied, but one degree of freedom must be subtracted for each independent parameter estimated. Test, at the 5% significance level, whether the 24 random integers of Exercise 6.147 can be fitted to a *normal* distribution whose mean and standard deviation are those you found in Exercise 6.149a. Choose 4 slots by dividing the interval from -0.5 to 999.5 at $\hat{\mu}_x - 0.6745\hat{\sigma}_x$, $\hat{\mu}_x$, and $\hat{\mu}_x + 0.6745\hat{\sigma}_x$; if the numbers were sampled from a normal distribution, each of these slots would be equally likely.

*160. My colleague and I have been teaching different sections of Chem. 1A. Over the past few years, the number of students receiving grades A, B, C, D, and E are

	A	B	C	D	E	Total
Colleague	110	190	470	190	40	1000
Me	30	90	160	90	30	400
Total	140	280	630	280	70	1400

Use the generalization of the chi-square test indicated in the previous exercise to determine, at the 5% significance level, whether my colleague and I use different grading curves. (*Hint:* There are only 4 independent parameters to be estimated, since $\hat{p}_A + \hat{p}_B + \hat{p}_C + \hat{p}_D + \hat{p}_E = 1$.) This sort of table, exhibiting two different modes of classification (here teacher and grades) is called a CONTINGENCY TABLE, and the chi-square test may be used to infer whether the two modes are related or independent.

Occasionally it is necessary to estimate a quantity μ_x on the basis of measurements of unequal reliability. For example, Avogadro's number can be determined in many ways, of various accuracies. Although an average of all the determinations would be more accurate than a single value, the mean of measurements of unequal reliability is not the best estimate. It is preferable to put a greater reliance on the more accurate values by using a WEIGHTED MEAN

$$\hat{\mu}_x = \bar{x}_w = \sum w(x_i)x_i / \sum w(x_i). \tag{6.46}$$

Here $w(x_i)$ is the weighting factor, chosen so that more reliable values are weighted more heavily. The best weighting scheme is to choose $w(x_i) = 1/\sigma_{x_i}{}^2$, where $\sigma_{x_i}{}^2$ is the variance of x_i. Furthermore, it can be shown that the error of the weighted mean is given by

$$1/\sigma_{\bar{x}_w}{}^2 = \sum (1/\sigma_{x_i}{}^2). \tag{6.47}$$

Exercise

*161. The best student in my laboratory has determined a rate constant to be $(3.2 \pm 0.6) \times 10^6\ \text{sec}^{-1}$. Two other students have obtained values of $(2.6 \pm 1.0) \times 10^6$ and $(3.7 \pm 1.2) \times 10^6\ \text{sec}^{-1}$. What value should we publish and what error should we admit to?

Next we consider the propagation of errors that results from calculating a function of quantities subject to error. Let the functional dependence of a desired quantity Q upon a set of other, measurable quantities $\{x, y, \ldots\}$ be given by $Q = f(x, y, \ldots)$. Let the estimates $\hat{x} = \bar{x}$, $\hat{y} = \bar{y}$, . . . , be independently determined experimental values, with errors $\sigma_{\bar{x}}$, $\sigma_{\bar{y}}$, . . . , respectively. Then it can be shown that

$$\hat{Q} = f(\hat{x}, \hat{y}, \ldots) \tag{6.48}$$

$$\sigma_{\hat{Q}}{}^2 = \left(\frac{\partial f}{\partial x}\right)^2 \sigma_{\bar{x}}{}^2 + \left(\frac{\partial f}{\partial y}\right)^2 \sigma_{\bar{y}}{}^2 + \cdots, \tag{6.49}$$

where the partial derivatives are evaluated at $x = \hat{x}$, $y = \hat{y}$, ...

Examples

64. Many times a desired quantity is the sum or difference of two experimental values. Then $(\widehat{x \pm y}) = \hat{x} \pm \hat{y}$, and it is readily shown that

$$\sigma_{\widehat{x \pm y}}{}^2 = \sigma_{\bar{x}}{}^2 + \sigma_{\bar{y}}{}^2.$$

Notice that the variances are additive for both sum and difference. The RELATIVE ERROR (error divided by the quantity of interest) is

$$\text{R.E.}_{x\hat{\pm}y} = (\sigma_{\bar{x}}^2 + \sigma_{\bar{y}}^2)^{1/2}/(\hat{x} \pm \hat{y}). \tag{6.50}$$

This result embodies the well-known fact that the difference of two large numbers is subject to a large relative error.

65. Many times a desired quantity is the product or ratio of two experimental values. Then $\widehat{(xy)} = \hat{x}\hat{y}$ and $\widehat{(x/y)} = \hat{x}/\hat{y}$. Notice that $\hat{x}\hat{y} = \bar{x}\bar{y} \neq \overline{xy}$ and $\hat{x}/\hat{y} = \bar{x}/\bar{y} \neq \overline{x/y}$. And it can be shown that

$$\text{R.E.}_{\widehat{xy}} = (\sigma_{\bar{x}}^2/\bar{x}^2 + \sigma_{\bar{y}}^2/\bar{y}^2)^{1/2} = \text{R.E.}_{\widehat{x/y}}. \tag{6.51}$$

The generalization of both these examples to more than two quantities is obvious.

Exercises

162. Derive both parts of Eq. 6.51 from Eq. 6.49.

163. The concentration of a substance is to be determined spectrophotometrically, according to O.D. $= \varepsilon lc$. The errors in the optical density, extinction coefficient, and path length, are ± 10, ± 6, and $\pm 2\%$, respectively. What is the resultant error in the concentration?

*164. Show that for most accurate spectrophotometric determination of concentration, a sample should be diluted so that its optical density is near $\log_{10} e \sim 0.43$. (*Hints:* Spectrophotometric measurements are subject to a constant error ΔT in the transmittance $T = I/I_0$. Assume Beer's law and minimize the relative error in the measured concentration.)

165. How accurate is the product (or quotient) of 9 measured values, each of which has been rounded off so that it is accurate to only $\pm 0.1\%$. Convince yourself that one extra significant figure should be carried throughout any calculation and rounded off only at the end.

*166. For C^{14} dating purposes, radioactive samples A and B were counted for 10 min and gave averages of 122 and 168 dpm, respectively. The background was also counted for 10 min and found to be 40 dpm.

(a) What is the activity, in dpm, of A alone (with background subtracted out)?
(b) Of B alone?
(c) What are the accuracies of these values?
(d) What is the ratio of A activity to B activity?
(e) What is the percent accuracy of this ratio?

*167. Let $Q = \mu_x - \mu_y$, where μ_x is the mean of the distribution from which the first 12 numbers of Exercise 6.147 were sampled and μ_y is the mean of the distribution from which the last 12 numbers were sampled. Of course, these two distributions are identical, but what if we didn't know that?

(a) Find $\hat{Q}$ and $\sigma_{\hat{Q}}^2$ and estimate a 68% confidence interval for Q. (*Hint:* Although a 12-point sample is not large, for simplicity you may substitute $\hat{\sigma}_{\bar{x}}$ as an estimate for $\sigma_{\bar{x}}$, the true error in $\bar{x}$. In practice, an adaptation of Student's t distribution should be used to determine confidence intervals for the difference of two means.)

(b) Can you conclude that these two samples came from distributions with different means? This problem arises in many guises. Another common problem is that of determining whether two samples have been taken from distributions with different variances. Should you encounter such a problem, investigate the so-called F distribution in a statistics text.

6.10 Curve Fitting. Method of Least Squares

A frequent problem is that of fitting a set of paired data, $\{x_i, y_i\}_{i=1}^{n}$, to a linear equation, $y = ax + b$. For example, the problem of finding $\Delta S^{\ddagger}$ and $\Delta H^{\ddagger}$ is the problem of fitting experimental data, $\{1/T_i, k_i/T_i\}$, to

$$\ln\left(\frac{k}{T}\right) = \frac{-\Delta H^{\ddagger}}{R}\frac{1}{T} + \left(\frac{\Delta S^{\ddagger}}{R} + \ln\frac{R}{Nh}\right).$$

If there are only two data points, x_1, y_1 and x_2, y_2, two equations may be solved for the two unknowns a and b. But if there are more than two data points, there are more equations than unknowns and no consistent solution since random errors intrude and a and b determined from the first two points do not necessarily satisfy $y_3 = ax_3 + b$. We would like to utilize all the data to find the "best" estimates of a and b. We may take the best estimates, $\hat{a}$ and $\hat{b}$, as those that minimize S, the sum of the squares of the deviations, d_i, of observed y_i from calculated $\hat{y}_i = \hat{a}x_i + \hat{b}$. To find these values, differentiate S with respect to both $\hat{a}$ and $\hat{b}$:

$$S = \sum d_i^2 = \sum (y_i - \hat{a}x_i - \hat{b})^2$$

$$\frac{\partial S}{\partial \hat{a}} = 2\sum (y_i - \hat{a}x_i - \hat{b})(-x_i) = 0,$$

$$\frac{\partial S}{\partial \hat{b}} = 2\sum (y_i - \hat{a}x_i - \hat{b})(-1) = 0.$$

Dividing by 2, separating the sums, and transposing, gives

$$\hat{a}\sum x_i^2 + \hat{b}\sum x_i = \sum x_i y_i, \qquad \hat{a}\sum x_i + \hat{b}n = \sum y_i.$$

Solving gives

$$\hat{a} = \frac{n\sum xy - (\sum x)(\sum y)}{n\sum x^2 - (\sum x)^2}, \qquad \hat{b} = \frac{(\sum x^2)(\sum y) - (\sum x)(\sum xy)}{n\sum x^2 - (\sum x)^2}. \tag{6.52}$$

A further quantity of interest is the sample CORRELATION COEFFICIENT

$$r \equiv \frac{n\sum xy - (\sum x)(\sum y)}{[n\sum x^2 - (\sum x)^2]^{1/2}[n\sum y^2 - (\sum y)^2]^{1/2}}, \tag{6.53}$$

which is a best estimate of ρ_{xy}, the correlation coefficient of x and y. This quantity always lies between -1 and 1, and its sign depends upon whether y tends to increase or decrease with x, as is clear on comparison with the numerator of $\hat{a}$. The magnitude $|r|$ is a criterion of the goodness of the linear correlation. If $|r| = 1$, all the points lie exactly on one line; if $r = 0$, no straight line is better than any other. Therefore the closer $|r|$ is to 1, the more confident we may be that the straight line $y = \hat{a}x + \hat{b}$ represents the data. How close $|r|$ must be to 1 depends upon context; however, in chemistry you run the risk of being laughed at if you publish with $|r| < 0.95$. •

Exercises

168. The following data are random numbers, arranged to be monotonically increasing. Fit them to $y = ax + b$ (3 significant figures) and calculate r:

x_i:	1	2	3	4	5	6	7	8
y_i:	9	10	18	20	29	49	53	54

169. In a $1M$ solution of pyridine in benzonitrile, k_1, the pseudo-first-order rate constant for reaction with methyl iodide, has been determined at five temperatures:

t, °C	0.0	25.0	39.9	60.0	80.0
$10^5 k$, sec^{-1}	3.59	30.4	91.8	340	1120

Fit these data to

$$\ln (k_1/T) = (\Delta S^{\ddagger}/R) + \ln (R/Nh) - (\Delta H^{\ddagger}/R)(1/T)$$

and thereby estimate $\Delta S^{\ddagger}$ and $\Delta H^{\ddagger}$. You may take the relative error in k_1 to be independent of T.

*170. Derive a formula for evaluating the best parameters to fit $y = ax^2 + bx + c$.

*171. Use the general formula for $\sigma_{\hat{Q}}^2$ to estimate the standard deviation $\sigma_{\hat{a}}$ of the slope which fits $y = ax + b$. Assume that the error in y is independent of x and may be estimated by $\hat{\sigma}_y^2 = (1/n) \sum d_i^2 = \overline{d^2}$. You will derive $\sigma_{\hat{a}}^2 = \overline{d^2}/n(\overline{x^2} - \bar{x}^2)$, but the proper answer replaces n by $(n - 2)$ since if there are only two points, $\sum d_i^2 = 0$, but $\sigma_{\hat{a}}^2$ should be indeterminate. Similarly, show that $\sigma_{\hat{b}}^2 = \overline{d^2}\, \overline{x^2}/(n - 2)(\overline{x^2} - \bar{x}^2)$. Again, these estimates are valid for estimating confidence intervals for a and b only when n is large. For small n, an adaptation of Student's t distribution should be applied.

*172. In fitting $y = ax + b$, show that $\sum d_i^2 = \sum y_i^2 - \hat{a} \sum x_i y_i - \hat{b} \sum y_i$.

Often we must fit a set of data $\{x_i, y_i\}$ to a nonlinear equation, $y = f(x)$. It is often possible to transform a nonlinear equation to a linear one by choosing new variables $u = F(x, y)$ and $v = G(x, y)$ such that the equation becomes $v = au + b$. Thus to find the best values of C_0 and k to fit a set of experimental data $\{t_i, C_i\}$ to a first-order rate equation, $C = C_0 e^{-kt}$, transform to the linear form $\ln C = -kt + \ln C_0$ and find the best values of $-k$ and $\ln C_0$ to fit the data $\{t_i, \ln C_i\}$.

There is an error inherent in this procedure for the usual situation in which the error in determining C is absolute. For example, if C is determined titrimetrically, there is a constant error σ_{C_i} due to our inability to read a buret to better than, say, 0.01 ml. Therefore the error in $\ln C$ increases as C becomes small, and not all points are equally reliable. Therefore a better procedure is to replace all sums in Eq. 6.52 by weighted sums (and n by $\sum w_i$). The points are to be weighted inversely to the variance associated with each point. Since the variance in $\ln C_i$ is given by $\sigma^2_{\ln C_i} = [(d \ln C_i)/dC_i)]^2 \sigma_{C_i}{}^2 = \sigma_{C_i}{}^2/C_i{}^2$, the weighting factor w_i should be C_i^2. A still better procedure is to minimize $\sum d_i^2 = \sum (C_i - C_0 e^{-kt_i})^2$, but the resulting equations are transcendental and can be solved for k only by computer.

Exercises

*173. Convert the problem of fitting $y = x/(a + bx^2)$ to the linear form $v = \alpha u + \beta$.

*174. Minimize $S = \sum w_i(y_i - \hat{a}x_i - \hat{b})^2$ to derive the analog of Eq. 6.52 that is applicable to unequal weighting of points.

*175. The second-order kinetics, $2A \xrightarrow{k} B$, follows the rate law $1/A = (1/A_0) + 2kt$. In order to follow the kinetics, A was determined as a function of time, to produce a set of data $\{t_i, A_i\}$. However, the error in the determination of A was absolute, subject to a constant uncertainty σ_A. Find a suitable expression for determining the best values of A_0 and k by a weighted linear least squares technique.

176. A common task in polarography is to fit a set of data $\{i_j, E_j\}$ to the equation $\ln [i/(i_d - i)] = (\alpha nF/RT)E - (\alpha nF/RT)E_{1/2}$. However, the error in i is constant, independent of E. Find a suitable expression for determining the best values of $(\alpha nF/RT)$ and $(\alpha nF/RT)E_{1/2}$ by a weighted linear least squares technique.

7

VECTORS

In this chapter we begin a discussion of some very useful topics in abstract mathematics—vectors, matrices, and groups. Our initial presentation of vectors is very abstract, so that we can carry over a generalized notion of vectors to the subsequent chapters. However, we immediately return to the familiar notion of a vector as something with magnitude and direction, and we devote the remainder of this chapter to developing the theory and applications of vectors in three dimensions. We begin this development with the geometric interpretations of vector sums and products, and we indicate the computational advantages of choosing an orthonormal basis (Cartesian axes). Short digressions on simultaneous equations and on construction of orthonormal bases are included. Then we demonstrate how vector notation simplifies problems of analytic geometry. The remainder of this chapter is devoted to various aspects of functions of vectors and functions whose values are vectors. We indicate how vector notation simplifies mechanics, with special attention to a classical-mechanical description of magnetic resonance. We describe the four differential operators and their geometric interpretations. We conclude the chapter with a discussion of line integrals and surface integrals, with special attention to line integrals independent of path. Exercises indicate the applicability of the topics of this chapter to orthogonal functions, molecular structure, conformational analysis, X-ray diffraction, mechanics, magnetism, electrostatics, quantum mechanics, fluid flow, and thermodynamics.

7.1 Abstract Vector Spaces

We begin our presentation of vectors by defining a vector space. This approach is more abstract than your experience would suggest, but we shall wish to consider other objects, such as functions, to be vectors in a more general sense. Therefore we abandon the concept of a vector as something with magnitude and direction in three-dimensional space, and we concentrate on the addition and multiplication properties of vectors. Then we return to the familiar concept of vectors and show that they satisfy the more abstract definition. The bulk of this chapter is then devoted to applications of ordinary three-dimensional vectors, and we return to the more abstract case in the following chapter.

Every vector space is associated with a set of SCALARS, which may be the real numbers or the complex numbers, with the usual rules for their addition, subtraction, multiplication, and division. Thus we shall speak of a vector space OVER the reals, or a vector space over the complex numbers. We shall represent the scalars as lower-case letters.

A VECTOR SPACE, $\mathcal{V}$, is a set of elements called VECTORS (which we now represent as capital letters) subject to the following axioms:

VECTOR ADDITION is a process by which two vectors, X and Y, may be added to obtain a vector symbolized by $X + Y$, *which is also in* $\mathcal{V}$. This property of vector addition is known as CLOSURE. As with addition of ordinary numbers, vector addition is:

1. COMMUTATIVE: $X + Y = Y + X$; the same element of $\mathcal{V}$ is obtained regardless of the order of addition.
2. ASSOCIATIVE: $(X + Y) + Z = X + (Y + Z)$, so that we may dispense with the parentheses.
3. There exists a unique element of $\mathcal{V}$, the ORIGIN, or ZERO VECTOR, 0, such that for any X in $\mathcal{V}$, $X + 0 = X$. Notice that we use the same symbol for the zero vector as we do for the familiar scalar zero; this usage should not cause confusion.
4. Every element X in $\mathcal{V}$ has an ADDITIVE INVERSE, $-X$, in $\mathcal{V}$, such that $X + -X = 0$.

SCALAR MULTIPLICATION is a process whereby a scalar, a, and a vector, X, may be multiplied to obtain a vector, symbolized by aX, which is also in $\mathcal{V}$. Thus scalar multiplication too is closed. Scalar multiplication is also:

5. ASSOCIATIVE: $a(bX) = (ab)X$. Notice that the multiplications on the left are this new kind of multiplication, whereas the multiplication of a by b on the right is just the familiar multiplication of two numbers. Therefore this new kind of multiplication is "really very like" ordinary multiplication.
6. DISTRIBUTIVE OVER ADDITION: $a(X + Y) = aX + aY$ and $(a + b)X = aX + bX$. Notice that in the second distributive rule the addition on the right is vector addition, whereas the addition on the left is just the familiar addition of two numbers. Thus, these two types of addition are also quite similar.

Examples

1. $\mathscr{P}_N$, the vector space of all Nth-order polynomials, P_N, with real or complex coefficients. To ensure closure, P_M, with $M < N$, must be considered as an Nth-order polynomial. If the coefficients are real, the scalars must be real to ensure closure under scalar multiplication.

2. $\mathcal{R}^n$, the vector space of all n-tuplets of real numbers, over the reals as scalars, where the sum of $(x_1, x_2, \ldots, x_n)$ and $(y_1, y_2, \ldots, y_n)$ is defined to be $(x_1 + y_1, x_2 + y_2, \ldots, x_n + y_n)$ and the scalar product of a and $(x_1, x_2, \ldots, x_n)$ is $(ax_1, ax_2, \ldots, ax_n)$.

3. $\mathcal{C}^n$, the vector space of all n-tuplets of complex numbers, usually over the complex numbers as scalars, although the reals are possible. Sums and scalar products are defined as in Example 7.2.

4. $\mathcal{F}$, the vector space of all complex-valued functions on a suitable interval, over the complex numbers. Indeed, the rationale for generalizing the concept of vectors is to enable us to apply the mathematical techniques for calculating with vectors to problems involving functions. We shall return to this vector space in Chapter 9.

Exercises

1. Convince yourself that each of the examples of vector spaces does indeed satisfy every axiom.
2. For any X, what is $0X$? What is $1X$? What is $a0$?

A SUBSPACE of a vector space $\mathcal{V}$ is a set of vectors in $\mathcal{V}$ which themselves form a vector space.

Examples

5. $\mathscr{P}_M (M \leqslant N)$ is a subspace of $\mathscr{P}_N$.
6. The set of triplets of the form $(x_1, x_2, 0)$ is a subspace of $\mathcal{R}^3$ or $\mathcal{C}^3$.
7. The set of all even functions is a subspace of $\mathcal{F}$.

The sum $\sum a_i X_i$ is said to be a LINEAR COMBINATION of the set of vectors $\{X_i\}$. If any vector in $\mathcal{V}$ can be expressed as a linear combination of a set of vectors $\{X_i\}$, then the set $\{X_i\}$ is said to SPAN the vector space.

A set of vectors $\{X_i\}$ is said to be LINEARLY DEPENDENT if and only if there exists a set of scalars $\{a_i\}$ not all zero, such that $\sum a_i X_i = 0$. If the only way whereby $\sum a_i X_i$ can equal the zero vector is that every a_i be zero, then the vectors are LINEARLY INDEPENDENT.

Examples

8. In $\mathscr{P}_2$, the polynomials $2 - 2t$, $t - t^2$, and $1 - t^2$ are linearly dependent, since $\frac{1}{2}(2 - 2t) + (t - t^2) - (1 - t^2) = 0$, the zero polynomial.

9. In $\mathcal{R}^2$, the vectors $(1, -2)$ and $(-2, 4)$ are linearly dependent, since $2(1, -2) - (-2, 4) = (0, 0)$.

10. In $\mathcal{C}^3$ over the complex numbers, the vectors $(1, i, 1 - i)$, $(2 + i, i, 2i)$, and $(1 - i, 1, i)$ are linearly dependent, since

$$i(1, i, 1 - i) - (2 + i, i, 2i) + (1 + i)(1 - i, 1, i) = (0, 0, 0).$$

However, in $\mathcal{C}^3$ over the reals, these three vectors are linearly independent.

Exercises

3. Vectors are drawn from the origin to each of the vertices of a regular tetrahedron centered at the origin. Convince yourself that these four vectors span three-dimensional space and that they are not linearly independent.
4. Convince yourself that any set of vectors that includes the zero vector is linearly dependent.
5. Convince yourself that if a set of vectors is linearly dependent, then there exists at least one of them which is a linear combination of the others. (Cf. Exercise 1.51.)

A BASIS SET, BASIS, or COORDINATE SYSTEM for the vector space $\mathcal{V}$ is a set $\{E_i\}$ of linearly independent vectors such that every X in $\mathcal{V}$ may be expanded as a linear combination of the BASIS VECTORS: $X = \sum x_i E_i$. In other words, $\{E_i\}$ spans $\mathcal{V}$, and the requirement of linear independence means that a basis set is a minimal set required to span a vector space. The DIMENSION of a vector space $\mathcal{V}$ is defined as the number of basis vectors, that is, the minimum number of vectors required to span $\mathcal{V}$. The dimension may be finite or infinite. Incidentally, a one-dimensional vector space is trivial, since it behaves exactly as a set of scalars.

Examples

11. In $\mathcal{P}_N$, a basis set is $\{t^n\}_{n=0}^N$ since any Nth-order polynomial, $P_N(t)$, may be expressed as a linear combination of the elements of this basis set. Since there are $(N + 1)$ basis vectors, the vector space $\mathcal{P}_N$ is of dimension $(N + 1)$. We may also generalize to the infinite-dimensional vector space, $\mathcal{P}_\infty$, composed of all functions expandable in Maclaurin series.

12. In $\mathcal{R}^3$, a basis set is the three triplets $(1, 0, 0)$, $(0, 1, 0)$, and $(0, 0, 1)$ since the arbitrary triplet (x_1, x_2, x_3) may be expanded as $x_1(1, 0, 0) + x_2(0, 1, 0) + x_3(0, 0, 1)$. This is a three-dimensional space. Furthermore, this basis set for $\mathcal{R}^3$ is also suitable for $\mathcal{C}^3$ over the complex numbers since $(z_1, z_2, z_3) = z_1(1, 0, 0) + z_2(0, 1, 0) + z_3(0, 0, 1)$.

13. If we restrict $\mathcal{F}$ to those functions expressible in Fourier series on the interval from $-\pi$ to π, then a basis set is $\{e^{in\phi}\}_{n=-\infty}^{\infty}$. This is an infinite-dimensional vector space.

The coefficient x_i in the expansion $X = \sum x_i E_i$ is called the COMPONENT of X along E_i. The set $\{x_i\}$ is known as the COORDINATES of X. Two N-dimensional vectors X_1 and X_2 are equal if and only if each of the N components of X_1 is equal to the corresponding component of X_2. Therefore the vector equation $X_1 = X_2$ is equivalent to N numerical equations.

Once a basis set is chosen, the components are determined uniquely. However, it is important to recognize that the choice of a basis set is arbitrary and not unique. Other basis sets in $\mathcal{P}_N$ are $\{P_l(t)\}_{l=0}^N$ and $\{H_n(t)\}_{n=0}^N$, the first $(N + 1)$ Legendre and Hermite polynomials, respectively. Another basis set in $\mathcal{R}^3$ or $\mathcal{C}^3$ is $(1, 1, 0)$, $(1, -1, 0)$, and $(0, 1, 2)$; indeed, any three linearly independent vectors are suitable.

Of course, different bases lead to different sets of components. However, the choice of a basis enables us to REPRESENT a vector as a set of numbers, which can be manipulated algebraically. A vector exists independently of the choice of basis, but such an abstract quantity is not suitable for computation. It is only by choosing a basis that computations involving vectors may be reduced to numerical computations.

Since the choice of basis set is arbitrary, it might as well be chosen for the sake of convenience. Some problems involve calculating a quantity independent of the choice of basis; therefore we should choose a basis which makes the calculation simple. Other quantities may not be independent of the choice of basis, but it may be possible to do the calculation in a basis set in which the problem is easy and then convert the answer to the form appropriate to the basis set in which it is to be expressed (See Exercise 8.110).

Exercises

*6. Let $\mathcal{C}^1$ be the vector space of all complex numbers, associated with the *complex numbers* as scalars. If addition of two vectors is just addition of complex numbers, and if scalar multiplication is just multiplication of two complex numbers, convince yourself that $\mathcal{C}^1$ is indeed a vector space.
(a) What is its dimension?
Next let $\mathcal{C}^1$ be the vector space of all complex numbers, associated with the *real numbers* as scalars. If addition is as above, and scalar multiplication is multiplication of a real number times a complex number, convince yourself that this $\mathcal{C}^1$ is also a vector space.
(b) What is its dimension?
(c) As which vector space should the complex numbers be considered if the parallelogram rule for their addition is to be stressed?
(d) Which should be considered if multiplication of two complex numbers to obtain a complex number is to be stressed?

*7. Convince yourself that the set of all functions defined at only a single argument forms a vector space. Of what dimension? Convince yourself that the set of all functions defined at only the particular arguments $\{x_i\}_{i=1}^N$ forms a vector space. Of what dimension?

*8. Use the fact that a basis $\{E_i\}$ must be linearly independent to show that the components in the expansion $X = \sum E_i x_i$ are unique. (*Hint:* Assume that it is also true that $X = \sum E_i x_i'$.)

*9. Express the arbitrary vector (x_1, x_2, x_3) as a linear combination of the basis vectors $(0, 1, 1)$, $(1, 0, 1)$, and $(1, 1, 0)$.

7.2 Real Three-Dimensional Vectors

The remainder of this chapter is restricted to a consideration of the familiar type of three-dimensional vectors. We shall emphasize the geometric idea of a vector as something with magnitude and direction, and return to the more general approach in the next chapter. Meanwhile, you should consider whether and how the concepts developed in this chapter may be extended to higher-dimensional vectors and to the vectors of function space, to which it is not so easy to attach a readily visualized geometrical interpretation.

We have shown that $\mathcal{R}^3$, the set of all triplets, (x_1, x_2, x_3), of real numbers, forms a three-dimensional vector space. The disadvantage of this formulation is that it specifies a vector in terms of an implied choice of basis vectors; namely, $(1, 0, 0)$, $(0, 1, 0)$, and $(0, 0, 1)$. Often we would like to avoid specifying a set of basis vectors and concentrate on relations independent of such a choice. There are no coordinate axes engraved on the universe, and yet we do specify distances and (relative) orientations of stars in galaxies and atoms in molecules.

Therefore we return to the idea of a vector as something with magnitude and direction, and which exists independently of any choice of basis. We shall denote such a vector by a boldface capital letter: **A**. We shall denote the magnitude (length) of the vector **A** by $|A|$. Of course, in order to specify a vector in terms of numbers or to perform calculations on vectors, it is necessary to choose a basis set. But in many problems most of the manipulations and reasoning may be carried out with the basis-free vectors, and the numerical result calculated only at the end.

We shall envision vectors as arrows, each with a tip and a tail. The locations of the tip and tail are irrelevant, and only the magnitude of the arrow and the direction from its tail to its tip are of significance. We shall consider two vectors to be equal if they are of the same length and point in the same direction. If the tails of two equal vectors coincide, then so do their tips. If the tails of two equal vectors do not coincide, then the vectors must be of the same length and situated parallel to each other. See Figure 7.1.

To relate this view of vectors to the abstract definition, we must define scalar multiplication and vector addition. By $a\mathbf{A}$ we shall mean a vector in the same direction as **A** and a times as long; if $a < 0$, then $a\mathbf{A}$ is in the opposite direction to **A**. The sum of two vectors is given by the TRIANGLE RULE (PARALLELOGRAM RULE): $\mathbf{A} + \mathbf{B}$ is the vector from the tail of **A** to the tip of **B**, when the tail of **B** is placed so as to coincide with the tip of **A**. See Figure 7.2.

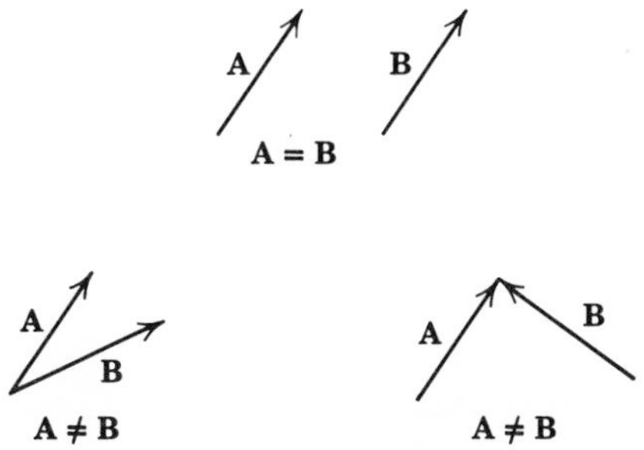

Figure 7.1

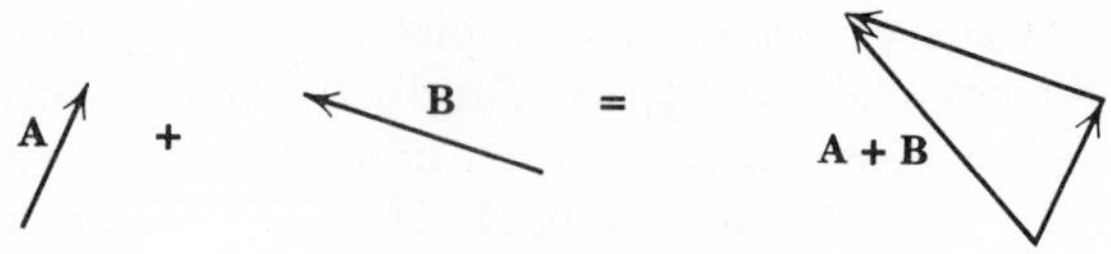

Figure 7.2

The zero vector (origin) is the vector with magnitude zero (and no direction). The additive inverse $-\mathbf{A}$ is a vector just as long as $\mathbf{A}$ but in the opposite direction. You will convince yourself that this definition of real three-dimensional vectors satisfies the abstract definition given in the previous section.

It is convenient to define VECTOR SUBTRACTION by $\mathbf{A} - \mathbf{B} \equiv \mathbf{A} + (-\mathbf{B})$. In words, $\mathbf{A} - \mathbf{B}$ is what must be added to $\mathbf{B}$ to produce $\mathbf{A}$. If the tails of $\mathbf{A}$ and $\mathbf{B}$ coincide, then $\mathbf{A} - \mathbf{B}$ is the vector from the tip of $\mathbf{B}$ to the tip of $\mathbf{A}$. Memorize this principle and Figure 7.3.

Exercises

10. Convince yourself that the set of real three-dimensional vectors, with magnitude and direction, does indeed satisfy every axiom defining an abstract vector space.
11. Convince yourself that two vectors that point in the same (or opposite) direction are linearly dependent. Convince yourself that three vectors lying entirely in the same plane are linearly dependent.

7.3 Vector Products

Thus far in our treatment the only multiplication involving vectors has been scalar multiplication, in which the product of a scalar and a vector is a vector. Next we consider how we may multiply two vectors together. Two types of multiplication may be distinguished, depending on whether the product is a scalar or another vector. The scalar product is readily generalized to other than three-dimensional vector spaces, although there are adjustments which must be made when we deal with vector spaces over the complex numbers. The vector product is unique to three-dimensional spaces.

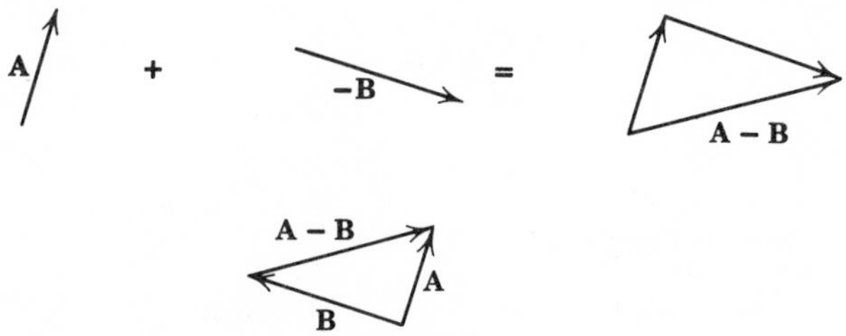

Figure 7.3

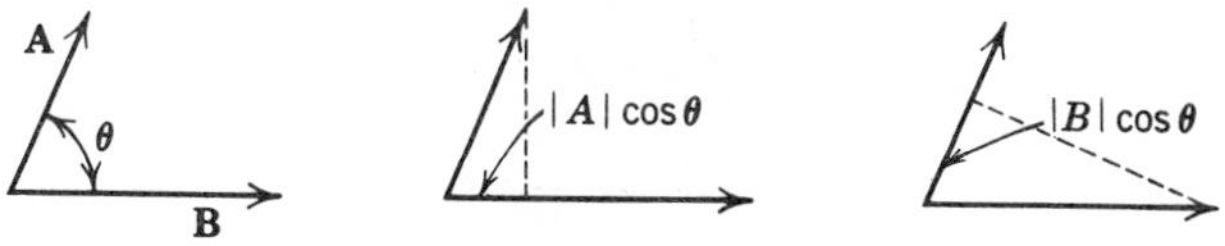

Figure 7.4

The SCALAR PRODUCT (DOT PRODUCT) of the vectors **A** and **B** is a scalar defined by

$$\mathbf{A} \cdot \mathbf{B} \equiv |A||B| \cos \theta \tag{7.1}$$

where θ is the angle between **A** and **B**. Since these quantities are independent of the choice of basis, the value of the scalar product is the same no matter what basis is chosen. Geometrically, $\mathbf{A} \cdot \mathbf{B} = |A| \times$ component of **B** in the direction of $\mathbf{A} = |B| \times$ component of **A** in the direction of **B** (See Figure 7.4). We may note further that $\mathbf{A} \cdot \mathbf{A} = |A|^2$, since $\theta = 0$ and $\cos \theta = 1$.

Since the angle between **A** and **B** is identical to the angle between **B** and **A**, it follows that the scalar product is commutative:

$$\mathbf{A} \cdot \mathbf{B} = \mathbf{B} \cdot \mathbf{A}. \tag{7.2}$$

It can be shown that the scalar product is distributive over vector addition:

$$\begin{aligned} \mathbf{A} \cdot (\mathbf{B} + \mathbf{C}) &= \mathbf{A} \cdot \mathbf{B} + \mathbf{A} \cdot \mathbf{C} \\ (\mathbf{A} + \mathbf{B}) \cdot \mathbf{C} &= \mathbf{A} \cdot \mathbf{C} + \mathbf{B} \cdot \mathbf{C}. \end{aligned} \tag{7.3}$$

Notice that the additions on the right are the familiar additions of two numbers. It is also readily shown that the scalar product is associative with scalar multiplication:

$$\begin{aligned} (a\mathbf{A}) \cdot \mathbf{B} &= a(\mathbf{A} \cdot \mathbf{B}) \\ \mathbf{A} \cdot (b\mathbf{B}) &= b(\mathbf{A} \cdot \mathbf{B}). \end{aligned} \tag{7.4}$$

Notice that the multiplications on the right are the familiar multiplications of two numbers.

So far the scalar product seems to obey all the familiar axioms of multiplication. However, the cancellation law of multiplication, which makes division by a number meaningful, does not hold: If $\mathbf{A} \cdot \mathbf{B} = 0$, we are not justified in concluding that either $\mathbf{A} = \mathbf{0}$ or $\mathbf{B} = \mathbf{0}$. It could be that **A** and **B** are both nonzero vectors, and that $\mathbf{A} \cdot \mathbf{B} = 0$ because $\theta = \pi/2$ and $\cos \theta = 0$. Two vectors **A** and **B** such that $\mathbf{A} \cdot \mathbf{B} = 0$ are said to be ORTHOGONAL.

Exercises

12. Does the associative law hold for the scalar product of vectors? For subtraction of real numbers?
13. For any nonzero vector **A** convince yourself that the vector $\mathbf{A}/|A|$ is a UNIT VECTOR (vector of magnitude unity).

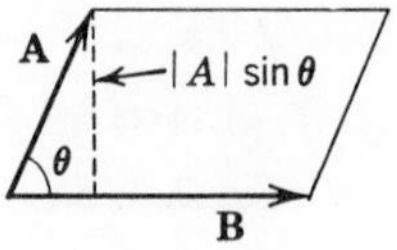

Figure 7.5

14. Show that for any two unit vectors **A** and **B**, **A** + **B** and **A** − **B** are orthogonal. What is the corresponding geometrical theorem? (*Hint:* Draw **A** − **B**.)
*15. Derive the law of cosines, relating the length of one side of a triangle to the lengths of the other two sides and the cosine of the opposite angle.
*16. Convince yourself that if we define the dot product of two functions (elements of the vector space $\mathcal{F}$) f and g by $f \cdot g = \int f(x)g(x)\,dx$, then this dot product satisfies all the properties of the dot product of two vectors, that is, it is commutative, distributive over addition, and associative with scalar multiplication, and the cancellation law of ordinary scalar multiplication does not hold. Notice that if $f \cdot g = 0$ the functions f and g are also said to be orthogonal (Exercise 1.69).

The VECTOR PRODUCT (CROSS PRODUCT) of **A** and **B** is a vector, **A** × **B** (sometimes written **A** ∧ **B**), whose magnitude is given by

$$|\mathbf{A} \times \mathbf{B}| \equiv |A||B| \sin \theta. \tag{7.5}$$

Since $|A| \sin \theta$ is the component of **A** in the direction of the perpendicular to **B**, $|\mathbf{A} \times \mathbf{B}|$ is equal to the area of the parallelogram whose edges are **A** and **B** (See Figure 7.5). Furthermore, we define the direction of **A** × **B** to be perpendicular to both **A** and **B** (perpendicular to the plane of **A** and **B**) and such that **A**, **B**, and **A** × **B** form a RIGHT-HANDED SYSTEM: A common screw or light bulb, twisted in the direction so as to carry **A** toward **B**, would be driven inward in the direction of **A** × **B**, as indicated in Figure 7.6. Notice especially that the vector product is therefore ANTICOMMUTATIVE:

$$\mathbf{B} \times \mathbf{A} = -\mathbf{A} \times \mathbf{B}. \tag{7.6}$$

From this it follows immediately that $\mathbf{A} \times \mathbf{A} = \mathbf{0}$. For our purposes we may take the vector product to be independent of the choice of basis.

It can be shown that the vector product is distributive over vector addition

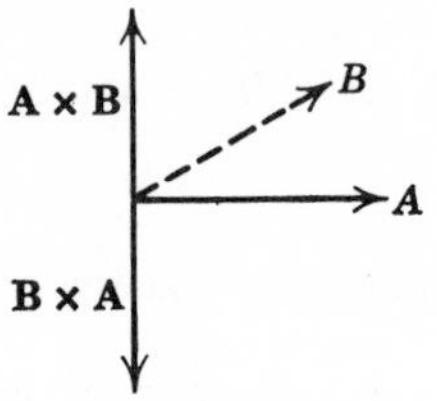

Figure 7.6

and associative with scalar multiplication

$$\begin{aligned}\mathbf{A} \times (\mathbf{B} + \mathbf{C}) &= \mathbf{A} \times \mathbf{B} + \mathbf{A} \times \mathbf{C} \\ (\mathbf{A} + \mathbf{B}) \times \mathbf{C} &= \mathbf{A} \times \mathbf{C} + \mathbf{B} \times \mathbf{C} \\ (a\mathbf{A}) \times \mathbf{B} &= a(\mathbf{A} \times \mathbf{B}) \\ \mathbf{A} \times (b\mathbf{B}) &= b(\mathbf{A} \times \mathbf{B}).\end{aligned} \tag{7.7}$$

However, the vector product is *not* associative:

$$(\mathbf{A} \times \mathbf{B}) \times \mathbf{C} \neq \mathbf{A} \times (\mathbf{B} \times \mathbf{C}).$$

That associativity does not hold is readily apparent in the special case that $\mathbf{A} = \mathbf{B}$, whereupon the left-hand side is automatically $\mathbf{0}$, but the right-hand side can be nonzero.

Exercises

*17. Screw in a few light bulbs for practice with the right-hand rule. To avoid confusion, do not unscrew any.

*18. Derive the law of sines. (*Hint:* Consider the quantity, twice the area of a triangle divided by the product of the lengths of the three sides.)

One further product is the (scalar) TRIPLE PRODUCT (BOX PRODUCT) of three vectors:

$$[ABC] \equiv \mathbf{A} \cdot \mathbf{B} \times \mathbf{C} \tag{7.8}$$

a scalar which results from a combination of the two types of vector multiplication. Notice that there can be no confusion about the order of performing the two types of multiplication, since $(\mathbf{A} \cdot \mathbf{B}) \times \mathbf{C}$ is meaningless.

Next we consider the geometrical significance of this triple product, $\mathbf{A} \cdot \mathbf{B} \times \mathbf{C} = |A||B \times C| \cos \theta$, where θ is the angle between $\mathbf{A}$ and $\mathbf{B} \times \mathbf{C}$. (We have drawn Figure 7.7 with θ acute and $\cos \theta > 0$; if θ is obtuse then the angle between $\mathbf{A}$ and $\mathbf{C} \times \mathbf{B} = -\mathbf{B} \times \mathbf{C}$ would be acute.) But $|A| \cos \theta$ is the component of $\mathbf{A}$ along $\mathbf{B} \times \mathbf{C}$, and therefore the length of the perpendicular from $\mathbf{A}$ to the plane of $\mathbf{B}$ and $\mathbf{C}$. But the volume of a parallelepiped is given by multiplying the area of one parallelogram face by the perpendicular height of the other edge. Therefore we conclude that the triple product, $\mathbf{A} \cdot \mathbf{B} \times \mathbf{C}$, is equal to the volume of the parallelepiped whose edges are $\mathbf{A}$, $\mathbf{B}$, and $\mathbf{C}$. Of course, if the angle between $\mathbf{A}$ and $\mathbf{B} \times \mathbf{C}$ is not acute, then

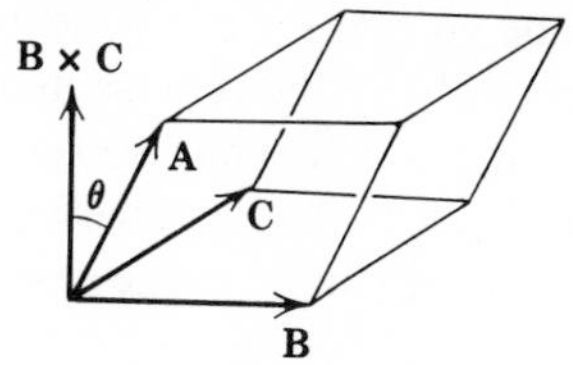

Figure 7.7

$\mathbf{A} \cdot \mathbf{B} \times \mathbf{C}$ equals $-\mathbf{A} \cdot \mathbf{C} \times \mathbf{B}$ and is therefore equal to the negative of the volume of the parallelepiped. Also, it is readily apparent that if the angle between $\mathbf{A}$ and $\mathbf{B} \times \mathbf{C}$ is acute, then so is the angle between $\mathbf{C}$ and $\mathbf{A} \times \mathbf{B}$. Since the volume of the parallelepiped does not depend upon which face is chosen as the parallelogram, we may conclude that $\mathbf{A} \cdot \mathbf{B} \times \mathbf{C} = \mathbf{C} \cdot \mathbf{A} \times \mathbf{B} = \mathbf{A} \times \mathbf{B} \cdot \mathbf{C}$. It is through this result that we are able to use the notation $[ABC]$ without confusion. Finally, we may apply the commutativity of the scalar product and the anticommutativity of the vector product to obtain the complete set of equalities:

$$[ABC] = -[ACB] = [BCA] = -[BAC] = [CAB] = -[CBA]. \quad (7.9)$$

7.4 Orthonormal Basis Sets

In our three-dimensional vector space, any set of three noncoplanar vectors is linearly independent and suitable for use as a basis set. The choice of a basis (and of origin) is arbitrary and may be made for convenience in computation. For example, to specify the points in a crystal it is often convenient to use as basis vectors the vectors along the edges of the unit cell; then the components of the corner of any unit cell take on only integral values.

For many purposes though, it is especially convenient to choose an ORTHONORMAL BASIS SET (CARTESIAN AXES, CARTESIAN COORDINATES), which we shall write as $\{\hat{\mathbf{i}}, \hat{\mathbf{j}}, \hat{\mathbf{k}}\}$, and which satisfies the following relations:

$$\hat{\mathbf{i}} \cdot \hat{\mathbf{i}} = \hat{\mathbf{j}} \cdot \hat{\mathbf{j}} = \hat{\mathbf{k}} \cdot \hat{\mathbf{k}} = 1 \quad (7.10)$$

$$\hat{\mathbf{i}} \cdot \hat{\mathbf{j}} = \hat{\mathbf{j}} \cdot \hat{\mathbf{k}} = \hat{\mathbf{k}} \cdot \hat{\mathbf{i}} = 0. \quad (7.11)$$

The first set of equalities states that the basis vectors are of length unity (NORMALIZED). Notice that in this context normal does not mean perpendicular. We shall use the caret to indicate a normalized vector. The second set of equalities states that each basis vector is orthogonal to the other basis vectors; since none is a zero vector, they must all be mutually perpendicular. By convention, $\hat{\mathbf{i}}$, $\hat{\mathbf{j}}$, and $\hat{\mathbf{k}}$ are taken to point along the x-, y-, and z-axes, respectively. There exist an infinite number of orthonormal basis sets, since orthonormal bases may be rotated and still remain orthonormal. There is no unique way of placing x-, y-, and z-axes in space, so the orientation of the basis vectors may still be chosen for convenience.

An arbitrary vector $\mathbf{A}$ may always be represented as a linear combination of basis vectors: $\mathbf{A} = a_x\hat{\mathbf{i}} + a_y\hat{\mathbf{j}} + a_z\hat{\mathbf{k}}$. Thus we return to an expression quite similar to the triplet (a_x, a_y, a_z). The geometrical interpretation of the components a_x, a_y, and a_z is quite simple in the case of an orthonormal basis. For example,

$$\mathbf{A} \cdot \hat{\mathbf{i}} = (a_x\hat{\mathbf{i}} + a_y\hat{\mathbf{j}} + a_z\hat{\mathbf{k}}) \cdot \hat{\mathbf{i}} = a_x\hat{\mathbf{i}} \cdot \hat{\mathbf{i}} = a_x.$$

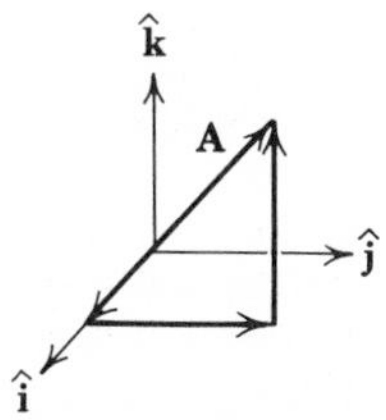

Figure 7.8

Also, $\mathbf{A} \cdot \hat{\mathbf{i}} = |A|\,|i| \cos\theta = |A| \cos\theta$, so a_x equals the component of **A** along the x-axis. Therefore we may rewrite the expansion of **A** as

$$\mathbf{A} = (\mathbf{A} \cdot \hat{\mathbf{i}})\hat{\mathbf{i}} + (\mathbf{A} \cdot \hat{\mathbf{j}})\hat{\mathbf{j}} + (\mathbf{A} \cdot \hat{\mathbf{k}})\hat{\mathbf{k}}. \tag{7.12}$$

The vector $(\mathbf{A} \cdot \hat{\mathbf{i}})\hat{\mathbf{i}}$ is known as the PROJECTION of **A** in the direction of $\hat{\mathbf{i}}$. (In contrast, a component of **A** is a scalar, and the coordinates of **A** are a set of scalars.) Equation 7.12 and Figure 7.8 show that a vector may be reconstituted by summing its projections. In terms of an orthonormal basis, the dot product takes on an especially simple form:

$$\mathbf{A} \cdot \mathbf{B} = (a_x\hat{\mathbf{i}} + a_y\hat{\mathbf{j}} + a_z\hat{\mathbf{k}}) \cdot (b_x\hat{\mathbf{i}} + b_y\hat{\mathbf{j}} + b_z\hat{\mathbf{k}}) = a_xb_x + a_yb_y + a_zb_z \tag{7.13}$$

since all cross-terms such as $a_xb_y\hat{\mathbf{i}} \cdot \hat{\mathbf{j}}$ disappear. Memorize this result.

Exercises

*19. A unit vector $\hat{\mathbf{u}}$ makes angles α, β, and γ with the x-, y-, and z-axes, respectively. Express $\hat{\mathbf{u}}$ as a linear combination of orthonormal basis vectors. The components are known as the DIRECTION COSINES of $\hat{\mathbf{u}}$. Show that the three angles are not independent.

20. Find $\mathbf{A} + \mathbf{B}$, $|\mathbf{A}|$, $|\mathbf{B}|$, $\mathbf{A} \cdot \mathbf{B}$, and $\cos\theta$ for the following vectors:
   (a) $\mathbf{A} = \hat{\mathbf{i}} + \hat{\mathbf{j}} + \hat{\mathbf{k}}$, $\mathbf{B} = \hat{\mathbf{i}} + 2\hat{\mathbf{j}} + 3\hat{\mathbf{k}}$.
   (b) $\mathbf{A} = 2\hat{\mathbf{i}} - 3\hat{\mathbf{j}} - \hat{\mathbf{k}}$, $\mathbf{B} = 2\hat{\mathbf{i}} - 5\hat{\mathbf{j}} + 3\hat{\mathbf{k}}$.
   (c) $\mathbf{A} = \hat{\mathbf{i}} + \hat{\mathbf{j}}$, $\mathbf{B} = \hat{\mathbf{i}} + \hat{\mathbf{k}}$.

21. A regular tetrahedron may be inscribed in a cube by placing the vertices of the tetrahedron at (1, 1, 1), (−1, −1, 1), (1, −1, −1), and (−1, 1, −1). (Draw it if you can't visualize it.) Find the tetrahedral bond angle.

22. Any sp^λ hybrid orbital may be written as $\psi = c_s s + c_x p_x + c_y p_y + c_z p_z$; it points in the direction $c_x\hat{\mathbf{i}} + c_y\hat{\mathbf{j}} + c_z\hat{\mathbf{k}}$. What is the angle between
   (a) $(\sqrt{2}s - p_x + \sqrt{3}p_y)/\sqrt{6}$ and $(\sqrt{2}s - p_x - \sqrt{3}p_y)/\sqrt{6}$?
   (b) $\frac{1}{2}(s + p_x + p_y + p_z)$ and $\frac{1}{2}(s + p_x - p_y - p_z)$?

23. The unit cell of a CaB_6 crystal is a cube 4.145 Å on an edge. If we choose orthonormal basis vectors along the three edges, then the Ca atom is at $\frac{1}{2}\hat{\mathbf{i}} + \frac{1}{2}\hat{\mathbf{j}} + \frac{1}{2}\hat{\mathbf{k}}$, in the middle of the cube. There is a B atom at $0.293\hat{\mathbf{i}}$, and two more at $0.707\hat{\mathbf{i}}$ and $0.293\hat{\mathbf{j}}$ (along with others we do not list). Find the distance, in Å, from the first B atom to:
   (a) The Ca atom.
   (b) The second B atom.
   (c) The third B atom.

*24. The dimensions of the unit cell of a glycine ($^+H_3NCH_2CO_2^-$) crystal are 5.10 Å × 11.96 Å × 5.45 Å. The angle between the long axis and either short axis is 90°, but the angle between the two short axes is 111° 38′. If we choose nonorthonormal basis vectors along the three edges, then the atomic coordinates are

$$\begin{aligned}
&\text{N:}\ 0.800\mathbf{e}_1 + 0.410\mathbf{e}_2 + 0.245\mathbf{e}_3\\
&\text{C:}\ 0.565\mathbf{e}_1 + 0.365\mathbf{e}_2 + 0.280\mathbf{e}_3\\
&\text{C:}\ 0.575\mathbf{e}_1 + 0.380\mathbf{e}_2 + 0.560\mathbf{e}_3\\
&\text{O:}\ 0.360\mathbf{e}_1 + 0.360\mathbf{e}_2 + 0.610\mathbf{e}_3\\
&\text{O}':\ 0.805\mathbf{e}_1 + 0.410\mathbf{e}_2 + 0.740\mathbf{e}_3
\end{aligned}$$

Find the NC, CC, CO, and CO′ bond lengths. Remember that the cross terms of Eq. 7.13 no longer disappear and that Eq. 7.10 does not hold.

25. How would you expect the dot product to be generalized to other than three-dimensional vectors: What is (1, 3) · (2, 4)? What is (1, 0, 4, 2) · (3, 1, −1, 0)? (Assume that these vectors are expressed in terms of an orthonormal basis.) How long is the N-dimensional vector whose every component is unity?

We shall further require that our basis set be a RIGHT-HANDED SYSTEM:

$$\hat{\mathbf{i}} \times \hat{\mathbf{j}} = \hat{\mathbf{k}} \qquad \hat{\mathbf{j}} \times \hat{\mathbf{k}} = \hat{\mathbf{i}} \qquad \hat{\mathbf{k}} \times \hat{\mathbf{i}} = \hat{\mathbf{j}}. \tag{7.14}$$

Then

$$\begin{aligned}
\mathbf{A} \times \mathbf{B} &= (a_x\hat{\mathbf{i}} + a_y\hat{\mathbf{j}} + a_z\hat{\mathbf{k}}) \times (b_x\hat{\mathbf{i}} + b_y\hat{\mathbf{j}} + b_z\hat{\mathbf{k}})\\
&= (a_yb_z - a_zb_y)\hat{\mathbf{i}} + (a_zb_x - a_xb_z)\hat{\mathbf{j}} + (a_xb_y - a_yb_x)\hat{\mathbf{k}}\\
&= \begin{vmatrix} \hat{\mathbf{i}} & \hat{\mathbf{j}} & \hat{\mathbf{k}} \\ a_x & a_y & a_z \\ b_x & b_y & b_z \end{vmatrix}.
\end{aligned} \tag{7.15}$$

The triple product $\mathbf{A} \cdot \mathbf{B} \times \mathbf{C} = a_x(B \times C)_x + a_y(B \times C)_y + a_z(B \times C)_z = a_x(b_yc_z - b_zc_y) + a_y(b_zc_x - b_xc_z) + a_z(b_xc_y - b_yc_x)$. In determinantal form

$$[ABC] = \begin{vmatrix} a_x & a_y & a_z \\ b_x & b_y & b_z \\ c_x & c_y & c_z \end{vmatrix}. \tag{7.16}$$

You will remember that this 3 × 3 determinant arises in connection with the problem of three simultaneous homogeneous linear equations in three unknowns, x, y, and z:

$$\begin{aligned}
a_xx + a_yy + a_zz &= 0\\
b_xx + b_yy + b_zz &= 0\\
c_xx + c_yy + c_zz &= 0.
\end{aligned}$$

If we let the three unknowns be represented by the vector $\mathbf{X} = x\hat{\mathbf{i}} + y\hat{\mathbf{j}} + z\hat{\mathbf{k}}$, then these three equations may be written as $\mathbf{A} \cdot \mathbf{X} = \mathbf{B} \cdot \mathbf{X} = \mathbf{C} \cdot \mathbf{X} = 0$. Thus the problem is to find a nonzero vector $\mathbf{X}$ which is orthogonal to all three vectors, **A**, **B**, and **C**. But if **A**, **B**, and **C** are linearly independent, there is no such **X**. The only way by which **X** can exist and be nonzero is if **A**, **B**, and **C** are coplanar. And if **A**, **B**, and **C** are coplanar, then the volume of the parallelepiped whose edges are **A**, **B**, and **C** must be zero. Thus we have provided a geometrical interpretation of the familiar rule that there exists a nontrivial solution to simultaneous homogeneous linear equations if and only if the determinant of coefficients is zero. We shall generalize this result to other than three dimensions and interpret an $N \times N$ determinant as the "volume" (except for sign) of an N-dimensional "parallelepiped" whose edges are N given vectors.

Exercises

26. (a) For the vectors of Exercise 7.20 find $\mathbf{A} \times \mathbf{B}$ and $|A \times B|$.
    (b) In Exercise 7.20c, if $\mathbf{C} = \hat{\mathbf{j}} + \hat{\mathbf{k}}$, what is $[ABC]$?
27. (a) What is the area of the parallelogram whose vertices are $(0, 0)$, (x_1, y_1), (x_2, y_2), and $(x_1 + x_2, y_1 + y_2)$?
    *(b) Express the result as a 2×2 determinant.
28. Find a nonzero vector orthogonal to $\hat{\mathbf{i}} + \hat{\mathbf{j}} + \hat{\mathbf{k}}$. Find another one.
29. Let $\mathbf{e}_1$, $\mathbf{e}_2$, $\mathbf{e}_3$ be basis vectors (not necessarily orthonormal) in a three-dimensional space. Consider three so-called RECIPROCAL VECTORS, $\boldsymbol{\epsilon}_1$, $\boldsymbol{\epsilon}_2$, and $\boldsymbol{\epsilon}_3$:

$$\boldsymbol{\epsilon}_1 \equiv \frac{\mathbf{e}_2 \times \mathbf{e}_3}{[e_1e_2e_3]} \qquad \boldsymbol{\epsilon}_2 \equiv \frac{\mathbf{e}_3 \times \mathbf{e}_1}{[e_1e_2e_3]} \qquad \boldsymbol{\epsilon}_3 \equiv \frac{\mathbf{e}_1 \times \mathbf{e}_2}{[e_1e_2e_3]}.$$

    (a) Convince yourself that $\mathbf{e}_i \cdot \boldsymbol{\epsilon}_j = \delta_{ij}$.
    (b) To help visualize these vectors, let $\mathbf{e}_3 = \hat{\mathbf{k}}$, and place $\mathbf{e}_1$ and $\mathbf{e}_2$ at an arbitrary angle in the xy-plane, with exaggeratedly different lengths. Now sketch the four vectors which lie in the xy-plane.
    Convince yourself that:
    (c) When a vector is expanded as a linear combination of nonorthonormal basis vectors—$\mathbf{X} = x_1\mathbf{e}_1 + x_2\mathbf{e}_2 + x_3\mathbf{e}_3$—$x_i$ is not equal to $\mathbf{e}_i \cdot \mathbf{X}$, the component of $\mathbf{X}$ on $\mathbf{e}_i$, but is equal to $\boldsymbol{\epsilon}_i \cdot \mathbf{X}$. Sketch these components.
    (d) If vectors are expanded in the basis set $\{\mathbf{e}_i\}$, $\mathbf{X} \cdot \mathbf{Y} \neq \sum x_i y_i$.
    (e) Such a simple form is obtained if $\mathbf{X}$ is expanded as a linear combination of reciprocal basis vectors.
*30. The functions $\{x^n\}_{n=0}^{\infty}$ are a basis in the vector space $\mathscr{P}_\infty$. If the dot product of two vectors in $\mathscr{P}_\infty$ is defined by $\mathbf{v}_i \cdot \mathbf{v}_j = \int_{-1}^{1} P_i(x)P_j(x)\,dx$, then $\{x^n\}$ is not an orthonormal basis. Use the result of Exercise 3.33 to show that the reciprocal basis is $\{(-1)^n D^n\delta(x)/n!\}_{n=0}^{\infty}$.
31. (a) Show that $\mathbf{A} \times (\mathbf{B} \times \mathbf{C}) = (\mathbf{A} \cdot \mathbf{C})\mathbf{B} - (\mathbf{A} \cdot \mathbf{B})\mathbf{C}$ by expanding both sides. (*Hint:* Since the cross product is independent of the choice of basis set, you may choose **B** to be along the x-axis and **C** to be in the xy-plane.)
    (b) Use Eq. 7.9 and this result to simplify $(\mathbf{A} \times \mathbf{B}) \cdot (\mathbf{C} \times \mathbf{D})$.
*32. Use Exercise 7.31a to show that $(\mathbf{A} \times \mathbf{B}) \times (\mathbf{C} \times \mathbf{D}) = [ABD]\mathbf{C} - [ABC]\mathbf{D}$.
*33. Simplify $\mathbf{A} \times [\mathbf{A} \times (\mathbf{A} \times \mathbf{B})]$ and $\mathbf{A} \times (\mathbf{B} \times \mathbf{C}) + \mathbf{B} \times (\mathbf{C} \times \mathbf{A}) + \mathbf{C} \times (\mathbf{A} \times \mathbf{B})$.

At this point we digress to indicate how any set of vectors $\{\mathbf{A}_i\}$ may be converted into a set of orthonormal vectors $\{\hat{\mathbf{a}}_i\}$. This procedure, the GRAM-SCHMIDT ORTHOGONALIZATION PROCEDURE, is applicable to any

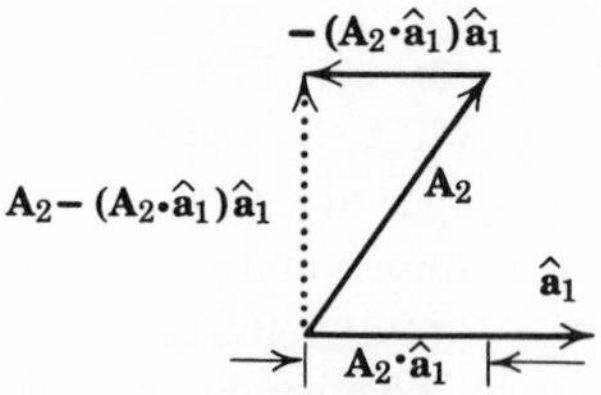

Figure 7.9

number of N-dimensional vectors, but we shall illustrate it with three three-dimensional vectors. The first vector of the orthonormal set is taken to be

$$\hat{\mathbf{a}}_1 = \mathbf{A}_1/|A_1| = \eta\mathbf{A}_1.$$

For notational convenience we shall use η as the symbol for a normalizing operator, which divides the vector upon which it acts by the length of that vector. To convert $\mathbf{A}_2$ into a vector that is orthogonal to $\hat{\mathbf{a}}_1$, we must subtract $(\mathbf{A}_2 \cdot \hat{\mathbf{a}}_1)\hat{\mathbf{a}}_1$, the projection of $\mathbf{A}_2$ on $\hat{\mathbf{a}}_1$ (Figure 7.9). To check, we note that

$$\begin{aligned}\mathbf{A}_1 \cdot [\mathbf{A}_2 - (\mathbf{A}_2 \cdot \hat{\mathbf{a}}_1)\hat{\mathbf{a}}_1] &= \mathbf{A}_1 \cdot \mathbf{A}_2 - (\mathbf{A}_2 \cdot \hat{\mathbf{a}}_1)(\mathbf{A}_1 \cdot \hat{\mathbf{a}}_1)\\ &= \mathbf{A}_1 \cdot \mathbf{A}_2 - (\mathbf{A}_2 \cdot \hat{\mathbf{a}}_1)|\mathbf{A}_1|\\ &= \mathbf{A}_1 \cdot \mathbf{A}_2 - \mathbf{A}_2 \cdot \mathbf{A}_1 = 0.\end{aligned}$$

Similarly, we must subtract from $\mathbf{A}_3$ its projections along $\hat{\mathbf{a}}_1$ and $\hat{\mathbf{a}}_2$:

$$\hat{\mathbf{a}}_3 = \eta[\mathbf{A}_3 - (\mathbf{A}_3 \cdot \hat{\mathbf{a}}_1)\hat{\mathbf{a}}_1 - (\mathbf{A}_3 \cdot \hat{\mathbf{a}}_2)\hat{\mathbf{a}}_2].$$

We may generalize this result to

$$\hat{\mathbf{a}}_{n+1} = \eta\left[\mathbf{A}_{n+1} - \sum_{i=1}^{n} (\mathbf{A}_{n+1} \cdot \hat{\mathbf{a}}_i)\hat{\mathbf{a}}_i\right]. \tag{7.17}$$

If the set $\{\mathbf{A}_i\}$ is linearly dependent, then some $\hat{\mathbf{a}}_i$s will be zero and should be ignored. It is readily apparent that the set $\{\hat{\mathbf{a}}_i\}$, composed only of nonzero vectors, is linearly independent, and that if the set $\{\mathbf{A}_i\}$ spans the vector space, the set $\{\hat{\mathbf{a}}_i\}$ is an orthonormal basis.

Exercises

34. The following vectors arise in the MO problem of benzene. They form a basis in six-dimensional space:

$$\begin{aligned}&\phi_1 = (1, 1, 1, 1, 1, 1) && \phi_4 = (2, -1, -1, 2, -1, -1)\\ &\phi_2 = (2, 1, -1, -2, -1, 1) && \phi_5' = (-1, 2, -1, -1, 2, -1)\\ &\phi_3' = (1, 2, 1, -1, -2, -1) && \phi_6 = (1, -1, 1, -1, 1, -1).\end{aligned}$$

(a) ϕ_1 and ϕ_6 are orthogonal to all others; normalize them.

(b) ϕ_2 and ϕ_3' are orthogonal to all others but not to each other. Normalize ϕ_2 and find ϕ_3, a linear combination of $\hat{\phi}_2$ and ϕ_3' which is orthogonal to $\hat{\phi}_2$. Then normalize this new vector.

(c) Do the same as (b) for ϕ_4 and ϕ_5'.

35. The functions $1, x, x^2, x^3, \ldots$, may be considered as basis vectors $\mathbf{e}_0, \mathbf{e}_1, \mathbf{e}_2, \mathbf{e}_3, \ldots$. If the dot product is defined as $\mathbf{v}_i \cdot \mathbf{v}_j = \int_{-1}^{1} v_i(x)v_j(x)\,dx$, these vectors are not orthonormal. Use the Gram-Schmidt procedure to make them so; you will thereby generate the first four normalized Legendre polynomials. (*Hint:* First evaluate all necessary integrals.)

*36. Do the same if the dot product is defined as $\mathbf{v}_i \cdot \mathbf{v}_j = \int_0^\infty v_i(x)v_j(x)e^{-x}\,dx$, and thereby generate the first four normalized Laguerre polynomials. (*Hint:* Remember the gamma function.)

7.5 Analytic Geometry

As a freshman you learned how to solve problems of geometry by numerical methods, in terms of coordinates. With vector notation such problems are considerably easier. We shall indicate how these problems may be set up in terms of vectors, and how numerical results may be obtained once a basis is chosen.

Although it doesn't matter where the tail of a vector is situated, in many contexts it is convenient to distinguish a vector whose tail is at the origin. Such a vector is known as a BOUND VECTOR. There is a one-to-one correspondence between bound vectors and points in three-dimensional space—to each bound vector there corresponds a unique point and vice versa. Therefore we may speak of bound vectors and points interchangeably. In the figures to follow, bound vectors are indicated as points rather than as arrows drawn from the origin. We shall use the symbol $\mathbf{X}_0 = x_0\hat{\mathbf{i}} + y_0\hat{\mathbf{j}} + z_0\hat{\mathbf{k}}$ to represent the bound vector which corresponds to the fixed point whose coordinates are (x_0, y_0, z_0). We shall use the symbol $\mathbf{X}$ to represent a variable bound vector, or arbitrary point. Many problems in analytic geometry involve specifying a set of points satisfying some criterion. Not every $\mathbf{X}$ in three-dimensional space satisfies the criterion, but the set may be defined by an equation of constraint limiting the possible values of $\mathbf{X}$. For example, the equation of the sphere of radius r and centered at the point $\mathbf{X}_0$ is $|\mathbf{X} - \mathbf{X}_0| = r$; a point $\mathbf{X}$ on the sphere is required to be a distance r from the fixed point $\mathbf{X}_0$. You are perhaps more familiar with this equation in the notation of analytic geometry, $[(x - x_0)^2 + (y - y_0)^2 + (z - z_0)^2]^{1/2} = r$, written in terms of coordinates.

First we shall find the equation of the plane passing through the point $\mathbf{X}_0$ and with the vector $\mathbf{N}$ normal (i.e., perpendicular) to the plane. If $\mathbf{X}$ is any point on the plane, then $\mathbf{X} - \mathbf{X}_0$ must be perpendicular to $\mathbf{N}$ (See Figure 7.10). Therefore the equation of this plane in vector notation is

$$(\mathbf{X} - \mathbf{X}_0) \cdot \mathbf{N} = 0.$$

Notice that there are three variables, x, y, and z, and one equation of constraint. Therefore there remain two degrees of freedom, as required for

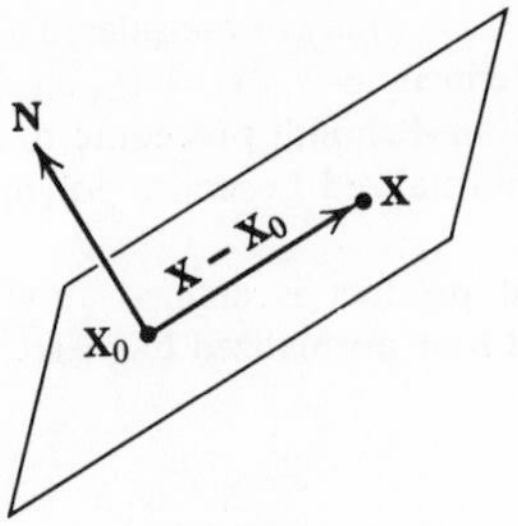

Figure 7.10

a surface. If the coordinates of **N** are given as (a, b, c), and if we define $d = \mathbf{X}_0 \cdot \mathbf{N} = ax_0 + by_0 + cz_0$, then this equation becomes $ax + by + cz = d$, the familiar form of the equation of a plane. However, in a different basis set (if the coordinate axes are rotated) this equation has different constants whereas the vector equation remains the same. We may further define the DIHEDRAL ANGLE θ between the two planes, $(\mathbf{X} - \mathbf{X}_0) \cdot \mathbf{N}_0 = 0$ and $(\mathbf{X} - \mathbf{X}_1) \cdot \mathbf{N}_1 = 0$ by

$$\cos \theta = \frac{|\mathbf{N}_0 \cdot \mathbf{N}_1|}{|N_0|\,|N_1|}$$

where taking the absolute value ensures choosing the smaller angle between the normals.

Next we consider how we might express the equation of a line passing through the point $\mathbf{X}_0$ and going in the direction of the vector **D**. If **X** is any point on the line, then $\mathbf{X} - \mathbf{X}_0$ must have the same direction as **D** (See

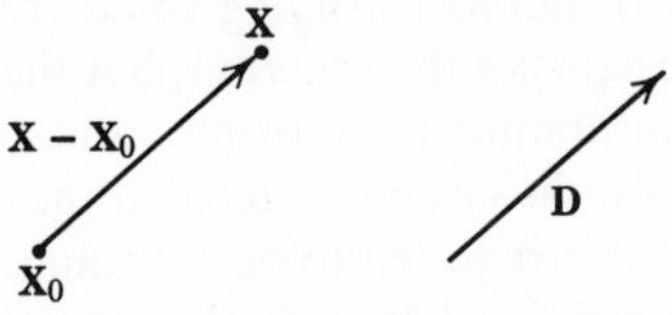

Figure 7.11

Figure 7.11). One way of expressing this constraint is by the vector equation

$$(\mathbf{X} - \mathbf{X}_0) \times \mathbf{D} = \mathbf{0}.$$

Another is by stating that $\mathbf{X} - \mathbf{X}_0$ must be a scalar multiple of **D**

$$\mathbf{X} - \mathbf{X}_0 = t\mathbf{D} \qquad \text{or} \qquad \mathbf{X} = \mathbf{X}_0 + t\mathbf{D}$$

where t is a parameter which runs from $-\infty$ to ∞; to each value of t there corresponds one point on the line. We may define the ANGLE θ BETWEEN

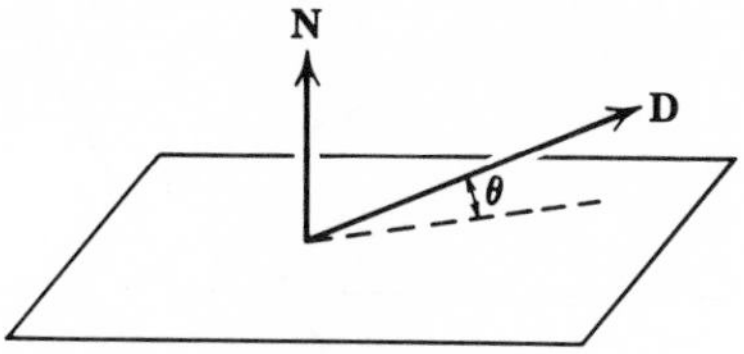

Figure 7.12

A LINE, $(\mathbf{X}-\mathbf{X}_0)\times\mathbf{D}=\mathbf{0}$, AND A PLANE, $(\mathbf{X}-\mathbf{P}_0)\cdot\mathbf{N}=0$ (see Figure 7.12), by

$$\sin\theta=\frac{|\mathbf{N}\cdot\mathbf{D}|}{|N|\,|D|}=\cos\left(\frac{\pi}{2}-\theta\right).$$

Examples of the Use of Vector Notation

14. What is the point of intersection $\mathbf{X}_I$ of the line $\mathbf{X}=\mathbf{X}_0+t\mathbf{D}$ and the plane $(\mathbf{X}-\mathbf{P}_0)\cdot\mathbf{N}=0$? Since $\mathbf{X}_I$ is common to both the line and the plane, there must be some t_I such that $\mathbf{X}_I=\mathbf{X}_0+t_I\mathbf{D}$ and $(\mathbf{X}_I-\mathbf{P}_0)\cdot\mathbf{N}=(\mathbf{X}_0+t_I\mathbf{D}-\mathbf{P}_0)\cdot\mathbf{N}=0$. Solving for t_I gives $t_I=(\mathbf{P}_0-\mathbf{X}_0)\cdot\mathbf{N}/\mathbf{D}\cdot\mathbf{N}$. Therefore the point of intersection is

$$\mathbf{X}_I=\mathbf{X}_0+\frac{(\mathbf{P}_0-\mathbf{X}_0)\cdot\mathbf{N}}{\mathbf{D}\cdot\mathbf{N}}\mathbf{D}.$$

15. What is the (minimum) distance d from the point $\mathbf{P}_0$ to the line $(\mathbf{X}-\mathbf{X}_0)\times\mathbf{D}=\mathbf{0}$? As Figure 7.13 shows, d is the perpendicular distance from $\mathbf{P}_0$ to the line, but we do not

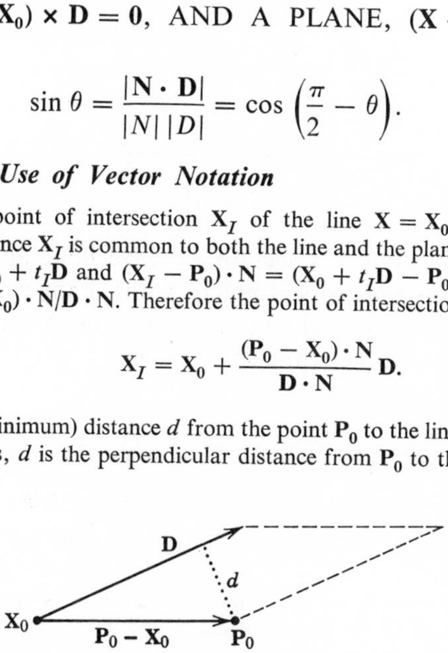

Figure 7.13

know where that perpendicular hits the line. However, the area of the parallelogram whose edges are $\mathbf{D}$ and $\mathbf{P}_0-\mathbf{X}_0$ is equal to $|D|$ times this perpendicular. But this area equals $|\mathbf{D}\times(\mathbf{P}_0-\mathbf{X}_0)|$. Therefore

$$d=\frac{|\mathbf{D}\times(\mathbf{P}_0-\mathbf{X}_0)|}{|D|}.$$

16. One of the most complicated problems of analytic geometry is that of finding the (minimum) distance d between two skew lines, $(\mathbf{X}-\mathbf{X}_0)\times\mathbf{D}_0=\mathbf{0}$ and $(\mathbf{X}-\mathbf{X}_1)\times\mathbf{D}_1=\mathbf{0}$. Notice first that $\mathbf{D}_0\times\mathbf{D}_1$ (dashed line in Fig. 7.14) is a vector perpendicular to both $\mathbf{D}_0$ and $\mathbf{D}_1$. From Figure 7.14 it is clear that d is equal to the distance between the two parallel planes, with common normal $\mathbf{D}_0\times\mathbf{D}_1$, each containing one of the two lines. But this distance is just the absolute value of the component of $\mathbf{X}_1-\mathbf{X}_0$ in the direction of $\mathbf{D}_0\times\mathbf{D}_1$. Therefore

$$d=\left|\frac{(\mathbf{X}_1-\mathbf{X}_0)\cdot\mathbf{D}_0\times\mathbf{D}_1}{|\mathbf{D}_0\times\mathbf{D}_1|}\right|.$$

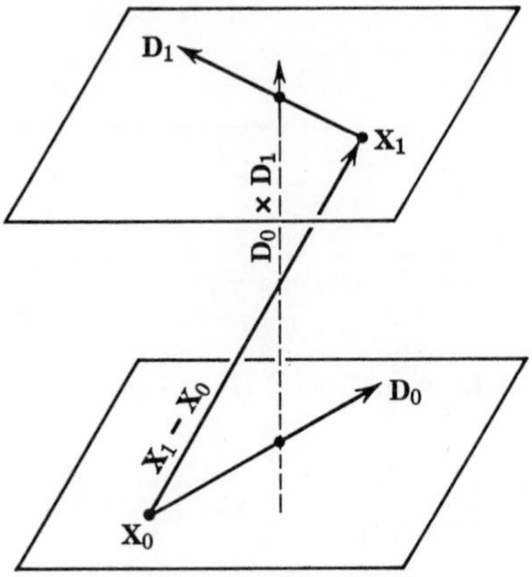

Figure 7.14

Notice that throughout this section we have done all our reasoning in terms of dot products and cross products, without any choice of basis. Only if we needed a numerical answer would we finally choose a basis and perform the indicated manipulations of the components.

Exercises

*37. Show that the line $\mathbf{X} - \mathbf{X}_0 = t\mathbf{D}$ may be written in the form

$$(x - x_0)/a = (y - y_0)/b = (z - z_0)/c.$$

38. (a) Find the equation of the plane through the point $\hat{\imath} + 3\hat{\jmath} - 2\hat{k}$ with normal $2\hat{\imath} - \hat{\jmath} - 2\hat{k}$.
    (b) Find $\mathbf{N}$ and some $\mathbf{X}_0$ for the plane whose equation is $4x + y - 3z = 5$.
39. Find the equation of the plane tangent to the sphere $|\mathbf{X} - \mathbf{A}_0| = r$ and passing through the point on the sphere $\mathbf{X}_0$.
40. Find the equation of the plane through the points $\mathbf{R}_1, \mathbf{R}_2, \mathbf{R}_3$ (not colinear).
41. Find the equation of the line through the points $\mathbf{X}_0$ and $\mathbf{X}_1$.
42. (a) Find the direction of the line of intersection of the (nonparallel) planes $(\mathbf{X} - \mathbf{X}_0) \cdot \mathbf{N}_0 = 0$ and $(\mathbf{X} - \mathbf{X}_1) \cdot \mathbf{N}_1 = 0$.
    *(b) Find the equation, in vector form, of the line of intersection of the planes $4x + y - 3z = 5$ and $2x - y + z = 1$.

*43. Show that the two nonparallel lines with directions $\mathbf{D}_0$ and $\mathbf{D}_1$ and through the points $\mathbf{X}_0$ and $\mathbf{X}_1$, respectively, intersect if and only if $(\mathbf{X}_1 - \mathbf{X}_0) \times \mathbf{D}_0$ is in the same direction as $\mathbf{D}_0 \times \mathbf{D}_1$.

44. Find the (shortest) distance from the point $\mathbf{P}_0$ to the plane $(\mathbf{X} - \mathbf{X}_0) \cdot \mathbf{N} = 0$. (*Hint:* Express this distance as a projection on the perpendicular from $\mathbf{P}_0$ to the plane.) Cf. Exercise 2.25.

*45. Find the point of intersection $\mathbf{R}$ of the three planes

$$(\mathbf{X}_1 - \mathbf{X}_1^0) \cdot \mathbf{n}_1 = 0, \qquad (\mathbf{X}_2 - \mathbf{X}_2^0) \cdot \mathbf{n}_2 = 0, \qquad (\mathbf{X}_3 - \mathbf{X}_3^0) \cdot \mathbf{n}_3 = 0.$$

(*Hint:* Expand $\mathbf{R}$ in the basis set of the vectors reciprocal to $\{\mathbf{n}_i\}$.)

46. Let the vectors from the origin to the carbon atoms C_1—C_6 of the chair form of cyclohexane be $\{X_i\}_{i=1}^6$. Assume that all C—C bond lengths are equal and that every CCC angle is tetrahedral. Use the results of Exercises 7.21 and 7.31b to find the dihedral angle between the $C_1C_2C_3$ and $C_2C_3C_4$ planes. (*Hint:* If the origin is placed at the center of the molecule, then $\mathbf{X}_{i\pm3} = -\mathbf{X}_i$.)

*47. What are the coordinates of the carbon atoms of cyclobutane, in terms of d, the distance between adjacent carbon atoms, and the dihedral angle ψ? What is the CCC angle? (*Note:* Cyclobutane is not planar, but is twisted somewhat. You will find it easy to allow for this twisting by placing the atoms in the xz- and yz-planes such that the atoms are alternately above and below the xy-plane by the same amount.)

*48. What is the "bowsprit-flagpole" C—C distance in the boat form of cyclohexane?

49. Describe the surface which satisfies the equation $\mathbf{X} \cdot \hat{\mathbf{n}} = a\,|X|$, where $\hat{\mathbf{n}}$ and a are constants.

*50. Describe the surface which satisfies the equation $|(\mathbf{X} - \mathbf{A}_0) \times \mathbf{D}| = r\,|D|$. (*Hint:* Choose the coordinate system so that $\mathbf{A}_0 = \mathbf{0}$ and $\mathbf{D} = \hat{\mathbf{k}}$, then generalize.)

7.6 Vector Functions

So far we have been primarily concerned with scalar functions, whose domain and range are sets of numbers. Now we consider functions whose domain and/or range is a vector space.

A SCALAR POINT FUNCTION is a rule that takes a vector into a number. We may write the value of the function f at the point $\mathbf{X}$ as $f(\mathbf{X})$. Yet now we are confident that you will not mistake a function for its value, so for convenience we shall henceforth be sloppy and refer to the scalar point function $f(\mathbf{X})$. Actually, we have already encountered $f(\mathbf{X})$ in the form of a function of three variables, since a function of x, y, and z, or of r, θ, and ϕ, assigns a number to the point whose coordinates are (x, y, z) or $(r \sin\theta \cos\phi, r \sin\theta \sin\phi, r \cos\theta)$. However, the notation $f(\mathbf{X})$ allows us to assign a number to the point $\mathbf{X}$ without choosing a basis or even committing ourselves to Cartesian coordinates or polar coordinates or whatever.

Examples

17. $f(\mathbf{X}) = |X|$, assigning to each point the magnitude of the corresponding bound vector.

18. $T(\mathbf{X})$, assigning to each point in a room the temperature at that point.

19. $z(\mathbf{X}) = \mathbf{X} \cdot \hat{\mathbf{k}}$, assigning to each point its height above, say, sea level.

20. $\psi(\mathbf{X})$, an electronic wave function such that the probability of finding an electron within the volume element dV about the point $\mathbf{X}$ is $[\psi(\mathbf{X})]^2\, dV$.

A VECTOR FUNCTION is a rule that takes a number into a vector: $\mathbf{X} = f(t)$, which we shall write simply as $\mathbf{X}(t)$. Such a function may be viewed as a composite of three ordinary functions, f_1, f_2, and f_3, since we may write $\mathbf{X}(t) = f_1(t)\hat{\mathbf{i}} + f_2(t)\hat{\mathbf{j}} + f_3(t)\hat{\mathbf{k}}$. To visualize $\mathbf{X}(t)$, identify t with time and $\mathbf{X}(t)$ with the position of a moving particle at time t. The graph of a vector function is a curve in space, labelled with the values of t at representative points.

Examples

21. $\mathbf{X}(t) = \mathbf{X}_0 + t\mathbf{D}$ corresponds to a particle initially at the point $\mathbf{X}_0$ and traveling in the direction $\mathbf{D}$ with speed $|D|$.

22. $\mathbf{v}(t) = \mathbf{D}$, the velocity of the above particle, such that the same vector $\mathbf{D}$ is assigned to every scalar t. Notice that it is not always the case that the vector corresponds to position.

23. $\mathbf{X}(t) = r\cos\omega t\hat{\mathbf{i}} + r\sin\omega t\hat{\mathbf{j}}$ corresponds to a particle traveling in a circle in the xy-plane with angular frequency ω.

24. $\mathbf{X}(t) = r\cos\omega t\hat{\mathbf{i}} + r\sin\omega t\hat{\mathbf{j}} + vt\hat{\mathbf{k}}$ corresponds to a particle traveling along a RIGHT-HANDED HELIX whose axis is the z-axis. See Figure 7.15.

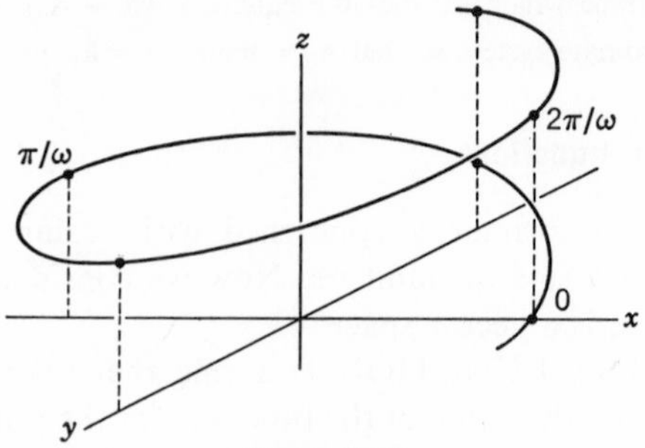

Figure 7.15

A VECTOR FIELD is a rule that takes vectors into vectors: $\mathbf{E} = f(\mathbf{X})$, which we shall write simply as $\mathbf{E}(\mathbf{X})$. Such a function may be viewed as a composite of three functions f_1, f_2, and f_3 of three variables since we may write $\mathbf{E}(\mathbf{X}) = f_1(x, y, z)\hat{\mathbf{i}} + f_2(x, y, z)\hat{\mathbf{j}} + f_3(x, y, z)\hat{\mathbf{k}}$. Another possibility is to express the vector field in cylindrical polar coordinates: $\mathbf{E} = g_1(r, \phi, z)\hat{\mathbf{e}}_r + g_2(r, \phi, z)\hat{\mathbf{e}}_\phi + g_3(r, \phi, z)\hat{\mathbf{e}}_z$, where the LOCAL COORDINATE SYSTEM $\{\hat{\mathbf{e}}_r, \hat{\mathbf{e}}_\phi, \hat{\mathbf{e}}_z\}$ is an orthonormal right-handed basis set whose orientation depends upon the argument, as shown in Figure 7.16a, with $\hat{\mathbf{e}}_r$ pointing

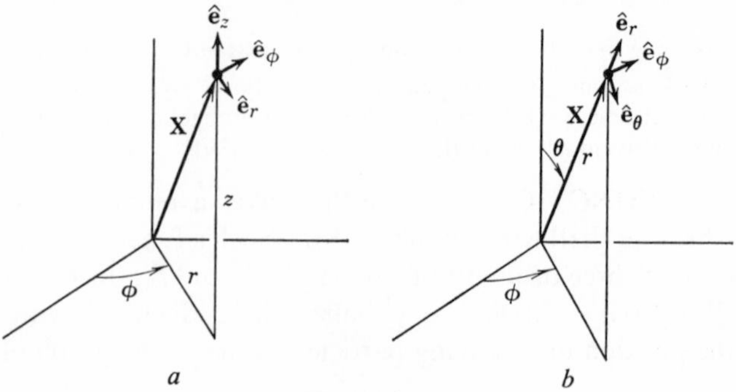

Figure 7.16

horizontally and away from the origin, $\hat{\mathbf{e}}_z$ parallel to the z-axis, and $\hat{\mathbf{e}}_\phi$ perpendicular to both $\hat{\mathbf{e}}_z$ and $\hat{\mathbf{e}}_r$ and pointing in the direction of increasing ϕ. Similarly, in spherical polar coordinates $\mathbf{E} = h_1(r, \theta, \phi)\hat{\mathbf{e}}_r + h_2(r, \theta, \phi)\hat{\mathbf{e}}_\theta + h_3(r, \theta, \phi)\hat{\mathbf{e}}_\phi$, where $\{\hat{\mathbf{e}}_r, \hat{\mathbf{e}}_\theta, \hat{\mathbf{e}}_\phi\}$ is also a local orthonormal right-handed system, as shown in Figure 7.16b, with $\hat{\mathbf{e}}_r$ directed radially outward, and $\hat{\mathbf{e}}_\theta$ and $\hat{\mathbf{e}}_\phi$ pointing along the surface of a sphere of radius r in the directions of increasing θ and ϕ, respectively. However, the notation $\mathbf{E}(\mathbf{X})$ allows us to assign the vector $\mathbf{E}$ to the point $\mathbf{X}$ without choosing a basis.

Examples

25. $\mathbf{E}(\mathbf{X}) = (\mathbf{X} \cdot \hat{\mathbf{k}})\hat{\mathbf{k}}$, assigning to each vector its projection on the z-axis.

26. $\mathbf{E}(\mathbf{R}) = \mathbf{R}/|R|^3$. This, the inverse-square field, is quite important: The gravitational force on a body of mass m at the point $\mathbf{R}$, due to a body of mass M at the origin, is $\mathbf{F} = -GmM\mathbf{R}/|R|^3$, where G is a constant and the minus sign means that the force is attractive. Similarly, the electrostatic force between objects of charges q_1 and q_2 is $\mathbf{F} = q_1q_2\mathbf{R}/|R|^3$.

A common technique for graphing a vector field is to indicate the vector assigned to each of certain representative points by drawing it with its tail

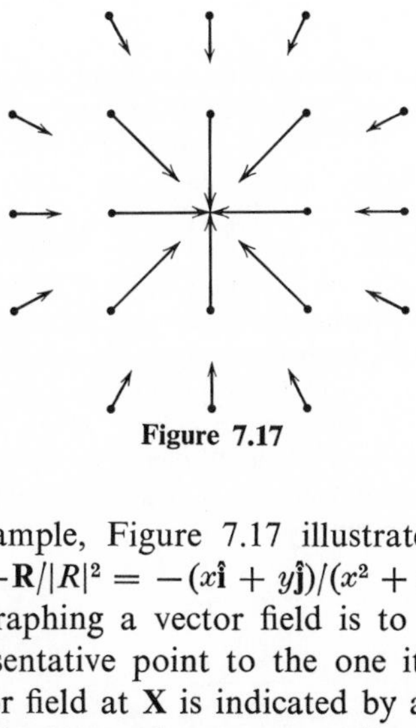

Figure 7.17

at that point. For example, Figure 7.17 illustrates the two-dimensional vector field, $\mathbf{E}(\mathbf{R}) = -\mathbf{R}/|R|^2 = -(x\hat{\mathbf{i}} + y\hat{\mathbf{j}})/(x^2 + y^2) = -\hat{\mathbf{e}}_r/r$. An alternative technique for graphing a vector field is to draw STREAMLINES connecting each representative point to the one its vector points at; the magnitude of the vector field at $\mathbf{X}$ is indicated by arranging the density of streamlines so that the distance to the next streamline is inversely proportional to $|\mathbf{E}(\mathbf{X})|$. The streamlines for an inverse-square field in three dimensions are straight lines emanating from the origin and uniformly spaced. It is clear that the lines are close together near the origin, where $|E|$ is large, and far apart at a distance from the origin, where $|E|$ is small, but the fact that the density of streamlines is indeed exactly proportional to $|E|$ will be apparent only when we consider the divergence of a vector field. Figure 7.18 shows a

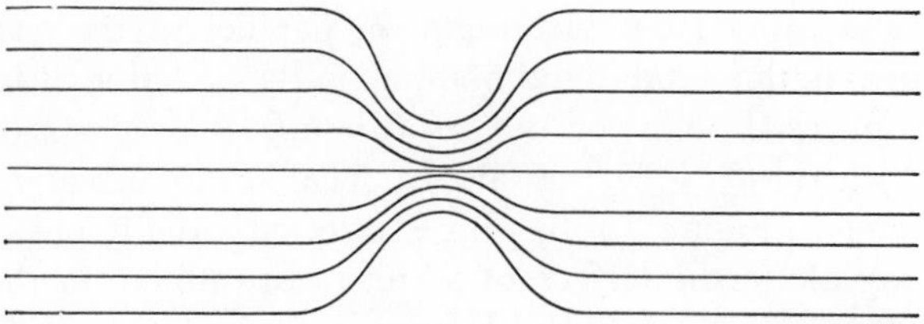

Figure 7.18

slice of the graph of the velocity field of a liquid flowing through a pipe with a constriction: $\mathbf{v}(\mathbf{X})$ is the velocity of the liquid flowing past the point $\mathbf{X}$. Notice that the streamlines are closer together in the constriction, where the speed of the liquid is greater.

Exercises

51. What kind of function is
    (a) $\mathbf{X}(t) \cdot \mathbf{Y}(t)$?
    (b) $f(\mathbf{X})\mathbf{E}(\mathbf{X})$?
    (c) $\mathbf{E}(\mathbf{X}) \times \mathbf{F}(\mathbf{X})$?
    (d) $\mathbf{E}(\mathbf{X}) \cdot \mathbf{F}(\mathbf{X})$?
    (e) $\mathbf{X}(t) \times \mathbf{Y}(t)$?
    (f) $f(t)\mathbf{X}(t)$?
52. Let $\mathbf{e}_1$, $\mathbf{e}_2$, and $\mathbf{e}_3$ be basis vectors, not necessarily orthonormal, in a three-dimensional space. Then $\mathbf{X} = x\mathbf{e}_1 + y\mathbf{e}_2 + z\mathbf{e}_3$ and $\Phi(\mathbf{X}) = f(x, y, z)$. Express the Fourier transform as an operator on $\Phi(\mathbf{X})$ which produces the function $G(\boldsymbol{\xi})$. In what basis set should $\boldsymbol{\xi}$ be expanded? (Cf. Exercises 2.46a and 7.29e.)

*53. When a crystal diffracts a beam of light there is a phase shift dependent on position, because different portions of the beam travel different distances. Let a parallel beam of light traveling in the direction $\hat{\mathbf{v}}_0$ be diffracted, thereby emerging in the direction $\hat{\mathbf{v}}$. What is the path-length difference Δd for the beam diffracted from $\mathbf{X}$, relative to that diffracted from the origin? What is the phase shift $\Delta\delta = 2\pi\Delta d/\lambda$ in terms of $\boldsymbol{\xi} = 2\pi(\hat{\mathbf{v}} - \hat{\mathbf{v}}_0)/\lambda$? If the probability that a beam will be diffracted is proportional to $\rho(\mathbf{X})\,dV$, the electron density within the infinitesimal volume dV about $\mathbf{X}$, then the amplitude of the diffracted wave, relative to the incident wave, is proportional to $\int \rho(\mathbf{X})e^{i\Delta\delta}\,dV$. Show that this amplitude is a Fourier transform of the electron density in the crystal.

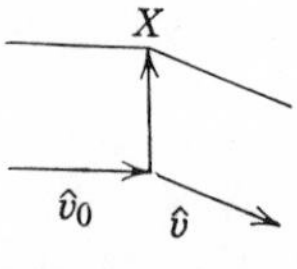

Figure 7.19

*54. Use the result of Exercise 5.20 to convince yourself that the Fourier transform of the lattice function $f(\mathbf{X}) = \sum \delta(\mathbf{X} - \mathbf{X}_{\text{lattice}})$, where $\mathbf{X}_{\text{lattice}} = n_1\mathbf{e}_1 + n_2\mathbf{e}_2 + n_3\mathbf{e}_3$ is proportional to $\sum \delta(\boldsymbol{\xi} - \boldsymbol{\xi}_{\text{lattice}})$, where $\boldsymbol{\xi}_{\text{lattice}} = 2\pi(h\boldsymbol{\epsilon}_1 + k\boldsymbol{\epsilon}_2 + l\boldsymbol{\epsilon}_3)$. The points $\{\boldsymbol{\xi}_{\text{lattice}}\}$ are known as the RECIPROCAL LATTICE; the integers h, k, l

are known as the MILLER INDICES. (The asymmetry of the factor of 2π may be removed by redefining the Fourier transform.) The electron density of an infinite crystal is the convolution $\rho(\mathbf{X}) * f(\mathbf{X})$, where $\rho(\mathbf{X})$ is the electron density in one unit cell whose edges are the vectors $\mathbf{e}_1$, $\mathbf{e}_2$, and $\mathbf{e}_3$. The diffracted wave is the Fourier transform of this convolution, which by Exercise 1.73d is the product of $G(\boldsymbol{\xi}) = F\rho(\mathbf{X})$ and $\sum \delta(\boldsymbol{\xi} - \boldsymbol{\xi}_{\text{lattice}})$. But by Rule 4 for delta functions

$$G(\boldsymbol{\xi}) \sum \delta(\boldsymbol{\xi} - \boldsymbol{\xi}_{\text{lattice}}) = \sum G(\boldsymbol{\xi}_{\text{lattice}})\delta(\boldsymbol{\xi} - \boldsymbol{\xi}_{\text{lattice}}),$$

so that the Fourier transform of the electron density of the crystal is sampled only at reciprocal lattice points. Thus the diffracted wave is seen only as spots. Unfortunately the phase of each $G(\boldsymbol{\xi}_{\text{lattice}})$ cannot be obtained, but only $|G(\boldsymbol{\xi}_{\text{lattice}})|$.

55. What is the vector function for a left-handed helix?

*56. Convince yourself that a right-handed helix remains right-handed "even when viewed from the other end." (Instead of rotating yourself relative to the coordinate axes, rotate the helix 180° about the x-axis: $x \to x$, $y \to -y$, $z \to -z$.)

*57. DNA is a double right-handed helix. There are two strands $\mathbf{X}_1(t)$ and $\mathbf{X}_2(t)$ such that a point on $\mathbf{X}_2$ is 180° around the z-axis from the corresponding point on $\mathbf{X}_1$. The diameter is 20 Å and it takes 34 Å along the z-axis to go 360° around the z-axis. Also, in this 34 Å height there are ten monomer units, whose positions may be marked by integral t. What are the $\mathbf{X}_1(t)$ and $\mathbf{X}_2(t)$?

58. Draw a graph of each of the following two-dimensional vector fields:
   (a) $\mathbf{E}(\mathbf{R}) = -\mathbf{R}$
   (b) $\mathbf{E}(\mathbf{X}) = \hat{\imath}$
   (c) $\mathbf{E}(\mathbf{X}) = (\mathbf{X} \cdot \hat{\imath})\hat{\imath}$
   (d) $\mathbf{E}(\mathbf{X}) = y^2\hat{\imath}$
   (e) $\mathbf{E}(\mathbf{X}) = -y\hat{\imath} + x\hat{\jmath}$ (*Hint:* $|E|$ is constant along the perimeter of any circle centered at the origin.)

   Notice that it is easy to draw streamlines, but that in (a) and (c) it is difficult to space them inversely to $|E|$.

59. Express the vector fields of Exercises 7.58a and 7.58e in polar coordinates.

*60. Sketch the streamlines of the velocity field of the light beam of Exercise 2.26. Be exact.

7.7 Differentiation and Integration of Vectors. Applications to Mechanics

We define the DERIVATIVE of the vector function $\mathbf{X}(t)$ by

$$\frac{d\mathbf{X}}{dt} \equiv \lim_{\Delta t \to 0} \frac{\mathbf{X}(t + \Delta t) - \mathbf{X}(t)}{\Delta t}. \tag{7.18}$$

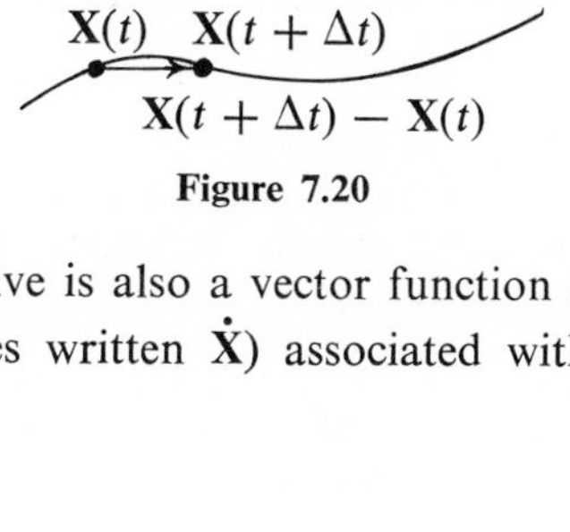

Figure 7.20

Notice that this derivative is also a vector function since there is a unique vector $d\mathbf{X}/dt$ (sometimes written $\dot{\mathbf{X}}$) associated with each t. In terms of components, if $\mathbf{X}(t) = f_1(t)\hat{\mathbf{i}} + f_2(t)\hat{\mathbf{j}} + f_3(t)\hat{\mathbf{k}}$, then $d\mathbf{X}/dt = f_1'(t)\hat{\mathbf{i}} + f_2'(t)\hat{\mathbf{j}} + f_3'(t)\hat{\mathbf{k}}$.

Exercises

61. Convince yourself that

$$\frac{d}{dt}[\mathbf{X}(t)\cdot\mathbf{Y}(t)] = \mathbf{X}\cdot\frac{d\mathbf{Y}}{dt} + \frac{d\mathbf{X}}{dt}\cdot\mathbf{Y},$$

$$\frac{d}{dt}[\mathbf{X}(t)\times\mathbf{Y}(t)] = \frac{d\mathbf{X}}{dt}\times\mathbf{Y} + \mathbf{X}\times\frac{d\mathbf{Y}}{dt}$$

and

$$\frac{d}{dt}[f(t)\mathbf{X}(t)] = f'(t)\mathbf{X}(t) + f(t)\frac{d\mathbf{X}}{dt}.$$

*62. Show that if $|X|$ is independent of time, then $\mathbf{X}(t)$ is orthogonal to $\mathbf{v}(t) = d\mathbf{X}/dt$.

63. What is the equation of the line tangent to the curve $\mathbf{X}(t)$ at $\mathbf{X}_0$?

64. (a) What is $d\mathbf{X}/dt$ for a right-handed helix?
(b) What is the PITCH ANGLE, the angle the tangent to the helix makes with a plane perpendicular to the axis?

*65. What is the pitch angle in DNA?

We are now is a position to define the ARC LENGTH S of the curve $\mathbf{X}(t)$ between the points $\mathbf{X}(t_0)$ and $\mathbf{X}(t_1)$ by

$$S \equiv \int_{t_0}^{t_1} \left|\frac{d\mathbf{X}}{dt}\right| dt \tag{7.19}$$

as is reasonable if we consider the arc length as the limit of the sum of lengths of line segments $\sum |\mathbf{X}(t+\Delta t) - \mathbf{X}(t)|$ (Fig. 7.20). If $t_0 > t_1$, it is customary to consider the arc length as negative. In terms of components, if $\mathbf{X}(t) = f_1(t)\hat{\mathbf{i}} + f_2(t)\hat{\mathbf{j}} + f_3(t)\hat{\mathbf{k}}$ or $x(t)\hat{\mathbf{i}} + y(t)\hat{\mathbf{j}} + z(t)\hat{\mathbf{k}}$, then

$$\begin{aligned} S &= \int_{t_0}^{t_1} \{[f_1'(t)]^2 + [f_2'(t)]^2 + [f_3'(t)]^2\}^{1/2}\, dt \\ &= \int_{t_0}^{t_1} \left\{\left(\frac{dx}{dt}\right)^2 + \left(\frac{dy}{dt}\right)^2 + \left(\frac{dz}{dt}\right)^2\right\}^{1/2} dt \\ &= \int_{x(t_0)}^{x(t_1)} \left\{1 + \left(\frac{dy}{dx}\right)^2 + \left(\frac{dz}{dx}\right)^2\right\}^{1/2} dx, \end{aligned} \tag{7.20}$$

where $dy/dx = (dy/dt)/(dx/dt)$, etc., as a function of x.

Exercises

66. Find the arc length of one turn of a helix.

*67. Find the arc length of the catenary $y = \cosh x$ from $x = 0$ to $x = x_0$.

*68. Show that the circumference of an ellipse, represented parametrically as $x = a\sin\theta$, $y = b\cos\theta$, in terms of complete elliptic integrals, is $4aE(\sqrt{a^2 - b^2}/a)$ or $4bE(\sqrt{b^2 - a^2}/b)$.

69. What is $\int_{t_0}^{t_1} (d\mathbf{X}/dt)\,dt$? (Notice that it is not arc length, but rather the definite integral of a vector function.)

*70. It is also possible to integrate a vector field to obtain a vector: If

$$\mathbf{E}(\mathbf{X}) = f_x(x, y, z)\hat{\mathbf{i}} + f_y(x, y, z)\hat{\mathbf{j}} + f_z(x, y, z)\hat{\mathbf{k}},$$

then

$$\iiint \mathbf{E}(\mathbf{X})\,dV \equiv \hat{\mathbf{i}}\iiint f_x\,dx\,dy\,dz + \hat{\mathbf{j}}\iiint f_y\,dx\,dy\,dz + \hat{\mathbf{k}}\iiint f_z\,dx\,dy\,dz.$$

The most common such integral is the center of mass of a solid object, which for particulate objects is $\sum m_i\mathbf{X}_i/\sum m_i$, where m_i is the mass of the ith particle. Convince yourself that this sum may be generalized to $\iiint \rho(\mathbf{X})\mathbf{X}\,dV/\iiint \rho(\mathbf{X})\,dV$, where $\rho(\mathbf{X})$ is the density of matter at $\mathbf{X}$ and the integration is over the volume of the object. (*Hint:* Express density as a derivative.)

*71. Find the center of mass of a solid osmium cone of height a and half-apical angle θ $(0 < \theta < \pi/2)$, with apex at the origin and axis along the positive z-axis.

If $\mathbf{X}(t)$ represents the position of a particle as a function of time, then $d\mathbf{X}/dt$ represents its VELOCITY, $\mathbf{v}$. The MOMENTUM of a particle of mass m is given by $\mathbf{p} = m\mathbf{v}$. The KINETIC ENERGY T is given by $T = \frac{1}{2}m\mathbf{v}\cdot\mathbf{v} = \mathbf{p}\cdot\mathbf{p}/2m$. Also, we may repeat the differentiation to obtain the ACCELERATION $\mathbf{a} \equiv d\mathbf{v}/dt = d^2\mathbf{X}/dt^2$. In vector notation Newton's second law is

$$\mathbf{F} = \frac{d\mathbf{p}}{dt} = \frac{d}{dt}(m\mathbf{v}) = m\mathbf{a}$$

where $\mathbf{F}$ is the force.

Next we consider rotatory motion. The position of a particle rotating about the z-axis in a circle of radius r at an angular frequency ω is $\mathbf{r}(t) = r\cos\omega t\hat{\mathbf{i}} + r\sin\omega t\hat{\mathbf{j}} + z_0\hat{\mathbf{k}}$. [We shall use $\mathbf{r}(t)$ instead of $\mathbf{X}(t)$ when we wish to emphasize the radial aspect of the position.] The velocity of this particle is $\mathbf{v}(t) = -r\omega\sin\omega t\hat{\mathbf{i}} + r\omega\cos\omega t\hat{\mathbf{j}}$. Let us define the ANGULAR VELOCITY $\boldsymbol{\omega}$ as a vector of magnitude ω perpendicular to the plane of the motion, such that a right-handed screw, rotated with the particle, would advance in the direction of $\boldsymbol{\omega}$. For the particle rotating about the z-axis, $\boldsymbol{\omega} = \omega\hat{\mathbf{k}}$. By expansion of the cross product, it becomes clear that

$$\boldsymbol{\omega} \times \mathbf{r} = \mathbf{v}$$

independent of z_0. Moreover, this result does not rest upon the fact that the rotation is about the z-axis, but is independent of basis, except that the origin must be on the axis of rotation.

We may define the ANGULAR MOMENTUM about the origin of a particle whose position is $\mathbf{r}(t)$ by

$$\mathbf{L} \equiv \mathbf{r} \times \mathbf{p} = m\mathbf{r} \times \mathbf{v} = m\mathbf{r} \times (\boldsymbol{\omega} \times \mathbf{r}) = m[(\mathbf{r}\cdot\mathbf{r})\boldsymbol{\omega} - (\mathbf{r}\cdot\boldsymbol{\omega})\mathbf{r}]$$

where we have invoked the result of Exercise 7.31a. This expression is most easily evaluated if we choose the origin so that $z_0 = 0$, whereby $\mathbf{r}\cdot\boldsymbol{\omega} = 0$.

Then $\mathbf{L} = mr^2\boldsymbol{\omega}$. If we define the MOMENT OF INERTIA about the axis of rotation by $I = mr^2$, then the angular momentum and the angular velocity are simply related by $\mathbf{L} = I\boldsymbol{\omega}$. For a collection of masses $\{m_i\}$ all rotating with angular velocity $\boldsymbol{\omega}$, the total angular momentum is $\mathbf{L} = \sum \mathbf{r}_i \times \mathbf{p}_i = \sum m_i[(\mathbf{r}_i \cdot \mathbf{r}_i)\boldsymbol{\omega} - (\mathbf{r}_i \cdot \boldsymbol{\omega})\mathbf{r}_i]$. However, unless all the masses are in the same plane it is not possible to choose the origin so that every $\mathbf{r}_i \cdot \boldsymbol{\omega} = 0$. Indeed, it is not even always possible to choose the origin so that $\sum m_i(\mathbf{r}_i \cdot \boldsymbol{\omega})\mathbf{r}_i = \mathbf{0}$; therefore it is possible that $\mathbf{L}$ and $\boldsymbol{\omega}$ will not be parallel (Exercise 8.33).

We may define the TORQUE, or MOMENT OF FORCE, about the origin produced by a force applied to a particle at $\mathbf{r}$ by

$$\mathbf{T} = \mathbf{r} \times \mathbf{F}.$$

Let us now consider how the angular momentum of the particle varies:

$$\frac{d\mathbf{L}}{dt} = \frac{d}{dt}(\mathbf{r} \times \mathbf{p}) = \frac{d\mathbf{r}}{dt} \times \mathbf{p} + \mathbf{r} \times \frac{d\mathbf{p}}{dt} = \mathbf{v} \times m\mathbf{v} + \mathbf{r} \times \mathbf{F} = \mathbf{T},$$

where we have used Newton's second law and the fact that $\mathbf{v} \times \mathbf{v} = \mathbf{0}$. Thus we may conclude that if there are no torques on the particle, its angular momentum is conserved.

Exercises

*72. Use Newton's second law to show that for a set of particles whose positions are given by $\{\mathbf{X}_i(t)\}$, $(d/dt)\sum_i \mathbf{p}_i(t) \cdot \mathbf{X}_i(t) = 2T + \sum \mathbf{F}_i(t) \cdot \mathbf{X}_i(t)$, where $T =$ total kinetic energy of all the particles. Then assuming that $|\sum \mathbf{p}_i(t) \cdot \mathbf{X}_i(t)|$ is always less than some maximum M, take the time average of both sides and derive the VIRIAL THEOREM $\langle T \rangle = -\frac{1}{2}\langle \sum \mathbf{F}_i \cdot \mathbf{X}_i \rangle$.

73. What is the acceleration of a particle traveling in a circle of radius r with an angular frequency ω? Relate $\mathbf{a}(t)$ to $\mathbf{X}(t)$.

74. Find the general solution to the differential equation $d^2\mathbf{X}/dt^2 = -g\hat{\mathbf{k}}$, which describes the trajectory of a particle subject to a downward gravitational force. Notice that the constants of integration are vectors since a vector differential equation is equivalent to three ordinary differential equations.

75. Solve the three-dimensional analog of Exercise 4.29: A particle at $\mathbf{X} = x\hat{\mathbf{i}} + y\hat{\mathbf{j}} + z\hat{\mathbf{k}}$ is attracted to the origin by a force $\mathbf{F} = -k_x x\hat{\mathbf{i}} - k_y y\hat{\mathbf{j}} - k_z z\hat{\mathbf{k}}$. (*Note:* This force law is appropriate for a hydrogen atom of benzene, which is attracted to its equilibrium position by three different force constants—the stiffest "spring" resists changing the C—H bond length and the floppiest holds the hydrogen atom in the molecular plane. Notice also that unlike the one-dimensional case, the force and the displacement are not necessarily in opposite directions.)
   (a) Find $\mathbf{X}(t)$, subject to the initial conditions $\mathbf{X}(0) = x_0\hat{\mathbf{i}} + y_0\hat{\mathbf{j}} + z_0\hat{\mathbf{k}}$ and $\mathbf{v}(0) = \mathbf{0}$, by solving the separate x, y, and z equations.
   (b) What are $T(t)$ and $T_{\max}$? (*Hint:* The absolute maximum occurs when every $\sin^2 = 1$.)
   (c) What are $V(t)$ and $V(\mathbf{X})$?

*76. So far we have considered only a fixed coordinate system with fixed origin. However, it is also possible to use a MOVING COORDINATE SYSTEM in which the origin moves in the direction $\mathbf{D}$ with speed $|D|$ relative to the fixed origin. For example, we usually refer measurements of planetary positions to coordinate axes moving with the sun, and it is more convenient to consider scattering problems (e.g., molecular beams) in a coordinate system traveling with the center of mass of the system. If $\mathbf{X}$ is a constant position vector (point) in the fixed system, it cannot be represented as a constant vector in the moving system since its components change as the origin moves. Similarly, if $\mathbf{v}$ represents the velocity of a moving particle as viewed from the fixed system, it cannot also represent the velocity relative to the moving system, as is obvious if $\mathbf{v} = \mathbf{D}$, whereupon the particle appears to be stationary as viewed from the moving system. Therefore we must distinguish $\mathbf{X}_f$ and $\mathbf{v}_f$, the representations of $\mathbf{X}$ and $\mathbf{v}$ in the fixed system, from $\mathbf{X}_m$ and $\mathbf{v}_m$, the representations in the moving system.

(a) Convince yourself that *relative* positions and *relative* velocities (differences between two position vectors or two velocity vectors) are nevertheless not expressed by different vectors in the two systems.

(b) Convince yourself that for position vectors $\mathbf{X}_m = \mathbf{X}_f - t\mathbf{D}$.

(c) Convince yourself that for velocities, $\mathbf{v}_m = \mathbf{v}_f - \mathbf{D}$.

(d) Convince yourself that an acceleration vector has the same representation in the two systems: $\mathbf{a}_m = \mathbf{a}_f$.

*77. It is also possible to use a ROTATING COORDINATE SYSTEM which is rotating about the origin of a fixed system with angular velocity $\boldsymbol{\omega}$. For example, we usually refer measurements of terrestrial positions to coordinate axes rotating with the earth. Again, it is necessary to distinguish $\mathbf{X}_f$, the representation of a vector (point) in the fixed system, from $\mathbf{X}_r$, the representation in the rotating system.

(a) Convince yourself that as the coordinate system rotates "out from under" a constant vector $\mathbf{X}_f$, an observer rotating with the coordinate system would see this vector rotate in the *opposite* direction to the direction of the rotation of the coordinates, that is, convince yourself that $\mathbf{X}_r$ is not constant but rotates with an angular velocity $-\boldsymbol{\omega}$, and therefore with an apparent linear velocity $-\boldsymbol{\omega} \times \mathbf{X}_r$.

(b) Convince yourself that if the particle is not stationary but moves with a velocity $\mathbf{v}_f = (d\mathbf{X}_f/dt)_f$ in the fixed system, then in the rotating system

$$\mathbf{v}_r = \left(\frac{d\mathbf{X}_r}{dt}\right)_r = \mathbf{v}_f - \boldsymbol{\omega} \times \mathbf{X}_r.$$

(c) Convince yourself that any vector $\mathbf{V}$ appears from the rotating system to have an additional backward rotation, as given by

$$\left(\frac{d\mathbf{V}}{dt}\right)_r = \left(\frac{d\mathbf{V}}{dt}\right)_f - \boldsymbol{\omega} \times \mathbf{V}.$$

*78. Apply the previous result to $\mathbf{v}_f$ to show that $\mathbf{a}_f \equiv (d\mathbf{v}_f/dt)_f = \mathbf{a}_r + 2(\boldsymbol{\omega} \times \mathbf{v}_r) + \boldsymbol{\omega} \times (\boldsymbol{\omega} \times \mathbf{X}_r)$. The significance of this result is as follows: In the fixed system Newton's second law is simply $\mathbf{F}_f = m\mathbf{a}_f$, where $\mathbf{F}_f$ is the applied force acting on a particle. But in the rotating system $m\mathbf{a}_r = \mathbf{F}_r = \mathbf{F}_f - 2m\boldsymbol{\omega} \times \mathbf{v}_r - m\boldsymbol{\omega} \times (\boldsymbol{\omega} \times \mathbf{X}_r)$, so that the effective force $\mathbf{F}_r$ operative in the rotating system is not the same as the "true" force $\mathbf{F}_f$. There are two extra, fictitious forces, the CORIOLIS FORCE and the CENTRIFUGAL FORCE, respectively, which arise as artifacts of the use of a rotating system. To clarify these forces, consider:

(a) What is the direction of the centrifugal force acting on a stationary object on the surface of the earth? (*Hint:* To determine the direction of $\boldsymbol{\omega}$, remember that the sun rises in the east.)

(b) What is the direction of the Coriolis force on a rocket that has been shot due north from Cape Kennedy?

(c) Straight up from Cape Kennedy?

These effects are also relevant to considerations of the vibrations of rotating molecules.

79. Show that the kinetic energy of a rotating particle is related to its angular momentum by $T = \frac{1}{2}\boldsymbol{\omega} \cdot \mathbf{L}$.

*80. Newton's third law is that if particle 1 exerts a force **F** on particle 2, then particle 2 exerts a force $-\mathbf{F}$ on particle 1.

(a) Use this law to convince yourself that if there are no external forces on the two particles, then the total momentum of the two particles is conserved no matter what the nature of the mutual force is.

(b) Show that if the mutual force is along the vector joining the two particles and if there are no external torques, then the total angular momentum is conserved.

81. Find the x-component of $\mathbf{L} = \sum m_i[(\mathbf{r}_i \cdot \mathbf{r}_i)\boldsymbol{\omega} - (\mathbf{r}_i \cdot \boldsymbol{\omega})\mathbf{r}_i]$ to see how each component of **L** is a linear combination of ω_x, ω_y, and ω_z.

*82. Apply Exercise 7.77c to **L**, and thus show that if the angular momentum and the angular velocity of a rotating body are not parallel, then even though there are no external torques there is an apparent torque as viewed from the rotating body. Try to imagine the motion of a spheroid rotating about some axis other than one of its principal axes; remember that in the fixed system the angular momentum is conserved, so that the motion is not merely rotation but also includes a precession of $\boldsymbol{\omega}$ about **L**. Viewed from the rotating system, the apparent torque causes **L** to precess about $\boldsymbol{\omega}$.

Next we demonstrate the economy and simplicity of vector notation by considering the precession of a magnetic moment in a magnetic field. Many nuclei have a magnetic moment $\boldsymbol{\mu}$ which may be considered classically to result from the rotation of a charged particle with an angular momentum **L**. These two vectors are in the same (or opposite) direction: $\boldsymbol{\mu} = \gamma\mathbf{L}$, where the scalar γ is known as the gyromagnetic ratio.

If there is a homogeneous magnetic field $\mathbf{H}_0$ at the nucleus, the nucleus will experience no *net* force **F**, but there will be a torque **T** given by $\mathbf{T} = \boldsymbol{\mu} \times \mathbf{H}_0$. This torque is in a direction so as to try to align $\boldsymbol{\mu}$ with $\mathbf{H}_0$. (In Fig. 7.21 **T** points out of the paper.) This situation is analogous to that of a

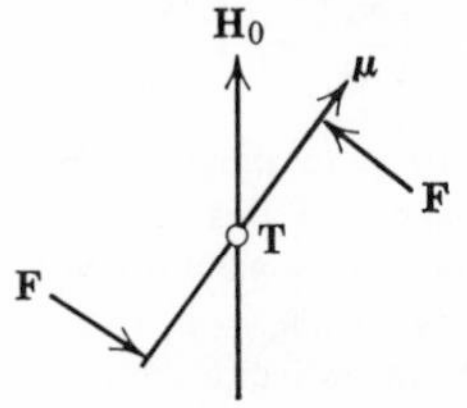

Figure 7.21

gyroscope balancing on a point and experiencing a torque due to the force of gravity acting through its center of mass.

But we have shown that $\mathbf{T} = d\mathbf{L}/dt$ or

$$\mathbf{T} = \frac{1}{\gamma}\frac{d\boldsymbol{\mu}}{dt} = \boldsymbol{\mu} \times \mathbf{H}_0$$

so that $d\boldsymbol{\mu}/dt$ is always perpendicular to both $\boldsymbol{\mu}$ and $\mathbf{H}_0$. (See Fig. 7.22, which

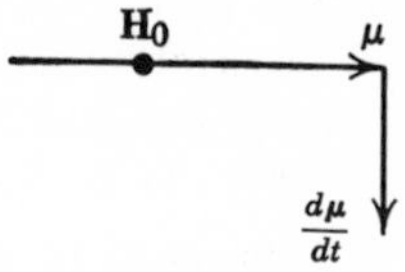

Figure 7.22

is a top view of Fig. 7.21; $\mathbf{H}_0$ points out of the plane of the paper and $\boldsymbol{\mu}$ slopes outward.) If $\mathbf{H}_0$ does not vary with time, then

$$\frac{d(\mathbf{H}_0 \cdot \boldsymbol{\mu})}{dt} = \mathbf{H}_0 \cdot \frac{d\boldsymbol{\mu}}{dt} = \gamma[H_0 \mu H_0] = 0$$

since the triple product vanishes when two factors are equal. This means that although the torque tries to align $\boldsymbol{\mu}$ with $\mathbf{H}_0$, the angle between $\boldsymbol{\mu}$ and $\mathbf{H}_0$ remains constant. The resultant motion is a PRECESSION of $\boldsymbol{\mu}$ about $\mathbf{H}_0$, that is, a rotation about $\mathbf{H}_0$ that maintains a constant angle between $\mathbf{H}_0$ and $\boldsymbol{\mu}$. Indeed, it can be shown (Exercise 7.84) that the tip of $\boldsymbol{\mu}$ describes a circle traversed with angular frequency

$$\omega = \gamma|\mathbf{H}_0|,$$

where ω is known as the LARMOR PRECESSION FREQUENCY.

Consider next what happens if a small magnetic field $\mathbf{H}_1$ is applied perpendicular to both $\mathbf{H}_0$ and $\boldsymbol{\mu}$. The additional torque due to $\mathbf{H}_1$ is given by $\mathbf{T}_1 = \boldsymbol{\mu} \times \mathbf{H}_1$. (See Fig. 7.23; $\mathbf{H}_1$ points into the paper.) Now this torque does

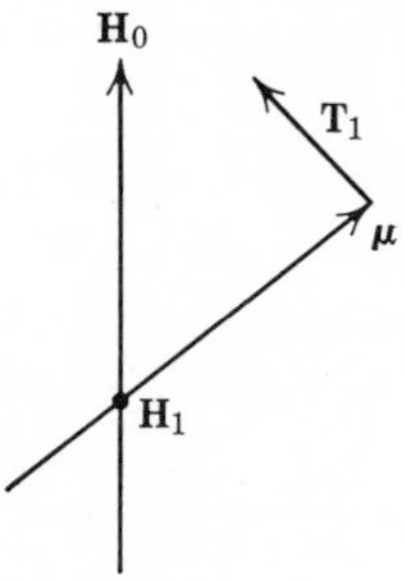

Figure 7.23

produce an additional $d\boldsymbol{\mu}/dt$ such as to move $\boldsymbol{\mu}$ toward $\mathbf{H}_0$. But if $\mathbf{H}_1$ is fixed, when the precession carries $\boldsymbol{\mu}$ around to the other side the additional torque will be such as to move $\boldsymbol{\mu}$ back away from $\mathbf{H}_0$. The net result would merely be a precession of $\boldsymbol{\mu}$ about $\mathbf{H}_0 + \mathbf{H}_1$.

But *if* it can be arranged that as $\boldsymbol{\mu}$ precesses, $\mathbf{H}_1$ also rotates with the frequency ω, so as to keep $\mathbf{H}_1$ always perpendicular to both $\boldsymbol{\mu}$ and $\mathbf{H}_0$, then the torque is always being applied in such a fashion that it really does align $\boldsymbol{\mu}$ with $\mathbf{H}_0$. Or if $\mathbf{H}_1$ is always in the opposite direction but still perpendicular to $\boldsymbol{\mu}$ and $\mathbf{H}_0$, the result of that $\mathbf{H}_1$ will be to align $\boldsymbol{\mu}$ with $-\mathbf{H}_0$.

This is a simplified classical description of nuclear magnetic resonance (and electron spin resonance). The quantum-mechanical description includes the fact that only certain values of $\boldsymbol{\mu} \cdot \mathbf{H}$ are permitted, but the rotating $\mathbf{H}_1$ will still cause transitions from one level to the next. Furthermore, since the energy of the system is given by $E = -\boldsymbol{\mu} \cdot \mathbf{H}$, these transitions correspond to absorption or emission of energy.

Exercises

*83. Show that magnetic torques cannot change the magnitude of the nuclear magnetic moment.

*84. "Solve" the differential equation $d\boldsymbol{\mu}/dt = \gamma\boldsymbol{\mu} \times \mathbf{H}$ by transforming to a rotating coordinate system in which $d\boldsymbol{\mu}/dt = \mathbf{0}$. What is the frequency at which the coordinate system must rotate? What is the frequency at which $\boldsymbol{\mu}$ precesses, as viewed from the fixed coordinate system?

7.8 Vector Differential Operators

Vector differential operators are operators involving differentiation which convert scalar point functions or vector fields into scalar point functions or vector fields. We shall consider four such operators, all of which can be written in terms of the same symbolic operator DEL, which is defined by

$$\nabla \equiv \hat{\mathbf{i}}\frac{\partial}{\partial x} + \hat{\mathbf{j}}\frac{\partial}{\partial y} + \hat{\mathbf{k}}\frac{\partial}{\partial z}. \tag{7.21}$$

The four operators differ from each other by the type of function they act on or the type of function they produce. The first operator, GRADIENT, symbolized by ∇ in juxtaposition, acts on a scalar point function ϕ to produce a vector field

$$\nabla\phi \equiv \frac{\partial\phi}{\partial x}\hat{\mathbf{i}} + \frac{\partial\phi}{\partial y}\hat{\mathbf{j}} + \frac{\partial\phi}{\partial z}\hat{\mathbf{k}}. \tag{7.22}$$

The second operator, DIVERGENCE, symbolized by $\nabla\cdot$, acts on a vector field $\mathbf{E}(\mathbf{X}) = E_x\hat{\mathbf{i}} + E_y\hat{\mathbf{j}} + E_z\hat{\mathbf{k}}$ to produce a scalar point function

$$\nabla \cdot \mathbf{E} \equiv \frac{\partial E_x}{\partial x} + \frac{\partial E_y}{\partial y} + \frac{\partial E_z}{\partial z}. \tag{7.23}$$

The third operator, LAPLACIAN, symbolized by ∇^2, acts on a scalar point function ϕ to produce a scalar point function

$$\nabla^2\phi \equiv \frac{\partial^2\phi}{\partial x^2} + \frac{\partial^2\phi}{\partial y^2} + \frac{\partial^2\phi}{\partial z^2}. \tag{7.24}$$

The fourth operator, CURL, symbolized by $\boldsymbol{\nabla}\times$, acts on a vector field $\mathbf{E}$ to produce a vector field

$$\boldsymbol{\nabla}\times\mathbf{E} \equiv \left(\frac{\partial E_z}{\partial y} - \frac{\partial E_y}{\partial z}\right)\hat{\mathbf{i}} + \left(\frac{\partial E_x}{\partial z} - \frac{\partial E_z}{\partial x}\right)\hat{\mathbf{j}} + \left(\frac{\partial E_y}{\partial x} - \frac{\partial E_x}{\partial y}\right)\hat{\mathbf{k}}. \tag{7.25}$$

These operators are defined in Cartesian coordinates, but they are intended to be invariant to a change of coordinates. For example, if $\phi(x, y, z) = f(r, \theta, \phi)$, then $\boldsymbol{\nabla}\phi$ as defined above is some vector field, and $\boldsymbol{\nabla}f$ must be the same vector field, but expressed in polar coordinates.

Exercises

85. Convince yourself that $\nabla^2 = \boldsymbol{\nabla}\cdot\boldsymbol{\nabla}$, that is, that Laplacian of = divergence of gradient of.
86. Convince yourself that all four vector differential operators are linear; you will need to generalize the definition of linearity to cover divergence and curl.
87. Find gradient, divergence, Laplacian, and curl (as many as applicable) for
    (a) xyz
    (b) $bz\hat{\mathbf{i}} + cx\hat{\mathbf{j}} + ay\hat{\mathbf{k}}$
    (c) $(x^2 + y^2)\hat{\mathbf{k}}$
    (d) $-y(x^2 + y^2)\hat{\mathbf{i}} + x(x^2 + y^2)\hat{\mathbf{j}} + z^3\hat{\mathbf{k}}$
    (e) $r \equiv (x^2 + y^2 + z^2)^{1/2}$
    (f) $\hat{\mathbf{r}} \equiv \mathbf{R}/|R|$
88. Show that $\mathbf{F}$ in Exercise 7.75 is given by $\mathbf{F} = -\boldsymbol{\nabla}V$. This is a general result applicable whenever a potential energy can be defined.

7.9 Gradient. The Oracle of Del Phi

The gradient of the scalar point function $\phi(x, y, z)$ is the vector field

$$\boldsymbol{\nabla}\phi = \frac{\partial\phi}{\partial x}\hat{\mathbf{i}} + \frac{\partial\phi}{\partial y}\hat{\mathbf{j}} + \frac{\partial\phi}{\partial z}\hat{\mathbf{k}}.$$

To understand the significance of this quantity, let us consider the total differential of ϕ:

$$d\phi = \frac{\partial\phi}{\partial x}\,dx + \frac{\partial\phi}{\partial y}\,dy + \frac{\partial\phi}{\partial z}\,dz$$

which represents the change in the value of ϕ on going from the point (x, y, z) to the neighboring point $(x + dx, y + dy, z + dz)$. In vector notation this becomes

$$d\phi = \nabla\phi \cdot \mathbf{dX},$$

where $\mathbf{dX} = dx\hat{\mathbf{i}} + dy\hat{\mathbf{j}} + dz\hat{\mathbf{k}}$. Therefore we may conclude that the change in ϕ on moving from the point $\mathbf{X}$ by a distance $|dX|$ in the direction $\mathbf{dX}$ is the dot product of $\mathbf{dX}$ with $\nabla\phi$. Just as $(\partial\phi/\partial x)\,dx$ represents the change in ϕ on changing x by the amount dx while holding y and z fixed, $\nabla\phi \cdot \mathbf{dX}$ represents the change in ϕ on changing $\mathbf{X}$ by the amount $\mathbf{dX}$. The partial derivative $\partial\phi/\partial x$ is the ratio of the change in ϕ to the change in x on changing x by the amount dx. The gradient $\nabla\phi$ is the ratio of the change in ϕ to the change in $\mathbf{X}$ on changing $\mathbf{X}$ by the amount $\mathbf{dX}$. Symbolically we *might* write

$$\nabla\phi = \frac{d\phi}{\mathbf{dX}}$$

except that division by a vector is undefined. However, if we define the unit vector $\hat{\mathbf{u}} = \mathbf{dX}/|dX|$, then

$$\nabla\phi \cdot \hat{\mathbf{u}} = \frac{\nabla\phi \cdot \mathbf{dX}}{|dX|} \equiv \frac{\partial\phi}{\partial u} \tag{7.26}$$

where $\partial\phi/\partial u$ is the symbolic notation for the DIRECTIONAL DERIVATIVE of ϕ in the direction of the unit vector $\hat{\mathbf{u}}$. It represents the ratio of the change in ϕ to the distance traveled in the direction $\hat{\mathbf{u}}$. The geometrical interpretation of the directional derivative is simple and readily demonstrated in two dimensions. Just as the partial derivative $\partial\phi/\partial y$ is the slope of the curve of intersection of $z = \phi(x, y)$ and a vertical plane parallel to the y-axis, the directional derivative is the slope of the curve of intersection of $z = \phi(x, y)$ and a vertical plane parallel to $\hat{\mathbf{u}}$, as illustrated in Figure 7.24.

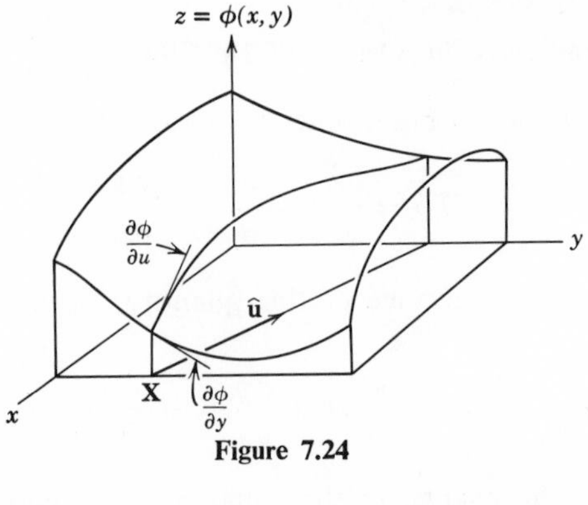

Figure 7.24

Now we are in a position to provide a geometrical interpretation of $\nabla\phi$. First let us consider the surface $\phi(x, y, z) = \phi_0$ and the tangent plane at a point $\mathbf{X}$ on the surface. (See Figure 7.25.) In the case where $\mathbf{dX}$ is a vector in

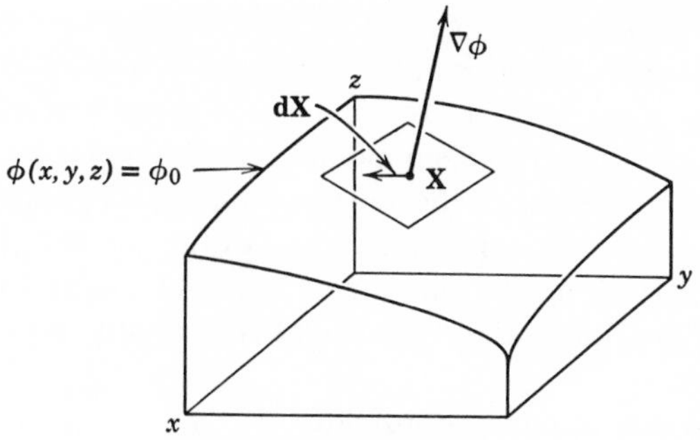

Figure 7.25

this plane, $d\phi$ must equal 0 because the value of ϕ does not change as long as we move only on the surface (and because for infinitesimally small displacements the surface and the tangent plane coincide). But $0 = d\phi = \nabla\phi \cdot \mathbf{dX}$ requires that at $\mathbf{X}$, $\nabla\phi$ must be orthogonal to every $\mathbf{dX}$ in the tangent plane. Therefore we conclude that $\nabla\phi$ is perpendicular to the tangent plane and to the surface $\phi(\mathbf{X}) = \phi_0$ at $\mathbf{X}$. Furthermore, if $\hat{\mathbf{n}}$ is a unit vector perpendicular to the surface and therefore parallel to $\nabla\phi$, then $\partial\phi/\partial n = \nabla\phi \cdot \hat{\mathbf{n}} = |\nabla\phi|$ is the maximum possible value of the directional derivative. Therefore in order to obtain the maximum possible increase in the value of ϕ, we should move in the direction $\nabla\phi$. The ratio of this increase in ϕ to the distance traveled in the direction $\nabla\phi$ is equal to $|\nabla\phi|$. An alternative viewpoint is to look from the point $\mathbf{X}$ on the surface $\phi(x, y, z) = \phi_0$ to the "adjacent" surface $\phi(x, y, z) = \phi_0 + \Delta\phi$. The shortest straight line from $\mathbf{X}$ to the adjacent surface is in the direction $\nabla\phi$, and the distance to the adjacent surface is $|\Delta X| = \Delta\phi/|\nabla\phi|$. Therefore we conclude that the surfaces of constant ϕ are everywhere perpendicular to $\nabla\phi$ and spaced inversely to $|\nabla\phi|$. These results are clearer in two dimensions, where the gradient $\nabla\phi = (\partial\phi/\partial x)\hat{\mathbf{i}} + (\partial\phi/\partial y)\hat{\mathbf{j}}$ is perpendicular to the curve $\phi(x, y) = \phi_0$. Figure 7.26 shows a contour diagram of a valley: The indentations, where the contour

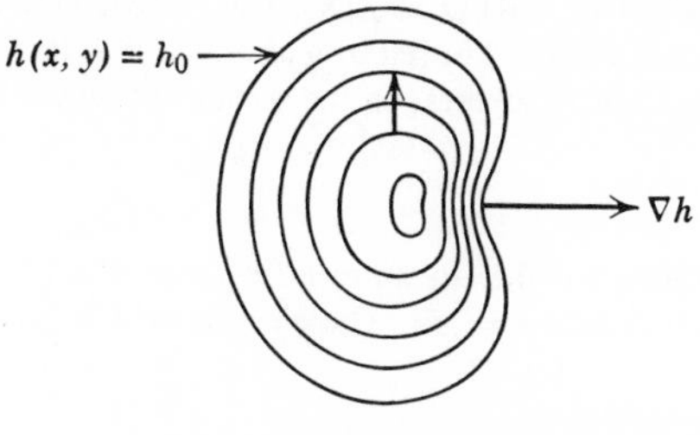

Figure 7.26

lines are closer together, correspond to a ridge running up one side of the valley. In any particular direction, the directional derivative is the slope of the ground in that direction. At every point the maximum slope is in the uphill direction, where the gradient points, perpendicular to the contour lines. If we wish to climb out of the valley as quickly as possible, of course we should travel uphill, that is, in the direction of the gradient. Furthermore, the ascent is steepest along the ridge, and we would ascend most quickly along the crest of the ridge, where the height is changing most rapidly with horizontal distance.

It can be shown that in polar coordinates

$$\nabla f = \frac{\partial f}{\partial r}\,\hat{\mathbf{e}}_r + \frac{1}{r}\frac{\partial f}{\partial \phi}\,\hat{\mathbf{e}}_\phi + \frac{\partial f}{\partial z}\,\hat{\mathbf{e}}_z$$

or

$$\nabla f = \frac{\partial f}{\partial r}\,\hat{\mathbf{e}}_r + \frac{1}{r}\frac{\partial f}{\partial \theta}\,\hat{\mathbf{e}}_\theta + \frac{1}{r\sin\theta}\frac{\partial f}{\partial \phi}\,\hat{\mathbf{e}}_\phi.$$

A vector field **E** that is equal to the gradient $\nabla\phi$ of a function is said to be a CONSERVATIVE FIELD. The function ϕ is said to be a SCALAR POTENTIAL for the vector field. The surfaces of constant ϕ are known as EQUIPOTENTIALS. When a conservative field is graphed by drawing streamlines, representative equipotentials are also often indicated. Since the surfaces of constant ϕ are perpendicular to the gradient $\nabla\phi$, we may conclude that any streamline must intersect any equipotential perpendicularly.

Examples

27. The potential energy of a particle of mass m at height z is $V(x, y, z) = mgz$. The gravitational force on the particle is $\mathbf{F} = -\nabla V = -mg\hat{\mathbf{k}}$. The gravitational field is the force per unit mass: $\mathbf{G} = \mathbf{F}/m = -g\hat{\mathbf{k}}$. The streamlines are all vertical and the equipotentials are horizontal planes.

28. The potential energy of a point charge Q (in appropriate units, statcoulombs) at the point **R** in the presence of a point charge q at the origin is $V(\mathbf{R}) = qQ/|R|$. The force is $\mathbf{F} = -\nabla V$. The electrostatic potential is the potential energy due to a charge q at the origin and a unit charge at **R**: $\phi(\mathbf{R}) = V(\mathbf{R})/Q = q/|R|$. The electric field at **R** due to the charge q at the origin is the force which would be felt by a unit charge at **R**: $\mathbf{E} = \mathbf{F}/Q = -\nabla(q/|R|)$. The equipotentials are concentric spheres, so the streamlines must be radial.

Exercises

89. The electrostatic potential due to an electric point charge of magnitude q at the origin is $\phi(\mathbf{R}) = q/r$, where $r = |R|$. The electric field is given by $\mathbf{E} = -\nabla\phi$. Find **E**:
   (a) In Cartesian coordinates.
   (b) In spherical polar coordinates.
   (c) Convince yourself that these are indeed the same vector field.

90. (a) Convince yourself that $\nabla(\phi\psi) = \phi\nabla\psi + \psi\nabla\phi$.
(b) Show that $\nabla(r^n) = nr^{n-2}\mathbf{r}$.
(c) Show that $\nabla(\mathbf{A}\cdot\mathbf{r}) = \mathbf{A}$, where $\mathbf{A}$ is a constant vector.

91. The electrostatic potential due to a point dipole of dipole moment $\boldsymbol{\mu}$ at the origin is $\phi = \boldsymbol{\mu}\cdot\mathbf{r}/r^3$. Use the above results to find $\mathbf{E} = -\nabla\phi$.

*92. Find the eigenvalues and corresponding eigenfunctions, $\psi(\mathbf{X})$, of the operator $i\nabla$, which in quantum mechanics corresponds to momentum. (*Note:* The eigenvalues are vectors, and *any* vector is an allowable eigenvalue. Try for a solution in which the variables x, y, and z are separable.)

93. If $\phi(x, y, z) = (x-1)^2 - 3y^2 + 2(z-2)^2$ and $\hat{\mathbf{u}} = \frac{2}{3}\hat{\mathbf{i}} - \frac{2}{3}\hat{\mathbf{j}} + \frac{1}{3}\hat{\mathbf{k}}$, what is $\partial\phi/\partial u$ at $(5, 1, 1)$?

94. What is the equation of the plane tangent to the surface $\phi(\mathbf{X}) = \phi_0$ at the point $\mathbf{X}_0$?

95. In what direction is the normal to the surface $z^2 = xy + 3x^2 + 2y^2 - 4$ at the point $\hat{\mathbf{i}} + 2\hat{\mathbf{j}} + 3\hat{\mathbf{k}}$?

96. In what direction is the normal to the plane tangent to the surface $z = F(x, y)$ at the point $[x_0, y_0, F(x_0, y_0)]$?

*97. Rationalize Lagrange's method of undetermined multipliers, for finding extrema of $F(x, y, z)$ subject to the constraint $G(x, y, z) = 0$:
(a) In what direction is the normal to the surface $G(x, y, z) = 0$ at the point (x, y, z)?
(b) If ∇F at (x, y, z) has a nonzero component perpendicular to this normal ("lying in the surface"), convince yourself that along the surface in the direction of this component the value of F must increase in one direction and decrease in the other, and that therefore F cannot have a constrained extremum at such a point.
(c) What is the necessary relation between ∇F and the normal to the surface in order for ∇F to have no such component? Compare this expression with Lagrange's formula (Eqs. 2.13 and 2.14).

*98. To rationalize the introduction of the Jacobian into the volume element on transforming from Cartesian coordinates to the generalized orthogonal coordinates (u, v, w), consider the rectangular box enclosed by the six surfaces, $u = u_0$, $u = u_0 + du$, $v = v_0$, $v = v_0 + dv$, $w = w_0$, $w = w_0 + dw$. The volume of the box, dV, is the product of the lengths of three mutually perpendicular edges. What is the distance between the "adjacent" surfaces $u = u_0$ and $u = u_0 + du$? What are the lengths of the other two edges? What is dV? If the coordinates are not necessarily orthogonal, this result becomes

$$dV = \frac{du\,dv\,dw}{[\nabla u\;\nabla v\;\nabla w]},$$

where the triple product

$$[\nabla u\;\nabla v\;\nabla w] = \frac{\partial(u, v, w)}{\partial(x, y, z)} = 1\bigg/\frac{\partial(x, y, z)}{\partial(u, v, w)}.$$

*99. How would you define the gradient of a function of n variables?

*100. What can you conclude about the direction of the gradient of a homogeneous function of order zero?

101. Convince yourself that a scalar potential is arbitrary to the extent of an additive constant function, that is, if ϕ is a scalar potential for $\mathbf{E}$, then so is $\phi + c$.

There is a further type of gradient, the field gradient, which should be mentioned here to avoid confusion. To introduce this quantity, let us consider an electric dipole $\boldsymbol{\mu}$ in an electric field $\mathbf{E}$. The potential energy is given by $V(\mathbf{X}) = -\boldsymbol{\mu} \cdot \mathbf{E}$, and the force is $\mathbf{F} = -\nabla V = \nabla(\boldsymbol{\mu} \cdot \mathbf{E}) = \nabla(\mu_x E_x + \mu_y E_y + \mu_z E_z) = \mu_x \nabla E_x + \mu_y \nabla E_y + \mu_z \nabla E_z$, since $\boldsymbol{\mu}$ is independent of position so ∇ acts only on $\mathbf{E}$. Therefore we may interchange $\boldsymbol{\mu}$ and ∇ and write symbolically

$$\nabla(\boldsymbol{\mu} \cdot \mathbf{E}) = \boldsymbol{\mu} \cdot \nabla\mathbf{E},$$

where $\nabla\mathbf{E}$ is something new, the field gradient. Likewise, the dot is something new, different from the familiar dot product. Whereas the gradient of a scalar point function is a vector field, the gradient of a vector field is a tensor, quite a different beast. Just as the gradient $\nabla\phi$ represents the change in ϕ on going from $\mathbf{X}$ to $\mathbf{X} + \mathbf{dX}$, the FIELD GRADIENT $\nabla\mathbf{E}$ represents the change in $\mathbf{E}$ on going from $\mathbf{X}$ to $\mathbf{X} + \mathbf{dX}$. Symbolically, we might define

$$\nabla\mathbf{E} \equiv \frac{\mathbf{dE}}{\mathbf{dX}}$$

if we knew how to divide two vectors. We shall return to this topic when we consider matrices, which can be used to represent tensors or ratios of vectors.

Exercises

*102. As an example of a formula involving the field gradient, expand in Cartesian components to convince yourself that

$$\nabla(\mathbf{E} \cdot \mathbf{F}) = \mathbf{E} \cdot \nabla\mathbf{F} + \mathbf{F} \cdot \nabla\mathbf{E} + \mathbf{E} \times (\nabla \times \mathbf{F}) + \mathbf{F} \times (\nabla \times \mathbf{E})$$

where

$$\mathbf{E} \cdot \nabla\mathbf{F} = (\mathbf{E} \cdot \nabla)\mathbf{F} = \left(E_x \frac{\partial}{\partial x} + E_y \frac{\partial}{\partial y} + E_z \frac{\partial}{\partial z}\right) \mathbf{F}.$$

*103. Express the Maclaurin series expansion of a function of three variables in terms of the operator $\mathbf{r} \cdot \nabla$.

7.10 Divergence

The divergence of the vector field $\mathbf{E}(x, y, z)$ is the scalar point function

$$\nabla \cdot \mathbf{E} = \left(\hat{\mathbf{i}} \frac{\partial}{\partial x} + \hat{\mathbf{j}} \frac{\partial}{\partial y} + \hat{\mathbf{k}} \frac{\partial}{\partial z}\right) \cdot (E_x\hat{\mathbf{i}} + E_y\hat{\mathbf{j}} + E_z\hat{\mathbf{k}}) = \frac{\partial E_x}{\partial x} + \frac{\partial E_y}{\partial y} + \frac{\partial E_z}{\partial z}.$$

To understand the significance of this quantity, let us consider the total flux of $\mathbf{E}$ out of the surface of an infinitesimal box of dimensions $dx \times dy \times dz$ at (x, y, z). See Fig. 7.27. The FLUX of a vector field over an infinitesimal surface of area dA is defined as $\mathbf{E} \cdot \hat{\mathbf{n}}\, dA$, where $\hat{\mathbf{n}}$ is a unit vector perpendicular to the surface and pointing outward. On the rear face $\hat{\mathbf{n}} = -\hat{\mathbf{i}}$ and $dA = dy\, dz$, so the flux out of the rear face is $-E_x(x, y + \eta\, dy, z + \zeta\, dz)\, dy\, dz$, where $0 \leqslant \eta, \zeta \leqslant 1$. On the front face $\hat{\mathbf{n}} = +\hat{\mathbf{i}}$ and $dA = dy\, dz$,

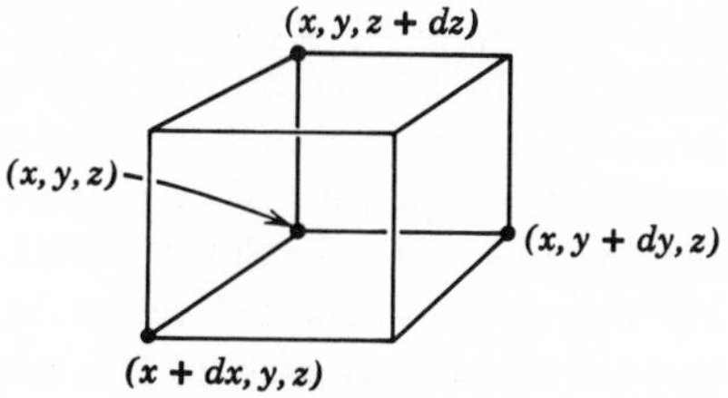

Figure 7.27

so the flux out of the front face is $E_x(x + dx, y + \eta' \, dy, z + \zeta' \, dz) \, dy \, dz$, where $0 \leqslant \eta', \zeta' \leqslant 1$. The total flux out the front and back faces is the sum of these. If we both multiply and divide this flux by dx, we obtain

$$\frac{E_x(x + dx, y + \eta' \, dy, z + \zeta' \, dz) - E_x(x, y + \eta \, dy, z + \zeta \, dz)}{dx} \, dx \, dy \, dz.$$

But the limit of the fraction as dx, dy, and dz approach zero is $\partial E_x/\partial x$. The flux out of the box through the other four faces may be calculated similarly. The total flux out of the box of volume dV is thus

$$\left(\frac{\partial E_x}{\partial x} + \frac{\partial E_y}{\partial y} + \frac{\partial E_z}{\partial z}\right) dx \, dy \, dz = \nabla \cdot \mathbf{E} \, dV.$$

Therefore we conclude that the divergence of **E** is equal to the total flux of **E** per unit volume.

Unfortunately this description only replaces one unfamiliar term, divergence, by another, flux. To gain an understanding of the significance of the flux, we consider the example of the velocity field **v**(**X**) of fluid flow. Let us find the rate at which fluid passes through an infinitesimal surface of area dA

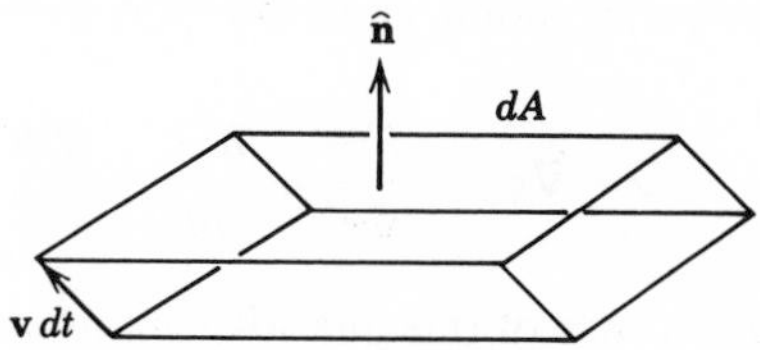

Figure 7.28

(See Figure 7.28). The only fluid that can pass through the surface during a time interval dt is that in a box whose face is the surface, whose edge is $\mathbf{v} \, dt$, and whose volume is thus $\mathbf{v} \cdot \hat{\mathbf{n}} \, dA \, dt$. Therefore the rate at which fluid passes through the surface is $\mathbf{v} \cdot \hat{\mathbf{n}} \, dA$ ml/sec, which is just the flux of the vector field **v** across the surface. The total flux of **v** out from a closed surface enclosing a volume dV is the net rate at which fluid flows out of the volume. But by the previous analysis this quantity equals $\nabla \cdot \mathbf{v}\, dV$. For a fluid whose density is constant there can be no net gain or loss of material from any volume element. Therefore we conclude that the velocity field of the flow of such a fluid must satisfy the equation $\nabla \cdot \mathbf{v} = 0$. A vector field whose divergence is everywhere zero is said to be SOLENOIDAL. In Exercise 7.105 you will discover that only for a solenoidal vector field is it easy to draw streamlines so that they are spaced inversely to $|E|$. But in any region where the divergence is nonzero there must be streamlines which disappear or begin out of nowhere.

It can be shown that in polar coordinates

$$\nabla \cdot \mathbf{E} = \frac{1}{r}\frac{\partial}{\partial r}(rE_r) + \frac{1}{r}\frac{\partial}{\partial \phi}E_\phi + \frac{\partial}{\partial z}E_z$$

or

$$\nabla \cdot \mathbf{E} = \frac{1}{r^2}\frac{\partial}{\partial r}(r^2E_r) + \frac{1}{r\sin\theta}\frac{\partial}{\partial \theta}(\sin\theta\, E_\theta) + \frac{1}{r\sin\theta}\frac{\partial}{\partial \phi}E_\phi.$$

Exercise

104. Expand in Cartesian coordinates to convince yourself that $\nabla \cdot (\phi\mathbf{E}) = \phi\nabla \cdot \mathbf{E} + \mathbf{E} \cdot \nabla\phi$.
105. Find the divergence of each vector field in Exercise 7.58. Notice that wherever $\nabla \cdot \mathbf{E} > 0$, streamlines must begin out of nowhere.
*106. Show that the divergence of the electric field due to a point charge at the origin is zero everywhere except at the origin.
*107. Consider the net rate of flow of fluid, in grams per second rather than milliliters per second, from a volume dV, to convince yourself that $\partial\rho/\partial t = -\nabla \cdot (\rho\mathbf{v})$, where ρ is the density of the fluid at $\mathbf{X}$. This equation is the fundamental partial differential equation governing hydrodynamic flow.

7.11 Laplacian

The Laplacian of the scalar point function $\phi(x, y, z)$ is the scalar point function

$$\nabla^2\phi = \nabla \cdot (\nabla\phi) = \frac{\partial^2\phi}{\partial x^2} + \frac{\partial^2\phi}{\partial y^2} + \frac{\partial^2\phi}{\partial z^2}.$$

To understand the significance of this quantity, let us consider how $\phi(x, y, z)$ differs from $\bar{\phi}$, the average of the values of ϕ at the six surrounding points on the axes a distance δ away:

$$6\bar{\phi}/\delta^2 = [\phi(x+\delta, y, z) + \phi(x-\delta, y, z) + \phi(x, y+\delta, z) + \phi(x, y-\delta, z) + \phi(x, y, z+\delta) + \phi(x, y, z-\delta)]/\delta^2.$$

Subtracting $6\phi(x, y, z)/\delta^2$, writing in appropriate fashion, and taking the limit as δ approaches zero gives

$$6(\bar{\phi} - \phi)/\delta^2 = \frac{1}{\delta}\left[\frac{\phi(x + \delta, y, z) - \phi(x, y, z)}{\delta}\right.$$

$$\left. - \frac{\phi(x, y, z) - \phi(x - \delta, y, z)}{\delta}\right] + \cdots$$

$$= \frac{1}{\delta}\left[\frac{\partial \phi}{\partial x}(x, y, z) - \frac{\partial \phi}{\partial x}(x - \delta, y, z)\right] + \cdots$$

$$= \frac{\partial^2 \phi}{\partial x^2} + \frac{\partial^2 \phi}{\partial y^2} + \frac{\partial^2 \phi}{\partial z^2} = \nabla^2 \phi.$$

Therefore we conclude that $\nabla^2\phi$ is a measure of how much the value of ϕ at (x, y, z) differs from the average of the values of ϕ at surrounding points.

A geometrical interpretation is possible in two dimensions, where we can draw the surface $z = \phi(x, y)$ and the tangent plane. For any point where $\nabla^2\phi = (\partial^2\phi/\partial x^2) + (\partial^2\phi/\partial y^2) = 0$, it is possible to find four surrounding points on the surface which also lie in the tangent plane. In two dimensions the Laplacian is then a measure of how much the surface deviates from the tangent plane, or how much "curvature" the surface has. In three dimensions there is no such surface or tangent plane, but we may still consider the Laplacian as a measure of the "curvature" of ϕ in the sense given in the previous paragraph (Cf. Exercise 5.5).

To take a particular example, let us consider the diffusion problem in three dimensions: How does $c(\mathbf{X}, t)$, the concentration of a substance, vary with time? Clearly, an even distribution of the substance throughout space is a stable situation, so the concentration will not change. What is required for mass transfer is an uneven distribution, so that material will pass from a region of higher concentration to one of lower concentration. The unevenness of the concentration is measured by the gradient ∇c. However, although a nonzero concentration gradient is sufficient for mass transfer, it does not necessarily lead to a change in the concentration of material at $\mathbf{X}$. It is entirely possible that the mass transfer is such that there is just as much material entering one side of a volume enclosing $\mathbf{X}$ as there is leaving the other side. Such would be the case if the concentration increased uniformly in one direction, so that material is always flowing from that direction, without ever changing $c(\mathbf{X}, t)$. What is required for $\partial c/\partial t$ to be nonzero is a nonuniform concentration gradient, as measured by $\nabla \cdot (\nabla c) = \nabla^2 c$. The situation in which $c(\mathbf{X}, t)$ will certainly change with time is if the concentration is a maximum or minimum at $\mathbf{X}$. Therefore we are again led to conclude that the Laplacian is a measure of the "curvature" of c, or the extent to which the concentration at a point differs from the average concentration in the neighborhood of the point.

In polar coordinates

$$\nabla^2 f = \frac{1}{r}\frac{\partial}{\partial r}\left(r\frac{\partial f}{\partial r}\right) + \frac{1}{r^2}\frac{\partial^2 f}{\partial \phi^2} + \frac{\partial^2 f}{\partial z^2}$$

or

$$\nabla^2 f = \frac{1}{r^2}\frac{\partial}{\partial r}\left(r^2\frac{\partial f}{\partial r}\right) + \frac{1}{r^2 \sin\theta}\frac{\partial}{\partial \theta}\left(\sin\theta\frac{\partial f}{\partial \theta}\right) + \frac{1}{r^2 \sin^2\theta}\frac{\partial^2 f}{\partial \phi^2}.$$

These results may of course be derived by repeated use of the chain rule for partial derivatives (Cf. Exercise 2.17). However, the method of "metric coefficients" is much less tedious and is also easily applicable to transformation of the other differential operators. Since Cartesian and polar coordinates are by far the most common in chemistry problems, we shall not pursue this topic.

Exercise

108. Use the results of Exercises 7.85, 7.90a, and 7.104 to simplify $\nabla^2(\phi\psi)$.

7.12 Curl

The curl of the vector field **E** is the vector field

$$\nabla \times \mathbf{E} = \left(\frac{\partial E_z}{\partial y} - \frac{\partial E_y}{\partial z}\right)\hat{\mathbf{i}} + \left(\frac{\partial E_x}{\partial z} - \frac{\partial E_z}{\partial x}\right)\hat{\mathbf{j}} + \left(\frac{\partial E_y}{\partial x} - \frac{\partial E_x}{\partial y}\right)\hat{\mathbf{k}}$$

$$= \begin{vmatrix} \hat{\mathbf{i}} & \hat{\mathbf{j}} & \hat{\mathbf{k}} \\ \partial/\partial x & \partial/\partial y & \partial/\partial z \\ E_x & E_y & E_z \end{vmatrix}.$$

To understand the significance of this quantity, let us consider the particular example of the vector field

$$\mathbf{v} = v_0\hat{\mathbf{k}}.$$

If this is taken to be a velocity field of fluid flow, it represents a stream of fluid flowing in the positive z-direction with uniform speed v_0. If we were to travel through the fluid, sometimes we would be traveling upstream, against the current, and sometimes we would be traveling downstream, with the current. However, the flow is such that along *any* closed path the "amount" (we shall be more precise in Example 7.31) of travel with the current exactly cancels the "amount" of travel against the current.

In contrast, the vector field

$$\mathbf{v} = -y\hat{\mathbf{i}} + x\hat{\mathbf{j}} + v_0\hat{\mathbf{k}}$$

represents a more complicated flow pattern. As you discovered in Exercise 7.58e, the horizontal components of the flow correspond to a swirling motion about the z-axis. In this case it is possible to traverse a closed path with net motion with the current. For example, if we were to traverse the unit circle in the xy-plane in the right-handed sense (with angular momentum in the $+z$ direction), we would always be traveling *with* the horizontal component of the current; traversal in the opposite sense would be against that current. This is a remarkable sort of flow: it would enable a particle initially at rest to acquire kinetic energy and return to its starting point without giving up that kinetic energy; a paddle-wheel in the stream would be turned continuously.

Of course, the crucial difference between these two fields is that for the former $\nabla \times \mathbf{v} = \mathbf{0}$, but for the latter

$$\nabla \times \mathbf{v} = \nabla \times (-y\hat{\mathbf{i}} + x\hat{\mathbf{j}} + v_0\hat{\mathbf{k}}) = 2\hat{\mathbf{k}}.$$

Any vector field whose curl is everywhere zero is said to be IRROTATIONAL, and the nonvanishing of the curl is always associated with a "swirling" motion. Among the vector fields of Exercise 7.87, only $\hat{\mathbf{r}}$ is irrotational; the swirling nature of $bz\hat{\mathbf{i}} + cx\hat{\mathbf{j}} + ay\hat{\mathbf{k}}$ and $-y(x^2 + y^2)\hat{\mathbf{i}} + x(x^2 + y^2)\hat{\mathbf{j}} + z^3\hat{\mathbf{k}}$ should be apparent. Although the vector field $(x^2 + y^2)\hat{\mathbf{k}}$ (similar to that of Exercise 7.58d) does not appear to swirl, the path from $\mathbf{0}$ to $\hat{\mathbf{j}}$ to $\hat{\mathbf{j}} + \hat{\mathbf{k}}$ to $\hat{\mathbf{k}}$ to $\mathbf{0}$ is an example of a closed path which is never against the current and sometimes with it, so the curl is nonzero.

Exercises

109. Expand in Cartesian coordinates to convince yourself that:
 (a) $\nabla \cdot (\mathbf{E} \times \mathbf{F}) = \mathbf{F} \cdot \nabla \times \mathbf{E} - \mathbf{E} \cdot \nabla \times \mathbf{F}$.
 (b) $\nabla \times (\phi\mathbf{E}) = \phi\nabla \times \mathbf{E} + \nabla\phi \times \mathbf{E}$.

*110. (a) Expand in Cartesian coordinates to find $\nabla \times \mathbf{R}$.
 (b) A CENTRAL-FORCE FIELD is one of the form $f(r)\mathbf{R}$. Use a result of the previous exercise, and others, to show that the curl of a central-force field is zero.

*111. Expand in Cartesian coordinates to show that if $\mathbf{v} = \boldsymbol{\omega} \times \mathbf{r}$, then $\nabla \times \mathbf{v} = 2\boldsymbol{\omega}$.

Next we shall investigate two important cases of repeated application of del—the curl of a gradient and the divergence of a curl. First we consider the curl of $\nabla\phi$:

$$\nabla \times \nabla\phi = \left[\frac{\partial}{\partial y}(\nabla\phi)_z - \frac{\partial}{\partial z}(\nabla\phi)_y\right]\hat{\mathbf{i}} + \cdots = \left[\frac{\partial}{\partial y}\frac{\partial\phi}{\partial z} - \frac{\partial}{\partial z}\frac{\partial\phi}{\partial y}\right]\hat{\mathbf{i}} + \cdots = \mathbf{0} \tag{7.27}$$

where we have invoked the result that if ϕ is a "reasonable" function, then $\partial^2\phi/\partial y\,\partial z = \partial^2\phi/\partial z\,\partial y$, etc. Therefore we conclude that the curl of a gradient is everywhere zero. Previously we defined a conservative field as one that is the gradient of some scalar point function. What we have just shown is that a conservative field is irrotational. It can be shown that the converse is also true: An irrotational field is conservative; if the curl of **E** is everywhere zero, then **E** is the gradient of some scalar point function. Therefore conservative and irrotational are synonymous.

Next we consider the divergence of $\nabla \times \mathbf{E}$:

$$\nabla \cdot \nabla \times \mathbf{E} = \frac{\partial}{\partial x}\left(\frac{\partial E_z}{\partial y} - \frac{\partial E_y}{\partial z}\right) + \frac{\partial}{\partial y}\left(\frac{\partial E_x}{\partial z} - \frac{\partial E_z}{\partial x}\right) + \frac{\partial}{\partial z}\left(\frac{\partial E_y}{\partial x} - \frac{\partial E_x}{\partial y}\right) = 0 \tag{7.28}$$

since the first term, $\partial^2 E_z/\partial x\, \partial y$, is exactly canceled by the fourth term, $-\partial^2 E_z/\partial y\, \partial x$, etc., again by the identity of the mixed second partial derivatives. Therefore we conclude that the divergence of a curl is everywhere zero, and that any vector field **B** that is expressible as the curl of another vector field **A** must be solenoidal. It can be shown that the converse is also true: A solenoidal vector field **B** may always be expressed as the curl of some other vector field **A**, which is called the VECTOR POTENTIAL of **B**.

Exercises

112. Show that a vector potential is arbitrary to the extent of an additive gradient.
113. Use the result of Exercise 7.109a to show that a vector field which is expressible as the cross product of two gradients is solenoidal.

7.13 Line Integrals

Originally we defined the definite integral $\int_a^b f(x)\, dx$ by choosing a set of points, $\{x_i\}$, on the interval from a to b, forming the sum $\sum f(x_i)\, \Delta x_i$, where Δx_i is the spacing between the adjacent points x_i and x_{i+1}, and taking the limit as the points become infinitely dense. However, we need not be restricted to points on the x-axis, for we can define the LINE INTEGRAL of a scalar point function over any path

$$\int \phi(\mathbf{X})\, ds \equiv \lim_{|\Delta \mathbf{X}_i| \to 0} \sum \phi(\mathbf{X}_i)\, |\Delta \mathbf{X}_i| \tag{7.29}$$

where $\{\mathbf{X}_i\}$ is a set of points on the path and $|\Delta \mathbf{X}_i|$ is the distance between the adjacent points $\mathbf{X}_i$ and $\mathbf{X}_{i+1}$. In general it is necessary to specify the path of integration, which is often symbolized by Γ. Two rules for dealing with line integrals are

$$\int_{\Gamma_1} \phi(\mathbf{X})\, ds + \int_{\Gamma_2} \phi(\mathbf{X})\, ds = \int_{\Gamma_1 + \Gamma_2} \phi(\mathbf{X})\, ds \tag{7.30}$$

$$\int_{-\Gamma} \phi(\mathbf{X})\, ds = -\int_{\Gamma} \phi(\mathbf{X})\, ds, \tag{7.31}$$

where $\Gamma_1 + \Gamma_2$ is the combination of paths Γ_1 and Γ_2, and $-\Gamma$ is path Γ traversed in the opposite direction.

A geometrical interpretation is possible in two dimensions. If we graph $z = \phi(x, y)$, then $\int_\Gamma \phi(x, y)\, ds$ represents the area of the vertical surface above the curve Γ and between the surface $z = \phi(x, y)$ and the xy-plane (See Figure 7.29).

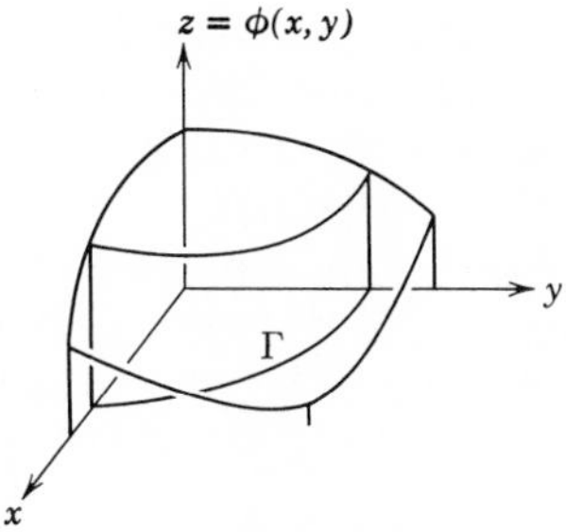

Figure 7.29

If the path is specified parametrically by $\Gamma = \mathbf{X}(t) = x(t)\hat{\mathbf{i}} + y(t)\hat{\mathbf{j}} + z(t)\hat{\mathbf{k}}$, $t_0 < t < t_1$, then the line integral may be reduced to an ordinary integral with respect to t:

$$\int_\Gamma \phi(\mathbf{X})\, ds = \int_{t_0}^{t_1} \phi[\mathbf{X}(t)] \left| \frac{d\mathbf{X}}{dt} \right| dt = \int_{t_0}^{t_1} f(t) \left[\left(\frac{dx}{dt}\right)^2 + \left(\frac{dy}{dt}\right)^2 + \left(\frac{dz}{dt}\right)^2 \right]^{1/2} dt. \tag{7.32}$$

More commonly, line integrals arise in such a fashion that $\phi(\mathbf{X})$ is the component of some vector field $\mathbf{F}(\mathbf{X})$ in the direction of the path at $\mathbf{X}$. Such a line integral has the form

$$\int_\Gamma \mathbf{F}(\mathbf{X}) \cdot d\mathbf{X} = \int_\Gamma F_x(x, y, z)\, dx + F_y(x, y, z)\, dy + F_z(x, y, z)\, dz,$$

the line integral of a differential form in three variables. Again, if the path is specified parametrically by $\Gamma = \mathbf{X}(t) = x(t)\hat{\mathbf{i}} + y(t)\hat{\mathbf{j}} + z(t)\hat{\mathbf{k}}$, then

$$\int_\Gamma \mathbf{F}(\mathbf{X}) \cdot d\mathbf{X} = \int_{t_0}^{t_1} \mathbf{F}[\mathbf{X}(t)] \cdot \frac{d\mathbf{X}}{dt}\, dt = \int_{t_0}^{t_1} \mathbf{G}(t) \cdot \frac{d\mathbf{X}}{dt}\, dt$$

$$= \int_{t_0}^{t_1} \left[G_x(t) \frac{dx}{dt} + G_y(t) \frac{dy}{dt} + G_z(t) \frac{dz}{dt} \right] dt, \tag{7.33}$$

where $G_x(t) = F_x[x(t), y(t), z(t)]$, etc. Alternatively, if the path is such that it is possible to solve for the y and z coordinates of points on the path as functions of x and thus express a point on the path by $[x, y(x), z(x)]$, then

$$\int_\Gamma F_x\,dx + F_y\,dy + F_z\,dz$$
$$= \int \left\{ F_x[x, y(x), z(x)] + F_y[x, y(x), z(x)]\frac{dy}{dx} + F_z[x, y(x), z(x)]\frac{dz}{dx} \right\} dx. \tag{7.34}$$

It should be pointed out that all the integrals of elementary thermodynamics are in fact line integrals in an eight-dimensional space. Two independent variables may be chosen from among T, P, V, S, E, H, A, and G. Then the space of accessible points is a surface, and one further constraint is needed to specify the path of integration.

For example, the heat absorbed on expansion of a gas may be written as a line integral in many ways. If E and V are taken as independent variables, and the expansion is from (E_0, V_0) to (E_1, V_1), then

$$\Delta q = \int dE + P\,dV = E_1 - E_0 + \int P\,dV.$$

To evaluate this integral it is necessary to know P as a function of E and V, and the relation between E and V on the particular expansion path chosen. In practice these functional relations are available only with difficulty. If S is taken as one of the independent variables, then simply

$$\Delta q = \int T\,dS,$$

but it is uncommon to know T as a function of S for every point on the path. However, the advantage of this expression is that for the particular case of an isentropic expansion path, $dS = 0$, so it is then easy to calculate $\Delta q = 0$. If T and V are taken as independent variables, then

$$\Delta q = \int C_V\,dT + T\left(\frac{\partial P}{\partial T}\right)_V dV$$

which is easy because the quantities C_V and $T(\partial P/\partial T)_V$ are often known as functions of T and V. Of course, it is necessary to specify the relationship between T and V along the path. Similarly, if P and V are taken as independent variables, then

$$\Delta q = \int C_V\left(\frac{\partial T}{\partial P}\right)_V dP + C_P\left(\frac{\partial T}{\partial V}\right)_P dV.$$

This form is especially convenient if the expansion is carried out at constant pressure.

Examples

To evaluate the line integral

$$\int_\Gamma (-y\hat{\mathbf{i}} + x\hat{\mathbf{j}} + v_0\hat{\mathbf{k}}) \cdot (dx\hat{\mathbf{i}} + dy\hat{\mathbf{j}} + dz\hat{\mathbf{k}}) = \int_\Gamma -y\,dx + x\,dy + v_0\,dz$$

29. where Γ goes from $(0, 0, 0)$ to $(3, 3, 0)$ along the straight line $y = x$, $z = 0$. This line integral may be transformed to an ordinary integral with respect to x, since anywhere along the path, $y = x$, $dy/dx = 1$, and $dz = 0$. Therefore by Eq. 7.34

$$\int_\Gamma -y\,dx + x\,dy = \int_0^3 (-x\,dx + x\,dx) = 0$$

as is obvious from the fact that the vector field $-y\hat{\mathbf{i}} + x\hat{\mathbf{j}} + v_0\hat{\mathbf{k}}$ is orthogonal to this straight-line path of integration.

30. where Γ goes from $(0, 0, 0)$ to $(3, 3, 0)$ along the parabola $y = x^2/3$. Again, we may transform to an ordinary integral with respect to x, since anywhere along the path, $y = x^2/3$, $dy/dx = 2x/3$, and $dz = 0$. Then

$$\int_\Gamma -y\,dx + x\,dy = \int_0^3 (-\tfrac{1}{3}x^2\,dx + \tfrac{2}{3}x^2\,dx) = \tfrac{1}{3}\int_0^3 x^2\,dx = 3.$$

Notice especially that although these two line integrals have the same integrand and the same endpoints, nevertheless the value of the line integral depends on the path.

31. where Γ is a circle of radius r centered at the origin, traveled once counterclockwise. We may parameterize the path by $\mathbf{X}(t) = r\cos t\hat{\mathbf{i}} + r\sin t\hat{\mathbf{j}}$, $0 \leqslant t \leqslant 2\pi$. Then by Eq. 7.33

$$\oint -y\,dx + x\,dy + v_0\,dz = \int_0^{2\pi} [-(r\sin t)(-r\sin t) + (r\cos t)(r\cos t)]\,dt$$
$$= r^2 \int_0^{2\pi} dt = 2\pi r^2.$$

The symbol $\oint$ is generally used to indicate a line integral over a closed path, conventionally traversed in a counterclockwise sense. Notice that if we again identify this vector field as a velocity field of a swirling, flowing liquid, the closed path is one that is always with the current, since all along the path $\mathbf{v} \cdot \mathbf{dX} > 0$. Only if $\oint \mathbf{v} \cdot \mathbf{dX} = 0$ would we conclude that the path is such that the amount of travel with the current exactly cancels the amount against it (Cf. Section 7.12). In the next section we consider the generality of such a cancellation.

Exercises

114. Convince yourself that if $\phi(\mathbf{X}) = 1$, then $\int_\Gamma \phi(\mathbf{X})\,ds$ is the arc length of Γ.

*115. Evaluate $\int_\Gamma \phi(\mathbf{X})\,ds$ if $\phi(\mathbf{X}) = |X|^2$ and Γ is the straight line from $\mathbf{X}_0$ to $\mathbf{X}_1$. Simplify your answer.

116. The gravitational force field is given by $\mathbf{F} = -mg\hat{\mathbf{k}}$, and the work done on a particle is given by the line integral $W = \int \mathbf{F} \cdot \mathbf{dX}$. Evaluate the work done on a particle of mass m when it moves from the point $\hat{\mathbf{i}} + \hat{\mathbf{j}} + \hat{\mathbf{k}}$ to the origin. You may choose any path, but specify what path you choose.

*117. Use Newton's second law to show that $W = \int \mathbf{F} \cdot \mathbf{dX}$ equals the increase in the kinetic energy of the particle.

118. Evaluate the line integral $\oint \mathbf{F} \cdot d\mathbf{X}$, where $\mathbf{F}$ is the vector field $(-y\hat{\mathbf{i}} + x\hat{\mathbf{j}})/(x^2 + y^2)$ and the path of integration is the unit circle centered at the origin, traveled once counterclockwise.

*119. Evaluate $\oint \mathbf{F} \cdot d\mathbf{X}$, where $\mathbf{F} = -ye^{-(x^2+y^2+z^2)}\hat{\mathbf{i}} + xe^{-(x^2+y^2+z^2)}\hat{\mathbf{j}}$ and the path of integration is the infinite helix $\hat{\mathbf{i}}r\cos\omega t + \hat{\mathbf{j}}r\sin\omega t + \hat{\mathbf{k}}vt$, $-\infty < t < \infty$.

120. The work done by an expanding gas is given by the line integral $W = \int P\,dV = \int P(\partial V/\partial T)_P\,dT + P(\partial V/\partial P)_T\,dP$. If the expansion of one mole of an ideal gas is carried out "reversibly," then at all points along the path $PV = RT$. Find the work done by 1 mole of an ideal gas expanding reversibly from (T_0, P_0) to (T_1, P_1) along each of the following paths:
(a) (T_0, P_0) to (T_0, P_1) at constant T, then (T_0, P_1) to (T_1, P_1) at constant P.
(b) The straight line in the T-P plane: $(T - T_0)/(P - P_0) = (T_1 - T_0)/(P_1 - P_0)$.

121. The entropy change accompanying the expansion of a substance is given by the line integral $\Delta S = \int (1/T)C_P\,dT - (\partial V/\partial T)_P\,dP$, where C_P is a constant. Find ΔS for the reversible expansion of 1 mole of an ideal gas from (T_0, P_0) to (T_1, P_1) along each of the paths of the previous exercise. Simplify your answers.

*122. It is also possible to form the line integral of a vector field:

$$\int_\Gamma \mathbf{E}(\mathbf{X})\,ds \equiv \lim \sum \mathbf{E}(\mathbf{X}_i)\,|\Delta\mathbf{X}_i| = \hat{\mathbf{i}}\int_\Gamma E_x(\mathbf{X})\,ds + \hat{\mathbf{j}}\int_\Gamma E_y(\mathbf{X})\,ds + \hat{\mathbf{k}}\int_\Gamma E_z(\mathbf{X})\,ds,$$

whose value is a vector. It is further possible to form the line integral of a vector-valued function of two vectors, $\int_\Gamma \mathbf{E}(\mathbf{X}_1, \mathbf{X}_2)\,ds_2$, whose value is a vector field. For example, the magnetic field at $\mathbf{R}$ due to a current I in a loop of wire is

$$\mathbf{H}(\mathbf{R}) = I\oint \frac{d\mathbf{X} \times (\mathbf{R} - \mathbf{X})}{|\mathbf{R} - \mathbf{X}|^3} = I \lim_{|\Delta\mathbf{X}_i| \to 0} \sum \frac{\Delta\mathbf{X}_i \times (\mathbf{R} - \mathbf{X}_i)}{|\mathbf{R} - \mathbf{X}_i|^3},$$

where $\mathbf{X}_i$ is a point on the wire. Show that $\mathbf{H}$ is solenoidal. (*Hints:* ∇ "commutes" with $\oint$ and acts only on $\mathbf{R}$. A zero curl is invariant to a change of origin.)

7.14 Line Integrals Independent of Path

Under what conditions will $\int_\Gamma \mathbf{F} \cdot d\mathbf{X}$ be independent of path and depend only on the endpoints? Geometric arguments (Fig. 7.30) show that this independence is equivalent to the vanishing of $\oint \mathbf{F} \cdot d\mathbf{X}$ for any closed path: If for any closed path, $\oint \mathbf{F} \cdot d\mathbf{X} = 0$, then for any arbitrary Γ_1 and Γ_2, both

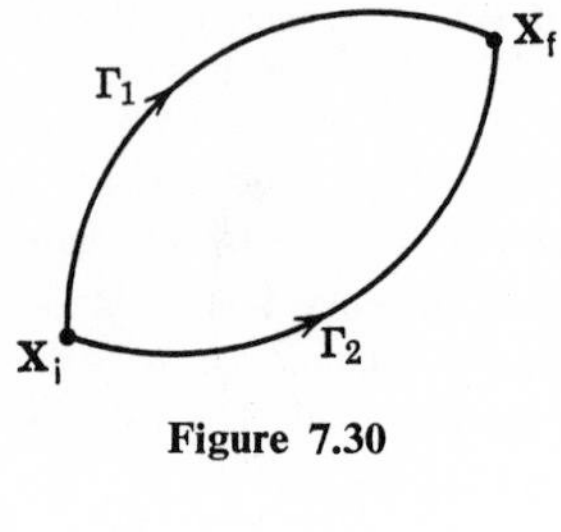

Figure 7.30

from $\mathbf{X}_i$ to $\mathbf{X}_f$, $0 = \oint_{\Gamma_1 - \Gamma_2} \mathbf{F} \cdot \mathbf{dX} = \int_{\Gamma_1} \mathbf{F} \cdot \mathbf{dX} - \int_{\Gamma_2} \mathbf{F} \cdot \mathbf{dX}$, or $\int_{\Gamma_1} \mathbf{F} \cdot \mathbf{dX} = \int_{\Gamma_2} \mathbf{F} \cdot \mathbf{dX}$. Similarly, if for all Γ_1 and Γ_2 connecting the same endpoints, $\int_{\Gamma_1} \mathbf{F} \cdot \mathbf{dX} = \int_{\Gamma_2} \mathbf{F} \cdot \mathbf{dX}$, then $\oint_{\Gamma_1 - \Gamma_2} \mathbf{F} \cdot \mathbf{dX} = 0$ for all closed paths.

Next we shall show that if $\mathbf{F}$ is a conservative field, then $\int_\Gamma \mathbf{F} \cdot \mathbf{dX}$ depends only upon the endpoints and $\oint \mathbf{F} \cdot \mathbf{dX} = 0$. Since $\mathbf{F}$ is conservative, it is the gradient of some scalar point function $\phi(\mathbf{X})$. Then

$$\int_\Gamma \mathbf{F} \cdot \mathbf{dX} = \int_\Gamma \nabla\phi \cdot \mathbf{dX} = \int_\Gamma d\phi = \phi(\mathbf{X}_f) - \phi(\mathbf{X}_i)$$

or

$$\oint \mathbf{F} \cdot \mathbf{dX} = \oint \nabla\phi \cdot \mathbf{dX} = \oint d\phi = \phi(\mathbf{X}_i) - \phi(\mathbf{X}_i) = 0.$$

It can be shown that the converse is also true: If for all closed paths $\oint \mathbf{F} \cdot \mathbf{dX} = 0$, then $\mathbf{F}$ is conservative. Also, since conservative and irrotational are synonymous, we conclude that $\oint \mathbf{F} \cdot \mathbf{dX} = 0$ for all paths if and only if $\nabla \times \mathbf{F} = \mathbf{0}$.

Finally, if $\mathbf{F}$ is a gradient, then $d\phi = \mathbf{F} \cdot \mathbf{dX} = F_x\,dx + F_y\,dy + F_z\,dz$ must be an exact differential. In Section 2.5 we stated that a necessary and sufficient condition for $F_x\,dx + F_y\,dy + F_z\,dz$ to be exact is

$$\frac{\partial F_x}{\partial y} = \frac{\partial F_y}{\partial x} \qquad \frac{\partial F_y}{\partial z} = \frac{\partial F_z}{\partial y} \qquad \frac{\partial F_z}{\partial x} = \frac{\partial F_x}{\partial z}$$

which is equivalent to the vector equation $\nabla \times \mathbf{F} = \mathbf{0}$.

We may list all the equivalent statements of this condition:

1. $\int_\Gamma \mathbf{F} \cdot \mathbf{dX}$ is independent of path and depends only upon endpoints.
2. $\oint \mathbf{F} \cdot \mathbf{dX} = 0$ for all closed paths.
3. $\mathbf{F}$ is conservative: $\mathbf{F} = \nabla\phi$.
4. $\mathbf{F}$ is irrotational: $\nabla \times \mathbf{F} = \mathbf{0}$.
5. $\mathbf{F} \cdot \mathbf{dX}$ is an exact differential.

In thermodynamics, if $\int_\Gamma \delta\phi$ is independent of path and depends only on endpoints, then ϕ is said to be a STATE FUNCTION. The value of a state function depends only on the state of the system and not on the process whereby that state was produced. In contrast, work and heat are not state functions since their differentials are inexact; the work done by a system on going from one state to another depends not only on the initial and final states of the system but also on the path between the states.

Exercises

123. A frictional force is proportional to the velocity $\mathbf{v}$ of a particle. Is this force conservative? Explain. (*Hint:* Consider $\oint \mathbf{F} \cdot \mathbf{dX}$.)

*124. In Exercise 7.118 you showed that $\oint (-y\,dx + x\,dy)/(x^2 + y^2) \neq 0$, even though $(-y\hat{\mathbf{i}} + x\hat{\mathbf{j}})/(x^2 + y^2) = \nabla \tan^{-1}(y/x)$. Explain why this vector field is not conservative.

125. In Section 2.5 we stated that an inexact differential form in three variables may not even be integrable. Show that if $\mathbf{F} \cdot d\mathbf{X}$ is integrable, then $\mathbf{F} \cdot \nabla \times \mathbf{F} = 0$. (*Hints:* Let μ be a nonzero integrating factor, such that $\mu\mathbf{F} \cdot d\mathbf{X}$ is exact. Then what is $\nabla \times (\mu\mathbf{F})$?) It can be shown that the converse is also true.

126. Is $-y\,dx + x\,dy + v_0\,dz$ integrable?

7.15 Surface Integrals

Originally we defined the double integral $\iint_A F(x, y)\,dx\,dy$ by dividing the region A into many tiny rectangles of area $\Delta A = \Delta x_i\,\Delta y_j$, forming the double sum $\sum F(x_i, y_j)\,\Delta A$, and taking the limit as the rectangles become infinitely small. However, we need not be restricted to surfaces in the xy-plane, for we can define the SURFACE INTEGRAL of a scalar point function over the surface Σ by

$$\iint_{\Sigma} F(\mathbf{X})\,dA \equiv \lim_{\Delta A_i \to 0} \sum F(\mathbf{X}_i)\,\Delta A_i, \tag{7.35}$$

where Σ has been subdivided so that the tiny area ΔA_i contains the point $\mathbf{X}_i$ on the surface. As expected

$$\iint_{\Sigma_1} F(\mathbf{X})\,dA + \iint_{\Sigma_2} F(\mathbf{X})\,dA = \iint_{\Sigma_1+\Sigma_2} F(\mathbf{X})\,dA. \tag{7.36}$$

More commonly, surface integrals arise in such fashion that $F(\mathbf{X})$ is the component of some vector field $\mathbf{E}(\mathbf{X})$ in the direction of the outward normal to the surface at $\mathbf{X}$. If the surface is specified by the constraint $\phi(\mathbf{X}) = 0$, then the unit normal to the surface is $\hat{\mathbf{n}} = \nabla\phi/|\nabla\phi|$; the outward direction is often arbitrary. Then the surface integral has the form

$$\iint_{\Sigma} \mathbf{E}(\mathbf{X}) \cdot \hat{\mathbf{n}}\,dA.$$

Examples

32. The area of Σ: $A = \iint_{\Sigma} dA$.

33. The FLUX of a vector field $\mathbf{E}$ over a closed surface: $\Phi = \iint \mathbf{E} \cdot \hat{\mathbf{n}}\,dA$.

34. The SOLID ANGLE, measured in STERADIANS, subtended at the origin by the surface Σ:

$$\Omega = \iint_{\Sigma} \frac{\mathbf{r} \cdot \hat{\mathbf{n}}}{|r|^3}\,dA.$$

To evaluate the surface integral $\iint_{\Sigma} F(x, y, z)\,dA$, it is necessary to reduce it to an ordinary double integral. Let us assume that it is possible to represent the surface Σ by $z = \phi(x, y)$. What we must do is relate dA to its projection ("shadow"), $dx\,dy$, on the xy-plane. Certainly $dA \geqslant dx\,dy$ to the extent that dA is tipped from the horizontal. If $\hat{\mathbf{n}} = \nabla(z - \phi)/|\nabla(z - \phi)| = \nabla(z - \phi)/[(\partial\phi/\partial x)^2 + (\partial\phi/\partial y)^2 + 1]^{1/2}$ is the unit normal to the surface (Cf. Exercise

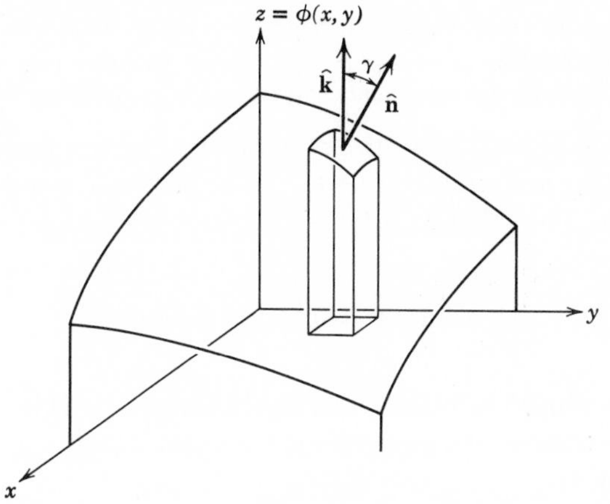

Figure 7.31

7.96), then $dx\,dy = \cos\gamma\, dA = \hat{\mathbf{k}}\cdot\hat{\mathbf{n}}\, dA$ (See Figure 7.31). Therefore

$$\iint F(x, y, z)\, dA = \iint_R F[x, y, \phi(x, y)]\frac{dx\,dy}{\hat{\mathbf{k}}\cdot\hat{\mathbf{n}}}$$
$$= \iint_R F[x, y, \phi(x, y)]\left[\left(\frac{\partial\phi}{\partial x}\right)^2 + \left(\frac{\partial\phi}{\partial y}\right)^2 + 1\right]^{1/2} dx\,dy \tag{7.37}$$

where R is the projection of Σ on the xy-plane.

Exercises

*127. Show that $\iint \mathbf{E}(\mathbf{X})\cdot\hat{\mathbf{n}}\, dA$ over the surface $z = \phi(x, y)$ may be reduced to the double integral

$$\iint\left\{-E_x[x, y, \phi(x, y)]\frac{\partial\phi}{\partial x} - E_y[x, y, \phi(x, y)]\frac{\partial\phi}{\partial y} + E_z[x, y, \phi(x, y)]\right\} dx\,dy.$$

128. Use Eq. 7.37 or the previous result to find the surface area of a sphere of radius r.

*129. Find the solid angle subtended by a cone whose apex is at the origin and whose apical angle is χ $(0 \leqslant \chi \leqslant 2\pi)$. (*Hint:* For all points on the cone except the apex, $\hat{\mathbf{r}}\cdot\hat{\mathbf{n}} = 0$; the integrand at the apex is infinite. Therefore, to evaluate the integral it is necessary to replace the cone with a portion of a sphere which subtends the same solid angle.)

*130. The electrostatic potential at the point $(0, 0, z_0)$ due to a charge $Q = 4\pi a^2\sigma$ distributed evenly over a sphere of radius a centered at the origin is

$$\phi(0, 0, z_0) = \sigma \oiint \frac{dA}{r},$$

where r is the distance from $(0, 0, z_0)$ to the surface element dA. Find ϕ. (*Caution:* Distinguish the cases $|z_0| < a$ and $|z_0| > a$. Also, depending upon the substitution you use to evaluate the integral, you may have to divide the sphere into two hemispheres, corresponding to $z = \pm(a^2 - x^2 - y^2)^{1/2}$. Compare your answer with the potential at $(0, 0, z_0)$ due to a point charge Q at the origin (Exercise 7.89).

Finally we state without proof two remarkable theorems that are especially handy for relating surface integrals to the more easily evaluated line and volume integrals. The DIVERGENCE THEOREM (GAUSS' THEOREM) states that

$$\oiint_{\Sigma} \mathbf{E} \cdot \hat{\mathbf{n}}\, dA = \iiint_{V} \nabla \cdot \mathbf{E}\, dx\, dy\, dz \tag{7.38}$$

where V is the volume enclosed by the closed surface Σ. This theorem is not unreasonable in view of the close relation between the flux of a vector field and its divergence.

STOKES' THEOREM states that

$$\oint_{\Gamma} \mathbf{E} \cdot \mathbf{dX} = \iint_{\Sigma} \nabla \times \mathbf{E} \cdot \hat{\mathbf{n}}\, dA, \tag{7.39}$$

where Σ is any surface whose boundary (edge) is the closed curve Γ; the choice of the outward side of Σ must be taken so that on the average, a right-handed screw rotated in the direction of Γ would advance in the direction of $\hat{\mathbf{n}}$ (See Figure 7.32).

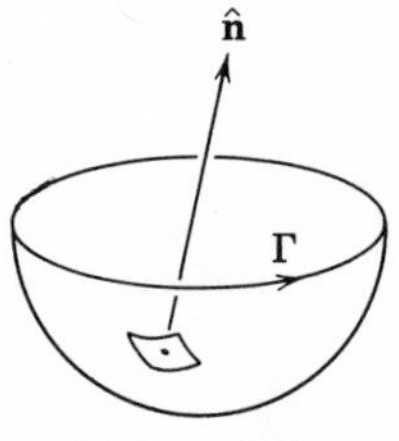

Figure 7.32

GREEN'S THEOREM is a corollary of Stokes' theorem for the special case that Γ and Σ are entirely in the xy-plane:

$$\oint_{\Gamma} E_x\, dx + E_y\, dy = \iint_{\Sigma} \left(\frac{\partial E_y}{\partial x} - \frac{\partial E_x}{\partial y}\right) dx\, dy. \tag{7.40}$$

This is especially handy for converting a line integral to an ordinary double integral.

Exercises

131. Show that $\iint d\Omega = \iint (\mathbf{r} \cdot \hat{\mathbf{n}}\, dA/r^3) = 4\pi$. (*Hint:* Use Gauss' theorem to relate this integral over an arbitrary closed surface to an easily evaluated surface integral over a small sphere of radius r_0 enclosing the origin.)
*132. Use Gauss' theorem to show that $\nabla \cdot (\mathbf{r}/r^3) = 4\pi\delta(\mathbf{r})$. What do you suppose this delta function means? (Cf. Exercise 7.106.)
*133. Derive Green's theorem from Stokes' theorem.
134. Use Green's theorem to provide a geometrical interpretation of $\oint_\Gamma x\, dy - y\, dx$.
*135. Evaluate $\oint (-\cosh x \sinh y\, dx + \sinh x \cosh y\, dy)$, where the path of integration is the unit square bounded by $x = \pm\frac{1}{2}$, $y = \pm\frac{1}{2}$. Then use Green's theorem to evaluate the same integral.
*136. Use Green's theorem to evaluate $\oint (e^{-x^2} \operatorname{erf} y\, dx - e^{-y^2} \operatorname{erf} x\, dy)$, where the path of integration is a circle of radius ρ centered at the origin and traveled once counterclockwise.
*137. Use the result of Exercise 1.43 to evaluate $\oint_\Gamma \mathbf{F} \cdot \mathbf{dX}$, where $\mathbf{F} = (-y\hat{\mathbf{i}} + x\hat{\mathbf{j}})/(x^2 + y^2)$, and Γ is a circle of radius r centered at $x = a$, traveled once counterclockwise. Then see what you can conclude from Green's theorem. Show that it is consistent to write $\nabla \times \mathbf{F} = 2\pi\delta(x)\delta(y)\hat{\mathbf{k}}$. (Cf. Exercise 7.124.)

8

MATRICES

It should be explained immediately that although matrix algebra may be totally unfamiliar, it is in fact nothing more than a convenient notational system, with no "complicated" mathematics involved. However, we must warn that this chapter tends to be more rigorous than previous ones. Since matrices do not obey all the familiar rules of arithmetic, we cannot thoughtlessly perform an operation on matrices just because it seems justified. Therefore we must learn definitions and theorems carefully so that we know what is correct and what is not. Also, we warn that this chapter tends to be rather abstract and lacking in obvious chemical applicability. Most chemical problems that are neatly solved by matrix methods involve chemistry that is too complicated to present. Toward the end of the chapter we shall see some applications to chemistry, but for now we shall have to trust that there exist chemical applications and wait for advanced courses, especially quantum mechanics, to appreciate most of these applications.

We begin this chapter with some rules for working with matrices. We go on to provide a somewhat abstract geometrical interpretation of matrices, involving n-dimensional vectors. And we also provide a very simple interpretation of tensors in terms of 3×3 matrices. We then proceed to two important scalar-valued functions of matrices—trace and determinant—with a digression on permutations along the way. More definitions are added in sections on partitioned matrices, direct sums and products, associated matrices, and special types of matrices; included are discussions of the extension of the scalar product to vectors over the complex numbers and the geometrical interpretations of unitary (orthogonal) and singular matrices. Then we discuss the familiar problem of simultaneous linear equations, with attention to both the inhomogeneous and homogeneous case. Also included is an elementary discussion of the general problem of simultaneous equations. So far there have been no subtle new concepts, but only a new system of notation. However, there are some subtle points introduced in Section 8.11, on linear transformation of basis, and the bewildered student is advised to concentrate only on Eq. 8.24. The chapter's most important section, on eigenvalue-eigenvector problems (diagonalization problems) follows. Included are a detailed explanation of the method of solution and, at last, an indication of problems of chemical interest solvable by matrix methods. We then proceed to some important theorems concerning diagonalization of Hermitian matrices and simultaneous diagonalization; the impatient student may skip the details of the proofs. We also indicate how the simultaneous diagonalizability of commuting matrices may simplify eigenvalue-eigenvector problems. Finally, we show how matrix-valued functions of a matrix may be evaluated and how they may be applied to problems of chemical interest. The numerous exercises of this chapter are primarily intended to provide practice, familiarity, and proficiency with matrices, rather than to demonstrate applications to chemistry. Nevertheless, there are problems involving matrix methods in quantum mechanics, electrostatics, Hückel molecular-orbital theory, spectrophotometry, balancing of chemical equations, chemical kinetics, magnetic resonance, and molecular vibrations.

A review of Appendix 3 may be helpful in connection with Section 8.12.

8.1 Definitions

We shall define a MATRIX as a rectangular array of numbers, enclosed in parentheses (sometimes in double straight lines), and abbreviated to a single boldface, sans-serif symbol

$$\left\| \begin{matrix} a_{11} & a_{12} & a_{13} & \cdots & a_{1n} \\ a_{21} & a_{22} & a_{23} & \cdots & a_{2n} \\ a_{31} & a_{32} & \cdot & \cdots & \\ \cdot & \cdot & \cdot & & \cdot \\ \cdot & \cdot & \cdot & & \cdot \\ \cdot & \cdot & & & \cdot \\ a_{m1} & a_{m2} & & \cdots & a_{mn} \end{matrix} \right\| = \mathbf{A}$$

whose properties are defined below. In particular this is an $m \times n$ matrix. The numbers $\{a_{ij}\}$ may be real or complex; for generality we shall always allow for complexity (pun intended). Notice the numbering convention: The first index tells which ROW of the matrix the number is in; the second index tells which COLUMN of the matrix the number is in. The *ij*th ELEMENT of the matrix $\mathbf{A}$ is the number a_{ij}, in the *i*th row and the *j*th column. Sometimes we may denote the *ij*th element of $\mathbf{A}$ by $(\mathbf{A})_{ij}$, and we also may denote the matrix $\mathbf{A}$ by (a_{ij}), read "the matrix whose elements are the numbers $\{a_{ij}\}$."

Notice that a $1 \times n$ matrix is only a single *row*; it is an ordered n-tuplet of numbers and therefore looks like an example of an n-dimensional vector. Likewise, an $m \times 1$ matrix is only a single *column* and looks like an m-dimensional vector. When we come to provide a geometrical interpretation of matrices, we shall identify ROW MATRICES and COLUMN MATRICES with vectors. Finally, a 1×1 matrix is only a single number, and looks like a scalar. Indeed, we shall not bother to distinguish between a 1×1 matrix and its single element, and we shall consider a 1×1 matrix to be a scalar, and vice versa.

We shall first consider only finite-dimensional matrices—those for which m and n are finite. In Chapter 9 we turn to infinite-dimensional matrices, in connection with the vector space of functions. However, as we proceed through finite-dimensional matrices, you should try to guess how the ideas introduced may be extended to infinite dimensions.

Two $m \times n$ matrices, $\mathbf{A}$ and $\mathbf{B}$, are equal if and only if for all i and j, $a_{ij} = b_{ij}$. Therefore the matrix equation, $\mathbf{A} = \mathbf{B}$, is equivalent to mn numerical equations.

We define SCALAR MULTIPLICATION of the matrix $\mathbf{A}$ by the scalar α to give the matrix

$$\alpha\mathbf{A} \equiv (\alpha a_{ij}), \tag{8.1}$$

that is, the matrix whose elements are $\{\alpha a_{ij}\}$. Alternatively, we could define this product by

$$(\alpha\mathbf{A})_{ij} \equiv \alpha a_{ij} = \alpha(\mathbf{A})_{ij},$$

specifying that the ijth element of the product matrix $\alpha\mathbf{A}$ is α times the ijth element of $\mathbf{A}$. It is obvious that

$$\alpha(\beta\mathbf{A}) = (\alpha\beta)\mathbf{A}$$

so that we may dispense with the parentheses.

We define the SUM of two $m \times n$ matrices, $\mathbf{A}$ and $\mathbf{B}$, by

$$\mathbf{A} + \mathbf{B} \equiv (a_{ij} + b_{ij}); \tag{8.2}$$

that is, add the matrices element by element. If the number of rows (columns) of $\mathbf{A}$ differs from the number of rows (columns) of $\mathbf{B}$, matrix addition is undefined. It is obvious that matrix addition is commutative and associative

$$\mathbf{A} + \mathbf{B} = \mathbf{B} + \mathbf{A}$$
$$\mathbf{A} + (\mathbf{B} + \mathbf{C}) = (\mathbf{A} + \mathbf{B}) + \mathbf{C}.$$

Also, there are the distributive rules

$$\alpha(\mathbf{A} + \mathbf{B}) = \alpha\mathbf{A} + \alpha\mathbf{B}$$
$$(\alpha + \beta)\mathbf{A} = \alpha\mathbf{A} + \beta\mathbf{A}.$$

We define a ZERO MATRIX by

$$(\mathbf{0})_{ij} \equiv 0,$$

a matrix whose every element is zero. Of course, there are an infinite number of zero matrices, one for each size array, but we shall use the same symbol for all of them. Then it follows that for any $\mathbf{A}$ there exists an appropriate $\mathbf{0}$ such that

$$\mathbf{A} + \mathbf{0} = \mathbf{A}.$$

If we also define the additive inverse $-\mathbf{A}$ by $(-\mathbf{A})_{ij} \equiv -(\mathbf{A})_{ij}$, we may show that the set of all $m \times n$ matrices forms a vector space, as we have already shown for $1 \times n$ and $m \times 1$ matrices in Examples 7.2 and 7.3.

Exercise

*1. Of what dimension is the *vector space* of $m \times n$ matrices of complex numbers, over the complex numbers as scalars?

However, there is more "structure" possible with matrices than with vectors, in that it is sometimes possible to multiply two matrices together to produce a matrix. (In general, two vectors can be multiplied together only to produce a scalar, with the cross product of two three-dimensional vectors being a special case which is not even associative.)

If $\mathbf{A}$ is an $m \times n$ matrix and $\mathbf{B}$ is an $n \times p$ matrix, we define the MATRIX PRODUCT $\mathbf{AB}$ as the $m \times p$ matrix whose elements are given by

$$(\mathbf{AB})_{ij} \equiv \sum_{k=1}^{n} a_{ik}b_{kj}. \tag{8.3}$$

Notice that the two matrices must be CONFORMABLE—the number of columns of the first matrix must equal the number of rows of the second. Then the product matrix has the same number of rows as the first matrix and the same number of columns as the second. As a mnemonic for these statements, the pictures of Figure 8.1 are helpful, and should be memorized. Notice that the elements $\{a_{ik}\}_{k=1}^{n}$ are distributed across the ith row, and that the elements $\{b_{kj}\}_{k=1}^{n}$ are distributed down the jth column. It is quite important to practice matrix multiplication in order to train your muscles so that your left arm moves horizontally while your right arm moves vertically.

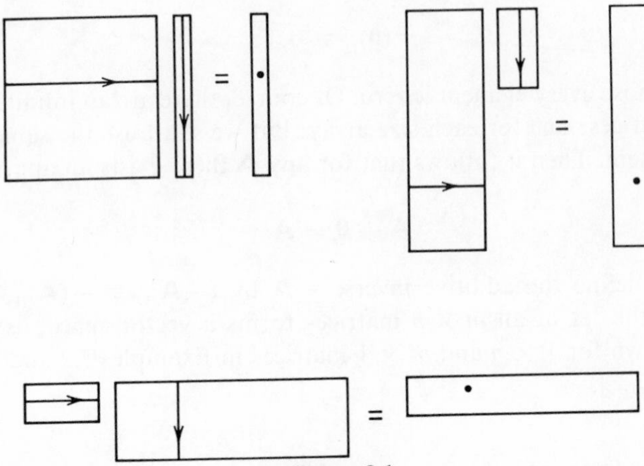

Figure 8.1

It is readily shown (cf. Exercise 8.2) that matrix multiplication is associative,

$$\mathbf{A}(\mathbf{BC}) = (\mathbf{AB})\mathbf{C}, \tag{8.4}$$

so that we may dispense with the parentheses. Matrix multiplication also has the further associative and distributive properties

$$\alpha(\mathbf{AB}) = (\alpha\mathbf{A})\mathbf{B} = \mathbf{A}(\alpha\mathbf{B}) \tag{8.5}$$

$$\mathbf{A}(\mathbf{B} + \mathbf{C}) = \mathbf{AB} + \mathbf{AC} \tag{8.6a}$$

$$(\mathbf{A} + \mathbf{B})\mathbf{C} = \mathbf{AC} + \mathbf{BC} \tag{8.6b}$$

but it is *not* commutative.

Examples

1. If

$$\mathbf{A} = \begin{pmatrix} 3 & i \\ 0 & 2 \end{pmatrix} \qquad \mathbf{B} = \begin{pmatrix} 3 & 0 \\ -i & 2 \end{pmatrix}$$

$$\mathbf{AB} = \begin{pmatrix} 3 \times 3 - i \times i & 3 \times 0 + i \times 2 \\ 0 \times 3 - 2 \times i & 0 \times 0 + 2 \times 2 \end{pmatrix} = \begin{pmatrix} 10 & 2i \\ -2i & 4 \end{pmatrix}$$

$$\mathbf{BA} = \begin{pmatrix} 3 \times 3 + 0 \times 0 & 3 \times i + 0 \times 2 \\ -i \times 3 + 2 \times 0 & -i \times i + 2 \times 2 \end{pmatrix} = \begin{pmatrix} 9 & 3i \\ -3i & 5 \end{pmatrix}.$$

2. If

$$\mathbf{A} = \begin{pmatrix} 1 & 3 & 2 \\ 2 & 1 & 3 \end{pmatrix} \qquad \mathbf{B} = \begin{pmatrix} 1 & 0 \\ 0 & 2 \\ 4 & 2 \end{pmatrix}$$

$$\mathbf{AB} = \begin{pmatrix} 1+0+8 & 0+6+4 \\ 2+0+12 & 0+2+6 \end{pmatrix} = \begin{pmatrix} 9 & 10 \\ 14 & 8 \end{pmatrix}$$

$$\mathbf{BA} = \begin{pmatrix} 1+0 & 3+0 & 2+0 \\ 0+4 & 0+2 & 0+6 \\ 4+4 & 12+2 & 8+6 \end{pmatrix} = \begin{pmatrix} 1 & 3 & 2 \\ 4 & 2 & 6 \\ 8 & 14 & 14 \end{pmatrix}.$$

Notice that in both these examples $\mathbf{AB} \neq \mathbf{BA}$. Indeed, in the second example the two products are not even the same size. We shall define the COMMUTATOR of two $n \times n$ matrices, $\mathbf{A}$ and $\mathbf{B}$, by

$$[\mathbf{A}, \mathbf{B}] \equiv \mathbf{AB} - \mathbf{BA}.$$

In the case that $\mathbf{AB} = \mathbf{BA}$, the commutator is zero, and the two matrices are said to COMMUTE.

Finally, it should be stated explicitly that in many respects the matrix equation $\mathbf{A} = \mathbf{B}$ may be manipulated as a numerical equation: Adding the same matrix $\mathbf{C}$ to both sides leads to an equality $\mathbf{A} + \mathbf{C} = \mathbf{B} + \mathbf{C}$. Multiplying both sides by a scalar α leads to the equality $\alpha\mathbf{A} = \alpha\mathbf{B}$. But since matrix multiplication is not commutative, multiplying both sides by the same matrix $\mathbf{C}$ must be performed either from the left, to give $\mathbf{CA} = \mathbf{CB}$, or from the right, to give $\mathbf{AC} = \mathbf{BC}$.

Exercises

2. What is the ijth element of the matrix $\mathbf{ABC}$? Convince yourself that matrix multiplication is associative (Eq. 8.4).

3. Find $\mathbf{AB}$ and $\mathbf{AB}$ if $\mathbf{A} = \begin{pmatrix} 1 & 1 \\ 0 & 1 \end{pmatrix}$ and $\mathbf{B} = \begin{pmatrix} 1 & 0 \\ 1 & 1 \end{pmatrix}$.

4. Find $\mathbf{AB}$ and $\mathbf{BA}$ if $\mathbf{A} = (1 \quad 1)$ and $\mathbf{B} = \begin{pmatrix} 2 & 3 & 1 \\ 1 & 2 & 3 \end{pmatrix}$.

5. (a) Perform the matrix multiplication

$$\begin{pmatrix} 0 & -u_z & u_y \\ u_z & 0 & -u_x \\ -u_y & u_x & 0 \end{pmatrix} \begin{pmatrix} v_x \\ v_y \\ v_z \end{pmatrix}.$$

(b) Does the result look familiar?

6. If $\mathbf{x} = (x_1 \quad x_2)$, $\mathbf{y} = \begin{pmatrix} y_1 \\ y_2 \end{pmatrix}$, $\mathbf{F} = \begin{pmatrix} f_{11} & f_{12} \\ f_{21} & f_{22} \end{pmatrix}$, find $\mathbf{xy}$, $\mathbf{xFy}$, and $\mathbf{yx}$.

7. (a) Perform the matrix multiplication

$$\begin{pmatrix} 2^{-1/2} & 0 & -2^{-1/2} \\ 0 & 1 & 0 \\ 2^{-1/2} & 0 & 2^{-1/2} \end{pmatrix} \begin{pmatrix} 0 & 1 & 0 \\ 1 & 0 & 1 \\ 0 & 1 & 0 \end{pmatrix} \begin{pmatrix} 2^{-1/2} & 0 & 2^{-1/2} \\ 0 & 1 & 0 \\ -2^{-1/2} & 0 & 2^{-1/2} \end{pmatrix}.$$

*(b) Check that this matrix multiplication is associative.

8. Find **IA**, **AI**, and **IAI** if

$$\mathbf{I} = \begin{pmatrix} 0 & 1 & 0 \\ 1 & 0 & 0 \\ 0 & 0 & 1 \end{pmatrix} \qquad \mathbf{A} = \begin{pmatrix} a_{11} & a_{12} & a_{13} \\ a_{21} & a_{22} & a_{23} \\ a_{31} & a_{32} & a_{33} \end{pmatrix}.$$

9. Generalize the previous result: What is the effect of premultiplication of an $n \times n$ matrix by the $n \times n$ matrix **I** whose only nonzero elements are 1 on the diagonal except for the four elements $(\mathbf{I})_{ij} = 1 = (\mathbf{I})_{ji}$ and $(\mathbf{I})_{ii} = 0 = (\mathbf{I})_{jj}$? What is the effect of postmultiplication by **I**?

*10. (a) What matrix multiplication will multiply the jth row of **A** by the scalar α?
(b) What matrix multiplication will replace the ith row of **A** by the sum (element-by-element) of the ith and jth rows?
(c) What matrix multiplication will replace the ith row of **A** by the sum of the ith row and α times the jth row? This operation is often called adding α times the jth row to the ith row.

*11. Perform the matrix multiplication

$$\begin{pmatrix} \cos\psi & \sin\psi & 0 \\ -\sin\psi & \cos\psi & 0 \\ 0 & 0 & 1 \end{pmatrix} \begin{pmatrix} \cos\theta & 0 & \sin\theta \\ 0 & 1 & 0 \\ -\sin\theta & 0 & \cos\theta \end{pmatrix} \begin{pmatrix} \cos\phi & \sin\phi & 0 \\ -\sin\phi & \cos\phi & 0 \\ 0 & 0 & 1 \end{pmatrix}$$

(The angles, ϕ, θ, ψ, are known as the EULERIAN ANGLES, and the resultant product matrix represents a rotation in three dimensions.)

12. Let **A** be an $n \times n$ matrix.
(a) Find a matrix **B** such that **BA** is a matrix whose first row is b_1 times the first row of **A**, whose second row is b_2 times the second row of **A**, etc.
(b) What matrix multiplication will convert **A** to another matrix whose first *column* is c_1 times the first column of **A**, whose second column is c_2 times the second column of **A**, etc.

13. What is $[\mathbf{A}, \mathbf{B}] + [\mathbf{B}, \mathbf{A}]$?

*14. The PAULI SPIN MATRICES are $\boldsymbol{\sigma}_x = \begin{pmatrix} 0 & 1 \\ 1 & 0 \end{pmatrix}$, $\boldsymbol{\sigma}_y = \begin{pmatrix} 0 & -i \\ i & 0 \end{pmatrix}$, and $\boldsymbol{\sigma}_z = \begin{pmatrix} 1 & 0 \\ 0 & -1 \end{pmatrix}$. Show that $\boldsymbol{\sigma}_i^2 = \begin{pmatrix} 1 & 0 \\ 0 & 1 \end{pmatrix}$. Find $[\boldsymbol{\sigma}_x, \boldsymbol{\sigma}_y]$, $[\boldsymbol{\sigma}_y, \boldsymbol{\sigma}_z]$, and $[\boldsymbol{\sigma}_z, \boldsymbol{\sigma}_x]$ in terms of the Pauli spin matrices.

15. (a) Find **AB** and **BA** if

$$\mathbf{A} = \begin{pmatrix} 1 & 1 & 0 \\ 1 & 0 & 0 \\ 0 & 1 & 0 \end{pmatrix} \qquad \mathbf{B} = \begin{pmatrix} 0 & 0 & 0 \\ 0 & 0 & 0 \\ 1 & 0 & 1 \end{pmatrix}.$$

(b) Find $\mathbf{A}^2 = \mathbf{AA}$ if

$$\mathbf{A} = \begin{pmatrix} 2 & i \\ 4i & -2 \end{pmatrix}.$$

(c) These two examples show that another familiar rule of numerical multiplication does not extend to matrix multiplication. Which rule?

8.2 Special Matrices

A SQUARE MATRIX is an $n \times n$ matrix. Henceforth we shall not consider any oblong matrices except row matrices and column matrices. We shall say that the DIMENSION of an $n \times n$ square matrix, a $1 \times n$ row matrix, or an $n \times 1$ column matrix is n.

We have already defined the n-dimensional zero matrix, which is readily shown to have the property of ANNIHILATING any conformable matrix $\mathbf{A}$:

$$\mathbf{0A} = \mathbf{0} \qquad \mathbf{A0} = \mathbf{0}. \tag{8.7}$$

The converse is an important theorem: The *only* n-dimensional matrix that annihilates *every* (conformable) matrix is the n-dimensional zero matrix.

We define the n-dimensional UNIT MATRIX by

$$(\mathbf{1})_{ij} \equiv \delta_{ij},$$

the matrix whose every DIAGONAL ELEMENT is unity and whose every off-diagonal element is zero. It is readily shown that for every conformable $\mathbf{A}$

$$\mathbf{1A} = \mathbf{A} \qquad \mathbf{A1} = \mathbf{A}. \tag{8.8}$$

We define an n-dimensional SCALAR MATRIX (CONSTANT MATRIX) by the scalar product

$$\mathbf{z} \equiv z\mathbf{1},$$

the matrix whose every diagonal element is the number z and whose every off-diagonal element is zero. It is readily shown that for every conformable $\mathbf{A}$

$$\mathbf{zA} = z\mathbf{A} \qquad \mathbf{Az} = z\mathbf{A}. \tag{8.9}$$

If $\mathbf{A}$ is square, then $\mathbf{zA} = \mathbf{Az}$. The converse of this last statement is an important theorem: The *only* matrices that commute with *every* $n \times n$ matrix are the n-dimensional scalar matrices.

A DIAGONAL MATRIX is one whose every off-diagonal element is zero. A diagonal matrix $\mathbf{D}$ whose diagonal elements are $\{d_i\}_{i=1}^n$ is often indicated by

$$\mathbf{D} = \text{diag}\,(d_1, d_2, \ldots, d_n).$$

It is readily shown that all diagonal matrices commute *with each other*, and that the product of two diagonal matrices is a diagonal matrix.

A TRIANGULAR MATRIX $\mathbf{T}$ is one for which $t_{ij} = 0$ for $i > j$ (or $t_{ij} = 0$ for $i < j$). All elements below (above) the diagonal are zero. We shall have little to do with triangular matrices, but they are of importance in computer programs.

We define the nth POWER of a matrix $\mathbf{A}$ as the matrix product

$$\mathbf{A}^n \equiv \mathbf{AA} \cdots \mathbf{A} \ (n \text{ times}).$$

A matrix $\mathbf{P}$ is said to be IDEMPOTENT if

$$\mathbf{P}^2 = \mathbf{P}.$$

A matrix $\mathbf{C}$ is said to be CYCLIC of order n if there exists an $n > 0$ such that

$$\mathbf{C}^n = \mathbf{1}.$$

Exercises

16. Perform the matrix multiplications to convince yourself that if $\mathbf{A}$ is a square matrix

$$\mathbf{0A} = \mathbf{0} = \mathbf{A0} \qquad \mathbf{1A} = \mathbf{A} = \mathbf{A1} \qquad \mathbf{zA} = z\mathbf{A} = \mathbf{Az}.$$

17. If $[\mathbf{A}, \mathbf{C}] = \mathbf{0}$ and $[\mathbf{B}, \mathbf{C}] = \mathbf{0}$, does $[\mathbf{A}, \mathbf{B}] = \mathbf{0}$? (*Hint:* Use Exercise 8.3 to construct a counterexample.)
18. Convince yourself that:
    (a) $\mathbf{0}$ and $\mathbf{1}$ are special cases of $\mathbf{z}$.
    (b) $\mathbf{z}$ is a special case of $\mathbf{D}$.
    (c) $\mathbf{D}$ is a special case of $\mathbf{T}$.

*19. Find a cyclic matrix among those given previously in the exercises.

20. Convince yourself that any two $n \times n$ diagonal matrices commute, and that their product is still diagonal.

*21. (a) How many N-letter words can be constructed with just the letters α and β?
    (b) Start from the left, treat α and β as noncommuting, and convince yourself that the sum of all these words is equal to the matrix product

$$(\alpha \quad \beta)\begin{pmatrix} \alpha & \beta \\ \alpha & \beta \end{pmatrix}^{N-1}\begin{pmatrix} 1 \\ 1 \end{pmatrix}.$$

    (c) There is no significance to summing all these words, but in the Ising model of ferromagnetism there arises the analogous problem of constructing the partition function Q for a sequence of N spins in a row. Each of these spins (of magnetic moment μ) can be either up or down (wave function α or β), and Q is a sum over all possible configurations (words): $Q = \sum e^{-E_i/kT}$, where E_i is the total energy of the ith configuration. According to the Ising model, the energy of a configuration in a magnetic field H may be calculated as follows: An α spin contributes μH to E_i and a β spin contributes $-\mu H$. A spin following an identical spin contributes J to E_i, and a spin following the opposite spin contributes $-J$. Build up the sequence of spins one at a time, calculate the contribution to E_i at each step, and convince yourself that

$$Q = (e^{-\mu H/kT} \quad e^{\mu H/kT})\begin{pmatrix} e^{-(\mu H+J)/kT} & e^{-(-\mu H-J)/kT} \\ e^{-(\mu H-J)/kT} & e^{-(-\mu H+J)/kT} \end{pmatrix}^{N-1}\begin{pmatrix} 1 \\ 1 \end{pmatrix}$$

    Although this one-dimensional model is too simple to account exactly for ferromagnetism in three-dimensional crystals, modifications of it do account for such phenomena as titration of linear polyelectrolytes, adsorption onto linear polymers, and helix-coil transitions in polypeptides and nucleic acids.

8.3 Geometrical Interpretation

So far we have only provided some definitions, without any hint as to the applicability of matrices. Next we consider, in geometrical terms, what square matrices do.

First we must try to imagine the existence of an n-dimensional vector space and a basis set $\{\mathbf{e}_i\}_{i=1}^n$. (Please forget that you ever heard that time is the fourth dimension. This notion is irrelevant to our purpose.) Then expanding an arbitrary n-dimensional vector $\mathbf{v}$ gives $\mathbf{v} = \sum_{i=1}^n v_i\mathbf{e}_i$. You should have no difficulty extending the notion of vectors beyond three dimensions, because once a basis is chosen, we need consider only the ordered n-tuplet $\{v_i\}$ of components. We shall never need to visualize any vector as something with magnitude and direction in n dimensions because we shall always reduce problems involving n-dimensional vectors to numerical problems involving components.

We may represent the vector $\mathbf{v}$ either as a row matrix whose elements are $\{v_i\}$ or as a column matrix whose elements are $\{v_i\}$. By convention we choose to REPRESENT every vector as a column matrix whose elements are the components of $\mathbf{v}$ in the basis set chosen.

$$\mathbf{v} \text{ is represented as } \mathbf{v} = \begin{pmatrix} v_1 \\ v_2 \\ \cdot \\ \cdot \\ \cdot \\ v_n \end{pmatrix}.$$

And in particular, the basis vectors are represented as the column matrices

$$\mathbf{e}_1 = \begin{pmatrix} 1 \\ 0 \\ \cdot \\ \cdot \\ \cdot \\ 0 \end{pmatrix} \quad \mathbf{e}_2 = \begin{pmatrix} 0 \\ 1 \\ \cdot \\ \cdot \\ \cdot \\ 0 \end{pmatrix} \quad \cdots \quad \mathbf{e}_n = \begin{pmatrix} 0 \\ 0 \\ \cdot \\ \cdot \\ \cdot \\ 1 \end{pmatrix}.$$

Next we investigate the significance of multiplying $\mathbf{v}$ by the $n \times n$ matrix $\mathbf{A}$. Only premultiplication by $\mathbf{A}$ is possible, and the product $\mathbf{Av}$ is also a column matrix

$$\mathbf{Av} = \begin{pmatrix} a_{11} & a_{12} & \cdots & a_{1n} \\ a_{21} & a_{22} & \cdots & a_{2n} \\ \cdot & \cdot & & \cdot \\ \cdot & \cdot & & \cdot \\ \cdot & \cdot & & \cdot \\ a_{n1} & a_{n2} & \cdots & a_{nn} \end{pmatrix} \begin{pmatrix} v_1 \\ v_2 \\ \cdot \\ \cdot \\ \cdot \\ v_n \end{pmatrix}$$

$$= \begin{pmatrix} a_{11}v_1 + a_{12}v_2 + \cdots + a_{1n}v_n \\ a_{21}v_1 + a_{22}v_2 + \cdots + a_{2n}v_n \\ \cdot \\ \cdot \\ \cdot \\ a_{n1}v_1 + a_{n2}v_2 + \cdots + a_{nn}v_n \end{pmatrix}.$$

Thus we conclude that the matrix $\mathbf{A}$ is something that acts on a vector to produce a vector. But the resultant vector $\mathbf{Av}$ is not necessarily of the same length as $\mathbf{v}$, or even in the same direction. Therefore we may view the matrix $\mathbf{A}$ as something that takes a vector, stretches (or squeezes) it, and twists it around to (in general) some other direction. However, by the distributive and associative properties of matrix multiplication (Eqs. 8.5 and 8.6), the process of applying $\mathbf{A}$ to vectors does satisfy the linearity condition

$$\mathbf{A}(c_1\mathbf{v}_1 + c_2\mathbf{v}_2) = c_1\mathbf{A}\mathbf{v}_1 + c_2\mathbf{A}\mathbf{v}_2. \tag{8.10}$$

The disadvantage of this formulation is that it relies upon the particular choice of $\{\mathbf{e}_i\}_{i=1}^n$ to represent the basis vectors. We saw that vectors may exist independent of any choice of basis, and that the choice of basis only serves to convert a vector problem into a numerical one. Similarly, it ought to be possible to imagine something that converts vectors into vectors, without choosing a basis. For example, the processes "Invert every vector through the origin" and "Halve every vector" may be specified without any reference to a basis set. In this context, something that assigns vectors to vectors is known as a TRANSFORMATION. We shall denote transformations by capital letters. However, we shall consider only LINEAR TRANSFORMATIONS, those for which $A(c_1\mathbf{v}_1 + c_2\mathbf{v}_2) = c_1A\mathbf{v}_1 + c_2A\mathbf{v}_2$, since matrix methods are applicable only to linear transformations.

Next we consider how the linear transformation A may be REPRESENTED as a set of numbers once a basis is chosen. First we note that if we wish to determine what A does to an arbitrary vector $\mathbf{v}$, we need only know what A does to each basis vector $\mathbf{e}_j$, since if $\mathbf{v} = \sum v_j\mathbf{e}_j$, then $A\mathbf{v} = \sum v_jA\mathbf{e}_j$. So let us focus on $\{A\mathbf{e}_j\}$ and expand each of these n vectors as a linear combination of basis vectors

$$A\mathbf{e}_j = \sum_i \mathbf{e}_i a_{ij}.$$

Thus we are led back again to a matrix $(a_{ij}) = \mathbf{A}$. Notice that we must sum over the *first* index on a_{ij} in order that the product of the matrix $\mathbf{A}$ with the column vector $\mathbf{e}_j$ be the column matrix whose elements are $\{a_{ij}\}_{i=1}^n$.

$$\begin{Vmatrix} a_{11} & a_{12} & \cdots & a_{1j} & \cdots & a_{1n} \\ a_{21} & a_{22} & \cdots & a_{2j} & \cdots & a_{2n} \\ \cdot & \cdot & & \cdot & & \cdot \\ \cdot & \cdot & & \cdot & & \cdot \\ \cdot & \cdot & & \cdot & & \cdot \\ a_{j1} & a_{j2} & \cdots & a_{jj} & \cdots & a_{jn} \\ \cdot & \cdot & & \cdot & & \cdot \\ \cdot & \cdot & & \cdot & & \cdot \\ \cdot & \cdot & & \cdot & & \cdot \\ a_{n1} & a_{n2} & \cdots & a_{nj} & \cdots & a_{nn} \end{Vmatrix} \begin{Vmatrix} 0 \\ 0 \\ \cdot \\ \cdot \\ \cdot \\ 1 \\ \cdot \\ \cdot \\ \cdot \\ 0 \end{Vmatrix} = \begin{Vmatrix} a_{1j} \\ a_{2j} \\ \cdot \\ \cdot \\ \cdot \\ a_{jj} \\ \cdot \\ \cdot \\ \cdot \\ a_{nj} \end{Vmatrix}.$$

Thus we may interpret the element a_{ij} (and *not* a_{ji}) as the component of $A\mathbf{e}_j$ along $\mathbf{e}_i$, and we may interpret the elements of the jth column of $\mathbf{A}$ as the components of the vector that results from applying the transformation A to the jth basis vector.

What if we apply the transformation A to a vector $\mathbf{v}_1$ and then apply the transformation B to the resultant vector, $\mathbf{v}_2$, to obtain $\mathbf{v}_3$? Clearly there must exist a single transformation which acts on $\mathbf{v}_1$ to produce $\mathbf{v}_3$ directly. Let us call this transformation C, so that $C\mathbf{v}_1 = B\mathbf{v}_2 = BA\mathbf{v}_1$. Since $\mathbf{v}_1$ is arbitrary, we may write $C = BA$. This product notation is consistent with our convention that operators act from the left. Again we need consider only what C does to the basis vectors:

$$C\mathbf{e}_j = \sum_i \mathbf{e}_i c_{ij}.$$

But also $A\mathbf{e}_j = \sum_k \mathbf{e}_k a_{kj}$ and $B\mathbf{e}_k = \sum_i \mathbf{e}_i b_{ik}$, so that if we apply B to $A\mathbf{e}_j$, to produce $C\mathbf{e}_j$, substitution and rearrangement leads to

$$BA\mathbf{e}_j = B\sum_k \mathbf{e}_k a_{kj} = \sum_k B\mathbf{e}_k a_{kj} = \sum_k \sum_i \mathbf{e}_i b_{ik} a_{kj} = \sum_i \mathbf{e}_i \sum_k b_{ik} a_{kj}.$$

Comparison with the previous equation shows that

$$c_{ij} = \sum_k b_{ik} a_{kj}.$$

Therefore we conclude that the transformation $C = BA$ is represented by the matrix $\mathbf{C} = (c_{ij})$, which is just the matrix product $\mathbf{BA}$. It is this result that is the motivation for the particular choice for the form of matrix multiplication.

Finally, it is desirable to suggest the relevance of these concepts to the vector space of functions. Originally we defined an operator as something that assigns functions to functions. However, if we are viewing a function as a vector, an operator becomes something that assigns vectors to vectors, that is, a transformation. If we are dealing only with linear operators, we need consider only linear transformations. Then once a basis set is chosen in the vector space of functions, a linear operator may be expressed as a matrix. And we shall eventually be able to solve operator problems, such as eigenvalue-eigenfunction differential equations, by the matrix methods that we are developing.

Exercises

22. Review the definition of an abstract vector space. Notice especially Example 7.3.
*23. Choose any orthonormal basis in three-dimensional space to represent the transformation, "Invert through the origin, then halve," as a matrix.

*24. A vector field is also something that assigns vectors to vectors, but in general a vector field cannot be represented as a matrix. Why not?

25. Convince yourself that the effect of the transformation A on the basis vectors $\{\mathbf{e}_i\}$ may also be described in matrix notation by $(\mathbf{e}_1\,\mathbf{e}_2\cdots\mathbf{e}_n)\mathbf{A} = (A\mathbf{e}_1\,A\mathbf{e}_2\cdots A\mathbf{e}_n)$, where the matrix $\mathbf{A} = (a_{ij})$ acts *from the right* on a *row* matrix. Notice that this is quite different from our usual approach, where a vector becomes a whole column matrix rather than a single element of a row matrix. However, this notation is often handy as a mnemonic for how to construct the matrix that represents a transformation, and it serves as a warning not to construct column matrices whose elements are basis vectors.

26. Express the definitions of Exercise 1.17 in matrix form, with $\alpha = \begin{pmatrix}1\\0\end{pmatrix}$ and $\beta = \begin{pmatrix}0\\1\end{pmatrix}$ taken as the basis set.

*27. Express the results of Exercises 2.5c and 2.5e in matrix form, representing S^2 and S_z as matrices that act on

$$\alpha\alpha = \begin{pmatrix}1\\0\\0\\0\end{pmatrix} \quad \alpha\beta = \begin{pmatrix}0\\1\\0\\0\end{pmatrix} \quad \beta\alpha = \begin{pmatrix}0\\0\\1\\0\end{pmatrix} \quad \beta\beta = \begin{pmatrix}0\\0\\0\\1\end{pmatrix}.$$

28. A PROJECTION MATRIX, $\mathbf{P}$, is one that acts on an arbitrary vector $\mathbf{v}$ in the vector space $\mathcal{V}$ to produce the projection of $\mathbf{v}$ in some subspace of $\mathcal{V}$. Convince yourself that $\mathbf{P}$ is idempotent. (*Hint:* $\mathbf{Pv}$ is already in the subspace, so acting $\mathbf{P}$ on it won't change it further.)

29. A special set of $n \times n$ projection matrices is the set $\{\mathbf{P}_i\}_{i=1}^n$, such that $\mathbf{P}_i$ acting on the arbitrary vector $\mathbf{v}$ produces the projection of $\mathbf{v}$ along $\mathbf{e}_i$, the ith basis vector.
(a) What is $\mathbf{P}_i$?
(b) Convince yourself that $\mathbf{P}_i\mathbf{P}_j = \delta_{ij}\mathbf{P}_j$.
(c) What is $\sum \mathbf{P}_i$?

*30. We may write the result of Exercise 6.32 as $P(A_j) = \sum P(A_j \mid E_i)P(E_i)$, where $\{E_i\}$ are mutually exclusive events that comprise a sample space, and $\{A_j\}$ likewise. We may express this result in matrix form: $\mathbf{p}_A = \mathbf{p}_E\mathbf{P}$, where $(\mathbf{P})_{ij} = P(A_j \mid E_i)$ and $\mathbf{p}_A$ and $\mathbf{p}_E$ are *row* matrices whose elements are $\{P(A_i)\}$ and $\{P(E_j)\}$, respectively. (*Warning:* The use of row matrices in this context is customary but opposite to our usual convention.) Notice that for any i, $\sum_j (\mathbf{P})_{ij} = \sum_j P(A_j \mid E_i) = 1$, since *some* A_j must result. Also, every $(\mathbf{P})_{ij} \geqslant 0$. Such a matrix, whose elements are all non-negative and whose row sums are all unity, is called a STOCHASTIC matrix. In this context, E_i and A_j are called STATES, and the matrix element $(\mathbf{P})_{ij}$ is called the TRANSITION PROBABILITY for transition from state E_i to state A_j. Furthermore, this situation, in which a set of final probabilities can be calculated in terms of a set of initial probabilities and a set of transition probabilities, is called a MARKOV CHAIN.

For example, let a box contain N molecules distributed between side A and side B. Let a molecule be chosen at random and be transferred to the other side. Let E_i = "there were i molecules in side A before the transfer" $(0 \leqslant i \leqslant N)$ and A_j = "there are now j molecules in side A after the transfer" $(0 \leqslant j \leqslant N)$.
(a) What is $P(A_{i+1} \mid E_i)$?
(b) What is $P(A_{i-1} \mid E_i)$?

(c) Convince yourself that **P** is the $(N + 1) \times (N + 1)$ matrix

$$\begin{Vmatrix} 0 & 1 & 0 & \cdots & & \\ 1/N & 0 & (N-1)/N & \cdots & & \\ 0 & 2/N & 0 & \cdots & & \\ \vdots & \vdots & \vdots & & & \\ & & & & & \vdots \\ & & & & & 1/N \\ & & & \cdots & 1 & 0 \end{Vmatrix}$$

(Notice that the upper left element is $(\mathbf{P})_{00}$, since we have included a zeroth row and a zeroth column for notational convenience.)

Furthermore, we may extend these results to account for a series of operations causing transitions between states:

(d) Convince yourself that if $\mathbf{P}'$ is the matrix of transition probabilities from the states $\{A_j\}$ to the states $\{B_k\}$, the matrix of transition probabilities from the states $\{E_i\}$ to the states $\{B_k\}$ is the matrix product $\mathbf{PP}'$. [*Hint:* Use Exercise 6.32 again to find $P(B_k)$.]

(e) Let a box contain N molecules, and let $p_0(r)$ be the probability that r molecules are initially in side A of the box. Let a molecule be chosen at random and be transferred to the other side, and let this process be carried out a total of n times. Express $p_n(s)$, the probability that there are then s molecules in side A, in terms of a matrix product. This result is the basis of Ehrenfest's model for diffusion through a membrane. Notice that the Markov-chain formulation permits consideration of not only the equilibrium distribution but also the rate and mode of approach to that equilibrium from any initial distribution of states.

*31. Show that the product of two stochastic matrices is still stochastic.

*32. Find the stochastic matrices for a single step of each of the following Markov chains:

(a) Random walk with absorbing barriers: There are $N + 1$ cells, and at each step a molecule in any interior cell has probability p of moving to the cell on the right and probability $q = 1 - p$ of moving to the cell on the left. A molecule in cell 0 or cell N is trapped there. (In gambling contexts interpret N as the total amount of money available to two players, and interpret each cell as a possible distribution of this money between the two players. Cells 0 and N then correspond to one player bankrupting the other.)

(b) Random walk with reflecting barriers: As (a), but a molecule in cell 0 or N necessarily moves to the adjacent cell.

(c) The chromatography column, with N theoretical plates (numbered from 0 to $N - 1$), of Example 6.41: The probability that a (V_m/N)-ml portion of mobile phase carries a molecule to the next theoretical plate is p. For convenience, take plate $\#(N - 1)$ as absorbing. Notice that this formulation permits a calculation of the distribution of material on the column in terms of *any* initial distribution, and is not limited to an initial distribution with all the material in the zeroth theoretical plate; for example, the column might be prepared under one set of conditions and then eluted further under another (as in temperature programming in vpc or changing the eluent in column chromatography).

(d) During a nanosecond time interval the probability that a given molecule will be excited from its ground state (S_0) to its second excited singlet state (S_2) is $10^{-9}k_I I$, where I is the light intensity. Whenever formed, S_2 immediately decays to the first excited singlet (S_1). During a nanosecond, S_1 might fluoresce and return to S_0 or it might cross to the triplet (T_1), with probabilities $10^{-9}k_F$ and $10^{-9}k_{isc}$, respectively. The probability that T_1 phosphoresces and returns to S_0 within a nanosecond is $10^{-9}k_P$.

8.4 Tensors

We have not as yet considered the ratio of two vectors. Let us investigate what can be said about $\mathbf{A}/\mathbf{B}$, that is, what it is that multiplies $\mathbf{B}$ to give $\mathbf{A}$.

In some special cases, when $\mathbf{A}$ and $\mathbf{B}$ are in the same direction, $\mathbf{A}/\mathbf{B}$ is a scalar. As examples, we may cite

$\mathbf{F} = m\mathbf{a}$ (force, mass, acceleration).

$\mathbf{D} = \varepsilon\mathbf{E}$ (electric displacement, dielectric constant, electric field).

$\boldsymbol{\mu} = \alpha\mathbf{E}$ (induced dipole moment, polarizability, electric field).

$\mathbf{M} = \chi\mathbf{H}$ (magnetization, magnetic susceptibility, magnetic field).

$\mathbf{L} = I\boldsymbol{\omega}$ (angular momentum, moment of inertia, angular velocity).

$\mathbf{v} = n\mathbf{v}_0$ (velocity of light, refractive index, velocity in a vacuum).

$\boldsymbol{\mu} = \gamma\mathbf{L}$ (nuclear magnetic moment, gyromagnetic ratio, angular momentum).

$\boldsymbol{\mu} = g\mathbf{L}$ (electronic magnetic moment, g-factor, angular momentum).

But if $\mathbf{A}$ and $\mathbf{B}$ are not in the same direction, then the entity which can multiply $\mathbf{B}$ to give $\mathbf{A}$ is more complicated. As an example, let us consider the magnetization $\mathbf{M}$ produced in a benzene molecule in the xy-plane in a magnetic field $\mathbf{H}$. When $\mathbf{H}$ is along the x-axis, the diamagnetic susceptibility results in a small magnetization in the opposite direction: $M_x = \chi_\perp H_x$ (see Figure 8.2a). Likewise $\mathbf{H}$ along the y-axis produces a magnetization $M_y = \chi_\perp H_y$. But when $\mathbf{H}$ is perpendicular to the plane, the ring currents lead to a greater magnetization, again in the opposite direction: $M_z = \chi_\parallel H_z$ (see Figure 8.2b). And if $\mathbf{H}$ is at some arbitrary angle to the plane, then $\mathbf{M}$ is at

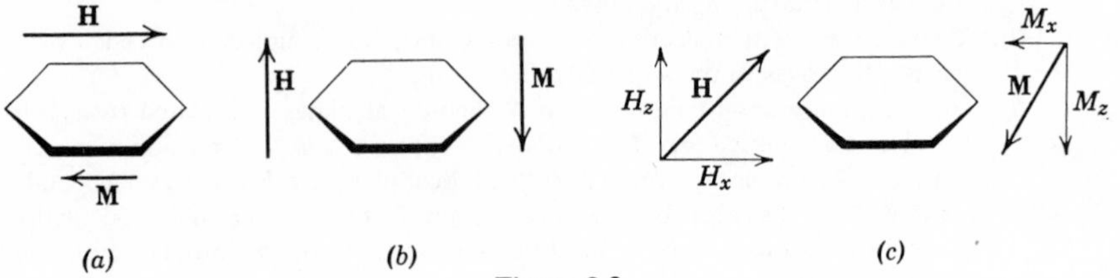

Figure 8.2

some different angle to the plane: $\mathbf{M} = M_x\hat{\mathbf{i}} + M_y\hat{\mathbf{j}} + M_z\hat{\mathbf{k}} = \chi_\perp H_x\hat{\mathbf{i}} + \chi_\perp H_y\hat{\mathbf{j}} + \chi_\parallel H_z\hat{\mathbf{k}} \neq \chi\mathbf{H}$ (see Figure 8.2c). Therefore we must express the relationship between **H** and **M** by $\mathbf{M} = \boldsymbol{\chi}\mathbf{H}$, where $\mathbf{M}$ and $\mathbf{H}$ are three-dimensional column vectors and $\boldsymbol{\chi} = (\partial M_i/\partial H_j)$ is a 3×3 matrix which represents the magnetic susceptibility:

$$\begin{pmatrix} M_x \\ M_y \\ M_z \end{pmatrix} = \begin{pmatrix} \chi_\perp & 0 & 0 \\ 0 & \chi_\perp & 0 \\ 0 & 0 & \chi_\parallel \end{pmatrix} \begin{pmatrix} H_x \\ H_y \\ H_z \end{pmatrix}.$$

In this context the ratio of two vectors is called a TENSOR although, strictly speaking, it is a second-order tensor. Just as a vector has significance independent of choice of basis, so does a tensor. However, we shall not define a tensor, but merely note that once a basis is chosen the tensor may be represented as a 3×3 matrix, which multiplies one column vector to produce another. In the special case that the two vectors are always proportional, the tensor is merely a multiple of the unit matrix and behaves like a scalar; such a tensor is called ISOTROPIC. In general, though, tensors are ANISOTROPIC, and a 3×3 matrix must be used. Indeed, under suitably anisotropic conditions, each of the above examples may become an equation involving tensors rather than scalars.

As another example of an anisotropic situation we might expect that the dipole moment induced in a Cl_2 molecule in an electric field will depend upon the orientation of the field relative to the molecular axis. The dipole moment produced when the field is perpendicular to the axis is smaller than that when the field is along the axis (because in the latter case the resonance form $Cl^+{=}Cl^-$ is stabilized). And for a random orientation, the induced moment will not even be in the same direction as the field. Therefore we must write $\boldsymbol{\mu} = \boldsymbol{\alpha}\mathbf{E}$, where $\boldsymbol{\alpha}$ is the polarizability tensor.

In the above examples the 3×3 matrices are diagonal. This fortunate circumstance arises solely from our choice of basis. We chose to express **H** in terms of its components parallel and perpendicular to the benzene plane, and by "symmetry" it was apparent that **H** along either of these directions produced an **M** in the opposite direction. On the other hand, if we had been so foolish as to choose a basis set inclined at arbitrary angles to the molecular plane, **H** along one of these basis vectors would produce an **M** in some different direction, so that $\boldsymbol{\chi}$ would not be diagonal. Likewise, the polarizability tensor of Cl_2 is diagonal when the basis vectors are chosen along "natural" axes. But in the case of *trans*-dichloroethylene, an electric field along the C—C bond induces a dipole moment at some angle to the bond. Therefore if the "natural" molecular axes are chosen as basis, the matrix $\boldsymbol{\alpha}$ is not diagonal.

Thus in general the given basis set may be such that the matrix representing a tensor is not diagonal. Nevertheless we shall see that it is always possible to find three directions such that a vector (**H**, **E**, etc.) along one of these directions produces a response vector (**M**, $\boldsymbol{\mu}$, etc.) which is proportional. Then choosing these directions as basis set leads to a diagonal matrix to represent the tensor. Such a revision of basis is known as a PRINCIPAL-AXIS TRANSFORMATION.

Exercises

*33. (a) Extend the result of Exercise 7.81 to find the y- and z-components of **L**, and express your result in the matrix form $\mathbf{L} = \mathbf{I}\boldsymbol{\omega}$, where $\mathbf{I}$ is the MOMENT OF INERTIA TENSOR.

(b) Find $\mathbf{I}$ for *trans*-FN=NF. Call the bond lengths d_{NF} and d_{NN} and the masses m_{N} and m_{F}; assume $<\mathrm{NNF} = 120°$. Also, it is customary to choose the origin at the center of mass. Notice that $\mathbf{I}$ is not diagonal. What this means physically is that viewed from the rotating system, an FN=NF molecule rotating about the N—N bond would experience a torque that would push the F atoms away from the axis of rotation (Cf. Exercise 7·82).

*34. Let us define the FIELD GRADIENT tensor, $\nabla\mathbf{E}$, as the matrix product

$$\nabla\mathbf{E} \equiv \begin{pmatrix} \dfrac{\partial}{\partial x} \\ \dfrac{\partial}{\partial y} \\ \dfrac{\partial}{\partial z} \end{pmatrix} (E_x \quad E_y \quad E_z).$$

(a) Show that if $\mathbf{dr} = (dx\ dy\ dz)$, then $\mathbf{dE} = \mathbf{dr} \cdot \nabla\mathbf{E}$, where the dot means matrix multiplication.

(b) Convince yourself that the expression for $\mathbf{E} \cdot \nabla\mathbf{F}$ given in Exercise 7.102 is a matrix product of **E** with $\nabla\mathbf{F}$.

(c) Show that $\nabla(\boldsymbol{\mu} \cdot \mathbf{E}) = \boldsymbol{\mu} \cdot \nabla\mathbf{E}$. (*Hint:* $\mathbf{E} = -\nabla\phi$.)

(d) Calculate the electric field gradient at the origin due to two like charges of magnitude q, at $\pm a$ along the z-axis.

The QUADRUPOLE MOMENT, $\mathbf{Q}$, of a set of point charges $\{q_k\}$ located at the points $\{x_k, y_k, z_k\}$ is a tensor which can be represented as a 3 × 3 matrix whose components are given by $Q_{\alpha\beta} = \sum_k q_k(3\alpha_k\beta_k - r_k^2\delta_{\alpha\beta})$, where the sum is over all point charges and α_k and β_k can be any of x_k, y_k, and z_k.

(e) Find the quadrupole moment of the (linear) CO_2 molecule if the C—O bond length is d and there is a charge of $-q$ on each oxygen, balanced by a charge of $+2q$ on the carbon.

(f) Do the same for *transf*-FN=NF, in terms of distances and charges. Assume $<\mathrm{NNF} = 120°$.

8.5 Scalar Functions of a Matrix

Next we consider two particular functions, trace and determinant, which assign scalars (numbers) to matrices.

The TRACE of a square matrix is the sum of its diagonal elements: $\mathrm{Tr}\,\mathbf{A} \equiv \sum_{i=1}^{n} a_{ii}$. This is certainly a "simple" function.

Exercise

35. Show that

$$\mathrm{Tr}\,(\mathbf{A} + \mathbf{B}) = \mathrm{Tr}\,\mathbf{A} + \mathrm{Tr}\,\mathbf{B}$$
$$\mathrm{Tr}\,(a\mathbf{A}) = a\,\mathrm{Tr}\,\mathbf{A}$$
$$\mathrm{Tr}\,(\mathbf{AB}) = \mathrm{Tr}\,(\mathbf{BA}).$$

Use this last result to convince yourself that $\mathrm{Tr}\,(\mathbf{ABC}) = \mathrm{Tr}\,(\mathbf{BCA}) = \mathrm{Tr}\,(\mathbf{CAB})$.

Before we consider determinants, we must first consider permutations. A PERMUTATION of the n objects $\{a, b, c, \ldots\}$ is a rearrangement of them. There are $n!$ permutations of n objects since any of n may be chosen for the first slot, any of $(n - 1)$ for the second slot, etc.

Example for $n = 3$

3. The six permutations of three objects may be symbolized by

$$\begin{pmatrix} a & b & c \\ a & b & c \end{pmatrix} \quad \begin{pmatrix} a & b & c \\ b & c & a \end{pmatrix} \quad \begin{pmatrix} a & b & c \\ c & a & b \end{pmatrix} \quad \begin{pmatrix} a & b & c \\ a & c & b \end{pmatrix} \quad \begin{pmatrix} a & b & c \\ c & b & a \end{pmatrix} \quad \begin{pmatrix} a & b & c \\ b & a & c \end{pmatrix}$$

or by

$$E \qquad (abc) \qquad (acb) \qquad (bc) \qquad (ac) \qquad (ab).$$

The letters a, b, and c merely stand for the objects on which the permutation operates, and it is the set of six letters, read downward in pairs, which tells what the permutation does. Thus the second symbol says that (*1*) wherever a occurs, replace it by b; (*2*) wherever b occurs, replace it by c; and (*3*) wherever c occurs, replace it by a. Permutations may be multiplied—applied successively—with the one at the right acting first; for example

$$\begin{pmatrix} abc \\ bca \end{pmatrix}\begin{pmatrix} abc \\ acb \end{pmatrix} (abc) = \begin{pmatrix} abc \\ bca \end{pmatrix} (acb) = (bac) = \begin{pmatrix} abc \\ bac \end{pmatrix} (abc),$$

or

$$\begin{pmatrix} abc \\ bca \end{pmatrix}\begin{pmatrix} abc \\ acb \end{pmatrix} = \begin{pmatrix} abc \\ bac \end{pmatrix}.$$

The second notation omits objects which remain unchanged, and lists CYCLES, enclosed in parentheses; the first element of a cycle is to be replaced by the second, the second by the third, etc., and the last by the first. The IDENTITY permutation, which "replaces" each object with itself, has no cycles and is usually symbolized by E (German *einheit*).

Further Examples of Cycle Notation

4. $\begin{pmatrix} a & b & c & d \\ c & d & a & b \end{pmatrix} = (ac)(bd)$, sometimes written as $P_{ac}P_b{}^p$.

5. $\begin{pmatrix} 1 & 2 & 3 & 4 & 5 & 6 & 7 \\ 5 & 7 & 3 & 6 & 4 & 1 & 2 \end{pmatrix} = (1546)(27).$

The cycle notation is more concise and more convenient for a discussion of the parity of a permutation; the built-up notation is more convenient for multiplying permutations.

A cycle is said to be EVEN if it contains an odd number of elements; a cycle is said to be ODD if it contains an even number of elements. The PARITY of a permutation is defined as follows: A permutation is said to be EVEN if it is composed of an even number of odd cycles; a permutation is said to be ODD if it is composed of an odd number of odd cycles. Notice that the even cycles are not counted. The first three permutations of three objects (including E) listed above are even, and the last three are odd. The permutations given in Examples 8.4 and 8.5 are both even, since each is composed of two odd cycles.

A cycle of two elements is called a (BINARY) INTERCHANGE or a TRANSPOSITION; it is the simplest kind of odd cycle. It can be shown that any permutation may be written as a "product" of binary interchanges, applied in succession. Although this "factorization" is not unique, it can be shown that the number of interchanges needed to effect a given permutation is either even or odd. An alternative and equivalent definition of the parity of a permutation is the parity of the number of binary interchanges needed to effect the permutation. However, the cycle method is easier to apply.

Exercises

36. Show that multiplication of permutations is not commutative:

$$\begin{pmatrix} abc \\ bca \end{pmatrix}\begin{pmatrix} abc \\ acb \end{pmatrix} \neq \begin{pmatrix} abc \\ acb \end{pmatrix}\begin{pmatrix} abc \\ bca \end{pmatrix}.$$

*37. Write down all 24 permutations of 4 objects.

*38. Convince yourself that for all n the identity permutation is indeed even. [*Hint:* What is $(ab)(ab)$?]

39. Is $\begin{pmatrix} 1 & 2 & 3 & 4 & 5 & 6 & 7 & 8 & 9 & 10 \\ 4 & 3 & 2 & 8 & 10 & 6 & 5 & 1 & 7 & 9 \end{pmatrix}$ even or odd?

*40. Express the permutation (1546)(27) as a product of successive binary interchanges. Check that there are indeed an even number of such interchanges.

The DETERMINANT is a scalar function of a square matrix, that is, a function which takes a square matrix into a number. The value of the function determinant, evaluated at the argument $\mathbf{A}$, is symbolized by $|\mathbf{A}|$ or det $\mathbf{A}$ and is defined by

$$|\mathbf{A}| \equiv \sum_{i=1}^{n!}(-1)^{p(P_i)}\prod_{j=1}^{n} a_{jP_i(j)} = \sum_{P_i}(-1)^{p(P_i)}a_{1P_i(1)}a_{2P_i(2)}\cdots a_{nP_i(n)}, \quad (8.11)$$

where the sum is over all $n!$ permutations $\{P_i\}$ of n objects, $p(P_i)$ is an integer that is even or odd according as the parity of the permutation P_i is even or odd, and $P_i(j)$ is that integer which is to replace integer j, according to the permutation P_i. Each term in the sum is a product of n factors. In each term there is one and only one factor from each row, and there is one and only one factor from each column. [If there were two factors from column k, then there would have to be l and m such that $P_i(l) = k = P_i(m)$, which is contrary to the notion of permutation.]

It is essential to keep in mind that a matrix is an array of numbers, whereas determinant is a scalar-valued function of a matrix. Determinant is only one such function of a matrix; trace is another. To clarify this notion, let us construct a table of examples:

Function	**Argument**	**Value**
Square	3	9
Square	x	x^2
Trace	$\begin{pmatrix} 1 & 0 \\ 3 & 2 \end{pmatrix}$	3
Determinant	$\begin{pmatrix} 1 & 0 \\ 3 & 2 \end{pmatrix}$	2
Determinant	$\mathbf{A}$	$\lvert\mathbf{A}\rvert$

You will often encounter the value of the determinant of the matrix $\begin{pmatrix} 1 & 0 \\ 3 & 2 \end{pmatrix}$ written as $\begin{vmatrix} 1 & 0 \\ 3 & 2 \end{vmatrix}$. However, this should be thought of as merely a longer way of writing the value 2. This long form, enclosed by a pair of vertical lines, is not a matrix but the determinant of a matrix—a number that must be evaluated by a rather elaborate procedure.

The following results follow readily from the definition:

$$\begin{aligned} |\mathbf{1}| &= 1 && \text{(unit matrix)} \\ |\mathbf{D}| &= \prod d_{ii} && \text{(diagonal matrix).} \\ |\mathbf{T}| &= \prod t_{ii} && \text{(triangular matrix).} \\ |\alpha\mathbf{A}| &= \alpha^n|\mathbf{A}| && \text{(scalar multiple of an } n \times n \text{ matrix).} \end{aligned}$$

A further theorem, which is not obvious and which we shall not prove, is one that we shall invoke frequently:

$$|\mathbf{AB}| = |\mathbf{A}|\,|\mathbf{B}| = |\mathbf{BA}|. \tag{8.12}$$

Exercises

*41. (a) Express the acceptable wave function of Exercise 4.62 as the difference of two 2×2 determinants. Neglect proportionality constants. Notice that applying the projection operator $1 - P_{12}$ to $\phi_1(x_1)\alpha(x_1)\phi_2(x_2)\beta(x_2)$ is equivalent to constructing the determinant of a matrix whose diagonal elements are $\phi_1(x_1)\alpha(x_1)$ and $\phi_2(x_2)\beta(x_2)$. A common short notation for this determinant is $|\phi_1\alpha\ \phi_2\beta|$.

(b) Express the acceptable wave function of Exercise 4.63 in determinantal form. (*Hint:* The operator P_{-1} of Exercise 4.61 means "construct into a 3×3 determinant.")

42. Convince yourself that the determinant of a triangular matrix is the product of the diagonal elements.

*43. Let $\mathbf{\Pi}$ be that product of matrices of the form given in Exercise 8.10c, such that $\mathbf{\Pi A}$ is a triangular matrix $\mathbf{T}$. Then $\mathbf{\Pi}$ represents the succession of operations on rows that converts $\mathbf{A}$ to triangular form. What is $|\mathbf{\Pi}|$? Use this result and Eq. 8.20 to convince yourself that $|\mathbf{A}| = |\mathbf{T}|$.

44. What is $|\mathbf{0}|$? $|\mathbf{z}|$?

*45. Use the rule for the derivative of an n-fold product to convince yourself that the derivative of the determinant of an $n \times n$ matrix $(a_{ij}(x))$ whose elements are functions of a variable, x, is a sum of n determinants, in the mth of which the mth column (row) contains $a'_{im}(x)$ [or $a'_{mj}(x)$].

The following properties of determinants also follow from the definition. They are stated in terms of rows, but they are all also true of columns.

1. If all the elements of a row are zero, (the value of) the determinant is zero (since every term of the determinant contains a zero factor).

2. Multiplying each element of a row by k multiplies the determinant by k (since exactly one factor of each term has been multiplied by k). Notice this means that if we consider the number $|\mathbf{A}|$ as the value of a function that depends upon only the n elements of the ith row, that is, if $|\mathbf{A}| = F(\{a_{ij}\}_{j=1}^{n})$, then F is a homogeneous function of degree unity.

3. Interchanging two rows changes the sign of the determinant (since each term in the new determinant is the same as a term in the old one, but with opposite sign because each permutation now has opposite parity).

4. If two rows are identical, the determinant is zero (as a corollary of the above property).

5. If the elements of one row are a multiple of those of another, the determinant is zero (as a corollary of properties 2 and 4).

6. If each element of the ith row is a sum of pairs—$a_{ij} = b_{ij} + c_{ij}$—then the determinant is a sum of two determinants, each of which is identical to the original determinant except that the ith row of one contains b_{ij} and the ith row of the other contains c_{ij} (since each term of the determinant has one factor which is a sum of two terms). This property is usually utilized to allow the sum of two determinants which differ only in one row to be combined into a single determinant.

7. If a multiple of one row is added element-by-element to another row, the determinant is unchanged (since by properties 2, 4, and 6 the resultant determinant is the sum of two determinants, one of which is zero).

Two simple cases may be easily evaluated according to the definition:

1. For $n = 2$, the two permutations are E and (12), the second of which is odd

$$\begin{vmatrix} a_{11} & a_{12} \\ a_{21} & a_{22} \end{vmatrix} = a_{11}a_{22} - a_{12}a_{21}.$$

2. For $n = 3$, the six permutations are given in Example 8.3, so

$$\begin{vmatrix} a_{11} & a_{12} & a_{13} \\ a_{21} & a_{22} & a_{23} \\ a_{31} & a_{32} & a_{33} \end{vmatrix} = \begin{aligned} &a_{11}a_{22}a_{33} + a_{12}a_{23}a_{31} + a_{13}a_{21}a_{32} \\ &- a_{11}a_{23}a_{32} - a_{13}a_{22}a_{31} - a_{12}a_{21}a_{33}. \end{aligned}$$

A handy mnemonic is to rewrite the first two columns, add the three downward diagonal products, and subtract the three upward diagonal products:

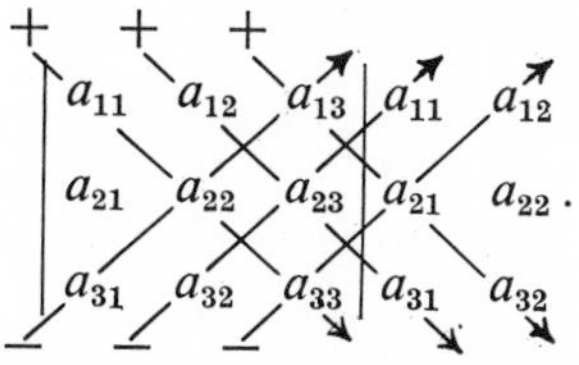

Such simple diagrams are not extendable to larger determinants, as is clear if we remember that a 4×4 determinant has 24 terms. The general method for evaluation of such determinants is EXPANSION IN COFACTORS (LAPLACE EXPANSION), whereby an $n \times n$ determinant is converted to a sum of n $(n-1) \times (n-1)$ determinants:

From property 2, that $|\mathbf{A}|$ is a homogeneous function of the elements of the ith row, it follows from Euler's theorem (Eq. 2.16) that

$$|\mathbf{A}| = \sum_j a_{ij} \frac{\partial |\mathbf{A}|}{\partial a_{ij}}.$$

But

$$\frac{\partial |\mathbf{A}|}{\partial a_{ij}} = \frac{\partial}{\partial a_{ij}} \sum_P^{n!} (-1)^{p(P)} \prod_{k=1}^{n} a_{kP(k)} = \sum_P^{(n-1)!} (-1)^{i+j}(-1)^{p(P)} \prod_{\substack{k=1 \\ k \neq i,\, P(k) \neq j}}^{n} a_{kP(k)}$$

since there are only $(n-1)!$ terms in $|\mathbf{A}|$ which contains a_{ij} as a factor, and none of the other factors in these terms can have an element from the ith row or from the jth column; the factor $(-1)^{i+j}$ is what is necessary to relate the permutations of $(n-1)$ objects to those of n objects. Although this expression looks formidable, it is just $(-1)^{i+j}$ times the $(n-1) \times (n-1)$ determinant of the matrix formed from $\mathbf{A}$ by deleting the ith row and the jth column. This $(n-1) \times (n-1)$ determinant formed by deleting the ith row and the jth column of $\mathbf{A}$ is called the MINOR of a_{ij} and is symbolized by $|A_{ij}|$. The COFACTOR of a_{ij} is the signed minor $(-1)^{i+j}|A_{ij}|$. Therefore the above partial derivative is just the cofactor of a_{ij}, and the expansion may be written

$$|\mathbf{A}| = \sum_{j=1}^{n} (-1)^{i+j} a_{ij} |A_{ij}|. \tag{8.13a}$$

This expansion is valid for any row i. Similar reasoning leads to the conclusion that $|\mathbf{A}|$ may be expanded in cofactors of any column j:

$$|\mathbf{A}| = \sum_{i=1}^{n} (-1)^{i+j} a_{ij} \, |A_{ij}|. \tag{8.13b}$$

In practice, it is convenient to choose a row or column with as many zeros as possible.

As an example, we shall indicate the expansion of a general 4×4 determinant in cofactors of its second row:

$$\begin{vmatrix} a_{11} & a_{12} & a_{13} & a_{14} \\ a_{21} & a_{22} & a_{23} & a_{24} \\ a_{31} & a_{32} & a_{33} & a_{34} \\ a_{41} & a_{42} & a_{43} & a_{44} \end{vmatrix} = -a_{21} \begin{vmatrix} a_{12} & a_{13} & a_{14} \\ a_{32} & a_{33} & a_{34} \\ a_{42} & a_{43} & a_{44} \end{vmatrix} + a_{22} \begin{vmatrix} a_{11} & a_{13} & a_{14} \\ a_{31} & a_{33} & a_{34} \\ a_{41} & a_{43} & a_{44} \end{vmatrix}$$

$$- a_{23} \begin{vmatrix} a_{11} & a_{12} & a_{14} \\ a_{31} & a_{32} & a_{34} \\ a_{41} & a_{42} & a_{44} \end{vmatrix} + a_{24} \begin{vmatrix} a_{11} & a_{12} & a_{13} \\ a_{31} & a_{32} & a_{33} \\ a_{41} & a_{42} & a_{43} \end{vmatrix}$$

An alternative procedure is REDUCTION TO TRIANGULAR FORM. An appropriate multiple of one row is subtracted from another row to convert one element to zero. Successive annihilations of off-diagonal elements convert the determinant to triangular form, but by property 7 none of these operations changes the value of the determinant. In the example below, a_{31}/a_{21} times the second row is subtracted from the third row in order to annihilate the 3-1 element. Then a_{21}/a_{11} times the first row is subtracted from the second row in order to annihilate the 2-1 element:

$$\begin{vmatrix} a_{11} & a_{12} & a_{13} \\ a_{21} & a_{22} & a_{23} \\ a_{31} & a_{32} & a_{33} \end{vmatrix} = \begin{vmatrix} a_{11} & a_{12} & a_{13} \\ a_{21} & a_{22} & a_{23} \\ 0 & a_{32} - a_{31}a_{22}/a_{21} & a_{33} - a_{31}a_{23}/a_{21} \end{vmatrix}$$

$$= \begin{vmatrix} a_{11} & a_{12} & a_{13} \\ 0 & a_{22} - a_{21}a_{12}/a_{11} & a_{23} - a_{21}a_{13}/a_{11} \\ 0 & a_{32} - a_{31}a_{22}/a_{21} & a_{33} - a_{31}a_{23}/a_{21} \end{vmatrix}$$

Next multiply the second row by $(a_{32} - a_{31}a_{22}/a_{21})/(a_{22} - a_{21}a_{12}/a_{11})$ and subtract from the third row to annihilate the 3-2 element and obtain the determinant of a triangular matrix. Although this method looks very cumbersome in symbols, it is very easy with numbers. Furthermore, notice that for an $n \times n$ determinant there are $n(n - 1)/2$ off-diagonal elements to annihilate, and each annihilation requires one division, $(n - 1)$ multiplications, and $(n - 1)$ subtractions. Therefore the labor involved for large n is proportional to only n^3 rather than to $n!$, as would be required for Laplace expansion.

Example

6. Let us follow the manipulations above in the same order to evaluate the determinant

$$\begin{vmatrix} 1 & 1 & -1 \\ -2 & \mathit{1} & \mathit{4} \\ 6 & 0 & 3 \end{vmatrix} = \begin{vmatrix} 1 & 1 & -1 \\ -2 & 1 & 4 \\ 6-6 & 0+3 & 3+12 \end{vmatrix} = \begin{vmatrix} \mathit{1} & \mathit{1} & \mathit{-1} \\ -2 & 1 & 4 \\ 0 & 3 & 15 \end{vmatrix}$$

$$= \begin{vmatrix} 1 & 1 & -1 \\ -2+2 & 1+2 & 4-2 \\ 0 & 3 & 15 \end{vmatrix} = \begin{vmatrix} 1 & 1 & -1 \\ \mathit{0} & \mathit{3} & \mathit{2} \\ 0 & 3 & 15 \end{vmatrix}$$

$$= \begin{vmatrix} 1 & 1 & -1 \\ 0 & 3 & 2 \\ 0 & 3-3 & 15-2 \end{vmatrix} = \begin{vmatrix} 1 & 1 & -1 \\ 0 & 3 & 2 \\ 0 & 0 & 13 \end{vmatrix} = 39$$

(For clarity, the row that is to be multiplied by an appropriate factor and subtracted from the one below it is indicated in italics.) This method is especially convenient on a blackboard or a computer, where it is not necessary to copy the other $n - 1$ rows at every manipulation.

A geometrical interpretion of the $n \times n$ determinant $|\mathbf{A}|$ is possible if all the elements of the matrix $\mathbf{A}$ are real. If we view the ith row of $\mathbf{A}$ as an n-dimensional vector $\mathbf{a}_i$, it can be shown that the absolute value of $|\mathbf{A}|$ is equal to the "volume" of the n-dimensional "parallelepiped" whose edges are the vectors $\{\mathbf{a}_i\}$. In Section 7.4 we demonstrated this identity for $n = 2$ and 3, but we shall not provide any proof for $n > 3$.

Exercises

46. Evaluate the determinant

$$\begin{vmatrix} 1 & 2 & 1 & 3 \\ 0 & 1 & 2 & 2 \\ 2 & 3 & 1 & 1 \\ 3 & 0 & 2 & 1 \end{vmatrix}.$$

(a) By expansion in cofactors.
(b) By reduction to triangular form. (Time yourself on each method. Notice that evaluation of a 5×5 determinant by Laplace expansion would take five times as long, whereas reduction to triangular form would take about twice as long. You should then be able to estimate which method is preferable whenever you must evaluate a determinant.)

*47. To test whether a set of functions $\{f_i\}_{i=1}^n$ is linearly dependent, form the $n \times n$ WRONSKIAN matrix $\mathbf{W}(x)$ defined by $(\mathbf{W}(x))_{ij} = D^{j-1}f_i(x)$. Show that if the set is linearly dependent, then $|\mathbf{W}(x)|$ is zero for all x. The converse is also true.

48. Use the previous result to test for linear independence:
(a) $\{1, \sin x, \sin 2x\}$.
(b) $\{1, \sin^2 x, \cos^2 x\}$.

*49. In the Hückel MO treatment of linear polyenes, there arises the problems of finding the roots of the nth-order polynomial $D_n(x)$ expressed as the $n \times n$ determinant

$$D_n(x) = \begin{vmatrix} x & 1 & 0 & & & \\ 1 & x & 1 & & & \\ 0 & 1 & x & & & \\ & & & \cdot & & \\ & & & & \cdot & 1 \\ & & & & 1 & x \end{vmatrix}$$

with x on the diagonal, 1 adjacent to the diagonal, and zeros elsewhere.

(a) Expand the determinant in cofactors of the first row to obtain a recursion formula relating $D_n(x)$ to $D_{n-1}(x)$ and $D_{n-2}(x)$. Check that it is consistent to define $D_0(x) = 1$.

(b) Let $G(x, t)$ be the generating function of the set $\{D_n\}$, that is, $G(x, t) = \sum_{n=0}^{\infty} D_n(x)t^n = 1 + xt + \sum_{n=2}^{\infty} D_n(x)t^n$. Multiply the recursion relationship by t^n and sum from $n = 2$ to ∞. Then change indices of summation so that each term may be related to $G(x, t)$. Finally, solve for $G(x, t)$.

(c) Let $x = 2\cos\theta$ and use the results of Exercise 1.74 to determine $D_n(2\cos\theta)$.

(d) Find the zeros of $D_n(x)$.

8.6 Partitioned Matrices. Direct Sums and Products

Any matrix may be PARTITIONED into SUBMATRICES for bookkeeping purposes or notational convenience. For example:

$$\mathbf{A} = \begin{pmatrix} a_{11} & a_{12} & a_{13} \\ a_{21} & a_{22} & a_{23} \\ a_{31} & a_{32} & a_{33} \end{pmatrix}$$

may be partitioned as

$$\left(\begin{array}{cc:c} a_{11} & a_{12} & a_{13} \\ a_{21} & a_{22} & a_{23} \\ \hdashline a_{31} & a_{32} & a_{33} \end{array}\right) = \begin{pmatrix} \mathbf{A}_{11} & \mathbf{A}_{12} \\ \mathbf{A}_{21} & \mathbf{A}_{22} \end{pmatrix} = (\mathbf{A}_{ij}).$$

The (dotted) partitions have no inherent significance except to indicate the partitioning. Here the matrix $\mathbf{A}$ is partitioned into the 2×2, 1×2, 2×1, and 1×1 submatrices, $\mathbf{A}_{11}$, $\mathbf{A}_{21}$, $\mathbf{A}_{12}$, and $\mathbf{A}_{22}$, respectively.

If two matrices $\mathbf{A}$ and $\mathbf{B}$ are partitioned identically, their sum, expressed as a matrix also partitioned in the same way, is

$$\mathbf{A} + \mathbf{B} = (\mathbf{A}_{ij} + \mathbf{B}_{ij}).$$

Thus, to add partitioned matrices, add corresponding submatrices.

If the partitioning of the columns of $\mathbf{A}$ matches the partitioning of the rows of $\mathbf{B}$, matrix multiplication proceeds row by column and each submatrix of the product is a matrix sum of matrix products. For example:

$$\mathbf{AB} = \begin{pmatrix} \mathbf{A}_{11} & \mathbf{A}_{12} & \mathbf{A}_{13} \\ \mathbf{A}_{21} & \mathbf{A}_{22} & \mathbf{A}_{23} \end{pmatrix} \begin{pmatrix} \mathbf{B}_{11} \\ \mathbf{B}_{21} \\ \mathbf{B}_{31} \end{pmatrix} = \begin{pmatrix} \mathbf{A}_{11}\mathbf{B}_{11} + \mathbf{A}_{12}\mathbf{B}_{21} + \mathbf{A}_{13}\mathbf{B}_{31} \\ \mathbf{A}_{21}\mathbf{B}_{11} + \mathbf{A}_{22}\mathbf{B}_{21} + \mathbf{A}_{23}\mathbf{B}_{31} \end{pmatrix}.$$

Partitioning is especially convenient for multiplication if there are zero submatrices. For example:

$$\mathbf{AB} = \begin{pmatrix} \mathbf{A}_{11} & \mathbf{0} \\ \mathbf{0} & \mathbf{A}_{22} \end{pmatrix} \begin{pmatrix} \mathbf{B}_{11} & \mathbf{0} \\ \mathbf{0} & \mathbf{B}_{22} \end{pmatrix} = \begin{pmatrix} \mathbf{A}_{11}\mathbf{B}_{11} & \mathbf{0} \\ \mathbf{0} & \mathbf{A}_{22}\mathbf{B}_{22} \end{pmatrix}.$$

The DIRECT SUM of two square matrices $\mathbf{A}$ and $\mathbf{B}$ is a partitioned matrix

$$\mathbf{A} \oplus \mathbf{B} \equiv \begin{pmatrix} \mathbf{A} & \mathbf{0} \\ \mathbf{0} & \mathbf{B} \end{pmatrix}.$$

The extension to more than two matrices is obvious. Notice that direct summation is associative but not commutative: $\mathbf{A} \oplus \mathbf{B} \neq \mathbf{B} \oplus \mathbf{A}$, but we shall find that these possible direct sums are equivalent, such that the non-commutativity never creates any complications. It is readily shown that the determinant of a direct sum simplifies: $|\mathbf{A} \oplus \mathbf{B}| = |\mathbf{A}||\mathbf{B}|$. This result is an example of the "independent behavior" of the diagonal submatrices of a direct sum. A matrix that can be written as the direct sum of submatrices is said to be in BLOCK-DIAGONAL FORM.

The direct product of two matrices is defined for any pair of matrices, without any requirement of conformability, but it most commonly involves square matrices, often of different sizes. The DIRECT PRODUCT of $\mathbf{A}$ and $\mathbf{B}$ is the matrix of all possible products of an element of $\mathbf{A}$ with an element of $\mathbf{B}$. Thus if $\mathbf{A}$ is an $m \times m$ matrix and $\mathbf{B}$ an $n \times n$ matrix, the direct product $\mathbf{A} \times \mathbf{B}$ is an $mn \times mn$ matrix which is most conveniently written as a partitioned matrix in which each submatrix is a scalar multiple of the matrix $\mathbf{B}$, with each scalar an appropriate element of $\mathbf{A}$:

$$\mathbf{A} \times \mathbf{B} \equiv \begin{pmatrix} a_{11}\mathbf{B} & a_{12}\mathbf{B} & \cdots & a_{1m}\mathbf{B} \\ a_{21}\mathbf{B} & a_{22}\mathbf{B} & \cdots & a_{2m}\mathbf{B} \\ \cdot & \cdot & & \cdot \\ \cdot & \cdot & & \cdot \\ \cdot & \cdot & & \cdot \\ a_{m1}\mathbf{B} & a_{m2}\mathbf{B} & \cdots & a_{mm}\mathbf{B} \end{pmatrix}$$

Notice that direct multiplication is not commutative.

Exercises

*50. Let $\mathcal{V}_2$ be a two-dimensional vector space, with basis vectors $\mathbf{e}_1$ and $\mathbf{e}_2$. Let $\mathcal{V}_3$ be a separate three-dimensional vector space, with basis vectors $\boldsymbol{\epsilon}_1$, $\boldsymbol{\epsilon}_2$, and $\boldsymbol{\epsilon}_3$. Let us form the set of all possible pairs of a vector in $\mathcal{V}_2$ with a vector in $\mathcal{V}_3$: If $\mathbf{x}$ is in $\mathcal{V}_2$ and $\mathbf{y}$ is in $\mathcal{V}_3$, then an element of this set is the pair, $(\mathbf{x}, \mathbf{y})$. Let us define the scalar multiple of the pair $(\mathbf{x}, \mathbf{y})$ by the scalar α to be the pair $(\alpha\mathbf{x}, \alpha\mathbf{y})$, and let us define the sum of the pairs $(\mathbf{x}_1, \mathbf{y}_1)$ and $(\mathbf{x}_2, \mathbf{y}_2)$ to be the pair $(\mathbf{x}_1 + \mathbf{x}_2, \mathbf{y}_1 + \mathbf{y}_2)$.
(a) Convince yourself that this set of all pairs forms a vector space. The space is known as the DIRECT PRODUCT of $\mathcal{V}_2$ and $\mathcal{V}_3$ and is symbolized by $\mathcal{V}_2 \times \mathcal{V}_3$.
(b) Convince yourself that $\{(\mathbf{e}_i, \boldsymbol{\epsilon}_j)\}$, the set of all pairs of a basis vector from $\mathcal{V}_2$ with a basis vector from $\mathcal{V}_3$, is a basis for $\mathcal{V}_2 \times \mathcal{V}_3$.
(c) Convince yourself that $\mathcal{V}_2 \times \mathcal{V}_3$ is six-dimensional, and that $\mathcal{V}_m \times \mathcal{V}_n$ is mn-dimensional. (*Note:* There is no geometrical interpretation to be attached to the pair $(\mathbf{x}, \mathbf{y})$. That $\mathcal{V}_2 \times \mathcal{V}_3$ is six-dimensional instead of five-dimensional shows that $\mathcal{V}_2 \times \mathcal{V}_3$ is not the vector space spanned by the five basis vectors.)

The direct product space is important in group theory and in quantum mechanics. For example, the set of all spatial wave functions for a single electron forms an infinite-dimensional vector space, and the set of all spin wave functions for a single electron forms a two-dimensional vector space. The total wave function is a product of a spatial part times a spin part, and the set of all these functions is a vector space that is the direct product of these two separate vector spaces.

*51. Let us arrange the six basis vectors of $\mathcal{V}_2 \times \mathcal{V}_3$ in the order $(\mathbf{e}_1, \boldsymbol{\epsilon}_1) = \mathbf{E}_1$, $(\mathbf{e}_1, \boldsymbol{\epsilon}_2) = \mathbf{E}_2$, $(\mathbf{e}_1, \boldsymbol{\epsilon}_3) = \mathbf{E}_3$, $(\mathbf{e}_2, \boldsymbol{\epsilon}_1) = \mathbf{E}_4$, $(\mathbf{e}_2, \boldsymbol{\epsilon}_2) = \mathbf{E}_5$, $(\mathbf{e}_2, \boldsymbol{\epsilon}_3) = \mathbf{E}_6$. Then

$$\begin{aligned}(\mathbf{x}, \mathbf{y}) &= (x_1\mathbf{e}_1 + x_2\mathbf{e}_2, y_1\boldsymbol{\epsilon}_1 + y_2\boldsymbol{\epsilon}_2 + y_3\boldsymbol{\epsilon}_3) \\ &= x_1y_1\mathbf{E}_1 + x_1y_2\mathbf{E}_2 + x_1y_3\mathbf{E}_3 + x_2y_1\mathbf{E}_4 + x_2y_2\mathbf{E}_5 + x_2y_3\mathbf{E}_6.\end{aligned}$$

(a) Convince yourself that if $\mathbf{x}$ and $\mathbf{y}$ are represented as column matrices x and y, $(\mathbf{x}, \mathbf{y})$ is represented as the direct product column matrix $\mathsf{x} \times \mathsf{y}$.
(b) Convince yourself that if the transformation A acts on $\mathbf{x}$ to produce $A\mathbf{x}$ and B on $\mathbf{y}$ to produce $B\mathbf{y}$, the transformation that acts on $(\mathbf{x}, \mathbf{y})$ to produce $(A\mathbf{x}, B\mathbf{y})$ is represented as the direct product matrix $\mathsf{A} \times \mathsf{B}$.

*52. Use general 2×2 matrices to convince yourself that $(\mathsf{A} \times \mathsf{B})(\mathsf{C} \times \mathsf{D}) = \mathsf{AC} \times \mathsf{BD}$.

*53. (a) Does $(\mathsf{A} \oplus \mathsf{B}) \oplus \mathsf{C} = \mathsf{A} \oplus (\mathsf{B} \oplus \mathsf{C})$?
(b) Does $(\mathsf{A} \times \mathsf{B}) \times \mathsf{C} = \mathsf{A} \times (\mathsf{B} \times \mathsf{C})$?
(c) Does $\mathsf{A} \times (\mathsf{B} \oplus \mathsf{C}) = \mathsf{A} \times \mathsf{B} \oplus \mathsf{A} \times \mathsf{C}$?
(d) Does $(\mathsf{A} \oplus \mathsf{B}) \times \mathsf{C} = (\mathsf{A} \times \mathsf{C}) \oplus (\mathsf{B} \times \mathsf{C})$?

54. Convince yourself that $\mathrm{Tr}\,(\mathsf{A} \oplus \mathsf{B}) = \mathrm{Tr}\,\mathsf{A} + \mathrm{Tr}\,\mathsf{B}$.

55. Convince yourself that $\mathrm{Tr}\,(\mathsf{A} \times \mathsf{B}) = (\mathrm{Tr}\,\mathsf{A})(\mathrm{Tr}\,\mathsf{B})$. This result is important in group theory.

*56. Convince yourself that $|\mathsf{A} \oplus \mathsf{B}| = |\mathsf{A}|\,|\mathsf{B}|$ for the particular case that A is 2×2.

57. If $\mathsf{1}$ is the two-dimensional unit matrix, and A is an arbitrary square matrix, what is $\mathsf{1} \times \mathsf{A}$?

8.7 Associated Matrices

Associated with any matrix A are three closely related matrices: its complex conjugate, its transpose, and its adjoint. Also, if A is square there may be a fourth associated matrix, its inverse. We now introduce these matrices.

We define the COMPLEX CONJUGATE, $\mathbf{A}^*$, of the matrix $\mathbf{A}$ by

$$(\mathbf{A}^*)_{ij} \equiv (\mathbf{A})_{ij}{}^*.$$

Thus, to form the complex conjugate of a matrix, take the complex conjugate of every element. It follows immediately that

$$\begin{aligned}
(\mathbf{A}^*)^* &= \mathbf{A} \\
(\mathbf{A} + \mathbf{B})^* &= \mathbf{A}^* + \mathbf{B}^* \\
(\mathbf{AB})^* &= \mathbf{A}^*\mathbf{B}^* \\
|\mathbf{A}^*| &= |\mathbf{A}|^* \\
(\alpha\mathbf{A})^* &= \alpha^*\mathbf{A}^*.
\end{aligned}$$

We define the TRANSPOSE, $\mathbf{A}^T$, of the matrix $\mathbf{A}$ by

$$(\mathbf{A}^T)_{ij} \equiv (\mathbf{A})_{ji}.$$

Thus the transpose of a column matrix is a row matrix and the transpose of a row matrix is a column matrix. To form the transpose of a square matrix, reflect it through its diagonal. It is obvious then that

$$\begin{aligned}
(\mathbf{A}^T)^T &= \mathbf{A} \\
(\mathbf{A} + \mathbf{B})^T &= \mathbf{A}^T + \mathbf{B}^T.
\end{aligned}$$

From the equivalence of rows and columns in the definition of determinant, it follows that

$$|\mathbf{A}^T| = |\mathbf{A}|.$$

Matrix multiplication to produce $\mathbf{AB}$ proceeds by multiplying the rows of $\mathbf{A}$ by the columns of $\mathbf{B}$. On forming transposes, rows are converted to columns and columns to rows, so that

$$(\mathbf{AB})^T = \mathbf{B}^T\mathbf{A}^T.$$

More important than either the complex conjugate or the transpose is the combination of the two, the ADJOINT

$$\mathbf{A}^\dagger \equiv \mathbf{A}^{T*} = \mathbf{A}^{*T} \qquad (\mathbf{A}^\dagger)_{ij} = (\mathbf{A})_{ji}{}^*.$$

Thus, to form the adjoint of a matrix, reflect it in the diagonal and take the complex conjugate of every element. (*Warning:* Many books, especially in pure mathematics, define the adjoint matrix as the transpose of the matrix of cofactors, and call our adjoint the conjugate transpose or the Hermitian adjoint.) It follows immediately that

$$\begin{aligned}
(\mathbf{A}^\dagger)^\dagger &= \mathbf{A} \\
(\mathbf{A} + \mathbf{B})^\dagger &= \mathbf{A}^\dagger + \mathbf{B}^\dagger \\
|\mathbf{A}^\dagger| &= |\mathbf{A}|^* \\
(\mathbf{AB})^\dagger &= \mathbf{B}^\dagger\mathbf{A}^\dagger \\
(\alpha\mathbf{A})^\dagger &= \alpha^*\mathbf{A}^\dagger.
\end{aligned}$$

Don't forget to apply these last two results in forming adjoints of matrix and scalar products! Notice also that if $\mathbf{v}$ and $\mathbf{Av}$ are column matrices, $\mathbf{v}^\dagger$ and $(\mathbf{Av})^\dagger = \mathbf{v}^\dagger\mathbf{A}^\dagger$ are row matrices.

Exercise

58. Convince yourself that $(\mathbf{AB})^T = \mathbf{B}^T\mathbf{A}^T$ and that $(\mathbf{AB})^\dagger = \mathbf{B}^\dagger\mathbf{A}^\dagger$.

We are now in a position to generalize the dot product to vectors over the complex numbers. If real n-dimensional vectors $\mathbf{x}$ and $\mathbf{y}$ are expanded in an orthonormal basis, the dot product is readily generalized to $\mathbf{x}\cdot\mathbf{y} = \sum x_i y_i$. However, we are now representing these vectors as column matrices $\mathbf{x}$ and $\mathbf{y}$, which are not conformable. In order to multiply two column matrices and obtain a scalar, the one on the left must be converted to a row matrix. But there are *two* possible row matrices that can be produced from $\mathbf{x}$, namely $\mathbf{x}^T$ and $\mathbf{x}^\dagger$; we can define the dot product as $\mathbf{x}^T\mathbf{y} = \sum x_i y_i$ or as $\mathbf{x}^\dagger\mathbf{y} = \sum x_i^* y_i$. But if we wish to retain the idea of $\mathbf{x}\cdot\mathbf{x}$ as the square of the length of $\mathbf{x}$, we require that this quantity be real and nonnegative. To ensure this result, we must define $\mathbf{x}\cdot\mathbf{x}$ as $\mathbf{x}^\dagger\mathbf{x} = \sum |x_i|^2$. Therefore we choose to generalize the scalar product of the two column vectors $\mathbf{x}$ and $\mathbf{y}$ by defining the HERMITIAN SCALAR PRODUCT, whereby

$$\mathbf{x}\cdot\mathbf{y} \qquad \text{becomes} \qquad \mathbf{x}^\dagger\mathbf{y} = \sum_{i=1}^{n} x_i^* y_i.$$

As a result, we must modify some of the rules for the scalar product. It is no longer necessarily true that $\mathbf{x}\cdot\mathbf{y} = \mathbf{y}\cdot\mathbf{x}$, since

$$\mathbf{x}^\dagger\mathbf{y} = (\mathbf{y}^\dagger\mathbf{x})^*. \tag{8.14}$$

This result follows from the fact that both sides are equal to $(\mathbf{y}^\dagger\mathbf{x})^\dagger$, the adjoint of a 1×1 matrix. Also, although $\mathbf{x}\cdot(\alpha\mathbf{y}) = \alpha(\mathbf{x}\cdot\mathbf{y})$ becomes $\mathbf{x}^\dagger(\alpha\mathbf{y}) = \alpha(\mathbf{x}^\dagger\mathbf{y})$, without change, it is no longer necessarily true that $(\alpha\mathbf{x})\cdot\mathbf{y} = \alpha(\mathbf{x}\cdot\mathbf{y})$, since

$$(\alpha\mathbf{x})^\dagger\mathbf{y} = \alpha^*(\mathbf{x}^\dagger\mathbf{y}). \tag{8.15}$$

Previously, when we expanded $\mathbf{x}$ as a linear combination of orthonormal basis vectors $\mathbf{x} = \sum \mathbf{e}_i x_i$, we identified x_i as the dot product of $\mathbf{x}$ with the basis vector $\mathbf{e}_i$. However, we must now recognize that this dot product is $\mathbf{e}_i^\dagger\mathbf{x}$, so that the expansion becomes

$$\mathbf{x} = \sum \mathbf{e}_i(\mathbf{e}_i^\dagger\mathbf{x}). \tag{8.16}$$

Of course, all these expressions reduce to the old rules if the vectors are real. And we may extend our definitions and say that if $\mathbf{x}\cdot\mathbf{y} = \mathbf{x}^\dagger\mathbf{y} = 0$, the vectors $\mathbf{x}$ and $\mathbf{y}$ are orthogonal, and that if $\mathbf{x}\cdot\mathbf{x} = \mathbf{x}^\dagger\mathbf{x} = 1$, the vector $\mathbf{x}$ is a unit vector.

Also, when we deal with vectors over the complex numbers, we must abandon the idea of the angle between two vectors. As an example of the paradoxes inherent in that idea, consider a unit vector $\mathbf{u}$ and its scalar multiple $i\mathbf{u}$, which is also of magnitude unity. Their dot product is $\mathbf{u}^\dagger(i\mathbf{u}) = i\mathbf{u}^\dagger\mathbf{u} = i$, which is an unusual value for the cosine of the angle between two vectors. Furthermore, the other dot product is $(i\mathbf{u})^\dagger\mathbf{u} = -i$, so that the angle between $\mathbf{u}$ and $i\mathbf{u}$ is not the same as the angle between $i\mathbf{u}$ and $\mathbf{u}$. To avoid this latest dilemma, it is common to define $\cos\theta = \operatorname{Re}(\mathbf{x}^\dagger\mathbf{y})/(\mathbf{x}^\dagger\mathbf{x})^{1/2}(\mathbf{y}^\dagger\mathbf{y})^{1/2}$. Then the cosine of the angle between $\mathbf{u}$ and $i\mathbf{u}$, or between $i\mathbf{u}$ and $\mathbf{u}$, is zero. But this is still an unusual value for the cosine of the angle between two vectors that are scalar multiples of each other, and that might therefore be presumed to be in the same direction. Indeed, the difficulties inherent in these considerations of angles suggest that it is not even very meaningful to attach the attribute direction to vectors over the complex numbers.

Exercises

59. Convince yourself that $\mathbf{x}\cdot\mathbf{x} = \mathbf{x}^\dagger\mathbf{x} = 0$ if and only if $\mathbf{x} = \mathbf{0}$.

*60. The SCHWARZ INEQUALITY

$$|\mathbf{x}\cdot\mathbf{y}| \leqslant |\mathbf{x}|\,|\mathbf{y}|$$

is "obviously" true for all vectors over the real numbers, since it is equivalent to $|\cos\theta| \leqslant 1$. However, for vectors over the complex numbers this theorem is not obviously true, since for complex numbers $|\cos z|$ may be greater than 1. Show that the Schwarz inequality does hold, even for vectors over the complex numbers, by considering $|(\mathbf{x}\cdot\mathbf{x})\mathbf{y} - (\mathbf{x}\cdot\mathbf{y})\mathbf{x}|^2/|\mathbf{x}|^2$.

61. Write out the Schwarz inequality in terms of components. Notice that it holds for all $\{x_i\}$ and $\{y_i\}$.

*62. Use the Schwarz inequality to prove that the TRIANGLE RULE, $|\mathbf{x}+\mathbf{y}| \leqslant |\mathbf{x}| + |\mathbf{y}|$, is true even for vectors over the complex numbers.

63. Express in matrix notation the dot product of $\mathbf{x}$ with the vector $A\mathbf{y}$, which results from applying the transformation A to the vector $\mathbf{y}$. What is the dot product of $A\mathbf{y}$ with $\mathbf{x}$?

*64. The dot product $\mathbf{x}\cdot\mathbf{y} = \mathbf{x}^\dagger\mathbf{y}$ takes on the simple form $\sum x_i^* y_i$ only in the special case that the basis set is orthonormal. How may the dot product be expressed in terms of components along a nonorthonormal basis set, $\{\mathbf{e}_i\}$. Express your answer in terms of the OVERLAP MATRIX, $\mathbf{S} = (\mathbf{e}_i\cdot\mathbf{e}_j)$. Check that your answer reduces to the simpler form when the basis set is orthonormal.

65. Generalize the formula for the Gram-Schmidt procedure so that it is correct for orthonormalizing a set of complex vectors. Check your answer. Then orthonormalize $(1, i, 1, i)^T$ and $(1, 1, i, i)^T$.

Finally, if $\mathbf{A}$ is square and $|\mathbf{A}| \neq 0$, we may define the INVERSE MATRIX

$$(\mathbf{A}^{-1})_{ij} \equiv \frac{(-1)^{i+j}|A_{ji}|}{|\mathbf{A}|}.$$

Thus, to invert a matrix, form the matrix of all cofactors, *then transpose*, and divide each element by $|\mathbf{A}|$. The inverse matrix satisfies

$$\mathbf{A}\mathbf{A}^{-1} = \mathbf{1} = \mathbf{A}^{-1}\mathbf{A}. \tag{8.17}$$

To prove the first equality, consider $(\mathbf{A}\mathbf{A}^{-1})_{ij} = \sum_k a_{ik}(\mathbf{A}^{-1})_{kj} = (1/|\mathbf{A}|) \sum_k (-1)^{k+j} a_{ik} |A_{jk}|$. If $i = j$, the sum is just the Laplace expansion of $|\mathbf{A}|$ in cofactors of the jth row. Therefore $(\mathbf{A}\mathbf{A}^{-1})_{jj} = 1$. If $i \neq j$, the sum is equal to the Laplace expansion of the determinant of a matrix whose ith row is identical to its jth row (Exercise 8.68), so that the determinant is zero. Therefore we conclude that $(\mathbf{A}\mathbf{A}^{-1})_{ij} = \delta_{ij}$; similar reasoning leads to $(\mathbf{A}^{-1}\mathbf{A})_{ij} = \delta_{ij}$. Further properties of the inverse include

$$(\mathbf{A}^{-1})^{-1} = \mathbf{A}$$

$$(\mathbf{A}^{\dagger})^{-1} = (\mathbf{A}^{-1})^{\dagger}$$

$$(\mathbf{A}\mathbf{B})^{-1} = \mathbf{B}^{-1}\mathbf{A}^{-1}.$$

This last result follows from $\mathbf{A}\mathbf{B}(\mathbf{B}^{-1}\mathbf{A}^{-1}) = \mathbf{A}\mathbf{1}\mathbf{A}^{-1} = \mathbf{1}$.

Example

7. Let

$$\mathbf{A} = \begin{pmatrix} 1 & i & 0 \\ 2 & 0 & -i \\ 0 & 1-i & -1 \end{pmatrix} \qquad \mathbf{x} = \begin{pmatrix} 0 \\ 1 \\ i \end{pmatrix} \qquad \mathbf{y} = \begin{pmatrix} i \\ -i \\ 2 \end{pmatrix}.$$

Then

$$\mathbf{A}^T = \begin{pmatrix} 1 & 2 & 0 \\ i & 0 & 1-i \\ 0 & -i & -1 \end{pmatrix} \qquad \mathbf{A}^* = \begin{pmatrix} 1 & -i & 0 \\ 2 & 0 & i \\ 0 & 1+i & -1 \end{pmatrix} \qquad \mathbf{A}^{\dagger} = \begin{pmatrix} 1 & 2 & 0 \\ -i & 0 & 1+i \\ 0 & i & -1 \end{pmatrix}$$

$$\mathbf{x}^{\dagger} = (0 \quad 1 \quad -i) \qquad \mathbf{y}^{\dagger} = (-i \quad i \quad 2)$$

$$\mathbf{x}^{\dagger}\mathbf{y} = (0 \quad 1 \quad -i)\begin{pmatrix} i \\ -i \\ 2 \end{pmatrix} = -3i \qquad \mathbf{y}^{\dagger}\mathbf{x} = (-i \quad i \quad 2)\begin{pmatrix} 0 \\ 1 \\ i \end{pmatrix} = 3i$$

$$\mathbf{x}^{\dagger}\mathbf{A}\mathbf{y} = (0 \quad 1 \quad -i)\begin{pmatrix} 1 & i & 0 \\ 2 & 0 & -i \\ 0 & 1-i & -1 \end{pmatrix}\begin{pmatrix} i \\ -i \\ 2 \end{pmatrix} = (0 \quad 1 \quad -i)\begin{pmatrix} 1+i \\ 0 \\ -3-i \end{pmatrix} = -1+3i$$

$$\mathbf{y}^{\dagger}\mathbf{A}^{\dagger}\mathbf{x} = (-i \quad i \quad 2)\begin{pmatrix} 1 & 2 & 0 \\ -i & 0 & 1+i \\ 0 & i & -1 \end{pmatrix}\begin{pmatrix} 0 \\ 1 \\ i \end{pmatrix} = (1-i \quad 0 \quad i-3)\begin{pmatrix} 0 \\ 1 \\ i \end{pmatrix} = -1-3i.$$

Also, $|\mathbf{A}| = i(1-i) + 2i = 1 + 3i$, and

$$\mathbf{A}^{-1} = \frac{1}{1+3i}\left\|\begin{matrix} \begin{vmatrix} 0 & -i \\ 1-i & -1 \end{vmatrix} & -\begin{vmatrix} 2 & -i \\ 0 & -1 \end{vmatrix} & \begin{vmatrix} 2 & 0 \\ 0 & 1-i \end{vmatrix} \\ -\begin{vmatrix} i & 0 \\ 1-i & -1 \end{vmatrix} & \begin{vmatrix} 1 & 0 \\ 0 & -1 \end{vmatrix} & -\begin{vmatrix} 1 & i \\ 0 & 1-i \end{vmatrix} \\ \begin{vmatrix} i & 0 \\ 0 & -i \end{vmatrix} & -\begin{vmatrix} 1 & 0 \\ 2 & -i \end{vmatrix} & \begin{vmatrix} 1 & i \\ 2 & 0 \end{vmatrix} \end{matrix}\right\|^T$$

$$= \frac{1}{1+3i}\begin{pmatrix} 1+i & i & 1 \\ 2 & -1 & i \\ 2-2i & -1+i & -2i \end{pmatrix}.$$

Exercises

*66. If A takes the vector $\mathbf{x}$ into the vector $\mathbf{y}$, let us assume that there exists a transformation A^{-1} that takes $\mathbf{y}$ back to $\mathbf{x}$. Use the rule for the matrix representation of the product transformation to convince yourself that if A is represented by the matrix $\mathbf{A}$, then A^{-1} is represented by the matrix $\mathbf{A}^{-1}$. (*Hint:* What transformation takes $\mathbf{x}$ into $\mathbf{x}$, and what must its matrix representation be?)

67. The reciprocal basis $\{\boldsymbol{\epsilon}_j\}$ may be generalized to n dimensions by requiring that $\mathbf{e}_i \cdot \boldsymbol{\epsilon}_j = \delta_{ij}$. Find the reciprocal basis vectors as linear combinations of original basis vectors: Let $\boldsymbol{\epsilon}_j = \sum \mathbf{e}_k t_{kj}$ and relate $\mathbf{T} = (t_{kj})$ to $\mathbf{S} = (\mathbf{e}_i \cdot \mathbf{e}_k)$.

*68. Convince yourself that $\sum_k (-1)^{j+k} a_{ik} |A_{jk}| = 0$ when $i \neq j$: The value of this sum does not depend upon the elements $\{a_{jk}\}_{k=1}^n$ of row j, since no a_{ik} depends on any a_{jk}, and since each minor $|A_{jk}|$ is the determinant of a matrix formed from $\mathbf{A}$ by deleting row j. Therefore the value of this sum is not changed if each a_{jk} of row j is replaced by the corresponding a_{ik} of row i. But this sum is then also the Laplace expansion, in cofactors of row j, of the determinant of a matrix whose ith row is identical with its jth row.

69. If $\mathbf{D} = \text{diag}\,(d_1, d_2, \ldots, d_n)$, what is $\mathbf{D}^{-1}$?

70. Use the rule for the determinant of a matrix product (Eq. 8.12) to convince yourself that $|\mathbf{A}^{-1}| = 1/|\mathbf{A}|$.

71. Convince yourself that $(\mathbf{A} \oplus \mathbf{B})^{-1} = \mathbf{A}^{-1} \oplus \mathbf{B}^{-1}$. Notice that this result means that the problem of inverting $\mathbf{A} \oplus \mathbf{B}$ is immediately simplified to the separate problems of inverting $\mathbf{A}$ and $\mathbf{B}$.

72. (a) Convince yourself that $(\mathbf{A}^n)^{-1} = (\mathbf{A}^{-1})^n$.
 *(b) Show that if $\mathbf{C}$ is cyclic of order n, then $\mathbf{C}^{-1}$ exists and is also cyclic of order n.

73. Show that if $\mathbf{P}$ is idempotent, either $\mathbf{P} = \mathbf{1}$ or $|\mathbf{P}| = 0$. (*Hint:* Assume $|\mathbf{P}| \neq 0$.)

74. Find the inverse of the matrix $\begin{pmatrix} 1 & i \\ i & 2 \end{pmatrix}$. Check your answer.

*75. Find the inverse of

$$\begin{pmatrix} 1/\sqrt{3} & 1/\sqrt{3} & 1/\sqrt{3} \\ -1/\sqrt{2} & 1/\sqrt{2} & 0 \\ -1/\sqrt{6} & -1/\sqrt{6} & 2/\sqrt{6} \end{pmatrix}.$$

8.8 Special Types of Matrices

We next define several special types of square matrices. We shall investigate the significance of these definitions in what follows.

A matrix $\mathbf{S}$ is said to be SINGULAR if $|\mathbf{S}| = 0$. Alternatively, a singular matrix is one with no inverse.

A matrix $\mathbf{R}$ is said to be REAL if $\mathbf{R}^* = \mathbf{R}$. A matrix $\mathbf{Y}$ is said to be SYMMETRIC if $\mathbf{Y}^T = \mathbf{Y}$. A matrix $\mathbf{A}$ is said to be ANTISYMMETRIC if $\mathbf{A}^T = -\mathbf{A}$. A matrix $\mathbf{H}$ is said to be HERMITIAN (also, SELF-ADJOINT) if $\mathbf{H}^\dagger = \mathbf{H}$. A matrix $\mathbf{K}$ is said to be SKEW-HERMITIAN if $\mathbf{K}^\dagger = -\mathbf{K}$.

A matrix $\mathbf{O}$ is said to be ORTHOGONAL if $\mathbf{O}^T = \mathbf{O}^{-1}$ or if $\mathbf{O}^T\mathbf{O} = \mathbf{1}$. A matrix $\mathbf{U}$ is said to be UNITARY if $\mathbf{U}^\dagger = \mathbf{U}^{-1}$ or if $\mathbf{U}^\dagger\mathbf{U} = \mathbf{1}$.

If we were to restrict ourselves to real matrices, there would be no distinction between Hermitian and symmetric, between skew-Hermitian and antisymmetric, or between unitary and orthogonal. However, since we do allow for matrices with complex elements, we shall concentrate on those matrices found to be most important—Hermitian, skew-Hermitian, and unitary. Since symmetric, antisymmetric, and orthogonal matrices with complex elements never occur in practical problems, any theorem about Hermitian (skew-Hermitian, unitary) matrices will also be true of those symmetric (antisymmetric, orthogonal) matrices encountered.

We summarize the definitions of the most important types of matrices:

SINGULAR	$\lvert\mathbf{S}\rvert = 0$
HERMITIAN	$\mathbf{H}^\dagger = \mathbf{H}$
UNITARY	$\mathbf{U}^\dagger = \mathbf{U}^{-1}$.

Exercises

76. Convince yourself that if $\mathbf{H}$ is Hermitian, then $\mathbf{H}^T = \mathbf{H}^*$.
77. Convince yourself that for any square matrix $\mathbf{M}$, the matrices $\mathbf{M} + \mathbf{M}^\dagger$ and $\mathbf{M}\mathbf{M}^\dagger$ are Hermitian.
*78. Show that any matrix may be expressed as the sum of a Hermitian matrix and a skew-Hermitian matrix. (Cf. Exercise 1.10.)
79. Convince yourself that if $\mathbf{A}$ and $\mathbf{B}$ are Hermitian, so are $\mathbf{ABA}$ and $\mathbf{BA}^2\mathbf{B}$. Convince yourself that if $\mathbf{U}$ is unitary, so are $\mathbf{U}^\dagger$ and $\mathbf{U}^{-1}$.
80. Show that for any Hermitian matrix $\mathbf{H}$ and any two vectors $\mathbf{x}$ and $\mathbf{y}$, $(\mathbf{x}^\dagger\mathbf{H}\mathbf{y})^* = \mathbf{y}^\dagger\mathbf{H}\mathbf{x}$. (*Hint:* Consider the triple product as a 1×1 matrix.) What can you conclude about $\mathbf{x}^\dagger\mathbf{H}\mathbf{x}$? What can you conclude about $\mathbf{x}^\dagger\mathbf{K}\mathbf{x}$ if $\mathbf{K}$ is skew-Hermitian?
81. Show that the commutator of two Hermitian matrices is skew-Hermitian.
82. Under what circumstance is the product of two Hermitian matrices Hermitian?
*83. Use the three previous results to show that if $\mathbf{F}$ and $\mathbf{G}$ are Hermitian, then for any $\mathbf{x}$, the quotient $(\mathbf{x}^\dagger[\mathbf{F}, \mathbf{G}]\mathbf{x})/(\mathbf{x}^\dagger\mathbf{G}^2\mathbf{x})$ is imaginary.
84. Is every unit matrix unitary? Is every unitary matrix a unit matrix? Do not confuse these terms.

85. Show that the product of two unitary matrices is unitary.
86. Show that if $\mathbf{H}$ is Hermitian and $\mathbf{U}$ unitary, then $\mathbf{U}^{-1}\mathbf{H}\mathbf{U}$ is also Hermitian.
*87. A matrix $\mathbf{D}$ is said to be POSITIVE DEFINITE if, for all $\mathbf{x}$, $\mathbf{x}^\dagger\mathbf{D}\mathbf{x} \geqslant 0$. Show that for any matrix $\mathbf{A}$, the matrix $\mathbf{A}^\dagger\mathbf{A}$ is positive definite. Then apply this result to $\mathbf{A} = \mathbf{F} + i\lambda\mathbf{G}$, with $\mathbf{F}$ and $\mathbf{G}$ Hermitian, and λ real (cf. Exercise 8.83) and equal to $(\mathbf{x}^\dagger[\mathbf{F}, \mathbf{G}]\mathbf{x})/2i(\mathbf{x}^\dagger\mathbf{G}^2\mathbf{x})$, to show that for all $\mathbf{x}$
$$(\mathbf{x}^\dagger\mathbf{F}^2\mathbf{x})(\mathbf{x}^\dagger\mathbf{G}^2\mathbf{x}) \geqslant -\tfrac{1}{4}(\mathbf{x}^\dagger[\mathbf{F}, \mathbf{G}]\mathbf{x})^2.$$
This expression is one form of the Heisenberg uncertainty principle.

We consider first the special properties of unitary matrices: If $\mathbf{U}$ is unitary, let us investigate $|\mathbf{U}|$. Firstly, since $\mathbf{U}^\dagger\mathbf{U} = \mathbf{1}$, $|\mathbf{U}^\dagger\mathbf{U}| = 1$. But $|\mathbf{U}^\dagger\mathbf{U}| = |\mathbf{U}^\dagger||\mathbf{U}|$ and $|\mathbf{U}^\dagger| = |\mathbf{U}|^*$. Therefore we conclude that
$$|\mathbf{U}|^*|\mathbf{U}| = 1 \tag{8.18}$$
or that $|\mathbf{U}|$ is of magnitude unity.

We next investigate the individual columns of $\mathbf{U}$. Let us consider the elements $\{(\mathbf{U})_{ki}\}_{k=1}^n$ of the ith column to be the components of an n-dimensional vector $\mathbf{u}_i$, and likewise for $\{(\mathbf{U})_{kj}\}_{k=1}^n$. Let us take the Hermitian scalar product of $\mathbf{u}_i$ and $\mathbf{u}_j$:
$$\mathbf{u}_i \cdot \mathbf{u}_j = \mathbf{u}_i^\dagger\mathbf{u}_j = \sum_{k=1}^n (\mathbf{U})_{ki}^*(\mathbf{U})_{kj} = \sum_k (\mathbf{U}^\dagger)_{ik}(\mathbf{U})_{kj} = (\mathbf{U}^\dagger\mathbf{U})_{ij} = (\mathbf{1})_{ij} = \delta_{ij}, \tag{8.19}$$
where we have used the definitions of the adjoint, of matrix multiplication, and of unitary matrices. Therefore we have shown that the column vectors $\{\mathbf{u}_i\}$ of the matrix $\mathbf{U}$ form an orthonormal set; each column of $\mathbf{U}$ represents a unit vector, and different columns are orthogonal. Similarly it can be shown that the same result holds for the rows of $\mathbf{U}$. And it can readily be shown that the converse is true: If the columns (rows) of a matrix form an orthonormal set of vectors, the matrix is unitary. For example, in Exercise 7.34 you used the Gram-Schmidt procedure to construct an orthonormal set of six six-dimensional row vectors; if these are stacked one above the other to form a 6×6 matrix, a unitary matrix results. One form of the general 2×2 unitary matrix is
$$\mathbf{U} = \begin{pmatrix} \cos\phi & \mp\sin\phi \\ \sin\phi & \pm\cos\phi \end{pmatrix}.$$

We may next consider the effect of transforming all column vectors by acting $\mathbf{U}$ on each. Then $\mathbf{x}$ becomes $\mathbf{U}\mathbf{x}$ and $\mathbf{y}$ becomes $\mathbf{U}\mathbf{y}$. And the dot product $\mathbf{x}^\dagger\mathbf{y}$ becomes $(\mathbf{U}\mathbf{x})^\dagger\mathbf{U}\mathbf{y} = \mathbf{x}^\dagger\mathbf{U}^\dagger\mathbf{U}\mathbf{y} = \mathbf{x}^\dagger\mathbf{1}\mathbf{y} = \mathbf{x}^\dagger\mathbf{y}$. What we have just shown is that the dot product is invariant to a unitary transformation of all vectors. And, in particular, $\mathbf{x}^\dagger\mathbf{x}$ is unchanged by a unitary transformation. Although a transformation generally twists vectors around and stretches (or squeezes) them, a unitary transformation preserves lengths. Furthermore, if we were considering only matrices with real elements, then we would be able to conclude that the invariance of the dot product also means that the angles between vectors are unchanged by a unitary (orthogonal) transformation. But the only kind of linear transformation that preserves lengths and angles is a rotation about some axis (possibly coupled with an inversion through the origin). When we consider matrices with complex elements, it is still profitable to consider a unitary transformation as a generalized rotation, but without allowing this interpretation to become too geometrical.

Finally, let us consider the effect of applying a unitary transformation U to a set $\{\mathbf{e}_i\}$ of orthonormal basis vectors, the ith of which is represented by the column matrix e_i, whose only nonzero element is the ith, which is unity. By the above property of unitary transformations, since $\{\mathbf{e}_i\}$ is an orthonormal set, so is $\{U\mathbf{e}_i\}$. And the column matrix that represents $U\mathbf{e}_i$ is Ue_i, which is just the ith column of U. Therefore we may conclude that the columns of U represent another orthonormal set of vectors that may be produced by applying the transformation U to the basis set.

Exercises

88. What can you conclude about the determinant of a real orthogonal matrix?
89. Convince yourself that the columns (rows) of $\begin{pmatrix} 1/\sqrt{2} & i/\sqrt{2} \\ i/\sqrt{2} & 1/\sqrt{2} \end{pmatrix}$ form an orthonormal set, and that this is therefore an example of a unitary matrix with complex elements.
90. Convince yourself that if U is unitary, so is the matrix formed from U by multiplying every element of a row (column) by a phase factor $e^{i\delta}$.
91. Convince yourself that if two rows (columns) of a unitary matrix are interchanged, the resultant matrix is still unitary.

*92. Find the 3×3 unitary matrix which effects each of the following transformations on the column vector $(x, y, z)^T$.
   (a) Inversion through the origin.
   (b) Reflection in the xy-plane.
   (c) Rotation by an angle ϕ about the z-axis.
   (d) Rotation by an angle ϕ about the x-axis.
   (e) The rotation of (c), followed by the reflection of (b). (*Hint:* Multiply the matrices, then check.)
   (f) The rotation of (c), followed by inversion. Notice that if $\phi = 180°$, this combination of operations is equivalent to the single operation, reflection (b).
   (g) Show that rotations (c) and (d) do not commute.
   (h) Show that rotation by an angle ϕ about the z-axis commutes with rotation by an angle θ about that axis.

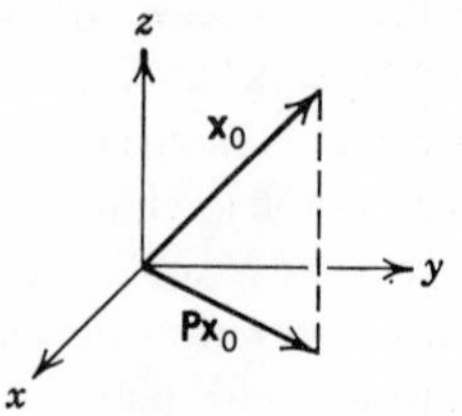

Figure 8.3

*93. Describe the transformation represented by each of the following unitary matrices:

(a) $\begin{pmatrix} 1 & 0 & 0 \\ 0 & -1 & 0 \\ 0 & 0 & 1 \end{pmatrix}$.

(b) $\begin{pmatrix} 0 & 1 & 0 \\ -1 & 0 & 0 \\ 0 & 0 & 1 \end{pmatrix}$.

(c) $\begin{pmatrix} 0 & 1 & 0 \\ 1 & 0 & 0 \\ 0 & 0 & 1 \end{pmatrix}$.

(d) $\begin{pmatrix} 0 & 1 & 0 \\ 0 & 0 & 1 \\ 1 & 0 & 0 \end{pmatrix}$.

Next we investigate the special properties of singular matrices. We may obtain a geometrical interpretation of a singular matrix by considering the particularly simple one

$$\mathbf{P} = \begin{pmatrix} 1 & 0 & 0 \\ 0 & 1 & 0 \\ 0 & 0 & 0 \end{pmatrix} \qquad |\mathbf{P}| = 0$$

(which is a projection matrix and thus not a general example of a singular matrix). Let us act $\mathbf{P}$ on the particular column vector $\mathbf{x}_0 = (x_0, y_0, z_0)^T$.

$$\mathbf{P}\mathbf{x}_0 = \begin{pmatrix} 1 & 0 & 0 \\ 0 & 1 & 0 \\ 0 & 0 & 0 \end{pmatrix} \begin{pmatrix} x_0 \\ y_0 \\ z_0 \end{pmatrix} = \begin{pmatrix} x_0 \\ y_0 \\ 0 \end{pmatrix}.$$

Thus we see that the result of operating $\mathbf{P}$ on $\mathbf{x}_0$ is a vector in the xy-plane (see Fig. 8.3). Indeed, the result of operating $\mathbf{P}$ on any column vector of the form $\mathbf{x} = (x_0, y_0, z)^T$, with z arbitrary, is still the vector $\mathbf{P}\mathbf{x}_0$. Given the result $(x_0, y_0, 0)^T$, we would not know which $\mathbf{x}$ it came from; there is a whole lineful of possibilities. In matrix terms, there is no matrix that will act on $\mathbf{P}\mathbf{x}$ to regenerate $\mathbf{x}$, so $\mathbf{P}^{-1}$ does not exist. We may also note that any vector $\mathbf{z} = (0, 0, z),^T$ along the z-axis, is annihilated by $\mathbf{P}$: $\mathbf{P}\mathbf{z} = \mathbf{0}$. And there certainly cannot exist any $\mathbf{P}^{-1}$ that could multiply a zero column matrix to produce a nonzero matrix. In terms of the three row vectors, $\mathbf{p}_1 = (1, 0, 0)$, $\mathbf{p}_2 = (0, 1, 0)$, and $\mathbf{p}_3 = (0, 0, 0)$, of $\mathbf{P}$, the matrix equation $\mathbf{P}\mathbf{z} = \mathbf{0}$ becomes the three numerical equations, $\mathbf{p}_1\mathbf{z} = 0$, $\mathbf{p}_2\mathbf{z} = 0$, and $\mathbf{p}_3\mathbf{z} = 0$. Geometrically, these equations mean that $\mathbf{z}$ is orthogonal to all three row vectors, $\mathbf{p}_1$, $\mathbf{p}_2$, and $\mathbf{p}_3$. (We have not taken the Hermitian scalar product because each $\mathbf{p}_i$ is already a row matrix.) The existence of a nonzero

z orthogonal to all three row vectors means that the three cannot span three-dimensional space, and that they must therefore be linearly dependent. That they in fact span only a two-dimensional space is obvious in this example. And we may see that $|\mathbf{P}|$, which is the volume of the parallelepiped whose edges are $\mathbf{p}_1$, $\mathbf{p}_2$, and $\mathbf{p}_3$, is zero.

By generalizing from this particular singular matrix, we may summarize all the properties of a singular $n \times n$ matrix: If **S** is singular, then

1. $|\mathbf{S}| = 0$.
2. The "volume" of the n-dimensional "parallelepiped" whose edges are the row vectors $\{\mathbf{s}_i\}$ of **S** is zero.
3. The row vectors $\{\mathbf{s}_i\}$ are linearly dependent.
4. The row vectors $\{\mathbf{s}_i\}$ do not span n-dimensional space.
5. There exists a nonzero vector **p** that is orthogonal to every $\mathbf{s}_i$: $\mathbf{s}_i\mathbf{p} = 0$.
6. There exists a nonzero vector **p** that is annihilated by **S**: $\mathbf{Sp} = \mathbf{0}$.
7. $\mathbf{S}^{-1}$ does not exist.
8. By the symmetry of rows and columns, properties 2–5 are also true of the column vectors of **S**.

Exercises

*94. Convince yourself that if the set $\{\mathbf{v}_i\}_{i=1}^n$ of n-dimensional vectors is linearly dependent, the overlap matrix $\mathbf{S} = (\mathbf{v}_i \cdot \mathbf{v}_j)$ is singular. In this context, **S** is known as the GRAMIAN and $|\mathbf{S}|$ as the GRAM DETERMINANT.

*95. Discuss the requirement for three planes, $(\mathbf{X}_1 - \mathbf{X}_1^0) \cdot \mathbf{n}_1 = 0$, $(\mathbf{X}_2 - \mathbf{X}_2^0) \cdot \mathbf{n}_2 = 0$, and $(\mathbf{X}_3 - \mathbf{X}_3^0) \cdot \mathbf{n}_3 = 0$, to intersect in a point? (*Hint:* What if $\mathbf{n}_1$, $\mathbf{n}_2$, and $\mathbf{n}_3$ are coplanar?)

8.9 Simultaneous Linear Equations

Next we consider how we may use matrix methods in connection with the problem of SIMULTANEOUS LINEAR EQUATIONS. We shall first consider the case in which there are n equations and n unknowns, $\{x_i\}$.

$$\begin{aligned} a_{11}x_1 + a_{12}x_2 + \cdots + a_{1n}x_n &= c_1 \\ a_{21}x_1 + a_{22}x_2 + \cdots + a_{2n}x_n &= c_2 \\ &\vdots \\ a_{n1}x_1 + a_{n2}x_2 + \cdots + a_{nn}x_n &= c_n. \end{aligned}$$

In matrix notation, we may write these equations concisely as

$$\mathbf{Ax} = \mathbf{c}, \tag{8.20}$$

where $\mathbf{A} = (a_{ij})$ is square and $\mathbf{x} = (x_i)$ and $\mathbf{c} = (c_i)$ are column matrices. If $|\mathbf{A}| \neq 0$, the solution to these equations may be obtained, at least symbolically, simply by multiplying both sides of Eq. 8.20 from the left by $\mathbf{A}^{-1}$, and invoking $\mathbf{A}^{-1}\mathbf{A}\mathbf{x} = \mathbf{1}\mathbf{x} = \mathbf{x}$:

$$\mathbf{x} = \mathbf{A}^{-1}\mathbf{c}.$$

Therefore the problem represented by the n equations is equivalent to inverting a matrix and performing a matrix multiplication. You may wish to show (Exercise 8.98) that this result is equivalent to the familiar Cramer's rule.

Exercises

96. Use the result of Exercise 8.74 to solve $x + iy = 3$ and $ix + 2y = 3i$.

*97. A common technique in chemistry is the spectrophotometric analysis of a multicomponent mixture. If the concentration of the ith component is c_i and its extinction coefficient at wavelength λ_j is ε_{ji}, the total absorbance at λ_j due to all components is given by $A_j = \sum_i \varepsilon_{ji} l c_i$, or in matrix form $\mathbf{A} = l\boldsymbol{\varepsilon}\mathbf{c}$, where $\mathbf{A}$ and $\mathbf{c}$ are column matrices, $\boldsymbol{\varepsilon}$ is assumed to be square, and l is a scalar, the pathlength of the cell. Then the solution to these simultaneous equations is simply $\mathbf{c} = (1/l)\boldsymbol{\varepsilon}^{-1}\mathbf{A}$, so that it is possible to determine each c_i by measuring the absorbances. Of course, for maximum accuracy the wavelengths should be chosen so that each is specific for a different component, with the ith component absorbing only λ_i. Then the matrix $\boldsymbol{\varepsilon}$ would be diagonal. In practice this is not possible, so that $\boldsymbol{\varepsilon}$ is only "almost diagonal." Let $\boldsymbol{\varepsilon} = \mathbf{D} + \boldsymbol{\delta}$ and find a two-term approximation for $\boldsymbol{\varepsilon}^{-1}$. (*Note:* It is not $\mathbf{D}^{-1} + \boldsymbol{\delta}^{-1}$.)

*98. Convince yourself that the solution of $\mathbf{A}\mathbf{x} = \mathbf{c}$, given by $\mathbf{x} = \mathbf{A}^{-1}\mathbf{c}$, is equivalent to the solution according to CRAMER'S RULE, $x_i = |\mathbf{A}_i|/|\mathbf{A}|$, where $\mathbf{A}_i$ is the matrix formed from $\mathbf{A}$ by replacing the ith column of $\mathbf{A}$ by $\mathbf{c}$. [*Hint:* What is $(\mathbf{A}^{-1}\mathbf{c})_i$]?

*99. A solution contains both Cl^- and Br^- ions. A 50.00-ml aliquot of the solution titrates 43.56 ml 0.1046*N* $AgNO_3$. Another 50.00-ml aliquot, treated with excess $AgNO_3$, gives 0.6873 g of mixed silver halide precipitate. Calculate $[Cl^-]$ and $[Br^-]$ in the solution.

An alternative method of solving n simultaneous linear equations in n unknowns is GAUSSIAN ELIMINATION, a systematic reduction to triangular form quite similar to that used to evaluate a determinant. Just as we reduced $|\mathbf{A}|$ to triangular form by repeatedly multiplying one row by an appropriate factor and adding to another row, so also may we solve the simultaneous linear equations $\mathbf{A}\mathbf{x} = \mathbf{c}$ by repeatedly multiplying one equation by an appropriate factor and adding it to another equation. Thus, multiplying the $(n - 1)$st equation above by $a_{n1}/a_{n-1,1}$ and subtracting from the nth equation annihilates a_{n1} and converts the nth equation to

$$0 + (a_{n2} - a_{n1}a_{n-1,2}/a_{n-1,1})x_2 + \cdots + (a_{nn} - a_{n1}a_{n-1,n}/a_{n-1,1})x_n = c_n - a_{n1}c_{n-1}/a_{n-1,1}.$$

Repeated application of this procedure converts the problem to the triangular form

$$\begin{aligned}
a_{11}x_1 + a_{12}x_2 + \cdots \qquad &+ \quad a_{1n}x_n = c_1 \\
a'_{22}x_2 + \cdots \qquad &+ \quad a'_{2n}x_n = c'_2 \\
&\vdots \\
a'_{n-1,n-1}x_{n-1} &+ a'_{n-1,n}x_n = c'_{n-1} \\
&\quad a'_{nn}x_n = c'_n.
\end{aligned}$$

Solving from the bottom up gives $x_n = c'_n/a'_{nn}$, $x_{n-1} = (c'_{n-1} - a'_{n-1,n}x_n)/a'_{n-1,n-1}$, etc. In practice, it is most convenient merely to write **c** next to **A** on a blackboard, and perform operations on this $n \times (n + 1)$ matrix. This reduction to triangular form is also applicable to matrix inversion by computer (Exercise 8.101).

Exercises

100. Solve by reduction to triangular form (cf. Exercise 8.46).

$$\begin{aligned}
x_1 + 2x_2 + x_3 + 3x_4 &= 2 \\
x_2 + 2x_3 + 2x_4 &= -1 \\
2x_1 + 3x_2 + x_3 + x_4 &= 4 \\
3x_1 \qquad + 2x_3 + x_4 &= -3.
\end{aligned}$$

*101. Let $\mathbf{\Pi}$ be the matrix of Exercise 8.43 such that $\mathbf{\Pi A}$ is a triangular matrix **T**. Reduce the problem of finding $\mathbf{A}^{-1}$ to the easier problem of finding $\mathbf{T}^{-1}$.

It is not necessary that **A** be square; in the general problem there may be more unknowns than equations or more equations than unknowns.

1. If there are more unknowns than equations, there are not sufficient constraints to determine the solution vector **x** uniquely. For example, the simplest case—one equation, $a_{11}x_1 + a_{12}x_2 = c_1$, in two unknowns—has a whole lineful of solutions, instead of a single point.

2. If there are more equations than unknowns, it is possible that the equations are inconsistent, so that there is no solution possible. The simplest example of this situation is the two equations, $x_1 = 0$ and $x_1 = 1$, in one unknown. Both these situations are covered in those mathematics texts where the "rank" of a matrix is discussed; neither occurs very frequently in practical problems.

3. Another possibility with more equations than unknowns is that the equations are not linearly independent, so that some of the extra equations are redundant. In some circumstances there does exist a unique solution. The simplest example of this situation is the two equations, $x_1 = 1$ and $2x_1 = 2$, in one unknown. The problem of n linearly dependent equations in $n - 1$ unknowns, with one equation redundant, is readily converted to the problem

of $n - 1$ linearly independent equations in $n - 1$ unknowns merely by discarding a redundant equation. We shall use this technique below, in a variant of this problem.

Exercise

102. Balance each of the following equations. (*Note:* This classical problem can be set up as a problem of simultaneous equations. The unknowns are the coefficients, and the equations are balance of charge and of each chemical element. What should you do about the fact that there are only $n - 1$ equations for n unknowns? For ease of solution, try to minimize the number of unknowns.)
(a) $Cr_2O_7^{2-} + Co(NO_2)_6^{3-} + H^+ = Cr^{3+} + Co^{2+} + NO_3^- + H_2O$.
*(b) $MnO_4^- + Fe(CN)_6^{4-} + H^+ = Mn^{2+} + Fe^{3+} + CO_2 + HNO_3 + H_2O$.

We next turn to the problem of n simultaneous linear HOMOGENEOUS equations in n unknowns

$$\mathbf{Ax} = \mathbf{0}.$$

Here $\mathbf{A}$ is square, but the constants on the right are all zero. First we note that there always exists the trivial solution $\mathbf{x} = \mathbf{0}$, which we shall ignore. Next we note that if $\mathbf{A}$ is nonsingular, the trivial solution is the only solution, since $\mathbf{A}^{-1}$ exists and $\mathbf{A}^{-1}\mathbf{Ax} = \mathbf{x} = \mathbf{0}$. Therefore we conclude that a necessary condition for a nontrivial solution is that $\mathbf{A}$ be singular. (From the following discussion, it may be apparent that this is also a sufficient condition.) We may summarize the various equivalent ways of recognizing the existence of a nontrivial solution:

1. There exists a nonzero vector $\mathbf{x}$ which is annihilated by $\mathbf{A}$: $\mathbf{Ax} = \mathbf{0}$.
2. There exists a nonzero vector $\mathbf{x}$ which is orthogonal to every row vector $\mathbf{a}_i$ of $\mathbf{A}$: $\mathbf{a}_i\mathbf{x} = 0$.
3. The row vectors $\{\mathbf{a}_i\}$ do not span n-dimensional space.
4. The row vectors $\{\mathbf{a}_i\}$ are linearly dependent; the equations $\{\mathbf{a}_i\mathbf{x} = 0\}$ are redundant.
5. The "volume" of the n-dimensional "parallelepiped" whose edges are the vectors $\{\mathbf{a}_i\}$ is zero.
6. $|\mathbf{A}| = 0$.
7. $\mathbf{A}$ is singular.
8. $\mathbf{A}^{-1}$ does not exist.

Before we find this nontrivial solution, we should note that any scalar multiple of a solution is also a solution, since if $\mathbf{Ax} = \mathbf{0}$, then $\mathbf{A}(\alpha\mathbf{x}) = \mathbf{0}$. Therefore there is a whole lineful of solutions, which goes through the origin. Furthermore, if the n row vectors $\{\mathbf{a}_i\}$ span a less-than-$(n - 1)$-dimensional space, then there is not merely a lineful of solutions, but a planeful, 3-spaceful, etc.; if $\mathbf{x}_1$ and $\mathbf{x}_2$ are two vectors, both orthogonal to every $\mathbf{a}_i$, then so is any linear combination of $\mathbf{x}_1$ and $\mathbf{x}_2$, since if $\mathbf{Ax}_1 = \mathbf{0}$ and $\mathbf{Ax}_2 = \mathbf{0}$, then $\mathbf{A}(\alpha_1\mathbf{x}_1 + \alpha_2\mathbf{x}_2) = \mathbf{0}$. And if $\mathbf{x}_1$ and $\mathbf{x}_2$ are linearly independent, the locus of all linear combinations of these two vectors is the plane they span.

We shall merely state that in the case where the n row vectors $\{\mathbf{a}_i\}$ span an $(n - 1)$-dimensional space, the solution to $\mathbf{Ax} = \mathbf{0}$ is given by

$$x_i = (-1)^{i+j} |A_{ji}| \tag{8.21}$$

or any multiple thereof. Here the choice of the jth row is arbitrary, except that it must be chosen so that not all the cofactors of its elements are zero; in practice it is convenient to choose the row whose elements are "most complicated." This solution arises in the following fashion: Let us assign a value to a particular unknown x_k and solve for the remaining variables in terms of this one. The value assigned to x_k may be chosen arbitrarily, and we choose this value to be $(-1)^{k+j} |A_{jk}|$ in order to preserve symmetry in the final answer. Next each constant term we have created is transposed across the equal sign. Then the problem becomes one of solving n linearly dependent equations in $n - 1$ unknowns. If we eliminate a redundant equation (the jth, corresponding to the choice of cofactor above), solution of the remaining $n - 1$ linearly independent inhomogeneous equations in $n - 1$ unknowns leads to Eq. 8.21 (see Exercise 8.103). If the remaining $n - 1$ equations happen still to be linearly dependent, all the cofactors will be zero and another row should be tried.

If the n row vectors $\{\mathbf{a}_i\}$ span only an $(n - 2)$-dimensional space, every cofactor of $\mathbf{A}$ will be found to be zero. The simplest remedy for such a situation is to add the small quantity δ to one arbitrarily chosen element of $\mathbf{A}$. Then evaluate the cofactors of some row not containing the δ; some of these cofactors are identically zero and some are proportional to δ. Finally, divide each cofactor by δ to obtain a set of cofactors, and a set of components, that are not all zero. Another solution vector may be obtained by repeating this procedure but with the δ added to another element of $\mathbf{A}$. Of course, you should check that these two solution vectors are linearly independent, and if not try again. In the rare case where there is more than a planeful of solutions, then several δs must be sprinkled through $\mathbf{A}$ in order to obtain a cofactor that is not identically zero. Although we have not provided any method of determining the dimension of the space spanned by $\{\mathbf{a}_i\}$, we shall see that in practice (in connection with degenerate eigenvalues) it is known how many linearly independent solutions vectors are to be found.

Exercises

*103. Convince yourself that the solution to $\mathbf{Ax} = \mathbf{0}$ is indeed $x_i = (-1)^{i+j} |A_{ji}|$: Let $\mathbf{A}$ be 3×3 and discard the third equation, then solve by Cramer's rule for x_1 and x_3 in terms of x_2, and finally set $x_2 = -|A_{32}|$.

104. Solve the homogeneous equations $\mathbf{Ax} = \mathbf{0}$, where

$$\mathbf{A} = \begin{pmatrix} 0 & -1 & 1 \\ 0 & 2 & -2 \\ 1 & 3 & 0 \end{pmatrix}.$$

First convince yourself that the minors $|A_{3i}|$ are all zero, since the first and second equations are still dependent. Then find a solution by evaluating the cofactors of some other row.

*105. Convince yourself that if the n row vectors of $\mathbf{A}$ span only an m-dimensional space, there is a whole $(n - m)$-dimensional-spaceful of solutions to $\mathbf{Ax} = \mathbf{0}$.

106. Solve the homogeneous equations $\mathbf{Ax} = \mathbf{0}$, where

$$\mathbf{A} = \begin{pmatrix} 1 & 2 & 3 \\ 2 & 4 & 6 \\ 3 & 6 & 9 \end{pmatrix}.$$

First convince yourself that the cofactor method fails because all $|A_{ji}|$ are zero. Then apply the delta method twice to obtain two linearly independent solution vectors. (In practice it is customary to go further and orthogonalize these vectors; do so.)

*107. What is the *general* solution of $\mathbf{Ax} = \mathbf{c}$ in the case that $|\mathbf{A}| = 0$? (*Hint:* Let $\mathbf{x}_0$ be a particular solution; such a solution does not necessarily exist.)

8.10 Digression on Simultaneous Equations

The simultaneous equations that we have solved by matrix techniques have all been linear, since there are no products of unknowns. Indeed the matrix techniques are applicable only to linear equations, and no general method exists for solving n simultaneous nonlinear equations in n unknowns. Nor is there even a general method of deciding whether the equations are inconsistent or dependent.

One possibility is to solve one equation for one of the unknowns, eliminate that unknown from the other equations, and thereby reduce a problem of n equations in n unknowns to one of $(n - 1)$ equations in $(n - 1)$ unknowns:

A Familiar Example

8. To find the unknowns x and y, given their sum $s = x + y$ and their ratio $r = x/y$: Solve the second equation for $x = ry$, substitute in the first, and solve for $y = s/(r + 1)$. Then $x = ry = rs/(r + 1)$.

A More Difficult Example

9. To solve $a_1x^2 + a_2xy + a_3y^2 = c_1$ and $b_1x^2 + b_2xy + b_3y^2 = c_2$ requires solving for one variable by the quadratic formula, substituting into the other equation, and solving the resulting quadratic (or worse) equation.

An "Impossible" Example

10. To solve $a_1xy + a_2e^x + a_3e^y = c_1$ and $b_1xy + b_2e^x + b_3e^y = c_2$ requires a trick, since neither equation can be solved for one variable as a function of the other. However, multiplying the first by b_1 and the second by a_1, subtracting, and transposing gives

$$(a_2b_1 - a_1b_2)e^x = (a_1b_3 - a_3b_1)e^y + (b_1c_1 - a_1c_2),$$

which can be solved for x as a function of y. This result may be substituted into either equation to produce a transcendental equation to be solved for y.

A Hopeless Example

11. There is not even a trick available to solve $a_1x^2y + a_2e^x + a_3e^y = c_1$ and $b_1xy^2 + b_2e^x + b_3e^y = c_2$. However, there does exist an iterative method which is a generalization of the Newton-Raphson method (Exercise 2.10).

One further feature of simultaneous equations deserves to be mentioned. Whenever you approach such a problem it is often helpful to decide how many unknowns there are and how many independent equations (data, constraints on the unknowns) are available.

Example

12. What is the pH of an $0.1M$ solution of NH_3? As stated, there are three unknowns, $[H^+]$, $[NH_3]$, and $[NH_4^+]$. One equation is $[NH_3] + [NH_4^+] = 0.1$, by conservation of nitrogen. Another is $K_a^{NH_4^+} = [H^+][NH_3]/[NH_4^+] = 10^{-9.25}$. Therefore we see that to solve the problem we must find another equation. Clearly, another comes from conservation of charge: $[H^+] + [NH_4^+] = [OH^-]$. However, we have introduced another unknown, $[OH^-]$. Now we have three equations in four unknowns, and we still need another independent equation, which is $[H^+][OH^-] = 10^{-14}$. Now there are sufficient equations to determine the unknowns, although the solution of these nonlinear equations is quite tedious without the usual approximations, $[NH_3] \gg [NH_4^+] \gg [H^+]$.

This sort of analysis of variables and constraints is not without intrinsic significance. As an example, let us derive the Gibbs phase rule.

In a thermodynamic system composed of c components and p phases in equilibrium, there are a total of $2 + cp$ variables, since a complete description of the situation requires specifying the temperature, the pressure, and the mole fraction of each component in each phase (but not the volume of each phase). Next let us enumerate the constraints.

1. For each phase the sum of the mole fractions must be unity. This condition imposes p constraints.

2. Each component must have the same chemical potential in every phase. This condition imposes $(p - 1)$ constraints for each component, since once the chemical potential of a given component is specified for one phase, it must have the same value in each of the other phases.

Therefore the total number of constraints is $p + c(p - 1)$. Subtracting the number of constraints from the number of variables leaves the number of independently variable degrees of freedom, $f = 2 + c - p$.

8.11 Linear Transformation of Basis Vectors

In the basis set $\{\mathbf{e}_j\}$ the vector $\mathbf{v}$ is represented by the column matrix $\mathbf{v}$. Next we consider what happens to $\mathbf{v}$ when we apply the transformation T to each member of the basis set, to obtain a new basis set. Let us call the jth member of this new basis set $\mathbf{e}'_j$, so that $\mathbf{e}'_j = T\mathbf{e}_j = \sum_i \mathbf{e}_i t_{ij}$. Then T is represented by the nonsingular matrix $\mathbf{T} = (t_{ij})$, whose jth column contains the components of the new basis vector $\mathbf{e}'_j$, expressed in the old basis set. Although the vector $\mathbf{v}$ was expanded in the old basis set as $\mathbf{v} = \sum \mathbf{e}_i v_i$, in the new basis the components of $\mathbf{v}$ are different, so that the expansion now becomes $\mathbf{v} = \sum \mathbf{e}'_j v'_j$. Thus the column matrix $\mathbf{v} = (v_i)$ is different from the column matrix $\mathbf{v}' = (v'_j)$. We now relate these two column matrices by equating the two expansions and substituting into the new one:

$$\sum_i \mathbf{e}_i v_i = \sum_j \mathbf{e}'_j v'_j = \sum_j T\mathbf{e}_j v'_j = \sum_{i,j} \mathbf{e}_i t_{ij} v'_j = \sum \mathbf{e}_i \sum_j t_{ij} v'_j.$$

Comparison of the components along each $\mathbf{e}_i$ shows that $v_i = \sum_j t_{ij} v'_j$, or

$$\mathbf{v} = \mathbf{T}\mathbf{v}'$$

and

$$\mathbf{v}' = \mathbf{T}^{-1}\mathbf{v}. \tag{8.22}$$

This result is to be compared with the way the transformation acts on basis vectors:

$$\mathbf{e}'_i = T\mathbf{e}_i = \sum_j \mathbf{e}_j (\mathbf{T})_{ji} \tag{8.23}$$

and

$$\mathbf{e}_j = T^{-1}\mathbf{e}'_j = \sum_i \mathbf{e}_i (\mathbf{T}^{-1})_{ij}.$$

This is a seemingly paradoxical result. It would seem as though operating on a basis vector should be the same as operating on some arbitrary vector $\mathbf{v}$. Yet we find that although $\mathbf{T}$ is the matrix that converts the old basis vector, $\mathbf{e}_i$, into the new one, $\mathbf{e}'_i$, it is not $\mathbf{T}$ but $\mathbf{T}^{-1}$ that converts the old vector, $\mathbf{v}$, into the new one, $\mathbf{v}'$.

The clue to this puzzle is that operating $\mathbf{T}$ on a vector $\mathbf{e}_i$ is indeed different from operating on $\mathbf{v}$, which is a column of numbers. These numbers express the orientation of the vector $\mathbf{v}$ relative to the basis vectors. If we twist the basis vectors around by acting T upon each, while we keep $\mathbf{v}$ fixed, this relative orientation changes as the basis vectors "rotate out from under $\mathbf{v}$." But we could achieve the same change of relative orientation by keeping the basis set fixed and "rotating $\mathbf{v}$ in the opposite way."

A Simple Example

13. Let $\mathbf{v} = \mathbf{e}_1 + \mathbf{e}_2$ (See Figure 8.4), or $\mathbf{v} = \begin{pmatrix} 1 \\ 1 \end{pmatrix}$. Let T rotate the basis vectors 90° clockwise. Then $\mathbf{e}'_1 = -\mathbf{e}_2$ ($t_{21} = -1$) and $\mathbf{e}'_2 = \mathbf{e}_1$ ($t_{12} = 1$), so $\mathbf{T} = \begin{pmatrix} 0 & 1 \\ -1 & 0 \end{pmatrix}$. And $\mathbf{v}' = \mathbf{T}^{-1}\mathbf{v} = \begin{pmatrix} 0 & -1 \\ 1 & 0 \end{pmatrix}\begin{pmatrix} 1 \\ 1 \end{pmatrix} = \begin{pmatrix} -1 \\ 1 \end{pmatrix}$, or $\mathbf{v} = -\mathbf{e}'_1 + \mathbf{e}'_2$, as is also apparent from Figure 8.4. We have changed the orientation of $\mathbf{v}$ relative to the basis by rotating the basis

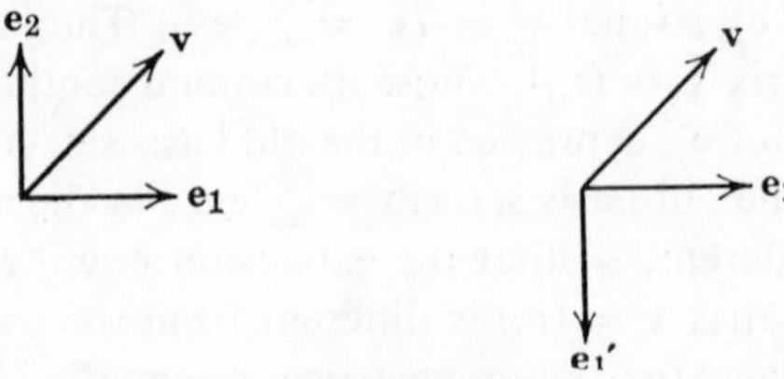

Figure 8.4

vectors 90° clockwise, but we could accomplish the same change of relative orientation by holding the basis vectors fixed and rotating **v** 90° counterclockwise.

By way of analogy to changing the basis, consider the length 3 cm. In inches, this length is 1.18 in., since 1 in. = 2.54 cm. This is still the same length, but the number specifying it is smaller because the unit of length in which it is expressed is now larger. For a fixed length the measure of length and the unit of length change in opposite directions. Likewise, for a fixed vector the "measure" of the vector (its components) and the reference (basis vectors) change in opposite directions.

Exercise

108. Let $\mathbf{E}$ be the row vector $(\mathbf{e}_1\,\mathbf{e}_2 \cdots \mathbf{e}_n)$, whose elements are basis vectors. (Cf. Exercise 8.25.) Then we may write the vector **v** as the matrix product $\mathbf{Ev}$, where $\mathbf{v} = (v_j)$, the column matrix of components of **v** in the basis $\{\mathbf{e}_j\}$. But $\mathbf{Ev} = \mathbf{ETT}^{-1}\mathbf{v}$. Use this simple result to convince yourself that on transforming to the new basis $\{\mathbf{e}_i'\}$, where $\mathbf{e}_i' = \sum_j \mathbf{e}_j(\mathbf{T})_{ji} = (\mathbf{ET})_i$, the column matrix **v** should be replaced by $\mathbf{T}^{-1}\mathbf{v}$.

Let us mix the two possible types of transformations. Let the matrix $\mathbf{A}$ convert the column matrix $\mathbf{v}$ into the column matrix $\mathbf{w}$: $\mathbf{Av} = \mathbf{w}$, where these are all referred to the basis set $\{\mathbf{e}_i\}$. Next we transform the basis vectors by the transformation T: $\mathbf{e}_i' = T\mathbf{e}_i$. Then **v** becomes $\mathbf{v}' = \mathbf{T}^{-1}\mathbf{v}$ and $\mathbf{w}$ becomes $\mathbf{w}' = \mathbf{T}^{-1}\mathbf{w}$. Substituting into $\mathbf{Av} = \mathbf{w}$ leads to $\mathbf{ATv}' = \mathbf{Tw}'$, or

$$\mathbf{T}^{-1}\mathbf{ATv}' = \mathbf{w}'. \tag{8.24}$$

Thus we see that the matrix representation of the transformation that produces the vector **w** from the vector **v** depends upon the choice of basis. The matrix appropriate for computation in the new coordinate system is no longer $\mathbf{A}$, but $\mathbf{T}^{-1}\mathbf{AT}$.

Exercises

109. The projection matrices $\{\mathbf{P}_i\}$ of Exercise 8.29 may be generalized to the set of operators $\{Q_i\}$ such that Q_i acting on a vector **v** produces the projection of **v** along the ith vector of an orthonormal set $\{\mathbf{e}_i'\}$. Let these vectors be given by $\mathbf{e}_i' = U\mathbf{e}_i = \sum \mathbf{e}_j u_{ji}$.
    (a) What is the matrix that represents Q_i in the basis set $\{\mathbf{e}_i'\}$? (*Hint:* In this basis the problem is especially easy.)
    (b) What is the matrix that represents Q_i in the basis set $\{\mathbf{e}_i\}$?

(c) Convince yourself that in either basis $\mathbf{Q}_i\mathbf{Q}_j = \delta_{ij}\mathbf{Q}_j$.

(d) What is $\sum \mathbf{Q}_i$ in the basis set $\{\mathbf{e}_i'\}$?

(e) What is $\sum \mathbf{Q}_i$ in the basis set $\{\mathbf{e}_i\}$?

*110. One form of the general three-dimensional rotation matrix is that given in Exercise 8.11. Another form is the matrix $\mathbf{O}$ which corresponds to a right-handed rotation by the angle ϕ about an axis in the direction of the unit vector, $\hat{\mathbf{u}} = \hat{\mathbf{i}}\cos\alpha + \hat{\mathbf{j}}\cos\beta + \hat{\mathbf{k}}\cos\gamma$. Show that

$$\mathbf{O} = \begin{pmatrix} \cos^2\alpha + \cos\phi\sin^2\alpha & (1-\cos\phi)\cos\alpha\cos\beta + \sin\phi\cos\gamma & (1-\cos\phi)\cos\alpha\cos\gamma - \sin\phi\cos\beta \\ (1-\cos\phi)\cos\alpha\cos\beta - \sin\phi\cos\gamma & \cos^2\beta + \cos\phi\sin^2\beta & (1-\cos\phi)\cos\beta\cos\gamma + \sin\phi\cos\alpha \\ (1-\cos\phi)\cos\alpha\cos\gamma + \sin\phi\cos\beta & (1-\cos\phi)\cos\beta\cos\gamma - \sin\phi\cos\alpha & \cos^2\gamma + \cos\phi\sin^2\gamma \end{pmatrix}$$

(*Hint:* Find two other unit vectors, $\hat{\mathbf{v}}$ and $\hat{\mathbf{w}}$, which with $\hat{\mathbf{u}}$ form an orthonormal, right-handed set. In this basis set the rotation matrix is especially simple. Then transform back to see what the matrix should be in the $\hat{\mathbf{i}}$, $\hat{\mathbf{j}}$, $\hat{\mathbf{k}}$ basis set.)

The operation of multiplying a matrix $\mathbf{A}$ from the right by $\mathbf{T}$ and from the left by $\mathbf{T}^{-1}$, to produce the matrix $\mathbf{T}^{-1}\mathbf{A}\mathbf{T}$, is known as a SIMILARITY TRANSFORMATION of $\mathbf{A}$ by $\mathbf{T}$. In the case that $\mathbf{T}$ is unitary (orthogonal), the similarity transformation is called a UNITARY (ORTHOGONAL) TRANSFORMATION. Matrices related by similarity transformation represent the same operation on vectors but expressed in different coordinate systems. We may expect that many problems are easier in a basis in which the transformation A is represented as a "simple" matrix. And if the given basis is such that $\mathbf{A}$ is "cumbersome," then a similarity transformation by some appropriate $\mathbf{T}$, to produce a "simpler" matrix $\mathbf{T}^{-1}\mathbf{A}\mathbf{T}$, corresponds to a transformation to a new basis in which the problem may be easier. In particular, it is especially convenient to deal with transformations represented by diagonal matrices, and we next investigate how to find the matrix $\mathbf{T}$ such that $\mathbf{T}^{-1}\mathbf{A}\mathbf{T}$ is diagonal.

Exercises

111. Use the result of Exercise 8.35 to convince yourself that the trace of a matrix is invariant to similarity transformation, that is, that $\mathrm{Tr}\,(\mathbf{T}^{-1}\mathbf{A}\mathbf{T}) = \mathrm{Tr}\,\mathbf{A}$.

112. Use the formula for the determinant of a matrix product (Eq. 8.12) to convince yourself that the determinant of a matrix is also invariant to similarity transformation.

113. Show that if two matrices commute, so do the corresponding matrices obtained by similarity transformation.

*114. Two matrices, $\mathbf{A}$ and $\mathbf{B}$, for which there exists a nonsingular matrix $\mathbf{T}$ such that $\mathbf{T}^{-1}\mathbf{A}\mathbf{T} = \mathbf{B}$, are said to be SIMILAR, written $\mathbf{A} \sim \mathbf{B}$.

(a) Show that $\mathbf{A} \sim \mathbf{A}$.

(b) Show that if $\mathbf{B} \sim \mathbf{A}$, then $\mathbf{A} \sim \mathbf{B}$.

(c) Show that if $\mathbf{A} \sim \mathbf{B}$ and $\mathbf{B} \sim \mathbf{C}$, then $\mathbf{A} \sim \mathbf{C}$.

8.12 Eigenvalue-Eigenvector Problems

The linear transformation A represents an operation that takes an n-dimensional vector $\mathbf{v}$ and twists it around and stretches (or squeezes) it to produce the vector $A\mathbf{v}$. Many problems can be stated in the form of an EIGENVALUE-EIGENVECTOR EQUATION (cf. Section 4.9)

$$A\mathbf{x} = \lambda\mathbf{x}.$$

"Which vectors are not twisted around by A, but are merely converted to scalar multiples of themselves?" Here $\mathbf{x}$ is the unknown EIGENVECTOR, and λ, the scalar multiplier, is the EIGENVALUE. The set of all eigenvalues of A is known as the SPECTRUM of A. Of course, there is always the trivial eigenvector $\mathbf{x} = \mathbf{0}$, which we shall ignore. Also, any scalar multiple of an eigenvector is also an eigenvector with the same eigenvalue since A is linear, whereby $A\mathbf{x} = \lambda\mathbf{x}$ implies $A(\alpha\mathbf{x}) = \lambda(\alpha\mathbf{x})$. It is possible to have more than one linearly independent eigenvector belonging to the same eigenvalue. Such an eigenvalue is said to be DEGENERATE, of a MULTIPLICITY equal to the number of linearly independent eigenvectors belonging to it. And any linear combination of eigenvectors belonging to a degenerate eigenvalue is also an eigenvector belonging to that eigenvalue, again since A is linear, so that $A\mathbf{x}_1 = \lambda\mathbf{x}_1$ and $A\mathbf{x}_2 = \lambda\mathbf{x}_2$ implies $A(\alpha_1\mathbf{x}_1 + \alpha_2\mathbf{x}_2) = \lambda(\alpha_1\mathbf{x}_1 + \alpha_2\mathbf{x}_2)$.

In order to reduce this eigenvalue-eigenvector problem to one involving numbers, it is necessary to choose a basis. If we choose an orthonormal basis, $\{\mathbf{e}_j\}$, then $\mathbf{x} = \sum \mathbf{e}_j x_j$ and $A\mathbf{e}_j = \sum_i \mathbf{e}_i a_{ij}$. Then $A\mathbf{x} = \lambda\mathbf{x}$ becomes

$$\sum A\mathbf{e}_j x_j = \sum_{i,j} \mathbf{e}_i a_{ij} x_j = \lambda \sum_j \mathbf{e}_j x_j.$$

Equating the components along each $\mathbf{e}_i$ gives $\sum_j a_{ij} x_j = \lambda x_i$, or in matrix notation

$$\mathbf{A}\mathbf{x} = \lambda\mathbf{x},$$

where $\mathbf{A} = (a_{ij})$ is $n \times n$ and $\mathbf{x} = (x_j)$ is an as yet unknown column matrix. This equation is also called an EIGENVALUE-EIGENVECTOR EQUATION. Transposing and factoring gives

$$(\mathbf{A} - \lambda\mathbf{1})\mathbf{x} = \mathbf{0}$$

which is just n simultaneous linear homogeneous equations in n unknowns, $\{x_j\}$. And a nontrivial solution to these equations exists if and only if

$$|\mathbf{A} - \lambda\mathbf{1}| = \begin{vmatrix} a_{11} - \lambda & a_{12} & \cdots & a_{1n} \\ a_{21} & a_{22} - \lambda & \cdots & a_{2n} \\ \cdot & \cdot & & \cdot \\ \cdot & \cdot & & \cdot \\ \cdot & \cdot & & \cdot \\ a_{n1} & a_{n2} & \cdots & a_{nn} - \lambda \end{vmatrix} = 0. \qquad (8.25)$$

This equation is known as the SECULAR EQUATION, and the determinant as the SECULAR DETERMINANT. Expansion of the secular determinant leads to an nth-order polynomial in λ, the CHARACTERISTIC POLYNOMIAL, $P_n(\lambda) = c_0 + c_1\lambda + \cdots + c_{n-1}\lambda^{n-1} + (-1)^n\lambda^n$, which has n roots, $\{\lambda_j\}$, not necessarily distinct, each satisfying $P_n(\lambda_j) = 0$. These are the n eigenvalues of $\mathbf{A}$. Notice that

$$|\mathbf{A}| = P_n(0) = c_0 = \prod \lambda_j \tag{8.26}$$

$$(-1)^{n-1}\,\mathrm{Tr}\,\mathbf{A} = c_{n-1} = (-1)^{n-1}\sum \lambda_j, \tag{8.27}$$

where the last equality in each case follows from theorems of algebra concerning sums and products of roots of polynomials (Appendix 3).

To find the eigenvector $\mathbf{x}_j$ belonging to the nondegenerate eigenvalue λ_j, we merely substitute into the simultaneous homogeneous equations to obtain

$$(\mathbf{A} - \lambda_j\mathbf{1})\mathbf{x} = \mathbf{0}.$$

But each λ_j has been chosen so that $\mathbf{A} - \lambda_j\mathbf{1}$, the matrix of coefficients, is singular. And from Eq. 8.21 we know that the solution vector $\mathbf{x}_j$ is given by any multiple of

$$x_{ij} = (-1)^{k+i}\,|(\mathbf{A} - \lambda_j\mathbf{1})_{ki}|, \tag{8.28}$$

where $(-1)^{k+i}\,|(\mathbf{A} - \lambda_j\mathbf{1})_{ki}|$ is the cofactor of the kith element of the matrix $\mathbf{A} - \lambda_j\mathbf{1}$, with the kth row chosen arbitrarily. Notice the subscripting: x_{ij} is the ith component of the jth eigenvector.

If λ_j is degenerate of multiplicity n_j, the cofactor of every element of $\mathbf{A} - \lambda_j\mathbf{1}$ is zero, and the delta method must be applied n_j times to find n_j linearly independent eigenvectors. In general it is not obvious what the multiplicity of λ_j is. However, in almost all practical situations, $\mathbf{A}$ commutes with $\mathbf{A}^\dagger$; it can then be shown that an eigenvalue λ_j of multiplicity n_j always contributes the factor $(\lambda - \lambda_j)^{n_j}$ to the characteristic polynomial. Therefore, in practice, the number of linearly independent eigenvectors to be found for each eigenvalue is determined immediately upon factorization of the characteristic polynomial.

Although any scalar multiple of an eigenvector is still an eigenvector, it is common to eliminate this arbitrariness and NORMALIZE each eigenvector so that $\mathbf{x}_j^\dagger\mathbf{x}_j = 1$. If the eigenvector $\mathbf{x}_j$ is not normalized, dividing it by $(\mathbf{x}_j^\dagger\mathbf{x}_j)^{1/2} = (\sum_i x_{ij}^* x_{ij})^{1/2}$ will normalize it. There is still an arbitrariness to even a normalized $\mathbf{x}_j$, since $e^{i\delta}\mathbf{x}_j$, with δ an arbitrary phase, is also a normalized eigenvector. This arbitrariness of phase is something to be remembered when you compare your eigenvectors with someone else's; they often differ by a factor of -1, but are in fact entirely equivalent.

Finally, let us set up the n column eigenvectors $\mathbf{x}_j$ side by side to form the square matrix

$$\mathbf{X} = (x_{ij}).$$

Then since the jth column of $\mathbf{X}$ is an eigenvector of $\mathbf{A}$ with eigenvalue λ_j, the product $\mathbf{AX}$ is an $n \times n$ matrix whose jth column is λ_j times $\mathbf{x}_j$. And if we let

$$\mathbf{\Lambda} = \text{diag}(\lambda_1, \lambda_2, \ldots, \lambda_n)$$

then, according to the result of Exercise 8.12b, the matrix whose jth column is λ_j times the jth column of $\mathbf{A}$ is simply $\mathbf{A\Lambda}$. Therefore we conclude that

$$\mathbf{AX} = \mathbf{X\Lambda}.$$

Furthermore, in almost all practical situations, $\mathbf{A}$ commutes with $\mathbf{A}^\dagger$; it can then be shown that there are n linearly independent eigenvectors of $\mathbf{A}$. Then $\mathbf{X}$ is nonsingular, $\mathbf{X}^{-1}$ exists, and

$$\mathbf{X}^{-1}\mathbf{AX} = \mathbf{\Lambda}. \tag{8.29}$$

What we have just shown is that the similarity transformation of $\mathbf{A}$ by $\mathbf{X}$ converts $\mathbf{A}$ to diagonal form (DIAGONALIZES $\mathbf{A}$). This similarity transformation corresponds to a transformation of the operator A to a new basis in which A has an especially simple representation; namely, as a diagonal matrix $\mathbf{\Lambda}$. This new basis is the set of eigenvectors of A, and the jth column of $\mathbf{X}$ contains the components of the new jth basis vector, expressed in the old basis. Therefore we see that an eigenvalue-eigenvector problem is equivalent to the problem of finding a basis in which a given operator is represented as a diagonal matrix. For this reason, the problem of finding the eigenvalues and eigenvectors of A is often called the problem of the DIAGONALIZATION of $\mathbf{A}$.

Example

14. To find the eigenvalues and corresponding normalized eigenvectors of $\mathbf{A} = \begin{pmatrix} 2 & 3 \\ 1 & 4 \end{pmatrix}$:

(a) Expansion of the secular determinant $\begin{vmatrix} 2-\lambda & 3 \\ 1 & 4-\lambda \end{vmatrix}$ gives the characteristic polynomial $P_2(\lambda) = 8 - 6\lambda + \lambda^2 - 3 = \lambda^2 - 6\lambda + 5$.

(b) $P_2(\lambda) = \lambda^2 - 6\lambda + 5$ factors to $(\lambda - 1)(\lambda - 5)$.

(c) The eigenvalues are $\lambda_1 = 1$ and $\lambda_2 = 5$.

(d) To find the eigenvectors, let us evaluate the cofactors of the first row of $\mathbf{A} - \lambda_j\mathbf{1}$:

	$\lambda_1 = 1$	$\lambda_2 = 5$
$x_{1j} = 4 - \lambda_j$	$x_{11} = 3$	$x_{12} = -1$
$x_{2j} = -1$	$x_{21} = -1$	$x_{22} = -1.$

(e) To normalize these eigenvectors, divide them by $\sqrt{3^2 + (-1)^2}$ and $\sqrt{(-1)^2 + (-1)^2}$, respectively, to obtain

$$\mathbf{x}_1 = \begin{pmatrix} 3/\sqrt{10} \\ -1/\sqrt{10} \end{pmatrix} \qquad \mathbf{x}_2 = \begin{pmatrix} -1/\sqrt{2} \\ -1/\sqrt{2} \end{pmatrix}.$$

Of course, $\mathbf{x}_2 = (1/\sqrt{2} \quad 1/\sqrt{2})^T$ would be just as suitable as an eigenvector.

(f) Let us check the eigenvectors:

$$\mathbf{A}\mathbf{x}_1 = \begin{pmatrix} 2 & 3 \\ 1 & 4 \end{pmatrix}\begin{pmatrix} 3/\sqrt{10} \\ -1/\sqrt{10} \end{pmatrix} = \begin{pmatrix} 3/\sqrt{10} \\ -1/\sqrt{10} \end{pmatrix} = 1\mathbf{x}_1 = \lambda_1\mathbf{x}_1$$

$$\mathbf{A}\mathbf{x}_2 = \begin{pmatrix} 2 & 3 \\ 1 & 4 \end{pmatrix}\begin{pmatrix} -1/\sqrt{2} \\ -1/\sqrt{2} \end{pmatrix} = \begin{pmatrix} -5/\sqrt{2} \\ -5/\sqrt{2} \end{pmatrix} = 5\mathbf{x}_2 = \lambda_2\mathbf{x}_2.$$

(g) Finally, we demonstrate how $\mathbf{A}$ is diagonalized:

$$\mathbf{X}^{-1}\mathbf{A}\mathbf{X} = \begin{pmatrix} \sqrt{5}/2\sqrt{2} & -\sqrt{5}/2\sqrt{2} \\ -1/2\sqrt{2} & -3/2\sqrt{2} \end{pmatrix}\begin{pmatrix} 2 & 3 \\ 1 & 4 \end{pmatrix}\begin{pmatrix} 3/\sqrt{10} & -1/\sqrt{2} \\ -1/\sqrt{10} & -1/\sqrt{2} \end{pmatrix} = \begin{pmatrix} 1 & 0 \\ 0 & 5 \end{pmatrix} = \mathbf{\Lambda}.$$

Notice that it is helpful to subscript eigenvalues and eigenvectors so as to manifest which eigenvector belongs to which eigenvalue. However, there is no standard ordering scheme for the eigenvalues and eigenvectors. We could just as well have taken $\lambda_1 = 5$ and $\lambda_2 = 1$, whereupon the columns of $\mathbf{X}$ would be interchanged and $\mathbf{\Lambda}$ would be diag (5, 1).

Next we indicate briefly some situations in which eigenvalue-eigenvector problems occur:

1. We have already seen how much more convenient it is to choose a coordinate system in which a tensor may be represented as a diagonal matrix. For example, we have seen that whenever **H** is either perpendicular to the molecular plane of benzene or parallel to it, **M** is a scalar multiple of **H**; it is obvious which coordinate axes to choose to make $\boldsymbol{\chi}$ diagonal. On the other hand, we have also seen (Exercise 14.33b) that with the "obvious" choice of coordinate axes, the moment of inertia of *trans*-FN=NF is not diagonal. But we have indicated how it is always possible to diagonalize a tensor by finding a new coordinate system. In this context diagonalization is called a PRINCIPAL-AXIS TRANSFORMATION.

2. In the LCAO-MO approximation, a molecular orbital ψ is taken to be a linear combination of n atomic orbitals $\{\phi_j\}$:

$$\psi = \sum c_j\phi_j.$$

If we define $W \equiv \int \psi^* H\psi \, dV$ and $S \equiv \int \psi^*\psi \, dV$, where H is the Hamiltonian operator, then one way of stating the quantum-mechanical problem is "choose the coefficients so as to minimize W subject to the constraint $S = 1$." Let us use one Lagrangian multiplier, $-\varepsilon$, and construct

$$u = W - \varepsilon(S - 1) = \sum_{i,j} c_i^* c_j \int \phi_i^* H\phi_j \, dV - \varepsilon\left(\sum_{i,j} c_i^* c_j \int \phi_i^*\phi_j \, dV - 1\right).$$

Setting $H_{ij} \equiv \int \phi_i^* H\phi_j \, dV$ and $S_{ij} \equiv \int \phi_i^*\phi_j \, dV$, and then taking the partial derivative of u with respect to each variable c_i^* gives

$$\sum_j H_{ij}c_j = \varepsilon \sum_j S_{ij}c_j,$$

or in matrix form

$$\mathbf{Hc} = \varepsilon\mathbf{Sc},$$

where $\mathbf{H} \equiv (H_{ij})$ and $\mathbf{S} \equiv (S_{ij})$ are $n \times n$ and $\mathbf{c} = (c_j)$ is an as yet unknown column matrix. In this form the problem is an eigenvalue-eigenvector problem in a nonorthonormal basis, to which we shall return later. However, in the HMO approximation, $\mathbf{S} = \mathbf{1}$, so that this is indeed the problem of finding eigenvalues and eigenvectors of the matrix $\mathbf{H}$. Diagonalization of $\mathbf{H}$ leads to n new basis vectors (the molecular orbitals), which are linear combinations of the old basis vectors (atomic orbitals). And the eigenvalues of $\mathbf{H}$ are the energies of the various molecular orbitals.

3. We have already encountered eigenvalue-eigen*function* problems, $\Omega\psi = \lambda\psi$. But we have also indicated that in the vector space of functions, a linear operator becomes a linear transformation. Therefore we may also view the problem $\Omega\psi = \lambda\psi$ as that of finding the eigenvalues and eigenvectors of the transformation Ω. If we choose a complete orthonormal basis, $\{\phi_i\}$, then the problem of finding ψ may be replaced by the problem of finding the coefficients in the expansion $\psi = \sum \phi_i x_i$. Also, the transformation Ω is determined by what it does to each basis vector, so that Ω may be represented by the matrix $\mathbf{\Omega}$, defined by $\Omega\phi_j = \sum_i \phi_i(\mathbf{\Omega})_{ij}$. Then the eigenvalue-eigenfunction problem $\Omega\psi = \lambda\psi$ is replaced by the problem, $\mathbf{\Omega x} = \lambda\mathbf{x}$, of finding the eigenvalues and eigenvectors of the matrix $\mathbf{\Omega}$. Thus we see that by choosing a basis, an eigenvalue-eigenfunction problem, which ordinarily involves solving differential equations, may be converted to a matrix-diagonalization problem, which involves only algebraic manipulations.

4. We shall see how problems of both simultaneous first-order kinetics and molecular vibrations may be expressed as eigenvalue-eigenvector problems.

Finally, we note that two basis vectors, $\mathbf{e}_i$ and $\mathbf{e}_j$, for which $\mathbf{e}_i \cdot A\mathbf{e}_j = a_{ij}$ is nonzero cannot be eigenvectors of A. These two vectors are then said to be COUPLED by the transformation A. Furthermore, since there then exist eigenvectors of A that are linear combinations of $\mathbf{e}_i$ and $\mathbf{e}_j$ (and perhaps others), these two vectors are also said to be MIXED by the transformation A.

Exercises

115. Convince yourself that the eigenvalues of a diagonal matrix are the n diagonal elements, and that the eigenvector belonging to the ith eigenvalue is the ith basis vector.
116. If $\mathbf{v}_i$ is a normalized eigenvector of $\mathbf{A}$ with eigenvalue λ_i, what is $\mathbf{v}_i^\dagger\mathbf{A}\mathbf{v}_i$?

*117. The rotational partition function of a nonlinear molecule contains the product of the three PRINCIPAL MOMENTS OF INERTIA, which are the diagonal elements of the tensor after it has been subjected to a principal-axis transformation. Use Eq. 8.26 and the result of Exercise 8.33b to find this product for *trans*-FN=NF.

*118. Show that every eigenvalue of a positive definite matrix is nonnegative.

119. What can you conclude about the eigenvectors and eigenvalues of $\mathbf{A}^{-1}$?

*120. What can you conclude about the eigenvalues of a cyclic matrix?

*121. (a) Show that an $N \times N$ matrix whose row sums are all zero is singular. (*Hint:* Add columns 2 through N to column 1.)

(b) Use this result to show that any stochastic matrix has 1 as an eigenvalue.

122. The HMO problem for conjugated hydrocarbons is that of finding eigenvalues and eigenvectors of $\mathbf{H} = \alpha\mathbf{1} + \beta\mathbf{B}$. Show that if $\mathbf{c}$ is an eigenvector of $\mathbf{H}$ with eigenvalue ε, it is also an eigenvector of $\mathbf{B}$. If $\mathbf{Bc} = m\mathbf{c}$, relate m to ε.

*123. The matrix $\begin{pmatrix} 1 & 1 \\ 0 & 1 \end{pmatrix}$ does not commute with its adjoint (Exercise 8.3). Convince yourself that although $P_2(\lambda) = (\lambda - 1)^2$, the eigenvalue $\lambda = 1$ is not doubly degenerate because the only linearly independent eigenvector is $\begin{pmatrix} 1 \\ 0 \end{pmatrix}$.

*124. Convince yourself that all Hermitian, skew-Hermitian, unitary, and diagonal matrices do commute with their adjoints.

125. Find the eigenvalues and *corresponding* normalized eigenvectors of:

(a) $\begin{pmatrix} 1 & -1 \\ -1 & 1 \end{pmatrix}$.

(b) $\begin{pmatrix} 3 & i \\ -i & 3 \end{pmatrix}$.

(c) $\begin{pmatrix} 2 & 1 \\ 1 & 0 \end{pmatrix}$.

(d) $\begin{pmatrix} 1 & 1 \\ 0 & 2 \end{pmatrix}$.

*(e) $\begin{pmatrix} 1 & 1 \\ -2 & 4 \end{pmatrix}$.

(f) $\begin{pmatrix} -1 & 2 & 0 \\ 2 & 0 & 2 \\ 0 & 2 & 1 \end{pmatrix}$.

*126. Find the eigenvalues and corresponding normalized eigenvectors of the matrix $\mathbf{S}^2$ of Exercise 8.27. Compare these answers with the answers to Exercise 4.57.

127. Let the N sp^2-hybridized carbon atoms of an unsaturated hydrocarbon be numbered from 1 to N. The HMO bond matrix $\mathbf{B}$ is defined by $b_{ij} = 1$ if atoms i and j are adjacent, $b_{ij} = 0$ otherwise (and, in particular, $b_{ii} = 0$). (Cf. Exercise 8.122 for comparison with the usual $\mathbf{H}$.) Find the eigenvalues and corresponding orthonormal eigenvectors of $\mathbf{B}$ for:

(a) Allyl, CH_2CHCH_2.

(b) Cyclopropenyl, $(CH)_3$.

*(c) Cyclobutadiene, $(CH)_4$.

*(d) Butadiene, $CH_2CHCHCH_2$. (*Hint:* $[(1 \pm \sqrt{2})/2]^2 = (3 \pm \sqrt{5})/2$.)

128. Express the answer to any one of the previous exercises in the form $\mathbf{X}^{-1}\mathbf{A}\mathbf{X} = \mathbf{\Lambda}$.
129. Find the eigenvalues and corresponding normalized eigenvectors for the general 2×2 unitary matrix $\begin{pmatrix} \cos\phi & \sin\phi \\ -\sin\phi & \cos\phi \end{pmatrix}$.
*130. For hydrocarbons that contain no odd-membered rings, the atoms can be numbered so that the HMO matrix $\mathbf{B}$ takes the partitioned form $\begin{pmatrix} \mathbf{0} & \mathbf{b} \\ \mathbf{b}^\dagger & \mathbf{0} \end{pmatrix}$. Show that if $\mathbf{x} = \begin{pmatrix} \mathbf{x}_* \\ \mathbf{x}_0 \end{pmatrix}$ is an eigenvector of $\mathbf{B}$ (partitioned as $\mathbf{B}$ is) with eigenvalue m, then $\begin{pmatrix} \mathbf{x}_* \\ -\mathbf{x}_0 \end{pmatrix}$ is also an eigenvector of $\mathbf{B}$, but with eigenvalue $-m$. This is the pairing theorem for alternant hydrocarbons.
*131. Find the eigenvalues and eigenvectors of

$$\mathbf{u}\times \equiv \begin{pmatrix} 0 & -u_z & u_y \\ u_z & 0 & -u_x \\ -u_y & u_x & 0 \end{pmatrix}.$$

How can $\mathbf{u} \times \mathbf{v} = \lambda\mathbf{v}$ if $\mathbf{u} \times \mathbf{v}$ is perpendicular to both $\mathbf{u}$ and $\mathbf{v}$?
132. Use a desk calculator to find the eigenvalues of

$$\begin{pmatrix} -2 & 0 & 1 \\ 0 & 1 & 2 \\ 1 & -2 & 2 \end{pmatrix}$$

to four decimal places.
*133. Sometimes a problem involves merely finding the largest (in absolute value) eigenvalue $\lambda_{\max}$ of a matrix $\mathbf{A}$. Let $\mathbf{x}_{\max}$ be the corresponding eigenvector.
(a) Convince yourself that if $\mathbf{u}_0$ is any unit vector not orthogonal to $\mathbf{x}_{\max}$, then as N gets larger and larger, the vector $\mathbf{A}^N\mathbf{u}_0$ approaches closer and closer to being a scalar multiple of $\mathbf{x}_{\max}$. (*Hint:* Expand $\mathbf{u}_0$ in the basis set of eigenvectors of $\mathbf{A}$.)
(b) A practical iterative procedure is to divide the largest component of $\mathbf{A}^{N+1}\mathbf{u}_0$ by the largest component of $\mathbf{A}^N\mathbf{u}_0$, and continue increasing N until this ratio approaches $\lambda_{\max}$ within the desired level of convergence. Apply this method to $\mathbf{u}_0 = (1, 0, 0)^T$ to find the largest eigenvalue (7 significant figures) of the matrix

$$\begin{pmatrix} 1.00 & 0 & 0.01 \\ 0 & 0.10 & 0.01 \\ 0.01 & 0.01 & 0.01 \end{pmatrix}.$$

Notice that the convergence to $\mathbf{x}_{\max}$ is much slower than convergence to $\lambda_{\max}$.
134. Sketch a graph of the eigenvalues of the matrix $\begin{pmatrix} 0 & x \\ x & d \end{pmatrix}$:
(a) As a function of x for fixed d.
(b) As a function of d for fixed x.
Convince yourself that "diagonal elements repel each other," especially when they are close together and there is a large off-diagonal element "coupling them."

*135. Given an $n \times n$ Hermitian matrix $\mathbf{A}$ whose eigenvalues are all nondegenerate, it is possible to define n PROJECTOR MATRICES which will act on an arbitrary vector $\mathbf{v}$ to produce the projection of $\mathbf{v}$ along a given eigenvector. (Cf. Exercise 4.55). Show that multiplying $\mathbf{v}$ by

$$\mathbf{P}_i \equiv \prod_{j \neq i}^{n} \frac{\mathbf{A} - \lambda_j \mathbf{1}}{\lambda_i - \lambda_j}$$

produces the projection of $\mathbf{v}$ along the ith eigenvector. (*Hint:* Express $\mathbf{v}$ as a linear combination of eigenvectors.) Then convince yourself that $\mathbf{P}_i\mathbf{P}_j = \delta_{ij}\mathbf{P}_j$ and $\sum \mathbf{P}_i = \mathbf{1}$. (*Note:* If $\mathbf{A}$ has degenerate eigenvalues, the product is over all distinct eigenvalues different from λ_i, but if λ_i is doubly degenerate, then $\mathbf{P}_i$ produces the projection in the planeful of eigenvectors belonging to λ_i.)

136. If $\mathbf{A} = \mathbf{\Sigma A}_i$, where $\mathbf{\Sigma}$ means direct sum, convince yourself that if $\mathbf{X}_i$ diagonalizes $\mathbf{A}_i$ to $\mathbf{\Lambda}_i$, then $\mathbf{X} = \mathbf{\Sigma X}_i$ diagonalizes $\mathbf{A}$ to $\mathbf{\Lambda} = \mathbf{\Sigma \Lambda}_i$. What this means is that diagonalizing a block-diagonal matrix requires only the separate diagonalization of each block.

137. Show that the spectrum of a matrix is invariant to similarity transformation. Note that this means further that each of the coefficients of the characteristic polynomial is invariant to similarity transformation. We have already shown that c_0 and c_{n-1} are invariant, since the determinant (product of the eigenvalues, according to Eq. 8.26) and the trace (sum of the eigenvalues, according to Eq. 8.27) are invariant (Exercises 8.111 and 8.112).

*138. A total of N molecules are distributed at random between side A and side B of a box.
(a) What is the probability that there are exactly r molecules $(0 \leqslant r \leqslant N)$ in side A? Construct these probabilities into an $(N + 1)$-dimensional row matrix $\mathbf{p}$.
(b) Convince yourself that after an infinite number of steps of the Markov chain for the Ehrenfest diffusion model (Exercise 8.30), the distribution of states is just this random distribution, no matter what the initial distribution of states was. In matrix notation, $\lim_{n\to\infty} \mathbf{p}_0\mathbf{P}^n = \mathbf{p}$, where $\mathbf{p}_0$ is the row matrix of initial probabilities.
(c) Use the fact that $\mathbf{p}$ is an equilibrium distribution of states to convince yourself that $\mathbf{pP} = \mathbf{p}$; $\mathbf{p}$ is said to be a LEFT EIGENVECTOR of $\mathbf{P}$, with eigenvalue 1 (cf. Exercise 8.121b).
(d) Check this result by matrix multiplication.

In the context of Markov chains, any eigenvector belonging to the eigenvalue 1 is called a STATIONARY DISTRIBUTION. Notice that $\mathbf{p}$ is normalized according to $\sum p_i = 1$ rather than $\sum |p_i|^2 = 1$.

8.13 Diagonalization of Hermitian Matrices

We now prove two important results concerning Hermitian matrices, $\mathbf{H}$. (*1*) The eigenvalues of $\mathbf{H}$ are real. (*2*) Eigenvectors of $\mathbf{H}$ belonging to different eigenvalues are orthogonal.

Let $\mathbf{x}$ be an eigenvector of $\mathbf{H}$ with eigenvalue λ. Then

$$\mathbf{Hx} = \lambda\mathbf{x}. \tag{8.30}$$

Let us take the adjoints of both sides of this equation, and invoke $(\mathbf{Hx})^\dagger = \mathbf{x}^\dagger\mathbf{H}^\dagger = \mathbf{x}^\dagger\mathbf{H}$

$$\mathbf{x}^\dagger\mathbf{H} = \lambda^*\mathbf{x}^\dagger. \tag{8.31}$$

Next we multiply Eq. 8.30 from the left by $\mathbf{x}^\dagger$ and Eq. 8.31 from the right by $\mathbf{x}$ and subtract, whereupon the left-hand side is zero:

$$0 = \mathbf{x}^\dagger\mathbf{H}\mathbf{x} - \mathbf{x}^\dagger\mathbf{H}\mathbf{x} = \lambda\mathbf{x}^\dagger\mathbf{x} - \lambda^*\mathbf{x}^\dagger\mathbf{x} = (\lambda - \lambda^*)\mathbf{x}^\dagger\mathbf{x}.$$

But the trivial eigenvector $\mathbf{x} = \mathbf{0}$ is not a solution, so $\mathbf{x}^\dagger\mathbf{x}$ cannot equal zero. Therefore we conclude that $(\lambda - \lambda^*) = 0$, or

$$\lambda = \lambda^* \tag{8.32}$$

and that λ must be real.

Let $\mathbf{x}_1$ and $\mathbf{x}_2$ be eigenvectors of $\mathbf{H}$ with distinct eigenvalues λ_1 and λ_2, respectively. Then

$$\mathbf{H}\mathbf{x}_1 = \lambda_1\mathbf{x}_1 \tag{8.33}$$

$$\mathbf{H}\mathbf{x}_2 = \lambda_2\mathbf{x}_2 \tag{8.34}$$

Let us take the adjoint of Eq. 8.33 and invoke $\lambda_1^* = \lambda_1$ (Eq. 8.32):

$$\mathbf{x}_1^\dagger\mathbf{H} = \lambda_1\mathbf{x}_1^\dagger.$$

Next we multiply this equation from the right by $\mathbf{x}_2$, and multiply Eq. 8.34 from the left by $\mathbf{x}_1^\dagger$, and subtract, whereupon the left-hand side is zero

$$0 = \mathbf{x}_1^\dagger\mathbf{H}\mathbf{x}_2 - \mathbf{x}_1^\dagger\mathbf{H}\mathbf{x}_2 = \lambda_1\mathbf{x}_1^\dagger\mathbf{x}_2 - \lambda_2\mathbf{x}_1^\dagger\mathbf{x}_2 = (\lambda_1 - \lambda_2)\mathbf{x}_1^\dagger\mathbf{x}_2.$$

But we have taken $\lambda_1 \neq \lambda_2$, so that we may conclude that

$$\mathbf{x}_1^\dagger\mathbf{x}_2 = 0 \tag{8.35}$$

or that eigenvectors belonging to different eigenvalues of a Hermitian matrix are orthogonal.

Next we consider the n_j eigenvectors belonging to a degenerate eigenvalue λ_j of multiplicity n_j. Since these n_j eigenvectors are linearly independent, application of the Gram-Schmidt procedure to them produces an orthonormal set of n_j new vectors. But since any linear combination of degenerate eigenvectors is still an eigenvector with the same eigenvalue, each of these new vectors is still an eigenvector belonging to λ_j. Thus it is always possible to arrange even for different eigenvectors belonging to the same eigenvalue to be orthogonal.

In summary, eigenvectors belonging to different eigenvalues are automatically orthogonal, and eigenvectors belonging to the same eigenvalue can be made orthogonal. Henceforth we shall assume that this has been done, along with the normalization. Then we may be sure that the n eigenvectors of an $n \times n$ Hermitian matrix form an orthonormal set and, furthermore, that the square matrix of eigenvectors, which diagonalizes $\mathbf{H}$, is unitary. Therefore we may conclude that a Hermitian matrix may be diagonalized by a unitary transformation

$$\mathbf{U}^\dagger\mathbf{H}\mathbf{U} = \mathbf{\Lambda}. \tag{8.36}$$

Exercises

139. Why must the determinant of a Hermitian matrix be real, and why must at least one eigenvalue of a singular matrix be zero?

140. A real QUADRATIC FORM in the n real variables $\{x_i\}$ is the double sum $\sum a_{ij}x_ix_j$, which in matrix notation becomes $\mathbf{x}^\dagger\mathbf{A}\mathbf{x}$.

   *(a) Convince yourself that without loss of generality, $\mathbf{A}$ may be taken as Hermitian. (*Hint:* Convince yourself that $\mathbf{x}^\dagger(\mathbf{A} + \mathbf{A}^\dagger)\mathbf{x} = 2\mathbf{x}^\dagger\mathbf{A}\mathbf{x}$.)

   (b) If $\mathbf{U}^\dagger\mathbf{A}\mathbf{U} = \mathbf{\Lambda}$, find a new set of variables $\{x_j'\}$ such that the quadratic form simplifies to the single sum $\sum \lambda_j x_j'^2$, without cross terms; express x_j' as a linear combination of $\{x_i\}$. (*Hint:* $\mathbf{x}^\dagger\mathbf{A}\mathbf{x} = \mathbf{x}^\dagger\mathbf{U}\mathbf{U}^\dagger\mathbf{A}\mathbf{U}\mathbf{U}^\dagger\mathbf{x}$.)

   (c) The equation

$$(x \quad y \quad z)\begin{pmatrix} 5 & 2 & -1 \\ 2 & 5 & -1 \\ -1 & -1 & 8 \end{pmatrix}\begin{pmatrix} x \\ y \\ z \end{pmatrix} = 1,$$

   of the form $\mathbf{x}^\dagger\mathbf{H}\mathbf{x} = 1$, is the equation of an ellipsoid centered at the origin. Find a transformation to a new set of variables, x', y', z', such that the equation takes the usual form $ax'^2 + by'^2 + cz'^2 = 1$, without cross terms. What are the new variables?

*141. Let $\mathbf{H}$ be a real symmetric matrix whose largest (in absolute value) off-diagonal element is $H_{ij} = H_{ji}$. Find a unitary matrix $\mathbf{U}$ whose only off-diagonal elements are U_{ij} and U_{ji} and such that $(\mathbf{U}^{-1}\mathbf{H}\mathbf{U})_{ij} = 0$. (*Hint:* Generalize the general 2×2 unitary matrix and solve for $\tan 2\phi$.) Repeated application of such transformations can be used to make *all* off-diagonal elements as small as desired. Also, accumulative multiplication of the transforming matrices produces the unitary matrix that diagonalizes $\mathbf{H}$. This technique is known as Jacobi's method, and it is the basis of computer programs for matrix diagonalization.

*142. Let $\mathbf{H} = \mathbf{R} + i\mathbf{I}$ be a complex $n \times n$ Hermitian matrix, whose eigenvalues λ_j and eigenvectors $\mathbf{v}_j$ are to be found by a computer than can do only real arithmetic. Let $\mathbf{v}_j = \mathbf{x}_j + i\mathbf{y}_j$, and show that this problem may be converted to that of diagonalizing the real $2n \times 2n$ matrix $\begin{pmatrix} \mathbf{R} & -\mathbf{I} \\ \mathbf{I} & \mathbf{R} \end{pmatrix}$. Is this matrix Hermitian? What about the fact that there are twice as many eigenvalues and twice as many eigenvectors as required? (*Hint:* If $\mathbf{v}_j$ is an eigenvector belonging to λ_j, then so is $i\mathbf{v}_j$.)

*143. Show that the eigenvalues of a skew-Hermitian matrix are imaginary, and that eigenvectors belonging to different eigenvalues are orthogonal.

*144. Show that the eigenvalues of a unitary matrix are of magnitude unity, and that eigenvectors belonging to different eigenvalues are orthogonal.

*145. Let $\mathbf{R}$ be a real $n \times n$ matrix, so that its characteristic polynomial has only real coefficients.

   (a) Convince yourself that if λ is an eigenvalue of $\mathbf{R}$, then so is λ^*.

   (b) If n is odd, convince yourself that at least one eigenvalue of $\mathbf{R}$ is real.

   (c) If also $\mathbf{R}$ is unitary, what values can this real eigenvalue take on?

   (d) If $\mathbf{v}$ is the eigenvector belonging to this eigenvalue, describe the transformation R (cf. Section 8.8).

8.14 Simultaneous Diagonalization

If $\mathbf{x}_j$ is an eigenvector of the matrix $\mathbf{A}$, it is not necessarily true that $\mathbf{x}_j$ is also an eigenvector of some other matrix $\mathbf{B}$. In general, the vector $\mathbf{B}\mathbf{x}_j$ is not a scalar multiple of $\mathbf{x}_j$ but must be expanded as a linear combination of all eigenvectors of $\mathbf{A}$:

$$\mathbf{Bx}_j = \sum_k \mathbf{x}_k \beta_{kj}.$$

(Here, and in what follows, we assume that $\{\mathbf{x}_j\}$ is linearly independent and therefore suitable as a basis.) If we form the column eigenvectors of $\mathbf{A}$ into the square matrix $\mathbf{X}$, then $\mathbf{X}^{-1}\mathbf{AX}=\mathbf{\Lambda}$ is diagonal, but $\mathbf{X}^{-1}\mathbf{BX} = \boldsymbol{\beta} = (\beta_{kj})$ is not.

We next show that if two matrices (operators) commute, they can be diagonalized simultaneously. Let $\mathbf{A}$ and $\mathbf{B}$ commute, and let $\mathbf{x}_j$ be an eigenvector of $\mathbf{A}$ with eigenvalue λ_j: $\mathbf{Ax}_j = \lambda_j \mathbf{x}_j$. Multiplying both sides of the equation from the left by $\mathbf{B}$, and invoking $\mathbf{BA} = \mathbf{AB}$, gives

$$\mathbf{A}(\mathbf{Bx}_j) = \lambda_j(\mathbf{Bx}_j).$$

We have inserted parentheses to point out that what we have just shown is that $\mathbf{Bx}_j$ is also an eigenvector of $\mathbf{A}$ with eigenvalue λ_j. Two cases must now be distinguished:

1. λ_j is nondegenerate. Then the only eigenvectors of $\mathbf{A}$ with eigenvalue λ_j are the lineful of scalar multiples of $\mathbf{x}_j$. Since $\mathbf{Bx}_j$ is such an eigenvector, it must be some scalar multiple of $\mathbf{x}_j$:

$$\mathbf{Bx}_j = \mu_j \mathbf{x}_j.$$

We have thus shown that $\mathbf{x}_j$ is also an eigenvector of $\mathbf{B}$ with eigenvalue μ_j (cf. Exercise 4.64).

2. λ_j is degenerate of multiplicity n_j. Then there exist n_j linearly independent eigenvectors of $\mathbf{A}$ belonging to λ_j. And if $\mathbf{x}_j$ is one of these, then it still follows that $\mathbf{Bx}_j$ is also an eigenvector of $\mathbf{A}$ with eigenvalue λ_j. But we can no longer conclude that $\mathbf{Bx}_j$ is a scalar multiple of $\mathbf{x}_j$, since the totality of eigenvectors of $\mathbf{A}$ belonging to λ_j is not merely a lineful, but an n_j-spaceful. Now all that we can conclude is that each $\mathbf{Bx}_j$ is a linear combination of eigenvectors belonging to λ_j:

$$\mathbf{Bx}_j = \sum_k \mathbf{x}_k \beta_{kj} \qquad \beta_{kj} = 0 \quad \text{if } \lambda_k \neq \lambda_j.$$

Notice that this differs from the general case in that this is not the most general linear combination of all eigenvectors in the basis set, but includes only those eigenvectors of $\mathbf{A}$ belonging to the degenerate eigenvalue λ_j. We may also note that any linear combination of these n_j vectors is still an eigenvector of $\mathbf{A}$ belonging to the degenerate eigenvalue λ_j. And we may find a set of linear combinations of these vectors such that each linear combination is also an eigenvector of $\mathbf{B}$.

The problem of finding these linear combinations is clarified by considering the matrices $\mathbf{X}^{-1}\mathbf{AX} = \mathbf{\Lambda}$ and $\mathbf{X}^{-1}\mathbf{BX} = \boldsymbol{\beta}$. Let us arrange the eigenvalues and eigenvectors of $\mathbf{A}$ so that identical eigenvalues are grouped together. Then

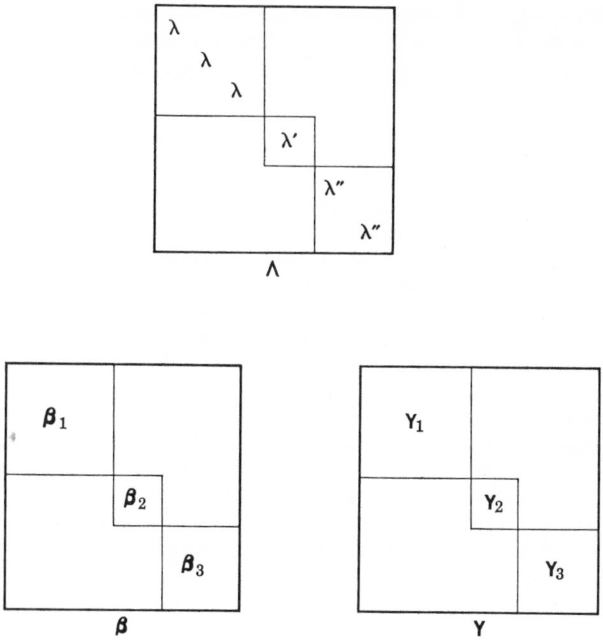

Figure 8.5

the diagonal matrix $\mathbf{\Lambda}$ is a direct sum of scalar matrices, each of dimension equal to the multiplicity of the corresponding eigenvalue: $\mathbf{\Lambda} = \Sigma\, \mathbf{\Lambda}_i = \Sigma\, \lambda_i \mathbf{1}$ (See Figure 8.5). Then, from the previous paragraph, the matrix $\boldsymbol{\beta}$ is in block-diagonal form, with one block for each distinct eigenvalue of $\mathbf{A}$: $\boldsymbol{\beta} = \Sigma\, \boldsymbol{\beta}_i$. Let $\mathbf{Y}_i$ be the matrix that diagonalizes $\boldsymbol{\beta}_i$. Then $\mathbf{Y} = \Sigma\, \mathbf{Y}_i$ diagonalizes $\boldsymbol{\beta}$: $\mathbf{Y}^{-1}\boldsymbol{\beta}\mathbf{Y} = \mathbf{M}$. But since the matrix $\mathbf{\Lambda}_i$ corresponding to $\boldsymbol{\beta}_i$ is a scalar matrix, it must commute with $\mathbf{Y}_i$, and $\mathbf{\Lambda}$ must commute with $\mathbf{Y}$. We summarize these results by

$$\mathbf{Y}^{-1}\mathbf{X}^{-1}\mathbf{A}\mathbf{X}\mathbf{Y} = \mathbf{Y}^{-1}\mathbf{\Lambda}\mathbf{Y} = \mathbf{\Lambda}$$

$$\mathbf{Y}^{-1}\mathbf{X}^{-1}\mathbf{B}\mathbf{X}\mathbf{Y} = \mathbf{Y}^{-1}\boldsymbol{\beta}\mathbf{Y} = \mathbf{M},$$

showing that $\mathbf{A}$ and $\mathbf{B}$ are simultaneously diagonalized by the matrix product $\mathbf{XY}$.

In summary, we conclude that if $\mathbf{A}$ and $\mathbf{B}$ commute, they may be diagonalized simultaneously. If $\mathbf{x}_j$ is an eigenvector of $\mathbf{A}$ belonging to a nondegenerate eigenvalue, it is automatically an eigenvector of $\mathbf{B}$. If $\mathbf{x}_j$ is an eigenvector of $\mathbf{A}$ belonging to a degenerate eigenvalue of multiplicity n_j, it is not automatically an eigenvector of $\mathbf{B}$. However, it is possible to construct n_j linearly independent linear combinations of the n_j eigenvectors belonging to that eigenvalue, such that each of the linear combinations is also an eigenvector of **B**. This result is readily extended to N commuting matrices: If each of them commutes with every other, then they may all be diagonalized simultaneously.

Example

15. For the hydrogen atom, the operators H, L^2, and L_z commute with each other. We shall choose the hydrogen-atom wave functions as basis, but consider only the first 5×5 submatrix of the matrix that represents each operator. The five wave functions, ψ_{1s}, ψ_{2s}, ψ_{2px}, ψ_{2py}, and ψ_{2pz}, are all eigenfunctions of H, with eigenvalues E_1, E_2, E_2, E_2, and E_2, respectively, where $E_n = -me^4/2n^2\hbar^2$. Therefore H is represented as a matrix that begins with the matrix **H** below. And since **H** is the direct sum of a 1×1 scalar matrix and a 4×4 scalar matrix, we may conclude that both L^2 and L_z are represented as matrices that begin as direct sums of a 1×1 matrix and a 4×4 matrix. This partitioning is indicated in the matrices $\mathbf{L}^2$ and $\mathbf{L}_z$ below. Since ψ_{1s} belongs to the nondegenerate eigenvalue E_1, ψ_{1s} must also be an eigenfunction of both L^2 and L_z; indeed it is, with eigenvalue zero for both. Since E_2 is a fourfold degenerate eigenvalue, we cannot automatically conclude that the four wave functions belonging to it are also eigenfunctions of L^2 and L_z. Fortuitously, the functions ψ_{2s}, ψ_{2px}, ψ_{2py}, and ψ_{2pz} do happen to be eigenfunctions of L^2, with eigenvalues 0, $2\hbar^2$, $2\hbar^2$, and $2\hbar^2$, respectively. Therefore, the representation of L^2 begins as the matrix $\mathbf{L}^2$, which is the direct sum of 1×1, 1×1, and 3×3 matrices.

$$\mathbf{H} = \begin{pmatrix} E_1 & & & & \\ & E_2 & 0 & 0 & 0 \\ & 0 & E_2 & 0 & 0 \\ & 0 & 0 & E_2 & 0 \\ & 0 & 0 & 0 & E_2 \end{pmatrix} \qquad \mathbf{L}^2 = \begin{pmatrix} 0 & & & & \\ & 0 & & & \\ & & 2\hbar^2 & 0 & 0 \\ & & 0 & 2\hbar^2 & 0 \\ & & 0 & 0 & 2\hbar^2 \end{pmatrix}$$

And we may conclude that the 4×4 submatrix of $\mathbf{L}_z$ must be the direct sum of a 1×1 matrix and a 3×3 matrix. This partitioning is also indicated in the matrix $\mathbf{L}_z$ below. Since ψ_{2s} is the only function of the four that is an eigenfunction of L^2 belonging to the eigenvalue zero, it must also be an eigenfunction of L_z; indeed it is, with eigenvalue zero. But since the eigenvalue $2\hbar^2$ is degenerate, we cannot conclude that ψ_{2px}, ψ_{2py}, and ψ_{2pz} are eigenfunctions of L_z. Fortuitously, ψ_{2pz} happens to be an eigenfunction of L_z, with eigenvalue zero. But ψ_{2px} and ψ_{2py} are not eigenfunctions of L_z, since $L_z\psi_{2px} = i\hbar\psi_{2py}$ and $L_z\psi_{2py} = -i\hbar\psi_{2px}$. Therefore the matrix representation of L_z begins as the matrix $\mathbf{L}_z$, below, which is indeed a direct sum of 1×1, 1×1, and 3×3 submatrices. If we diagonalize the 3×3 submatrix by transforming with $\mathbf{Y}_3$,

$$\mathbf{L}_z = \begin{pmatrix} 0 & & & & \\ & 0 & & & \\ & & 0 & -i\hbar & 0 \\ & & i\hbar & 0 & 0 \\ & & 0 & 0 & 0 \end{pmatrix} \qquad \mathbf{Y}_3 = \begin{pmatrix} 1/\sqrt{2} & 1/\sqrt{2} & 0 \\ i/\sqrt{2} & -i/\sqrt{2} & 0 \\ 0 & 0 & 1 \end{pmatrix},$$

we arrive at the linear combinations, $(\psi_{2px} + i\psi_{2py})/\sqrt{2}$, $(\psi_{2px} - i\psi_{2py})/\sqrt{2}$, and ψ_{2pz}, which are eigenfunctions of L_z with eigenvalues $\hbar$, $-\hbar$, and 0, respectively. Furthermore, $\mathbf{Y} = \mathbf{1} \oplus \mathbf{1} \oplus \mathbf{Y}_3$ commutes with **H** and $\mathbf{L}^2$, so that these new linear combinations are still eigenfunctions of $\mathbf{H}$ with eigenvalue E_2 and of L^2 with eigenvalue $2\hbar^2$. Therefore, in this new basis, H and L^2 are still represented by matrices beginning with $\mathbf{H}$ and $\mathbf{L}^2$, but L_z is represented as the diagonal matrix that begins

$$\mathbf{Y}^{-1}\mathbf{L}_z\mathbf{Y} = \begin{pmatrix} 0 & & & & \\ & 0 & & & \\ & & \hbar & 0 & 0 \\ & & 0 & -\hbar & 0 \\ & & 0 & 0 & 0 \end{pmatrix}.$$

Exercises

*146. Use Exercise 8.113 to show that two matrices that can be diagonalized simultaneously commute.

147. Let $\mathbf{C}_n$ be an $n \times n$ matrix with zeros everywhere except $(\mathbf{C}_n)_{i,i+1} = 1 = (\mathbf{C}_n)_{n1}$.
  (a) What does $\mathbf{C}_n$ do to the vector $\mathbf{x} = (x_1, x_2, \ldots, x_n)^T$? What does $\mathbf{C}_n{}^2$ do to $\mathbf{x}$? What does $\mathbf{C}_n{}^n$ do to $\mathbf{x}$?
  (b) Without evaluating a secular determinant, use these results to find all eigenvalues and eigenvectors of $\mathbf{C}_n$. (*Hint:* Let $x_1 = 1$ and remember the nth roots of unity. Then normalize each eigenvector.)
  (c) What is $\mathbf{C}_n{}^{-1}$?
  (d) The HMO problem for cyclic polyenes is to diagonalize the matrix, $\mathbf{C}_n + \mathbf{C}_n{}^{-1}$, which commutes with $\mathbf{C}_n$. But you have already found its eigenvectors "by inspection." What is the eigenvalue associated with each eigenvector; what is its multiplicity? (*Caution:* Distinguish n even from n odd.)

Now let us consider the problem of factoring secular equations, and how the feature of simultaneous diagonalizability may be applied to this problem. To diagonalize an $n \times n$ matrix $\mathbf{B}$, it is necessary to evaluate an $n \times n$ determinant, find the roots of an nth-order polynomial, and solve for n n-dimensional eigenvectors. This task becomes more and more enormous as n increases. It would certainly be a considerable simplification if we could easily find a matrix $\mathbf{X}$ such that $\mathbf{X}^{-1}\mathbf{B}\mathbf{X}$ is in block-diagonal form: $\mathbf{X}^{-1}\mathbf{B}\mathbf{X} = \boldsymbol{\beta} = \boldsymbol{\Sigma}\,\boldsymbol{\beta}_i$. Then we would merely need to diagonalize each separate submatrix $\boldsymbol{\beta}_i$. If similarity transformation by $\mathbf{Y}_i$ diagonalizes $\boldsymbol{\beta}_i$ to $\mathbf{M}_i$, similarity transformation by $\mathbf{Y} = \boldsymbol{\Sigma}\,\mathbf{Y}_i$ diagonalizes $\boldsymbol{\beta}$ to $\mathbf{M} = \boldsymbol{\Sigma}\,\mathbf{M}_i$ (Exercise 8.136). Then since $\mathbf{M}\boldsymbol{\beta} = \mathbf{Y}^{-1}\mathbf{Y} = \mathbf{Y}^{-1}\mathbf{X}^{-1}\mathbf{B}\mathbf{X}\mathbf{Y} = (\mathbf{X}\mathbf{Y})^{-1}\mathbf{B}(\mathbf{X}\mathbf{Y})$, we may conclude that similarity transformation by the matrix $\mathbf{X}\mathbf{Y}$ diagonalizes $\mathbf{B}$ to $\mathbf{M}$. Thus we may be able to simplify the diagonalization of $\mathbf{B}$ by doing it in two steps.

Of course, there remains the problem of finding the matrix $\mathbf{X}$ that effects the preliminary simplification. The clue to finding $\mathbf{X}$ lies in the discussion of simultaneous diagonalization. Let $\mathbf{A}$ be a Hermitian or unitary matrix that commutes with $\mathbf{B}$. Then we have seen that transformation to a new orthonormal basis set composed of eigenvectors of $\mathbf{A}$ converts $\mathbf{B}$ to block-diagonal form, with one submatrix for each distinct eigenvalue of $\mathbf{A}$. Therefore we conclude that we should choose $\mathbf{X}$ to be a unitary matrix that diagonalizes

A. If $\mathbf{x}_j$ is an eigenvector of **A** belonging to a nondegenerate eigenvalue, it is automatically an eigenvector of **B** also. If $\mathbf{x}_j$ is an eigenvector of **A** belonging to a degenerate eigenvalue λ_j of multiplicity n_j, it is not necessarily an eigenvector of **B**. However, it is possible to find new linear combinations of the n_j orthonormal eigenvectors of **A** belonging to λ_j such that each of these new linear combinations is an eigenvector of **B**. And this remaining problem is one of diagonalizing only an $n_j \times n_j$ matrix.

It is instructive to consider the block diagonalization from a somewhat different viewpoint. Since $\boldsymbol{\beta} = \mathbf{X}^{-1}\mathbf{B}\mathbf{X}$ is in block-diagonal form, with one submatrix for each distinct eigenvalue of **A**, we may conclude that $\beta_{ij} = 0$ if β_{ii} and β_{jj} are in different submatrices. Or if $\mathbf{x}_i$ and $\mathbf{x}_j$ are eigenvectors of **A** belonging to different eigenvalues, then $\mathbf{x}_i^\dagger\mathbf{B}\mathbf{x}_j = \beta_{ij} = 0$, that is, eigenvectors of A belonging to different eigenvalues cannot be coupled by the transformation B. Therefore we may conclude that if A and B commute, we may simplify the problem of finding eigenvalues and eigenvectors of B by transforming to a basis set composed of eigenvectors of A.

Example

16. Let us diagonalize the Hermitian matrix

$$\mathbf{B} = \begin{pmatrix} 0 & 1 & 0 \\ 1 & 0 & 1 \\ 0 & 1 & 0 \end{pmatrix}.$$

Matrix multiplication shows that **B** commutes with

$$\mathbf{A} = \begin{pmatrix} 0 & 0 & 1 \\ 0 & 1 & 0 \\ 1 & 0 & 0 \end{pmatrix}.$$

Alternatively, the fact that $\mathbf{AB} = \mathbf{BA}$ is obvious on inspection, since premultiplication of **B** by **A** interchanges rows 1 and 3, and postmultiplication by **A** interchanges columns 1 and 3 (cf. Exercise 8.9). The matrix of eigenvectors of **A** is readily determined to be

$$\mathbf{X} = \begin{pmatrix} 1/\sqrt{2} & 0 & 1/\sqrt{2} \\ 0 & 1 & 0 \\ -1/\sqrt{2} & 0 & 1/\sqrt{2} \end{pmatrix}.$$

Then from Exercise 8.7

$$\mathbf{X}^{-1}\mathbf{B}\mathbf{X} = \boldsymbol{\beta} = \left(\begin{array}{c|cc} 0 & 0 & 0 \\ \hline 0 & 0 & \sqrt{2} \\ 0 & \sqrt{2} & 0 \end{array}\right).$$

One eigenvector of this partitioned matrix is immediately seen to be $(1, 0, 0)^T$, with eigenvalue zero. The other two are readily determined to be $(0, 1/\sqrt{2}, -1/\sqrt{2})^T$ and $(0, 1/\sqrt{2}, 1/\sqrt{2})^T$, with eigenvalues $-\sqrt{2}$ and $\sqrt{2}$, respectively. Therefore the matrix that diagonalizes

$\boldsymbol{\beta}$ to $\mathbf{M} = \text{diag}\,(0, -\sqrt{2}, \sqrt{2})$ is

$$\mathbf{Y} = \begin{pmatrix} 1 & 0 & 0 \\ 0 & 1/\sqrt{2} & 1/\sqrt{2} \\ 0 & -1/\sqrt{2} & 1/\sqrt{2} \end{pmatrix},$$

and the matrix that diagonalizes $\mathbf{B}$ to $\mathbf{M}$ is the matrix product

$$\mathbf{XY} = \begin{pmatrix} 1/\sqrt{2} & -1/2 & 1/2 \\ 0 & 1/\sqrt{2} & 1/\sqrt{2} \\ -1/\sqrt{2} & -1/2 & 1/2 \end{pmatrix}.$$

Admittedly, this two-step process does not seem to offer any promise of simplification. However, we have included unnecessary steps in order to demonstrate the principle involved. In practice, it is often possible to write down the matrix $\boldsymbol{\beta}$ directly without most of the preliminary labor presented above. In particular, the matrix $\mathbf{B}$ of this example may be considered to have arisen in the HMO problem of allyl, CH_2CHCH_2 (cf. Exercise 8.127a). If we take as basis ϕ_1, ϕ_2, and ϕ_3, the $2p_z$ atomic orbitals on the three carbon

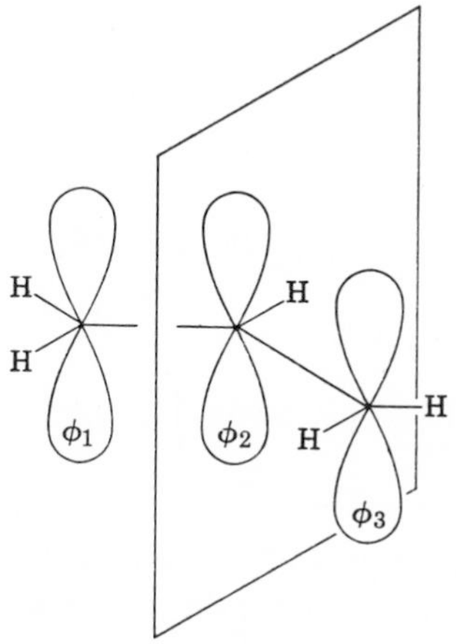

Figure 8.6

atoms (see Figure 8.6), then $\mathbf{B} = (b_{ij}) = (\int \phi_i^* B \phi_j \, dV)$. Here $B = (H - \alpha 1)/\beta$, with H the usual HMO Hamiltonian operator, in terms of the Coulomb integral α and the resonance integral β. (Cf. Exercise 8.122 for the justification for our diagonalizing $\mathbf{B}$, rather than $\mathbf{H}$.) However, we note that the two CH_2 groups are equivalent, so that reflecting the molecule in a vertical plane cannot affect the electrons or the Hamiltonian. Mathematically, this reflection corresponds to transformation to a new basis in which ϕ_1 and ϕ_3 have been interchanged. This interchange is effected by similarity transformation by the matrix **A** above, and the result $\mathbf{A}^{-1}\mathbf{B}\mathbf{A} = \mathbf{B}$ follows from matrix multiplication or, more obviously, from the fact that the molecule is symmetric about the vertical plane. Next we must find the eigenvectors of this transformation. But it is trivial to note that the linear combination $(\phi_1 - \phi_3)/\sqrt{2} \equiv \chi_1$ is antisymmetric with respect to reflection in the plane, whereas both $\phi_2 \equiv \chi_2$ and $(\phi_1 + \phi_3)/\sqrt{2} \equiv \chi_3$ are symmetric with respect to this reflection. Next we must find $\boldsymbol{\beta} = (\beta_{ij})$, the matrix representation of B in this new basis. We know immediately that $\beta_{12} = \beta_{21} = 0 = \beta_{13} = \beta_{31}$, because χ_1 is antisymmetric with respect to reflection (eigenvalue -1 for reflection) and χ_2 and χ_3 are symmetric (eigenvalue $+1$ for reflection). Indeed, since χ_1 is the only member of the basis that is antisymmetric with respect to reflection (since χ_1 belongs to the nondegenerate eigenvalue -1 of **A**), it must itself be an eigenfunction of B. And it belongs to the eigenvalue

$$\int \chi_1^* B \chi_1 \, dV = \frac{1}{2}\left(\int \phi_1^* B \phi_1 \, dV + \int \phi_3^* B \phi_3 \, dV - \int \phi_1^* B \phi_3 \, dV - \int \phi_3^* B \phi_1 \, dV\right)$$
$$= \tfrac{1}{2}(b_{11} + b_{33} - b_{13} - b_{31}) = 0.$$

Evaluation of the remaining four matrix elements is easy:

$$\beta_{22} = \int \chi_2^* B \chi_2 \, dV = \int \phi_2^* B \phi_2 \, dV = b_{22} = 0$$

$$\beta_{33} = \int \chi_3^* B \chi_3 \, dV$$
$$= \frac{1}{2}\left(\int \phi_1^* B \phi_1 \, dV + \int \phi_3^* B \phi_3 \, dV + \int \phi_1^* B \phi_3 \, dV + \int \phi_3^* B \phi_1 \, dV\right)$$
$$= \tfrac{1}{2}(b_{11} + b_{33} + b_{13} + b_{31}) = 0$$

$$\beta_{32} = \int \chi_3^* B \chi_2 \, dV = \left(\int \phi_1^* B \phi_2 \, dV + \int \phi_3^* B \phi_2 \, dV\right)\Big/\sqrt{2}$$
$$= (b_{12} + b_{32})/\sqrt{2} = \sqrt{2} = \beta_{23}.$$

Thus we see that it is indeed possible to construct the matrix $\boldsymbol{\beta}$ directly. It was, in fact, not necessary to construct the matrix **A** or to diagonalize it. All that was necessary was to recognize the symmetry inherent in the chemical problem, determine the new basis set by inspection, and evaluate the elements of $\boldsymbol{\beta}$ in terms of the given elements of **B**. In cases of more elaborate symmetry, the new basis set may not be obvious, but group theory provides a simple recipe for constructing it.

Exercise

148. Find the eigenvalues and corresponding eigenvectors of the HMO bond matrix **B** for bicyclohexatriene (Fig 8.7). Choose a new LCAO basis, $\{\chi_i\}_{i=1}^{6}$, with each χ_i

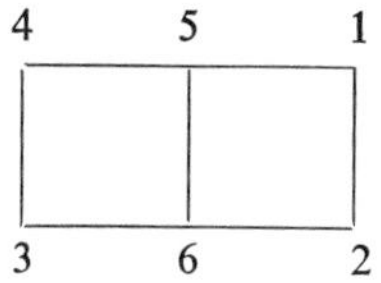

Figure 8.7

symmetric or antisymmetric to each of the two vertical planes perpendicular to the plane of the molecule (a total of four possibilities). Then construct the matrix $\boldsymbol{\beta} = (\int \chi_i^* B \chi_j \, dV)$, which is a direct sum of two 2×2 matrices and two 1×1 matrices. It is then sufficient to find the eigenvalues and corresponding eigenvectors of $\boldsymbol{\beta}$, if you indicate clearly how $\{\chi_i\}$ is defined.

8.15 Matrix Functions

Next we consider functions that take square matrices into square matrices. Notice that such functions are quite different from the functions det and Tr, whose values are numbers. We shall write the value of the function f at the argument $\mathbf{A}$ as $\mathbf{f}(\mathbf{A})$.

Examples

17. $f =$ nth power: $\mathbf{f}(\mathbf{A}) = \mathbf{A}^n$, with $\mathbf{A}^0 \equiv \mathbf{1}$.
18. $f =$ Nth-order polynomial:

$$\mathbf{f}(\mathbf{A}) = \mathbf{P}_N(\mathbf{A}) = c_0\mathbf{1} + c_1\mathbf{A} + \cdots + c_N\mathbf{A}^N = \sum_{n=0}^{N} c_n\mathbf{A}^n.$$

19. $f =$ reciprocal (multiplicative inverse): $\mathbf{f}(\mathbf{A}) = \mathbf{A}^{-1}$. But just as the domain of the function reciprocal excludes zero, so does the domain of this function exclude all singular matrices.

20. $f =$ square root: $\mathbf{f}(\mathbf{A}) = \mathbf{A}^{1/2}$, such that $(\mathbf{A}^{1/2})^2 = \mathbf{A}$. But just as there are uniqueness (sign) problems with $f(x) = \sqrt{x}$, so also are there several matrices whose squares are $\mathbf{A}$. Below we shall see how the square root of a matrix can be calculated.

21. $f =$ exponential: $\mathbf{f}(\mathbf{A}) = e^{\mathbf{A}} \equiv \sum_{n=0}^{\infty} \mathbf{A}^n/n!$, if the series converges; this one does converge for every $\mathbf{A}$.

22. $f =$ cosine:

$$\mathbf{f}(\mathbf{A}) = \cos \mathbf{A} \equiv \tfrac{1}{2}(e^{i\mathbf{A}} + e^{-i\mathbf{A}}) = \sum_{n=0}^{\infty} (-1)^n\mathbf{A}^{2n}/(2n)!.$$

Likewise

$$\sin \mathbf{A} \equiv \frac{1}{2i}(e^{i\mathbf{A}} - e^{-i\mathbf{A}}) = \sum_{n=0}^{\infty} (-1)^n\mathbf{A}^{2n+1}/(2n+1)!.$$

Exercises

*149. (a) Find a power series expansion for $(\mathbf{1} + \mathbf{A})^n$, where n is a positive integer.
(b) Find a power series expansion for $(\mathbf{1} + \mathbf{A})^{-1}$. Check your answer.
(c) Find a power series expansion for $(\mathbf{1} + \mathbf{A})^{-1/2}$ (*Hint:* cf. Exercise 1.59).
What would you suppose to be a necessary condition for convergence of the infinite series?

150. Convince yourself that if $\mathbf{D} = \text{diag}(d_1, d_2, \ldots, d_n)$ and if f has a power series expansion, then $\mathbf{f}(\mathbf{D}) = \text{diag}(f(d_1), f(d_2), \ldots, f(d_n))$.

*151. (a) Find four different diagonal matrices which solve $\mathbf{A}^2 = \begin{pmatrix} 1 & 0 \\ 0 & 1 \end{pmatrix}$.

(b) Find the most general form of the nondiagonal $\begin{pmatrix} 1 & 0 \\ 0 & 1 \end{pmatrix}^{1/2}$.

*152. Under what circumstance does $e^{\mathbf{A}}e^{\mathbf{B}} = e^{\mathbf{A}+\mathbf{B}}$?

*153. Solve:

(a) $e^{\mathbf{M}} = \mathbf{1}$,

(b) $e^{\mathbf{M}} = e\mathbf{1}$.

(c) What is the requirement on $\mathbf{M}$ to ensure that $e^{\mathbf{M}} = \mathbf{1} + \mathbf{M}$?

154. Convince yourself that if $\mathbf{H}$ is Hermitian, $e^{i\mathbf{H}}$ is unitary. (*Hint:* What is the series expansion of $(e^{i\mathbf{H}})^\dagger$?)

*155. (a) Show that $\cos^2 \mathbf{A} + \sin^2 \mathbf{A} = \mathbf{1}$.

(b) What is $\sin 2\mathbf{A}$?

(c) Does $\sin(\mathbf{A} + \mathbf{B}) = \sin \mathbf{A} \cos \mathbf{B} + \cos \mathbf{A} \sin \mathbf{B}$?

Except for the simplest functions, the definitions do not provide any convenient method for evaluating $\mathbf{f}(\mathbf{A})$. It is not very convenient to find $\mathbf{A}^{1/2}$ by equating the square of the most general $n \times n$ matrix to $\mathbf{A}$ and solving the n^2 simultaneous quadratic equations that result; it is not very convenient to evaluate an infinite power series in $\mathbf{A}$ by summing terms until a desired level of accuracy is reached. Next we consider how we may reduce the evaluation of $\mathbf{f}(\mathbf{A})$ to a diagonalization problem.

Let us find the eigenvalues and eigenvectors of $\mathbf{f}(\mathbf{A})$. First we notice that if f has a power series expansion $\mathbf{f}(\mathbf{A}) = \sum c_n \mathbf{A}^n$, then $\mathbf{f}(\mathbf{A})$ commutes with $\mathbf{A}$. Therefore $\mathbf{A}$ and $\mathbf{f}(\mathbf{A})$ can be diagonalized simultaneously: If $\mathbf{X}^{-1}\mathbf{A}\mathbf{X} = \mathbf{\Lambda} = \operatorname{diag}(\lambda_1, \lambda_2, \ldots, \lambda_n)$, then $\mathbf{X}^{-1}\mathbf{f}(\mathbf{A})\mathbf{X} = \mathbf{F}$, which is also diagonal. To find $\mathbf{F}$, we proceed as follows. Let $\mathbf{A}\mathbf{x} = \lambda\mathbf{x}$, and apply $\mathbf{f}(\mathbf{A})$ to $\mathbf{x}$

$$\mathbf{f}(\mathbf{A})\mathbf{x} = \sum c_n \mathbf{A}^n \mathbf{x} = \sum c_n \lambda^n \mathbf{x} = f(\lambda)\mathbf{x}, \tag{8.37}$$

since $\mathbf{A}^n\mathbf{x} = \lambda\mathbf{A}^{n-1}\mathbf{x} = \cdots = \lambda^n\mathbf{x}$. Therefore we conclude that if $\mathbf{x}$ is an eigenvector of $\mathbf{A}$ with eigenvalue λ, it is also an eigenvector of $\mathbf{f}(\mathbf{A})$, but with eigenvalue $f(\lambda)$. And therefore

$$\mathbf{X}^{-1}\mathbf{f}(\mathbf{A})\mathbf{X} = \mathbf{F} = \operatorname{diag}(f(\lambda_1), f(\lambda_2), \ldots, f(\lambda_n)) = \mathbf{f}(\mathbf{\Lambda}).$$

This result now provides a method for calculating $\mathbf{f}(\mathbf{A})$ if we know the matrix that diagonalizes $\mathbf{A}$:

$$\mathbf{f}(\mathbf{A}) = \mathbf{X}\mathbf{F}\mathbf{X}^{-1} = \mathbf{X}\mathbf{f}(\mathbf{\Lambda})\mathbf{X}^{-1}. \tag{8.38}$$

Some other methods are suggested in the exercises.

Exercises

156. Evaluate Q for the one-dimensional Ising model (Exercise 8.21) in the special case $H = 0$. (*Hint:* Here $\mathbf{f}(\mathbf{A}) = \mathbf{A}^{N-1}$.)

*157. Expansion of the secular determinant $|\mathbf{A} - \lambda\mathbf{1}|$ gives the characteristic polynomial $P_n(\lambda) = c_0 + c_1\lambda + \cdots + c_n\lambda^n$. The CAYLEY-HAMILTON THEOREM states that $\mathbf{P}_n(\mathbf{A}) = \mathbf{0}$. Show this theorem to be true for an $n \times n$ matrix that has n linearly independent eigenvectors. (*Note:* There exists a method for determining the coefficients without expanding the determinant; for details, look up "Newton's sums," the sum of powers of roots of a polynomial, in an appropriate reference.)

158. Use the Cayley-Hamilton theorem to find the coefficients in the expansion of $\mathbf{A}^{-1}$ as an $(n - 1)$st-order polynomial in $\mathbf{A}$.

*159. Show that $\sum f(\lambda_j)\mathbf{P}_j = \mathbf{f}(\mathbf{A})$, where $\mathbf{P}_j$ is the projector matrix of Exercise 8.135. This is an alternative method for computing a function of a matrix.

Next we consider the applications of matrix functions to (*1*) simultaneous first-order chemical kinetics, (*2*) eigenvalue-eigenvector problems in a nonorthonormal basis, and (*3*) molecular vibrations.

The most general case of simultaneous first-order chemical kinetics leads to the coupled linear first-order differential equations

$$-\frac{dc_1}{dt} = k_{11}c_1 + k_{12}c_2 + \cdots + k_{1n}c_n$$

$$-\frac{dc_2}{dt} = k_{21}c_1 + k_{22}c_2 + \cdots + k_{2n}c_n$$

$$\vdots$$

$$-\frac{dc_n}{dt} = k_{n1}c_1 + k_{n2}c_2 + \cdots + k_{nn}c_n,$$

where $c_j(t)$ is the concentration of the jth species as a function of time, k_{jj} is the first-order rate constant for the disappearance of the jth species, and $-k_{ij}c_j$ is the rate at which the jth species is converted into the ith species. In matrix notation, this becomes

$$-\frac{d\mathbf{C}}{dt} = \mathbf{KC}, \tag{8.39}$$

where $\mathbf{C} = (c_j(t))$ and $d\mathbf{C}/dt = (dc_i/dt)$ are column matrices and $\mathbf{K} = (k_{ij})$ is $n \times n$. For example, for the kinetics $\mathrm{A} \xrightarrow{k_1} \mathrm{B} \xrightarrow{k_2} \mathrm{C}$, which we have solved previously by other methods, these matrices are

$$\mathbf{C} = \begin{pmatrix} \mathrm{A}(t) \\ \mathrm{B}(t) \\ \mathrm{C}(t) \end{pmatrix} \qquad \frac{d\mathbf{C}}{dt} = \begin{pmatrix} d\mathrm{A}/dt \\ d\mathrm{B}/dt \\ d\mathrm{C}/dt \end{pmatrix} \qquad \mathbf{K} = \begin{pmatrix} k_1 & 0 & 0 \\ -k_1 & k_2 & 0 \\ 0 & -k_2 & 0 \end{pmatrix}.$$

For help in finding the solution, let us pretend that we have, not a matrix equation, but the ordinary differential equation $-dc/dt = kc$, whose solution is simply $c = c_0e^{-kt}$. This result suggests that we try the solution

$$\mathbf{C} = e^{-t\mathbf{K}}\mathbf{C}_0,$$

where $\mathbf{C}_0 = (c_j(0))$ is a column matrix of initial concentrations. (This order of the factors is required for conformability.) Let us check this answer by substituting into Eq. 8.39:

$$\frac{d\mathbf{C}}{dt} = \frac{d}{dt} e^{-t\mathbf{K}}\mathbf{C}_0 = \frac{d}{dt}\sum_{n=0}^{\infty}(-1)^n \frac{t^n\mathbf{K}^n}{n!}\mathbf{C}_0 = \sum_{n=1}^{\infty}(-1)^n \frac{t^{n-1}\mathbf{K}^n}{(n-1)!}\mathbf{C}_0$$

$$= \mathbf{K}\sum_{n=1}^{\infty}(-1)^n \frac{t^{n-1}\mathbf{K}^{n-1}}{(n-1)!}\mathbf{C}_0 = \mathbf{K}\sum_{n=0}^{\infty}(-1)^{n+1}\frac{t^n\mathbf{K}^n}{n!}\mathbf{C}_0 = -\mathbf{KC}$$

as required. Furthermore, at $t = 0$, $\mathbf{C} = e^{\mathbf{0}}\mathbf{C}_0 = \mathbf{1C}_0 = \mathbf{C}_0$. Therefore we conclude that we have indeed found a formal solution to the coupled differential equations. Of course it still remains to evaluate $e^{-t\mathbf{K}}$. But this problem may be reduced to the problem of diagonalizing $\mathbf{K}$ to $\mathbf{\Lambda} = \text{diag}\,(\lambda_1, \lambda_2, \ldots \lambda_n)$ since if $\mathbf{X}^{-1}\mathbf{KX} = \mathbf{\Lambda}$, then $e^{-t\mathbf{K}} = \mathbf{X}e^{-t\mathbf{\Lambda}}\mathbf{X}^{-1}$, and

$$\mathbf{C} = \mathbf{X}e^{-t\mathbf{\Lambda}}\mathbf{X}^{-1}\mathbf{C}_0,$$

where $e^{-t\mathbf{\Lambda}} = \text{diag}\,(e^{-\lambda_1 t}, e^{-\lambda_2 t}, \ldots, e^{-\lambda_n t})$. By way of warning it should be pointed out that in general $\mathbf{K}$ is not Hermitian.

Exercises

160. (a) Show that $(d/dt)e^{\pm it\omega} = \pm i\omega e^{\pm it\omega}$.
(b) Then evaluate $(d^2/dt^2)e^{\pm it\omega}$, $(d^2/dt^2)\cos(t\boldsymbol{\omega})$, and $(d^2/dt^2)\sin(t\boldsymbol{\omega})$.

*161. Let $\boldsymbol{\psi}(t)$ be a column vector whose components depend upon the variable t, and let $d\boldsymbol{\psi}/dt$ be the column vector whose components are the time derivatives of the components of $\boldsymbol{\psi}(t)$. Let $\boldsymbol{\psi}$ satisfy the differential equation, $i\hbar(d\boldsymbol{\psi}/dt) = \mathbf{H}\boldsymbol{\psi}$, with $\mathbf{H}$ a Hermitian matrix and $\boldsymbol{\psi}(0) = \boldsymbol{\psi}_0$.
(a) Show that $(d/dt)(\boldsymbol{\psi}^\dagger\boldsymbol{\psi}) = 0$.
(b) Solve the differential equation. (*Note:* This equation is the Schrödinger equation including the time dependence, and the unitary matrix $e^{-it\mathbf{H}/\hbar}$ is known as the time-evolution operator.)

162. Use the matrix method to solve the kinetics $A \xrightarrow{k_1} B \xrightarrow{k_2} C$, $A(0) = A_0$, $B(0) = 0 = C(0)$. (*Hint:* This need be only a two-dimensional problem, since $C(t)$ may be determined from the stoichiometry.)

163. The magnetization $\mathbf{M}$ of a sample of spins is the vector sum of the magnetic moments in a unit volume. The Bloch equations, describing the time dependence of $\mathbf{M}$, are basic to magnetic resonance: (*1*) In any magnetic field $\mathbf{H}$, $d\mathbf{M}/dt = -\gamma\mathbf{H}\times\mathbf{M}$, since $d\mathbf{L}/dt = -\gamma\mathbf{H}\times\mathbf{L}$, and $\mathbf{M} = \sum\boldsymbol{\mu}_i = \sum\gamma\mathbf{L}_i$. (*2*) In a strong static magnetic field along the z-axis, there is an equilibrium $M_z = M_0$ (cf. Exercise 2.52 for the net electric dipole moment in the direction of the electric field). If M_z is displaced from the value M_0, then M_z approaches M_0 with a LONGITUDINAL RELAXATION TIME T_1. (*3*) If any magnetization is produced perpendicular to the field, this magnetization decays to zero with a TRANSVERSE RELAXATION TIME T_2.
(a) Let $\mathbf{M} = (M_x, M_y, M_z)^T$, and express these statements in the form $-d\mathbf{M}/dt = \mathbf{KM} - \mathbf{K}_0\mathbf{M}_0$. Let $\mathbf{H}$ have the particular form $H_1\cos\omega t\hat{\mathbf{i}} - H_1\sin\omega t\hat{\mathbf{j}} + H_0\hat{\mathbf{k}}$. Then the matrix differential equation is difficult to solve in this form because $\mathbf{K}$ depends upon t.

(b) Let us transform to a coordinate system rotating at the same frequency as does **H** by defining

$$\mathbf{M}' = \begin{pmatrix} u \\ v \\ M_z \end{pmatrix} = \begin{pmatrix} \cos \omega t & -\sin \omega t & 0 \\ \sin \omega t & \cos \omega t & 0 \\ 0 & 0 & 1 \end{pmatrix} \begin{pmatrix} M_x \\ M_y \\ M_z \end{pmatrix} = \mathbf{U}^\dagger \mathbf{M}$$

and $\mathbf{M}_0' = \mathbf{U}^\dagger \mathbf{M}_0$. Then $-d\mathbf{M}/dt = \mathbf{K}\mathbf{M} - \mathbf{K}_0\mathbf{M}_0$ becomes $-d\mathbf{M}'/dt = \mathbf{K}'\mathbf{M}' - \mathbf{K}_0'\mathbf{M}_0'$, with $\mathbf{K}'$ now time-independent. But $\mathbf{K}'$ is not simply equal to $\mathbf{U}^\dagger\mathbf{K}\mathbf{U}$, since there is a further effect to consider on transforming to a rotating coordinate system: H_0 must be replaced by $H_z^{\text{eff}} = H_0 - \omega/\gamma$ since there would appear to be no applied H_z if we were rotating at the Larmor frequency, $\omega_0 = \gamma H_0$. Find $\mathbf{K}'$.

(c) The resultant equations are simply three coupled linear differential equations with constant coefficients (as in chemical kinetics, but inhomogeneous) and could be solved by diagonalizing $\mathbf{K}'$ and evaluating $e^{-t\mathbf{K}'}$ (although the eigenvalues and eigenvectors are messy). But for many purposes it is sufficient to determine the value of $\mathbf{M}'$ for which $d\mathbf{M}'/dt = \mathbf{0}$; indeed, it is the steady-state $v(\omega)$ that is ordinarily observed in nmr spectrometers, and $dv/d\omega$ in esr. Show that this steady-state $\mathbf{M}'$ is equal to

$$M_0[1 + T_2^2(\omega_0 - \omega)^2 + \gamma^2 H_1^2 T_1 T_2]^{-1}$$
$$(\gamma H_1 T_2^2(\omega_0 - \omega,) \quad \gamma H_1 T_2, \quad 1 + T_2^2(\omega_0 - \omega)^2)^T.$$

In a nonorthonormal basis set $\{\mathbf{e}_j\}$ an eigenvalue-eigenvector problem

$$A\mathbf{x} = \lambda \mathbf{x}$$

is slightly more complicated. We may still expand $\mathbf{x} = \sum_j \mathbf{e}_j x_j$ in order to represent the vector $\mathbf{x}$ as the column matrix $\mathbf{x}$. However, we do not define $\mathbf{A}$ as the matrix of coefficients in the expansion $A\mathbf{e}_j = \sum_i \mathbf{e}_i \alpha_{ij}$, because we do not ordinarily know $\alpha_{ij} = \boldsymbol{\epsilon}_i \cdot A\mathbf{e}_j$, where $\{\boldsymbol{\epsilon}_i\}$ is the reciprocal basis. In practice it is $\mathbf{e}_i \cdot A\mathbf{e}_j$ that we know, and we therefore represent A as the matrix $\mathbf{A} \equiv (\mathbf{e}_i \cdot A\mathbf{e}_j)$. Substituting the expansion of $\mathbf{x}$ into the eigenvalue-eigenvector equation, and taking the dot product with $\mathbf{e}_i$ from the left, gives

$$\sum_j (\mathbf{e}_i \cdot A\mathbf{e}_j)\mathbf{x}_j = \lambda \sum_j (\mathbf{e}_i \cdot \mathbf{e}_j)\mathbf{x}_j$$

If we also define the matrix $\mathbf{S} \equiv (\mathbf{e}_i \cdot \mathbf{e}_j)$, we may write this result in matrix form

$$\mathbf{A}\mathbf{x} = \lambda \mathbf{S}\mathbf{x}$$

which is not in the form of an eigenvalue-eigenvector equation as previously presented, although it can be solved by the same technique as previously: Find the roots of the secular equation $|\mathbf{A} - \lambda\mathbf{S}| = 0$ (*Warning*: Remember that λ occurs off the diagonal), and solve the resulting linear homogeneous equations for the solution vector $\mathbf{x}$.

One way of converting this equation into a true eigenvalue-eigenvector equation is to multiply from the left by $\mathbf{S}^{-1}$:

$$\mathbf{S}^{-1}\mathbf{A}\mathbf{x} = \lambda \mathbf{x}$$

which is just the problem of finding eigenvalues and eigenvectors of the matrix $\mathbf{S}^{-1}\mathbf{A}$. Another way is to let $\mathbf{y} = \mathbf{S}\mathbf{x}$ and $\mathbf{x} = \mathbf{S}^{-1}\mathbf{y}$, so that the equation becomes

$$\mathbf{A}\mathbf{S}^{-1}\mathbf{y} = \lambda\mathbf{y}$$

which is the problem of finding eigenvalues and eigenvectors of the matrix $\mathbf{A}\mathbf{S}^{-1}$. The disadvantage of both these methods is that in most practical situations $\mathbf{A}$ and $\mathbf{S}$ are Hermitian but $\mathbf{S}^{-1}\mathbf{A}$ and $\mathbf{A}\mathbf{S}^{-1}$ are not. It is often more convenient to deal only with Hermitian matrices. Therefore let us set $\mathbf{y} = \mathbf{S}^{1/2}\mathbf{x}$ and $\mathbf{x} = \mathbf{S}^{-1/2}\mathbf{y}$. Then $\mathbf{A}\mathbf{x} = \lambda\mathbf{S}\mathbf{x}$ becomes $\mathbf{A}\mathbf{S}^{-1/2}\mathbf{y} = \lambda\mathbf{S}^{1/2}\mathbf{y}$. And if we multiply from the left by $\mathbf{S}^{-1/2}$, we obtain

$$\mathbf{S}^{-1/2}\mathbf{A}\mathbf{S}^{-1/2}\mathbf{y} = \lambda\mathbf{y}$$

which is the problem of finding eigenvalues and eigenvectors of the matrix $\mathbf{S}^{-1/2}\mathbf{A}\mathbf{S}^{-1/2}$, which is Hermitian if $\mathbf{A}$ and $\mathbf{S}$ are Hermitian (Exercise 8.164).

Of course there still remains the problem of finding $\mathbf{S}^{-1/2}$. But if we diagonalize $\mathbf{S}$ according to $\mathbf{T}^{-1}\mathbf{S}\mathbf{T} = \mathbf{\Sigma} = \text{diag}\,(\sigma_1, \sigma_2, \ldots, \sigma_n)$, then $\mathbf{S}^{-1/2} = \mathbf{T}\mathbf{\Sigma}^{-1/2}\mathbf{T}^{-1}$, where $\mathbf{\Sigma}^{-1/2} = \text{diag}\,(\sigma_1^{-1/2}, \sigma_2^{-1/2}, \ldots, \sigma_n^{-1/2})$. Therefore the problem of finding the eigenvalues and eigenvectors of a Hermitian operator in a non-orthonormal basis may be converted to that of diagonalizing *two* Hermitian matrices by the usual techniques.

Exercises

*164. Let $\{\mathbf{e}_i\}$ be a set of n linearly independent n-dimensional vectors. Then the matrix $\mathbf{S} \equiv (\mathbf{e}_i \cdot \mathbf{e}_j)$ is nonsingular (cf. Exercise 8.94).
(a) Show that $\mathbf{S}$ is Hermitian.
(b) Use the result of Exercise 8.64 to show that $\mathbf{S}$ is positive definite.
(c) Show that there exists an $\mathbf{S}^{-1/2}$ that is also Hermitian.
(d) Show that if $\mathbf{H}$ is Hermitian, so is $\mathbf{S}^{-1/2}\mathbf{H}\mathbf{S}^{-1/2}$.

165. If $\mathbf{S} = \begin{pmatrix} 1 & \frac{1}{4} \\ \frac{1}{4} & 1 \end{pmatrix}$, diagonalize $\mathbf{S}$ and find $\mathbf{S}^{-1/2}$. Check your result: Does $(\mathbf{S}^{-1/2})^2 = \mathbf{S}^{-1}$?

*166. (a) Use Exercise 8.157 to convince yourself that if $\mathbf{A}$ is $n \times n$ and if f has a power-series expansion, then $\mathbf{f}(\mathbf{A})$ can be reduced to an $(n - 1)$st-order polynomial. [*Hints:* Solve $\mathbf{P}_n(\mathbf{A}) = \mathbf{0}$ for $\mathbf{A}^n$, then multiply this result by $\mathbf{A}$.]
(b) Let $\mathbf{f}(\mathbf{A}) = \sum_{j=1}^{n} b_j\mathbf{A}^{j-1}$, a finite sum whose coefficients are as yet unknown. Show that if all eigenvalues $\{\lambda_i\}$ of $\mathbf{A}$ are distinct, the solution for the coefficients in this expansion is given by $\mathbf{b} = \mathbf{\Pi}^{-1}(f(\lambda_1), f(\lambda_2), \ldots, f(\lambda_n))^T$, where $\mathbf{b} = (b_j)$ and $(\mathbf{\Pi})_{ij} = \lambda_i^{j-1}$. [*Hint:* If $\mathbf{A}\mathbf{x}_i = \lambda_i\mathbf{x}_i$, what is $\mathbf{f}(\mathbf{A})\mathbf{x}_i$?]
(c) Find $\begin{pmatrix} 1 & S \\ S & 1 \end{pmatrix}^{-1/2}$ by this method.

To describe the motion of an N-atomic molecule, it is necessary to specify $3N$ coordinates, each as a function of time. For simplicity, let us choose cartesian coordinates, $\{X_i(t)\}_{i=1}^{3N} = \{x_1, y_1, z_1, x_2, y_2, \ldots, x_N, y_N, z_N\}$. The potential energy of the molecule depends on these coordinates, but there is a configuration $\{X_i^0\}$ at which the potential energy is minimum and which is the equilibrium position for the molecule. Let us define new coordinates $\xi_i \equiv X_i - X_i^0$, representing the displacements from equilibrium positions. And let us expand the potential energy as a Maclaurin series in these displacements (Exercise 2.9):

$$V = V_{\min} + \tfrac{1}{2}\sum_{i,j} k_{ij}\xi_i\xi_j + \cdots,$$

where the linear term vanishes because $\xi_i = 0$ is the equilibrium position. We shall neglect terms higher than quadratic. And we may choose the zero of energy as $V_{\min}$. In cartesian coordinates the kinetic energy is given by

$$T = \tfrac{1}{2}\sum_i m_i\dot{\xi}_i^2,$$

where $\dot{\xi}_i \equiv d\xi_i/dt = dX_i/dt$, an x-, y-, or z-component of the velocity of some atom, and m_i is the mass of that atom. If we define the $3N$-dimensional column matrices $\boldsymbol{\xi} \equiv (\xi_i(t))$, $\dot{\boldsymbol{\xi}} \equiv (d\xi_i/dt)$, and $\ddot{\boldsymbol{\xi}} \equiv (d^2\xi_i/dt^2)$, and the $3N \times 3N$ matrices $\mathbf{F} \equiv (\partial^2 V/\partial\xi_i\,\partial\xi_j) = (k_{ij})$ and $\mathbf{G}^{-1} \equiv (\partial^2 T/\partial\dot{\xi}_i\,\partial\dot{\xi}_j) = \text{diag}\,(m_1, m_2, \ldots, m_{3N})$, then the quadratic forms V and T become $\frac{1}{2}\boldsymbol{\xi}^\dagger\mathbf{F}\boldsymbol{\xi}$ and $\frac{1}{2}\dot{\boldsymbol{\xi}}^\dagger\mathbf{G}^{-1}\dot{\boldsymbol{\xi}}$, respectively.

According to Lagrange's formulation of classical mechanics, Newton's laws of motion become the matrix differential equation

$$\mathbf{G}^{-1}\ddot{\boldsymbol{\xi}} + \mathbf{F}\boldsymbol{\xi} = \mathbf{0} \qquad \text{or} \qquad \ddot{\boldsymbol{\xi}} + \mathbf{GF}\boldsymbol{\xi} = \mathbf{0}. \tag{8.40}$$

If there were only one variable, this equation would become the familiar differential equation for simple harmonic motion, force $= ma = -kx$, or

$$\ddot{x} + (k/m)x = 0$$

whose solution, subject to the initial conditions $x(0) = x_0$ and $\dot{x}(0) = \dot{x}_0$, is

$$x(t) = x_0 \cos \omega t + (\dot{x}_0/\omega) \sin \omega t,$$

where $\omega = \sqrt{k/m}$.

This result suggests that the solution to the matrix equation is

$$\boldsymbol{\xi} = \cos(t\boldsymbol{\omega})\boldsymbol{\xi}_0 + \boldsymbol{\omega}^{-1}\sin(t\boldsymbol{\omega})\dot{\boldsymbol{\xi}}_0, \tag{8.41}$$

where $\boldsymbol{\omega} \equiv (\mathbf{GF})^{1/2}$, $\cos(t\boldsymbol{\omega}) \equiv \frac{1}{2}(e^{it\boldsymbol{\omega}} + e^{-it\boldsymbol{\omega}})$, and $\sin(t\boldsymbol{\omega}) \equiv (1/2i)(e^{it\boldsymbol{\omega}} - e^{-it\boldsymbol{\omega}})$, and $\boldsymbol{\xi}_0$ and $\dot{\boldsymbol{\xi}}_0$ are column matrices of initial displacements and velocities. For help in checking this solution let us note (Exercise 8.160) that

$$\frac{d^2}{dt^2}\cos(t\boldsymbol{\omega}) = -\boldsymbol{\omega}^2\cos(t\boldsymbol{\omega}) \qquad \frac{d^2}{dt^2}\sin(t\boldsymbol{\omega}) = -\boldsymbol{\omega}^2\sin(t\boldsymbol{\omega}).$$

Therefore, if we take the second derivative of the suggested solution, $\boldsymbol{\xi}$, with respect to t, we obtain

$$\ddot{\boldsymbol{\xi}} = -\boldsymbol{\omega}^2 \cos(t\boldsymbol{\omega})\boldsymbol{\xi}_0 - \boldsymbol{\omega}^2\boldsymbol{\omega}^{-1}\sin(t\boldsymbol{\omega})\dot{\boldsymbol{\xi}}_0 = -\boldsymbol{\omega}^2\boldsymbol{\xi} \qquad \text{or} \qquad \ddot{\boldsymbol{\xi}} + \boldsymbol{\omega}^2\boldsymbol{\xi} = \mathbf{0}$$

which is indeed the original differential equation. And it is readily shown that $\boldsymbol{\xi}$ does satisfy the initial conditions.

There still remains the problem of finding $\cos(t\boldsymbol{\omega})$, $\sin(t\boldsymbol{\omega})$, and $\boldsymbol{\omega}^{-1}$. Let $\mathbf{C} = (c_{ij})$ be the $3N \times 3N$ matrix that diagonalizes $\boldsymbol{\omega}^2 = \mathbf{GF}$:

$$\mathbf{C}^{-1}\mathbf{GFC} = \boldsymbol{\Lambda} \equiv \operatorname{diag}(\lambda_1, \lambda_2, \ldots, \lambda_{3N}).$$

Then

$$\cos(t\boldsymbol{\omega}) = \mathbf{C}\cos(t\boldsymbol{\Lambda}^{1/2})\mathbf{C}^{-1} \qquad \sin(t\boldsymbol{\omega}) = \mathbf{C}\sin(t\boldsymbol{\Lambda}^{1/2})\mathbf{C}^{-1}$$
$$\boldsymbol{\omega}^{-1} = \mathbf{C}\boldsymbol{\Lambda}^{-1/2}\mathbf{C}^{-1},$$

where $\cos(t\boldsymbol{\Lambda}^{1/2}) = \operatorname{diag}(\cos\lambda_1^{1/2}t, \cos\lambda_2^{1/2}t, \ldots, \cos\lambda_{3N}^{1/2}t)$, $\sin(t\boldsymbol{\Lambda}^{1/2}) = \operatorname{diag}(\sin\lambda_1^{1/2}t, \sin\lambda_2^{1/2}t, \ldots, \sin\lambda_{3N}^{1/2}t)$, and $\boldsymbol{\Lambda}^{-1/2} = \operatorname{diag}(\lambda_1^{-1/2}, \lambda_2^{-1/2}, \ldots, \lambda_{3N}^{-1/2})$. Substituting these expressions for the matrix functions into Eq. 8.41 gives

$$\boldsymbol{\xi} = \mathbf{C}\cos(t\boldsymbol{\Lambda}^{1/2})\mathbf{C}^{-1}\boldsymbol{\xi}_0 + \mathbf{C}\boldsymbol{\Lambda}^{-1/2}\sin(t\boldsymbol{\Lambda}^{1/2})\mathbf{C}^{-1}\dot{\boldsymbol{\xi}}_0.$$

Furthermore, let us multiply from the left by $\mathbf{C}^{-1}$, and set $\mathbf{Q} = \mathbf{C}^{-1}\boldsymbol{\xi}$, $\dot{\mathbf{Q}}_0 = \mathbf{C}^{-1}\boldsymbol{\xi}_0$, and $\dot{\mathbf{Q}}_0 = \mathbf{C}^{-1}\dot{\boldsymbol{\xi}}_0$, to obtain

$$\mathbf{Q} = \cos(t\boldsymbol{\Lambda}^{1/2})\mathbf{Q}_0 + \boldsymbol{\Lambda}^{-1/2}\sin(t\boldsymbol{\Lambda}^{1/2})\dot{\mathbf{Q}}_0. \tag{8.42}$$

But the square matrices here are all diagonal, so that in terms of components this equation becomes

$$Q_i = Q_i(0)\cos\lambda_i^{1/2}t + \dot{Q}_i(0)\lambda_i^{-1/2}\sin\lambda_i^{1/2}t \tag{8.43}$$

which shows that the coordinate Q_i executes simple harmonic motion without interaction with other coordinates. Alternatively, let us substitute $\boldsymbol{\xi} = \mathbf{CQ}$ into the original matrix differential equation (Eq. 8.40) and multiply from the left by $\mathbf{C}^{-1}$ to obtain

$$\ddot{\mathbf{Q}} + \mathbf{C}^{-1}\mathbf{GFCQ} = \ddot{\mathbf{Q}} + \boldsymbol{\Lambda}\mathbf{Q} = \mathbf{0}$$

which in terms of components is the set of uncoupled differential equations

$$\ddot{Q}_i + \lambda_i Q_i = 0$$

whose solution is indeed Eqs. 8.42 and 8.43. Thus we see that the transformation from cartesian coordinates to the new set of coordinates $\{Q_i\}$ provides an important simplification toward solving the equations of molecular motion. These new coordinates, in which the differential equations are uncoupled, are known as NORMAL COORDINATES or NORMAL MODES.

In summary, we see that solving the equations of motion is essentially the problem of finding the normal coordinates. To find these, it is necessary to find the matrix $\mathbf{C}$ that diagonalizes the matrix $\mathbf{GF}$. Then the ith normal coordinate is given by the linear combination $\sum_j (\mathbf{C}^{-1})_{ij}\xi_j$ of cartesian coordinates. Furthermore, the eigenvalue λ_i of $\mathbf{GF}$ is the square of the angular frequency with which Q_i executes harmonic motion: $\lambda_i = \omega_i^2 = (2\pi\nu_i)^2$.

To diagonalize $\mathbf{GF}$ it is necessary to solve the eigenvalue-eigenvector equation $\mathbf{GF\xi} = \lambda\mathbf{\xi}$. However, although $\mathbf{G}$ and $\mathbf{F}$ are Hermitian, they do not ordinarily commute, so that $\mathbf{GF}$ is not Hermitian, nor is the matrix $\mathbf{C}$ that diagonalizes $\mathbf{GF}$ unitary. However, if we multiply both sides of this equation from the left by $\mathbf{G}^{-1/2} = \text{diag}\,(m_1^{1/2}, m_2^{1/2}, \ldots, m_{3N}^{1/2})$, and let $\mathbf{q} = \mathbf{G}^{-1/2}\mathbf{\xi}$, we obtain

$$\mathbf{G}^{1/2}\mathbf{F}\mathbf{G}^{1/2}\mathbf{q} = \lambda\mathbf{q}$$

which is the problem of diagonalizing the Hermitian matrix $\mathbf{G}^{1/2}\mathbf{F}\mathbf{G}^{1/2}$. Since $q_i = m_i^{1/2}\xi_i$, the coordinates $\{q_i\}$ are known as MASS-WEIGHTED CARTESIAN DISPLACEMENTS. Also, since $\mathbf{C}^{-1}\mathbf{GFC} = \mathbf{\Lambda}$, it must be $\mathbf{G}^{-1/2}\mathbf{C}$ that diagonalizes $\mathbf{G}^{1/2}\mathbf{F}\mathbf{G}^{1/2}$. And from Eq. 8.36, $\mathbf{G}^{-1/2}\mathbf{C}$ must be a unitary matrix $\mathbf{U}$. Then the motion of the molecule is given in terms of mass-weighted coordinates by

$$\mathbf{q} = \mathbf{U}\cos(t\mathbf{\Lambda}^{1/2})\mathbf{U}^{-1}\mathbf{q}_0 + \mathbf{U}\mathbf{\Lambda}^{-1/2}\sin(t\mathbf{\Lambda}^{1/2})\mathbf{U}^{-1}\dot{\mathbf{q}}_0.$$

Let us next investigate the forms that the potential and kinetic energies take in these various coordinates. In the original cartesian displacements $V = \frac{1}{2}\mathbf{\xi}^\dagger\mathbf{F}\mathbf{\xi}$ and $T = \frac{1}{2}\dot{\mathbf{\xi}}^\dagger\mathbf{G}^{-1}\dot{\mathbf{\xi}}$. If we transform to mass-weighted cartesian displacement coordinates by $\mathbf{\xi} = \mathbf{G}^{1/2}\mathbf{q}$, we obtain $V = \frac{1}{2}\mathbf{q}^\dagger\mathbf{G}^{1/2}\mathbf{F}\mathbf{G}^{1/2}\mathbf{q}$ and $T = \frac{1}{2}\dot{\mathbf{q}}^\dagger\mathbf{G}^{1/2}\mathbf{G}^{-1}\mathbf{G}^{1/2}\dot{\mathbf{q}} = \frac{1}{2}\dot{\mathbf{q}}^\dagger\mathbf{1}\dot{\mathbf{q}}$. The transformation from mass-weighted coordinates to normal coordinates is given by $\mathbf{q} = \mathbf{UQ}$. Therefore V becomes $\frac{1}{2}\mathbf{Q}^\dagger\mathbf{U}^\dagger\mathbf{G}^{1/2}\mathbf{F}\mathbf{G}^{1/2}\mathbf{UQ} = \frac{1}{2}\mathbf{Q}^\dagger\mathbf{\Lambda}\mathbf{Q}$ and T becomes $\frac{1}{2}\dot{\mathbf{Q}}^\dagger\mathbf{U}^\dagger\mathbf{U}\dot{\mathbf{Q}} = \frac{1}{2}\dot{\mathbf{Q}}^\dagger\mathbf{1}\dot{\mathbf{Q}}$. Thus we may conclude that the transformation to normal coordinates represents a transformation of the quadratic forms T and V to $\frac{1}{2}\sum\dot{Q}_i^2$ and $\frac{1}{2}\sum\lambda_i Q_i^2$, respectively, without cross-terms.

Finally, let us look at some special cases of the molecular motion. First let us investigate the solution subject to the initial conditions $\mathbf{\xi}_0 = (A, 0, 0, \ldots, 0)^T$, $\dot{\mathbf{\xi}}_0 = \mathbf{0}$, corresponding to displacing the first atom along the x-axis a distance A from its equilibrium position, and then releasing it. The resulting motion of the molecule is given by

$$\begin{aligned}\xi_i(t) &= [\cos(t\mathbf{\omega})\mathbf{\xi}_0]_i = A[\cos(t\mathbf{\omega})]_{i1}\\ &= A[\mathbf{C}\cos(t\mathbf{\Lambda}^{1/2})\mathbf{C}^{-1}]_{i1} = A\textstyle\sum_j(\mathbf{C})_{ij}\cos\lambda_j^{1/2}t(\mathbf{C}^{-1})_{j1}\end{aligned}$$

which is not simple because the first atom is coupled to the other atoms, which are set in motion. And in general each of the frequencies of vibration is represented. For contrast, let us investigate the solution subject to the initial conditions $\mathbf{Q}_0 = (A, 0, 0, \ldots, 0)^T$, $\dot{\mathbf{Q}}_0 = \mathbf{0}$. Since $\mathbf{\xi}_0 = \mathbf{CQ}_0$, or

$\xi_i(0) = Ac_{i1}$, these initial conditions correspond to releasing the molecule from a configuration in which each atom has been pulled away from its equilibrium position to an extent and in a direction specified by the coefficients $\{c_{i1}\}_{i=1}^{3N}$. Then the motion of the molecule is given by

$$\xi_i(t) = (\mathbf{CQ})_i = Ac_{i1} \cos \lambda_1^{1/2} t,$$

showing that the molecule vibrates in only this first normal mode. This is a much simpler motion because all atoms move in unison with angular frequency $\lambda_1^{1/2}$, each passing through its equilibrium position at the same time.

Example

23. Let us find the longitudinal normal modes and frequencies of a linear symmetrical AB_2 molecule. We neglect motion perpendicular to the molecular axis and take $\boldsymbol{\xi} = (x_1, x_A, x_2)^T$.

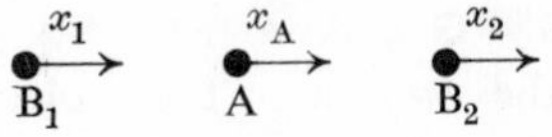

Then $\mathbf{G}^{-1} = \text{diag}\,(m_B, m_A, m_B)$. We shall choose a force field with two force constants: the force constant k opposing stretching an A—B bond and the interaction force constant k' opposing stretching both bonds simultaneously. (For example, stretching one bond of CO_2 tends to shorten the other, so that k' is positive.) Then the potential energy is a quadratic form in the two A—B bond length changes:

$$\begin{aligned} V &= \tfrac{1}{2}k\,\Delta d_{AB_1}^2 + \tfrac{1}{2}k\,\Delta d_{AB_2}^2 + k'\,\Delta d_{AB_1}\,\Delta d_{AB_2} \\ &= \tfrac{1}{2}k(x_1 - x_A)^2 + \tfrac{1}{2}k(x_2 - x_A)^2 + k'(x_1 - x_A)(x_2 - x_A) \end{aligned}$$

and

$$\mathbf{F} = (\partial^2 V/\partial \xi_i\,\partial \xi_j) = \begin{pmatrix} k & -(k+k') & k' \\ -(k+k') & 2(k+k') & -(k+k') \\ k' & -(k+k') & k \end{pmatrix}.$$

If we transform to mass-weighted coordinates, then $\mathbf{q} = \mathbf{G}^{-1/2}\boldsymbol{\xi} = (m_B^{1/2}x_1, m_A^{1/2}x_A, m_B^{1/2}x_2)^T$, and $\mathbf{G}^{1/2}\mathbf{F}\mathbf{G}^{1/2}$ equals the Hermitian matrix

$$\begin{pmatrix} m_B^{-1}k & -m_A^{-1/2}m_B^{-1/2}(k+k') & m_B^{-1}k' \\ -m_A^{-1/2}m_B^{-1/2}(k+k') & 2m_A^{-1}(k+k') & -m_A^{-1/2}m_B^{-1/2}(k+k') \\ m_B^{-1}k' & -m_A^{-1/2}m_B^{-1/2}(k+k') & m_B^{-1}k \end{pmatrix}.$$

The eigenvalues of this matrix are $\lambda_1 = 4\pi^2\nu_1^2 = 0$, $\lambda_2 = 4\pi^2\nu_2^2 = m_B^{-1}(k - k')$, and $\lambda_3 = 4\pi^2\nu_3^2 = (2m_A^{-1} + m_B^{-1})(k + k')$. The unitary matrix of normalized eigenvectors, which diagonalizes $\mathbf{G}^{1/2}\mathbf{G}\mathbf{F}^{1/2}$, is

$$\mathbf{U} = \mathbf{G}^{-1/2}\mathbf{C} = \begin{pmatrix} (m_B/M)^{1/2} & -1/\sqrt{2} & (m_A/2M)^{1/2} \\ (m_A/M)^{1/2} & 0 & -2(m_B/2M)^{1/2} \\ (m_B/M)^{1/2} & 1/\sqrt{2} & (m_A/2M)^{1/2} \end{pmatrix},$$

where $M = m_A + 2m_B$. And the normal modes are given by $\mathbf{Q} = \mathbf{U}^\dagger \mathbf{q} = \mathbf{U}^\dagger \mathbf{G}^{-1/2}\boldsymbol{\xi}$, or

$$\begin{aligned} M^{1/2}Q_1 &= m_B^{1/2}q_1 + m_A^{1/2}q_A + m_B^{1/2}q_2 = m_B x_1 + m_A x_A + m_B x_2 \\ 2^{1/2}Q_2 &= -q_1 + q_2 = -m_B^{1/2}x_1 + m_B^{1/2}x_2 \\ 2M^{1/2}Q_3 &= m^{1/2}q_1 - 2m_B^{1/2}q_A + m_A^{1/2}q_2 = x_1 - 2x_A + x_2. \end{aligned}$$

Finally, let us solve for the cartesian coordinates in terms of the normal coordinates: $\boldsymbol{\xi} = \mathbf{G}^{1/2}\mathbf{q} = \mathbf{G}^{1/2}(\mathbf{G}^{-1/2}\mathbf{C})\mathbf{Q} = \mathbf{C}\mathbf{Q}$, or

$$\begin{aligned} x_1 &= M^{-1/2}Q_1 - (2m_B)^{-1/2}Q_2 + (m_A/2m_BM)^{1/2}Q_3 \\ x_A &= M^{-1/2}Q_1 - 2(m_B/2m_AM)^{1/2}Q_3 \\ x_2 &= M^{-1/2}Q_1 + (2m_B)^{-1/2}Q_2 + (m_A/2m_BM)^{1/2}Q_3. \end{aligned}$$

To show what these normal modes "look like," it is customary to illustrate each Q_j by drawing an arrow at each atom proportional to the coefficient c_{ij} in the jth column of $\mathbf{C}$ (the coefficient of Q_j in the expansion of x_i). An example of this, with $m_A = \frac{3}{4}m_B$, is shown in Figure 8.8.

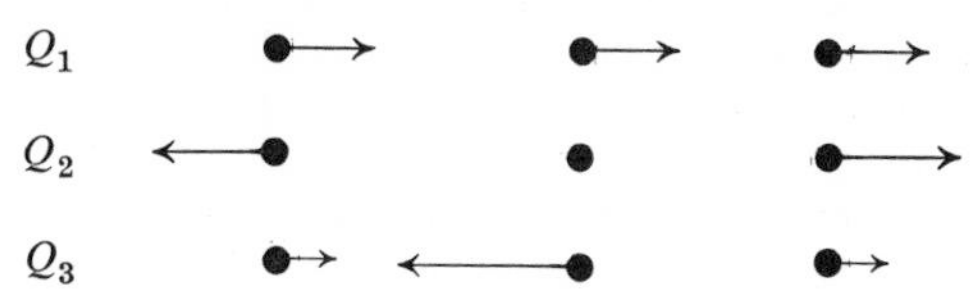

Figure 8.8

To demonstrate the significance of these pictures, let us distort the AB_2 molecule along each of the normal modes in turn: Each such distortion corresponds to displacing each atom from its equilibrium position to the tip of the arrow at that atom.

1. Let us set $Q_1 = Q_1^0$ and $Q_2 = 0 = Q_3$. Then $x_1 = x_A = x_2 = M^{-1/2}Q_1^0$, so that all three atoms have been displaced the identical distance $M^{-1/2}Q_1^0$ from their equilibrium positions. But this distortion along Q_1 corresponds to a uniform translation of the molecule. And since the restoring force opposing translation is zero, we can understand why ν_1, the frequency of this "vibration," is zero.

2. Let us set $Q_2 = Q_2^0$ and $Q_1 = 0 = Q_3$. Then $x_2 = (2m_B)^{-1/2}Q_2^0 = -x_1$ and $x_A = 0$. This distortion along the normal mode Q_2 corresponds to displacing atom B_2 a distance $(2m_B)^{-1/2}Q_2^0$ in the positive x-direction (away from atom A), displacing atom B_1 an equal distance in the negative x-direction (also away from A), and holding atom A fixed. Alternatively, we may say that the atoms have been displaced to the tips of the arrows drawn to illustrate Q_2. If the atoms are released from these distorted positions, the resultant motion is given by $Q_2(t) = Q_2^0 \cos(2\pi\nu_2 t)$: The two B atoms oscillate in unison about their equilibrium positions, with frequency ν_2, and the two A—B bonds alternately lengthen and shorten, in phase.

3. Let us set $Q_3 = Q_3^0$ and $Q_1 = 0 = Q_2$. This distortion along the normal mode Q_3 corresponds to displacing both B atoms in the positive x-direction, while the A atom is displaced in the opposite direction, thereby shortening the A—B_1 bond and lengthening the A—B_2 bond. This distortion is also illustrated by the arrows. If the atoms are released from these distorted positions, the resultant motion is given by $Q_3(t) = Q_3^0 \cos(2\pi\nu_3 t)$. The two B atoms oscillate in phase with each other and 180° out of phase with the A atom, and each A—B bond alternately shortens and lengthens, but out of phase with the other.

Exercises

*167. Show that the problem $\mathbf{GF}\boldsymbol{\xi} = \lambda\boldsymbol{\xi}$ may lead to any of the secular determinants, $|\mathbf{GF} - \lambda\mathbf{1}|$, $|\mathbf{F} - \lambda\mathbf{G}^{-1}|$, $|\mathbf{G} - \lambda\mathbf{F}^{-1}|$, or $|\mathbf{FG} - \lambda\mathbf{1}|$. (*Hint:* Let $\boldsymbol{\xi}' = \mathbf{F}\boldsymbol{\xi}$.) The choice of which secular equation to solve depends upon which matrices are most readily calculated. We chose cartesian coordinates as our basis so that the quadratic form T and the matrix $\mathbf{G}$ were simple. However, this is not the only choice of basis. For example, there are various sets of internal coordinates $\{S_i\}$, involving only interatomic distances and angles, that may be more convenient. But then $\mathbf{G}^{-1} \equiv (\partial^2 T/\partial \dot{S}_i\, \partial \dot{S}_j)$ and $\mathbf{F} \equiv (\partial^2 V/\partial S_i\, \partial S_j)$ are different.

*168. For the linear AB_2 molecule, what does the constraint $Q_1 = 0$ say about the center of mass of the molecule?

*169. An alternative approach to normal modes of motion is the problem of the simultaneous diagonalization of two noncommuting $n \times n$ Hermitian matrices $\mathbf{F} = (\partial^2 V/\partial S_i\, \partial S_j)$ and $\mathbf{G}^{-1} = (\partial^2 T/\partial \dot{S}_i\, \partial \dot{S}_j)$. Of course, since these matrices do not commute, there is no similarity transformation that can diagonalize them simultaneously. Therefore we shall seek a transformation by the matrix $\mathbf{C}$ such that $\mathbf{C}^\dagger\mathbf{G}^{-1}\mathbf{C} = \mathbf{1}$ and $\mathbf{C}^\dagger\mathbf{FC} = \mathbf{\Lambda}$; this sort of transformation, involving premultiplication by $\mathbf{C}^\dagger$ instead of $\mathbf{C}^{-1}$, is known as a CONGRUENT TRANSFORMATION. Let us carry out the transformation in three steps: (*1*) Let $\mathbf{U}$ diagonalize $\mathbf{G}^{-1}$: $\mathbf{U}^\dagger\mathbf{G}^{-1}\mathbf{U} = \mathbf{M} = \text{diag}\,(\mu_1, \mu_2, \ldots, \mu_n)$. But $\mathbf{U}^\dagger\mathbf{FU}$ is not diagonal. (*2*) By the positive-definite nature of kinetic energy, every eigenvalue of $\mathbf{G}^{-1}$ is necessarily positive, so that it is possible to construct $\mathbf{M}^{-1/2} = \text{diag}\,(\mu_1^{-1/2}, \mu_2^{-1/2}, \ldots, \mu_n^{-1/2}) = (\mathbf{M}^{-1/2})^\dagger$. Then $(\mathbf{M}^{-1/2})^\dagger\mathbf{M}\mathbf{M}^{-1/2} = \mathbf{1}$, which now does commute with

$$(\mathbf{M}^{-1/2})^\dagger\mathbf{U}^\dagger\mathbf{FUM}^{-1/2},$$

which is still Hermitian. (3) Let $\mathbf{W}$ diagonalize this matrix:

$$\mathbf{W}^\dagger(\mathbf{M}^{-1/2})^\dagger\mathbf{U}^\dagger\mathbf{FUM}^{-1/2}\mathbf{W} = \mathbf{\Lambda}.$$

But also $\mathbf{W}^\dagger\mathbf{1}\mathbf{W} = \mathbf{1}$. Therefore congruent transformation by $\mathbf{C} = \mathbf{UM}^{-1/2}\mathbf{W}$ does indeed simultaneously convert $\mathbf{G}^{-1}$ to $\mathbf{1}$ and $\mathbf{F}$ to $\mathbf{\Lambda}$. Now *you* show how the problem of finding $\mathbf{C}$ may be reduced to the problem of diagonalizing $\mathbf{GF}$. (*Hint:* This is easy: Solve for $\mathbf{C}^\dagger$.)

170. The equations of motion for two identical coupled pendulums (Fig. 8.9) are $m\ddot{x}_1 = -kx_1 - k'x_2$, $m\ddot{x}_2 = -k'x_1 - kx_2$, where $k = mg/L$. Show that the particular solution to these equations, subject to the initial conditions $x_1(0) = A$, $\dot{x}_1(0) = 0 = x_2(0) = \dot{x}_2(0)$, is

$$\begin{pmatrix} x_1(t) \\ x_2(t) \end{pmatrix} = \tfrac{1}{2}A \begin{pmatrix} \cos\omega_+ t + \cos\omega_- t \\ \cos\omega_+ t - \cos\omega_- t \end{pmatrix}$$

where $\omega_\pm^2 = (k \pm k')/m$. (*Note:* This is a special case, since $\mathbf{G}^{-1}$ is a scalar matrix

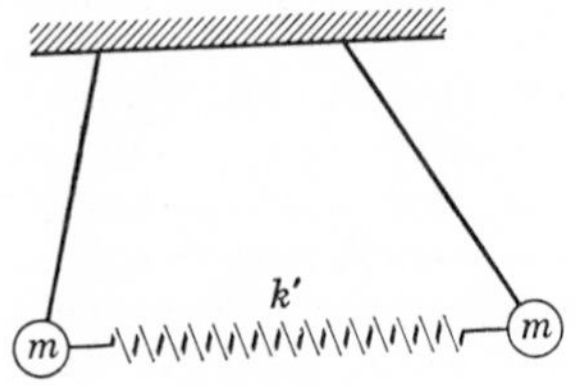

Figure 8.9

and therefore commutes with **F**. However, only in the very simplest cases is it possible to solve for the normal modes explicitly.)

*171. Find the in-plane normal modes of the H_2O molecule:

(a) Show that a force field that involves a force constant k (in dyne/cm), opposing changing O—H bond lengths, and a force constant k' (in dyne/radian), opposing changing the H—O—H angle (see Figure 8.10), leads to

$$V = \tfrac{1}{2}k(x_1 + x_0 \cos\phi + y_0 \sin\phi)^2 + \tfrac{1}{2}k(x_2 + x_0\cos\phi - y_0 \sin\phi)^2 + \tfrac{1}{2}k' d_0^{-2}(y_1 + y_2 - 2x_0 \sin\phi)^2,$$

where d_0 is the equilibrium O—H bond length. (*Hint:* What are $\partial d_{OH_1}/\partial x_1$, $\partial d_{OH_1}/\partial x_0$, $\partial(2\phi)/\partial y_1$, $\partial(2\phi)/\partial x_0$, etc., at the equilibrium positions?)

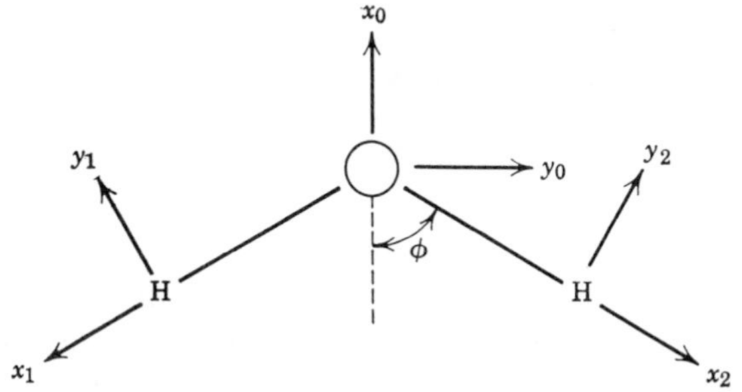

Figure 8.10

(b) Let $\boldsymbol{\xi} = (x_0, x_1, x_2, y_1, y_2, y_0)^T$ and $\mathbf{G}^{-1} = \text{diag}\,(16, 1, 1, 1, 1, 16)$, and construct the 6×6 matrices **F** and $\mathbf{G}^{1/2}\mathbf{F}\mathbf{G}^{1/2}$.

(c) Block-diagonalize $\mathbf{G}^{1/2}\mathbf{F}\mathbf{G}^{1/2}$ by transforming by

$$\mathbf{U} = \left\|\begin{matrix} 1 & 0 & 0 & 0 & 0 & 0 \\ 0 & 1/\sqrt{2} & 0 & 0 & 1/\sqrt{2} & 0 \\ 0 & 1/\sqrt{2} & 0 & 0 & -1/\sqrt{2} & 0 \\ 0 & 0 & 1/\sqrt{2} & 1/\sqrt{2} & 0 & 0 \\ 0 & 0 & 1/\sqrt{2} & -1/\sqrt{2} & 0 & 0 \\ 0 & 0 & 0 & 0 & 0 & 1 \end{matrix}\right\|.$$

Convince yourself that this transformation corresponds to choosing new coordinates that are either symmetric or antisymmetric to the plane bisecting the H—O—H angle.

(b) Show that each diagonal submatrix of $\mathbf{U}^\dagger\mathbf{G}^{1/2}\mathbf{F}\mathbf{G}^{1/2}\mathbf{U}$ has a zero eigenvalue.

(e) Find the nonzero eigenvalue of the 2×2 submatrix.

(f) Find and sketch the normal mode corresponding to this eigenvalue.

(g) Find the normal mode corresponding to the zero eigenvalue of the 3×3 submatrix. Sketch this normal mode, and convince yourself that it is a translation.

9

EXTENSION TO INFINITE DIMENSIONS

In this chapter we pursue the previously indicated analogy between functions and vectors. Indeed, there is little really new material in this chapter, except for a new notation and the approximation methods at the end. Nevertheless, this notation is standard in the research literature and should be learned. We begin by generalizing vectors, basis sets, operators, and eigenvalue-eigenvector problems to infinite dimensions, and we show how these are equivalent to functions, complete sets of functions, operators, and eigenvalue-eigenfunction problems, respectively. We then show how quantum mechanics may be formulated in these terms and how this formulation is related to the familiar approach. Finally, we discuss two approximation methods that are very important in chemical applications—the variation principle and perturbation theory. The exercises of this chapter are primarily aimed at providing practice and familiarity with some very abstract ideas, but some exercises involve applications to various aspects of quantum chemistry.

9.1 Vectors and Functions

In quantum mechanics the state of a physical system is described by an eigenfunction of an appropriate operator. However, it is often convenient to treat functions and operators in a more general fashion; namely, in terms of vectors. The fundamental notion of DIRAC NOTATION (BRACKET NOTATION) is the correspondence between functions and vectors in an infinite-dimensional space. We shall represent each such vector as a KET VECTOR, written $|\phi\rangle$.

The definition of a vector space is the same as in Section 7.1: Vector addition, to give $|\phi\rangle + |\psi\rangle$, is closed, commutative, and associative; there exists a zero ket vector, and there exist additive inverses, $-|\phi\rangle$. Scalar multiplication, to give $a|\phi\rangle$, is closed, associative, and distributive. What is important now is that functions also form a vector space over the complex numbers, as indicated in Example 7.4. The advantage of denoting a function by $|\phi\rangle$ is that we need not specify what the independent variables are. For example, the electronic state of a hydrogen atom may be specified (*1*) as a function of x, y, and z; (*2*) as a function of r, θ, and ϕ; or (*3*) as a function of E (energy), L^2, and L_z (various angular momenta); these are different functions ("rules") which correspond to the same physical state and which are to be represented by the same ket vector.

A basis set is still a set, $\{|i\rangle\}$, of linearly independent vectors such that any $|\phi\rangle$ in the vector space may be expanded as a linear combination of the basis vectors: $|\phi\rangle = \sum \phi_i|i\rangle$, where ϕ_i is the ith component of $|\phi\rangle$. For the vector space of functions, the number of basis vectors is infinite. Notice that each basis vector is itself a function.

The choice of basis is still arbitrary and not unique, and the vector (function) exists even if a basis set is not chosen. However, once a basis is chosen, the vector $|\phi\rangle$, or the function f, may be REPRESENTED as an infinite set of numbers, that is, as a column matrix, $\boldsymbol{\phi}$, whose components are ϕ_i. Calculations involving vectors or functions may thereby be transformed to calculations involving matrices and numbers. Indeed, most real problems are not so abstract as to be solvable purely in terms of vectors; ultimately calculations with numbers must be performed, and the basis may as well be chosen so as to simplify the calculations.

Examples

1. A basis for the vector space of all functions expressible in Maclaurin series is $\{x^n\}$. We replace x^n by the basis ket $|n\rangle$, so that the expansion $f(x) = \sum c_n x^n$ becomes $|\phi\rangle = \sum c_n|n\rangle$, with $c_n = D^n f(0)/n!$.

2. The Fourier expansion $f(x) = \sum c_n e^{inx}/\sqrt{2\pi}$ becomes $|\phi\rangle = \sum c_n|n\rangle$, where $c_n = \int_{-\pi}^{\pi} e^{-inx} f(x)\,dx/\sqrt{2\pi}$.

3. Any "reasonable" function may be expanded as a linear combination of Hermite functions: $f(x) = \sum c_n \hat{H}_n(x)e^{-x^2/2}$ becomes $|\phi\rangle = \sum c_n|n\rangle$.

4. Any function of θ and ϕ may be expanded as a linear combination of spherical harmonics: $F(\theta, \phi) = \sum c_{lm} Y_{lm}(\theta, \phi)$ becomes $|\Phi\rangle = \sum c_{lm}|lm\rangle$. Notice that there are two indices on each basis vector, but $\boldsymbol{\Phi}$ is still a column matrix

$$(c_{00}, c_{11}, c_{10}, c_{1-1}, c_{22}, c_{21}, \ldots)^T.$$

5. Any function f may be represented as an infinite set of numbers merely by indicating the value of the function at every argument x. However, these components and the basis vectors are indexed, not by an integer, but by a continuous variable. Therefore the column matrix which represents $|\phi\rangle$ in this choice of basis does not have discrete entries, but is

continuous. This column matrix cannot be displayed in tabular form, but must be graphed:

Column Matrix of Components

In expansion $|\phi\rangle = \sum \phi_i|i\rangle$ In expansion $|\phi\rangle = \int f(x)\,|x\rangle\,dx$

The expansion sum must then be replaced by an integral: $|\phi\rangle = \int f(x)\,|x\rangle\,dx$. To obtain an analogous equation in terms of functions, we must identify the xth basis vector $|x\rangle$ with the function $\delta(\xi - x)$ since $f(\xi) = \int f(x)\delta(\xi - x)\,dx$ (cf. Exercise 3.32). This basis set, $\{|x\rangle\}$ or $\{\delta(\xi - x)\}$, is known as the CONTINUOUS BASIS; the other bases are DISCRETE.

Remember that the choice of basis is arbitrary. It is the vector $|\phi\rangle$ that is fundamental, and not the set of components that may be obtained once a basis is chosen. In particular, the function f no longer occupies a central position because the set of values $\{f(x)\}$ is only one of many possible sets of components that may be obtained. Likewise, our representation of, say, the nth member of the Hermite basis set as $\hat{H}_n(x)e^{-x^2/2}$ is only one of many possible representations; a more "natural" representation of the basis vector $|n\rangle$ is as $(0, 0, \ldots, 1, \ldots)^T$, the column matrix whose only nonzero element is the nth, which is unity. And our identification of $|x\rangle$ with $\delta(\xi - x)$ is in fact the particular representation of kets by functions of ξ.

Notice that there exist functions which cannot be expanded as a linear combination of some of these basis sets. Some functions cannot be expanded in Maclaurin series, or Fourier series, or Hermite series. Whenever we are interested in a particular class of functions, we must choose a basis set which does span the space of functions of interest. Such a set is said to be COMPLETE with respect to that class of functions.

Exercises

1. An ANTISYMMETRIC FUNCTION, F, of n variables is a function such that for all i and j $F(x_1, x_2, \ldots, x_j, \ldots, x_i, \ldots, x_n) = -F(x_1, x_2, \ldots, x_i, \ldots, x_j, \ldots, x_n)$. Convince yourself that the set of all such functions forms a subspace of the vector space of functions of n variables.
2. Use the Fourier basis set $\{(2\pi)^{-1/2}e^{inx}\}_{n=-\infty}^{\infty}$ to express the function $f(x) = x$ $(-\pi \leqslant x \leqslant \pi)$ as a column matrix.

In connection with the dot product of two N-dimensional vectors, we associated to every column matrix $\mathbf{x}$ its adjoint $\mathbf{x}^\dagger$, a row matrix, with the property that $(a\mathbf{x} + b\mathbf{y})^\dagger = a^*\mathbf{x}^\dagger + b^*\mathbf{y}^\dagger$. In analogous fashion we associate to every ket vector $|\phi\rangle$ a BRA VECTOR $\langle\phi|$, in a different vector space, such that $\langle x|$ is also identified with $\delta(\xi - x)$ and such that the bra vector associated with $a|\phi\rangle + b|\psi\rangle$ is $a^*\langle\phi| + b^*\langle\psi|$. Thus, if we expand $|\phi\rangle = \sum \phi_i|i\rangle$, the corresponding expansion is $\langle\phi| = \sum \phi_i^*\langle i|$; if $|\phi\rangle$ is represented as a column matrix $\boldsymbol{\phi}$, $\langle\phi|$ is represented as the adjoint row matrix $\boldsymbol{\phi}^\dagger$.

To investigate what function represents $\langle\phi|$, expand $\langle\phi|$ in the continuous basis: If $|\phi\rangle = \int f(x)\,|x\rangle\,dx$, $\langle\phi| = \int f^*(x)\langle x|\,dx$ is represented by $\int f^*(x)\,\delta(\xi - x)\,dx = f^*(\xi)$. As might be expected, the bra vector corresponds to the complex conjugate function f^*.

We now define the dot product of the vectors $\langle\phi|$ and $|\psi\rangle$ as a number, $\langle\phi|\cdot|\psi\rangle$ or $\langle\phi\,|\,\psi\rangle$ (read "ϕ dot ψ"), independent of the choice of basis, with the following properties:

$$\langle\phi|\cdot(a|\chi\rangle + b|\psi\rangle) = a\langle\phi\,|\,\chi\rangle + b\langle\phi\,|\,\psi\rangle \tag{9.1a}$$

$$(a\langle\phi| + b\langle\chi|)\cdot|\psi\rangle = a\langle\phi\,|\,\psi\rangle + b\langle\chi\,|\,\psi\rangle \tag{9.1b}$$

$$\langle x\,|\,x'\rangle = \delta(x - x'). \tag{9.2}$$

Do not forget that the dot product is formed from one of each type of vector, *not* two kets.

To relate the dot product to something involving functions, expand $\langle\phi\,|\,\psi\rangle$ in the continuous basis: $\langle\phi\,|\,\psi\rangle = \int f^*(x)\langle x|\,dx \int g(x')\,|x'\rangle\,dx' = \iint f^*(x)g(x')\langle x\,|\,x'\rangle\,dx\,dx' = \iint f^*(x)g(x')\,\delta(x - x')\,dx\,dx' = \int f^*(x)g(x)\,dx$, or

$$\langle\phi\,|\,\psi\rangle = \int f^*(x)g(x)\,dx \tag{9.3}$$

(cf. Exercise 7.16, where the complex conjugate was omitted for simplicity). From this result, we conclude that for any nonzero vector $|\phi\rangle$,

$$\langle\phi\,|\,\phi\rangle > 0. \tag{9.4}$$

Thus we may still attach a geometrical significance to the dot product whenever it is real. The dot product of $|\phi\rangle$ with its corresponding bra is somehow the square of the "size" of the function f, and the dot product of $\langle\phi|$ with $|\psi\rangle$ reflects the extent to which the functions f and g "overlap" or "point in the same direction." A vector $|\phi\rangle$ such that $\langle\phi\,|\,\phi\rangle = 1$ is said to be NORMALIZED. Two vectors, $|\phi\rangle$ and $|\psi\rangle$,such that $\langle\phi\,|\,\psi\rangle = 0$ are said to be ORTHOGONAL. Furthermore, since $\int g^*(x)f(x)\,dx = [\int f^*(x)g(x)\,dx]^*$,

$$\langle\psi\,|\,\phi\rangle = \langle\phi\,|\,\psi\rangle^*, \tag{9.5}$$

as we found (Eq. 8.14) for dot products in vector spaces over the complex numbers.

A basis set $\{|i\rangle\}$, such that $\langle i\,|\,j\rangle = \delta_{ij}$, is again said to be ORTHONORMAL. By referring to Eq. 9.3, we may see that it was indeed appropriate to use this geometric term in Section 5.7 in connection with complete sets of functions. Expansion of the dot product $\langle\phi\,|\,\psi\rangle$ in a discrete basis gives

$\sum \phi_i^* \langle i | \sum \psi_j | j \rangle = \sum \phi_i^* \psi_j \langle i | j \rangle$, which simplifies to $\langle \phi | \psi \rangle = \sum \phi_i^* \psi_i$ in an orthonormal basis. Henceforth we shall assume that all basis sets are orthonormal, so that the dot product will have this simple form.

Next let us investigate the geometric significance of the components in the expansion $|\phi\rangle = \sum \phi_j |j\rangle$. Take the dot product of both sides of this equation with $\langle i|$: $\langle i | \phi \rangle = \sum \phi_j \langle i | j \rangle = \phi_i$. Thus we see that the component ϕ_i is the dot product of $|\phi\rangle$ with the ith basis vector, or the component of $|\phi\rangle$ in the direction of $|i\rangle$. Now we may rewrite the expansion as $|\phi\rangle = \sum \langle i | \phi \rangle |i\rangle$, although for esthetic reasons it is common to write

$$|\phi\rangle = \sum |i\rangle\langle i | \phi\rangle \tag{9.6a}$$

$$\langle\phi| = \sum \langle\phi | i\rangle\langle i|. \tag{9.6b}$$

Similarly, in the expansion $|\phi\rangle = \int f(x)\,|x\rangle\,dx$, the component $f(x) = \langle x | \phi\rangle$, so that the expansions become $|\phi\rangle = \int |x\rangle\langle x | \phi\rangle\,dx$ and $\langle\phi| = \int \langle\phi | x\rangle\langle x|\,dx$. If we now substitute this form of $|\phi\rangle$ into the right-hand side of Eq. 9.6a, we obtain $|\phi\rangle = \sum |i\rangle\langle i | \phi\rangle = \sum |i\rangle \int \langle i | x\rangle\langle x | \phi\rangle\,dx = \sum |i\rangle \int \langle x | i\rangle^*\langle x | \phi\rangle\,dx$. Comparing components along $|i\rangle$ shows that $\langle i | \phi\rangle = \int \langle x | i\rangle^*\langle x | \phi\rangle\,dx$. But $\langle x | \phi\rangle = f(x)$ and $\langle x | i\rangle = \phi_i(x)$, the ith basis *function*, so that the coefficient $\langle i | \phi\rangle$ does have the familiar form $\int \phi_i^*(x) f(x)\,dx$ (cf. Eq. 5.21).

Exercises

3. Show that if $|\phi\rangle$ is a normalized vector, so is $e^{i\theta}\,|\phi\rangle$.
4. If $\{|i\rangle\}$ and $\{|j\rangle\}$ are both orthonormal basis sets, show that the matrix $\mathbf{U} = (\langle i | j\rangle)$ is unitary.
5. A function f is defined on the interval from -1 to 1. In the Fourier basis $\{2^{-1/2}e^{in\pi x}\}_{n=-\infty}^{\infty}$, f is represented by the column matrix $\boldsymbol{\phi}$. In the Legendre basis $\{\hat{P}_l(x)\}_{l=0}^{\infty}$, f is represented by the column matrix $\boldsymbol{\phi}'$. Find a matrix $\mathbf{T}$ such that $\mathbf{T}\boldsymbol{\phi} = \boldsymbol{\phi}'$, and express its elements as integrals.
*6. In the vector space of all functions of two variables, the expansion of $|\Phi\rangle$ in the continuous basis is $|\Phi\rangle = \iint |xy\rangle\langle xy | \Phi\rangle\,dx\,dy$, where $\langle xy | \Phi\rangle = F(x, y)$. But there is another continuous basis, $\{|r\phi\rangle\}$, in which the $r\phi$th component of $|\Phi\rangle$ is $\langle r\phi | \Phi\rangle = G(r, \phi)$. To relate these components, dot the expansion in the xy basis with $\langle r\phi|$ to obtain $\langle r\phi | \Phi\rangle = \iint \langle r\phi | xy\rangle\langle xy | \Phi\rangle\,dx\,dy$. What is $\langle r\phi | xy\rangle$? (*Hint:* It is a number that depends upon r, ϕ, x, and y.)
*7. Convince yourself that another possible basis is that in which a function f is represented by the set of all its moments: $|\phi\rangle = \sum c_n|n\rangle$, where $c_n = \int x^n f(x)\,dx$. Use the result of Exercise 7.30 to describe this nonorthonormal basis.
8. Rewrite the Schwarz inequality in terms of functions and integrals. Notice that this result is not as obvious as $|\cos\theta| \leqslant 1$.

9.2 Operators

We have already defined an operator as something which takes functions into functions.

Examples

6. Thrice: the function f is taken into the function Tf, such that $(Tf)(x) = 3f(x)$.
7. Multiply by x: the function f is taken into the function g, such that $g(x) = xf(x)$.
8. Differentiate: $Df = f'$.
9. Square: $Sf(x) = [f(x)]^2$.
10. Complex conjugate: $*f(x) = f^*(x) = [f(x)]^*$.
11. Displace: $\Delta_c f(x) = f(x - c)$.
12. Invert: $\mathrm{I}f(x, y, z) = f(-x, -y, -z)$.
13. Rotate by $2\pi/n$: $C_n f(r, \theta, \phi) = f(r, \theta, \phi - 2\pi/n)$.
14. Interchange: $P_{12}F(x_1, x_2) = F(x_2, x_1)$.
15. Permute: $P_{123}F(x_1, x_2, x_3) = F(x_3, x_1, x_2)$, etc., or in general $P_j F(\{x_i\}) = F(\{x_{P_j i}\})$.
16. Antisymmetrize: $A_2 F(x_1, x_2) = F(x_1, x_2) - F(x_2, x_1) = F(x_1, x_2) - P_{12}F(x_1, x_2)$, or $A_n F(\{x_i\}_{i=1}^n) = \sum_{j=1}^{n!} (-1)^{p(P_j)} P_j F(\{x_i\}_{i=1}^n)$, where the sum is over all permutations.

But in the context of the vector space of functions, an operator becomes a transformation—something which takes vectors into vectors, something which twists a vector around to some other direction and changes its magnitude. Just as we indicated the application of the operator Ω to the function f by juxtaposition, we shall write the result of applying the transformation Ω to the vector $|\phi\rangle$ as $\Omega|\phi\rangle$. Furthermore, the result of applying a transformation to a vector is a vector, so that we may write $\Omega|\phi\rangle = |\psi\rangle$. We shall not distinguish operators from transformations, and we shall call something that converts functions into functions, or vectors in function-space into vectors in function-space, an operator. Furthermore, we shall restrict ourselves to LINEAR OPERATORS, those such that $\Omega(a|\phi\rangle + b|\psi\rangle) = a\Omega|\phi\rangle + b\Omega|\psi\rangle$.

We may define the sum of two operators Ω_1 and Ω_2 by $(\Omega_1 + \Omega_2)|\phi\rangle = \Omega_1|\phi\rangle + \Omega_2|\phi\rangle$; notice that the addition on the right is the addition of two vectors, whereas the addition on the left is the addition of two operators. We may also define the product of two operators by $(\Omega_1\Omega_2)|\phi\rangle = \Omega_1(\Omega_2|\phi\rangle)$, such that $\Omega_1\Omega_2$ is an operator whose result is the same as that produced by applying first Ω_2 and then applying Ω_1. This multiplication is associative: $\Omega_1(\Omega_2\Omega_3) = (\Omega_1\Omega_2)\Omega_3$, so that we may omit the parentheses. Again, it is not necessarily true that $\Omega_1\Omega_2 = \Omega_2\Omega_1$. The COMMUTATOR of two operators is $[\Omega_1, \Omega_2] \equiv \Omega_1\Omega_2 - \Omega_2\Omega_1$. Two operators whose products are the same regardless of order are said to COMMUTE.

We have already encountered one special operator in Eq. 9.6. Since $\sum |i\rangle\langle i|$ (or $\int |x\rangle\langle x| dx$) acts on $|\phi\rangle$ merely to produce $|\phi\rangle$ again,

$$\sum |i\rangle\langle i| = 1 = \int |x\rangle\langle x| dx. \tag{9.7}$$

Since these are the identity operator, they may be inserted anywhere.

Notice that although $\langle\psi \mid \psi\rangle$ is a number, the product of two vectors in the reverse order, $|\psi\rangle\langle\psi|$, is an operator since it acts on the arbitrary vector $|\phi\rangle$

to produce $|\psi\rangle\langle\psi \mid \phi\rangle$, a vector in the direction $|\psi\rangle$. Indeed, this is an example of a PROJECTION OPERATOR, which acts on an arbitrary vector to produce its projection in some subspace of the vector space.

If $\Omega|\chi\rangle = |\phi\rangle$, it is not necessarily true that the same Ω is appropriate for converting $\langle\chi|$ into $\langle\phi|$. The operator that is appropriate is called the ADJOINT operator, symbolized by $\Omega^\dagger$: $\langle\chi|\Omega^\dagger = \langle\phi|$. Taking the dot product with $\langle\psi|$ or $|\psi\rangle$, and invoking Eq. 9.5, we obtain

$$\langle\psi| \, \Omega \, |\chi\rangle = \langle\chi| \, \Omega^\dagger \, |\psi\rangle^*. \tag{9.8}$$

If $\Omega^\dagger = \Omega$, then Ω is said to be HERMITIAN, or SELF-ADJOINT.

Exercises

9. Which of the operators given in the examples are not linear?

*10. The commutation rules for angular momentum operators are $[L_x, L_y] = i\hbar L_z$, $[L_y, L_z] = i\hbar L_x$, and $[L_z, L_x] = i\hbar L_y$. It is possible to combine these operators into a three-dimensional VECTOR OPERATOR, $\mathbf{L} = \hat{\mathbf{i}}L_x + \hat{\mathbf{j}}L_y + \hat{\mathbf{k}}L_z$, which acts on a function (infinite-dimensional vector) to produce a three-dimensional vector whose components are functions. What is $\mathbf{L} \times \mathbf{L}$?

*11. For a set $\{|i\rangle\}$ of orthonormal vectors which is not complete, $\sum |i\rangle\langle i| \neq 1$. We might expect a relation like $\sum |i\rangle\langle i| < 1$ to be true, except that we cannot compare operators in this fashion. Provide an inequality which is true. This is known as BESSEL'S INEQUALITY.

12. Show that if $|\phi\rangle$ is a normalized vector, $|\phi\rangle\langle\phi|$ is idempotent.

13. Find a projection operator that will act on an arbitrary vector to produce its projection in the plane spanned by the two particular basis vectors $|j\rangle$ and $|k\rangle$.

14. Convince yourself that the antisymmetrizing operator of Example 9.16 is a projection operator. What is the dimension of the subspace into which all vectors are projected?

15. What is $|\phi\rangle\langle\psi|^\dagger$?

16. Rewrite Eq. 9.8 in terms of functions and integrals. Then integrate by parts to find $D^\dagger$; assume that all functions vanish at the limits of integration.

17. Convince yourself that $(a_1\Omega_1 + a_2\Omega_2)^\dagger = a_1^*\Omega_1^\dagger + a_2^*\Omega_2^\dagger$ and that $(\Omega_1\Omega_2)^\dagger = \Omega_2^\dagger\Omega_1^\dagger$.

If $|\phi\rangle$ is expanded in a basis set, then $\Omega|\phi\rangle = \Omega \sum |j\rangle\langle j \mid \phi\rangle = \sum \Omega|j\rangle\langle j \mid \phi\rangle$ so that the result of operating Ω on the arbitrary vector $|\phi\rangle$ is determined both by what Ω does to each basis vector and by the component of $|\phi\rangle$ along each basis vector. Furthermore, the projection of $\Omega|\phi\rangle$ along the basis vector $|i\rangle$ is $\langle i| \, \Omega \, |\phi\rangle = \sum \langle i| \, \Omega \mid j\rangle\langle j \mid \phi\rangle$, and the expansion of the vector $\Omega|\phi\rangle$ is $\sum |i\rangle\langle i| \, \Omega \mid j\rangle\langle j \mid \phi\rangle$. Therefore the result of operating Ω on the arbitrary vector $|\phi\rangle$ is determined by the set of numbers $\{\langle i| \, \Omega \mid j\rangle\}$ and the set of components $\{\langle j \mid \phi\rangle\}$, which are independent of Ω. Just as we found that by choosing a basis we could express a function as a set of numbers, we now find that by choosing a basis we may also express an operator as a set of numbers, but with two indices. Just as a ket vector may be represented as a column matrix, and a bra vector as a row matrix, an operator may be REPRESENTED as an infinite square matrix $\mathbf{\Omega}$, such that $(\mathbf{\Omega})_{ij} = \langle i|\,\Omega\,|j\rangle$:

$$\mathbf{\Omega} = \left\| \begin{matrix} \langle 1|\,\Omega\,|1\rangle & \langle 1|\,\Omega\,|2\rangle & \langle 1|\,\Omega\,|3\rangle & \cdots \\ \langle 2|\,\Omega\,|1\rangle & \langle 2|\,\Omega\,|2\rangle & \cdots & \\ \vdots & \vdots & & \end{matrix} \right\|$$

Exercises

18. In the basis set $\{|i\rangle\}$ the operator Ω is represented by the matrix $\mathbf{\Omega} = (\langle i|\,\Omega\,|i'\rangle)$. In a different basis set, $\{|j\rangle\}$, Ω is represented by a different matrix, $\mathbf{\Omega}' = (\langle j|\,\Omega\,|j'\rangle)$. What matrix multiplication will convert $\mathbf{\Omega}$ into $\mathbf{\Omega}'$?
19. Show that a Hermitian operator is represented by a Hermitian matrix no matter what basis set is chosen.
20. What is the matrix representation of each of the following *operators* in the Fourier basis set:
    (a) 1.
    (b) D.
    (c) x.

*21. What is the matrix representation of the Fourier transform operator in the Fourier basis set?

The representative of an operator in a continuous basis is still a matrix, but one that is difficult to write down as a square array. The xx'th element of such a matrix is $\langle x|\,\Omega\,|x'\rangle$, but the indices are continuous variables and the matrix is called a CONTINUOUS MATRIX. Thus, in the continuous basis, the vector equation $\Omega|\phi\rangle = |\psi\rangle$ becomes $(\mathbf{\Omega}\boldsymbol{\phi})_x = \int \langle x|\,\Omega\,|x'\rangle\langle x' \mid \phi\rangle\,dx' = \langle x \mid \psi\rangle = (\boldsymbol{\psi})_x$, where the integration corresponds to the row-by-column summation of ordinary matrix multiplication. But $\langle x' \mid \phi\rangle = f(x')$ and $\langle x \mid \psi\rangle = g(x)$, so that the matrix multiplication of $\boldsymbol{\phi}$ by $\mathbf{\Omega}$ to produce $\boldsymbol{\psi}$ corresponds to the familiar application of the operator Ω to the function f to produce the function g. The continuous matrix $\mathbf{\Omega} = (\langle x|\,\Omega\,|x'\rangle)$ is in fact the familiar operator Ω, written so as to convert a function of x' into a function of x. It should not be mistaken for a function of two variables. For example, the Fourier transform operator F is defined by $g(x) = F[f(x')] = (2\pi)^{-1/2} \int e^{-ixx'} f(x')\,dx'$, so that $\langle x|\,F\,|x'\rangle = (2\pi)^{-1/2} e^{-ixx'}$. The operator x (multiply by x) is represented by the matrix whose xx'th element is $x\delta(x' - x)$; it converts $f(x')$ into $g(x) = xf(x)$.

Exercises

22. The identity operator 1 is defined by $1 = \sum |i\rangle\langle i|$ or by $1|\phi\rangle = |\phi\rangle$. In any discrete basis 1 is represented as the unit matrix $\mathbf{1}$, for which $(\mathbf{1})_{ij} = \delta_{ij}$. What is the matrix that represents 1, or $\int |x\rangle\langle x|\,dx$, in the continuous basis?

23. For a system described by the normalized vector $|\Psi\rangle$, the quantum-mechanical DENSITY MATRIX in the basis set $\{|i\rangle\}$ is defined by $(\boldsymbol{\rho})_{ij} = \langle i \mid \Psi\rangle\langle\Psi \mid j\rangle$.
(a) What is Tr $\boldsymbol{\rho}$?
(b) What is the density matrix in the continuous basis $\{|x\rangle\}$? Notice that this density matrix is a continuous matrix, with $(\boldsymbol{\rho})_{x,x'} = \rho(x, x')$.
(c) How would you define the trace of this matrix?
(d) Show that $\langle\Psi|\,\Omega\,|\Psi\rangle = \text{Tr}\,(\boldsymbol{\rho\Omega})$.

*24. In many situations it is not possible to know what state a physical system is in. For example, in proton magnetic resonance experiments, it is not possible to know whether a given proton has α spin or β spin, and the most complete description is limited to a specification of the probability of finding each state. In these situations, if the probability that the system is described by $|\Psi_n\rangle$ is P_n, the density matrix becomes $\boldsymbol{\rho} = (\sum_n P_n\langle i \mid \Psi_n\rangle\langle\Psi_n \mid j\rangle)$, the average of the density matrices for the pure states.
(a) Show that the choice of the $\{|\Psi_n\rangle\}$ basis diagonalizes $\boldsymbol{\rho}$. (*Hint:* The basis is orthonormal.)
(b) Find the eigenvalues of $\boldsymbol{\rho}$ and thereby convince yourself that if p is an eigenvalue of $\boldsymbol{\rho}$, then $0 \leqslant p \leqslant 1$.
(c) Under what circumstance can $\boldsymbol{\rho}^2 = \boldsymbol{\rho}$? Notice that since idempotence is invariant to similarity transformation, this result does not depend on the choice of basis.

9.3 Eigenvalue Problems

Originally we indicated that many differential equations arise in the form of an eigenvalue-eigenfunction problem $\Omega\psi = \lambda\psi$. In the context of the vector space of functions, such a problem becomes the eigenvalue-eigenvector problem $\Omega|\psi\rangle = \lambda|\psi\rangle$. Choosing an orthonormal basis leads to the simultaneous equations

$$\sum_j \langle i|\,\Omega\,|j\rangle\langle j \mid \psi\rangle = \lambda \sum_j \langle i \mid j\rangle\langle j \mid \psi\rangle = \lambda\langle i \mid \psi\rangle.$$

Here the unknowns are the expansion coefficients $\{\langle j \mid \psi\rangle\}$, and there is one equation for each component of $\Omega|\psi\rangle$. But if we define the column matrix $\boldsymbol{\psi} = (\langle j \mid \psi\rangle)$ and the square matrix $\boldsymbol{\Omega} = (\langle i|\,\Omega\,|j\rangle)$, this problem is equivalent to the matrix equation $\boldsymbol{\Omega\psi} = \lambda\boldsymbol{\psi}$, which is the problem of finding the eigenvalues and eigenvectors of $\boldsymbol{\Omega}$. Thus by choosing a discrete basis we have transformed a differential equation involving functions into a problem involving numbers.

In order to have a nontrivial solution to the simultaneous homogeneous equations, $(\boldsymbol{\Omega} - \lambda\mathbf{1})\boldsymbol{\psi} = \mathbf{0}$, it is necessary that the secular determinant be zero:

$$|(\langle i|\,\Omega\,|j\rangle - \lambda\delta_{ij})| = |\boldsymbol{\Omega} - \lambda\mathbf{1}| = 0.$$

We shall neglect the difficulties inherent in expanding infinite-dimensional determinants, finding the roots of an infinite-order polynomial, solving for the infinity of coefficients, and normalizing the infinite-dimensional eigenvectors. We shall maintain the geometrical interpretation of an eigenvalue-eigenvector problem as the problem of finding the transformation to a new basis set in which the operator Ω is represented as an especially simple matrix; namely, a diagonal one. Furthermore, we shall assume that all the theorems concerning finite matrices also hold for these infinite-dimensional ones. In particular, the eigenvalues of an infinite Hermitian matrix are real, and eigenvectors belonging to different eigenvalues are orthogonal; an infinite Hermitian matrix may be diagonalized by an infinite unitary matrix. Also, two matrices that commute may be diagonalized simultaneously.

A further feature of Dirac notation is to label the eigenvector by its eigenvalue: $\Omega|\lambda\rangle = \lambda|\lambda\rangle$, or by an index related to the eigenvalue: $L^2|l\rangle = l(l+1)\hbar^2|l\rangle$. A simultaneous eigenvector is labeled by an index related to each eigenvalue. For a two-dimensional harmonic oscillator:

$$H|nm\rangle = (n+1)h\nu|nm\rangle, \qquad L_z|nm\rangle = m\hbar|nm\rangle.$$

Exercises

25. Find the eigenvalues and corresponding normalized eigenvectors of the operator $|\phi\rangle\langle\phi|$.
26. Show that the eigenvalues of a self-adjoint operator are real and that eigenvectors belonging to different eigenvalues are orthogonal. Use Dirac notation throughout.
27. Show that if Ω_1 and Ω_2 commute, and if $|\lambda\rangle$ is an eigenvector of Ω_1 belonging to the nondegenerate eigenvalue λ, then $|\lambda\rangle$ is also an eigenvector of Ω_2. Use Dirac notation throughout.
28. The commutation rules for the angular momentum operators are given in Exercise 9.10. If $|l_z\rangle$ is an eigenvector of L_z with eigenvalue $l_z\hbar$, and if the raising and lowering operators are defined by $L_\pm = L_x \pm iL_y$, show that the vectors $L_+|l_z\rangle$ and $L_-|l_z\rangle$ are also eigenvectors of L_z. (*Hint:* Evaluate the commutator $[L_z, L_\pm]$ and solve for $L_\pm L_z$.) What is the eigenvalue to which $L_+|l_z\rangle$ belongs? $L_-|l_z\rangle$?

*29. The quantum-mechanical commutation rule for the position operator q and the momentum operator p is $[p, q] = -(ih/2\pi)1$. The Hamiltonian operator for the harmonic oscillator is $H = \frac{1}{2}p^2 + \frac{1}{2}(2\pi\nu)^2q^2$. Define the raising operator $R = (2h\nu)^{-1/2}(p + 2\pi i\nu q)$. The operators p, q, and H are Hermitian, but R is not.
   (a) What is $R^\dagger$?
   (b) Use the commutation rule to relate $R^\dagger R$ and $RR^\dagger$ to H.
   (c) Multiply the equation involving $R^\dagger R$ by R from the left and multiply the other equation by R from the right, then subtract to find $[H, R]$.
   (d) If $|E\rangle$ is an eigenvector of H with eigenvalue E, show that $R|E\rangle$ is also an eigenvector of H.
   (e) What is its eigenvalue?
   (f) If $E_0 = \frac{1}{2}h\nu$ is the lowest possible eigenvalue of H, what is the general form for the eigenvalues $\{E_n\}$ of H?

9.4 General Formulation of Quantum Mechanics

It is helpful to indicate briefly the physical significance of such an abstract entity as an infinite-dimensional vector. We list five postulates of quantum mechanics:

1. The state of a physical system is described by a normalized vector, $|\phi\rangle$, in an appropriate space.

2. To each observable quantity (such as momentum, position, energy, angular momentum) there corresponds a linear, Hermitian operator.
3. The Correspondence Principle: For "large" systems, quantum mechanics must reduce to classical mechanics. This postulate provides a systematic procedure for constructing the operator appropriate for each observable, but we shall not pursue such details further.
4. When an experiment is performed to determine the value of the observable corresponding to the operator Ω, the only possible values that can be measured are the eigenvalues of Ω.
5. When an experiment is performed on a system which is in the state $|\phi\rangle$, the probability of measuring the particular eigenvalue λ_0 is $P(\lambda_0) = |\langle\lambda_0 \mid \phi\rangle|^2$.

Now let us see how this formulation is connected with the elementary approach to quantum mechanics. The usual problem in quantum chemistry is that of finding the eigenvalues and eigenstates of the Hamiltonian operator, H, which corresponds to the observable, energy. We may write the solution to this problem as $H|E\rangle = E|E\rangle$. Next let us express this solution in the continuous basis: $\int \langle x'| H |x\rangle\langle x \mid E\rangle\, dx = E \int \langle x' \mid x\rangle\langle x \mid E\rangle\, dx$. Let us investigate the significance of $\langle x \mid E\rangle$. Firstly, since this is a number that depends upon x, it is the value of some function evaluated at the arbitrary argument x. Secondly, since $|x\rangle$ is identified with the function $\delta(\xi - x)$, the significance of the vector $|x\rangle$ is that it is an eigenket of the position operator ξ with eigenvalue x: $\xi|x\rangle = x|x\rangle$ or $\xi\delta(\xi - x) = x\delta(\xi - x)$. If we had a system in the particular state $|x\rangle$, a measurement of position (an experiment to determine where the object of interest is) would give the particular value x. But, according to postulate 5, the quantity $|\langle x \mid E\rangle|^2$ is the probability of observing the value x in a measurement of position on a system which is in the state $|E\rangle$. The familiar notation for this probability is $|\psi(x)|^2$, and we may therefore write the quantity $\langle x \mid E\rangle$ in more familiar form as the WAVE FUNCTION, $\psi(x)$. In this context, the expansion in the $\{|x\rangle\}$ basis is known as the SCHRÖDINGER REPRESENTATION, and corresponds to converting the eigenvalue-eigenvector equation $H|E\rangle = E|E\rangle$ back to the eigenvalue-eigenfunction equation $H\psi = E\psi$. The extension to more than one variable should be obvious.

Exercises

30. A system is in the particular eigenstate $|\lambda_0\rangle$. What is the probability of observing the particular eigenvalue λ_0? What is the probability of observing some other eigenvalue?
31. Show that $\sum P(\lambda) = 1$, that is, that the probability of measuring some eigenvalue is unity, as required.
32. Show that the average value (averaged over many experiments) of an observable Ω for a system in the state $|\phi\rangle$ is $\langle\phi| \Omega |\phi\rangle$.

*33. For a system describable only by the density matrix $\boldsymbol{\rho}$, what is the average value of the observable Ω? Notice that there are two types of averaging involved, one due to inherent quantum-mechanical limitations, the other due to lack of complete knowledge about the state of the system.

34. The quantum-mechanical momentum operator p is $(\hbar/i)(d/dx)$. Use the result of Example 4.5 to express the solution to $p|k\rangle = k|k\rangle$ in the Schrödinger representation, that is, find $\langle x \mid k\rangle$, not normalized. Find the probability of finding the value k_0 in a measurement of momentum on a system in the state $|E\rangle$. Express this probability in terms of a Fourier transform of $\psi(x) = \langle x \mid E\rangle$ (cf. Section 5.6).

9.5 The Variation Principle

Almost never in quantum chemistry is it possible to solve an eigenvalue problem exactly. Therefore it is necessary to have recourse to some approximation methods. One of the most powerful of these is the Variation Principle.

Let the problem be that of finding eigenvalues $\{E\}$ and normalized eigenvectors $\{|E\rangle\}$ of the self-adjoint operator H. Assume that it is possible to know that there exists a lowest eigenvalue E_0, such that E_0 is less than or equal to every other E. Consider the arbitrary, normalized vector $|\phi\rangle$, and form the quantity $W \equiv \langle\phi|\, H\, |\phi\rangle$. Then if we expand $|\phi\rangle$ in the basis $\{|E\rangle\}$

$$\begin{aligned} W - E_0 &= \langle\phi|\, H\, |\phi\rangle - \langle\phi \mid \phi\rangle E_0 \\ &= \sum_{E',E} \langle\phi \mid E'\rangle\langle E'|\, H\, |E\rangle\langle E \mid \phi\rangle - \sum_{E} \langle\phi \mid E\rangle\langle E \mid \phi\rangle E_0 \\ &= \sum_{E} \langle\phi \mid E\rangle\langle E \mid \phi\rangle E - \sum_{E} \langle\phi \mid E\rangle\langle E \mid \phi\rangle E_0 \\ &= \sum_{E} \langle\phi \mid E\rangle\langle E \mid \phi\rangle(E - E_0), \end{aligned}$$

where we have invoked $\langle E'|\, H\, |E\rangle = E\langle E' \mid E\rangle = E\delta_{E'E}$. But $\langle\phi \mid E\rangle\langle E \mid \phi\rangle = |\langle\phi \mid E\rangle|^2$ and is necessarily nonnegative, and $E - E_0$ is nonnegative by hypothesis. Therefore we conclude that

$$W \equiv \langle\phi|\, H\, |\phi\rangle \geqslant E_0, \tag{9.9}$$

the equality holding if and only if $|\phi\rangle$ is an eigenvector of H with eigenvalue E_0.

The significance of this result is that it provides a scheme for approaching a true eigenvector (eigenfunction) of H. What we should do is construct a normalized TRIAL VECTOR $|\phi_{\{\alpha\}}\rangle$ or TRIAL FUNCTION $f_{\{\alpha\}}$ which includes as-yet unspecified parameters $\{\alpha\}$. Then we calculate

$$W(\{\alpha\}) = \langle\phi_{\{\alpha\}}|\, H\, |\phi_{\{\alpha\}}\rangle = \int f^*_{\{\alpha\}}(x) H f_{\{\alpha\}}(x)\, dx$$

which depends upon the parameters. Finally we minimize $W(\{\alpha\})$ with respect to all the parameters. By the Variation Principle, we are guaranteed that even this minimum, $W_{\min}$, cannot be less than E_0, so that $W_{\min}$ is the closest we may approach E_0 with a trial function of the form chosen. Furthermore, we may expect that as W approaches closer and closer to E_0, the trial function approaches the true eigenfunction.

Example

17. To find the lowest eigenvalue of the quartic oscillator, whose Hamiltonian operator is $H = -\frac{1}{2}D^2 + kx^4$, we shall use a variation function of the form $e^{-\alpha^2x^2/2}$. Normalization of this function on the interval from $-\infty$ to ∞ leads to the trial function

$$f_\alpha(x) = \alpha^{1/2}\pi^{-1/4}e^{-\alpha^2x^2/2}.$$

Then

$$W = \langle\phi|\, H\,|\phi\rangle = \frac{\alpha}{\pi^{1/2}}\int_{-\infty}^{\infty} e^{-\alpha^2x^2/2}[-\tfrac{1}{2}D^2 + kx^4]e^{-\alpha^2x^2/2}\,dx$$

$$= \frac{\alpha}{\pi^{1/2}}\int_{-\infty}^{\infty} e^{-\alpha^2x^2}[\tfrac{1}{2}\alpha^2 - \tfrac{1}{2}\alpha^4x^2 + kx^4]\,dx$$

$$= \frac{\alpha}{\pi^{1/2}}\,[\tfrac{1}{2}\alpha\pi^{1/2} - \tfrac{1}{4}\alpha\pi^{1/2} + \tfrac{3}{4}k\alpha^{-5}\pi^{1/2}] = \frac{\alpha^2}{4} + \frac{3}{4}\frac{k}{\alpha^4}$$

$$\frac{dW}{d\alpha} = \frac{\alpha}{2} - 3\frac{k}{\alpha^5} = 0 \qquad \text{or} \qquad \alpha = (6k)^{1/6}$$

$$W_{\min} = \tfrac{3}{8}(6k)^{1/3} \sim .681k^{1/3}.$$

Although the trial function is appropriate only for a harmonic oscillator, it gives an answer quite close to the true value $E_0 \sim .668k^{1/3}$.

In general, the variation method cannot be extended to other eigenvalues. However, if we could construct a trial function $|\phi_1\rangle$ orthogonal to $|E_0\rangle$, we could conclude $W = \langle\phi_1|\, H\,|\phi_1\rangle \geqslant E_1$. The difficulty is that we don't usually know $|E_0\rangle$, so that it is not possible to ensure its orthogonality to $|\phi_1\rangle$. The theorem does not hold if we merely orthogonalize $|\phi_1\rangle$ to an approximate $|\phi_0\rangle$. However, it is possible to extend the variation method rigorously to other states which are known to be orthogonal to $|E_0\rangle$ on the basis of symmetry. Thus, if L is a Hermitian operator that commutes with H, so that $L|E_0\rangle = l_0|E_0\rangle$, and if $|E_i\rangle$ is any eigenvector of both H and L that belongs to an eigenvalue of L different from l_0, then $\langle E_0 \mid E_i\rangle = 0$, as explained in Section 8.14. For example, the ground state of helium is a singlet, but there exist states which are triplets, so that it is possible to calculate the energies of both the lowest singlet state and the lowest triplet of helium.

Exercises

35. Convince yourself that if $|\phi\rangle$ is not normalized, the appropriate W to consider in connection with the variation principle is $\langle\phi|\, H\,|\phi\rangle/\langle\phi \mid \phi\rangle$.

36. If a trial $|\phi\rangle$ differs from the true $|E_0\rangle$ by the vector $\varepsilon|\chi\rangle$, where ε is small, show that W differs from E_0 by order ε^2. In words, a reasonable approximation to $|E_0\rangle$ gives quite a good approximation to E_0; a good approximation to E_0 does not guarantee that $|\phi\rangle$ is a good approximation to $|E_0\rangle$.

*37. Starting from $W - E_0 = \sum_{E \neq E_0} \langle\phi \mid E\rangle\langle E \mid \phi\rangle(E - E_0)$, show that

$$1 - \langle\phi \mid E_0\rangle\langle E_0 \mid \phi\rangle \leqslant (W - E_0)/(E_1 - E_0),$$

where E_1 is the next lowest eigenvalue. The quantity on the left represents the extent to which $|\phi\rangle$ is not identical with $|E_0\rangle$, so that this result demonstrates that as W approaches E_0, the trial function must approach the true eigenfunction.

*38. Convince yourself that the $|\phi\rangle$ which minimizes $W = \langle\phi| H |\phi\rangle$ is not ordinarily the same as that which maximizes $|\langle\phi \mid E_0\rangle|^2$.

39. Convince yourself that further elaboration of a trial function by including yet another parameter cannot increase W, but must either decrease W or leave it unchanged. Therefore, "The more work you do, the closer the approximate answer is to the true one."

40. (a) Use a trial function of the form $xe^{-\beta^2 x^2/2}$ to obtain an approximation to the next lowest eigenvalue of the quartic oscillator.
 (b) Why is this procedure applicable to this state?
 (c) On the basis of the variation principle, what can you conclude about $W_1 - W_0$, which is an estimate of the excitation energy of the quartic oscillator?

*41. Use a trial function of the form $e^{-\alpha r^2}$ to obtain an approximation to the lowest energy level of the hydrogen atom, whose Hamiltonian is $-\frac{1}{2}\nabla^2 - (1/r)$. (*Note:* In these units, the true eigenvalue is $E = -\frac{1}{2}$.)

9.6 Linear Variation Principle

So far we have considered a general trial $|\phi\rangle$ in which the parameters may occur in any fashion. The simultaneous equations that result on setting each partial derivative equal to zero are in general nonlinear, but a convenient special case arises if $|\phi\rangle$ is linear in the parameters.

Let us use a trial vector of the form

$$|\phi\rangle = \sum_{j=1}^{N} c_j |j\rangle = \sum_{j=1}^{N} |j\rangle\langle j \mid \phi\rangle, \tag{9.10}$$

where $\{|j\rangle\}_{j=1}^{N}$ is a set of linearly independent vectors, not necessarily orthonormal, and not complete. Our problem is to find those values of the coefficients $\langle j | \phi\rangle$ which minimize $W = \langle\phi| H |\phi\rangle = \sum_{i,j}^{N} \langle\phi \mid i\rangle\langle i| H |j\rangle\langle j| \phi\rangle$, subject to the constraint $\langle\phi \mid \phi\rangle = \sum_{i,j}^{N} \langle\phi \mid i\rangle\langle i \mid j\rangle\langle j| \phi\rangle = 1$. According to Lagrange's method of undetermined multipliers, we should construct the function $\langle\phi| H |\phi\rangle - \lambda(\langle\phi \mid \phi\rangle - 1)$, and set the partial derivative with respect to each $\langle\phi \mid i\rangle$ equal to zero. Such a procedure leads to the N simultaneous homogeneous linear equations

$$\sum_{j=1}^{N} (\langle i| H |j\rangle - \lambda\langle i \mid j\rangle)\langle j \mid \phi\rangle = 0$$

in the N unknowns $\{\langle j \mid \phi\rangle\}$. In matrix notation these equations become the matrix equation $(\mathbf{H} - \lambda\mathbf{S})\boldsymbol{\phi} = \mathbf{0}$, where $\mathbf{H} = (\langle i|\, H\,|j\rangle)$, $\mathbf{S} = (\langle i \mid j\rangle)$, and $\boldsymbol{\phi} = (\langle j \mid \phi\rangle)$ are N-dimensional. For a nontrivial solution the $N \times N$ determinant of coefficients must equal zero:

$$|(\langle i|\, H\,|j\rangle - \lambda\langle i \mid j\rangle)| = |\mathbf{H} - \lambda\mathbf{S}| = 0. \tag{9.11}$$

(Notice that if the set $\{|j\rangle\}$ is made complete, then the problem again becomes that of finding the eigenvalues and eigenvectors of the infinite matrix that represents the operator H in the basis chosen.) Expansion of the determinant leads to an Nth-order polynomial in λ, which has N roots, λ_i. Each of these roots may be substituted into the N simultaneous linear equations to determine the coefficients of the normalized solution vector $|\phi_i\rangle$. Furthermore, it is easy to show that $\lambda_i = \langle\phi_i|\, H\,|\phi_i\rangle$, so that by the variation principle we may conclude that no eigenvalue of the finite matrix $\mathbf{H}$ can be less than the lowest eigenvalue, E_0, of H, and that the lowest eigenvalue of $\mathbf{H}$ is the closest approximation to E_0 possible with the type of trial vector chosen. It may further be mentioned that each λ_i has as its lower bound *some* eigenvalue of H, but this result is useful only for lower eigenvalues (Exercise 9.49).

Previously we included in the definition of a basis set the requirement that the set be complete. We shall relax this definition and call the finite set $\{|j\rangle\}_{j=1}^{N}$ a basis set, although the proper name is TRUNCATED BASIS SET or LIMITED BASIS SET. Of course, as the set approaches completeness, the trial vector $|\phi\rangle = \sum_{j=1}^{N} |j\rangle\langle j \mid \phi\rangle$ approaches the true eigenvector $|\psi\rangle = \sum_{j=1}^{\infty} |j\rangle\langle j \mid \psi\rangle$, and the approximate eigenvalues λ_i approach closer and closer to the true ones. What we would like is a very rapid convergence, so that we can obtain adequate accuracy with only a small number of basis vectors.

Much of the art of quantum chemistry consists in the choice of basis. A convenient basis set is one that permits a good approximation to the true eigenvalues and eigenvectors without too much labor. Often it is possible to choose as basis set the set of solutions to a similar but simpler problem. This choice has the added advantage that it is then possible to visualize a complicated solution in terms of a simpler one.

Examples

18. To find the energy levels and wave functions for the helium atom, the set of appropriate products of hydrogen-atom wave functions (perhaps about a nucleus with an adjustable charge Ze) is convenient. Indeed, the designation of the configurations of the helium atom as $(1s)^2$, $1s2s$, $1s2p$, etc., corresponds to the approximation that a single basis vector is an adequate description of a wave function that can be written down only as an infinite series.

19. To find the energy levels and wave functions of the hydrogen molecule, various sets of products of hydrogen-atom wave functions are convenient. Let us take as basis $|1\rangle = 1s_A(1)1s_B(2)$, $|2\rangle = 1s_B(1)1s_A(2)$, $|3\rangle = 1s_A(1)1s_A(2)$, and $|4\rangle = 1s_B(1)1s_B(2)$, where $1s_A(1)$ is the wave function with the first electron in a $1s$ orbital on atom A, etc.

Then simple LCAO-MO theory approximates the lowest-energy wave function of H_2 by the finite linear combination $(|1\rangle + |2\rangle + |3\rangle + |4\rangle)/2$. Simple VB theory approximates this wavefunction by $(|1\rangle + |2\rangle)/\sqrt{2}$. A better approximation is obtained by finding the best linear combination of the four basis functions. A still better approximation is obtained by mixing in products involving $2p$ orbitals, corresponding to hybridizing the orbitals so as to concentrate electron density between the atoms.

20. To find the pi molecular orbitals of an aromatic hydrocarbon, we may choose atomic orbitals on each carbon atom as basis, so that each molecular orbital is a linear combination of atomic orbitals. The exact solution is an infinite linear combination of atomic orbitals, but the finite set of only $2p_z$ orbitals is a truncated basis that is adequate for most purposes.

Exercises

*42. Show that if every $\langle j \mid \phi\rangle$ is real, then $\partial[\langle\phi| H |\phi\rangle - \lambda(\langle\phi \mid \phi\rangle - 1)]/\partial\langle\phi \mid i\rangle = 2\sum_j \langle i| H |j\rangle\langle j \mid \phi\rangle - 2\lambda \sum_j \langle i \mid j\rangle\langle j \mid \phi\rangle$.

*43. Show that λ_k does indeed equal $\langle\phi_k| H |\phi_k\rangle$.

44. Use the linear variation principle to approximate the lowest eigenvalue of the operator $-D^2 + V(x)$, where $V(x) = V_0$ for $0 < x < L/2$, $V(x) = 0$ for $L/2 < x < L$, and $V(x) = \infty$ elsewhere. Use as basis set only the first two solutions to the problem with $V_0 = 0$, namely $(2/L)^{1/2} \sin(\pi x/L)$ and $(2/L)^{1/2} \sin(2\pi x/L)$. (Cf. Exercise 4.52.)

45. Use the linear variation principle to approximate the lowest eigenvalue of the operator $-D^2 + V(\phi)$, where $V(\phi) = V_0$ for $-\pi/6 < \phi < \pi/6$ and $V(\phi) = 0$ elsewhere. Use as basis set only the first three solutions to the problem with $V_0 = 0$; namely, $(2\pi)^{-1/2}$, $(2\pi)^{-1/2}e^{i\phi}$, and $(2\pi)^{-1/2}e^{-i\phi}$. [*Hint:* One root of the cubic is $1 + \frac{1}{6}V_0 - V_0\sqrt{3}/4\pi$.] (Cf. Exercise 4.53. Just as the problem with $V_0 = 0$ is a convenient model for cyclic polyenes, this problem is a convenient model for pyridine.)

9.7 Perturbation Theory

Frequently the eigenvalue-eigenvector problem $H|E_i\rangle = E_i|E_i\rangle$ is too difficult to solve exactly. However, if H is a Hermitian operator that is "approximately equal to" an operator H° whose eigenvalues $\{E_i^\circ\}$ and complete orthonormal set of eigenvectors $\{|E_i^\circ\rangle\}$ are known exactly, it is possible to obtain approximations to each E_i and $|E_i\rangle$. Let $H = H^\circ + \varepsilon H'$, where H' is a PERTURBATION OPERATOR and ε is a small scalar quantity. (Sometimes a variable parameter measuring the perturbation arises naturally in the problem, such as when the perturbation is an applied electric or magnetic field. Sometimes the perturbation is of fixed magnitude, with $H = H^\circ + H'$, but it is convenient to introduce ε and then to evaluate the resulting answer at $\varepsilon = 1$.) Now let us expand the perturbed eigenvalues and eigenvectors as a power series in ε:

$$\begin{aligned} E_i &= E_i^\circ + \varepsilon E_i' + \varepsilon^2 E_i'' + \cdots \\ |E_i\rangle &= |E_i^\circ\rangle + \varepsilon|E_i'\rangle + \cdots. \end{aligned} \tag{9.12}$$

It can then be shown that

$$E_i' = \langle E_i^\circ | \, H' \, | E_i^\circ \rangle \tag{9.13a}$$

$$| E_i' \rangle = - \sum_{j \neq i} \frac{| E_j^\circ \rangle \langle E_j^\circ | \, H' \, | E_i^\circ \rangle}{E_j^\circ - E_i^\circ} \tag{9.13b}$$

$$E_i'' = - \sum_{j \neq i} \frac{\langle E_i^\circ | \, H' \, | E_j^\circ \rangle \langle E_j^\circ | \, H' \, | E_i^\circ \rangle}{E_j^\circ - E_i^\circ}. \tag{9.13c}$$

Notice especially that the perturbed answer is expressed entirely in terms of the unperturbed solution. Let us consider each of these equations in turn.

Since $\langle E_i^\circ | \, H^\circ + \varepsilon H' \, | E_i^\circ \rangle = E_i^\circ + \varepsilon E_i'$, the eigenvalue, correct to first order, is simply the average value of H, averaged over the unperturbed eigenvector (cf. Exercise 9.32). Notice that the eigenvalue is obtained correct to first order with only zeroth-order eigenvectors. This is another manifestation of the general phenomenon that an approximate eigenvector can give quite a close approximation to the true eigenvalue.

The first-order correction to the eigenvector is expressed in the basis $\{| E_i^\circ \rangle\}$, and only those basis vectors that are coupled to $| E_i^\circ \rangle$ by the perturbation are included in $| E_i' \rangle$. Notice especially the quantity $E_j^\circ - E_i^\circ$ in the denominator. If $E_j^\circ = E_i^\circ$, with $\langle E_j^\circ | \, H' \, | E_i^\circ \rangle \neq 0$, then a term in the sum is infinite. The occurrence of such an infinity contradicts our initial supposition that we could expand E_i and $| E_i \rangle$ as power series in ε. Let there be n_i degenerate eigenvectors $\{| E_{i,k}^\circ \rangle\}_{k=1}^{n_i}$ of H° belonging to E_i°. To eliminate this infinity it is then necessary and sufficient to find n_i new linear combinations of these eigenvectors such that none is coupled to any other by H'. This problem is equivalent to diagonalizing the $n_i \times n_i$ matrix $(\langle E_{i,k}^\circ | \, H' \, | E_{i,l}^\circ \rangle)$. The eigenvalues of this matrix are then the correct first-order corrections to the eigenvalues, and the corresponding eigenvectors are the PROPER ZEROTH-ORDER EIGENVECTORS. These new eigenvectors are still eigenvectors of H° belonging to E_i°, but the first-order correction may SPLIT (remove) the degeneracy.

The second-order correction to the ith eigenvalue is a sum over all basis vectors (unperturbed eigenvectors) except the ith. Alternatively, we may write $E_i'' = \langle E_i^\circ | \, H' \, | E_i' \rangle$, suggesting that E_i'' arises from the way the ith basis vector "adapts" to the perturbation. Directing this viewpoint at $| E_0^\circ \rangle$, the basis vector belonging to the lowest unperturbed eigenvalue E_0°, provides a mnemonic for the signs in E_i'': Coupled basis vectors repel, so that the effect of mixing of $| E_0^\circ \rangle$ with all other basis vectors, whose eigenvalues are greater than E_0°, is to decrease the resulting lowest eigenvalue. Therefore E_0'' must be negative, as is consistent with the fact that every $\langle E_0^\circ | \, H' \, | E_j^\circ \rangle \langle E_j^\circ | \, H' \, | E_0^\circ \rangle = |\langle E_j^\circ | \, H' \, | E_0^\circ \rangle|^2$ and every $E_j^\circ - E_0^\circ$ are positive. We also note that just as the zeroth-order eigenvectors provide eigenvalues correct to first order, so do eigenvectors correct to first order provide eigenvalues correct to second order. And we note that unless proper zeroth-order eigenvectors are chosen, there is an infinity in E_i'', and thus a wrong E_i'.

Example

21. We shall treat the effect of a magnetic field H_z along the z-axis as a perturbation on the energy levels of a hydrogen atom. For simplicity, we neglect spin. In the basis set $|1s\rangle$, $|2s\rangle$, $|2p_x\rangle$, $|2p_y\rangle$, $|2p_z\rangle, \ldots$, of unperturbed solutions, the operators H° and H' are represented as matrices that begin

$$\mathbf{H}^\circ = \begin{pmatrix} E_1 & 0 & 0 & 0 & 0 \\ 0 & E_2 & 0 & 0 & 0 \\ 0 & 0 & E_2 & 0 & 0 \\ 0 & 0 & 0 & E_2 & 0 \\ 0 & 0 & 0 & 0 & E_2 \end{pmatrix}, \qquad \mathbf{H}' = H_z \begin{pmatrix} 0 & 0 & 0 & 0 & 0 \\ 0 & 0 & 0 & 0 & 0 \\ 0 & 0 & 0 & i\beta & 0 \\ 0 & 0 & -i\beta & 0 & 0 \\ 0 & 0 & 0 & 0 & 0 \end{pmatrix},$$

where β is the Bohr magneton, $e\hbar/2mc$. Here the expansion parameter is naturally H_z. Since the diagonal elements of $\mathbf{H}'$ are all zero, it would seem as though all $n = 2$ states remain degenerate in first order. But the states $|2p_x\rangle$ and $|2p_y\rangle$, which are degenerate with respect to H°, are coupled by H'. Therefore it is necessary to diagonalize the 4×4 submatrix of $\mathbf{H}$ corresponding to the basis vectors belonging to the fourfold-degenerate eigenvalue E_2 of H°. The proper zeroth-order eigenvectors that result (cf. Example 8.15) are $|2s\rangle$, $|2p_+\rangle$, $|2p_-\rangle$, and $|2p_z\rangle$, with first-order eigenvalues E_2, $E_2 + \beta H_z$, $E_2 - \beta H_z$, and E_2, respectively. There are then no off-diagonal terms, so the second-order correction vanishes. Indeed, this example is rather trivial, because H' commutes with H°, but it is intended to illustrate how the choice of proper zeroth-order eigenvectors is essential for a correct evaluation of first-order eigenvalues.

Exercises

*46. Derive Eqs. 9.13a and 9.13b by the following procedure: (*1*) Substitute Eq. 9.12 into $(H^\circ + \varepsilon H')|E_i\rangle = E_i|E_i\rangle$. Since this result must hold for all ε, the coefficient of every power of ε must be zero. (*2*) Set the coefficient of ε equal to zero. (*3*) Expand $|E_i'\rangle$ in the basis set $\{|E_j^\circ\rangle\}$. (*4*) Multiply the resulting equation:
(a) By $\langle E_i^\circ|$ and solve for E_i'. (*Hint:* Remember that $\{|E_j^\circ\rangle\}$ is orthonormal.)
(b) By $\langle E_j^\circ|$ and solve for $\langle E_j^\circ \mid E_i'\rangle$.

*47. Show that if $H = H^\circ + \varepsilon H' + \varepsilon^2 H''$, there is an additional term $\langle E_i^\circ|\, H''\, |E_i^\circ\rangle$ in E_i''.

48. Rewrite the formulas of perturbation theory in terms of functions and integrals.

*49. The linear variation principle often gives close approximations to the lowest eigenvalues of H, but higher eigenvalues of the finite matrix $\mathbf{H}$ are poor approximations to higher eigenvalues of H. Why? (*Hint:* Which eigenvalues will be most strongly affected on going to a complete basis set?)

50. Solve the problem of Exercise 9.44 by perturbation theory, correct to second order.

51. Solve the problem of Exercise 9.45 by perturbation theory, correct to second order.

52. Find, to first order, *all* eigenvalues of the operator $-D^2 + V(\phi)$ of Exercise 9.45.

*53. Find the eigenvalues of the matrix

$$\begin{pmatrix} \lambda_1^\circ & 0 & \kappa \\ 0 & \lambda_1^\circ & \kappa \\ \kappa & \kappa & \lambda_2^\circ \end{pmatrix}$$

by second-order perturbation theory. Then find the eigenvalues exactly. Why is there a discrepancy? (*Hint:* In the fourth-order correction to the ith eigenvalue, there are terms of the form

$$\frac{\langle E_i^\circ | H' | E_j^\circ \rangle \langle E_j^\circ | H' | E_k^\circ \rangle \langle E_k^\circ | H' | E_l^\circ \rangle \langle E_l^\circ | H' | E_i^\circ \rangle}{(E_i^\circ - E_j^\circ)(E_i^\circ - E_k^\circ)(E_i^\circ - E_l^\circ)} .$$

10

GROUP THEORY

We begin this chapter by defining an abstract group and giving some examples. We describe symmetry operations on molecules and we indicate how the set of all symmetry operations that leave a molecule unchanged form a group. The various point groups (molecular symmetries) are catalogued. We quickly introduce the concepts of conjugate elements, classes, and subgroups. Then we introduce the two central concepts of applied group theory—group representations and bases for representations. We discuss the related topics, reduction of representations and transformation of basis, and we focus on irreducible representations. We indicate in detail how group theory may be used to simplify eigenvalue problems. Then we provide both a catalog and a visual description of the irreducible representations of the point groups. Some important theorems concerning group representations follow. Then we show how we may simplify application of group theory by working with characters (matrix traces). We show how the reduction formula may be used to determine the extent to which group theory can help to simplify an eigenvalue problem, and we provide a recipe for adapting basis sets to take maximum advantage of symmetry. And we conclude with an explanation of how we may use group theory to show that certain integrals are equal to zero.

Again, the exercises are intended to provide practice, familiarity, and proficiency with some very abstract concepts. Most applications of group theory will become apparent only on taking advanced courses, especially quantum mechanics and spectroscopy, and on reading the research literature. However, exercises are included that demonstrate applications to molecular structure, statistical mechanics, molecular vibrations, molecular-orbital theory, ligand-field theory, and molecular spectroscopy. The exercises, and the examples in the body of the text, refer primarily to organic and inorganic molecules since it is easier to learn the principles of group theory by particularizing them with familiar topics. Therefore we have omitted applications of group theory to crystal structure and to atomic structure since these topics are less familiar and involve more complicated aspects of group theory. However, the principles are the same, and only the notations are different, so a student should have little difficulty in applying what he has learned to other areas.

10.1 Definitions

A BINARY OPERATION is a rule for taking two objects and obtaining a unique result. For example, subtraction is a binary operation which takes two numbers and produces a number, their difference. Another example, which does not involve numbers, is "relation," which takes two members of a family and produces their relation—"son," "aunt," "third cousin once removed," etc. In order to specify the binary operation, it is necessary to specify the result of combining every possible ordered pair of elements under consideration. The operation should be considered as a generalized multiplication, and we shall indicate the "product" of two elements, A and B, by juxtaposition: AB.

A GROUP $\mathcal{G}$ is a set of elements, A, B, C, . . . , and a binary operation, satisfying the following rules:

1. CLOSURE: AB is a member of $\mathcal{G}$.
2. ASSOCIATIVITY: $(AB)C = A(BC)$, so that we may drop the parentheses.
3. IDENTITY ELEMENT: There exists a unique element E (German *einheit*) in $\mathcal{G}$, such that for every element A in $\mathcal{G}$, $EA = A = AE$.
4. INVERSES: For every A in $\mathcal{G}$, there exists a unique element A^{-1} (read "A inverse") in $\mathcal{G}$, such that $AA^{-1} = E = A^{-1}A$.

Notice that it is not required that $AB = BA$; it is quite possible that the order of multiplication is important. However, if $AB = BA$, then A and B are said to COMMUTE. According to rules 3 and 4, E always commutes with every element, and every element commutes with its inverse. A group is said to be ABELIAN when all of its elements commute.

The number of elements in the group is called the ORDER of the group, and is usually denoted by g.

Examples of Groups

1. The set of all integers, with addition as the binary operation (abbreviated "integers under addition"). Here $E = 0$ and $A^{-1} = -A$.
2. The positive real numbers under multiplication; $E = 1$, $A^{-1} = 1/A$.
3. The set of Nth-order polynomials, $P_N(x)$, under addition; $E = 0$, $P_N(x)^{-1} = -P_N(x)$. Notice that in order to ensure closure, $P_M(x)$, with $M < N$, must be considered as an Nth-order polynomial.

4. The set of all $n \times n$ unitary matrices under matrix multiplication. Here $E = \mathbf{1}$ and inverses are inverse matrices. Closure is guaranteed by the result of Exercise 8.85. This group is not Abelian.

The remaining examples are groups whose elements are themselves operations, in particular, operations which rearrange objects. In these examples, the binary operation is "carry out in succession," and by convention the product, O_2O_1, is the operation, "first perform O_1, then perform O_2." Closure is guaranteed by the nature of the operations—the product operation is always equivalent to some single operation. Associativity may not be obvious, but does hold. The identity element is always the operation, "leave the objects alone." Inverses must be worked out for each case, but are usually obvious from the context.

5. The set of symmetry operations which leave the NH_3 molecule unchanged. Some of the elements are "rotate by $2\pi/3$ about the axis of the nitrogen lone pair," "reflect in a plane containing the lone-pair axis and an N—H bond," and the identity "neither rotate nor reflect." The inverse of a rotation by $2\pi/3$ is a rotation by $4\pi/3$, and a reflection is its own inverse.

6. The set of symmetry operations which leave an infinite crystal unchanged. Besides rotations and reflections, translations are also included in the group, along with other operations which we shall not consider.

7. $\mathscr{P}_n$, the set of permutations of n objects. For example, with $n = 3$, we may call the objects a, b, and c, and one of the permutations is written $\begin{pmatrix} abc \\ acb \end{pmatrix}$. To apply this permutation to any arrangement of a, b, and c, leave a alone and replace b by c and c by b; thus $\begin{pmatrix} abc \\ acb \end{pmatrix}(abc) = (acb)$. If next we apply the permutation $\begin{pmatrix} abc \\ bca \end{pmatrix}$, the final result is (bac), which could have been achieved by the single permutation $\begin{pmatrix} abc \\ bac \end{pmatrix}$. Therefore we write $\begin{pmatrix} abc \\ bca \end{pmatrix}\begin{pmatrix} abc \\ acb \end{pmatrix}(abc) = (bac) = \begin{pmatrix} abc \\ bac \end{pmatrix}(abc)$, and we may note that the product

$$\begin{pmatrix} abc \\ bca \end{pmatrix}\begin{pmatrix} abc \\ acb \end{pmatrix} = \begin{pmatrix} abc \\ bac \end{pmatrix}$$

does not depend upon the initial arrangement of the objects. See also Example 8.3. For $n > 2$ these groups are not Abelian (Exercise 8.36).

Exercises

1. Which of the following are groups? For each that is not, indicate a rule which is not satisfied.
   (a) Integers under multiplication.
   (b) All 3×3 matrices under matrix multiplication.
   (c) Binary interchanges of n elements.
   (d) Even integers under addition.
   (e) Odd integers under addition.
   (f) Three-dimensional vectors under addition.
   (g) Three-dimensional vectors under multiplication (both types).
   (h) $N \times N$ Hermitian matrices under matrix addition.
   (i) Real numbers under subtraction.
   (j) Nonnegative reals under "greater of."

2. What is E^{-1}? What is $(AB)^{-1}$?
3. Of what order is the group $\mathcal{P}_n$?
4. Let $\mathcal{G}_1 = \{R_i\}_{i=1}^{g_1}$ and $\mathcal{G}_2 = \{S_j\}_{j=1}^{g_2}$ be groups such that every R_i commutes with every S_j. Define the DIRECT PRODUCT $\mathcal{G}_1 \times \mathcal{G}_2 = \{R_iS_j\}$, the set of all possible products, with a total of g_1g_2 distinct elements. Convince yourself that if we define the product of R_iS_j and R_kS_l as $(R_iR_k)(S_jS_l)$, then $\mathcal{G}_1 \times \mathcal{G}_2$ is also a group.

10.2 Group Multiplication Tables

We may specify all products of group elements by constructing a multiplication table for the group. The product AB is entered at the intersection of row A and column B. (Notice the order.)

The group of order one is trivial, since it contains only the identity element.

The multiplication tables for $g = 2$ (the group $\mathcal{C}_2$) and $g = 3$ (the group $\mathcal{C}_3$) are

$\mathcal{C}_2$	E	A
E	E	A
A	A	E

$\mathcal{C}_3$	E	A	B
E	E	A	B
A	A	B	E
B	B	E	A

or

$\mathcal{C}_3$	A	B
A	B	E
B	E	A

.

In this last form the E column and the E row are omitted, since such products are determined by group postulate 3. A group $\mathcal{G}$ is said to be CYCLIC if there exists a GENERATOR G in $\mathcal{G}$ such that every element of $\mathcal{G}$ is some power of G, where by G^n is meant $GG \cdots G$ (n times). Both the above groups are cyclic: in $\mathcal{C}_2$, $A^1 = A$, $A^2 = E$; in $\mathcal{C}_3$, $A^1 = A$, $A^2 = B$, $A^3 = E$.

There are two "distinctly different" groups of order four:

$\mathcal{C}_4$	A	B	C
A	B	C	E
B	C	E	A
C	E	A	B

$\mathcal{V}$	A	B	C
A	E	C	B
B	C	E	A
C	B	A	E

.

Notice that $\mathcal{C}_4$ is cyclic but $\mathcal{V}$ is not.

The five groups detailed above are all Abelian. Next we consider a non-Abelian group of order six.

	A	B	C	D	F
A	B	E	D	F	C
B	E	A	F	C	D
C	F	D	E	B	A
D	C	F	A	E	B
F	D	C	B	A	E

.

As examples of the non-Abelian character, notice that $AC = D \neq F = CA$ and $CD = B \neq A = DC$. Since this is the simplest non-Abelian group, it is very useful as an example for particularizing the theorems of group theory when they seem too general and abstract.

Exercises

5. Convince yourself that the multiplication table for an Abelian group is symmetric across the main diagonal.
6. Prove the "Latin-square" property of group multiplication tables, that is, in any column (row) every element of the group appears once and only once. (*Hint:* Assume some element X appears twice in the column under A.)
7. Use the Latin-square property to convince yourself that if S is an element of the group $\mathcal{G}$ of order g, then a sum over all the g products of S with elements of $\mathcal{G}$ is the same as a sum over the g elements of $\mathcal{G}$: $\sum_{RS} = \sum_{R}$.
8. Construct a multiplication table for a group of order five. Is it Abelian?
9. To spot-check associativity for the group of order six given, show that $(AC)B = A(CB)$, $(AC)D = A(CD)$, $(CD)F = C(DF)$.
10. Rotations of a rectangular parallelepiped (length, width, and height all unequal) which leave the figure unchanged form a group of order four. Construct the multiplication table for this group. (*Hint:* Practice with a book.) If you call the rotations E, A, B, and C appropriately, you will obtain the same multiplication table as that given for one of the groups of order four. Which one? These two groups are said to be ISOMORPHIC, the same abstract group, but with the elements labelled by different names and with different significance.
11. Construct the multiplication table for $\mathscr{P}_2$. To what group given is it isomorphic?
*12. Construct the multiplication table for $\mathscr{P}_3$. Find an isomorphism between $\mathscr{P}_3$ and the group of order six given, that is, find how the elements should be renamed to make the multiplication tables identical.
*13. Construct a multiplication table for a group of order nine. (*Hint:* There is always a cyclic group of order n.) Construct a multiplication table for another (non-isomorphic) group of order nine. (*Hint:* Choose two different elements, A and B, such that $A^3 = E = B^3$.)

10.3 Symmetry Operations on Molecules

We have already implied that symmetry operations which transform a molecule or an infinite crystal into itself form a group. We shall not consider crystals, but we shall concentrate on symmetry operations which leave molecules unchanged. These form the so-called "point groups."

Since every group must have an identity element, we must include as a symmetry operation the identity operation, which leaves the molecule alone. This operation is still represented by the symbol E.

An n-fold ROTATION AXIS, denoted by C_n, is the operation rotate (in a right-handed sense) by an angle $2\pi/n$ about some axis, usually the z-axis. Since $C_1 = E$, this is not considered as a rotation. Examples of molecules with such axes are water ($n = 2$), ammonia ($n = 3$), square-pyramidal B_5H_9

($n = 4$), ferrocene ($C_5H_5FeC_5H_5$, $n = 5$), benzene ($n = 6$), and hydrogen fluoride ($n = \infty$). No notational distinction need be made between the operator C_n and the symmetry axis C_n.

Remember that if C_n is a symmetry element of a molecule, then so is C_n^k, by the closure property. Also, notice abbreviations such as $C_6^2 = C_3$ and $C_6^3 = C_2$.

One way of describing symmetry operations is to specify what they do to the point (x, y, z). If the axis of rotation is the z-axis, C_n sends (x, y, z) into $(x \cos 2\pi/n - y \sin 2\pi/n, x \sin 2\pi/n + y \cos 2\pi/n, z)$. In matrix notation

$$C_n \begin{pmatrix} x \\ y \\ z \end{pmatrix} = \begin{pmatrix} \cos 2\pi/n & -\sin 2\pi/n & 0 \\ \sin 2\pi/n & \cos 2\pi/n & 0 \\ 0 & 0 & 1 \end{pmatrix} \begin{pmatrix} x \\ y \\ z \end{pmatrix}.$$

It is possible to have more than one axis of symmetry. A twofold axis perpendicular to a C_n is denoted by C_2', and if there are two "different types of" perpendicular twofold axes, they are distinguished as C_2' and C_2''. For example, ethylene has a C_2 perpendicular to the molecular plane and both a C_2' and a C_2'' perpendicular to C_2. Boron trifluoride has three C_2's perpendicular to the C_3. The square planar $PtCl_4^{2-}$ ion has two C_2's along Pt—Cl bonds and two C_2''s bisecting Cl-Pt-Cl angles (Fig. 10.1a). Benzene has three C_2's and

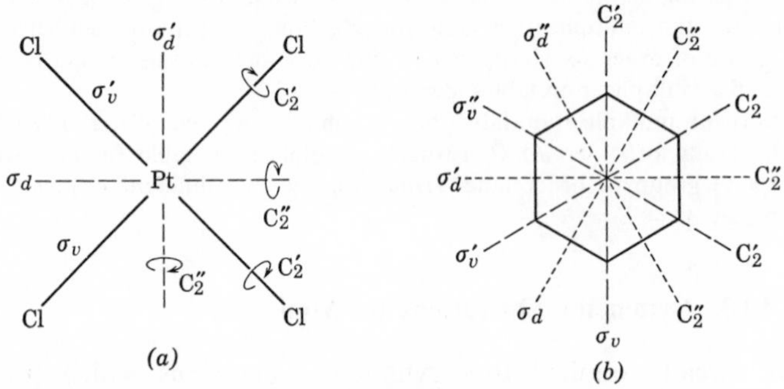

Figure 10.1

three C_2''s perpendicular to the C_6 (Fig. 10.1b). The hydrogen molecule has an infinity of C_2's perpendicular to the C_∞. More difficulty visualized C_2's are found in ethane (staggered configuration) and allene (Fig. 10.2).

Usually there is a single major axis, C_n ($n > 2$), and other C_2's, as in most of the above examples. Molecules of high symmetry have several C_ns ($n > 2$). In a tetrahedral molecule there are four C_3s. In a cubical or octahedral molecule there are three perpendicular C_4s and four C_3s. The regular dodecahedron and icosahedron have six C_5s and ten C_3s.

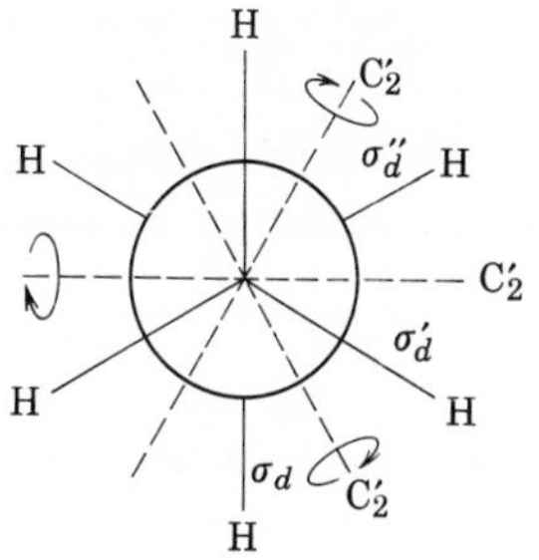

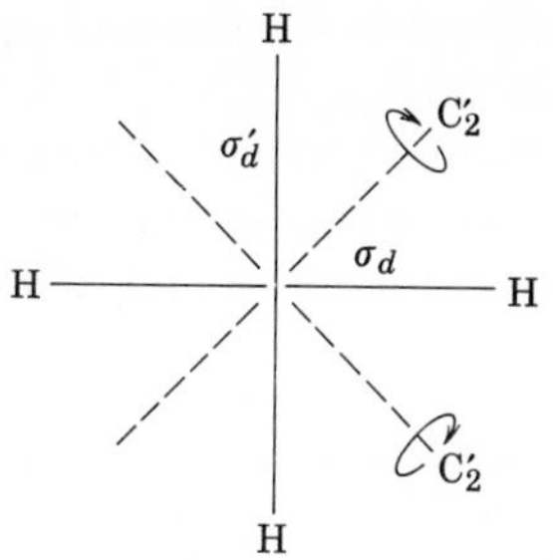

Figure 10.2

Next we consider MIRROR PLANES, denoted by σ. These are of two types, horizontal (σ_h, perpendicular to the major axis) and vertical (σ_v and σ_d, containing the major axis).

Examples of species with a σ_h are ethylene, cyclopropane, $PtCl_4^{2-}$, benzene, and H_2. If the major axis is along the z-axis, σ_h takes (x, y, z) into $(x, y, -z)$.

Before we go on to the vertical planes, let us consider the INVERSION CENTER. The operation $\sigma_h C_2$ takes the point (x, y, z) into $(-x, -y, -z)$. This operation is an inversion through the origin and is denoted by i.

Ethylene has both a C_2 and a σ_h, and therefore it must have i. Likewise, $PtCl_4^{2-}$ and benzene have a C_4 and a C_6, respectively, and therefore each must have a C_2; since each has a σ_h, each must also have a center of inversion. However, the inversion center may exist in the absence of both C_2 and σ_h. Examples of molecules with this feature are cyclohexane (chair form) and ferrocene (staggered conformation).

The more common type of vertical plane is σ_v. The σ_d (diagonal) symbol is reserved to denote vertical planes which bisect the angles between C_2's. Both NH_3 and BF_3 have three σ_vs. But $PtCl_4^{2-}$ has two σ_vs, containing the C_2's, and two σ_ds, straddled by the C_2's and containing the C_2''s (Fig. 10.1a). Likewise, benzene has three σ_vs and three σ_ds (Fig. 10.1b). The vertical planes of ethane and allene are σ_ds, since they are straddled by the C_2's (Fig. 10.2). The several σ_vs and σ_ds are distinguished by primes; thus NH_3 has σ_v, σ_v', and σ_v''. A vertical plane containing the x- and z-axes takes the point (x, y, z) into $(x, -y, z)$.

The last symmetry element is the n-fold ROTATION-REFLECTION AXIS, denoted by S_n. This operation is simply $\sigma_h C_n$, rotation by $2\pi/n$, followed by reflection in a plane perpendicular to the axis of rotation. (*Caution:* Some authors, especially in the crystallographic literature, call $S_n = iC_n$.)

The species BF_3, $PtCl_4^{2-}$, $C_5H_5^-$, benzene, and H_2 have C_n and σ_h, and therefore S_n. But if n is even, it is possible for S_n to be present without C_n or σ_h. Examples of molecules with this feature are methane and allene (S_4), the chair form of cyclohexane and the staggered conformation of ethane (S_6), and ferrocene (S_{10}).

Notice that σ, i, and S_n all involve reflection through a plane. A molecule which has any of these operations is superimposable on its mirror image, and it must therefore be optically inactive.

Exercises

*14. Convince yourself that i or a C_2' is sufficient to guarantee the vanishing of the dipole moment of a molecule.

15. Do Exercises 8.92 and 8.93.

16. An alternative way of describing symmetry operators is to specify what they do to $f(x, y, z)$. For example, we may take $C_n f(\mathbf{X})$ as a function whose graph is the same as that of $f(\mathbf{X})$ but rotated counterclockwise by $2\pi/n$ about the z-axis. However, just as basis vectors and column matrices "go in opposite directions," so also do arguments and functions (values). We have already seen (Exercise 5.2) that displacing the argument in the positive x-direction from x to $x + ct$ displaces the graph of f a distance ct in the negative x-direction. Use this example and that of C_4 acting on $f(\mathbf{X}) = x$ to convince yourself that although Ω takes the point $\mathbf{X}$ into the point $\Omega\mathbf{X}$, the operator Ω takes $f(\mathbf{X})$ into $f(\Omega^{-1}\mathbf{X})$. In practice, this feature will cause no confusion, but since we shall be applying symmetry operators to functions, it is advisable to define this process.

17. Let $\psi(x, y, z)$ be an eigenfunction of σ_v. What can you conclude about the eigenvalue to which ψ belongs? (*Hint:* What is σ_v^2?) What can you conclude about the eigenvalues of C_n? Of S_n? (*Caution:* Distinguish n even from n odd.)

18. Use the matrix formulation to show that $C_n\sigma_h$ takes (x, y, z) into the same point as does $\sigma_h C_n$, and that therefore these operations commute.

*19. What is the matrix form of a σ_v oriented at an angle ϕ to the x-axis? (*Hint:* Transform to a new coordinate system in which the matrix is especially simple, and then transform back.)

20. What are the more commonly used symbols for S_1 and S_2?

21. Show that if n is odd, the existence of S_n implies the existence of both σ_h and C_n. (*Hint:* S_n^n must be in the group.)

22. Does the existence of S_n imply the existence of C_{2n} or of $C_{n/2}$?

23. Visualize each of the above examples of S_n without C_n.

10.4 Stereographic Projections

To visualize symmetry operations of molecules, it is helpful to draw a stereographic projection, which indicates what a symmetry operation does to an object placed "at random" (off all symmetry elements). We shall indicate the initial position by $\times_1$ and the resultant position by $\times_2$. The major axis is perpendicular to the plane of the paper, and a point above this plane is an $\times$, a point below the plane an $\circ$.

A C_n is indicated by a solid n-sided polygon (a lune for $n = 2$). A C_2' is indicated by a lune and a dotted line. A σ_h is indicated by a solid circle, and a vertical plane is indicated by a solid line. An S_n in the absence of C_n is indicated by an open polygon which includes a solid polygon of $n/2$ sides for the $C_{n/2}$. Figure 10.3 illustrates each of these operations.

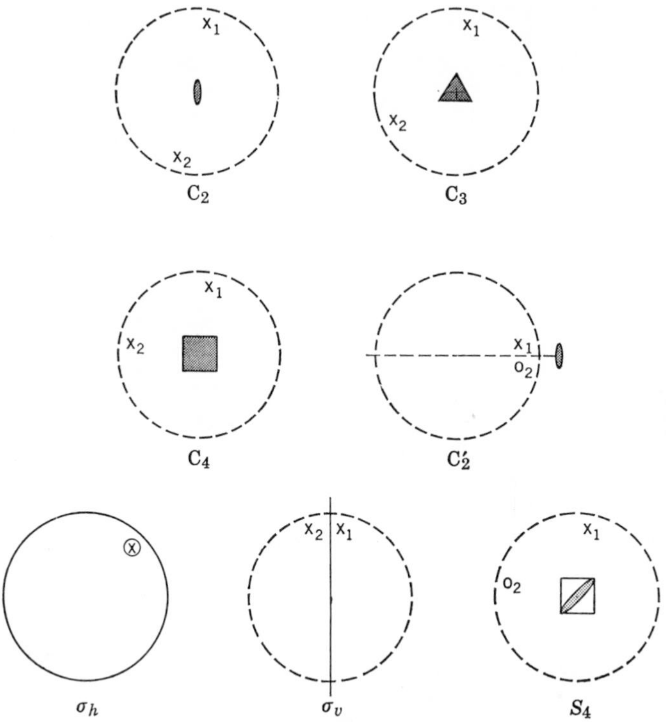

Figure 10.3

Let's practice using stereographic projections to evaluate products of symmetry operations. We shall label the points sequentially. Figure 10.4a shows that $\sigma_v'\sigma_v$ (90° apart) $= C_2$. Figure 10.4b shows that $\sigma_d\sigma_v$ (45° apart) $=$ C_4, and Figure 10.4c shows that $\sigma_v\sigma_d = C_4^3$, so that these two operations do not commute. Figure 10.4d shows that $\sigma_d S_4 = C_2'$ at 45° from the σ_d.

Exercises

24. Evaluate each of the following products:
    (a) $C_2'C_2$.
    (b) $C_2'\sigma_v$ (coincident).
    (c) $C_3\sigma_v$.
    (d) $\sigma_v C_3$.
    (e) $\sigma_d C_2'$ (30° apart).
    (f) $C_3^2 i$.

*25. (a) Evaluate $C_2'C_3$ and $C_2'C_4$.
    (b) Use these results to evaluate $C_2'C_3^2$ and $C_2'C_4^2$.
    (c) If a group contains a C_n and a C_2', how many perpendicular C_2's must there be in all?
    (d) Similarly, convince yourself that if a group contains a C_n and a vertical plane, it must contain a total of *n* vertical planes.

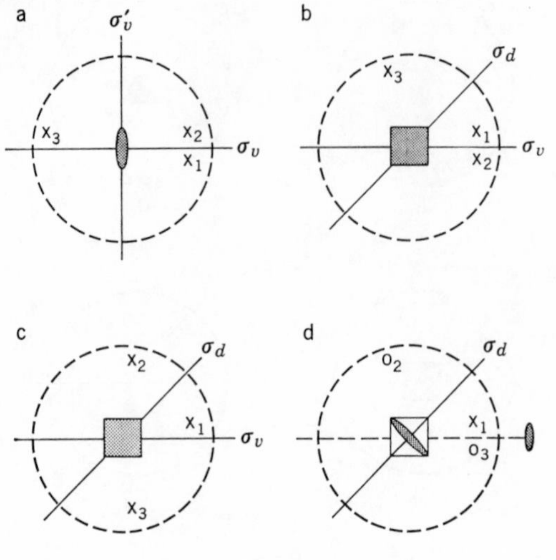

Figure 10.4

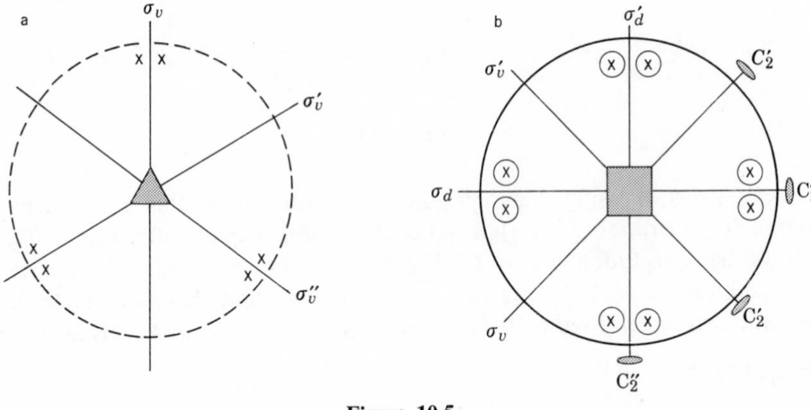

Figure 10.5

Most commonly, stereographic projection is used to illustrate all the symmetry elements that make up a point group. We may dispense with the subscripts, and use the xs and os to indicate all the equivalent points that are taken into one another by the group operations. Figure 10.5a shows the symmetry elements of NH_3: (E), C_3, (C_3^2), σ_v, σ_v', σ_v''. Figure 10.5b shows the symmetry elements of $PtCl_4^{2-}$: (E), C_4, (C_2), (C_4^3), σ_v, σ_v', σ_d, σ_d', σ_h, C_2', C_2', C_2'', C_2'', (i), (S_4), (S_4^3). The elements in parentheses are not indicated, but are understood to be present.

Exercises

26. Illustrate in stereographic projection all the symmetry elements of
   (a) Naphthalene: (E), C_2, C_2', C_2'', (i), σ_h, σ_v, σ_d.
   (b) Allene: (E), S_4, C_2, (S_4^3), C_2', C_2', σ_d, σ_d'.
   In each case show all equivalent points.
27. What is the total number of xs and os in the stereographic projection of a group?

10.5 A Catalog of the Point Groups

The symmetry of a molecule may be described by listing the symmetry operations that leave it unchanged. But rather than list all the operations, it is sufficient to specify only a few since, by the closure property, multiplication generates the entire group. Thus the existence of C_n guarantees the existence of C_n^2, C_n^3, etc.; C_n and σ_h imply S_n. If n is odd, C_n and C_2' imply $(n - 1)$ other C_2's, and C_n and σ_v imply $(n - 1)$ other σ_vs. If n is even, C_n and C_2' imply $(\frac{1}{2}n - 1)$ other C_2's and $\frac{1}{2}n$ C_2''s, and C_n and σ_v imply $(\frac{1}{2}n - 1)$ other σ_vs and $\frac{1}{2}n$ σ_ds. (Cf. Exercise 10.25.)

A concise notational scheme has been devised to indicate the symmetry operations which generate each of the molecular point groups. Examples of species which belong to the symmetries described below are listed in the Examples of Molecular Symmetries (page 323). Study each example until you see its symmetry; make a model if necessary.

$\mathcal{C}_n$ represents the groups whose only symmetry element is C_n. A molecule with "no" symmetry belongs to $\mathcal{C}_1$. $\mathcal{C}_2$ is the group whose only elements are E and C_2. Molecules with only C_n $(n > 2)$ and no other symmetry element are rarely encountered.

$\mathcal{S}_n$ represents the groups whose only symmetry element is an S_n. $\mathcal{S}_2$ is often called $\mathcal{C}_i$; its only elements are E and i. $\mathcal{S}_4 = \{E, S_4, C_2, S_4^3\}$.

$\mathcal{C}_{nh}$ represents the groups whose elements are E, C_n, σ_h, and all possible products. All these groups—$\mathcal{C}_n$, $\mathcal{S}_n$, $\mathcal{C}_{nh}$—are Abelian. $\mathcal{C}_{1h}$ is usually called $\mathcal{C}_s$, whose only elements are E and σ. $\mathcal{C}_{2h} = \{E, C_2, \sigma_h, i\}$. The elements of $\mathcal{C}_{3h}$ are E, C_3, C_3^2, σ_h, S_3, and S_3^5 (not S_3^2, which is C_3^2).

$\mathcal{C}_{nv}$ represents the groups whose elements are E, C_n, σ_v, and all possible products. $\mathcal{C}_{2v} = \{E, C_2, \sigma_v, \sigma_v'\}$; this symmetry is fairly common. $\mathcal{C}_{3v} = \{E, C_3, C_3^2, \sigma_v, \sigma_v', \sigma_v''\}$. The elements of $\mathcal{C}_{4v}$ are E, C_4, C_2, C_4^3, and four σ_vs. If n is greater than 2, $\mathcal{C}_{nv}$ is non-Abelian. If n is even, alternate σ_vs are sometimes called σ_d, even though there are no C_2's. The symmetry elements of $\mathcal{C}_{\infty v}$ are E, C_∞ and all its powers (rotation by any angle) and an infinite number of σ_vs.

$\mathcal{D}_n$ (dihedral) represents the groups whose elements are E, C_n, C_2', and all possible products. By the closure property, there are a total of n C_2's, although for n even these are sometimes divided into two "types," $\frac{1}{2}n$ C_2's and $\frac{1}{2}n$ C_2''s, alternately. $\mathcal{D}_2$ has three perpendicular C_2s, and the choice of one of them as the major axis is arbitrary. For $n > 2$, $\mathcal{D}_n$ is non-Abelian; these groups are rarely encountered.

The groups $\mathcal{D}_{nh}$ are formed by introducing σ_h into the elements of $\mathcal{C}_{nv}$ and forming all possible products. Since σ_h (and E) commutes with every element of $\mathcal{C}_{nv}$, we may write $\mathcal{D}_{nh}$ as the direct product $\mathcal{C}_{nv} \times \mathcal{C}_s$, although it is more informative to write $\mathcal{D}_{nh} = \mathcal{C}_{nv} \times \sigma_h$. $\mathcal{D}_{2h}$ has E, three perpendicular C_2s, three perpendicular σs, and i; the choice of major axis is arbitrary. The elements of $\mathcal{D}_{3h}$ are E, C_3, C_3^2, three σ_vs, σ_h, S_3, S_3^5, and three C_2's. $\mathcal{D}_{4h} = \{E, C_4, C_2, C_4^3, \sigma_v, \sigma_v', \sigma_d, \sigma_d', \sigma_h, S_4, i, S_4^3, C_2', C_2', C_2'', C_2''\}$. For $n > 2$, $\mathcal{D}_{nh}$ is non-Abelian.

$\mathcal{D}_{nd}$ represents the group whose elements are those of $\mathcal{D}_n$ along with a σ_d and all possible products. The clue to determining whether a molecule has this symmetry is to look for the S_{2n} which must be present. (See Exercise 10.34.) None of these groups is Abelian.

$\mathcal{T}$ is the group appropriate to species whose symmetry elements are four C_3s at tetrahedral angles (109.5°) to each other, along with three perpendicular C_2s. More important is $\mathcal{T}_d$, the symmetry of the regular tetrahedron, which includes the elements of $\mathcal{T}$ plus six σ_ds which bisect the tetrahedral angles and convert the C_2s into S_4s. $\mathcal{T}_h = \mathcal{T} \times i$; this is very rare, but can be identified by the presence of the four C_3s and i (which converts each C_3 to an S_6), but the absence of a C_4.

$\mathcal{O}$ is the group appropriate to molecules with three perpendicular C_4s, four tetrahedrally oriented C_3s, and six other C_2's. More common is $\mathcal{O}_h = \mathcal{O} \times i$, the symmetry of the cube and the regular octahedron.

$\mathcal{I}$ is the group appropriate to molecules with six C_5s, ten C_3s, and fifteen C_2s. More important is $\mathcal{I}_h = \mathcal{I} \times i$, the symmetry of the regular dodecahedron and icosahedron.

$\mathcal{R}^{\pm}(3)$ is the group with an infinity of C_∞s and of σs, passing through the origin in every direction in three-dimensional space. This is the symmetry of the sphere.

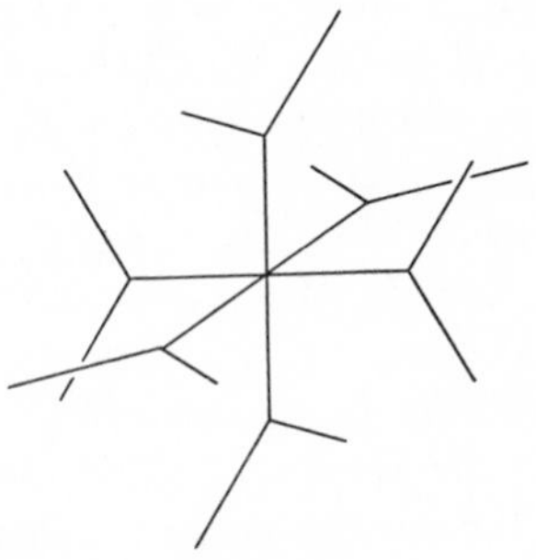

Figure 10.6

Examples of Molecular Symmetries

$\mathcal{C}_1$	HCFClBr, cholesterol.
$\mathcal{C}_2$	H_2O_2, *cis*-decalin.
$\mathcal{C}_3$	H_3CCCl_3, twisted slightly about C—C bond.
$\mathcal{C}_4$	H_8B_5—B_5D_8, twisted slightly about apical B—B bond.
$\mathcal{C}_i$	*meso*-1,2-Dibromo-1,2-dichloroethane (*trans* conformation).
$\mathcal{S}_4$	1,3,5,7-Tetrabromocyclooctatetraene (tub conformation).
$\mathcal{C}_s$	HOCl, HN_3.
$\mathcal{C}_{2h}$	*trans*-1,2-Difluoroethylene, *trans*-decalin.
$\mathcal{C}_{3h}$	1,4-Dibromo-2,5-dichlorobenzene.
$\mathcal{C}_{4h}$	Zinc etioporphyrin I (planar, neglecting hydrogens).
$\mathcal{C}_{6h}$	Hexavinylbenzene (planar).
$\mathcal{C}_{2v}$	H_2O, *cis*-difluoroethylene, 1,1-difluoroethylene, H_2CCl_2.
$\mathcal{C}_{3v}$	NH_3, $HCCl_3$, H_3CCl, $OPCl_3$.
$\mathcal{C}_{4v}$	B_5H_9, $ClSbF_5^-$.
$\mathcal{C}_{\infty v}$	HF, HCN, NO.
$\mathcal{D}_2$	Biphenyl (dihedral angle between benzene rings ca. 40°).
$\mathcal{D}_3$	Triphenylmethyl (propellor form).
$\mathcal{D}_4$	$H_8B_5B_5H_8$, twisted slightly about apical B—B bond.
$\mathcal{D}_{2h}$	Ethylene, naphthalene, *trans*-$Br_2PtCl_2^{2-}$.
$\mathcal{D}_{3h}$	BF_3, cyclopropane.
$\mathcal{D}_{4h}$	$PtCl_4^{2-}$, copper phthalocyanine.
$\mathcal{D}_{5h}$	$C_5H_5^-$, nickelocene (pentagonal prism).
$\mathcal{D}_{6h}$	Benzene.
$\mathcal{D}_{7h}$	Tropylium ion ($C_7H_7^+$).
$\mathcal{D}_{\infty h}$	H_2, CO_2, C_2H_2.
$\mathcal{D}_{2d}$	Allene, cyclooctatetraene (tub form), cyclobutane.
$\mathcal{D}_{3d}$	Ethane (staggered conformation), cyclohexane (chair form).
$\mathcal{D}_{4d}$	$B_{10}H_{10}^{2-}$ (two square pyramids, base to base, twisted 45°).
$\mathcal{D}_{5d}$	Ferrocene (staggered, "pentagonal antiprism").
$\mathcal{T}$	$(H_3C)_4C$ [conformation with each CH_3 group twisted slightly, to the same extent, relative to the $(H_3C)_3C$ group to which it is attached].
$\mathcal{T}_d$	CH_4, adamantane, P_4O_{10}, SO_4^{2-}.
$\mathcal{T}_h$	$Fe(OH_2)_6^{3+}$ in the conformation of Fig. 10.6, pretending $FeOH_2$ group is planar.
$\mathcal{O}_h$	SF_6, cubane, $Fe(CN)_6^{4-}$.
$\mathcal{I}_h$	$B_{12}H_{12}^{2-}$
$\mathcal{R}^{\pm}(3)$	All atoms.

Exercises

28. What is the more commonly used symbol for the group that might be expressed as
    (a) $\mathcal{S}_1$?
    (b) $\mathcal{S}_3$?
    (c) $\mathcal{C}_{1v}$?
    (d) $\mathcal{D}_1$?
    (e) $\mathcal{D}_{1d}$?
    (f) $\mathcal{D}_{1h}$?
29. Which of the groups of order four—$\mathcal{S}_4$, $\mathcal{C}_{2h}$, $\mathcal{C}_{2v}$, and $\mathcal{D}_2$—are isomorphic to $\mathcal{C}_4$ and which are isomorphic to $\mathcal{V}$?
30. Which of the groups of order six—$\mathcal{C}_6$, $\mathcal{S}_6$, $\mathcal{C}_{3h}$, $\mathcal{C}_{3v}$, and $\mathcal{D}_3$—are isomorphic to $\mathscr{P}_3$? Which are cyclic?

*31. Is $\mathcal{D}_{nh} = \mathcal{D}_n \times \sigma_h$? Is $\mathcal{C}_{nh} = \mathcal{C}_n \times \sigma_h$? Is $\mathcal{C}_{nv} = \mathcal{C}_n \times \sigma_v$?

32. List the elements of $\mathcal{D}_3$. Of $\mathcal{D}_4$.
33. List the elements of $\mathcal{D}_{6h}$. Draw the stereographic projection for this group. Show all equivalent points.
34. In $\mathcal{D}_{nd}$, with n C_2's and n σ_ds, what is the spacing in radians between successive C_2's? What is the angle between a C_2' and its nearest σ_d? Convince yourself that the product of C_2' and σ_d, oriented in this fashion, is S_{2n} (or S_{2n}^{2n-1}).
35. Be sure to check over the examples of molecules of $\mathcal{D}_{nd}$ symmetry in order to convince yourself that S_{2n}, n σ_ds, and n C_2's are all present.
36. Draw a cube and indicate one C_4, one S_6, one σ_h, one σ_d, and one C_2'.

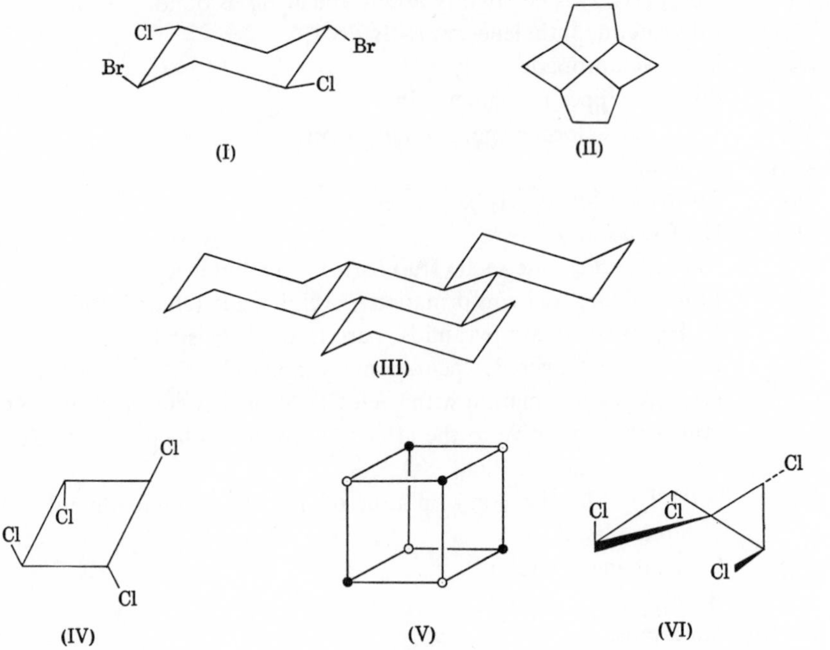

Figure 10.7

37. Draw a tetrahedron and indicate one C_3, one S_4, and one σ_d. Convince yourself that $\mathcal{T}_d$ is isomorphic to $\mathcal{P}_4$.

*38. Stare at a regular icosahedron and find its six S_{10}s, ten S_6s, and fifteen C_2s.

39. What is the order of $\mathcal{T}$? (*Hint:* It is necessary only to add up the distinct elements associated with each axis: C_n, C_n^2, . . . , C_n^{n-1}, and, finally, E.)

*40. What is the order of $\mathcal{O}$? Of $\mathcal{I}$?

41. What is the symmetry of each of the following species?
   (a) *m*-Dinitrobenzene.
   (b) *p*-Dinitrobenzene.
   (c) *m*-Chloronitrobenzene.
   (d) *p*-Chloronitrobenzene.
   (e) PCl_5.
   (f) Diphenylacetylene.
   (g) Cyanogen (C_2N_2).
   (h) 1,3-Dichloroallene.
   (i) S_8 ("crown" form).
   (j) *trans*-1,4-Dichlorocyclohexane.
   (k) Trivinylboron.
   (l) Chlorocubane.
   (m) IF_5 (square pyramid).
   (n) Cyclooctatetraene dianion ($C_8H_8^{2-}$, aromatic).
   (o) *trans*-$Cl_2SiF_4^{2-}$.
   (p) Fe(acac)$_3$: acac$^-$ = $H_3CCOCHCOCH_3^-$ (neglect hydrogens).
   (q) Dibromodichlorocyclohexane isomer (I in Figure 10.7).
   (r) "Twistane" (II).
   (s) All *trans* perhydrotriphenylene (III).
   (t) All *trans* tetrachlorocyclobutane (IV).
   (u) NaCl unit cell (V).
   (v) Tetrachlorospiropentane isomer (VI).

10.6 Conjugate Elements and Classes

Two elements, A and B, in a group $\mathcal{G}$, are said to be CONJUGATE if there exists an element X in $\mathcal{G}$ such that $X^{-1}AX = B$. The set of all elements which are conjugate to each other is called a CLASS. For example, in $\mathcal{C}_{3v}$ the classes are $\{E\}$, $\{C_3, C_3^2\}$, and $\{\sigma_v, \sigma_v', \sigma_v''\}$. In $\mathcal{D}_{4h}$ the classes are $\{E\}$, $\{C_4, C_4^3\}$, $\{C_2\}$, $\{\sigma_v, \sigma_v'\}$, $\{\sigma_d, \sigma_d'\}$, $\{\sigma_h\}$, $\{S_4, S_4^3\}$, $\{i\}$, $\{C_2', C_2'\}$, and $\{C_2'', C_2''\}$.

The physical significance of a class is that symmetry elements of a class are somehow the same sort of operation, but expressed with respect to different coordinate systems. This is what we had in mind when we distinguished C_2's from C_2''s and σ_vs from σ_ds, as "different types" of operations.

We state without proof (Exercise 10.49) the following helpful theorem: The number of elements in any class must be an integral divisor of the order of the group.

Exercises

*42. Convince yourself that $\{C_A\}$, the class to which A belongs, may be generated by forming the products $X^{-1}AX$ with every X in $\mathcal{G}$.

*43. Use the multiplication table for the non-Abelian group of order six to show that the classes are $\{E\}$, $\{A, B\}$, and $\{C, D, F\}$.
44. Show that in any group E forms a class by itself.
45. Show that for an Abelian group of order n there are n classes, each of one element.
46. To understand why conjugate elements are really quite similar, consider the elements of $\mathcal{G}$ as operators that act on bra vectors, $\langle\phi|, \langle\chi|, \ldots$, such that $\langle\phi|A = \langle\chi|$. If all the vectors are subjected to a transformation by X in $\mathcal{G}$, then A is no longer appropriate for converting $\langle\phi|X$ to $\langle\chi|X$. Find the operator that is appropriate.
*47. Use your intuition to convince yourself that the classes of $\mathcal{C}_{5v}$ are $\{E\}$, $\{C_5, C_5^4\}$, $\{C_5^2, C_5^3\}$, and the class of the five σ_vs.
48. Reason by analogy to $\mathcal{D}_{4h}$ to assign elements of $\mathcal{D}_{6h}$ to classes.
*49. Let there be n_A elements of $\mathcal{G}$ which commute with A; call them $\{X_\alpha\}$. Then when forming all products $\{X^{-1}AX\}$ generates $\{C_A\}$, forming the products $\{X_\alpha^{-1}AX_\alpha\}$ generates A exactly n_A times. Show that if B is in $\{C_A\}$, there are exactly n_A elements of $\mathcal{G}$ such that $X^{-1}AX = B$. Then convince yourself that if there are c_A distinct elements in $\{C_A\}$, $c_A = g/n_A$.

10.7 Subgroups

A SUBGROUP of the group $\mathcal{G}$ is a set of elements of $\mathcal{G}$ which themselves form a group with the same binary operation. Obvious and trivial subgroups of any $\mathcal{G}$ are $\mathcal{G}$ itself and the group whose only element is E.

For example, $\mathcal{C}_2 = \{E, C_2\}$ is a subgroup of $\mathcal{C}_4 = \{E, C_4, C_2, C_4^3\}$, but $\{E, C_4\}$ is not, because it is not closed. Subgroups of $\mathcal{D}_2 = \{E, C_2, C_2', C_2''\}$ are three "equivalent" $\mathcal{C}_2$s. Subgroups of $\mathcal{C}_{3v}$ are $\mathcal{C}_3$ and three "equivalent" $\mathcal{C}_s$s. Subgroups of $\mathcal{D}_{3h}$ are $\mathcal{C}_{3v}$, $\mathcal{D}_3$, and their subgroups. Subgroups of $\mathcal{O}_h$ include $\mathcal{O}$, $\mathcal{T}_d$, $\mathcal{D}_{4h}$, and $\mathcal{D}_{3d}$. All the point groups are subgroups of $\mathcal{R}^{\pm}(3)$.

For chemical purposes, the chief significance of subgroups is that they represent the result of a reduction in symmetry when a molecule is distorted or substituted. Thus the symmetry of an octahedral metal complex is reduced to $\mathcal{D}_{4h}$ if two *trans* ligands are changed, either by changing the bond length or changing the ligands.

Also, the SYMMETRY NUMBER, which is involved in the rotational partition function of a molecule, is the order of the PURE-ROTATION SUBGROUP of the point group of the molecule, that is, the subgroup obtained by deleting the reflection operators i, σ, and S_n. (But since rotation about a C_∞ can never be detected, such an axis does not count toward the symmetry number.) Thus the pure-rotation subgroup for benzene is $\mathcal{D}_6$, whose order is 12. The pure-rotation subgroup for methane is $\mathcal{T}$, whose order is 12. The appropriate subgroup of H_2 is $\mathcal{C}_2$, of order 2.

Exercises

*50. Convince yourself that if $\mathcal{G} = \{R_iS_j\} = \mathcal{G}_1 \times \mathcal{G}_2$, then the set $\{R_iE\}$ is a subgroup of $\mathcal{G}$ that is isomorphic to $\mathcal{G}_1$.
51. List the nontrivial subgroups of $\mathscr{P}_3$.

52. Half the elements of $\mathscr{P}_n$ are even permutations, and the other half are odd. Convince yourself that the set of even permutations is a subgroup of $\mathscr{P}_n$. This group is known as the ALTERNATING GROUP.

*53. Let the vectors along the unit cell of an infinite crystal be $\mathbf{t}_1$, $\mathbf{t}_2$, and $\mathbf{t}_3$, and let $\mathbf{t}_1$ also be the operator "displace a distance $|\mathbf{t}_1|$ in the direction $\mathbf{t}_1$," etc. In this context it is conventional to indicate repeated application of operators as addition, rather than as multiplication, so that $\mathbf{t}_2 + \mathbf{t}_1$ is the operator "first displace by $\mathbf{t}_1$ and then displace by $\mathbf{t}_2$" and $2\mathbf{t}_1 = \mathbf{t}_1 + \mathbf{t}_1$, etc.
   (a) Convince yourself that the TRANSLATION GROUP, $\{n_1\mathbf{t}_1 + n_2\mathbf{t}_2 + n_3\mathbf{t}_3\}$, composed of all LATTICE TRANSLATIONS, is a subgroup of the group of all symmetry operators that leave an infinite crystal unchanged.
   (b) Is this subgroup Abelian?
   (c) To avoid dealing with infinite groups, let us impose the fiction that for all $\{n_1, n_2, n_3\}$

$$(n_1 + N_1)\mathbf{t}_1 + (n_2 + N_2)\mathbf{t}_2 + (n_3 + N_3)\mathbf{t}_3 = n_1\mathbf{t}_1 + n_2\mathbf{t}_2 + n_3\mathbf{t}_3,$$

   where N_1, N_2, and N_3 are large but finite integers. This, the BORN-VON KARMAN CYCLIC BOUNDARY CONDITION, corresponds to replacing the infinite crystal by a set of identical finite crystals, each containing $N_1N_2N_3$ unit cells. Convince yourself that $\{n_1\mathbf{t}_1 + n_2\mathbf{t}_2 + n_3\mathbf{t}_3\}$ is now an Abelian group of order $N_1N_2N_3$.

*54. A subgroup $\mathcal{H}$ of $\mathcal{G}$ is said to be INVARIANT if for all R in $\mathcal{G}$ and all S in $\mathcal{H}$, $R^{-1}SR$ is still in $\mathcal{H}$. Convince yourself that $\mathcal{C}_3$ is an invariant subgroup of $\mathcal{C}_{3v}$ but the "equivalent" subgroups $\{E, \sigma_v\}$, $\{E, \sigma_v'\}$, and $\{E, \sigma_v''\}$ are not.

*55. Construct the multiplication table for $\mathcal{C}_{4v}$ by the following procedure: $\mathcal{C}_4$ and two equivalent (not invariant) $\mathcal{C}_{2v}$s are subgroups of $\mathcal{C}_{4v}$; evaluate products within each subgroup and enter them into the $\mathcal{C}_{4v}$ table. From Figure 10.4b $\sigma_d\sigma_v = C_4$; this equation, of the form $AB = C$, may be solved for A, A^{-1}, B, B^{-1}, and C^{-1}. (Pay attention to order. This group is not Abelian!) to yield a total of six more entries. Then use the Latin-square property to complete the table.

*56. The eigenvalue-eigenvector problem for angular momentum leads to solutions, $\Phi_m(\phi) = (2\pi)^{-1/2}e^{im\phi}$. Although the usual boundary condition $\Phi_m(\phi + 2\pi) = \Phi_m(\phi)$ restricts m to integers, half-integral values are allowable in the theory of spinning electrons. Then a rotation of 2π takes $\Phi_m(\phi)$ into $\Phi_m(\phi + 2\pi) = (-1)^{2m}\Phi_m(\phi)$, so that this is not necessarily the same as the identity operation. To deal with this weird situation, it is necessary to work with the so-called DOUBLE GROUP, $\mathcal{G}'$, associated with the point group $\mathcal{G}$. The double group is constructed by adding to $\mathcal{G}$ a new element, R, representing rotation by 2π, and forming all possible products.
   (a) Show that $R^2 = E$.
   (b) What is the order of $\mathcal{G}'$?
   (c) Construct the multiplication table for $\mathcal{D}_2'$: $C_2'C_2' = C_2''C_2'' = C_2C_2 = R$; $C_2C_2' = C_2''$, $C_2'C_2'' = C_2$, $C_2''C_2 = C_2'$; new elements $S = RC_2$, $T = RC_2'$, $U = RC_2''$. (*Caution:* This group is not Abelian!)
   (d) Note that $\mathcal{D}_2'$ is not isomorphic to any of the point groups. (*Hint:* How many elements of $\mathcal{D}_2'$ are equal to their inverses? How many of $\mathcal{C}_{4v}$?)

57. What is the symmetry number of
   (a) NH_3?
   (b) Naphthalene?
   (c) HCN?
   (d) Cyclohexane?
   (e) Methane?

10.8 Representations

Consider a set of (linearly independent) "objects" $\{\phi_j\}_{j=1}^{n}$, each of which may be acted upon by each element R of a group $\mathcal{G}$ to produce a linear combination of the objects: $R\phi_j = \sum_i \phi_i r_{ij}$. The objects are said to form a BASIS for a representation of $\mathcal{G}$. Furthermore, we may arrange the objects as a *row* vector, $\mathbf{\Phi}$, and the objects after application of R as another row vector, $\mathbf{\Phi}_R$, such that to each R in $\mathcal{G}$, we associate the $n \times n$ matrix, $\mathbf{R} = (r_{ij})$, which specifies what R does to $\mathbf{\Phi}$: $\mathbf{\Phi R} = \mathbf{\Phi}_R$ (see Exercise 8.25 for the justification for using row vectors).

Examples of Bases

8. The xs and os of the stereographic projection of $\mathcal{G}$ provide a basis for a representation of $\mathcal{G}$. Each symmetry operation rearranges the objects. Thus for $\mathcal{D}_2$ we may form the row vector $(\mathsf{x}_1 \ \ \mathsf{x}_2 \ \ \mathsf{o}_1 \ \ \mathsf{o}_2)$, which is transformed as follows (Fig. 10.8):

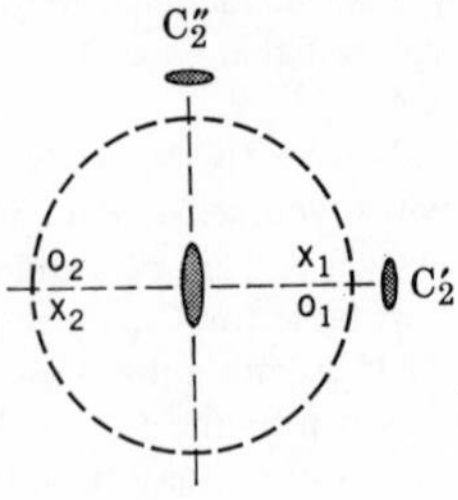

Figure 10.8

$$
\begin{aligned}
(\mathsf{x}_1 \ \ \mathsf{x}_2 \ \ \mathsf{o}_1 \ \ \mathsf{o}_2)E &= (\mathsf{x}_1 \ \ \mathsf{x}_2 \ \ \mathsf{o}_1 \ \ \mathsf{o}_2)\\
(\mathsf{x}_1 \ \ \mathsf{x}_2 \ \ \mathsf{o}_1 \ \ \mathsf{o}_2)C_2 &= (\mathsf{x}_2 \ \ \mathsf{x}_1 \ \ \mathsf{o}_2 \ \ \mathsf{o}_1)\\
(\mathsf{x}_1 \ \ \mathsf{x}_2 \ \ \mathsf{o}_1 \ \ \mathsf{o}_2)C_2' &= (\mathsf{o}_1 \ \ \mathsf{o}_2 \ \ \mathsf{x}_1 \ \ \mathsf{x}_2)\\
(\mathsf{x}_1 \ \ \mathsf{x}_2 \ \ \mathsf{o}_1 \ \ \mathsf{o}_2)C_2'' &= (\mathsf{o}_2 \ \ \mathsf{o}_1 \ \ \mathsf{x}_2 \ \ \mathsf{x}_1)
\end{aligned}
$$

Or we may express the application of each group operation as a 4×4 matrix:

$$
E \text{ as } \begin{pmatrix} 1 & 0 & 0 & 0 \\ 0 & 1 & 0 & 0 \\ 0 & 0 & 1 & 0 \\ 0 & 0 & 0 & 1 \end{pmatrix} \qquad C_2 \text{ as } \begin{pmatrix} 0 & 1 & 0 & 0 \\ 1 & 0 & 0 & 0 \\ 0 & 0 & 0 & 1 \\ 0 & 0 & 1 & 0 \end{pmatrix}
$$

$$
C_2' \text{ as } \begin{pmatrix} 0 & 0 & 1 & 0 \\ 0 & 0 & 0 & 1 \\ 1 & 0 & 0 & 0 \\ 0 & 1 & 0 & 0 \end{pmatrix} \qquad C_2'' \text{ as } \begin{pmatrix} 0 & 0 & 0 & 1 \\ 0 & 0 & 1 & 0 \\ 0 & 1 & 0 & 0 \\ 1 & 0 & 0 & 0 \end{pmatrix}.
$$

9. The two objects, "sum of all the xs" and "sum of all the os," are a basis for a representation of $\mathcal{G}$, since some of the elements of $\mathcal{G}$ exchange xs for os and some do not. For

$\mathcal{D}_2$ we may form the row vector $(x_1 + x_2, o_1 + o_2)$ and express the application of each group operation as a 2×2 matrix:

$$E \text{ as } \begin{pmatrix} 1 & 0 \\ 0 & 1 \end{pmatrix} \quad C_2 \text{ as } \begin{pmatrix} 1 & 0 \\ 0 & 1 \end{pmatrix} \quad C_2' \text{ as } \begin{pmatrix} 0 & 1 \\ 1 & 0 \end{pmatrix} \quad C_2'' \text{ as } \begin{pmatrix} 0 & 1 \\ 1 & 0 \end{pmatrix}.$$

10. A basis for a representation of $\mathcal{C}_{2v}$ is formed by the $2p$ orbitals of an oxygen atom and the $1s$ orbitals of two hydrogen atoms, arranged in a water molecule (Fig. 10.9).

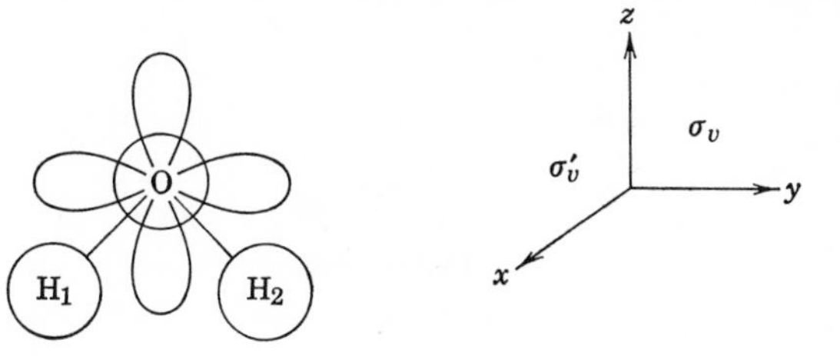

Figure 10.9

The two hydrogen orbitals are interchanged by C_2 and σ_v'. The $2p_z$ orbital of the oxygen is invariant to all operations of $\mathcal{C}_{2v}$. The $2p_x$ orbital is converted to its negative by C_2 and σ_v, and the $2p_y$ orbital is converted to its negative by C_2 and σ_v'. Therefore the row vector, $(s_1 \ s_2 \ p_x \ p_y \ p_z)$, is transformed as follows:

$$\begin{aligned}
(s_1 \ s_2 \ p_x \ p_y \ p_z)E &= (s_1 \ s_2 \ p_x \ p_y \ p_z) \\
(s_1 \ s_2 \ p_x \ p_y \ p_z)C_2 &= (s_2 \ s_1 \ -p_x \ -p_y \ p_z) \\
(s_1 \ s_2 \ p_x \ p_y \ p_z)\sigma_v &= (s_1 \ s_2 \ -p_x \ p_y \ p_z) \\
(s_1 \ s_2 \ p_x \ p_y \ p_z)\sigma_v' &= (s_2 \ s_1 \ p_x \ -p_y \ p_z)
\end{aligned}$$

We may express the application of each group operation as a 5×5 matrix:

$$E \text{ as } \begin{pmatrix} 1 & 0 & 0 & 0 & 0 \\ 0 & 1 & 0 & 0 & 0 \\ 0 & 0 & 1 & 0 & 0 \\ 0 & 0 & 0 & 1 & 0 \\ 0 & 0 & 0 & 0 & 1 \end{pmatrix} \quad C_2 \text{ as } \begin{pmatrix} 0 & 1 & 0 & 0 & 0 \\ 1 & 0 & 0 & 0 & 0 \\ 0 & 0 & -1 & 0 & 0 \\ 0 & 0 & 0 & -1 & 0 \\ 0 & 0 & 0 & 0 & 1 \end{pmatrix}$$

$$\sigma_v \text{ as } \begin{pmatrix} 1 & 0 & 0 & 0 & 0 \\ 0 & 1 & 0 & 0 & 0 \\ 0 & 0 & -1 & 0 & 0 \\ 0 & 0 & 0 & 1 & 0 \\ 0 & 0 & 0 & 0 & 1 \end{pmatrix} \quad \sigma_v' \text{ as } \begin{pmatrix} 0 & 1 & 0 & 0 & 0 \\ 1 & 0 & 0 & 0 & 0 \\ 0 & 0 & 1 & 0 & 0 \\ 0 & 0 & 0 & -1 & 0 \\ 0 & 0 & 0 & 0 & 1 \end{pmatrix}.$$

Thus we see that if a basis is chosen, it is possible to associate matrices to group operations. We next focus on the matrices and define group representations.

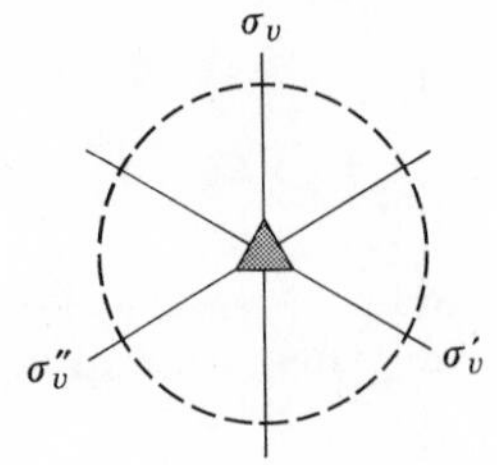

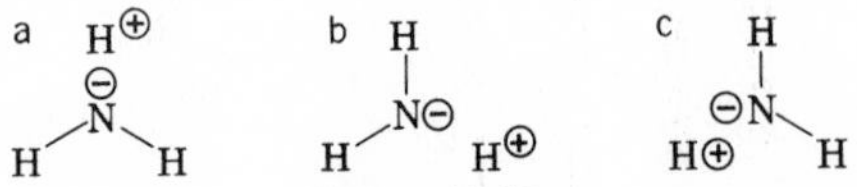

Figure 10.10

A REPRESENTATION of the group $\mathcal{G}$ is a function Γ whose domain is $\mathcal{G}$ and whose range is a set of unitary matrices $\{\mathbf{R}_i\}$ such that if $R_iR_j = R_k$, then $\Gamma(R_i)\Gamma(R_j) = \Gamma(R_k)$, or $\mathbf{R}_i\mathbf{R}_j = \mathbf{R}_k$. Notice that multiplication of the matrices must "mirror" the multiplication of the group elements; the matrices $\{\mathbf{R}_i\}$ are said to FORM a representation of $\mathcal{G}$. This definition differs from the usually accepted one, where the set $\{\mathbf{R}_i\}$ is the representation, but it is less confusing to consider the matrix $\mathbf{R}_i$ as the value of the function Γ at the argument R_i. It is clear, or will be, that there are an infinite number of representations of $\mathcal{G}$; different representations may be distinguished by subscripts. The DIMENSION of a representation is equal to the dimension of the matrices.

For every $\mathcal{G}$ there is always an obvious representation; namely that which associates 1, or the one-dimensional unit matrix, to every element of the group. This representation is denoted by Γ_1 or by A (maybe with subscripts, depending upon $\mathcal{G}$), and it is called the TOTALLY SYMMETRIC representation.

For further examples, we shall consider $\mathcal{C}_{3v}$. A second one-dimensional representation Γ_2 associates 1 with E, C_3, and C_3^2, and associates -1 with σ_v, σ_v', and σ_v''. For a three-dimensional representation, let us use as basis three functions which represent three ionic valence-bond structures for NH_3 (Fig. 10.10). We may form the row vector (a, b, c) and see what each operation of $\mathcal{C}_{3v}$ does to the vector:

$$(a\ b\ c)E = (a\ b\ c) \qquad \text{or} \qquad \mathbf{E} = \begin{pmatrix} 1 & 0 & 0 \\ 0 & 1 & 0 \\ 0 & 0 & 1 \end{pmatrix}$$

$$(a\,b\,c)C_3 = (c\,a\,b) \quad \text{or} \quad \mathbf{C}_3 = \begin{pmatrix} 0 & 1 & 0 \\ 0 & 0 & 1 \\ 1 & 0 & 0 \end{pmatrix}$$

$$(a\,b\,c)C_3^2 = (b\,c\,a) \quad \text{or} \quad \mathbf{C}_3^2 = \begin{pmatrix} 0 & 0 & 1 \\ 1 & 0 & 0 \\ 0 & 1 & 0 \end{pmatrix}$$

$$(a\,b\,c)\sigma_v = (a\,c\,b) \quad \text{or} \quad \boldsymbol{\sigma}_v = \begin{pmatrix} 1 & 0 & 0 \\ 0 & 0 & 1 \\ 0 & 1 & 0 \end{pmatrix}$$

$$(a\,b\,c)\sigma_v' = (c\,b\,a) \quad \text{or} \quad \boldsymbol{\sigma}_v' = \begin{pmatrix} 0 & 0 & 1 \\ 0 & 1 & 0 \\ 1 & 0 & 0 \end{pmatrix}$$

$$(a\,b\,c)\sigma_v'' = (b\,a\,c) \quad \text{or} \quad \boldsymbol{\sigma}_v'' = \begin{pmatrix} 0 & 1 & 0 \\ 1 & 0 & 0 \\ 0 & 0 & 1 \end{pmatrix}.$$

As a check, $\mathbf{C}_3\boldsymbol{\sigma}_v = \boldsymbol{\sigma}_v'$, as required (Exercise 10.24c).

Exercises

58. An arbitrary even function e and an arbitrary odd function o form a basis for a representation of $\mathcal{C}_i$, where i is the operation such that $if(x) = f(-x)$. Find the 2×2 matrix associated with each element of $\mathcal{C}_i$.
59. The two functions x and y form a basis for a representation of $\mathcal{C}_{4v}$ since each operation of $\mathcal{C}_{4v}$ takes each function into some linear combination of the functions. For example, $C_4x = y$ and $C_4y = -x$, so $(x \quad y)C_4 = (y \quad -x)$. Find the eight 2×2 matrices which this representation associates with the elements of $\mathcal{C}_{4v}$. Check that $\boldsymbol{\sigma}_d\boldsymbol{\sigma}_v = \mathbf{C}_4$ (cf. Fig. 10.4b). Find a pair of matrices which do not commute.
60. The three functions x^2, y^2, and xy also form a basis for a representation of $\mathcal{C}_{4v}$. For example, $(x^2 \quad y^2 \quad xy)C_4 = (y^2 \quad x^2 \quad -xy)$. Find the 3×3 matrix which this basis associates with each element of $\mathcal{C}_{4v}$. Can you find a pair of matrices which do not commute?
61. Show that for any representation, $\Gamma(E) = \mathbf{1}$.
62. Convince yourself that the representation Γ has an inverse function if and only if every matrix (value of Γ) is different. Such a representation is said to be FAITHFUL.
63. Convince yourself that if Γ is a representation of $\mathcal{G}$, so is Γ^*, where $\Gamma^*(R) = \Gamma(R)^*$
64. The REGULAR REPRESENTATION is that representation which uses the xs and os of the stereographic projection as basis. What is the regular representation for $\mathcal{C}_{3v}$?

65. Show that if every matrix of a representation of $\mathcal{G}$ is subjected to the same similarity transformation, the set of transformed matrices also forms a representation of $\mathcal{G}$. These two representations are said to be EQUIVALENT.
66. The DIRECT SUM $\Gamma_\alpha + \Gamma_\beta$ of two representations Γ_α and Γ_β of $\mathcal{G}$ is defined by $(\Gamma_\alpha + \Gamma_\beta)(R_i) \equiv \Gamma_\alpha(R_i) \oplus \Gamma_\beta(R_i)$, the direct sum of the matrices $\Gamma_\alpha(R_i)$ and $\Gamma_\beta(R_i)$. Convince yourself that $\Gamma_\alpha + \Gamma_\beta$ is also a representation of $\mathcal{G}$. Convince yourself that the converse is also true: If $\Gamma_\alpha + \Gamma_\beta$ is a representation, so are Γ_α and Γ_β individually.
67. Infinitesimal displacements of the chlorine atoms of $PtCl_4^{2-}$ form a basis for a representation of $\mathcal{D}_{4h}$. These may be drawn as little arrows, three per chlorine, one pointing radially, one pointing tangentially, and one pointing vertically; arrows on the same atom, but pointing in opposite directions, are the negative of each other. Find the representation. (*Hint:* To avoid writing sixteen 12×12 matrices, consider this twelve-dimensional representation as the direct sum of three four-dimensional ones, one for radial displacements, one for tangential displacements, and one for out-of-plane displacements. Also, you need only write the matrices $\mathbf{C}_4$, $\boldsymbol{\sigma}_v$, and $\boldsymbol{\sigma}_h$ since then matrix multiplication gives the remaining matrices.)

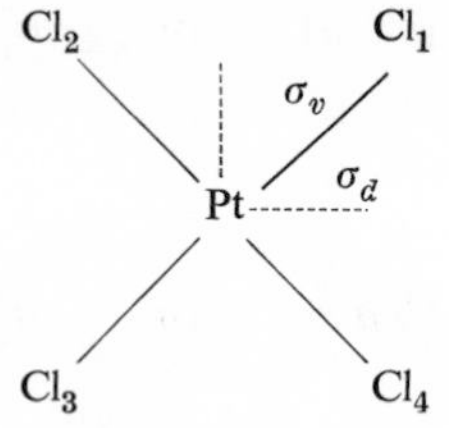

Figure 10.11

68. The DIRECT PRODUCT $\Gamma_\alpha \times \Gamma_\beta$ of two representations Γ_α and Γ_β of $\mathcal{G}$ is defined by $(\Gamma_\alpha \times \Gamma_\beta)(R_i) \equiv \Gamma_\alpha(R_i) \times \Gamma_\beta(R_i)$, the direct product of the matrices $\Gamma_\alpha(R_i)$ and $\Gamma_\beta(R_i)$.
    (a) Use the result of Exercise 8.52 to convince yourself that $\Gamma_\alpha \times \Gamma_\beta$ is also a representation of $\mathcal{G}$.
    (b) Convince yourself that if $\{\phi_i\}$ is a basis for Γ_α and $\{\chi_j\}$ is a basis for Γ_β, and if the set $\{\phi_i\chi_j\}$ of all products is linearly independent, then $\{\phi_i\chi_j\}$ is a basis for $\Gamma_\alpha \times \Gamma_\beta$.
69. For any group $\mathcal{G}$, the single object "sum of all the xs and os of the stereographic projection," is a basis for a representation of $\mathcal{G}$. Which representation?
*70. Convince yourself that the Γ_2 given for $\mathcal{C}_{3v}$ is indeed a representation. Find a basis for this representation.
71. Find two one-dimensional representations of $\mathscr{P}_n$.

Since it is less convenient to work with matrices than with scalars, let us try to diagonalize these matrices so as to obtain three one-dimensional representations (cf. Exercises 10.65 and 10.66). Observe, though, that this is a non-Abelian group and that the matrices do not all commute. Therefore it is not possible to diagonalize all six matrices simultaneously. The best we can do is to convert the matrices to block-diagonal form. There is no unique block diagonalization, so we choose, arbitrarily, to transform by the unitary matrix

$$\mathbf{X} = \begin{Vmatrix} \frac{1}{\sqrt{3}} & 0 & \frac{2}{\sqrt{6}} \\ \frac{1}{\sqrt{3}} & \frac{1}{\sqrt{2}} & -\frac{1}{\sqrt{6}} \\ \frac{1}{\sqrt{3}} & -\frac{1}{\sqrt{2}} & -\frac{1}{\sqrt{6}} \end{Vmatrix}.$$

Then by matrix multiplication the transformed matrices are

$$\mathbf{X}^{-1}\mathbf{E}\mathbf{X} = \begin{pmatrix} 1 & 0 & 0 \\ 0 & 1 & 0 \\ 0 & 0 & 1 \end{pmatrix} \qquad \mathbf{X}^{-1}\boldsymbol{\sigma}_v\mathbf{X} = \begin{pmatrix} 1 & 0 & 0 \\ 0 & -1 & 0 \\ 0 & 0 & 1 \end{pmatrix}$$

$$\mathbf{X}^{-1}\mathbf{C}_3\mathbf{X} = \begin{pmatrix} 1 & 0 & 0 \\ 0 & -\frac{1}{2} & -\frac{\sqrt{3}}{2} \\ 0 & \frac{\sqrt{3}}{2} & -\frac{1}{2} \end{pmatrix} \qquad \mathbf{X}^{-1}\mathbf{C}_3^2\mathbf{X} = \begin{pmatrix} 1 & 0 & 0 \\ 0 & -\frac{1}{2} & \frac{\sqrt{3}}{2} \\ 0 & -\frac{\sqrt{3}}{2} & -\frac{1}{2} \end{pmatrix}$$

$$\mathbf{X}^{-1}\boldsymbol{\sigma}_v'\mathbf{X} = \begin{pmatrix} 1 & 0 & 0 \\ 0 & \frac{1}{2} & -\frac{\sqrt{3}}{2} \\ 0 & -\frac{\sqrt{3}}{2} & -\frac{1}{2} \end{pmatrix} \qquad \mathbf{X}^{-1}\boldsymbol{\sigma}_v''\mathbf{X} = \begin{pmatrix} 1 & 0 & 0 \\ 0 & \frac{1}{2} & \frac{\sqrt{3}}{2} \\ 0 & \frac{\sqrt{3}}{2} & -\frac{1}{2} \end{pmatrix}.$$

Thus the three-dimensional representation has been block diagonalized to a one-dimensional representation, which is Γ_1 again, and a new two-dimensional representation Γ_3.

The process of simultaneous diagonalization or block diagonalization of the matrices of a representation is known as REDUCTION. The three-dimensional representation is REDUCIBLE. A one-dimensional representation is necessarily IRREDUCIBLE. The two-dimensional representation, Γ_3, is also irreducible. In some contexts an irreducible representation is known as a SYMMETRY SPECIES.

For future reference we list all three irreducible representations of $\mathcal{C}_{3v}$:

	E	C_3	C_3^2	σ_v	σ_v'	σ_v''
Γ_1	1	1	1	1	1	1
Γ_2	1	1	1	-1	-1	-1
Γ_3	$\begin{pmatrix}1 & 0\\0 & 1\end{pmatrix}$	$\begin{pmatrix}\frac{-1}{2} & \frac{-\sqrt{3}}{2}\\ \frac{\sqrt{3}}{2} & \frac{-1}{2}\end{pmatrix}$	$\begin{pmatrix}\frac{-1}{2} & \frac{\sqrt{3}}{2}\\ \frac{-\sqrt{3}}{2} & \frac{-1}{2}\end{pmatrix}$	$\begin{pmatrix}-1 & 0\\0 & 1\end{pmatrix}$	$\begin{pmatrix}\frac{1}{2} & \frac{-\sqrt{3}}{2}\\ \frac{-\sqrt{3}}{2} & \frac{-1}{2}\end{pmatrix}$	$\begin{pmatrix}\frac{1}{2} & \frac{\sqrt{3}}{2}\\ \frac{\sqrt{3}}{2} & \frac{-1}{2}\end{pmatrix}$.

It can be shown that these are all the irreducible representations there are for $\mathcal{C}_{3v}$. Of course, other forms of Γ_3 may be obtained by applying a similarity transformation to each matrix. But such forms do not constitute a different irreducible representation, and are therefore considered to be equivalent.

Exercises

*72. "Irreducible representation" is much too long a name for something that is used so frequently in practical applications of group theory.
(a) Think up a shorter name.
(b) Get it accepted universally.

73. Convince yourself that if $\{\phi_i\}$ is a basis for the representation Γ_α of $\mathcal{G}_1$ and $\{\chi_j\}$ is a basis for the representation Γ_β of $\mathcal{G}_2$, the set $\{\phi_i\chi_j\}$ of all possible products is a basis for a representation Γ of $\mathcal{G}_1 \times \mathcal{G}_2$, where $\Gamma(R_iS_j) = \Gamma_\alpha(R_i) \times \Gamma_\beta(S_j)$. This result may be used to construct the irreducible representations of the groups $\mathcal{D}_{nh}$ $\mathcal{T}_h$, $\mathcal{O}_h$, and $\mathcal{I}_h$. (This is a different situation from that of Exercise 10.68, where there is only one group.)

74. Show that all irreducible representations of an Abelian group are one-dimensional.

75. Show that no faithful representation of a non-Abelian group can be completely diagonalized to a direct sum of only one-dimensional representations.

*76. Convince yourself that the irreducible representations of the finite translation group of Exercise 10.53 are all one-dimensional and given by

$$\Gamma_{\mathbf{k}}(\mathbf{T}) = \exp(i\mathbf{k} \cdot \mathbf{T}),$$

where $\mathbf{T} = n_1\mathbf{t}_1 + n_2\mathbf{t}_2 + n_3\mathbf{t}_3$ and $\mathbf{k} = 2\pi[(K_1/N_1)\boldsymbol{\tau}_1 + (K_2/N_2)\boldsymbol{\tau}_2 + (K_3/N_3)\boldsymbol{\tau}_3]$, with $K_i = 0, \pm 1, \pm 2, \ldots, \pm(N_i - 1)/2, N_i/2$. Notice that since lattice vectors of crystals are not necessarily orthonormal, we must expand $\mathbf{k}$ in the basis $\{\boldsymbol{\tau}_i\}$, reciprocal to $\{\mathbf{t}_i\}$.

77. Reduce the four-dimensional representation of $\mathcal{D}_2$ given in Example 10.8 to a direct sum of four one-dimensional representations.

*78. Obtain another block diagonalization of the three-dimensional representation of $\mathcal{C}_{3v}$ by noting that since $\mathcal{C}_3$ is an Abelian subgroup of $\mathcal{C}_{3v}$, $\mathbf{E}$, $\mathbf{C}_3$, and $\mathbf{C}_3^2$ may be diagonalized simultaneously. Find the transformation that converts $\mathbf{C}_3$ to diag $(1, \omega, \omega^2)$, where $\omega = \exp(2\pi i/3)$. Then apply this transformation to the other five matrices to obtain Γ_1 and an equivalent form of Γ_3.

79. Reduce the three-dimensional representation of $\mathcal{C}_{4v}$ found in Exercise 10.60.

10.9 Transformation of Basis

Let $\{\phi_j\}$ be a set of functions which form a basis for a representation Γ of $\mathcal{G}$, which associates the unitary matrix $\mathbf{R}$ with group element R: $R\phi_j = \sum_i \phi_i(\mathbf{R})_{ij}$. (If what follows seems too abstract, use the *abc* basis for the reducible three-dimensional representation of $\mathcal{C}_{3v}$ as an example.) We have seen that in general the operation R mixes the functions: $\mathbf{\Phi R} = \mathbf{\Phi}_R$. And we may next ask the question, "Which linear combination of these functions is not mixed, but is merely multiplied by a constant?" This is equivalent to the problem of simultaneously diagonalizing the $\mathbf{R}$ matrices. But if $\mathcal{G}$ is non-Abelian, the matrices may not all commute, so that only a block diagonalization may be possible.

Let $\mathbf{X}$ be a matrix that maximally diagonalizes the $\mathbf{R}$ matrices: $\mathbf{X}^{-1}\mathbf{RX} = \mathbf{R}'$ and $\mathbf{\Phi X R}' = \mathbf{\Phi}_R\mathbf{X}$. Then $(\mathbf{\Phi X})_j$ is a new basis function ϕ_j', and each column of $\mathbf{X}$ is the set of coefficients of a new basis function as a linear combination of the old ones. [For example, for the three-dimensional representation of $\mathcal{C}_{3v}$, the new basis functions are $(a + b + c)/\sqrt{3}$, $(b - c)/\sqrt{2}$, and $(2a - b - c)/\sqrt{6}$.] Thus by a transformation of basis, we may reduce the representation Γ to a new representation Γ' which associates the block-diagonal matrix $\mathbf{R}'$ to group element R. And since the matrices have been maximally diagonalized, Γ' must be a direct sum of irreducible representations, each of which associates to R some diagonal submatrix of $\mathbf{R}'$.

Let us first concentrate on a one-dimensional representation. Let $(\mathbf{R}')_{jj}$ be a 1×1 diagonal submatrix of $\mathbf{R}'$. Then the set of such numbers forms a one-dimensional representation of $\mathcal{G}$. Furthermore $(\mathbf{R}')_{jj}$ is an eigenvalue of the unitary matrix $\mathbf{R}$ and must be of magnitude unity. And the jth column of $\mathbf{X}$ is an eigenvector of $\mathbf{R}$ belonging to that eigenvalue. Therefore we may conclude that the new basis function $\phi_j' = (\mathbf{\Phi X})_j$ must be an eigenfunction of R such that operating R on ϕ_j' merely multiplies it by a phase factor. Indeed, this linear combination of the old basis functions is an eigenfunction of *every* R in $\mathcal{G}$, with eigenvalue $(\mathbf{R}')_{jj}$, despite the possible non-Abelian character of $\mathcal{G}$. Such an eigenfunction is said to BELONG to that irreducible representation which assigns $(\mathbf{R}')_{jj}$ to R, or to TRANSFORM AS that irreducible representation. [The linear combination $(a + b + c)/\sqrt{3}$ is an eigenfunction of every operation of $\mathcal{C}_{3v}$, with eigenvalue one. It is said to belong to Γ_1 or to transform as Γ_1.]

Multidimensional irreducible representations are less simple. They arise because it is not always possible for every new basis function to be a simultaneous eigenfunction of every R in $\mathcal{G}$. An m-dimensional irreducible representation corresponds to a set of m new basis functions such that application of R to any of them produces a linear combination of those m functions. [Thus, in the $\mathcal{C}_{3v}$ example, $(b - c)/\sqrt{2} = b'$ and $(2a - b - c)/\sqrt{6} = c'$ are eigenfunctions of E with eigenvalue 1 and eigenfunctions of σ_v with eigenvalue -1 and 1, respectively. Each is converted to a linear combination of the pair by every other operation of $\mathcal{C}_{3v}$, as specified by the 2×2 matrices of Γ_3. For example, $(b', c')\sigma_v' = ((b - a)/\sqrt{2}, (2c - b - a)/\sqrt{6}) = (\frac{1}{2}b' - (\sqrt{3}/2)c', -(\sqrt{3}/2)b' - \frac{1}{2}c')$, as indicated in the 2×2 matrix $\Gamma_3(\sigma_v)$. But no operation of $\mathcal{C}_{3v}$ mixes either b' or c' with $(a + b + c)/\sqrt{3}$. The pair is said to belong to Γ_3, or to transform as Γ_3.]

In summary, it is sometimes possible to find a linear combination of the basis functions which is an eigenfunction of every R in $\mathcal{G}$, with eigenvalues given by a one-dimensional irreducible representation. The linear combination is said to transform as that representation. And it is sometimes necessary to find sets of linear combinations of the basis functions which are transformed among themselves by the elements of $\mathcal{G}$, but are not transformed into other linear combinations. The transformation properties are given by the matrices of a multidimensional irreducible representation, and the set of linear combinations is said to transform as that irreducible representation.

Exercises

80. Show that if $|\phi\rangle$ transforms as some one-dimensional representation of $\mathcal{G}$, and if R in $\mathcal{G}$ is such that $R = R^{-1}$, then either $R|\phi\rangle = |\phi\rangle$ or $R|\phi\rangle = -|\phi\rangle$. Under these circumstances $|\phi\rangle$ is said to be SYMMETRIC or ANTISYMMETRIC, respectively, under R. (cf. Exercise 10.17.)
81. What linear combinations of the xs and os of the stereographic projection of $\mathcal{D}_2$ are "eigenobjects" of every symmetry operation of $\mathcal{D}_2$?
82. What is the result of operating C_3 on $(b - c)/\sqrt{2}$? What linear combination of $(b - c)/\sqrt{2}$ and $(2a - b - c)/\sqrt{6}$ is this? Check your result with the 2×2 matrix $\Gamma_3(C_3)$.
83. Convince yourself that if ϕ_1 and ϕ_2 both belong to irreducible representation Γ, so does any linear combination of ϕ_1 and ϕ_2.
*84. If ϕ transforms as the irreducible representation Γ, as what irreducible representation does ϕ^* transform? (*Note:* These irreducible representations are different only if $\mathcal{G}$ is cyclic of order greater than 2.)

10.10 Factorization of Secular Equations

We wish to show generally how ideas of symmetry may be used to simplify the eigenvalue problem $H|E\rangle = E|E\rangle$ or $H\psi = E\psi$. If we choose an orthonormal basis $\{|i\rangle\}$ or $\{\phi_i\}$, we are led to the matrix equation $\mathbf{H}\boldsymbol{\psi} = E\boldsymbol{\psi}$, where $(\mathbf{H})_{ij} = \langle i|\,H\,|j\rangle$ or $\int \phi_i^*(x)H\phi_j(x)\,dx$ and $(\boldsymbol{\psi})_i = \langle i \mid E\rangle$ or $\int \phi_i^*(x)\psi(x)\,dx$. What we wish to show now is that if there are some operators that commute with H, then it is possible to find a basis set such that $\mathbf{H}$ is in block-diagonal form.

Let us first investigate the nature of an operator that commutes with H. Let R be an operator that acts on a function merely by transforming the argument: $Rf(x) = f(x')$, where f is an arbitrary function. (Examples 9.11 through 9.16 are of this general type.) Furthermore, let H be the same whether it is written in terms of the old variable x or of the new variable x': $H(x) = H(x')$. Then $H(x)Rf(x) = H(x')Rf(x) = H(x')f(x') = RH(x)f(x)$. But since f is arbitrary, we may conclude that H and R commute.

Examples

11. The operator "multiply by x^2" commutes with the inversion operator I, for which $If(x) = f(-x)$, since the operator x^2 is the same even when written in terms of the new variable $-x$: $x^2 = (-x)^2$, or since $x^2If(x) = x^2f(-x) = (-x)^2f(-x) = Ix^2f(x)$.

12. The differentiation operator D commutes with the displacement operator Δ_c, defined by $\Delta_c f(x) = f(x - c)$, since $d/dx = d/d(x - c)$, or since $D\Delta_c f(x) = Df(x - c) = f'(x - c) = \Delta_c f'(x) = \Delta_c Df(x)$.

13. The Hamiltonian $H = -(\hbar^2/2m)\nabla^2 - (e^2/r)$ for the hydrogen atom commutes with the rotation operator C_n about the z-axis since ϕ enters into ∇^2 only as $\partial^2/\partial\phi^2$, and ∇^2 is thus unchanged upon replacing ϕ by $\phi - (2\pi/n)$. Indeed, H must commute with every element of the group $\mathcal{R}^{\pm}(3)$.

14. The Hamiltonian for the helium atom commutes with the permutation operator P_{12}, which interchanges the coordinates of the two electrons:

$$P_{12}F(r_1, \theta_1, \phi_1, r_2, \theta_2, \phi_2) = F(r_2, \theta_2, \phi_2, r_1, \theta_1, \phi_1).$$

Indeed, the Hamiltonian for an n-electron atom commutes with every element of $\mathcal{S}_n$ since H is a sum over all electrons plus a sum over all pairs of electrons, and any permutation merely rearranges the terms in the sums.

15. The electronic Hamiltonian for the NH_3 molecule commutes with the operator C_3, which rotates the nuclei 120° about the lone-pair axis, since the electrons still see the same potential after the rotation. Indeed, this Hamiltonian commutes with every element of $\mathcal{C}_{3v}$ applied to the nuclei.

16. The total Hamiltonian for the NH_3 molecule commutes not only with every element of $\mathcal{C}_{3v}$ applied to the nuclei, but also with every element of $\mathscr{P}_{10}$ permuting the electrons. Furthermore, this Hamiltonian commutes with every product of an element of $\mathcal{C}_{3v}$ and an element of $\mathscr{P}_{10}$, so that it must commute with every element of the direct-product group $\mathcal{C}_{3v} \times \mathscr{P}_{10}$.

17. Since rotating the entire NH_3 molecule does not change the interactions of the various particles, the total Hamiltonian for NH_3 also commutes with every element of $\mathcal{R}^{\pm}(3)$, so it must commute with every element of $\mathcal{C}_{3v} \times \mathscr{P}_{10} \times \mathcal{R}^{\pm}(3)$.

In summary, if H remains unchanged when it is written in terms of a new set of variables, then H commutes with the operator that effects the transformation to those variables. And if two operators R_1 and R_2 commute with H, so must their products R_1R_2 and R_2R_1. Thus we are led to the idea of a group of operators, each of which commutes with H (but not necessarily with each other).

Previously (Section 8.14) we concluded that the basis set which block-diagonalizes H is one composed of eigenvectors of an operator R that commutes with H. (Actually, we demonstrated this rule only for matrices and transformations, but the extension to operators should be obvious.) The reason why this approach works is that eigenvectors of R belonging to different eigenvalues cannot be coupled by H, so that the matrix $\mathbf{H}$ contains one diagonal submatrix for each distinct eigenvalue of R. The only new feature is that now there is a whole group $\mathcal{G}$ of operators, each of which commutes with H. We may still conclude that if R is any member of $\mathcal{G}$, eigenvectors of R belonging to different eigenvalues cannot be coupled by H. What we would like to do is construct basis vectors that are eigenvectors of every R in $\mathcal{G}$: $R_j|i\rangle = r_{ij}|i\rangle$. Then two of these basis vectors that belong to different eigenvalues of even one element of $\mathcal{G}$ cannot be coupled by H, and the matrix $\mathbf{H}$ would contain one diagonal submatrix for each distinct set $\{r_{ij}\}_{j=1}^{g}$ of eigenvalues. Unfortunately, $\mathcal{G}$ is not necessarily Abelian, so it is not always possible to construct basis vectors that are eigenvectors of every R in $\mathcal{G}$. Although some basis vectors may be constructed to transform as one-dimensional irreducible representations of $\mathcal{G}$, there are other basis vectors that will transform as multidimensional irreducible representations of $\mathcal{G}$. However, it is possible to show that basis vectors that transform as different irreducible representations cannot be coupled by H: Let $|\alpha\rangle$ transform as Γ_α and $|\beta\rangle$ as $\Gamma_\beta \neq \Gamma_\alpha$. If Γ_α and Γ_β are one-dimensional, for all R we may conclude that $R|\alpha\rangle = \Gamma_\alpha(R)|\alpha\rangle$ and $R|\beta\rangle = \Gamma_\beta(R)|\beta\rangle$. But there must exist at least one R in $\mathcal{G}$ such that $\Gamma_\alpha(R) \neq \Gamma_\beta(R)$, for if there were none, Γ_α and Γ_β would be identical. Therefore we conclude that $|\alpha\rangle$ and $|\beta\rangle$ are eigenvectors of this R, but with different eigenvalues, and they therefore cannot be coupled by H. It can further be shown that basis vectors that transform as different multidimensional irreducible representations cannot be coupled by H, but the general proof is tedious. Let us instead note that in the *abc* basis for NH_3, the function $a' = (a + b + c)/\sqrt{3}$ cannot be coupled with $b' = (b - c)/\sqrt{2}$, since a' and b' are eigenfunctions of σ_v with different eigenvalues $+1$ and -1, respectively. Next let us write $c' = (a - b)/\sqrt{6} + (a - c)/\sqrt{6}$. Since a' is symmetric with respect to both σ_v'' and σ_v', whereas $(a - b)/\sqrt{6}$ and $(a - c)/\sqrt{6}$ are antisymmetric with respect to σ_v'' or σ_v', respectively, we may conclude that neither $(a - b)/\sqrt{6}$ nor $(a - c)/\sqrt{6}$, nor their sum c', can be coupled with a'. Notice also that b' and c' are respectively antisymmetric and symmetric with respect to σ_v, so they cannot be coupled to each other. In Example 10.23 we shall consider the coupling of different components of the same irreducible representation.

Thus we may conclude that vectors that transform as different irreducible representations of $\mathcal{G}$ cannot be coupled by any operator H that commutes with every element of $\mathcal{G}$. Therefore we should construct our basis vectors to transform as irreducible representations of $\mathcal{G}$, so that the matrix $\mathbf{H}$ will be in block-diagonal form, with one diagonal submatrix for each irreducible representation of $\mathcal{G}$. In Section 10.17 we present a simple method for constructing such basis vectors without the necessity of diagonalizing representation matrices.

Finally, let us return to the eigenvectors of H rather than the convenient basis vectors. The theorem underlying the previous discussion is one concerning simultaneous diagonalization of commuting operators. Although H commutes with every element of $\mathcal{G}$, we cannot automatically conclude that an eigenvector $|E\rangle$ of H is simultaneously an eigenvector of every R in $\mathcal{G}$, since $\mathcal{G}$ is not necessarily Abelian. What we can conclude is that $|E\rangle$ must transform as some irreducible representation of $\mathcal{G}$. This is a very important result. Even though E and $|E\rangle$ may be known only approximately, it is sometimes possible to draw conclusions solely on the basis of knowing the irreducible representation to which $|E\rangle$ belongs. Indeed, arguments based upon symmetry are among the most powerful of all. We next turn to one such.

Exercises

85. In the NH_3 example, if $\langle a| H |a\rangle = Q$ and $\langle a| H |b\rangle = J$, what is the matrix **H** in the *abc* basis? What is **H** in the new basis set of functions transforming as irreducible representations of $\mathcal{C}_{3v}$?
86. What linear combinations of 1*s* orbitals of the hydrogens in H_2O are eigenfunctions of every operation of $\mathcal{C}_{2v}$? With which oxygen orbital(s) can each linear combination be coupled by the Hamiltonian?

10.11 Degeneracy

Let $|E_1\rangle$ and $|E_2\rangle$ be a pair of eigenvectors of H which transform as some two-dimensional irreducible representation Γ of $\mathcal{G}$. We wish to show that the eigenvalues E_1 and E_2 are identical.

Since Γ is two-dimensional and irreducible, there must be some R in $\mathcal{G}$ which mixes $|E_1\rangle$ and $|E_2\rangle$: $R|E_1\rangle = r_{11}|E_1\rangle + r_{12}|E_2\rangle$, with $r_{12} \neq 0$. Apply R to $H|E_1\rangle = E_1|E_1\rangle$, invoke $RH = HR$, and substitute for $R|E_1\rangle$:

$$HR|E_1\rangle = E_1R|E_1\rangle$$

$$H(r_{11}|E_1\rangle + r_{12}|E_2\rangle) = E_1(r_{11}|E_1\rangle + r_{12}|E_2\rangle).$$

Collecting terms gives

$$r_{12}(H|E_2\rangle - E_1|E_2\rangle) = -r_{11}(H|E_1\rangle - E_1|E_1\rangle).$$

But the right-hand side is zero and $r_{12} \neq 0$, so we conclude that

$$H|E_2\rangle = E_1|E_2\rangle.$$

Thus $|E_2\rangle$ is also an eigenvector of H with eigenvalue E_1, so $|E_1\rangle$ and $|E_2\rangle$

are indeed degenerate. By similar reasoning, three eigenvectors which transform as some three-dimensional irreducible representation of $\mathcal{G}$ are triply degenerate, etc.

Note the chain of reasoning: If $\mathcal{G}$ is non-Abelian, there must exist multidimensional irreducible representations. If so, there exist sets of degenerate eigenvectors of $\mathbf{H}$.

One necessary feature of a non-Abelian point group is a threefold or higher axis. However, even in the Abelian groups, $\mathcal{C}_n$, $\mathcal{S}_n$, $\mathcal{C}_{nh}$, when $n > 2$ and $H = H^*$, there are pairs of eigenfunctions which are complex conjugates of each other, and therefore degenerate. Therefore we may conclude that a C_n or S_n with $n > 2$ is a sufficient feature for the existence of some degenerate eigenvalues.

Exercises

87. Label the atomic orbitals of cyclobutadiene sequentially as χ_1, χ_2, χ_3, and χ_4; the linear combinations $\phi_1 = (\chi_1 - \chi_3)/\sqrt{2}$ and $\phi_2 = (\chi_2 - \chi_4)/\sqrt{2}$ form a basis for a two-dimensional irreducible representation of $\mathcal{D}_{4h}$. Let ϕ_1 be an eigenfunction of some operator H. Find an element of $\mathcal{D}_{4h}$ which mixes ϕ_1 and ϕ_2 and show that ϕ_1 and ϕ_2 are degenerate.

*88. Show that if ψ is an eigenfunction of a *real* Hermitian operator, ψ^* is degenerate with ψ.

10.12 Designation of Irreducible Representations

Rather than list irreducible representations as $\Gamma_1, \Gamma_2, \ldots$, in some arbitrary order, a more descriptive notation has been developed for designating irreducible representations of the point groups.

One-dimensional representations are designated by A or B. Designations E, F (or T), G, and H are used for two-, three-, four-, and five-dimensional irreducible representations, respectively. The Abelian groups $\mathcal{C}_n$, $\mathcal{S}_n$, and $\mathcal{C}_{nh}$ have some irreducible representations which occur in pairs that are complex conjugates of each other (Exercise 10.89); these pairs are often called E, as though they were two-dimensional irreducible representations.

The choice of A or B depends upon the (one-dimensional) matrix which represents the major C_n (or S_n). If $\mathbf{C}_n$ (or $\mathbf{S}_n$) is 1, the representation is A; if it is -1, the representation is B.

In $\mathcal{C}_{nv}$ a subscript 1 or 2 is added, according as $\boldsymbol{\sigma}_v = 1$ or -1. In $\mathcal{D}_n$, $\mathcal{D}_{nd}$, and $\mathcal{D}_{nh}$ a subscript 1 or 2 is added, according as $\mathbf{C}_2' = 1$ or -1. Since there is no major C_2 in $\mathcal{D}_2$ (or $\mathcal{D}_{2h}$), the representations are often called A, B_1, B_2, and B_3, rather than A_1, A_2, B_1, and B_2.

Multidimensional irreducible representations are labeled with a subscript that corresponds loosely to an angular momentum quantum number. If $\mathbf{C}_n^{n/m} = \mathbf{1}$, the irreducible representation is E_m, F_m, etc. (cf. Exercise 10.90).

In $\mathcal{C}_{\infty v}$ and $\mathcal{D}_{\infty h}$, A becomes Σ, and $E_1, E_2, \ldots,$ become Π, Δ, Φ, etc. You are already familiar with the notation for $\mathcal{R}^{\pm}(3)$: the irreducible representations S, P, D, and F are respectively one-, three-, five-, and seven-dimensional.

In groups which contain i, a subscript g or u is added, according as $\mathbf{i} = \mathbf{1}$ or $-\mathbf{1}$. In $\mathcal{D}_{nh}$ and $\mathcal{C}_{nh}$, with n odd, a prime or double prime is added, according as $\boldsymbol{\sigma}_h = \mathbf{1}$ or $-\mathbf{1}$.

Exercises

*89. All the irreducible representations of $\mathcal{C}_n$ are one-dimensional since $\mathcal{C}_n$ is Abelian. Use the fact that $C_n^n = E$ to find $\mathbf{C}_n = \Gamma_j(C_n)$ in the jth irreducible representation. What is $\mathbf{C}_n^k$ in the jth irreducible representation? (*Note:* The usual notation is $j = 0$, $\pm 1, \pm 2, \ldots,$ up to $\pm(n-1)/2$ if n is odd, or up to $+n/2$ if n is even. Then $\Gamma_{-j} = \Gamma_j^*$.)

*90. The functions $x^2 - y^2$ and $2xy$ form a basis for an irreducible representation of $\mathcal{C}_{5v}$. Find $\mathbf{C}_5 = E_2(C_5)$. (*Hint:* Use the 3×3 matrix form for C_n to find what C_5 does to x and y.) Show that $\mathbf{C}_5^{(5/2)} = \mathbf{1}$; thus these functions transform as E_2, not E_1.

*91. Let ϕ_1 and ϕ_2 be the positions of two electrons around the molecular axis of HD. The function $\phi_1 - \phi_2$ forms a basis for a one-dimensional representation of $\mathcal{C}_{\infty v}$. What is its eigenvalue for every rotation? What is its eigenvalue for every σ_v? This representation is denoted by Σ^-.

92. Use the result of Exercise 10.73 to find the irreducible representations of $\mathcal{D}_{2h} = \mathcal{D}_2 \times i$ and $\mathcal{D}_{3h} = \mathcal{C}_{3v} \times \sigma_h$. Designate them properly, according to the above rules. (*Note:* The irreducible representations of $\mathcal{D}_2$ are collected in Section 10.14.)

10.13 "Appearance" of the Irreducible Representations

Each irreducible representation of $\mathcal{G}$ may be "illustrated" by drawing an object which has $\mathcal{G}$ symmetry and indicating the effect of each symmetry operation in $\mathcal{G}$. This is especially easy for one-dimensional representations whose elements are ± 1. For example, the irreducible representations of $\mathcal{C}_{2v}$ may be indicated on the top view of a puptent (Fig. 10.12). A function which

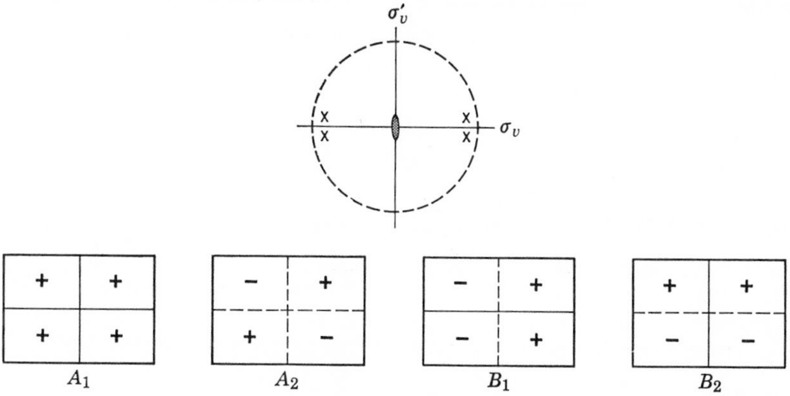

Figure 10.12

transforms as A_1 is symmetric with respect to every element of $\mathcal{C}_{2v}$. A function which transforms as A_2 is symmetric with respect to E and C_2 and antisymmetric with respect to σ_v and σ_v', a function which transforms as B_1 is antisymmetric with respect to C_2 and σ_v'; etc. Alternatively, the $\pm$ signs tell what linear combinations of the xs and os of the stereographic projection to take in order to obtain an object that transforms as a specific irreducible representation.

Illustration of multidimensional irreducible representations is less revealing, since a function that transforms as such a representation is not an eigenfunction of every symmetry element of $\mathcal{G}$. Also, the representation matrices are arbitrary and depend upon the matrix used to effect the block diagonalization. The remedy is to choose the transformation matrix so as to obtain one simple linear combination that is an eigenfunction of the elements of some subgroup of $\mathcal{G}$. (In $\mathcal{C}_{3v}$, the linear combination $b/\sqrt{2} - c/\sqrt{2}$ is an eigenfunction of the subgroup $\{E, \sigma_v\}$, and is readily illustrated.) The other component is sometimes difficult to draw since it may have coefficients other than ± 1. (In $\mathcal{C}_{3v}$ this is the linear combination $2a/\sqrt{6} - b/\sqrt{6} - c/\sqrt{6}$, which is symmetric with respect to the elements of $\{E, \sigma_v\}$.) One method of illustration is to draw the plane or axis of antisymmetry, which is not necessarily an element of the group, but is orthogonal to the σ_v or C_2' of antisymmetry of the first component. Figure 10.13 shows this first method of illustrating the irreducible representations of $\mathcal{C}_{3v}$.

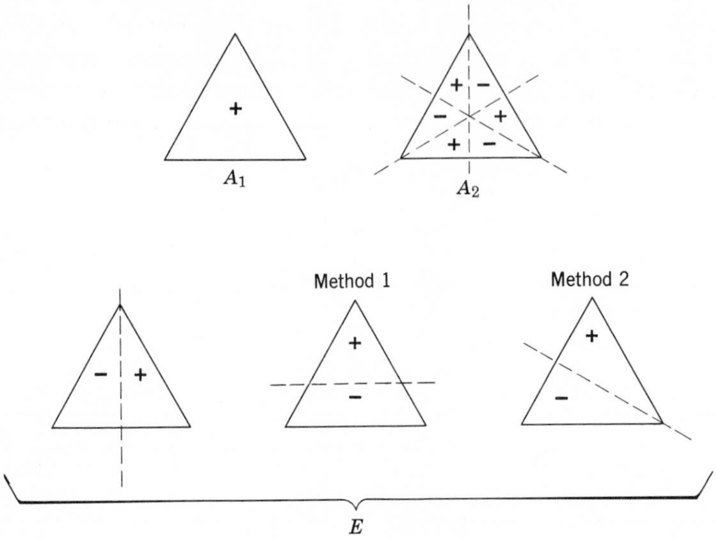

Figure 10.13

An alternative is to abandon orthogonality and draw (linearly independent) linear combinations that correspond to equivalent representations related by a similarity transformation. This corresponds to finding a second component which is an eigenfunction of a different subgroup. In Figure 10.13 the first component of the E irreducible representation for $\mathcal{C}_{3v}$ is still illustrated by $b/\sqrt{2} - c/\sqrt{2}$, an eigenfunction of the elements of $\{E, \sigma_v\}$, but the non-orthogonal $a/\sqrt{2} - c/\sqrt{2}$, an eigenfunction of the elements of $\{E, \sigma_v'\}$, illustrates the other component.

The irreducible representations of $\mathcal{D}_{6h}$ are illustrated in *Quantum Chemistry*, by W. Kauzmann, p. 467. Notice that E_1 is illustrated by Method 1, and E_2 by Method 2. Also, notice that both components of E_1 are antisymmetric with respect to C_2, whereas both components of E_2 are symmetric; this is a graphic illustration of the significance of the subscript.

Some important irreducible representations of $\mathcal{O}_h$ are illustrated in Figure 10.14. For clarity, shading represents minus.

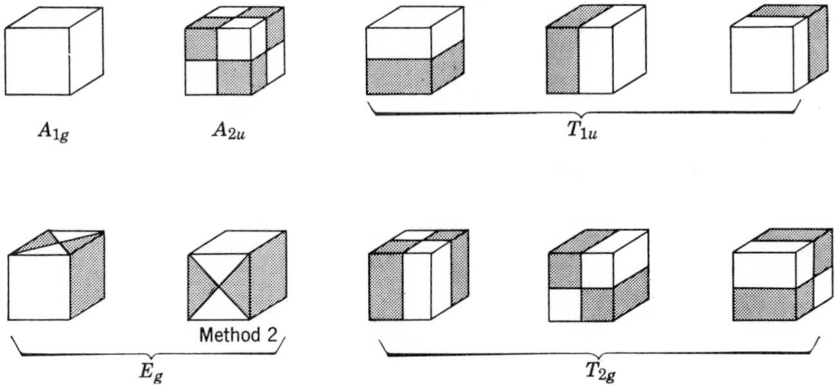

Figure 10.14

Exercises

93. Try your hand at illustrating the irreducible representations—A_1, A_2, B_1, B_2, E— of $\mathcal{C}_{4v}$ on the top view of a square pyramid.

*94. Some of the normal modes of ethylene are illustrated in Figure 10.15. Assign each to an irreducible representation of $\mathcal{D}_{2h}$.

95. The five 3d orbitals may be illustrated as in Figure 10.16. (The last three are not linearly independent since their sum is zero, and it is customary to define $d_{z^2} = d_{z^2-x^2} - d_{y^2-z^2}$, but this illustration renders their transformation properties more apparent.)
   (a) As which irreducible representations of $\mathcal{O}_h$ do these orbitals transform?
   (b) As which irreducible representations of $\mathcal{C}_{4v}$ do they transform if the z-axis is the C_4? Notice that reducing the symmetry splits some degeneracies.

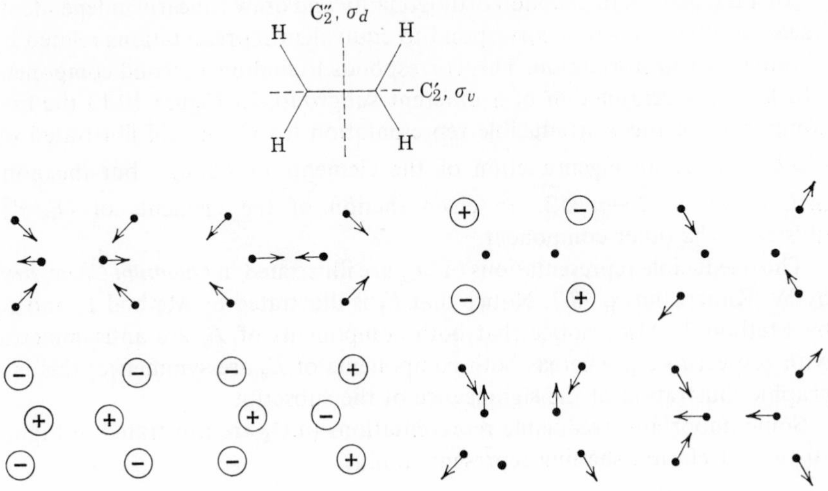

Figure 10.15

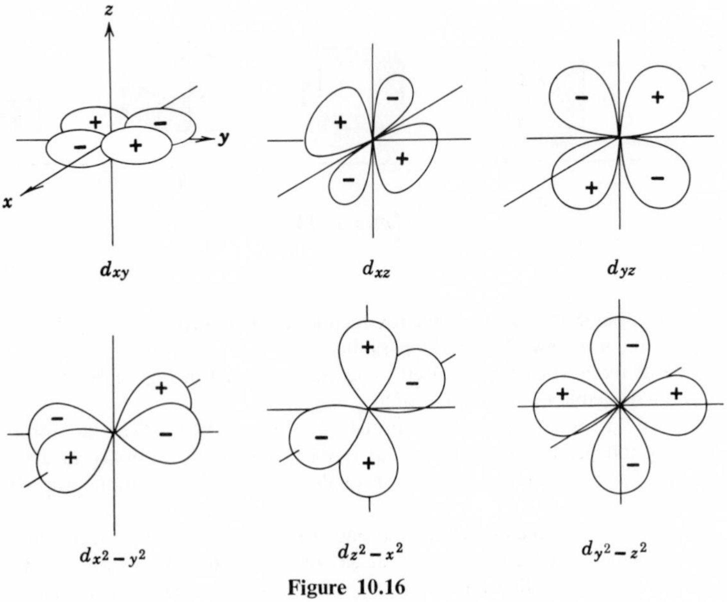

Figure 10.16

10.14 Some Remarkable Theorems

We state, without proof, the fundamental orthogonality theorem of group representations: If Γ_α and Γ_β are irreducible representations of dimension d_α and d_β, respectively, of a group $\mathcal{G}$ of order g, such that Γ_α assigns matrix $\mathbf{R}_\alpha$ to group element R and Γ_β assigns matrix $\mathbf{R}_\beta$ to group element R, then the sum over group elements

$$\sum_R (\mathbf{R}_\alpha)^*_{ij}(\mathbf{R}_\beta)_{i'j'} = \frac{g}{d_\alpha^{1/2} d_\beta^{1/2}} \delta_{\alpha\beta}\delta_{ii'}\delta_{jj'}. \tag{10.1}$$

(Here the $\delta_{\alpha\beta}$ factor must be interpreted to be zero if Γ_α and Γ_β are different irreducible representations, unity if Γ_α and Γ_β are identical irreducible representations, and undefined if Γ_α and Γ_β are equivalent irreducible representations, related by a similarity transformation.) This result means that if we take corresponding elements from each matrix of an irreducible representation, to form a g-dimensional vector whose components (one for each R) are $\{(\mathbf{R}_\alpha)_{ij}\}$, then vectors formed from different irreducible representations, or even from different matrix elements of the same representation, are orthogonal, and each vector formed from corresponding elements of Γ_α is of length $(g/d_\alpha)^{1/2}$.

Examples

18. In Exercise 10.77 you found all irreducible representations of $\mathcal{D}_2$. We list them

	E	C_2	C_2'	C_2''
A	1	1	1	1
B_1	1	1	−1	−1
B_2	1	−1	1	−1
B_3	1	−1	−1	1

The *rows* may be considered as four four-dimensional vectors. Clearly different vectors are orthogonal, and each is of magnitude $\sqrt{4}$.

19. The irreducible representations of $\mathcal{C}_{3v}$ are given in Section 10.8. We see that there are six orthogonal six-dimensional vectors

$$\mathbf{R}_1 = (1, 1, 1, 1, 1, 1)$$

$$\mathbf{R}_2 = (1, 1, 1, -1, -1, -1)$$

$$(\mathbf{R}_3)_{11} = (1, -\tfrac{1}{2}, -\tfrac{1}{2}, -1, \tfrac{1}{2}, \tfrac{1}{2})$$

$$(\mathbf{R}_3)_{12} = \left(0, -\frac{\sqrt{3}}{2}, \frac{\sqrt{3}}{2}, 0, -\frac{\sqrt{3}}{2}, \frac{\sqrt{3}}{2}\right)$$

$$(\mathbf{R}_3)_{21} = \left(0, \frac{\sqrt{3}}{2}, -\frac{\sqrt{3}}{2}, 0, -\frac{\sqrt{3}}{2}, \frac{\sqrt{3}}{2}\right)$$

$$(\mathbf{R}_3)_{22} = (1, -\tfrac{1}{2}, -\tfrac{1}{2}, 1, -\tfrac{1}{2}, -\tfrac{1}{2}).$$

That there are exactly four or six orthogonal vectors is no accident: We shall see (Exercise 10.107) that the total number of matrix elements in all the distinct irreducible representations is given by $\sum_\alpha d_\alpha^2 = g$: For $\mathfrak{D}_2$, $\sum d_\alpha^2 = 1 + 1 + 1 + 1 = 4$; for $\mathcal{C}_{3v}$, $\sum d_\alpha^2 = 1 + 1 + 2^2 = 6$. Thus we conclude that these vectors are not only orthogonal, but they span g-dimensional space.

A further helpful theorem is that the number of different irreducible representations is equal to the number of classes. For example, in $\mathcal{C}_{3v}$ there are three classes and three irreducible representations.

Exercises

96. In Exercises 10.59 and 10.79 you found four of the five irreducible representations of $\mathcal{C}_{4v}$. List the seven orthogonal eight-dimensional vectors included therein. Find an eighth vector orthogonal to each of the seven, and thereby find the remaining irreducible representation.

*97. Find the classes of the double group $\mathfrak{D}_2'$. How many classes are there? How many three-dimensional irreducible representations can there be? Two-dimensional? One-dimensional?

*98. What are the one-dimensional representations of the double group $\mathfrak{D}_2'$? (*Hint:* When the quantum number m is integral, R really is E, so the one-dimensional representations must be "the same as" those of $\mathfrak{D}_2$.) Find Γ_5 if $\mathbf{C}_2 = \begin{pmatrix} i & 0 \\ 0 & -i \end{pmatrix}$ and $\mathbf{C}_2' = \begin{pmatrix} 0 & i \\ i & 0 \end{pmatrix}$. Check that your answer satisfies the orthogonality theorem.

10.15 Characters

If dealing with the matrices of a representation has overwhelmed you, you may relax now. Happily, we may dispense with the matrices for most computational purposes and deal instead with only the trace of each matrix: $\operatorname{Tr} \mathbf{R} = \sum r_{ii}$. In this context, the trace is called the CHARACTER, χ_R^Γ or $\chi^\Gamma(R)$, of group element R in the representation Γ.

Fortunately, the characters for all the irreducible representations of the important point groups have already been evaluated and are available in texts. An especially extensive set of CHARACTER TABLES is given in *Molecular Vibrations*, by Wilson, Decius, and Cross.

For $\mathfrak{D}_2$ the irreducible representations are all one-dimensional, so each matrix is equal to its character, and the character table is the same as the table of irreducible representations given in Example 10.18. For $\mathcal{C}_{3v}$ the character table is

$\mathcal{C}_{3v}$	E	C_3	C_3^2	σ_v	σ_v'	σ_v''
A_1	1	1	1	1	1	1
A_2	1	1	1	−1	−1	−1
E	2	−1	−1	0	0	0

A shortened version of a character table is obtained by listing only classes, since elements of the same class must have the same character (Exercise 10.102). The columns are headed by the number of elements in the class and a representative member of the class. Thus the character table for $\mathcal{C}_{3v}$ is usually written

$\mathcal{C}_{3v}$	E	$2C_3$	$3\sigma_v$
A_1	1	1	1
A_2	1	1	-1
E	2	-1	0

We include two more character tables that we shall refer to:

$\mathcal{C}_{6v}$	E	$2C_6$	$2C_3$	C_2	$3\sigma_v$	$3\sigma_d$	
A_1	1	1	1	1	1	1	T_z
A_2	1	1	1	1	-1	-1	R_z
B_1	1	-1	1	-1	1	-1	
B_2	1	-1	1	-1	-1	1	
E_1	2	1	-1	-2	0	0	$T_{x,y}$; $R_{x,}$
E_2	2	-1	-1	2	0	0	

$\mathcal{O}$	E	$6C_4$	$3C_2$	$6C_2'$	$8C_3$	
$\mathcal{T}_d$	E	$6S_4$	$3C_2$	$6\sigma_d$	$8C_3$	
A_1	1	1	1	1	1	
A_2	1	-1	1	-1	1	
E	2	0	2	0	-1	
T_1	3	1	-1	-1	0	**T** in $\mathcal{O}$; **R**
T_2	3	-1	-1	1	0	**T** in $\mathcal{T}_d$

Exercises

99. For any representation Γ, what is χ_E^Γ?
100. What is χ_E for the regular representation? What is the character of each other group operation?
101. Why do equivalent representations have the same characters: $\chi_R^\Gamma = \chi_R^{\Gamma'}$?
102. Show that if R and S belong to the same class, $\chi_R = \chi_S$.
103. (a) Use the results of Exercise 10.92 to construct the character table for $\mathcal{D}_{3h}$.
     (b) Extend the result of Exercise 10.73 to convince yourself that the characters of the representation Γ of the direct-product group $\mathcal{G}_1 \times \mathcal{G}_2$ are given by $\chi_R^\Gamma = \chi_R^{\Gamma_\alpha} \chi_R^{\Gamma_\beta}$.
     (c) Check that using this method to construct the character table of $\mathcal{D}_{3h} = \mathcal{C}_{3v} \times \sigma_h$ leads to the same character table as you obtained in part (a).

104. Exercises 10.59, 10.79, and 10.96 led to all irreducible representations of $\mathcal{C}_{4v}$. What is the character table for $\mathcal{C}_{4v}$?

In terms of characters the orthogonality theorem becomes

$$\sum_R (\chi_R{}^{\Gamma_\alpha})^* \chi_R{}^{\Gamma_\beta} = g\delta_{\alpha\beta}, \tag{10.2a}$$

which may also be written as a sum over classes

$$\sum_C n_C (\chi_C{}^{\Gamma_\alpha})^* \chi_C{}^{\Gamma_\beta} = g\delta_{\alpha\beta}, \tag{10.2b}$$

where n_C is the number of elements in class C. If there are c classes, there are c irreducible representations, as stated without proof in Section 10.14. Eq. 10.2a means that the *rows* of characters (in the long form of the table) are c orthogonal vectors in a g-dimensional space.

Exercise

105. Derive Eq. 10.2a from Eq. 10.1.

10.16 Reduction of Reducible Representations

Let $\mathcal{G}$ be a group whose character table is known and Γ any representation of $\mathcal{G}$, with $\Gamma(R) = \mathbf{R}_\Gamma$. This Γ is not necessarily irreducible, but may be reduced by some transformation $\mathbf{X}$ to a direct sum of irreducible representations, some of which may occur more than once:

$$\mathbf{X}^{-1}\mathbf{R}_\Gamma\mathbf{X} = \mathbf{\Sigma}\, n_\alpha \mathbf{R}_\alpha. \tag{10.3}$$

Here the $\mathbf{\Sigma}$ notation means direct sum, and $n_\alpha \mathbf{R}_\alpha$ means $\mathbf{R}_\alpha \oplus \mathbf{R}_\alpha \oplus \cdots \oplus \mathbf{R}_\alpha$ (n_α times). Symbolically, this is usually written $\Gamma = \sum n_\alpha \Gamma_\alpha$, read "$\Gamma$ CONTAINS the irreducible representation Γ_α n_α times."

To obtain the REDUCTION FORMULA, take the trace of both sides of Eq. 10.3:

$$\chi_R^\Gamma = \sum_\alpha n_\alpha \chi_R^{\Gamma_\alpha} \tag{10.4}$$

since the trace is invariant to similarity transformation and since the trace of a direct sum is the sum of the traces of the submatrices. Next multiply both sides by $(\chi_R{}^{\Gamma_\beta})^*$, sum over R, and use Eq. 10.2:

$$\sum_R (\chi_R{}^{\Gamma_\beta})^* \chi_R^\Gamma = \sum_{R,\alpha} n_\alpha (\chi_R{}^{\Gamma_\beta})^* \chi_R{}^{\Gamma_\alpha} = \sum_\alpha n_\alpha g \delta_{\alpha\beta} = n_\beta g$$

or

$$n_\beta = \frac{1}{g} \sum_R (\chi_R{}^{\Gamma_\beta})^* \chi_R^\Gamma = \frac{1}{g} \sum_C n_C (\chi_C^{\Gamma_\beta})^* \chi_C^\Gamma. \tag{10.5}$$

Therefore, in order to determine how many times the irreducible representation Γ_α is contained (OCCURS) in the representation Γ, find the characters of Γ, take the "dot product" with the characters of Γ_α, and divide by g.

Exercises

106. Show that the regular representation contains each irreducible representation, Γ_α, d_α times. (This result shows that in principle it is possible to find all the irreducible representations of a group; "all" that is necessary is simultaneous block diagonalization of g $g \times g$ matrices.)
107. Use the previous result and those of Exercises 10.99 and 10.100 to show that $\sum d_\alpha^2 = g$.
108. Use the character table for $\mathcal{C}_{3v}$ and the characters of the matrices of the *abc* representation to find out, without diagonalization, how many times each irreducible representation occurs.
109. Why does $\chi_R^{\Gamma_\alpha \times \Gamma_\beta} = \chi_R^{\Gamma_\alpha} \chi_R^{\Gamma_\beta}$?
110. (a) Use the previous result to find, in $\mathcal{D}_2$, the characters of each of the direct-product representations $A \times A$, $A \times B_1$, $A \times B_2$, $A \times B_3$, $B_1 \times B_2$, $B_1 \times B_3$, and $B_2 \times B_3$.
     (b) Compare each set of characters with those in the character table to determine what irreducible representation each of these direct-product representations is.
     (c) Construct a multiplication table for the irreducible representations of $\mathcal{D}_2$. Notice the sense of multiplication of representations.
111. (a) Do the same as in Exercise 10.110a and b for $A_1 \times A_1$, $A_2 \times A_2$, $A_1 \times A_2$, $A_1 \times E$, and $A_2 \times E$ in $\mathcal{C}_{3v}$.
     (b) What are the characters of the reducible four-dimensional representation $E \times E$?
     (c) What may $E \times E$ be reduced to?
     (d) Construct a multiplication table for the irreducible representations of $\mathcal{C}_{3v}$. Notice the sense of multiplication and addition of representations.
112. Show that $\Gamma_\alpha \times \Gamma_1 = \Gamma_\alpha$.
*113. Reduce $E \times T_1 \times T_2$ in $\mathcal{O}$.

In real situations it is usually not Γ which is given, but rather a basis $\{\phi_i\}$ for the representation Γ (for example, again, the *abc* basis for $\mathcal{C}_{3v}$). If Γ is reducible, the basis functions are not appropriate for maximum utilization of symmetry, and we would wish to construct new basis functions, linear combinations of the old ones, which transform as irreducible representations. Before we set out to construct such linear combinations, it is helpful to know how many linear combinations there are that transform as Γ_α. The desired number is n_α, given by the reduction formula (Eq.10.5).

Exercises

114. In $\mathcal{C}_{4v}$, what may $E \times E$ be reduced to? Just as the functions x and y form a basis for the E irreducible representation, so do the functions x^3 and y^3. Therefore the functions x^4, x^3y, xy^3, and y^4 form a basis for $E \times E$. Find linear combinations of these four functions which transform as each of the irreducible representations contained in $E \times E$. (*Hint:* If the linear combinations are not obvious, diagonalize $E(C_4) \times E(C_4)$.)

115. The two functions $e^{\pm i\phi}$ transform as E_1 in $\mathcal{C}_{5v}$ since $C_5 e^{\pm i\phi} = e^{\pm 2\pi i/5} e^{\pm i\phi}$ and $\sigma_v e^{\pm i\phi} = e^{\mp i\phi}$, or $\chi_{C_5} = 2\cos(2\pi/5)$ and $\chi_{\sigma_v} = 0$. Similarly, the two functions $e^{\pm 2i\phi}$ transform as E_2. What are the transformation properties of the four possible products of $e^{\pm i\phi}$ with $e^{\pm 2i\phi}$; to what irreducible representation does each product belong? Check your answer by evaluating $E_1 \times E_2$.
116. The functions $e^{\pm i\phi}$ and $e^{\pm 2i\phi}$ transform as E_1 and E_2, respectively, in $\mathcal{C}_{6v}$. Form the four possible products into appropriate linear combinations which transform as irreducible representations of $\mathcal{C}_{6v}$. Check your answer by evaluating $E_1 \times E_2$
*117. Just as the functions $e^{\pm i\phi}$ transform as E_1 in $\mathcal{C}_{6v}$, so do the functions $e^{\pm 5i\phi}$. Form the four possible products into appropriate linear combinations which transform as irreducible representations of $\mathcal{C}_{6v}$. Check your answer by evaluating $E_1 \times E_1$.
*118. In $\mathcal{C}_{4v}$ the functions x and y form a basis for the E irreducible representation, but the set $\{x^2, xy, yx, y^2\}$ of all possible products of $\{x, y\}$ and $\{x, y\}$ does *not* form a basis for the direct-product representation $E \times E$ because there are only three linearly independent product functions. Therefore it is necessary to define another type of product, the SYMMETRIC DIRECT PRODUCT: If $\{\phi_i\}$ forms a basis for the d_α-dimensional irreducible representation Γ_α, then the set $\{\phi_i\phi_j\}$ of all *distinct* products forms a basis for a $\frac{1}{2}d_\alpha(d_\alpha + 1)$-dimensional representation $\Gamma_\alpha \cdot \Gamma_\alpha$, or $\Gamma_\alpha \times \Gamma_\alpha$(sym).
   (a) In Exercise 10.60 you used the three distinct functions, x^2, y^2, and xy, to find the representation $E \cdot E$ in $\mathcal{C}_{4v}$.
   (b) Find the characters of this representation.
   (c) Evaluate how many times each irreducible representation of $\mathcal{C}_{4v}$ occurs in $E \cdot E$; compare this result with that for $E \times E$ in Exercise 10.114.
   (d) It can be shown that if Γ_α is two-dimensional, then the characters of $\Gamma_\alpha^{[n]} = \Gamma_\alpha \cdot \Gamma_\alpha \cdots \Gamma_\alpha$ (*n* times) are given by the recursion formula

$$\chi_R^{\Gamma_\alpha^{[n]}} = \tfrac{1}{2}\left[\chi_R^{\Gamma_\alpha}\chi_R^{\Gamma_\alpha^{[n-1]}} + \chi_{(R^n)}^{\Gamma_\alpha}\right].$$

   Find the characters of $E \cdot E \cdot E$ in $\mathcal{C}_{4v}$ and evaluate how many times each irreducible representation of $\mathcal{C}_{4v}$ occurs in $E \cdot E \cdot E$. [*Note:* The symmetric direct product arises whenever identical particles are assigned to degenerate levels. For example, there are only three (not four) singlet states of cyclobutadiene that arise from distributing two electrons into the degenerate pair of nonbonding orbitals, and there are only six (not nine) singlet states that arise from assigning two electrons among three *p* orbitals. There are only three (not four) overtone states that result from distributing two quanta to the two components of a doubly degenerate vibration.]

It still remains to show how easy it is to find the characters of Γ without writing down the representation matrices. For each R in $\mathcal{G}$ we seek $\sum r_{ii}$, the trace of the matrix **R** defined by $R\phi_j = \sum \phi_i r_{ij}$. The simplest situation occurs when $R\phi_j = \phi_i \neq \phi_j$, that is, when R converts the jth basis function into a different basis function; then that ϕ_j contributes zero to χ_R^Γ. Other simple situations occur when $R\phi_j = \phi_j$ or $R\phi_j = -\phi_j$; then ϕ_j contributes $+1$ or -1, respectively, to χ_R^Γ. (In the *abc* example, C_3 converts each basis function into a different basis function, so $\chi_{C_3}^\Gamma = 0$. Only *a* is left unchanged by σ_v, so $\chi_{\sigma_v}^\Gamma = 1$.) If ϕ_j is a "little arrow" or a *p* orbital attached to an atom on a C_n and oriented perpendicular to the C_n, it can be shown (Exercises 10.119 and 10.120) that ϕ_j contributes $\cos(2\pi k/n)$ to $\chi^\Gamma(C_n^k)$ and $\chi^\Gamma(S_n^k)$.

Examples

20. The reducible representation of $\mathcal{C}_{2v}$ formed by $1s$ orbitals of two hydrogens and three $2p$ orbitals of an oxygen in H_2O was given in Example 10.10. But the characters of this representation may be obtained without the matrices: $\chi_E^\Gamma = 5$, the dimension of the representation, since all five basis functions are unchanged by E. The contributions to $\chi_{C_2}^\Gamma$ are zero from the hydrogen orbitals, which are interchanged, -1 from each of $2p_x$ and $2p_y$, which are converted to their negatives by C_2, and $+1$ from $2p_z$, which is left unchanged; therefore $\chi_{C_2}^\Gamma = -1$. The contributions to $\chi_{\sigma_v}^\Gamma$ are $+1$ from each of $1s_1$, $1s_2$, $2p_y$, and $2p_z$, and -1 from $2p_x$, which is converted to its negative by σ_v; therefore $\chi_{\sigma_v}^\Gamma = 3$. The contributions to $\chi^\Gamma(\sigma_v')$ are $+1$ from $2p_x$ and $2p_z$, -1 from $2p_y$, and zero from $1s_1$ and $1s_2$; therefore $\chi^\Gamma(\sigma_v') = 1$. We write the character table for $\mathcal{C}_{2v}$, and the characters, χ^Γ:

$\mathcal{C}_{2v}$	E	C_2	σ_v	σ_v'
A_1	1	1	1	1
A_2	1	1	-1	-1
B_1	1	-1	1	-1
B_2	1	-1	-1	1
Γ	5	-1	3	1

By the reduction formula

$$n_{A_1} = (1 \times 5 + 1 \times -1 + 1 \times 3 + 1 \times 1)/4 = \tfrac{8}{4} = 2$$
$$n_{A_2} = (1 \times 5 + 1 \times -1 - 1 \times 3 - 1 \times 1)/4 = 0$$
$$n_{B_1} = (1 \times 5 - 1 \times -1 + 1 \times 3 - 1 \times 1)/4 = \tfrac{8}{4} = 2$$
$$n_{B_2} = (1 \times 5 - 1 \times -1 - 1 \times 3 + 1 \times 1)/4 = \tfrac{4}{4} = 1.$$

In the symbolic notation $\Gamma = 2A_1 + 2B_1 + B_2$; there are two linear combinations which transform as A_1, two as B_1, and one as B_2.

21. Orthogonal infinitesimal displacements of the nitrogen and three hydrogens of NH_3 form a basis for a twelve-dimensional representation of $\mathcal{C}_{3v}$. [Since we are interested only in matrix traces, which are invariant to orthogonal transformation, we may choose a local coordinate system at each hydrogen, with one arrow along the N—H bond (not shown), one arrow in the plane of the three hydrogens, and the third arrow perpendicular to the other two, rather than pointing in x, y, z directions. This simplifies the evaluation

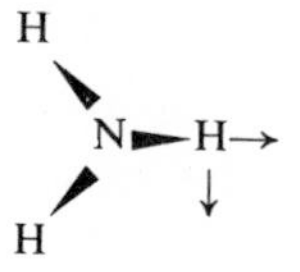

of χ_σ^Γ.] Necessarily $\chi_E^\Gamma = 12$. The hydrogen displacements contribute zero to $\chi_{C_3}^\Gamma$. The nitrogen displacement along the z-axis is unchanged by C_3, and contributes $+1$ to $\chi_{C_3}^\Gamma$; the other two nitrogen displacements each contribute $\cos(2\pi/3)$. Therefore $\chi_{C_3}^\Gamma = 0$. Only the nitrogen and one hydrogen can contribute to $\chi_{\sigma_v}^\Gamma$. Since two arrows at each atom are left unchanged by the reflection and the third is converted to its negative, $\chi_{\sigma_v}^\Gamma = 2$.

Being careful to sum all the elements in each class, we use the reduction formula to find

$$n_{A_1} = (1 \times 12 + 2 \times 1 \times 0 + 3 \times 1 \times 2)/6 = 3$$
$$n_{A_2} = (1 \times 12 + 2 \times 1 \times 0 + 3 \times -1 \times 2)/6 = 1$$
$$n_E = (2 \times 12 + 2 \times 1 \times 0 + 3 \times 0 \times 2)/6 = 4.$$

Thus we see that $\Gamma = 3A_1 + A_2 + 4E$; there are three linear combinations which transform as A_1, one which transforms as A_2, and four *pairs* of linear combinations which transform as E. Notice that we obtain just as many linear combinations—12—as were put in.

If we are interested only in normal modes of NH_3, infinitesimal translations and rotations of the whole molecule do not count, but have been included in the basis. The transformation properties of translations and rotations may be evaluated by considering them as little arrows and current loops (Fig. 10.17), but the irreducible representations to which they

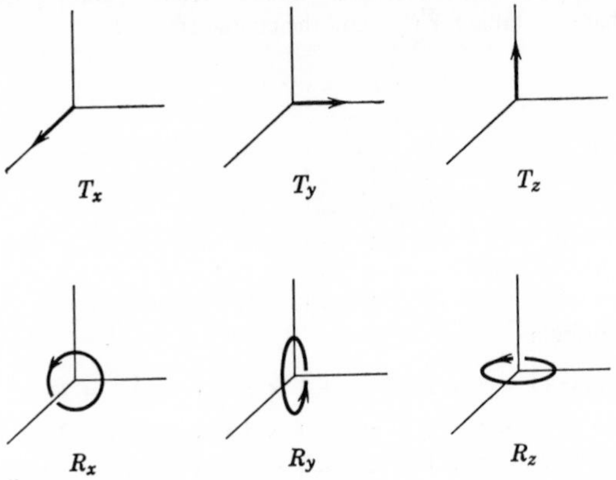

Figure 10.17

belong are conveniently indicated in character tables by T_x, T_y, T_z, R_x, R_y, R_z. In $\mathfrak{C}_{3v}$ **T** transforms as $A_1 + E$ and **R** transforms as $A_2 + E$. Therefore there remain two normal modes of A_1 symmetry and two (doubly degenerate) normal modes of E symmetry.

Exercises

119. (a) Show that if an atom is situated on a C_n, the set of its three mutually orthogonal p orbitals (or atomic displacements) contributes a total of -1 to χ_{C_2} and a total of $+1$ to χ_{C_4}. (*Hint:* Since the trace is invariant to any unitary transformation of the orbitals, we may for convenience choose one orbital to be along the C_n.)

(b) Show that this set contributes zero to χ_{C_3}. (*Hint:* Choose the coordinate axes so that the C_3 is along $\hat{\mathbf{i}} + \hat{\mathbf{j}} + \hat{\mathbf{k}}$, and choose the orbitals to be along the coordinate axes.)

*120. Show that an atomic displacement or a p orbital oriented perpendicular to a C_n passing through the atom contributes $\cos(2\pi k/n)$ to $\chi(C_n^k)$.

121. The four π atomic orbitals (antisymmetric under σ_h) of BF_3 form a basis for a reducible representation Γ of $\mathfrak{D}_{3h}$. What are the characters χ_R^Γ? How many times does each irreducible representation of $\mathfrak{D}_{3h}$ occur in Γ? (Cf. Exercise 10.103a.)

122. Since the four π atomic orbitals of BF_3 are all antisymmetric under σ_h, it is not necessary to use the full $\mathfrak{D}_{3h}$ symmetry. If σ orbitals (symmetric with respect to σ_h)

are not under consideration, it is sufficient to consider the π orbitals as forming a basis for a representation Γ of $\mathcal{C}_{3v}$. How many times does each irreducible representation of $\mathcal{C}_{3v}$ occur in Γ?

123. How many times does each irreducible representation of $\mathcal{C}_{2v}$ occur in the reducible representation formed by the ten π AOs of naphthalene as basis (Fig. 10.18)?

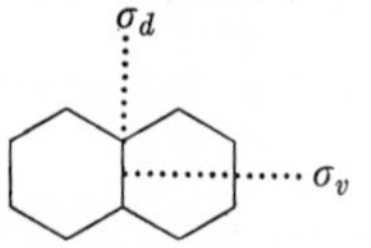

Figure 10.18

124. The 1s orbitals of the six hydrogen atoms of cyclopropane form a basis for a reducible representation of $\mathcal{D}_{3h}$. How many linear combinations of these orbitals are there which transform as each irreducible representation of $\mathcal{D}_{3h}$? Do the same for the 2s and three 2p orbitals of each of the carbons.

*125. The set of one 3s, three 3p, and five 3d orbitals of sulfur and the three 2p orbitals of each of the six fluorines of SF_6 form a basis for a 27-dimensional representation of the group $\mathcal{O}_h = \mathcal{O} \times i$. Find how many times each irreducible representation occurs. (Cf. Exercises 10.95, 10.103b, and 10.119.)

*126. Do the same for the 20-dimensional representation of the group $\mathcal{T}_d$ formed by all 2s and 2p orbitals of CF_4.

*127. Convince yourself that in $\mathcal{C}_{3v}$ the translations transform as $A_1 + E$ and the rotations as $A_2 + E$. (*Hint:* Don't puzzle too long over χ_{C_3} for $T_{x,y}$ and $R_{x,y}$, but recognize that no linear combination of these can be an eigenfunction of every operation of $\mathcal{C}_{3v}$.)

128. The set of 15 infinitesimal displacements of the atoms of $PtCl_4^{2-}$ forms a basis for a reducible representation of $\mathcal{D}_{4h} = \mathcal{C}_{4v} \times i$. Find the number of times each irreducible representation occurs. (*Hint:* Use the result of Exercise 10.103b to find the character table of $\mathcal{D}_{4h}$. Remember that the subscripts now refer to the behavior under C_2', not σ_v.) Translations and rotations transform according to the following symmetry species: $T_{x,y}$ as E_u, T_z as A_{2u}, $R_{x,y}$ as E_g, R_z as A_{2g}. Find the number of normal modes of each symmetry species.

*129. Infinitesimal radial, tangential, and vertical displacements of the vertices of a regular hexagon each form a basis for six-dimensional representations of $\mathcal{D}_{6h} = \mathcal{C}_{6v} \times i$. Find the total number of normal modes of each symmetry species. Take σ_v and C_2' through vertices, and remember that the subscripts now refer to the behavior under C_2', not σ_v. (*Hint:* See page 347 for $\Gamma_{\mathbf{T}}$ and $\Gamma_{\mathbf{R}}$.)

10.17 Construction of Eigenvectors of Symmetry Operators

In order to take advantage of symmetry in solving for eigenvalues and eigenvectors of H, we must construct our basis vectors so that they transform according to irreducible representations of the group $\mathcal{G}$ of symmetry operators that commute with H. Given a basis $\{\phi_i\}$ we can determine by the reduction formula how many linear combinations of the basis vectors there are that transform as each irreducible representation Γ_α of $\mathcal{G}$. These linear combinations are often called SYMMETRY COORDINATES, although the name varies with context. One way of finding them is to reduce the representation of $\mathcal{G}$ for which $\{\phi_i\}$ form the basis. However, there is an easier way that does not require diagonalization of any matrices. Next we consider this method.

To construct a linear combination of basis vectors which transforms as Γ_α, the recipe is to apply the projection operator

$$P_{\Gamma_\alpha} = \sum_R ({\chi_R}^{\Gamma_\alpha})^* R \tag{10.6}$$

to an arbitrary basis vector, ϕ_i. (For examples, in $\mathcal{C}_{2v}$, $P_{A_2} = E + C_2 - \sigma_v - \sigma_v'$, and in $\mathcal{C}_{3v}$, $P_E = 2E - C_3 - C_3^2$.)

If Γ_α is one-dimensional, it is easy to show that $P_{\Gamma_\alpha}\phi_i$ transforms as Γ_α. Let us find what group element S does to $P_{\Gamma_\alpha}\phi_i$:

$$SP_{\Gamma_\alpha}\phi_i = \sum_R ({\chi_R}^{\Gamma_\alpha})^* SR\phi_i.$$

Multiply by $1 = \chi_S^{\Gamma_\alpha}(\chi_S^{\Gamma_\alpha})^*$:

$$SP_{\Gamma_\alpha}\phi_i = {\chi_S}^{\Gamma_\alpha} \sum_R ({\chi_S}^{\Gamma_\alpha})^*({\chi_R}^{\Gamma_\alpha})^* SR\phi_i.$$

But since these are characters of a one-dimensional representation, ${\chi_S}^{\Gamma_\alpha}{\chi_R}^{\Gamma_\alpha} = {\chi_{SR}}^{\Gamma_\alpha}$. Also, by the result of Exercise 10.7, $\sum_R = \sum_{(SR)}$. Then

$$SP_{\Gamma_\alpha}\phi_i = {\chi_S}^{\Gamma_\alpha} \sum_{(SR)} ({\chi_{(SR)}}^{\Gamma_\alpha})^*(SR)\phi_i = {\chi_S}^{\Gamma_\alpha} P_{\Gamma_\alpha}\phi_i, \tag{10.7}$$

where we have used the fact that the summation index is a dummy variable.

What we have shown is that $P_{\Gamma_\alpha}\phi_i$ is an eigenvector of every S in $\mathcal{G}$, with eigenvalue ${\chi_S}^{\Gamma_\alpha}$; in other words, the vector $P_{\Gamma_\alpha}\phi_i$ transforms as Γ_α. Further application of P_{Γ_α} to other basis vectors will produce the remaining vectors which transform as Γ_α.

This projection operator is also appropriate for constructing a set of vectors which transforms as a multidimensional irreducible representation. (See Exercise 10.131.) Operating P_{Γ_α} on ϕ_i will give one component of the set, and operating P_{Γ_α} on a different vector, ϕ_j, will, in general, give a different component of the set. Alternatively, since $P_{\Gamma_\alpha}\phi_i$ is not an eigenvector of every element of $\mathcal{G}$, other components of the set may be obtained by operating a suitable element of $\mathcal{G}$ on $P_{\Gamma_\alpha}\phi_i$.

Example

22. Construction of appropriate linear combinations of the (abc) basis set: Obviously, $P_{A_1}a = (E + C_3 + C_3^2 + \sigma_v + \sigma_v' + \sigma_v'')a = 2a + 2b + 2c$, which transforms as A_1. To obtain a pair which transforms as E, first operate P_E on a: $P_E a = (2E - C_3 - C_3^2)a = 2a - b - c$. To obtain another component, apply P_E to b: $P_E b = 2b - c - a$. Alternatively, apply C_3 to $P_E a$: $C_3(2a - b - c) = 2c - a - b$. These components are not orthogonal, and they should be orthogonalized before proceeding to the eigenvalue-eigenvector problem.

Exercises

*130. Show that if Γ_α and Γ_β are one-dimensional representations, $P_{\Gamma_\alpha}P_{\Gamma_\beta} = g\delta_{\alpha\beta}P_{\Gamma_\alpha}$. (*Hint:* Multiply the double sum by $(\chi_R^{\Gamma_\beta})^*\chi_R^{\Gamma_\beta} = 1$.) This result holds for all irreducible representations and shows that P_{Γ_α} is indeed a projection operator, except for the factor g.

*131. Even if Γ_α is a multidimensional irreducible representation, show that for any arbitrary i and j the operator, $P_{\Gamma_\alpha,ij} \equiv \sum_R (\mathbf{R}_\alpha)^*_{ij}R$, acts on an arbitrary function ϕ to produce a function which transforms as Γ_α: $SP_{\Gamma_\alpha,ij}\phi = \sum_k P_{\Gamma_\alpha,kj}\phi(\mathbf{S}_\alpha)_{kl}$. (*Hint:* Since $\mathbf{S}_\alpha$ is unitary, $\delta_{li} = \sum_k (\mathbf{S}_\alpha^{-1})^*_{lk}(\mathbf{S}_\alpha)^*_{ki}$.) Then use the result of Exercise 10.83 to show that $\sum_R (\chi_R^{\Gamma_\alpha})^* R\phi$ also transforms as Γ_α.

132. In the (abc) basis, what is $P_{A_2}a$? Why?

133. Why are the *three* functions $P_E a$, $P_E b$, and $P_E c$ not suitable as a basis?

134. Let $\Phi(x_1, x_2)$ be a function of the two variables x_1 and x_2, and let P_{12} be the operator which interchanges the two variables: $P_{12}\Phi(x_1, x_2) = \Phi(x_2, x_1)$. This P_{12} and the identity operator comprise the permutation group $\mathscr{P}_2$.
   (a) Construct the operator P_{Γ_1}.
   (b) What is $P_{\Gamma_1}\Phi$?
   (c) Construct P_{Γ_2}.
   (d) What is $P_{\Gamma_2}\Phi$? (Cf. Exercises 4.60 and 9.14.)
   (e) For the special case, $\Phi(x_1, x_2) = \phi_1(x_1)\phi_2(x_2)$, express $P_{\Gamma_2}\Phi$ as a Slater determinant.

135. Let $\Phi(x_1, x_2, x_3)$ be a function of three variables and $\mathscr{P}_3$ the group of operators which permute the variables.
   (a) Construct P_{Γ_1} and P_{Γ_2}.
   (b) For the special case, $\Phi(x_1, x_2, x_3) = \phi_1(x_1)\phi_2(x_2)\phi_3(x_3)$, express $P_{\Gamma_2}\Phi$ as a Slater determinant.

136. A general statement of the Pauli Exclusion Principle is that the wave function for a system of n identical fermions (bosons) must transform as Γ_2 (Γ_1) of $\mathscr{P}_n$.
   (a) Use a result of Exercise 10.71 to construct a suitable wave function for an n-electron system, based on the simple product $\prod_{i=1}^n \phi_i(\mathbf{X}_i, s_i)$, where $\mathbf{X}_i$ is the position of the ith electron and s_i its spin.
   (b) Convince yourself that if two one-electron functions are identical (if two electrons are in the same spinorbital), the wave function is zero.

137. Construct orthonormal molecular orbitals from the four C—H bond orbitals of methane, that is, find a transformation from localized bond orbitals to orbitals that transform as irreducible representations of $\mathscr{T}_d$. (*Hint:* To avoid calculating many zeros, first find each n_α.)

138. Do the same for the six S—F bond orbitals of SF_6, but for $\mathscr{O}_h$.

139. (a) Construct an operator $P_\mathbf{k}$ that will operate on an arbitrary function $\phi(\mathbf{X})$ to produce a function that transforms as the $\mathbf{k}$th irreducible representation of the finite translation group of Exercises 10.53 and 10.76.
   (b) Construct the function $P_\mathbf{k}\phi(\mathbf{X})$. [*Hint:* Generalize the displacement operator of Example 9.11 to determine what $\mathbf{T}\phi(\mathbf{X})$ is.] In the case that $\phi(\mathbf{X})$ is a function localized within a single unit cell, the delocalized function $P_\mathbf{k}\phi(\mathbf{X})$ is often known as a BLOCH FUNCTION.

With care, further factorization of the submatrix associated with a multidimensional irreducible representation Γ_α may be effected. For definiteness, assume that Γ_α is two-dimensional and occurs n_α times in the original basis set. Since there are n_α pairs of new basis vectors, $\{\chi_{i1}, \chi_{i2}\}_{i=1}^{n_\alpha}$, which transform as Γ_α, the submatrix we wish to diagonalize is a $2n_\alpha \times 2n_\alpha$ one. However, we already know that every eigenvalue is doubly degenerate, so that it ought to be possible to diagonalize only an $n_\alpha \times n_\alpha$ matrix. To do so it is necessary to work with the "same component" of every pair: If the set $\{\chi_{i1}\}$ is chosen so that H mixes the elements of the set, but does not mix any χ_{j2} with any χ_{i1}, then the $2n_\alpha \times 2n_\alpha$ matrix is factored into the direct sum of two (identical) $n_\alpha \times n_\alpha$ matrices, one associated with $\{\chi_{i1}\}$ and the other with $\{\chi_{i2}\}$. The trick for finding such nonmixing sets is to choose, arbitrarily, some subgroup of $\mathcal{G}$ and form linear combinations which transform as different irreducible representations of the subgroup.

Example

23. Construction of linear combinations of π atomic orbitals $\{\phi_i\}$ of perinaphthyl which transform as one component of the E irreducible representation of $\mathcal{C}_{3v}$ (Fig. 10.19): By the reduction formula there are four pairs of linear combinations which transform as E. Operating $P_E = (2E - C_3 - C_3^2)$ on ϕ_{1-6}, ϕ_{10}, ϕ_{11} and orthogonalizing $P_E\phi_4$ to $P_E\phi_1$, $P_E\phi_5$ to $P_E\phi_2$, $P_E\phi_6$ to $P_E\phi_3$, and $P_E\phi_{11}$ to $P_E\phi_{10}$ leads to the linear combinations

$$\begin{aligned} P_E\phi_1 &= 2\phi_1 - \phi_4 - \phi_7 & \quad \mathcal{O}P_E\phi_4 &= \phi_4 - \phi_7 \\ P_E\phi_2 &= 2\phi_2 - \phi_5 - \phi_8 & \mathcal{O}P_E\phi_5 &= \phi_5 - \phi_8 \\ P_E\phi_3 &= 2\phi_3 - \phi_6 - \phi_9 & \mathcal{O}P_E\phi_6 &= \phi_6 - \phi_9 \\ P_E\phi_{10} &= 2\phi_{10} - \phi_{11} - \phi_{12} & \mathcal{O}P_E\phi_{11} &= \phi_{11} - \phi_{12}. \end{aligned}$$

where

$$\mathcal{O}P_E\phi_4 = P_E\phi_4 - (P_E\phi_1 \cdot P_E\phi_4)P_E\phi_1/(P_E\phi_1)^2, \text{ etc.}$$

By inspection, these may be combined into linear combinations which are either symmetric or antisymmetric to σ_v (transforming as A' or A'' in $\mathcal{C}_s = \{E, \sigma_v\}$)

$$A'$$

$$\begin{aligned} P_E\phi_1 + P_E\phi_3 &= 2\phi_1 + 2\phi_3 - \phi_4 - \phi_6 - \phi_7 - \phi_9 \\ P_E\phi_2 &= 2\phi_2 - \phi_5 - \phi_8 \\ P_E\phi_{10} &= 2\phi_{10} - \phi_{11} - \phi_{12} \\ \mathcal{O}P_E\phi_4 - \mathcal{O}P_E\phi_6 &= \phi_4 - \phi_6 - \phi_7 + \phi_9 \end{aligned}$$

$$A''$$

$$\begin{aligned} P_E\phi_1 - P_E\phi_3 &= 2\phi_1 - 2\phi_3 - \phi_4 + \phi_6 - \phi_7 + \phi_9 \\ \mathcal{O}P_E\phi_5 &= \phi_5 - \phi_8 \\ \mathcal{O}P_E\phi_{11} &= \phi_{11} - \phi_{12} \\ \mathcal{O}P_E\phi_4 + \mathcal{O}P_E\phi_6 &= \phi_4 + \phi_6 - \phi_7 - \phi_9. \end{aligned}$$

Since H commutes with σ_v, the A' linear combinations cannot mix with the A'' linear combinations.

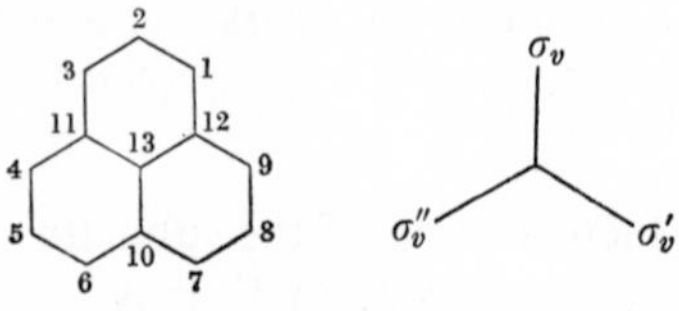

Figure 10.19

To complete the procedure, normalize each of the symmetry "coordinates" and check orthogonality. Finally, write down the block-diagonal matrix which represents H in this new basis and diagonalize each submatrix. (Remember that eigenvectors will be expressed as linear combinations of the new basis vectors, and that further transformation would be necessary to express them in terms of the original basis.) Notice that is it not necessary to write down H in terms of the old basis and apply a similarity transformation to obtain H in terms of the new basis. (See also Section 8.14.)

In summary, the steps involved in using symmetry to factor an eigenvalue-eigenvector problem are: (*1*) Find the group $\mathcal{G}$ of symmetry operators which commute with H, the operator to be diagonalized. (*2*) Evaluate how many times each Γ_α occurs in the representation of $\mathcal{G}$ formed by the given basis. (*3*) Construct orthonormal symmetry "coordinates." (*4*) Evaluate matrix elements of H in this new basis. (5) Diagonalize each submatrix.

Exercises

*140. Do Exercise 8.148 by the projection-operator method.

141. Construct orthonormal SYMMETRY ORBITALS from the ten π atomic orbitals of naphthalene (assumed orthonormal). (Cf. Exercise 10.123.) Find the eigenvalues of B if the only nonvanishing matrix elements are unity between adjacent atomic orbitals. (Cf. Exercise 8.127.)

142. Do the same for tetramethylenecyclobutane, $(C{=}CH_2)_4$ (Fig. 10.20), and find the eigenvectors but do not bother to normalize them.

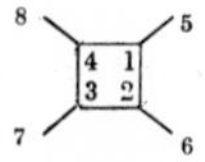

Figure 10.20

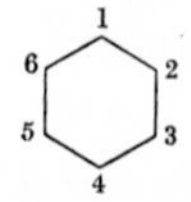

Figure 10.21

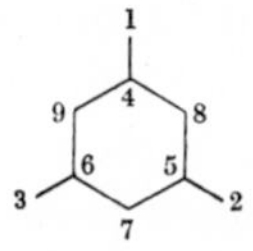

Figure 10.22

143. Do the same for benzene (Fig. 10.21). (Cf. Exercise 7.34.)

144. Do the same for 1,3,5-trimethylenebenzene, $C_6H_3(CH_2)_3$ (Fig. 10.22).

*145. Find the characteristic polynomial $P_{n\alpha}(m)$ for each irreducible representation involved in solving for the eigenvalues of the matrix **B** for coronene. (If you can do this, you may be confident that you understand how to find the "same component" for each pair.)

146. Construct symmetry coordinates from the 15 infinitesimal displacements of $PtCl_4^{2-}$. (Cf. Exercise 10.128.) Group those which mix with each other; separate the two components of E vibrations.

147. Construct symmetry orbitals from the 1*s* orbitals of the hydrogens and the 2*s*, $2p_x$, $2p_y$, and $2p_z$ orbitals of the carbons of cyclopropane. (Cf. Exercise 10.124.) Group the symmetry orbitals which mix with each other.

10.18 Vanishing of Integrals

We have often invoked the principle that the integral of an odd function over a symmetric interval is zero. Group theory enables us to generalize this principle to integrals over any region of integration that is unchanged by every symmetry operation of a group. (Usually the region of integration is all space, which is invariant to every operation of every point group.)

Let ϕ be a function that transforms as some irreducible representation Γ_α of a group $\mathcal{G}$. Then we may take advantage of symmetry to determine whether $\int \phi \, d\tau$ vanishes. Indeed, unless Γ_α is the totally symmetric irreducible representation, $\int \phi \, d\tau$ must be zero by the following reasoning: Since $\int \phi \, d\tau$ is a number independent of the coordinates, it must be invariant to a transformation of coordinates by any R in $\mathcal{G}$: $\int \phi \, d\tau = R \int \phi \, d\tau$. But if Γ_α is one-dimensional, $R\phi = \chi_R^{\Gamma_\alpha}\phi$, so $\int \phi \, d\tau = \chi_R^{\Gamma_\alpha} \int \phi \, d\tau$. And if $\Gamma_\alpha \neq \Gamma_1$, there exists an R for which $\chi_R^{\Gamma_\alpha} \neq 1$. Therefore we may conclude that $\int \phi \, d\tau = 0$. Furthermore, if ϕ transforms as some multidimensional irreducible representation, then ϕ can always be written as a linear combination of functions, each of which is converted to some different multiple of itself by some R. Therefore we may conclude that the only situation whereby it is possible that $\int \phi \, d\tau$ be different from zero is if ϕ transforms as Γ_1. Notice that group theory can never tell us the value of this integral, but only whether it is necessarily zero by symmetry.

This technique is also applicable to integrals of products. If the set of functions $\{\phi_i\}$ transforms as Γ_α and the set $\{\chi_j\}$ as Γ_β, the set of all possible products transforms as $\Gamma_\alpha \times \Gamma_\beta$. If $\Gamma_\alpha \times \Gamma_\beta$ is irreducible and different from Γ_1, every $\int \phi_i \chi_j \, d\tau = 0$. If $\Gamma_\alpha \times \Gamma_\beta$ is reducible to $\sum n_\gamma \Gamma_\gamma$, there are n_γ linear combinations of the products which transform as Γ_γ and only n_1 integrals which do not necessarily vanish.

The extension to triple products is obvious. Such integrals arise in the evaluation of transition probabilities. The probability that a system in state ψ_α will absorb or emit a photon and attain state ψ_β is proportional to the square of the integral $\int \psi_\alpha^* \Omega \psi_\beta \, d\tau$, where Ω may be any of several operators. For the usual electric-dipole transitions, Ω is $\mathbf{T} = x\hat{\mathbf{i}} + y\hat{\mathbf{j}} + z\hat{\mathbf{k}}$, but it may be $\mathbf{R}$ (for magnetic-dipole transitions) or $\boldsymbol{\alpha}$ (polarizability, for Raman transitions). Fortunately, the transformation properties of $\mathbf{T}$, $\mathbf{R}$, and $\boldsymbol{\alpha}$ are listed in character tables. Also, it can be shown that $(\Gamma_\alpha \times \Gamma_\beta) \times \Gamma_\gamma = \Gamma_\alpha \times (\Gamma_\beta \times \Gamma_\gamma)$ and that $\Gamma_\alpha \times (\Gamma_\beta + \Gamma_\gamma) = \Gamma_\alpha \times \Gamma_\beta + \Gamma_\alpha \times \Gamma_\gamma$.

If $\Gamma_\alpha^* \times \Gamma_{\mathrm{T}} \times \Gamma_\beta$ does not contain Γ_1, the integrals $\int \psi_\alpha^* \mathbf{T} \psi_\beta \, d\tau$ vanish and the transition is said to be FORBIDDEN. If some $\Gamma_\alpha^* \times \Gamma_{\mathrm{T}} \times \Gamma_\beta$ does contain Γ_1, the transition is said to be ALLOWED and x-, y-, or z-POLARIZED, depending upon which direct product contains Γ_1. Likewise, transitions may be magnetic-dipole allowed or Raman active.

Although it is possible to evaluate the irreducible representations contained in a direct product by multiplication of characters and use of the reduction formula, many simple rules for these products are conveniently given in Wilson, Decius, and Cross, p. 331. Further simplification results from the fact that the ground state of most molecules is Γ_1. The vibrational state with one quantum of excitation in a normal mode of symmetry Γ_α transforms as Γ_α.

Exercises

148. Use a geometrical argument to convince yourself that if ϕ is antisymmetric with respect to any operation R, then $\iiint \phi(\mathbf{X})\, dV = 0$: Convince yourself that for every volume element that contributes positively to the integral, there is a corresponding one that contributes an equal amount negatively.
149. Show that if $\Gamma_\alpha \neq \Gamma_\beta$, $\Gamma_\alpha^* \times \Gamma_\beta$ does not contain Γ_1. Show that if $\Gamma_\alpha = \Gamma_\beta$, $\Gamma_\alpha^* \times \Gamma_\beta$ contains Γ_1 once.
150. The UV bands of benzene correspond to excitation of an electron from an e_{1g} orbital to an e_{2u} orbital; the symmetries of the excited states are given by $E_{1g} \times E_{2u}$. What is the symmetry of each state? Which transition from the A_{1g} ground state is allowed? What is the polarization? (See page 347 for $\Gamma_{\mathbf{T}}$.)
151. The ground state of benzene radical anion is of E_{2u} symmetry. What electronic transitions are allowed?
152. Which normal modes of $PtCl_4^{2-}$ are IR-active? Which are Raman-active?

*153. Which normal modes of a regular hexagon are IR-active? Which are Raman-active?

*154. In a molecule with a center of symmetry, each component of $\mathbf{T}$ transforms as some Γ_u and each component of $\boldsymbol{\alpha}$ transforms as some Γ_g. Show that no normal mode of a centrosymmetric molecule can be both IR- and Raman-active.

APPENDIX 1 THE TERMINOLOGY OF LOGIC

This brief appendix is an explicit statement of some common-sense ideas that are often understood too vaguely. The rudiments of logic are not readily accessible to chemists, so we here present them, especially as they apply to the way the definitions and theorems of mathematics are stated. The student who has difficulty in grasping the precise meaning of mathematical statements may find this discussion helpful.

Let p and q represent statements that are capable of being judged true or false, such as "$2 + 2 = 4$," "All men are mortal," "J. S. Bach is the brand name of a filter-tip cigarette," or "If the sum of the interior angles of a plane polygon is 180°, the polygon is a square." If we simply write p, we mean "The statement p is true." The opposite, "p is false," will be written $\sim p$. It is clear that a statement and its opposite cannot both be true.

The following statements (whose components are themselves statements) are all equivalent ways of making the *same* statement!

1a. If p is true, then q is true. (Or, by the above convention, "If p, then q.")
1b. (The truth of) p implies (the truth of) q (written $p \rightarrow q$).
1c. If q is false, then p is false (written $\sim q \rightarrow \sim p$).
1d. Either p is false or q is true (or both).
1e. It is not possible that both p and $\sim q$ be true (that p be true and q false).

For example, $x = 3 \rightarrow x^2 = 9$ and $x^2 \neq 9 \rightarrow x \neq 3$. Form c is known as the CONTRAPOSITIVE form. It is often the easier statement to prove—assume q is false, and then prove that p is false. If $p \rightarrow q$, then (the truth of) p is a SUFFICIENT CONDITION for (the truth of) q. If $p \rightarrow q$, then (the truth of) q is a NECESSARY CONDITION for (the truth of) p. Most mathematical theorems are stated in the form of an implication. If the theorem

$p \to q$ can be proved, and if p is known to be true, the truth of q may be deduced.

The opposite of the above statements may be stated in several ways:

2a. It is not true that p implies q [written $\sim(p \to q)$].
2b. Even if p is true, q may be false.

The CONVERSE of statement $p \to q$ is $q \to p$. It is not necessarily true that a statement and its converse be true. For example, $x = 3 \to x^2 = 9$, but it is not true that $x^2 = 9 \to x = 3$.

But it is *possible* that a statement and its converse both be true. We abbreviate the pair of statements $p \to q$ and $q \to p$ by $p \leftrightarrow q$: "p is true if and only if q is true." The truth of p is then both a necessary and sufficient condition for the truth of q. Also, "p is EQUIVALENT to q."

Definitions are always stated in an if-and-only-if form. Thus "A (plane) triangle is a figure bounded by three straight lines" means both "If x is a figure bounded by three straight lines, x is a triangle," and "If x is a triangle, x is a figure bounded by three straight lines."

Most of the confusion about logical terminology arises in the usage of such words as "all" and "some." In mathematical usage, the words "all," "every" "each," and "any" are synonymous. Thus the following are all equivalent statements:

3a. All squares are quadrilaterals.
3b. Every square is a quadrilateral.
3c. Each square is a quadrilateral.
3d. For all x (No matter what x you choose) if x is a square, x is a quadrilateral.
3e. Any square is a quadrilateral.

"Any" is the most difficult of the lot, because in ordinary usage "any" is often used to mean "any, even one." Compare "If every juror votes Not Guilty, the defendant will be acquitted" with "If any juror votes Not Guilty, the defendant will be acquitted." In ordinary usage the former means that a unanimous vote for acquittal results in acquittal, and the latter means that a unanimous vote of Guilty is required for conviction. In mathematical usage, "any" has the sense of "any arbitrary," and the logical meaning of "Any square is a quadrilateral" is "No matter what square you might pick, it is a quadrilateral."

The words "some" and "there exist(s)" are likewise synonymous. The following are equivalent:

4a. Some quadrilateral(s) is a (are) square(s).
4b. There exists (at least) one quadrilateral which is a square.
4c. There is an x which is both a quadrilateral and a square.

Here the plural should be avoided since it suggests more than one example, whereas logically we are usually content with a single example.

The *opposite* of statement 3 may be stated in any of the following ways:

5a. Not all squares are quadrilaterals.
5b. Not every square is a quadrilateral.
5c. Some square is not a quadrilateral.
5d. There exists at least one square which is not a quadrilateral.
5e. There exists a nonquadrilateral which is a square.

The *opposite* of statement 4 may be stated:

6a. No quadrilaterals are squares.
6b. No quadrilateral is a square.
6c. All squares are nonquadrilaterals.
6d. All quadrilaterals are nonsquares.
6e. Nothing is both a square and a quadrilateral.

All forms of statements 5 and 6 are false, so that it is quite difficult to detect the equivalence of the various forms. You may convert each to a true statement by replacing "quadrilateral" by "triangle," whereupon the truth and the equivalences will be obvious. Notice especially that "all" and "not all" are opposites, as are "some" and "no," but the opposite of "all" is *not* "none."

Exercises

1. Is $(p \to q) \leftrightarrow (\sim q \to \sim p)$ true?

*2. To which statement above is "p only if q" equivalent?

3. Theory A predicts that experiment e leads to result r, abbreviated $A \to (e \to r)$. Theory B predicts the opposite result: $B \to (e \to \sim r)$. When the experiment is performed, result $\sim r$ is obtained. What can you conclude about the truth of A? Of B?

*4. Let $p \,\&\, q$ mean both p and q are true. Let $p \circ q$ mean p is true or q is true or both are true.
(a) Is $\sim(p \,\&\, q) \leftrightarrow \sim p \,\&\, \sim q$ true?
(b) $\sim(p \,\&\, q) \leftrightarrow \sim p \circ \sim q$?
(c) $\sim(p \circ q) \leftrightarrow \sim p \,\&\, \sim q$?
(d) $\sim(p \,\&\, q) \leftrightarrow p \to \sim q$?
(e) $p \circ q \leftrightarrow \sim p \to q$?
(f) $p \to (q \circ r) \leftrightarrow \sim q \to (p \to r)$?

*5. For each of the following statements, in as many ways as possible, (*1*) state its contrapositive (or paraphrase it), (*2*) state its opposite:
(a) No man is an island.
(b) Everything in the world is good for something.
(c) If everyone does his work carefully, no one will have to stay after 5 P.M.
(d) You cannot fool all of the people all of the time.
(e) Only the perfect martini gin makes any quinine water a treat.

APPENDIX 2
CONVERSION OF UNITS

Almost all chemists learn early how to convert units, but for those who never learned the simple trick of canceling units we include it here. We also include a brief introduction to the powerful technique of dimensional analysis, which may be used to derive the form of physical laws without detailed computation.

Often it is necessary to express a quantity in units other than those given. For example, we might want to know a length in centimeters when it is given in inches, or we might want to know how many moles there are in 5 lb of some material.

Two "tricks" are helpful in such conversions. The first is to maintain dimensions throughout and "cancel" them from numerator and denominator where applicable. The second is to multiply by "unity," which will not change the quantity of interest. For example, since 1 in. = 2.54 cm, 2.54 cm/1 in. = unity, so we can multiply a length expressed in inches by this "unity" in order to convert to centimeters. The choice of which number goes into the numerator of the "unity" and which into the denominator is made so as to "cancel" dimensions.

Examples

1. 3 cm = ? in:

$$3\ \text{cm} \times \frac{1\ \text{in}}{2.54\ \text{cm}} = 1.18\ \text{in.}$$

2. 0.25 lb I_2 = ? moles I_2:

$$0.25\ \text{lb}\ I_2 \times \frac{453.6\ \text{g}}{\text{lb}} \times \frac{\text{mol}\ I_2}{253.8\ \text{g}\ I_2} = 0.446\ \text{mol}\ I_2.$$

3. $R = 1.9865$ cal/mol-deg. $=$? l-atm/mol-deg.

$$1.9865\,\frac{\text{cal}}{\text{mol-deg}} \times \frac{4.185\text{ J}}{\text{cal}} \times \frac{10^7\text{ erg}}{\text{J}} \times \frac{1\text{ dyn-cm}}{\text{erg}} \times \frac{\text{mm Hg}}{1333.22\text{ dyn/cm}^2}$$

$$\times \frac{\text{atm}}{760\text{ mm Hg}} \times \frac{1}{10^3\text{ cc}} = 8.207 \times 10^{-2}\,\frac{\text{l-atm}}{\text{mol-deg}}.$$

4. The reciprocal of the wavelength of a monochromatic beam of light is expressed in cm^{-1}. Since $E = h\nu$, and $\nu\lambda = c = 2.998 \times 10^{10}$ cm/sec, these units are directly related to energy per photon. How many cm^{-1} correspond to 1 kcal/mole? To 1 eV? To kT at 20°C?

a. $$1\,\frac{\text{kcal}}{\text{mol}} \times \frac{10^3\text{ cal}}{\text{kcal}} \times \frac{\text{mol}}{6.023 \times 10^{23}\text{ ``quanta''}} \times \frac{4.185\text{ J}}{\text{cal}} \times \frac{10^7\text{ erg}}{\text{J}}$$

$$\times \frac{10^{27}}{6.625\text{ erg-sec}} \times \frac{\text{sec}}{2.998 \times 10^{10}\text{ cm}} = 350\,\frac{\text{cm}^{-1}}{\text{quantum}}.$$

b. $$1\,\frac{\text{eV}}{\text{quantum}} \times \frac{F}{6.023 \times 10^{23}\text{ electron}} \times \frac{96{,}500\text{ coulomb}}{F} \times \frac{1\text{ W-sec}}{\text{coulomb-V}}$$

$$\times \frac{1\text{ J}}{\text{W-sec}} \times \frac{10^7\text{ erg}}{\text{J}} \times \frac{10^{27}}{6.625\text{ erg-sec}} \times \frac{\text{sec}}{2.998 \times 10^{10}\text{ cm}} = 8067\,\frac{\text{cm}^{-1}}{\text{quantum}}.$$

c. $$kT = 1.38 \times 10^{-16}\,\frac{\text{erg}}{\text{quantum-°K}} \times 293.16\text{°K}$$

$$\times \frac{10^{27}}{6.625\text{ erg-sec}} \times \frac{\text{sec}}{2.998 \times 10^{10}\text{ cm}} = 203.7\,\frac{\text{cm}^{-1}}{\text{quantum}}.$$

Exercises

1. The distance from the earth to the sun is 9.3×10^7 miles. What is the velocity of the earth in its orbit, in km/sec?
*2. What is $kT = RT/N$ at 0°C, in electron volts?
*3. A 100-ohm resistor is connected to a 6V battery. How much power is produced, in cal/min?
*4. Aniline ($C_6H_5NH_2$) reacts with 3 moles bromine to give tribromoaniline + $3H^+$ + $3Br^-$. How many ml Br_2 should be taken to react with 10 ml aniline?
*5. How many coulombs of electricity are required to reduce 2 liters chlorine gas (25°C, 1 atm) to Cl^-?

This sort of analysis is not without intrinsic significance. DIMENSIONAL ANALYSIS is a technique for determining physical laws in the form of a proportionality. The underlying principle is merely the requirement that both sides of an equation must be expressed in the same units. Of course, since only dimensions are involved, the technique cannot specify dimensionless constants; therefore it is limited to proportionalities. However, the advantage is that the only scientific input is a guess at what physical constants and parameters might be involved, and no detailed calculations are required.

Example

5. The quantum-mechanical energy of the ground state of the hydrogen atom: The quantities we may expect are m, the mass of the electron in grams; e, the charge on the electron in statcoulombs (dyne$^{1/2}$-cm); and h, Planck's constant in erg-sec. Let us try for a relationship of the form $E \propto m^x e^y h^z$. Dimensionally this requires erg $=$ g^x (dyne$^{1/2}$ cm)^y (erg-sec)^z, or g cm^2 sec^{-2} $=$ g^x (g$^{1/2 y}$ cm$^{1/2 y}$ sec^{-y} cm^y) (g^z cm^{2z} sec^{-z}). Therefore $1 = x + \frac{1}{2}y + z$, $2 = \frac{3}{2}y + 2z$, and $-2 = -y - z$, whose solution is $x = 1$, $y = 4$, and $z = -2$. Thus we find very simply that $E \propto me^4/h^2$. A genuine quantum-mechanical calculation shows that the proportionality constant is $2\pi^2$.

Exercises

6. What is the angular momentum, in g cm^2/sec, of an electron in a hydrogen atom?
7. What is the Bohr radius, the "average" distance between the proton and the electron in a hydrogen atom?
*8. What is the period of a pendulum, in seconds, in terms of its mass, its length, and the acceleration of gravity?
*9. Find an expression for the viscosity of a gas, in poise (g/cm sec), in terms of the gas constant, the absolute temperature, the mass of a molecule (g/molecule), Avogadro's number, the density of the gas, and the cross-sectional area of the molecule (cm^2/molecule).

APPENDIX 3
SOME THEOREMS OF ALGEBRA CONCERNING POLYNOMIALS

Throughout this text we invoke various theorems of algebra concerning polynomials. We collect them here, without proof.

Let the Nth-order polynomial $P_N(x) = c_0 + c_1 x + \cdots + c_{N-1}x^{N-1} + c_N x^N = \sum_{j=0}^{N} c_j x^j$. Then

1. $P_N(x)$ may be factored to $c_N(x - \lambda_1)(x - \lambda_2) \cdots (x - \lambda_N) = c_N \prod_{j=1}^{N} (x - \lambda_j)$. It is not necessary that each λ_j be distinct or that each λ_j be real.
2. The equation $P_N(x) = 0$ is satisfied by each λ_j: $P_N(\lambda_j) = 0$. Each λ_j is said to be a ROOT of P_N. And if $P_N(\lambda) = 0$, then $(x - \lambda)$ is a factor of $P_N(x)$.
3. If every c_j is real, complex roots occur in pairs that are complex conjugates of each other. If $\lambda = a + ib$ is a root, so is $\lambda^* = a - ib$. And if N is odd, there must be at least one real root.
4. $P_N(0) = c_0 = (-1)^N c_N \prod_{j=1}^{N} \lambda_j$.
5. $c_{N-1} = -c_N \sum_{j=1}^{N} \lambda_j$.
6. The roots of $P_2(x)$ are given by the quadratic formula

$$\lambda_{\pm} = \frac{-c_1 \pm \sqrt{c_1^2 - 4c_0c_2}}{2c_2}.$$

There exist formulas for the roots of a cubic or a quartic, but no formula exists for the roots of higher-order polynomials. However, if λ is known to be a root of $P_N(x)$, polynomial long division of $P_N(x)$ by $(x - \lambda)$ converts the problem of finding roots of an Nth-order polynomial to that of finding roots of an $(N-1)$st-order polynomial.

$$
\begin{array}{r|l}
 & c_N x^{N-1} + (c_{N-1} + \lambda c_N)x^{N-2} + \cdots \\
\cline{2-2}
x - \lambda & c_N x^N \qquad\qquad + c_{N-1}x^{N-1} + c_{N-2}x^{N-2} \\
 & c_N x^N \qquad\qquad - \lambda c_N\, x^{N-1} \\
\cline{2-2}
 & \qquad\qquad (c_{N-1} + \lambda c_N)x^{N-1} + c_{N-2}x^{N-2} \\
 & \qquad\qquad (c_{N-1} + \lambda c_N)x^{N-1} \ \text{etc.}
\end{array}
$$

APPENDIX 4
COMPLEX NUMBERS

Throughout this text we do not restrict ourselves to real numbers. Therefore the student who is unfamiliar with complex numbers is advised to read this appendix before beginning Chapter 1. Here we define the arithmetic of complex numbers, present a geometrical representation for them, and show how their powers and roots may be computed.

A4.1 Real Numbers

By REAL NUMBER is meant any number which has a decimal representation, not necessarily finite or even repeating. Thus the set of real numbers includes integers (positive and negative), rational numbers (fractions), and many more numbers. Examples of real numbers are:

$$\begin{array}{ll} 42 & 1.4142\cdots = \sqrt{2} \\ -0.125 & 3.14159\cdots = \pi \\ 1.666\cdots = \frac{5}{3} & 2.718\cdots = e. \end{array}$$

We may add, subtract, multiply, and divide (except by 0) real numbers and obtain a real number. We then say that the real numbers are CLOSED under addition, subtraction, multiplication, and division.

Furthermore, the real numbers may be put into a ONE-TO-ONE CORRESPONDENCE with the points on the real axis; to each number there corresponds a unique point, and vice versa. And we may speak of one real number as being greater or less than another.

A4.2 Complex Numbers

The square root of -1 is not a real number since no real number times itself equals -1. We shall write

$$i = \sqrt{-1} \qquad i^2 = -1$$

and represent a complex number in the form

$$a + ib$$

where a and b are real numbers. A real number is a special case of a complex number with $b = 0$; an IMAGINARY number is a special case of a complex number with $a = 0$. The complex numbers are also closed under addition, subtraction, multiplication, and division (except by $0 = 0 + i0$) since

$$(a + ib) + (c + id) \equiv (a + c) + i(b + d)$$

$$(a + ib) - (c + id) \equiv (a - c) + i(b - d)$$

$$(a + ib)(c + id) \equiv (ac - bd) + i(ad + bc)$$

$$\frac{a + ib}{c + id} \equiv \frac{ac + bd}{c^2 + d^2} + i\,\frac{bc - ad}{c^2 + d^2}\,.$$

Notice this last—complex numbers are defined to have the form $a + ib$, and any complex number with i in the denominator is understood to be convertible to standard form.

To each complex number z we associate a complex number z^*, known as its COMPLEX CONJUGATE, according to the rule: Replace i by $-i$.

$$z = a + ib \qquad \leftrightarrow \qquad z^* = (a + ib)^* = a - ib.$$

Obviously a real number, and *only* a real number, is its own complex conjugate. Other simple rules include

a. $(z^*)^* = z$.
b. $(z + w)^* = z^* + w^*$.
c. $(zw)^* = z^*w^*$.

From (b) and (a), it follows that for any z

$$(z + z^*)^* = z^* + (z^*)^* = z^* + z$$

so that $z + z^*$ must be real. From (c) and (a), it follows that for any z

$$(zz^*)^* = z^*(z^*)^* = z^*z$$

so that zz^* must also be real.

We define the REAL AND IMAGINARY PARTS of a complex number:

$$\text{Re}\,(z) \equiv \tfrac{1}{2}(z + z^*) \qquad \text{Re}\,(a + ib) = \tfrac{1}{2}[(a + ib) + (a - ib)] = a$$

$$\text{Im}\,(z) \equiv \frac{1}{2i}(z - z^*) \qquad \text{Im}\,(a + ib) = \frac{1}{2i}[(a + ib) - (a - ib)] = b.$$

It follows that any z may be reconstituted from its real and imaginary parts:

$$z = \text{Re}\,(z) + i\,\text{Im}\,(z).$$

To each complex number z, we may associate a nonnegative real number $|z|$, known as its ABSOLUTE VALUE, according to the rule

$$|z| \equiv +\sqrt{zz^*} \qquad |a + ib| = \sqrt{(a + ib)(a - ib)} = \sqrt{a^2 + b^2}.$$

Notice that although $|z|$ is nonnegative and may be considered as a "magnitude" of z, we cannot speak of one complex number as being greater than another.

A4.3 Geometrical Representation

We may put the complex numbers into one-to-one correspondence with the points in a plane. If $z = a + ib$, let z correspond to the point whose coordinates are (a, b) (Fig. A4.1). Notice that zero corresponds to the

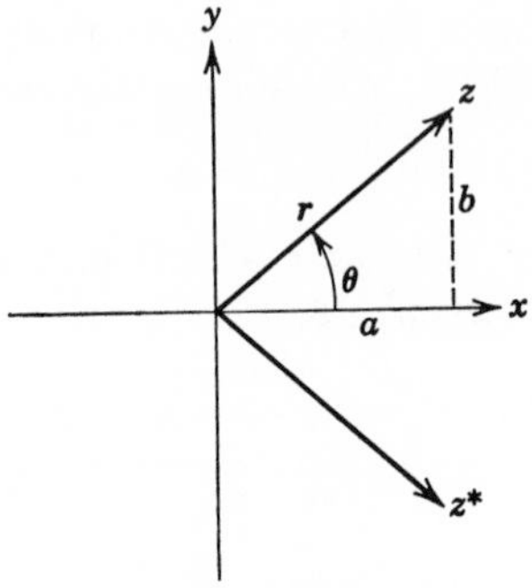

Figure A4.1

origin. Also, we see that addition of two complex numbers looks like the addition of two two-dimensional vectors since we add the real and imaginary components separately. And the point corresponding to z^* is related to the point corresponding to z by reflection through the x-axis.

We may also transform to polar coordinates r and θ, where θ is measured counterclockwise from the x-axis and runs from 0 to 2π.

$$a + ib = r(\cos\theta + i\sin\theta)$$
$$r = \sqrt{a^2 + b^2} = |a + ib|$$
$$\tan\theta = \frac{b}{a}$$
$$(a + ib)^* = r(\cos\theta - i\sin\theta)$$

Thus, a complex number may be specified either by its real and imaginary parts, or by the two parameters r and θ.

Exercises

1. Use a geometrical argument to show that for all z_1, z_2

$$|z_1 + z_2| \leqslant |z_1| + |z_2|$$
$$|z_1 - z_2| \geqslant |z_1| - |z_2|.$$

Then use the fact that Re $(z_1^* z_2) \leqslant |z_1^* z_2|$ to prove the same results.

*2. If $z_1 = r_1(\cos\theta_1 + i\sin\theta_1)$ and $z_2 = r_2(\cos\theta_2 + i\sin\theta_2)$, show that

$$z_1 z_2 = r_1 r_2[\cos(\theta_1 + \theta_2) + i\sin(\theta_1 + \theta_2)].$$

A4.4 Powers and Roots

In Exercise A4.2 you showed that to multiply two complex numbers you multiply their absolute values and add their angles. By repeated application of this product rule, we may obtain a simple expression for the nth power of a complex number

$$z^n = [r(\cos\theta + i\sin\theta)]^n = r^n(\cos n\theta + i\sin n\theta).$$

Next we consider the nth roots of a complex number, $Z = R(\cos\theta + i\sin\theta)$. Let $Z^{1/n} = r(\cos\phi + i\sin\phi)$, with r and ϕ to be determined. Substituting, we obtain

$$Z = R(\cos\theta + i\sin\theta) = r^n(\cos n\phi + i\sin n\phi) = (Z^{1/n})^n.$$

Clearly, it is necessary to solve the equations $R = r^n$, $\cos\theta = \cos n\phi$, and $\sin\theta = \sin n\phi$. The obvious solution to the first equation is

$$r = R^{1/n}.$$

One obvious solution to the last pair of equations is $\theta = n\phi$, or $\phi = \theta/n$. But another solution is $\theta = n\phi \pm 2\pi$, or $\phi = (\theta \pm 2\pi)/n$, since $\cos(n\phi \pm 2\pi) = \cos n\phi$ and $\sin(n\phi \pm 2\pi) = \sin n\phi$. Indeed, the totality of distinct solutions is

$$\phi = (\theta + 2\pi k)/n \qquad k = 0, 1, 2, \ldots, n - 1.$$

Notice that there are n nth roots in all.

Example

Find $(-3)^{1/4}$. The four roots lie on a circle of radius $3^{1/4}$ in the complex plane. Since

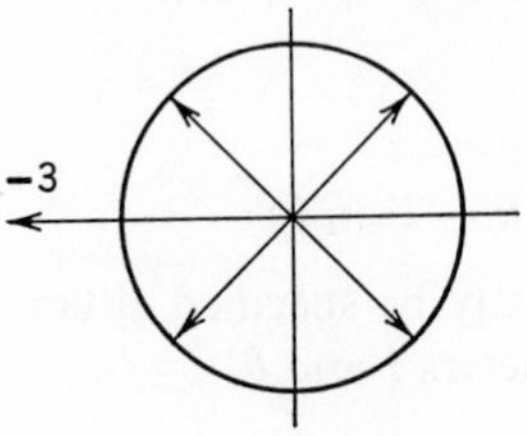

Figure A4.2

$-3 = 3\cos\pi$, the angular coordinates of the roots are $\pi/4$, $3\pi/4$, $5\pi/4$, and $7\pi/4$ (see Figure A4.2), so

$$
\begin{aligned}
(-3)^{1/4} &= \frac{3^{1/4}(1+i)}{\sqrt{2}} \\
&= \frac{3^{1/4}(-1+i)}{\sqrt{2}} \\
&= \frac{3^{1/4}(-1-i)}{\sqrt{2}} \\
&= \frac{3^{1/4}(1-i)}{\sqrt{2}} .
\end{aligned}
$$

Exercises

3. Find the cube roots of $-i$.
4. Use a geometrical argument to convince yourself that the sum of the nth roots of unity is zero ($n \geqslant 2$).

APPENDIX 5
LINE INTEGRALS IN THE COMPLEX PLANE

We present here only a brief introduction to complex-variable theory since a more extensive account is of interest only to the more physical physical chemists. We begin with a discussion of complex-valued functions of a complex variable, especially those expandable in Taylor series. We then consider line integrals of complex functions in the complex plane, and we show how these considerations may be applied to the evaluation of ordinary definite integrals. We do not proceed to discuss conformal mapping, but the student who encounters partial differential equations and boundary-value problems is advised to pursue this topic in advanced texts. It is advisable not to take up this appendix until after Chapter 7 is completed.

A5.1 Analytic Functions

The functions to which we turn our attention are complex-valued functions of a complex variable. Such a function takes a complex number z and unambiguously produces a complex number $w = f(z)$:

$$z = x + iy$$
$$w = f(z) = u(x, y) + iv(x, y).$$

Here u and v are real-valued functions of the two real variables, x and y.

Consider the derivative of f, evaluated at $z = z_0$:

$$f'(z_0) \equiv \lim_{z \to z_0} \frac{f(z) - f(z_0)}{z - z_0} = \lim_{\Delta x, \Delta y \to 0} \frac{\Delta u + i\Delta v}{\Delta x + i\Delta y}.$$

For this limit to exist, the approach of z to z_0 must give the same limit no matter what the path of approach is. In particular, let $\Delta x \to 0$ first; then

$$f'(z_0) = \lim_{\Delta y \to 0} \frac{\Delta u + i\Delta v}{i\Delta y} = \frac{\partial v}{\partial y} - i\frac{\partial u}{\partial y}.$$

If, on the other hand, Δy approaches zero first, then

$$f'(z_0) = \lim_{\Delta x \to 0} \frac{\Delta u + i\Delta v}{\Delta x} = \frac{\partial u}{\partial x} + i\frac{\partial v}{\partial x}.$$

Comparing real and imaginary parts of these two expressions for $f'(z_0)$, we see that necessary (and also sufficient, if u and v are "reasonable") conditions that $f'(z_0)$ exist, independent of the way z approaches z_0, are

$$\frac{\partial v}{\partial y} = \frac{\partial u}{\partial x} \quad \text{and} \quad -\frac{\partial u}{\partial y} = \frac{\partial v}{\partial x}. \tag{A5.1}$$

These equations are known as the CAUCHY-RIEMANN CONDITIONS.

Examples

1. $f(z) = e^z = e^{x+iy} = e^x \cos y + ie^x \sin y$

$$\frac{\partial v}{\partial y} = e^x \cos y = \frac{\partial u}{\partial x} \qquad -\frac{\partial u}{\partial y} = e^x \sin y = \frac{\partial v}{\partial x}$$

so the derivative $f'(z_0)$ exists at all points in the complex plane.

2. $f(z) = z^* = x - iy$

$$\frac{\partial v}{\partial y} = -1 \neq 1 = \frac{\partial u}{\partial x} \qquad -\frac{\partial u}{\partial y} = 0 = \frac{\partial v}{\partial x}.$$

Since the first condition is satisfied nowhere, the derivative exists nowhere.

3.

$$f(z) = \frac{1}{z} = \frac{1}{x+iy} = \frac{x-iy}{x^2+y^2}$$

$$\frac{\partial v}{\partial y} = \frac{y^2 - x^2}{(x^2+y^2)^2} = \frac{\partial u}{\partial x} \qquad -\frac{\partial u}{\partial y} = \frac{2xy}{(x^2+y^2)^2} = \frac{\partial v}{\partial x}.$$

These conditions are satisfied everywhere except at $z = 0$, where the quantities are infinite. Indeed, the function is not even continuous at $z = 0$.

A (single-valued) function f for which $f'(z)$ exists for every z "near" z_0 is said to be ANALYTIC at z_0. And by extension we may speak of a function as being analytic in a region of the complex plane. The functions in the above examples are, respectively, analytic everywhere, analytic nowhere, and analytic everywhere except at $z = 0$. What we have shown is that a necessary (and sufficient, too) condition that the function f be analytic in a region is that the Cauchy-Riemann conditions hold for all points in that region.

Furthermore, it can be shown that if f is analytic in a region, not only does $f'(z_0)$ exist at every point in that region, but also *all* derivatives exist; that is to say, $f(z)$ may be expanded in Taylor series about any z_0 in the region. Also, those points of a region where $f'(z)$ does not exist are called SINGULAR POINTS or SINGULARITIES.

It is readily shown that all the familiar rules of differentiation hold for complex functions. Furthermore, the derivative of a complex function is the "same" as the derivative of the corresponding real function:

$$\begin{aligned} f(z) &= z & f'(z) &= 1 \\ f(z) &= z^2 & f'(z) &= 2z \\ f(z) &= z^n & f'(z) &= nz^{n-1} \\ f(z) &= e^z & f'(z) &= e^z \\ f(z) &= \sin z & f'(z) &= \cos z \\ f(z) &= \cos z & f'(z) &= -\sin z. \end{aligned}$$

Therefore the series expansion of an analytic function of a complex variable must be the familiar series expansion.

Exercises

1. Test for analyticity and find any singularities of:
   (a) $f(z) = z^2$.
   (b) $f(z) = \text{Re}\,(z)$.
   (c) $f(z) = \cos z$.
   (d) $f(z) = 1/(1 + z^2)$.

*2. Test for analyticity and find any singularities of:
   (a) $f(z) = z^n$ (n is a positive integer).
   (b) $f(z) = z^{1/2}$.
   (c) $f(z) = \ln(1 + z)$. [*Hint:* $w = \frac{1}{2}(w + w^*) + \frac{1}{2}(w - w^*)$.]

3. Convince yourself that a finite linear combination of analytic functions is also analytic.

*4. Show that if f is analytic, then $u(x, y) = \text{Re}\,f(z)$ and $v(x, y) = \text{Im}\,f(z)$ both satisfy Laplace's equation in two variables.

5. It is also possible to express a complex function in polar coordinates: $f(z) = f(re^{i\theta}) = Re^{i\Theta}$, where R and Θ are real functions of the real variables r and θ. Express the following functions in polar coordinates:
   (a) $f(z) = z^2$.
   (b) $f(z) = z^{1/2}$.
   (c) $f(z) = z^*$.
   (d) $f(z) = |z|$.
   (e) $f(z) = 1/z$.

A5.2 Line Integrals of Complex Functions

Next we define a LINE INTEGRAL over some path Γ in the complex plane:

$$\int_\Gamma f(z)\,dz \equiv \lim_{\Delta z_i \to 0} \sum f(z_i)\,\Delta z_i,$$

where $\{z_i\}$ is a set of points on Γ and $\Delta z_i = z_{i+1} - z_i$. Rewriting the integral in terms of real functions leads to

$$\int f(z)\,dz = \int (u - iv)(dx + i\,dy) = \int (u\,dx - v\,dy) + i \int (v\,dx + u\,dy).$$

From our considerations of line integrals in the xy-plane (Section 7.14), the requirement that $\int_\Gamma f(z)\,dz$ be independent of path is that

$$\frac{\partial u}{\partial y} = -\frac{\partial v}{\partial x} \quad \text{and} \quad \frac{\partial v}{\partial y} = \frac{\partial u}{\partial x}. \tag{A5.2}$$

Notice that what we have just shown is that for $\int_\Gamma f(z)\,dz$ to be independent of the path we choose, the Cauchy-Riemann equations must hold, that is, f must be analytic in the region of all possible paths. Alternatively, we may state this result as the CAUCHY INTEGRAL THEOREM. If f is analytic at all points within and on a closed curve (CONTOUR), then

$$\oint f(z)\,dz = 0. \tag{A5.3}$$

By convention a contour is traversed in a counterclockwise direction.

Exercises

6. Convince yourself that $\int_{\Gamma_1} f(z)\,dz + \int_{\Gamma_2} f(z)\,dz = \int_{\Gamma_1+\Gamma_2} f(z)\,dz$ and $\int_{-\Gamma} f(z)\,dz = -\int_\Gamma f(z)\,dz$, where $-\Gamma$ is path Γ traversed in the opposite direction.
7. Generalize the fact that $|z_1 + z_2| \leqslant |z_1| + |z_2|$ (Exercise A4.1) to convince yourself that $|\int f(z)\,dz| \leqslant \int |f(z)|\,|dz|$.

*8. If $f(z) = f(re^{i\theta}) = R(r, \theta)e^{i\Theta(r,\theta)}$, convince yourself that $\int_\Gamma f(z)\,dz$ may also be reduced to the line integral $\int R(r, \theta)e^{i\theta}e^{i\Theta(r,\theta)}\,dr + irR(r, \theta)e^{i\theta}e^{i\Theta(r,\theta)}\,d\theta$.

Examples

To evaluate $\oint_C f(z)\,dz$, where C is the unit circle, $|z| = 1$:

4. If $f(z) = z^2$, then $\oint_C z^2\,dz = \oint d(\frac{1}{3}z^3) = 0$, as was also clear from the fact that $f(z) = z^2$ is analytic everywhere.

5. If $f(z) = z^*$, let us transform to polar coordinates: Since $r = 1$ everywhere on C, $z = e^{i\theta}$, $z^* = e^{-i\theta}$, and $dz = ie^{i\theta}\,d\theta$. Then $\oint z^*\,dz = \int_0^{2\pi} e^{-i\theta}ie^{i\theta}\,d\theta = 2\pi i$. Since $f(z) = z^*$ is not analytic, this contour integral is not necessarily zero.

6. If $f(z) = z^{1/2}$, transformation to polar coordinates gives $\oint z^{1/2}\,dz = \int_0^{2\pi} e^{i\theta/2}ie^{i\theta}\,d\theta = \frac{2}{3}[e^{3i\theta/2}]_0^{2\pi} = -\frac{4}{3}$. However, if the unit circle were traversed twice instead of once, the contour integral would vanish. The "nastiness" of $z^{1/2}$ is not so much that it is not analytic at $z = 0$, but rather that it is not even single-valued. On the unit circle at $\theta = 0$, $z^{1/2} = 1$, but after one traversal of the circle, at $\theta = 2\pi$, $z^{1/2} = -1$. This means that by traversing the circle once, we have not been able to restrict ourselves to the "principal value" of the square root. Similar dilemmas arise with most inverse functions, and the troubled reader is warned to look into the subjects of "Riemann surfaces" and "branch points."

Exercises

9. Evaluate $\oint \cos z \, dz$ and $\oint dz/(4 - z^2)$, where the contour is the unit circle in the complex plane.

*10. Apply the Cauchy integral theorem to $\oint_C \exp(-\frac{1}{2}\pi z^2)\, dz$, where C is the contour of Figure A5.1, and show that the Fresnel integrals $C(\infty) = \int_0^\infty \cos(\frac{1}{2}\pi t^2)\, dt$ and

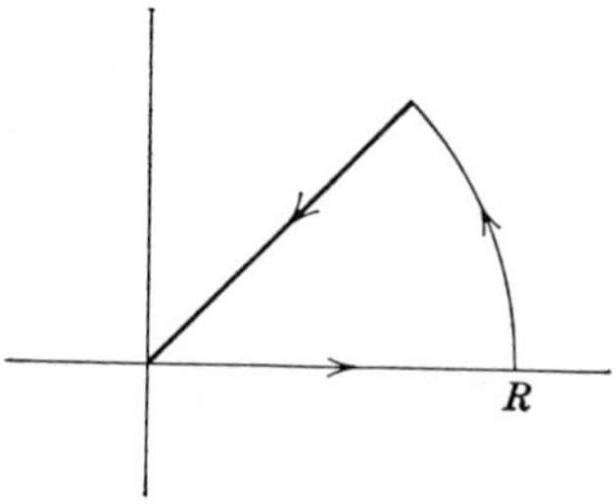

Figure A5.1

$S(\infty) = \int_0^\infty \sin(\frac{1}{2}\pi t^2)\, dt$ are both equal to $\frac{1}{2}$. (*Hints:* Along the x-axis, let $z = x$. Along the circular portion of the contour let $z = Re^{i\theta}$, and use the result of Exercise A5.7 to help convince yourself that the integral along this path approaches zero for large R. Along the diagonal path, let $z = 2^{-1/2}t(1 + i)$.)

11. Transform to polar coordinates to evaluate $\oint dz/z$, where the contour is a circle of radius r in the complex plane.

*12. Evaluate $\oint z^2 \ln z \, dz$, where the contour is the unit circle in the complex plane. Notice that although $f'(z) = z + 2z \ln z$ exists everywhere, $f(z) = z^2 \ln z$ is not analytic because ln is not single-valued in the complex plane. (Cf. Section 1.7.)

Next we consider changing the contour. First we notice that if f is analytic

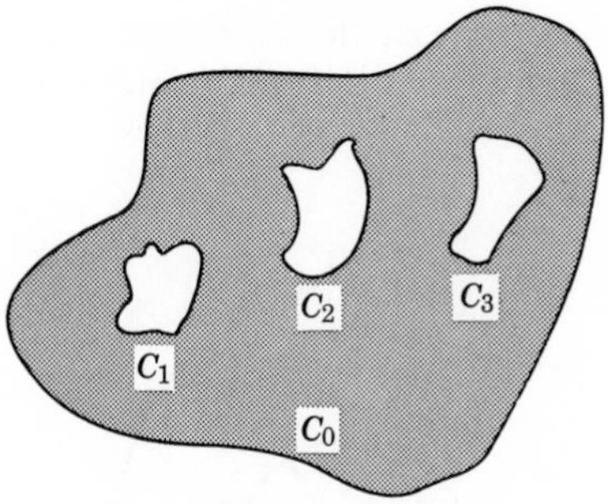

Figure A5.2

in the shaded region of Figure A5.2 and on its boundary, then the integral around that shaded region is zero:

$$\oint_{\text{shaded}} f(z)\, dz = \int_\Gamma f(z)\, dz + \int_{-C_2} f(z)\, dz + \int_{-\Gamma} f(z)\, dz + \int_{C_1} f(z)\, dz = 0.$$

Since the integrals over $-\Gamma$ and $-C_2$ are the negatives of the integrals over Γ and C_2, respectively, we may conclude that

$$\oint_{C_1} f(z)\, dz = \oint_{C_2} f(z)\, dz.$$

Thus we may deform the contour at will through a region of analyticity, without changing the value of the contour integral. Notice that it is not necessary that f be analytic within C_2; the only restriction is that we must not cross a singularity in deforming the contour. Furthermore, we may readily extend this result to more than one "hole:"

$$\oint_{C_0} f(z)\,dz = \oint_{C_1} f(z)\,dz + \oint_{C_2} f(z)\,dz + \oint_{C_3} f(z)\,dz$$

if f is analytic in the shaded region (Figure A5.3).

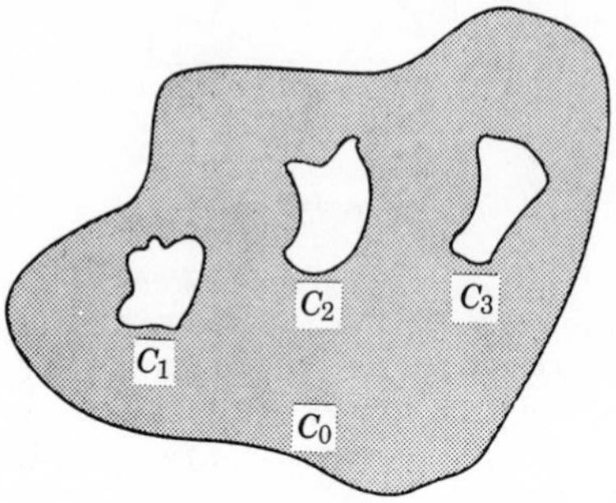

Figure A5.3

A5.3 Contour Integrals around Singularities

We next consider the contour integral $\oint_C dz/(z - z_0)^n$, where n is a positive integer and C encloses z_0 (otherwise the integrand would be analytic and the integral would be zero). But we may deform C into γ, a tiny circle of radius ρ and centered at z_0 (Figure A5.4). If z is on γ, then $z - z_0 = \rho e^{i\theta}$, $0 \leqslant \theta < 2\pi$,

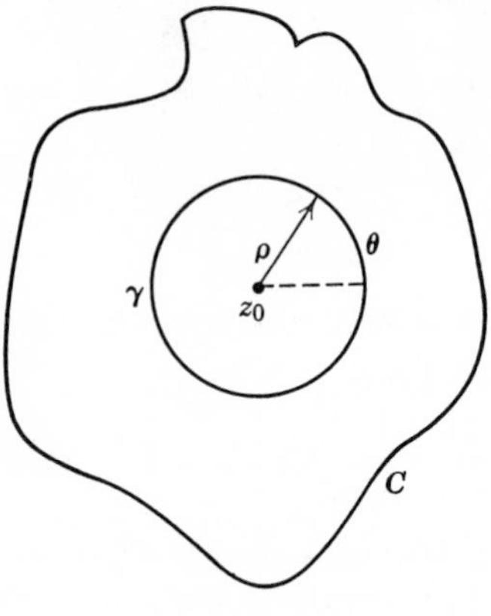

Figure A5.4

and

$$\oint_C \frac{dz}{(z-z_0)^n} = \oint_\gamma \frac{dz}{(z-z_0)^n} = \oint \frac{i\rho e^{i\theta}\,d\theta}{\rho^n e^{in\theta}} = \frac{i}{\rho^{n-1}} \int_0^{2\pi} e^{(1-n)i\theta}\,d\theta = 2\pi i \delta_{n1}. \tag{A5.4}$$

Thus we see that even though $(z - z_0)^{-n}$ is not analytic at z_0, a contour integral around z_0 vanishes unless $n = 1$.

Next we consider a more general integral, $\oint_C f(z)/(z - z_0)$, where f is analytic within and on C. By deformation this may be reduced to an integral around γ, a circle so tiny that $f(z)$ is constant on the circle and as close as desired to $f(z_0)$ (Figure A5.4). Then

$$\oint_C \frac{f(z)\,dz}{z-z_0} = \oint_\gamma \frac{f(z)\,dz}{z-z_0} = \oint_\gamma \frac{f(z_0)\,dz}{z-z_0} = f(z_0) \oint \frac{dz}{z-z_0} = 2\pi i f(z_0). \tag{A5.5}$$

This result is known as the CAUCHY INTEGRAL FORMULA, and we now show how it may be applied to the evaluation of some definite integrals.

Examples

7. To evaluate $I = \int_0^{2\pi} d\theta/(1 + a\sin\theta)$, $0 \leqslant a < 1$, let $z = e^{i\theta}$, so that $dz = ie^{i\theta}\,d\theta$ and $\sin\theta = (1/2i)(e^{i\theta} - e^{-i\theta}) = (1/2i)(z - 1/z)$. Then

$$I = \oint \frac{dz}{iz} \frac{1}{1 + (a/2i)(z - 1/z)} = \oint \frac{dz}{iz + (a/2)z^2 - a/2} = \frac{2}{a} \oint \frac{dz}{z^2 + (2i/a)z - 1},$$

where the contour is the unit circle, $|z| = 1$. The integrand is an analytic function except where the denominator equals zero. By the quadratic formula the zeros are at

$$z = -\frac{i}{a} \pm \left(-\frac{1}{a^2} + 1\right)^{1/2} = -\frac{i}{a} \pm \frac{i}{a}(1 - a^2)^{1/2}.$$

Call these roots z_+ and z_-, so that $(z - z_+)(z - z_-) = z^2 + (2i/a)z - 1$. But only z_+ (the root with the $+$ sign) lies within the unit circle. By the Cauchy integral formula, with $f(z) = 1/(z - z_-)$, we find

$$I = \frac{2}{a} \oint \frac{1}{z - z_-} \frac{dz}{z - z_+} = \frac{2}{a} \frac{2\pi i}{z_+ - z_-} = \frac{2}{a} \frac{2\pi i}{(2i/a)(1 - a^2)^{1/2}} = \frac{2\pi}{(1 - a^2)^{1/2}}.$$

8. To evaluate $I = \int_0^\infty (\cos mx\,dx)/(x^2 + 1)$, $m \geqslant 0$,

$$I = \int_0^\infty \frac{\cos mx\,dx}{x^2+1} = \frac{1}{2}\int_{-\infty}^\infty \frac{\cos mx\,dx}{x^2+1} = \frac{1}{2}\int_{-\infty}^\infty \frac{\mathrm{Re}\,e^{imx}\,dx}{x^2+1} = \frac{1}{2}\lim_{R\to\infty} \int_{-R}^R \frac{\mathrm{Re}\,e^{imx}\,dx}{x^2+1}.$$

This is still not an integral over a closed path, so our procedure will be to complete the contour by adding an integral over a large semicircle S in the $+i$ half of the plane (Figure A5.5). First we must show that the integral over S approaches zero as R becomes large. Then we may evaluate our definite integral as a contour integral around a region whose

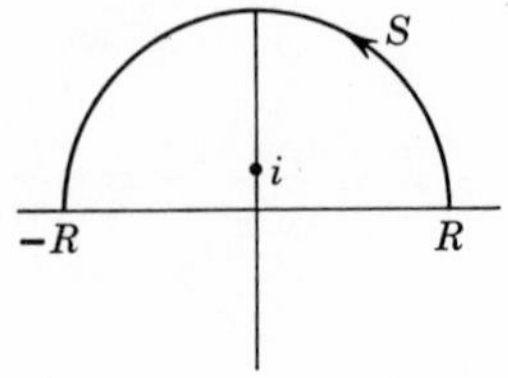

Figure A5.5

only singularity is at $z = i$.

$$I = \tfrac{1}{2}\text{Re}\oint \frac{e^{imz}\,dz}{z^2+1} - \tfrac{1}{2}\lim_{R\to\infty}\text{Re}\int_S \frac{e^{imz}\,dz}{z^2+1}.$$

But

$$\left|\int_S \frac{e^{imz}\,dz}{z^2+1}\right| \leqslant \int_S \frac{|e^{imz}|\,|dz|}{|z^2+1|} = \int_S \frac{|e^{imx}|\,e^{-my}\,|dz|}{|R^2e^{2i\theta}-(-1)|}$$

$$\leqslant \int_S \frac{|dz|}{R^2-1} = \int_0^\pi \frac{R\,d\theta}{R^2-1} = \frac{R\pi}{R^2-1} \to 0.$$

Here we have invoked (*1*) the result of Exercise A5.7, (*2*) $|z_1z_2| = |z_1|\,|z_2|$, (*3*) $|e^{imx}| = 1$, (*4*) $e^{-my} < 1$ for $y > 0$, and (*5*) $|z_1 - z_2| \geqslant |z_1| - |z_2|$. Indeed, it can be shown that a line integral $\int f(z)\,dz/g(z)$ over this infinite semicircular path always vanishes as long as $\lim_{z\to\infty} z^2\,|f(z)|/g(z)$ is finite.

Therefore, by the Cauchy integral formula, with $f(z) = e^{imz}/(z+i)$,

$$I = \tfrac{1}{2}\text{Re}\oint \frac{e^{imz}\,dz}{(z+i)(z-i)} = \tfrac{1}{2}\text{Re}\left[2\pi i\frac{e^{i2m}}{2i}\right] = \tfrac{1}{2}\text{Re}[\pi e^{-m}] = \frac{\pi}{2}e^{-m}.$$

9. To evaluate

$$I = \int_{-\infty}^{\infty} \frac{x\sin x\,dx}{(x^2+1)(x^2+4)} = \text{Im}\int_{-\infty}^{\infty} \frac{ze^{iz}\,dz}{(z^2+1)(z^2+4)} = \text{Im}\oint \frac{ze^{iz}\,dz}{(z^2+1)(z^2+4)}.$$

As in the previous example, we have converted the given integral to one around a semicircular contour, and we have used the fact that the integral along the semicircular path vanishes. The only singularities within this contour are at $z = i$ and $z = 2i$, so that the contour integral is equal to the sum of two contour integrals, one around each singularity:

$$I = \text{Im}\oint \frac{ze^{iz}}{(z+i)(z^2+4)}\frac{dz}{z-i} + \text{Im}\oint \frac{ze^{iz}}{(z^2+1)(z+2i)}\frac{dz}{z-2i}$$

$$= \text{Im}\left\{2\pi i\left[\frac{ie^{-1}}{2i(-1+4)} + \frac{2ie^{-2}}{(-4+1)4i}\right]\right\} = \frac{\pi}{3}(e^{-1}-e^{-2}).$$

Exercises

*13. A set of functions $\{f_n\}$ is defined in terms of the generating function $G(x, z) = \sum f_n(x)z^n$. Multiply by $z^{-(m+1)}$ and integrate around the origin to obtain a representation of $f_n(x)$ as a contour integral. Then use this result and the generating function for the Bessel functions

$$e^{\frac{1}{2}x(z-z^{-1})} = \sum_{n=-\infty}^{\infty} J_n(x)z^n$$

to show that $J_n(x) = (2\pi)^{-1}\int_{-\pi}^{\pi} e^{-in\theta}e^{ix\sin\theta}\,d\theta$.

*14. A distribution of N "objects" among s energy levels is specified by the set $\{n_i\}_{i=1}^s$ of occupation numbers, where $\sum n_i = N$. The number of ways of achieving a distribution with $\{n_i\}$ is given by the multinomial coefficient $M_{\{n_i\}}^N = N!/\prod n_i!$. In statistical mechanics it is necessary to evaluate such quantities as $\sum^* M_{\{n_i\}}^N$, where the asterisk means that the sum is over only those distributions for which $\sum n_i\varepsilon_i = E$. Use Eq. A5.4 (take all exponents as integers) to show that this sum is given by

$$\sum{}^* M_{\{n_i\}}^N = \frac{1}{2\pi i}\oint \frac{(\sum z^{\varepsilon_i})^N}{z^E}\frac{dz}{z}$$

where the contour encloses the origin. (*Hint:* Remember that the multinomial coefficients arise in the expansion $(\sum x_i)^N = \sum M_{\{n_i\}}^N \prod x_i^{n_i}$, and let $x_i = z^{\varepsilon_i}$.) This result is the basis for Fowler's approach to statistical mechanics, although the integrals must be approximated by the "method of steepest descent," which we shall not investigate.

*15. If f is analytic within and on a contour C enclosing $z = z_0$, show that if $n \neq 1$, $\oint_C f(z)\,dz/(z - z_0)^n = 0$.

16. Evaluate

(a) $$\int_0^\infty \frac{dx}{1 + x^4}.$$

(b) $$\int_{-\infty}^\infty \frac{x^2 + 1}{x^4 + 5x^2 + 6}\,dx.$$

(c) $$\int_0^\infty \frac{\cos x\,dx}{(x^2 + a^2)(x^2 + b^2)} \qquad (a \neq b).$$

(d) $$\int_0^{2\pi} \frac{d\theta}{a + b\cos\theta + c\sin\theta} \qquad (a^2 > b^2 + c^2).$$

(e) $$\oint_{|z|=1} \cot z\,dz = \oint \frac{z\cos z}{\sin z}\frac{dz}{z}.$$

A5.4 Series Expansions and Residues

We have already stated that a function which is analytic at z_0 may be expanded in a Taylor series about z_0. A function which is not analytic at z_0 cannot be so expanded, but may be expanded in a more general series, with negative powers, known as a LAURENT SERIES:

$$f(z) = \sum_{n=0}^{\infty} a_n(z - z_0)^n + \sum_{n=1}^{N} \frac{a_{-n}}{(z - z_0)^n} = \sum_{n=-N}^{\infty} a_n(z - z_0)^n. \qquad \text{(A5.6)}$$

The type of singularity at $z = z_0$ is characterized by N, the largest negative exponent for which $a_n \neq 0$. We shall deal only with situations in which N is finite, in which case z_0 is said to be a POLE OF ORDER N. Alternatively, a pole of order N may be defined by noting that $(z - z_0)^N f(z)$ is analytic at $z = z_0$, but $(z - z_0)^{N-1} f(z)$ is not. A pole of order unity is generally called a SIMPLE POLE; we have already seen how to evaluate a contour integral around a simple pole by the Cauchy integral formula.

We may use Eq. A5.4 to simplify an integral around a closed contour containing a pole of order N at z_0:

$$\oint f(z)\,dz = \sum_{n=-N}^{\infty} a_n \oint (z - z_0)^n\,dz = \sum_{n=-N}^{\infty} a_n \cdot 2\pi i \delta_{n,-1} = 2\pi i a_{-1}. \qquad \text{(A5.7)}$$

The coefficient a_{-1} thus plays a special role in this treatment. It is known as the RESIDUE of f at $z = z_0$, and we shall write it as $R(z_0)$.

This result may be generalized to the RESIDUE THEOREM. An integral around a contour that encloses several singularities, $\{z_j\}$, is equal to $2\pi i$ times the sum of the residues at the enclosed singularities:

$$\oint f(z)\,dz = 2\pi i \sum_j R(z_j), \qquad \text{(A5.8)}$$

where $R(z_j)$ is the residue of f at $z = z_j$, that is, the coefficient of $(z - z_j)^{-1}$ in the expansion of f about $z = z_j$. Also, it can be shown that if z_j is a simple pole *on* the contour, it contributes only half as much, $\pi i R(z_j)$, to the "principal value" of the integral.

Exercises

17. Previously we defined the Laplace transform by $L\{f(t)\} = F(s) = \int_0^\infty e^{-st} f(t)\,dt$. It can be shown that the Laplace transform may be inverted, according to

$$f(t) = L^{-1}\{F(s)\} = \frac{1}{2\pi i} \int_{c-i\infty}^{c+i\infty} e^{st} F(s)\,ds.$$

Notice that this is an integral along a path parallel to the imaginary axis; c is a constant such that Re c is greater than the real part of any singularity of F, that is, in the complex plane, c is to the right of all singularities of F. If $F(s)$ is sufficiently small in the negative half-plane, the line integral over a semicircular path vanishes so that the integral becomes a contour integral around a semicircle enclosing all singularities $\{z_j\}$ of F:

$$f(t) = \frac{1}{2\pi i} \oint e^{zt} F(z)\,dz = \sum R(z_j).$$

Find $f(t)$ for

(a) $F(s) = \dfrac{1}{s - a}$.

(b) $F(s) = \dfrac{1}{s^2 + a^2}$.

(c) $F(s) = \dfrac{s}{s^2 + a^2}$

(d) $F(s) = \dfrac{1}{s^2 - a^2}$

(e) $F(s) = \dfrac{s}{s^2 - a^2}$.

(f) $F(s) = \dfrac{s - a}{(s - a)^2 + b^2}$

18. Solve the set of simultaneous differential equations

$$-\frac{dA}{dt} = k_1 A \qquad A(0) = A_0$$

$$\frac{dB}{dt} = k_1 A - k_2 B \qquad B(0) = 0$$

by taking Laplace transforms of both sides, solving for LA and LB, and then inverting the Laplace transforms.

*19. Find $L^{-1}[s^3/(s^4 - a^4)]$ and $L^{-1}[s^3/(s^4 + 4)]$. Simplify your answers.

20. It can be shown that if $\lim_{z\to\infty} z\,|f(z)|/|g(z)|$ is finite, and $m > 0$, then $\int f(z)e^{imz}\,dz/g(z)$ also vanishes over an infinite semicircular path in the top half of the complex plane. Use this result to evaluate $\int_{-\infty}^{\infty} x \sin mx\, dx/(x^2 + a^2)$.

21. Find the principal value of

(a) $\displaystyle\int_{-\infty}^{\infty} dx/(1 - x^4).$

(b) $\displaystyle\int_{0}^{\infty} \cos mx\, dx/(x^2 - 1).$

*22. Find the principal value of $Si(\infty) = \int_0^\infty \sin t\, dt/t$.

We still have the problem of finding a_{-1}. In the next example the entire Laurent expansion is available, and we return to the general problem of finding residues after considering this special situation.

Example

10. According to Exercise 1.84

$$I = \int_0^{2\pi} [\cos 3\theta \sin(\cos\theta)\cosh(\sin\theta) + \sin 3\theta \cos(\cos\theta)\sinh(\sin\theta)]\, d\theta.$$

$$= \operatorname{Re} \int_0^{2\pi} e^{-3i\theta} \sin(e^{i\theta})\, d\theta.$$

Let $z = e^{i\theta}$, $dz = ie^{i\theta}\, d\theta$, and thereby transform to a contour integral around the unit circle, whose only enclosed singularity is at $z = 0$:

$$I = \operatorname{Re} \frac{1}{i} \oint \frac{\sin z\, dz}{z^4} = \operatorname{Re} \frac{1}{i} \oint \frac{1}{z^4}\left[z - \frac{z^3}{3!} + \frac{z^5}{5!} - \cdots\right] dz$$

$$= \operatorname{Re} \frac{1}{i} \oint \left[\frac{1}{z^3} - \frac{1}{3!z} + \frac{z}{5!} - \cdots\right] dz = \operatorname{Re} \frac{1}{i}\,(2\pi i)(-\tfrac{1}{6}) = -\frac{\pi}{3}.$$

To determine the residue in the general case, we return to the Laurent expansion and multiply by $(z - z_0)^N$ to obtain

$$(z - z_0)^N f(z) = \sum_{n=-N}^{\infty} a_n (z - z_0)^{n+N} = a_{-1}(z - z_0)^{N-1} + \sum_{\substack{n=-N \\ n\neq -1}}^{\infty} a_n (z - z_0)^{n+N}.$$

But this is an analytic function and may be differentiated $(N-1)$ times and evaluated at $z = z_0$. Then

$$(N-1)!a_{-1} = D^{N-1}[(z-z_0)^N f(z)]_{z=z_0}$$

$$a_{-1} = R(z_0) = \frac{1}{(N-1)!}\frac{d^{N-1}}{dz^{N-1}}[(z-z_0)^N f(z)]\bigg|_{z=z_0}. \tag{A5.9}$$

Example

11.

$$I = \int_{-\infty}^{\infty}\frac{x^2\,dx}{(x^2+a^2)^3} = \int_{-\infty}^{\infty}\frac{z^2\,dz}{(z^2+a^2)^3} = \oint\frac{z^2\,dz}{(z^2+a^2)^3} = \oint\frac{z^2}{(z+ia)^3}\frac{dz}{(z-ia)^3}.$$

The only singularity within the semicircular contour is a pole of order three at $z = ia$. The residue is given by

$$R(ia) = \frac{1}{2!}\frac{d^2}{dz^2}\left[\frac{z^2}{(z+ia)^3}\right]_{z=ia}$$

$$= \left[\frac{1}{(z+ia)^3} - \frac{6z}{(z+ia)^4} + \frac{6z^2}{(z+ia)^5}\right]_{z=ia}$$

$$= \frac{1}{8i^3a^3} - \frac{6ia}{16i^4a^4} + \frac{6i^2a^2}{32i^5a^5} = \frac{i}{8a^3}[1 - 3 + \tfrac{3}{2}] = -\frac{i}{16a^3}$$

so

$$I = 2\pi i\left(-\frac{i}{16a^3}\right) = \frac{\pi}{8a^3}.$$

Exercises

23. Evaluate $\oint \cos z\,dz/z^3$, where the contour encloses the origin.
24. Evaluate

(a) $\displaystyle\oint_{|z|=1}\frac{1+z}{z^2}\,dz.$

(b) $\displaystyle\int_{-\infty}^{\infty}\frac{dx}{(x^2+a^2)^2(x^2+b^2)^2}$

(c) $\displaystyle\int_{-\infty}^{\infty}\frac{dx}{(1+x^2)^n}.$

25. Find $L^{-1}[s/(s^2+a^2)^2]$.

APPENDIX 6
USES OF A COMPUTER

The following discussion is not intended to teach you computer programming, but only to outline the basic features of digital and analog computers. In order for you to decide whether a computer is applicable to the problem at hand, you must recognize the capabilities of computers. Although digital computers are by far the more widely applicable, they here receive only a very brief discussion since students frequently receive an introduction to them. Instead, analog computers receive a disproportionate share of this appendix since they are less familiar and an introduction to their capabilities is probably less readily accessible.

A6.1 Digital Computers

Digital computers are capable only of addition, subtraction, multiplication, division, and comparison of two numbers to decide whether one is greater than, equal to, or less than the other. All mathematical manipulations must be reduced to these. However, the operations are performed extremely rapidly, that is, in about a microsecond. Therefore all computations may be modified to utilize iterative or repetitive techniques.

For example, a computer cannot take the square root of N directly but rather finds $N^{1/2}$ by an iterative Newton-Raphson solution of $x^2 - N = 0$. Likewise the definite integral $\int_a^b f(x)\,dx$ may be evaluated by dividing the range of integration into equal intervals, Δx, and evaluating $\sum f(x_i)\,\Delta x$. The MESH SIZE Δx may then be decreased and the process repeated until the limit is reached within the accuracy desired. (There exist higher levels of approximation, and it may also be possible to decide in advance what mesh size is required.) Similarly, to find the particular solution to the first-order differential equation $y' = f(x, y)$, subject to the boundary condition $y(x_0) = y_0$, choose a mesh size Δx and approximate $y(x_0 + \Delta x)$, $y(x_0 + 2\,\Delta x)$, etc:

$$\begin{aligned} y(x_0 + \Delta x) &\simeq y_0 + y'\,\Delta x = y_0 + f(x_0, y_0)\,\Delta x \\ y(x_0 + 2\,\Delta x) &\simeq y(x_0 + \Delta x) + y'\,\Delta x \\ &\simeq y_0 + f(x_0, y_0)\,\Delta x + f(x_0 + \Delta x, y_0 + f(x_0, y_0)\,\Delta x)\,\Delta x \end{aligned}$$

until the entire range of x has been covered. Then decrease the mesh size and repeat the procedure until the solution converges to the desired accuracy.

There still remains the problem of PROGRAMMING, telling the computer *exactly* what is to be done. Fortunately, there exists a language, Fortran, which is sufficiently close to mathematical symbolism and which the machine translates into its own language for computation.

Example of Fortran Symbolism

1. To evaluate the product, **AB** = **C**, of two $N \times N$ matrices: $c_{ij} = \sum_k a_{ik}b_{kj}$ requires the following five statements:

```
  DO 8  I = 1, N
  DO 8  J = 1, N
  C(I, J) = 0.0
  DO 8  K = 1, N
8 C(I, J) = C(I, J) + A(I, K)*B(K, J).
```

The first two statements arrange to have the machine repeat the following commands (up to and including the statement labelled 8) for all i and j from 1 to N. The third statement sets the contents of the memory register allocated to c_{ij} equal to zero, preparatory to the summation. The next statement arranges that the following statement be repeated for all k from 1 to N. The last statement (labelled 8) multiplies a_{ik} by b_{kj}, adds the product to the sum so-far accumulated in the memory location allocated to c_{ij}, and stores the resultant sum back in that memory location. Only after all the N^2 elements of **C** have been calculated will the machine move on to the next command. Notice that multiplication of two $N \times N$ matrices requires N^3 multiplications and N^3 additions, but these may all be effected by only five Fortran statements. Indeed, operations on matrices are especially easy to do on a digital computer because of the repetitive nature of the computations.

A further advantage of Fortran is the applicability of subroutines. If in the course of a problem it is necessary to perform many matrix multiplications, it is not necessary to reprogram each time. Instead it is possible to preface the above five instructions with SUBROUTINE MATMULT (N, A, B, C). Then when it is time to multiply two matrices, only one further statement is necessary. Thus, to multiply the two 20 × 20 matrices TRANSF times COEFF and store the product in the memory locations allocated to the matrix MATR, merely invoke CALL MATMULT (20, TRANSF, COEFF, MATR). Thus it is possible to break up a complex problem into its logical parts and call upon each part as needed.

If it is an advantage of Fortran subroutines that such subroutines need be programmed only once, it is an even greater advantage that some subroutines need not be programmed even once. Many such subroutines have already been programmed and are available ready for use. Among these are operations on matrices (determinant, product, inverse, eigenvalue-eigenvector), evaluation of functions (roots, exponential, trigonometric, gamma, Bessel, many more), evaluation of definite integrals, solution of differential equations, and curve-fitting.

Exercises

1. Write a Fortran program to fit a set of data $\{x_i, y_i\}$ to $y = ax + b$. (Don't worry about correct Fortran symbolism if you don't know it.)
2. Write a Fortran program to evaluate the $N \times N$ determinant $|\mathbf{A}|$ by reduction to triangular form. If you know how, check that you aren't dividing by zero.
3. Use the above result to write a Fortran program to evaluate $\mathbf{A}^{-1}$ in terms of cofactors.

*4. Solve $\mathbf{Ax} = \mathbf{c}$ by reduction to triangular form.

A6.2 Analog Computers

An analog device works by representing one set of physical quantities by another. The most familiar analog computer is the slide rule, which represents, say, molarity, by length. An electronic analog computer uses a voltage as analog for the quantity of interest. The accuracy of such machines is generally better than 0.1 %, which is adequate for most chemical purposes.

The possibility of an efficient electronic analog computer depends upon the existence of high-gain, high-impedance DC amplifiers. Circuits involving such amplifiers may be devised to (*1*) multiply voltages by a constant, (*2*) add voltages, or (*3*) integrate voltages with respect to time. We shall not describe such circuits, but merely note that they exist. We shall symbolize these circuits as indicated in Figures A6.1–A6.3:

1. For an inverting amplifier (multiplying amplifier) (Figure A6.1)

$$V_{\text{out}} = -kV_{\text{in}}.$$

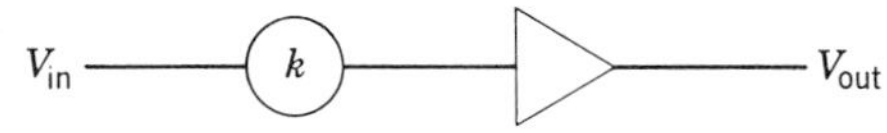

Figure A6.1

2. For a summing amplifier (Figure A6.2)

$$V_{\text{out}} = -(k_1V_1 + k_2V_2 + k_3V_3).$$

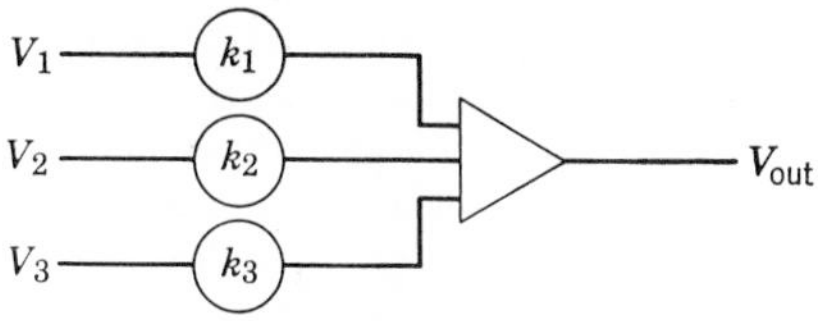

Figure A6.2

3. For an integrating amplifier (Figure A6.3)

$$V_{out} = -[V_0 + \int_0^t (k_1V_1 + k_2V_2)\,dt].$$

Notice the sign reversal in each of these circuits and notice also that the negative of the initial output of an integrating amplifier must be applied.

Examples

2. First-order chemical kinetics, $A \xrightarrow{k} B$, or $-dA/dt = kA$, with $A(0) = A_0$, $B(0) = 0$: The analog circuit for this problem requires integrating $-dA/dt = -\dot{A}$ through amplifier #1 in Figure A6.4, to produce A, and setting kA equal to $\dot{A}$. The voltage source for this circuit is the voltage, $-A_0$, applied to amplifier #1. (Here is where the scale of the analogy must be specified—what voltage corresponds to A_0 moles/liter.) The feedback in the circuit then guarantees that the output voltage will be $A = A_0e^{-kt}$. Amplifier #2 (with unit multipliers, omitted) adds A and $-A_0$ and changes the sign of the sum, to produce $A_0 - A = B$. Either output may be plotted on a recorder and the multiplier k adjusted until the computer trace fits experimental data. The setting of the multiplier may then be converted to a rate constant if the scale and chart speed of the recorder are taken into account.

3. Damped harmonic oscillator, $y'' + ky' + \omega^2y = 0$, with $y(0) = y_0$, $y'(0) = y_0'$: Integrating y'' twice produces both $-y'$ and y, and setting $y'' = -(ky' + \omega^2y)$ leads to the circuit diagram of Figure A6.5.

The above examples are linear differential equations whose solutions can be written out in analytic form. Therefore an analog computer offers little advantage here, except that it does present functional relationships in visual form. The principal utility of analog computers is for solving nonlinear differential equations. To do this it is necessary to be able to multiply together two variable voltages. There do exist such circuits, which we shall symbolize as shown in Figure A6.6.

4. For a multiplier

$$V_{out} = V_1V_2.$$

(Actually $V_{out} = V_1V_2/V_0$, where V_0 is some voltage inherent in the machine, but we shall neglect this complication. Also, on some machines, the output voltage is changed in sign.)

Examples

4. Second-order chemical kinetics, $2A \xrightarrow{k} B$, or $-dA/dt = 2kA^2$, with $A(0) = A_0$, $B(0) = 0$. Passing $-dA/dt$ through an integrating amplifier produces A, which may be multiplied by itself, then by the constant $2k$, and set equal to $-dA/dt$ (Fig. A6.7). Again these kinetics may be solved in closed form.

5. Competitive, consecutive second-order chemical kinetics: $A + B \xrightarrow{k_1} C$, $A + C \xrightarrow{k_2} D$, with $A(0) = A_0$, $B(0) = B_0$, $C(0) = 0 = D(0)$. Setting up integrating amplifiers to integrate $-dA/dt$, $-dB/dt$, and $-dC/dt$, and completing the feedback loops so that $-dA/dt = k_1AB + k_2AC$, $-dB/dt = k_1AB$, and $-dC/dt = -k_1AB + k_2AC$ yields the schematic shown in Figure A6.8. These kinetics cannot be solved in terms of familiar functions, and they arise rather often. The saponification of a diester is an example ($A = OH^-$, B = diester, C = monoester, D = diacid) in which it is possible to determine

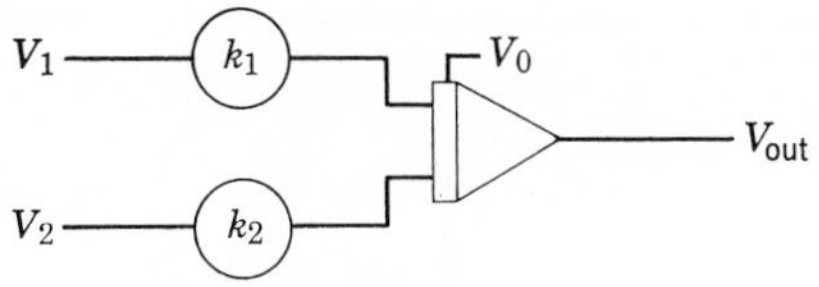

Figure A6.3

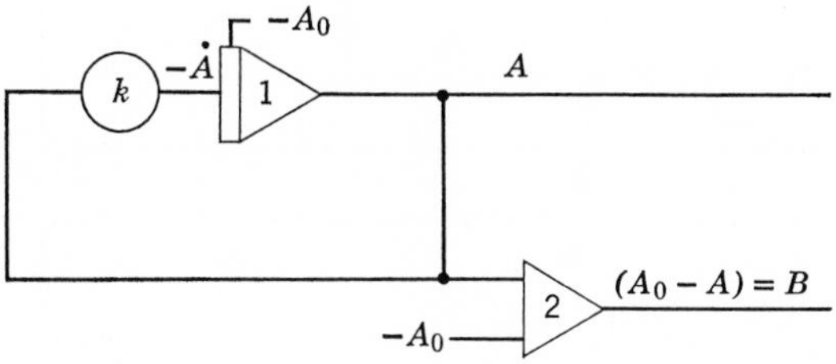

Figure A6.4

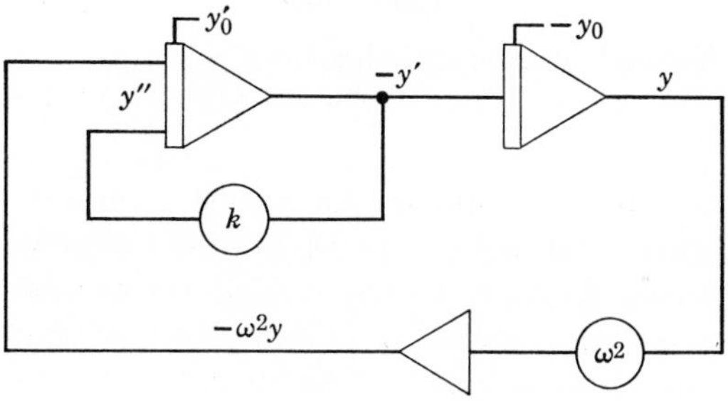

Figure A6.5

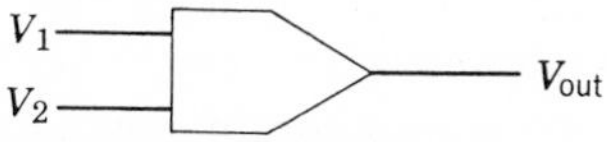

Figure A6.6

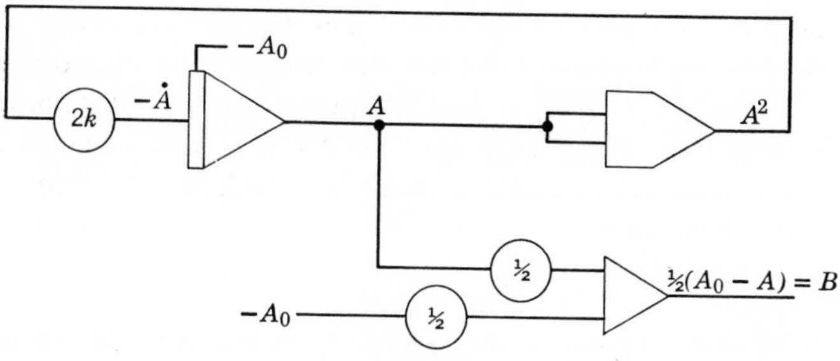

Figure A6.7

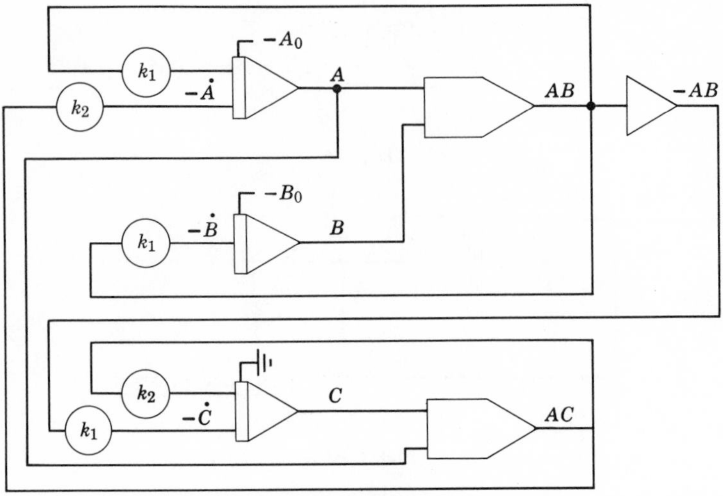

Figure A6.8

k_2 independently from the kinetics of saponification of pure C. If k_2 (or k_2/k_1) is known, it is possible to determine k_1 by adjusting the computer controls until the machine reproduces the experimental data.

The actual procedure for utilizing an analog computer is quite simple. First plot the experimental data on paper to fit the attached recorder. Next sketch the schematic diagram, choose a scale for the analogy, and then connect the components as indicated. (This is often accomplished merely by inserting wires into a special board which fits onto the computer during the actual run.) Set the multiplying constants on the machine to trial values and turn the machine on. Readjust the multipliers and rerun until the machine reproduces the plotted data. With a little familiarity, fitting a kinetic run requires only about five minutes.

The advantages of programming for analog computers are that it requires less than a day to learn, an hour to carry out, and little debugging. Furthermore, you are right there, "plugged into the machine." The major disadvantages are the limited capabilities and the unavailability of analog computers. Although nearly everyone has access to a megadollar digital computer, the cost of a useful analog computer (around $\$10^4$) paradoxically renders it rare. (However, there exist subroutines which enable a digital computer to pretend she is an analog computer which accepts circuit diagrams translated into Fortran.)

Exercises

5. Sketch a schematic diagram of an analog computer circuit to simulate the kinetics $A \xrightarrow{k_1} B \xrightarrow{k_2} C$.

6. The kinetics of isomerization via dissociation-recombination are solvable only in terms of Bessel functions. Sketch a schematic diagram of an analog computer circuit to simulate the kinetics $A \xrightarrow{k_1} B + C \xrightarrow{k_2} A'$, with $A(0) = A_0$, $B(0) = C(0) = A'(0) = 0$.

*7. The equation of motion for a pendulum of length l is $d^2\theta/dt^2 = -(g/l)\sin\theta$, and this nonlinear differential equation cannot be solved except in terms of elliptic integrals. Sketch a schematic diagram of an analog computer circuit to simulate this motion, subject to the initial conditions $\theta(0) = 0$, $\dot{\theta}(0) = \omega_0$. (*Hint:* It is necessary to generate both $\sin\theta$ and $\cos\theta$ in the circuit, but remember that the computer integrates only with respect to t, so that it is necessary to use $d(\sin\theta)/dt = \cos\theta(d\theta/dt)$.)

BIBLIOGRAPHY

This bibliography is intended to guide the student to books that will provide further discussion, proofs, and more advanced material. I have included only what I consider to be the best possible sources, so that this list is not complete. Books that are especially highly recommended are indicated with a plus and comments. Books that are considerably more advanced than the level of this book are preceded by an asterisk; these are generally standard references of whose existence you should be aware.

+ Abramowitz, M., and I. A. Stegun, *Handbook of Mathematical Functions, with Formulas, Graphs, and Mathematical Tables*, Dover, New York, 1965. (An extraordinary compendium.)

Aitken, A. C., *Determinants and Matrices*, 9th ed., Oliver and Boyd, Edinburgh, 1956.

Amundson, N. R., *Mathematical Methods in Chemical Engineering: Matrices and Their Application*, Prentice-Hall, Englewood Cliffs, N.J., 1966.

Anderson, J. M., *Mathematics for Quantum Chemistry*, Benjamin, New York, 1966.

Arfken, G., *Mathematical Methods for Physicists*, Academic Press, New York, 1966.

Atkin, R. H., *Mathematics and Wave Mechanics*, Wiley, New York, 1959.

+ Bak, T. A., and J. Lichtenberg, *Mathematics for Scientists*, Benjamin, New York, 1966. (A useful exposition of elementary material.)

Bennett, C. A., and N. L. Franklin, *Statistical Analysis in Chemistry and the Chemical Industry*, Wiley, New York, 1954.

+ Bierens de Haan, D., *Nouvelles Tables d'intégrales définies*, 1867 edition, corrected, Hafner, New York, 1957. (The standard table of definite integrals.)

+ Boas, M. L., *Mathematical Methods in the Physical Sciences*, Wiley, New York, 1966. (An eminently clear and readable text, with many examples and exercises.)

Bracewell, R., *The Fourier Transform and Its Applications*, McGraw-Hill, New York, 1965.

Bronwell, A. B., *Advanced Mathematics in Physics and Engineering*, 2nd ed., McGraw-Hill, New York, 1953.

Brookes, C. J., I. G. Betteley, and S. M. Loxston, *Mathematics and Statistics for Students of Chemistry, Chemical Engineering, Chemical Technology, and Allied Subjects*, Wiley, New York, 1966.

Brownlee, K. A., *Statistical Theory and Methodology in Science and Engineering*, Wiley, New York, 1960.

Brunk, H. D., *An Introduction to Mathematical Statistics*, 2nd ed., Blaisdell, New York, 1965.

Burington, R. S., *Handbook of Mathematical Tables and Formulas*, 4th ed., McGraw-Hill, New York, 1965.

Chirgwin, B. H., and C. Plumpton, *A Course of Mathematics for Engineers and Scientists*, Vols. I–V, Pergamon, London, 1961–1964.

Chisholm, J. S. R., and R. M. Morris, *Mathematical Methods in Physics*, Saunders, Philadelphia, 1965.

* Churchill, R. V., *Complex Variables and Applications*, 2nd ed., McGraw-Hill, New York, 1960.

\+ Cotton, F. A., *Chemical Applications of Group Theory*, Wiley-Interscience, New York, 1963. (A very readable account of molecular symmetry.)

Courant, R., *Differential and Integral Calculus*, Vols. I–II, E. F. McShane, transl., Wiley, New York, 1936–1937.

* Courant, R., and D. Hilbert, *Methods of Mathematical Physics*, Vols. I–II, Wiley-Interscience, New York, 1953, 1962.

Cramér, H., *The Elements of Probability Theory and Some of Its Applications*, Wiley, New York, 1955.

\+ *CRC Handbook of Chemistry and Physics*, *CRC Handbook of Tables for Mathematics*, and *CRC Handbook of Tables for Probability and Statistics*, The Chemical Rubber Co., Cleveland. (Standard references.)

Davis, P. J., *The Mathematics of Matrices*, Blaisdell, New York, 1965.

* Dennery, P., and A. Krzywicki, *Mathematics for Physicists*, Harper & Row, New York, 1967.

Dickson, T. R., *The Computer and Chemistry*, Freeman, San Francisco, 1968.

Dwight, H. B., *Tables of Integrals and Other Mathematical Data*, 4th ed., Macmillan, New York, 1961.

Eyring, H., J. Walter, and G. E. Kimball, *Quantum Chemistry*, Wiley, New York, 1944.

* Faddeev, D. K., and V. N. Faddeeva, *Computational Methods of Linear Algebra*, R. C Williams, transl., Freeman, San Francisco, 1963.

Fadell, A. G., *Vector Calculus and Differential Equations*, Van Nostrand, Princeton, N.J., 1968.

\+ Feller, W., *An Introduction to Probability Theory and Its Applications*, 3rd ed., Vol. I, Wiley, New York, 1968. (An excellent text, with many examples and exercises.)

Ferraro, J. R., and J. S. Ziomek, *Introductory Group Theory and Its Application to Molecular Structure*, Plenum, New York, 1969.

* Fletcher, A., J. C. P. Miller, L. Rosenhead, and L. J. Comrie, *An Index of Mathematical Tables*, 2nd ed., Vols. I–II, Addison-Wesley, Reading, Mass., 1962.

* Frazer, R. A., W. J. Duncan, and A. R. Collar, *Elementary Matrices and Some Applications to Dynamics and Differential Equations*, Cambridge, 1938.

Fuller, L. E., *Basic Matrix Theory*, Prentice-Hall, Englewood Cliffs, N.J., 1962.

* Goertzel, G., and N. Tralli, *Some Mathematical Methods of Physics*, McGraw-Hill, New York, 1960.

Gradshteyn, I. S., and I. M. Ryshik, *Tables of Integrals, Series, and Products*, 4th ed., A. Jeffrey, transl., Academic Press, New York, 1965.

* Hamermesh, M., *Group Theory and Its Applications to Physical Problems*, Addison-Wesley, Reading, Mass., 1962.

Heine, V., *Group Theory in Quantum Mechanics*, Pergamon, London, 1960.

Herzberg, G., *Molecular Spectra and Molecular Structure. III. Infrared and Raman Spectra of Polyatomic Molecules*, Van Nostrand, Princeton, N.J., 1945.

Hildebrand, F. B., *Advanced Calculus for Applications*, Prentice-Hall, Englewood Cliffs, N.J., 1962.

Hoel, P. G., *Introduction to Mathematical Statistics*, 3rd ed., Wiley, New York, 1962.

* Ince, E. L., *Ordinary Differential Equations*, Dover, New York, 1926.

Irving, J., and N. Mullineux, *Mathematics in Physics and Engineering*, Academic Press, New York, 1959.

Jackson, J. D., *Mathematics for Quantum Mechanics*, Benjamin, New York, 1962.

Jaffé, H. H., and M. Orchin, *Symmetry in Chemistry*, Wiley, New York, 1965.

Jahnke, E., and F. Emde, *Tables of Functions*, 4th ed., Dover, New York, 1945.

Jeffreys, H., and B. S. Jeffreys, *Methods of Mathematical Physics*, 3rd ed., Cambridge, 1956.

Jenson, V. G., and G. V. Jeffreys, *Mathematical Methods in Chemical Engineering*, Academic Press, New York, 1963.

* Johnson, D. E., and J. R. Johnson, *Mathematical Methods in Engineering and Physics: Special Functions and Boundary-Value Problems*, Ronald Press, New York, 1965.

Kaplan, W., *Advanced Calculus*, Addison-Wesley, Reading, Mass., 1952.

Kauzmann, W., *Quantum Chemistry*, Academic Press, New York, 1957.

Kemeny, J. G., J. L. Snell, and G. L. Thompson, *Introduction to Finite Mathematics*, 2nd ed., Prentice-Hall, Englewood Cliffs, N.J., 1966.

* Korn, G. A., and T. M. Korn, *Mathematical Handbook for Scientists and Engineers: Definitions, Theorems, and Formulas for Reference and Review*, 2nd ed., McGraw-Hill, New York, 1968.

Kreyszig, E., *Advanced Engineering Mathematics*, 2nd ed., Wiley, New York, 1967.

Kynch, G. J., *Mathematics for the Chemist*, Plenum, London, 1962.

Lark, P. D., B. R. Craven, and R. C. L. Bosworth, *The Handling of Chemical Data*, Pergamon, Oxford, 1968.

* Lebedev, A. V., and R. M. Fedorova, *A Guide to Mathematical Tables*, D. G. Fry, transl., Pergamon, London, 1960.

McLachlan, D., Jr., *Statistical Mechanical Analogies*, Prentice-Hall, Englewood Cliffs, N.J., 1968.

McWeeny, R., *Symmetry: An Introduction to Group Theory and Its Applications*, Pergamon, Oxford, 1963.

* Magnus, W., and F. Oberhettinger, *Formulas and Theorems for the Special Functions of Mathematical Physics*, Chelsea, New York, 1949.

* Margenau, H., and G. M. Murphy, *The Mathematics of Physics and Chemistry*, 2nd ed., Vol. I, Van Nostrand, Princeton, N.J., 1956.

* Mathews, J., and R. L. Walker, *Mathematical Methods of Physics*, Benjamin, New York, 1964.

Mellor, J. W., *Higher Mathematics for Students of Chemistry and Physics*, 4th ed., Dover, New York, 1955.

Menzel, D. H., *Fundamental Formulas of Physics*, Vols. I–II, Dover, New York, 1960.

* Morse, P. M., and H. Feshbach, *Methods of Theoretical Physics*, McGraw-Hill, New York, 1953.

Mosteller, F., R. E. K. Rourke, and G. B. Thomas, Jr., *Probability and Statistics*, Addison-Wesley, Reading, Mass., 1961.

Murdoch, D. C., *Linear Algebra for Undergraduates*, Wiley, New York, 1957.

+ Murphy, G. M., *Ordinary Differential Equations and Their Solutions*, Van Nostrand, Princeton, N.J., 1960. (Comparable to integral tables, but for differential equations.)

Nicholson, M. M., *Fundamentals and Techniques of Mathematics for Scientists*, Wiley, New York, 1961.

Parke, N. G., III, *Guide to the Literature of Mathematics and Physics*, 2nd ed., Dover, New York, 1958.

Pauling, L., and E. B. Wilson, Jr., *Introduction to Quantum Mechanics, with Applications to Chemistry*, McGraw-Hill, New York, 1935.

Peirce, B. O., *A Short Table of Integrals*, 4th ed., Ginn, Boston, 1957.

Perlis, S., *Theory of Matrices*, Addison-Wesley, Reading, Mass., 1952.

Petit Bois, G., *Tables of Indefinite Integrals*, Dover, New York, 1961.

Pipes, L. A., *Applied Mathematics for Engineers and Physicists*, 2nd ed., McGraw-Hill, New York, 1958.

Polya, G., *How to Solve It: A New Aspect of Mathematical Method*, 2nd ed., Doubleday Anchor, Garden City, N.Y., 1957.

* Ralston, A., and H. S. Wilf, *Mathematical Methods for Digital Computers*, Wiley, New York, 1960.

\+ Schaum Outline Series (many topics), Schaum, New York. (Texts on such topics as statistics, differential equations, Laplace transforms, vector analysis, and matrices, valuable primarily for the many examples and exercises.)

Schelkunoff, S. A., *Applied Mathematics for Engineers and Scientists*, 2nd ed., Van Nostrand, Princeton, N.J., 1965.

Schwartz, M., S. Green, and W. A. Rutledge, *Vector Analysis, with Applications to Geometry and Physics*, Harper, New York, 1960.

Searle, S. R., *Matrix Algebra for the Biological Sciences*, Wiley, New York, 1966.

Sneddon, I. N., *Special Functions of Mathematical Physics and Chemistry*, Oliver and Boyd, Edinburgh, 1961.

Snedecor, G. W., and W. G. Cochran, *Statistical Methods*, 6th ed., Iowa State University Press, Ames, 1967.

\+ Sokolnikoff, I. S., and R. M. Redheffer, *Mathematics of Physics and Modern Engineering*, 2nd ed., McGraw-Hill, New York, 1966. (A detailed text, with many examples and exercises.)

Stephenson, G., *Mathematical Methods for Science Students*, Wiley, New York, 1961.

Stice, J. E., and B. S. Swanson, *Electronic Analog Computer Primer*, Blaisdell, New York, 1965.

Taylor, A. E., *Advanced Calculus*, Ginn, Boston, 1955.

* Tinkham, M., *Group Theory and Quantum Mechanics*, McGraw-Hill, New York, 1963.

* Whittaker, E. T., and G. N. Watson, *A Course of Modern Analysis*, 4th ed., Cambridge, 1927.

* Wiberg, K. B., *Computer Programming for Chemists*, Benjamin, New York, 1965.

Widder, D. V., *Advanced Calculus*, 2nd ed., Prentice-Hall, Englewood Cliffs, N.J., 1961.

* Wigner, E. P., *Group Theory and Its Application to the Quantum Mechanics of Atomic Spectra*, J. J. Griffin, transl., Academic Press, New York, 1959.

Wilf, H. S., *Mathematics for the Physical Sciences*, Wiley, New York, 1962.

\+ Wilson, E. B., Jr., *An Introduction to Scientific Research*, McGraw-Hill, New York, 1952. (Includes much mathematical material not readily available to chemists and many handy hints.)

Wilson, E. B., Jr., J. C. Decius, and P. C. Cross, *Molecular Vibrations: The Theory of Infrared and Raman Vibrational Spectra*, McGraw-Hill, New York, 1955.

Wylie, C. R., Jr., *Advanced Engineering Mathematics*, 3rd ed., McGraw-Hill, New York, 1966.

Young, H. D., *Statistical Treatment of Experimental Data*, McGraw-Hill, New York, 1962.

ANSWERS TO EXERCISES

Chapter 1

2. (a) $(f+g)(x) \equiv f(x) + g(x)$. (b) Yes. (c) $f^*(x) \equiv [f(x)]^*$.
3. (a) $E = h\nu$. (b) $k = (RT/Nh)e^{\Delta S^{\ddagger}/R}e^{-\Delta H^{\ddagger}/RT}$ or $k = Ae^{-E_A/RT}$.
   (c) $E = E^\circ + \dfrac{RT}{F} \ln [\mathrm{Ag}^+]$. (d) $f = [nV_W/(nV_W + KV_E)]^n$.
7. (a) $f^{-1}(y) = \ln (1 + y)$. (b) $f^{-1}(y) = \tan^{-1} y$. (c) $f^{-1}(y) = e^{\cos^{-1} y}$.
   (d) $f^{-1}(y) = [-b \pm (b^2 - 4ac + 4ay)^{1/2}]/2a$.
   (e) $f^{-1}(y) = \ln [(1 + y)/(1 - y)]^{1/2}$. (*Note:* $f = \tanh$ and $f^{-1} \tanh^{-1}$.
   (f) $f^{-1}(z) = z^*$. Uniqueness problems with (b), (c), and (d).
10. $f(x) = \frac{1}{2}[f(x) + f(-x)] + \frac{1}{2}[f(x) - f(-x)]$.
11. (a) $(1 - x^2)^{-1/2}$. (b) $-\frac{1}{2}[\cos (e^{1/2 x^2} - 1)]^{-1/2} \sin (e^{1/2 x^2} - 1)xe^{1/2 x^2}$.
    (c) $e^{2x} \sin x \cos x\, (1 + \ln x)[2 + \cot x - \tan x + (x + x \ln x)^{-1}]$.
12. (a) $-(12 + 16\pi)/\pi^2$. (b) $-(1 + 2\pi)/\pi^2$.
13. (a) $(\Omega_1 + \Omega_2)f \equiv \Omega_1 f + \Omega_2 f$. (b) Yes. (c) $(\Omega_1 - \Omega_2)f \equiv \Omega_1 f - \Omega_2 f$.
15. (a) Yes. (c) $(DX - XD)f = DXf - XDf = xf' + f - xf' = f$.
16. $2D$ (meaning $D + D$).
17. (a) $0, \alpha, \beta, 0$. (b) $\frac{3}{4}\alpha, \frac{3}{4}\beta$. (c) is_z, is_x, is_y.
    (d) $s_x^2 + s_y^2 + i[s_y, s_z], s_x^2 + s_y^2 + s_z$. (e) $s_x^2 + s_y^2 + i[s_x, s_y], s_x^2 + s_y^2 - s_z$.
    (f) $s_- s_+ + s_z^2 + s_z, s_+ s_- + s_z^2 - s_z$. (g) $2s_z$.
18. Odd, even.
19. $2a^2 D_e$.
20. (a) $\nu_0, 1/\sigma\sqrt{2\pi}, 1/\sigma\pi\sqrt{3}$. (b) $\nu_0 \pm \sigma$. (c) $2\sigma\sqrt{\ln 4}, 2\sigma\sqrt{3}$.
21. $d \log (I^\circ/I)/d(lc) = (\varepsilon_1 I_1^\circ 10^{-\varepsilon_1 lc} + \varepsilon_2 I_2^\circ 10^{-\varepsilon_2 lc})/(I_1^\circ 10^{-\varepsilon_1 lc} + I_2^\circ 10^{-\varepsilon_2 lc})$.
22. $(f^{-1})'' = -f''/(f')^3$.
23. $K_W/[\mathrm{H}^+]^2 + M_0 K([\mathrm{H}^+] - K)/([\mathrm{H}^+] + K)^3 = 1$.
24. $(1.023340 - 1.022146)/(2 \times 1.022740) = 5.84 \times 10^{-4}\ \mathrm{deg}^{-1}$.
25. (a) $16.61 + 3.230n^{1/2}$. (b) 19.84 ml.
    (c) $d\Delta H/dn = 923 + 714n^{1/2} - 1452n^2 + 608n^{3/2}$. (d) 793 cal.
    (e) $793/6.025 \times 10^{23}$ cal.
26. $V(1\%) = V(0.2526m) = 1000.7$ ml, $V_{\mathrm{NaOH}}(1\%) = -3.5$ ml.
27. 1.4142136. Newton-Raphson formula becomes $x_{i+1} = (x_i + 2/x_i)/2$.
28. 0.443.
29. $2.821k/h$; 2.821 is the root of $(3 - x)e^x - 3 = 0$.
30. $1.634 \times 10^{-4}M$. (If $x = 10^4[\mathrm{Ca}^{2+}]$, then x is the solution of $x^4 - 1.74x^2 - 1.98x + 0.756 = 0$.)
31. (a) $3.58 \times 10^{-2}M$. (If $x = 100[\mathrm{H}^+]$, then x is the solution of $x^2 = 2(10 - x)$.)
    (b) $7.29 \times 10^{-2}M$. (If $x = 100[\mathrm{H}^+]$, then x is the solution of $2 \log x - \log (10 - x) - 0.2(10 + 2x)^{1/2} - \log 2 = 0$.)
32. Let $F(x) = f(x) - x$, and solve by Newton-Raphson method. Let $f(x) = 3(1 - e^{-x})$.

33. $\int_0^\infty \exp(-h^2x^2/8mL^2k)\,dx$, $\int_0^\infty (2Tx+1)\exp[-Tx(Tx+1)h^2/2IkT]\,dx$.
34. A careful estimate will be within 0.2% of the exact value, $\ln 2 \sim 0.693$.
36. Even, since $f(x) - f(-x) = \int_{-x}^{x} o(x)\,dx = 0$.
37. $\Theta^4/T^5[1 - \exp(\Theta/T)]$.
38. 431.7 kcal/mole, 11.49 e.u.

39. $E(T) = \left[\dfrac{8\pi k^4}{h^3c^3}\displaystyle\int_0^\infty \dfrac{x^3\,dx}{e^x - 1}\right]T^4$.

40. (a) $\frac{1}{2}x^2[(\ln x)^2 - \ln x + \frac{1}{2}]$. (b) $\sin x[\ln(\sin x) - 1]$. (c) $\dfrac{1}{\lambda_1 - \lambda_2}\ln\dfrac{x - \lambda_1}{x - \lambda_2}$.

(d) $\dfrac{1}{2b}\ln\dfrac{a+b}{a}$. (e) $\ln 3$. (f) $e^a\left(\dfrac{1}{a} - \dfrac{2}{a^2} + \dfrac{2}{a^3}\right) - \dfrac{2}{a^3}$.

(g) $\frac{1}{2}e^x(\sin x + \cos x)$.
41. $18.069 - 0.1496n_{\mathrm{NaCl}}^2 - 0.0233n_{\mathrm{NaCl}}^{5/2}$.
42. (a) $\int \sec\theta\,d\theta$.
(c) Derivative $= \sec\theta\tan\theta + \sec^2\theta$, so integral is $\ln(\sec\theta + \tan\theta)$.
(d) $\ln[x + (1 + x^2)^{1/2}]$.
43. $0(k < 1)$, $\pi/2(k = 1)$, $\pi(k > 1)$.
45. $(N-1)!/N! = 1/N$.
46. $(2n)!/2^{2n}n!^2$.

47. (a) $\dfrac{1}{N}\displaystyle\sum_{i=1}^{N} a_i$. (b) $\left[\displaystyle\prod_{i=1}^{N} a_i\right]^{1/N}$. (c) $\displaystyle\sum_{r=0}^{n}\dfrac{n!}{r!(n-r)!}(D^rf)(D^{n-r}g)$.

48. (a) $\displaystyle\sum_{n=0}^{\infty}(-1)^n(n+1)x^n$. (b) $\displaystyle\sum_{n=0}^{\infty}(-1)^n\dfrac{2^{2n}}{(2n)!}x^{2n}$. (c) $\displaystyle\sum_{n=0}^{\infty}(-1)^n\dfrac{(2n)!}{2^{2n}n!^2}\dfrac{x^{2n+1}}{2n+1}$.

50. (a) As $n \to \infty$, $n!/x^n$ approaches ∞ for *all* x.
(c) Ten terms gives $0.091542e^{-10} = 4.1560 \times 10^{-6}$, correct within 0.004%.
53. (a) 1. (b) 1. (c) 0.
54. $e^{-\frac{1}{2}h\nu/kT}/(1 - e^{-h\nu/kT})$.

55. Binomial formula, $\displaystyle\sum_{r=0}^{n}\dfrac{n!}{r!(n-r)!}x^r$.

56. $x\ln x - x - (t - x)^2/2x$.
58. (a) $\sum_{n=0}^{\infty} x^{N+n}$ or $\sum_{n=N}^{\infty} x^n$. (b) $(1 - x^N)/(1 - x)$.

59. $1 - \dfrac{1}{2}x + \dfrac{1\cdot 3}{2\cdot 2}\dfrac{x^2}{2!} - \dfrac{1\cdot 3\cdot 5}{2\cdot 2\cdot 2}\dfrac{x^3}{3!} + \cdots = \displaystyle\sum_{n=0}^{\infty}(-1)^n\dfrac{(2n)!}{2^{2n}n!^2}x^n$, $\displaystyle\sum_{n=0}^{\infty}\dfrac{(2n)!}{2^{2n}n!^2}x^n$.

60. $\displaystyle\sum_{n=0}^{\infty}\dfrac{(2n)!}{2^{2n}n!^2}\dfrac{x^{2n+1}}{2n+1}$.

61. (a) $\displaystyle\sum_{n=0}^{\infty}(-1)^n\dfrac{x^{2n+1}}{n!(2n+1)}$. (b) $\displaystyle\sum_{n=0}^{\infty}(-1)^{n+1}\dfrac{x^n}{n!n}$.

62. $-\dfrac{c}{b}\left(1 + \dfrac{ac}{b^2}\right)$ by all methods.

63. $1, -\frac{1}{2}, \frac{1}{6}, 0, -\frac{1}{30}$.
64. $f(x + h)$.
65. (a) $e^{x\ln a}[\cos(y\ln a) + i\sin(y\ln a)]$. (b) $e^{-\pi/2}e^{2\pi n}$ (n integral).

66. $\cos 5\theta = \cos^5\theta - 10\cos^3\theta\sin^2\theta + 5\cos\theta\sin^4\theta$,
$\sin 5\theta = 5\cos^4\theta\sin\theta - 10\cos^2\theta\sin^3\theta + \sin^5\theta$.
67. (a) $[\cos\frac{1}{2}\theta - \cos(n+\frac{1}{2})\theta]/2\sin\frac{1}{2}\theta$, $-\frac{1}{2}+\frac{1}{2}\sin(n+\frac{1}{2})\theta/\sin\frac{1}{2}\theta$.
(b) $(4n+2)\alpha + 4\beta\csc\pi/(4n+2)$.
68. (a) π, π. (b) 0, $(-1)^n$, $n\pi$, $\dfrac{\pi}{2}+2n\pi$ (n = any integer).
71. (b) $F[f'(x)] = iyF[f(x)]$. (c) $(2/\pi)^{1/2}a/(a^2+y^2)$. (d) $(2/\pi)^{1/2}(\sin\pi y)/y$.
72. $F[f(x)] = -i(2/\pi)^{1/2}\int_0^\infty f(x)\sin xy\,dx$.
73. (d) $F\{f_1 * f_2\}(y) = g_1(y)g_2(y)$.
74. $\{f_n(x)\} = \{\sin[(n+1)x]\}$.
75. (a) n. (b) $\frac{3}{2}$.
76. $$\sum_{n=0}^{\infty}(-1)^n\frac{x^{2n+1}}{(2n+1)!(2n+1)}$$
77. (a) Four terms give 0.4794255. (b) Term #1000 = .001.
78. e^x is neither even nor odd.
79. $\sinh 2x = 2\sinh x\cosh x$, $\cosh 2x = \cosh^2 x + \sinh^2 x$.
80. $$\theta = \left(\frac{\pi}{2}+n\pi\right) + i\sinh^{-1}[(-1)^{n+1}] = \left(\frac{\pi}{2}+n\pi\right) + i\ln[(-1)^{n+1}+\sqrt{2}] =$$
$$\left(\frac{\pi}{2}+n\pi\right) + (-1)^{n+1}i\ln(1+\sqrt{2}).$$ "Principal value" is $\dfrac{\pi}{2} - i\ln(1+\sqrt{2})$.
81. (a) $1 = \cosh y(dy/dx)$, or $D\sinh^{-1}x = 1/\cosh y$. (b) $e^y = x + (1+x^2)^{1/2}$.
(c) $(1+x^2)^{-1/2}$.
82. (a) $s/(s^2+\omega^2)$. (b) $\omega/(s^2+\omega^2)$. (c) $s/(s^2-a^2)$. (d) $a/(s^2-a^2)$.
85. $\frac{1}{2}\operatorname{csch}(\frac{1}{2}h\nu/kT)$.

Chapter 2

1. (b) See Figure 2.5.
3. $1s$ orbital, $2p_z$ orbital.
4. (b) $-L_z$. (c) $-L_x$, $-L_y$.
5. (a) iS_z. (c) $\alpha\alpha$, 0, 0, $-\beta\beta$.
(d) 0, $\alpha\alpha$, $\alpha\alpha$, $\alpha\beta+\beta\alpha$, $\alpha\beta+\beta\alpha$, $\beta\beta$, $\beta\beta$, 0. (e) $2\alpha\alpha$, $\alpha\beta+\beta\alpha$, $\alpha\beta+\beta\alpha$, $2\beta\beta$.
(f) $\frac{15}{4}\alpha\alpha\alpha$, $\frac{7}{4}\alpha\alpha\beta+\alpha\beta\alpha+\beta\alpha\alpha$, $\frac{7}{4}\alpha\beta\beta+\beta\alpha\beta+\beta\beta\alpha$, $\frac{15}{4}\beta\beta\beta$.
6. $$\left(\cos^2\phi - \frac{n^2}{x^2}\right)\cos(x\sin\phi - n\phi) - \frac{\sin\phi}{x}\sin(x\sin\phi - n\phi).$$
7. $$\frac{\partial^2 F}{\partial x\,\partial y} =$$
$$\lim_{\Delta x\to 0}\lim_{\Delta y\to 0}\left[\frac{F(x+\Delta x, y+\Delta y) - F(x+\Delta x, y) - F(x, y+\Delta y) + F(x,y)}{\Delta x\Delta y}\right],$$
and $\dfrac{\partial^2 F}{\partial y\,\partial x}$ is the same, except the order in which the limits are taken is reversed.

8. (a) $2x, -2y, 2, -2, 0$. (b) $y, x, 0, 0, 1$.
(d) In (a) $D_x{}^2F > 0$ but $D_y{}^2F < 0$. In (b) $(D_x{}^2F)(D_y{}^2F) = 0 < 1 = (D_xD_yF)^2$.
9. (a) $F(x + h, y + k) = F(x, y) + hD_xF + kD_yF + \frac{1}{2}h^2D_x{}^2F + hkD_xD_yF + \frac{1}{2}k^2D_y{}^2F$.
(b) Because expansion is about a minimum, where first derivatives are all zero.
(c) $\displaystyle \frac{1}{6}\sum_{i=1}^{3N}\sum_{j=1}^{3N}\sum_{k=1}^{3N} \left.\frac{\partial^3 V}{\partial X_i \partial X_j \partial X_k}\right|_{\{X_i{}^\circ\}} (X_i - X_i^\circ)(X_j - X_j^\circ)(X_k - X_k^\circ)$.

10. $\displaystyle x_{i+1} = x_i - \left(F\frac{\partial G}{\partial y} - G\frac{\partial F}{\partial y}\right)\Big/\Delta,\ y_{i+1} = y_i - \left(G\frac{\partial F}{\partial x} - F\frac{\partial G}{\partial x}\right)\Big/\Delta,$

where $\displaystyle \Delta = \left(\frac{\partial F}{\partial x}\right)\left(\frac{\partial G}{\partial y}\right) - \left(\frac{\partial F}{\partial y}\right)\left(\frac{\partial G}{\partial x}\right)$ and all functions and derivatives are evaluated at (x_i, y_i).

11. (a) $-V/P,\ R/P$. (b) $-(V - b)/P,\ R/P$.
(c) $-(V - b)/[P + a/V^2 - 2a(V - b)/V^3],\ R/[P + a/V^2 - 2a(V - b)/V^3]$.
12. $(\partial P/\partial V)_T = 2a/V^3 - RT/(V - b)^2$, $(\partial^2 P/\partial V^2)_T = -6a/V^4 + 2RT/(V - b)^3$. Now substitute the answers given.
14. (a) $-(1/T^2)(\partial T/\partial E)_V < 0$, $(1/T)(\partial P/\partial V)_E - (P/T^2)(\partial T/\partial V)_E < 0$,
$-(1/T^2)(\partial T/\partial E)_V[(1/T)(\partial P/\partial V)_E - (P/T^2)(\partial T/\partial V)_E] > (1/T^4)(\partial T/\partial V)_E{}^2$.
(b) Multiply first inequality by $-T^2$, and multiply third inequality by $-T^4$.

15. $\displaystyle D_x = \cos\phi\ D_r - \frac{1}{r}\sin\phi\ D_\phi,\ D_y = \sin\phi\ D_r + \frac{1}{r}\cos\phi\ D_\phi$.

16. $\displaystyle \left(\frac{\partial u}{\partial r}\right)_\phi = \cos\phi\left(\frac{\partial v}{\partial x}\right)_y + \sin\phi\left(\frac{\partial v}{\partial y}\right)_x,\ \left(\frac{\partial u}{\partial \phi}\right)_r = -r\sin\phi\left(\frac{\partial v}{\partial x}\right)_y + r\cos\phi\left(\frac{\partial v}{\partial y}\right)_x.$

17. $\displaystyle \frac{\partial^2 u}{\partial r^2} + \frac{1}{r}\frac{\partial u}{\partial r} + \frac{1}{r^2}\frac{\partial^2 u}{\partial \phi^2}.$

18. $\displaystyle \frac{d}{dx}\left[F - y'\frac{\partial F}{\partial y'}\right] = 0.$

19. (a) $\displaystyle \left(\frac{\partial F}{\partial x}\right)_y = \left(\frac{\partial G}{\partial x}\right)_z + \left(\frac{\partial G}{\partial z}\right)_x\left(\frac{\partial f}{\partial x}\right)_y.$ (c) $\displaystyle \left(\frac{\partial P}{\partial V}\right)_E = \left(\frac{\partial P}{\partial V}\right)_T + \left(\frac{\partial P}{\partial T}\right)_V\left(\frac{\partial T}{\partial V}\right)_E.$

20. $\displaystyle \left(\frac{\partial G}{\partial x}\right)_y = \left(\frac{\partial F}{\partial x}\right)_{y,z} + \left(\frac{\partial F}{\partial z}\right)_{x,y}\left(\frac{\partial f}{\partial x}\right)_y.$

21. $\frac{9}{16}$.
22. $(\partial P/\partial T)_V = \alpha/\beta$.
24. (a) $(\partial P/\partial V)_S = \gamma(\partial P/\partial V)_T$. (b) $-\gamma/V$.
(c) $P_1/P_0 = (V_0/V_1)^\gamma$. (d) $P_0[V_1(V_0/V_1)^\gamma - V_0]/(1 - \gamma)$.
25. $|ax_0 + by_0 + cz_0 - d|/(a^2 + b^2 + c^2)^{1/2}$.
26. Constraint is $a \tan\theta_A + b \tan\theta_B$ = component of A-B distance parallel to the phase boundary.
27. $R \ln N$.
28. Maximum $c_1c_2 = \frac{1}{2}$.
29. (a) $r = 3$.
(b) $r = 3c_N/c_H$, where c_N and c_H are the costs per mole of N_2 and H_2, respectively.
33. $\mu^\circ_{Tl} + 2RT\alpha\{[\beta + (1 - x)/x]^{-1} - \frac{1}{2}\beta[\beta + (1 - x)/x]^{-2}\} - RT\alpha/\beta$.
34. $V = V(n_A, n_B) \rightarrow dV = V_A\,dn_A + V_B\,dn_B$. But by Euler's theorem $V = n_AV_A + n_BV_B$, so $dV = n_A\,dV_A + V_A\,dn_A + n_B\,dV_B + V_B\,dn_B$. Subtracting gives the desired equation.

35. Concentric about origin and equally spaced.

36. (a) Via $\left(\dfrac{\partial E}{\partial V}\right)_T = T\left(\dfrac{\partial S}{\partial V}\right)_T - P$ and $\left(\dfrac{\partial S}{\partial V}\right)_T = -\dfrac{\partial^2 A}{\partial V \partial T} = \left(\dfrac{\partial P}{\partial T}\right)_V$.

(b) Via $\left(\dfrac{\partial T}{\partial V}\right)_E = -\left(\dfrac{\partial E}{\partial V}\right)_T \Big/ \left(\dfrac{\partial E}{\partial T}\right)_V$.

37. $\left(\dfrac{\partial T}{\partial P}\right)_H = -\left(\dfrac{\partial H}{\partial P}\right)_T \Big/ \left(\dfrac{\partial H}{\partial T}\right)_P = -\left[T\left(\dfrac{\partial S}{\partial P}\right)_T + V\right] \Big/ C_P = \left[T\left(\dfrac{\partial V}{\partial T}\right)_P - V\right] \Big/ C_P.$

39. Via $\left\{T\left(\dfrac{\partial P}{\partial V}\right)_T + \left(\dfrac{\partial T}{\partial V}\right)_E \left[T\left(\dfrac{\partial P}{\partial T}\right)_V - P\right]\right\} \Big/ \left(\dfrac{\partial E}{\partial T}\right)_V + \left(\dfrac{\partial T}{\partial V}\right)_E^2 < 0.$

40. (a) Exact. (b) $\mu = 1/PV$ or $1/T$.
42. Since $(\partial S/\partial E)_V = 1/T$ and $(\partial S/\partial V)_E = P/T$, $T(E, V) = 1/(\partial S/\partial E)_V$ and $P(E, V) = T(\partial S/\partial V)_E$. Then $H = E + PV$, $A = E - TS$, and $G = E + PV - TS$. The equation of state is $P(V, T) = P[V, T(E, V)]$.
45. Regular tetrahedron, with each apex corresponding to a pure component.
46. (a) $F\{f(x, y, z)\} = g(\xi, \eta, \zeta)$
$$\equiv (2\pi)^{-3/2} \iiint f(x, y, z) \exp\left[-(x\xi + y\eta + z\zeta)\right] dx\, dy\, dz.$$
(Both these integrals are over all space.)
(b) $f_1 * f_2(x, y, z) \equiv (2\pi)^{-3/2} \iiint f_1(\xi, \eta, \zeta) f_2(x - \xi, y - \eta, z - \zeta)\, d\xi\, d\eta\, d\zeta$.
47. Area of A, volume of V.

48. (a) $\int_0^1 \int_0^{ax}$, $\int_0^a \int_0^{b-(b/a)x}$, $\int_{-a}^a \int_0^{\sqrt{a^2-x^2}}$, $\int_0^{1/\sqrt{2}} \int_0^x + \int_{1/\sqrt{2}}^1 \int_0^{\sqrt{1-x^2}}$

(b) $\int_0^a \int_{y/a}^1$, $\int_0^b \int_0^{a-(a/b)y}$, $\int_0^a \int_{-\sqrt{a^2-y^2}}^{\sqrt{a^2-y^2}}$, $\int_0^{1/\sqrt{2}} \int_y^{\sqrt{1-y^2}}$.

49. (a) $abc/6$. (b) $a/4$, $b/4$, $c/4$.
50. $d \log (I^\circ/I)/d(lc) = \int \varepsilon(\lambda) I^\circ(\lambda) 10^{-\varepsilon(\lambda)lc}\, d\lambda / \int I^\circ(\lambda) 10^{-\varepsilon(\lambda)lc}\, d\lambda$.

51. (a) $\dfrac{1 \cdot 3 \cdot 5 \cdots (2n-1)}{2^n}\left[\dfrac{\pi}{a^{2n+1}}\right]^{1/2} = \dfrac{(2n)!}{2^{2n} n!}\left[\dfrac{\pi}{a^{2n+1}}\right]^{1/2}$.

(b) $\int_0^\infty x^{2n+1} e^{-ax^2}\, dx = \dfrac{1}{2n! a^{n+1}}$.

52. (a) $\dfrac{e^x + e^{-x}}{e^x - e^{-x}} - \dfrac{1}{x}\left(\text{or } \coth x - \dfrac{1}{x}\right) = \displaystyle\int_{-1}^1 y e^{xy}\, dy \Big/ \int_{-1}^1 e^{xy}\, dy.$

(b) $\mu\left[\coth x - \dfrac{1}{x}\right]$. (c) $\mu x/3$.

53. r, $r^2 \sin \theta$.
54. $\pm a^3(\lambda^2 - \mu^2)$.
56. If first integration is with respect to r: $\int_0^{\tan^{-1} a} \int_1^{\sec \phi}$, $\int_0^{\pi/2} \int_a^{((\cos\phi/a) + (\sin\phi/b))^{-1}}$, $\int_0^\pi \int_0^a$, $\int_0^{\pi/4} \int_0^1$.
57. $\frac{1}{2}\sqrt{\pi}$.
58. $(2\pi m L^2 k/h^2)^{1/2}$, $2Ik/h^2$; therefore $Q_{\text{trans}}^{[1]} \sim (2\pi m L^2 kT/h^2)^{1/2}$, $Q_{\text{rot}} \sim 2IkT/h^2$.
59. (a) $e^{-k\tau}/2k$. (b) $\pi \cos n\tau$. (c) $\pi^{1/2}\sigma \exp\left[-\tau^2/4\sigma^2\right]$.
(d) $G(\tau) = a^2[1 - |\tau|/2t_0]$, $|\tau| < 2t_0$, $G(\tau) = 0$ elsewhere.
60. $4\pi \int_{r_{\min}}^{r_{\max}} r^2 G(r)\, dr$.

61. $4\pi \int_0^r r^2\, dr = \frac{4}{3}\pi r^3$.
62. $3a_0/2$.

Chapter 3

1. $\pi^{1/2} \operatorname{erf}(x^{1/2})$.
2. $\Phi(x) = \frac{1}{2} \operatorname{erf}(x/\sqrt{2})$. $\Phi(\infty) = \frac{1}{2}$.
3. $\pi^{-1/2} e^{-x^2}\left[\frac{1}{x} - \frac{1}{2}\cdot\frac{1}{x^3} + \frac{1\cdot 3}{2\cdot 2}\cdot\frac{1}{x^5} - \cdots\right]$.
4. $s^{-1} \exp(s^2/4) \operatorname{erfc}(s/2)$.
5. $s^{-1} \exp(-2s^{1/2})$, $cs^{-1} \exp[-2(s/a)^{1/2}]$.
8. (b) $s^{-1} \ln(1 + s)$.
9. $\operatorname{li}(x) = \operatorname{Ei}(\ln x)$.
10. $F(k, \sin^{-1} x)$, $E(k, \sin^{-1} x)$.
11. $\sqrt{\pi}$, $\frac{1}{2}\sqrt{\pi}$. $\dfrac{1\cdot 3\cdots(2n-1)}{2^n}\Gamma(\tfrac{1}{2}) = \dfrac{(2n)!}{2^{2n}n!}\sqrt{\pi}$.
12. $n!/s^{n+1}$. $\Gamma(n+1)/s^{n+1}$.
13. $x\Gamma(x) = \Gamma(x+1)$.
14. $(-1)^n n!$.
15. $\Gamma(4)\sum_{n=1}^{\infty}\dfrac{1}{n^4}$.
16. $I_n = \pi^{n/2} = \frac{1}{2}nc_n\Gamma(n/2)$, $c_n = 2\pi^{n/2}/n\Gamma(n/2)$; $V_2 = \pi r^2$, $V_3 = \frac{4}{3}\pi r^3$.
17. (b, c) One is $\int_0^\infty \int_0^\infty x^n e^{-x^2} e^{-y^2}\, dx\, dy = \frac{1}{2}\pi^{1/2} \int_0^\infty x^n e^{-x^2}\, dx$.
    (d) Substitute $t = r^2$ and $t = x^2$ or y^2.
18. $\displaystyle\sum_{n=0}^{\infty} \frac{\pi^{1/2}}{2} \frac{(2n)!}{2^{2n}n!^2} \frac{\Gamma\left(\frac{n+1}{2}\right)}{\Gamma\left(\frac{n}{2}+1\right)} k^{2n}$. Since n is integral, the gamma functions can be further reduced to factorials by Equation 3.1 and Exercise 3.11.
19. (b) $4 \int_0^\infty \int_0^\infty e^{-u^2} e^{-v^2} u^{2x-1} v^{2y-1}\, du\, dv$
    $= 4 \int_0^\infty e^{-r^2} r^{2(x+y)-1}\, dr \int_0^{\pi/2} \cos^{2x-1}\theta \sin^{2y-1}\theta\, d\theta$.
20. $\mathrm{B}(a, b-a) = \Gamma(a)\Gamma(b-a)/\Gamma(b)$.
21. $-\operatorname{Ei}(-x) = \gamma(0, x)$, $\operatorname{erfc}(x) = \pi^{-1/2}\gamma(\frac{1}{2}, x^2)$, where $\gamma(x, y) \equiv \int_y^\infty e^{-t} t^{x-1}\, dt$.
22. $1/e$.
23. $1/\sqrt{\pi}$.
26. A single term gives 4.0235×10^{-2567}, correct to 1:12000, or 0.008%.
27. (a) 1.0352316. (b) 1.6050. (c) 0.10158, 1.00515. (d) 0.8427.
    (e) −2.27172 (requires quadratic interpolation). (f) 1.6858. (g) 1.5869.
    (h) 0.7824. (i) 0.5522 (requires two-dimensional interpolation).
    (j) 6.2539.
28. $\int_{-\pi}^{\pi} \sin mx \cos nx\, dx = 0$, $\int_{-\pi}^{\pi} \sin mx \sin nx = \pi\delta_{mn}(1 - \delta_{n0})$,
    $\int_{-\pi}^{\pi} \cos mx \cos nx\, dx = \pi\delta_{mn}(1 + \delta_{n0})$.
31. (a) $\delta(a - b)$ or $\delta(b - a)$. (b) $\delta(0) = \infty$.
32. $f(x) = \int_{-\infty}^{\infty} f(x')\delta(x - x')\, dx'$.

34. (a) $(2\pi)^{-1/2}f(x-a)$.
(b) Sum of two Gaussians, one centered at $x=-1$, the other at $x=+1$.
(d) $(2\pi)^{-1/2}\sum c_i f(x-a_i)$. (e) $\frac{1}{2}\exp(-\frac{1}{4}x^2)$, a Gaussian $\sqrt{2}$ times as wide.
35. Electron density within the entire crystal (neglecting "nuisance" factors of $\sqrt{2\pi}$).
36. $e^{-as}(a>0)$, $\frac{1}{2}(a=0)$, $0(a<0)$, although conventions differ for $L\{\delta(t)\}$. We obtain $\frac{1}{2}$ since we took δ to be an even function.

Chapter 4

1. $y''+\omega^2 y=0$.
2. $y''-y=0$.
4. $y''+2ky'+(k^2+\omega^2)y=0$.
5. $(1-x^2)y''-xy'-y=0$.
7. (a) $x^3+y^3-xy=C$.
(b) $\mu(x,y)=\sinh(x+y)$ leads to $xy\cosh(x+y)=C$.
8. (a) $G=$ constant (Example 2.6).
(b) $\mu(P,V,T)=1/PV$ leads to $PV=$ constant $\times\ T$.
9. $\ln(I^\circ/I)=\varepsilon\ln 10\ Lc$.
10. (a) $y=\ln(x+C)$. (b) $y=e^{Cx}$.
11. (a) $A(t)=A_0e^{-k_1t}$. (b) $A(t)=A_0e^{-t/\tau}$. (c) $A(t)=A_0[1+2k_2A_0t]^{-1}$.
(d) $\tau_{1/2}=(\ln 2)/k_1$, $\tau_{1/2}=1/2k_2A_0$.
12. $\dfrac{1}{A_0-B_0}\ln\dfrac{AB_0}{A_0B}=kt$.
13. $x=F(k,\sin^{-1}y)$.
14. $g(y)=\sigma\exp[-\sigma^2y^2/2]$, solution of $g'(y)=-\sigma^2yg(y)$.
15. $y=y_0+a\ln\dfrac{\cos(x/a)}{\cos(d/a)}$, $\theta=d/a$.
16. $(a^2-y^2)^{1/2}=x+C$.
17. (a) $\int du/[H(1,u)-u]=C+\int dx/x$. (b) $x^2=2y^2\ln(Cy)$.
19. (a) $y=Ce^{-\sin x}+2\sin x-2$. (b) $y=C(1+x)^{-2}+(1+x)^4/6$.
(c) $y=Ce^x+\sin x$.
20. $z=\left(\dfrac{v_0}{k}+\dfrac{g}{k^2}\right)(1-e^{-kt})$; $v\to -g/k$.
21. $y=(1-x^2)^{-1/2}\sin^{-1}x$, solution of $(1-x^2)y'=1+xy$.
22. $\mathrm{A}(t)=\mathrm{A}_0[1+k_1\mathrm{A}_0t]^{-1}=\mathrm{B}(t)$.

$$\mathrm{C}(t)=\mathrm{A}_0e^{-k_2t}-\frac{\mathrm{A}_0}{1+k_1\mathrm{A}_0t}+\frac{k_2}{k_1}\exp\left[-\frac{k_2}{k_1\mathrm{A}_0}(1+k_1\mathrm{A}_0t)\right]\times\left[\mathrm{Ei}\left(\frac{k_2}{k_1\mathrm{A}_0}+k_2t\right)-\mathrm{Ei}\left(\frac{k_2}{k_1\mathrm{A}_0}\right)\right].$$

23. $L[\mathrm{A}]=\mathrm{A}_0/(s+k_1)$, $L[\mathrm{B}]=k_1\mathrm{A}_0/(s+k_1)(s+k_2)$.
25. $y=y_0\left\{C_2+C_1\displaystyle\int\exp\left[-\int\left(p+2\frac{y_0'}{y_0}\right)dx\right]dx\right\}$. $y=Cx\sin x$.
27. (a) $y=A\sin 2x+B\cos 2x=A'e^{2ix}+B'e^{-2ix}$. (b) $y=e^{2x}[A\sin x+B\cos x]$.
(c) $y=Ae^x+Be^{-2x}$. (d) $y=Ae^{-x}+Bxe^{-x}$.

28. (a) $y = \exp\left[\frac{-1+\sqrt{5}}{2}x\right]$. (b) $y = e^x(e^{-\pi/2}\sin x + \cos x)$.

29. (a) $x(t) = x_0 \cos \omega t$, where $\omega = \sqrt{k/m}$.
(b) $T(t) = \frac{1}{2}m\omega^2 x_0^2 \sin^2 \omega t = \frac{1}{2}kx_0^2 \sin^2 \omega t$. $T_{max} = \frac{1}{2}kx_0^2$.
(c) $V(t) = \frac{1}{2}kx_0^2 \cos^2 \omega t$. (d) $V(x) = \frac{1}{2}kx^2$.
30. $y = (A + Bx + Cx^2)e^{rx}$.
31. (a) General solution of $y' - y = 0$ is $y = Ce^x$; particular solution to inhomogeneous equation is $y = \sin x$.
(b) $y = \cos x + e^{x/\sqrt{2}}[A \cos (x/\sqrt{2}) + B \sin (x/\sqrt{2})]$
$+ e^{-x/\sqrt{2}}[C \cos (x/\sqrt{2}) + D \sin (x/\sqrt{2})]$.

32. $x(t) = F_0(\omega^2 - \omega^2)^{-1}[\sin \omega t - (\omega/\omega_0) \sin \omega_0 t]$, resonance when $\omega = \omega_0$.
33. $x(t) = (F_0/2\omega) \sin \omega t$.
34. $y_1'c_1' + y_2'c_2' = r$ leads to $c_1 = -\int (ry_2\,dx/W(x))$, $c_2 = \int (ry_1\,dx/W(x))$.
35. $d\theta/dt = (2g/l)^{1/2}(\cos \theta - \cos \theta_0)^{1/2}$,
$E(\frac{1}{2}\theta, \csc \frac{1}{2}\theta) - E(\frac{1}{2}\theta_0, \csc \frac{1}{2}\theta_0) = (g/l)^{1/2}(\sin \frac{1}{2}\theta_0)t$.
36. $a_n = (-1)^{n+1}/n$.
37. $a_n = -a_{n-2}/n^2$ means $y = a_0 \sum_{n=0}^{\infty} (-1)^n \frac{x^{2n}}{2^{2n}n!^2}$.
38. (a) $1 - x^2 = \frac{2}{3}P_0(x) - \frac{2}{3}P_2(x)$. (b) $\sin^2 \theta = \frac{2}{3}P_0(\cos \theta) - \frac{2}{3}P_2(\cos \theta)$.

39. Via $(1 - t)^{-1} = \sum_{l=0}^{\infty} t^l$.

40. Via $[1 - 2(-x)t + t^2]^{-1/2} = [1 - 2x(-t) + (-t)^2]^{-1/2}$.
41. $(35x^4 - 30x^2 + 3)/8$.

42. $\frac{1}{r_{12}} = \frac{1}{r_1}\sum_{l=0}^{\infty}\left(\frac{r_2}{r_1}\right)^l P_l(\cos \theta)$.

43. $\int_0^{\pi} P_l(\cos \theta)P_m(\cos \theta) \sin \theta\, d\theta = 0$ if $l \neq m$.
44. $\sin \theta e^{i\phi}$, $\sin \theta e^{-i\phi}$, $\sin \theta \cos \theta e^{i\phi}$, $\sin \theta \cos \theta e^{-i\phi}$, $\sin^2 \theta e^{2i\phi}$, $\sin^2 \theta e^{-2i\phi}$, $(5 \cos^3 \theta - 3 \cos \theta)$. Nodal cones for $Y_{30}(\theta, \phi)$ at $\theta = \cos^{-1}\sqrt{3/5}$; nodal plane at $\theta = \pi/2$ (xy-plane). For $Y_{32} - Y_{3-2}$, the three nodal planes are the xy, yz, and xz planes; $Y_{32} + Y_{3-2}$ looks the same, but rotated 45° about the z-axis.
46. $y = x^{-1/2} \sin x$ (continuous at $x = 0$) and $y = x^{-1/2} \cos x$.
47. Indicial equation gives $r = 0$. Recursion formula is $a_{k+1} = -(n - k)a_k/(k + 1)^2$.

Solution is $L_n(x) = a_0 \sum_{k=0}^{n} (-1)^k \frac{n!}{k!^2(n-k)!} x^k$; by convention $a_0 \equiv n!$.

48. Second-order. (c) $[(1 - x^2)u']' + [l(l + 1)]u = 0$. (d) $[xu']' - \left(\frac{n^2}{x} - x\right)u = 0$.

49. (a) $[(x^2 - 1)D^2 + 2xD]\psi = l(l + 1)\psi$. (b) $[x^2D^2 + xD + x^2]\psi = n^2\psi$.
50. $s^2\alpha = \frac{3}{4}\alpha$, $s^2\beta = \frac{3}{4}\beta$, $s_z\alpha = \frac{1}{2}\alpha$, $s_z\beta = -\frac{1}{2}\beta$.
52. $\lambda = n^2\pi^2/L^2$, $n = 1, 2, 3, \ldots$; $\psi_\lambda(x) = (2/L)^{1/2} \sin (n\pi x/L)$.
53. $\lambda = n^2$, $n = 0, 1, 2, \ldots$; $\psi_\lambda = (2\pi)^{-1/2}e^{\pm in\theta}$ or $\psi_0 = (2\pi)^{-1/2}$, $\psi_\lambda = \pi^{-1/2} \cos n\theta$ and $\pi^{-1/2} \sin n\theta$. Notice that all eigenvalues except $\lambda = 0$ are doubly degenerate.
54. $H'' - 2xH' + (\lambda - 1)H = 0$ leads to recursion formula $a_{n+2} = (2n + 1 - \lambda)a_n/$

$(n + 1)(n + 2)$ and $\lambda = 2n + 1$. Then $H_n(x) = \sum_{r=0}^{n/2} (-1)^r 2^{n-2r} \frac{n!}{r!(n-2r)!} x^{n-2r}$.

55. (a) $(\Omega - \lambda_i 1)\psi_i = \Omega\psi_i - \lambda_i\psi_i = 0$. (b) If $\phi = \sum c_j\psi_j$, then $P_i\phi = c_i\psi_i$.
(c, d) $P_iP_j\phi = c_i\psi_i\delta_{ij}$.
56. (a) If $I\psi = \lambda\psi$, then $I^2\psi = \lambda I\psi$. But $I^2\psi(x) = I\psi(-x) = \psi(x)$, so $\psi = \lambda I\psi$, or $I\psi = (1/\lambda)\psi$. Therefore $\lambda = 1/\lambda$, or $\lambda = \pm 1$. Any *even* function is an eigenfunction of I with eigenvalue $+1$; any *odd* function is an eigenfunction of I with eigenvalue -1.
(b) $P_{+1} = \frac{1}{2}(1 + I)$, $P_{-1} = \frac{1}{2}(1 - I)$. To check, notice that $P_{+1}f(x) = \frac{1}{2}[f(x) + f(-x)]$, which is even, and $P_{-1}f(x) = \frac{1}{2}[f(x) - f(-x)]$, which is odd. Notice further that $P_{+1} + P_{-1} = 1$.
57. $S^2\alpha\alpha = 2\alpha\alpha$, $S^2(\alpha\beta + \beta\alpha) = 2(\alpha\beta + \beta\alpha)$, $S^2(\alpha\beta - \beta\alpha) = 0$, $S^2\beta\beta = 2\beta\beta$; $S_z\alpha\alpha = \alpha\alpha$, $S_z\alpha\beta = 0$, $S_z\beta\alpha = 0$, $S_z\beta\beta = -\beta\beta$. Note that any linear combination of $\alpha\beta$ and $\beta\alpha$ is an eigenfunction of S_z with eigenvalue 0, but only the linear combinations $\alpha\beta + \beta\alpha$ and $\alpha\beta - \beta\alpha$ are eigenfunctions of S^2.
58. $P_2 = -\frac{1}{3}S^2 + \frac{5}{4}1$, $P_4 = \frac{1}{3}S^2 - \frac{1}{4}1$, $P_2\alpha\alpha\beta = \frac{2}{3}\alpha\alpha\beta - \frac{1}{3}\alpha\beta\alpha - \frac{1}{3}\beta\alpha\alpha$, $P_4\alpha\alpha\beta = \frac{1}{3}\alpha\alpha\beta + \frac{1}{3}\alpha\beta\alpha + \frac{1}{3}\beta\alpha\alpha$.
59. $P_1 = (S^2 - 6)(S^2 - 2)/12$,
$P_1\alpha\beta\alpha\beta = \frac{1}{6}(2\alpha\beta\alpha\beta + 2\beta\alpha\beta\alpha - \beta\alpha\alpha\beta - \beta\beta\alpha\alpha - \alpha\alpha\beta\beta - \alpha\beta\beta\alpha)$.
60. (a) $P_{ij}P_{jk}\Phi(x_i, x_j, x_k) = \Phi(x_j, x_k, x_i)$, but $P_{jk}P_{ij}\Phi(x_i, x_j, x_k) = \Phi(x_k, x_i, x_j)$.
(b) $P_{ij}^2 = 1$. (c) 1. (d) ± 1.
(e) $P_{+1} = \frac{1}{2}(1 + P_{12})$, $P_{-1} = \frac{1}{2}(1 - P_{12})$. The functions $P_{+1}\Phi(x_1, x_2) = \frac{1}{2}[\Phi(x_1, x_2) + \Phi(x_2, x_1)]$ and $P_{-1}\Phi(x_1, x_2) = \frac{1}{2}[\Phi(x_1, x_2) - \Phi(x_2, x_1)]$ are eigenfunctions of P_{12} belonging to the eigenvalues $+1$ and -1, respectively.
61. $P_{-1} = \frac{1}{8}(1 - P_{12})(1 - P_{13})(1 - P_{23})$.
62. $P = \frac{1}{4}(2 - S^2)(1 - P_{12})$; $P[\phi_1\alpha\phi_2\beta] = \frac{1}{4}[\phi_1\alpha\phi_2\beta - \phi_2\beta\phi_1\alpha - \phi_1\beta\phi_2\alpha + \phi_2\alpha\phi_1\beta]$.
63. The trial function $\phi_1(x_1)\alpha(x_1)\phi_2(x_2)\alpha(x_2)\phi_3(x_3)\beta(x_3)$ is an eigenfunction of S_z with eigenvalue 1/2. Desired operator is $(-\frac{1}{3}S^2 + \frac{5}{4}1)P_{-1}$, where P_{-1} is that given in Exercise 4.61. The acceptable function has a total of 18 terms; we shall find that determinants provide a less cumbersome way to write such a function.
64. (a) $-f''(y) + y^2f(y)$.
(b) Via $F[f''(x)] = -y^2F[f(x)]$, $g''(y) = -F[x^2f(x)]$, and
$\Omega Ff(x) = F[x^2f(x)] + y^2F[f(x)] = F\Omega f(x)$.
(c) Apply F to both sides of $\Omega\psi_n = \lambda_n\psi_n$.

Chapter 5

1. $u(x, y) = f(x + y)$, with f arbitrary.
3. If $v(x, s) = L[u(x, t)]$, then $\partial^2 v/\partial x^2 = sv/D$, or $v(x, s) = A \exp[(s/D)^{1/2}x] + B \exp[-(s/D)^{1/2}x]$, with $A = 0$ and $B = C_0/s$. Therefore $u(x, t) = C_0 \operatorname{erfc}[x/2(Dt)^{1/2}]$ and $c(x, t) = C_0 \operatorname{erf}[x/2(Dt)^{1/2}]$.
4. A particular solution is $R(r)\Phi(\phi) = r^{n^2}e^{in\phi}$, $n = 0, \pm 1, \pm 2, \ldots$. The general solution is $u(r, \phi) = \sum_{n=-\infty}^{\infty} c_n r^{n^2} e^{in\phi}$.
5. To have a maximum, $\partial^2\phi/\partial x^2$, $\partial^2\phi/\partial y^2$, and $\partial^2\phi/\partial z^2$ must all be negative; for a minimum, all must be positive (cf. Exercise 2.8c).
6. $$\phi(x, y, z) = \sum_{\lambda,\mu,\nu} [a_{\lambda,\mu,\nu}e^{\lambda x} + a_{-\lambda,\mu,\nu}e^{-\lambda x}][b_{\lambda,\mu,\nu}e^{\mu y} + b_{\lambda,-\mu,\nu}e^{-\mu y}] \times [c_{\lambda,\mu,\nu}e^{\nu z} + c_{\lambda,\mu,-\nu}e^{-\nu z}],$$
where sum is over all $\{\lambda, \mu, \nu\}$ satisfying $\lambda^2 + \mu^2 + \nu^2 = 0$.
7. $\Phi''(\phi) = -m^2\Phi(\phi)$.
$\frac{1}{\sin\theta}\frac{d}{d\theta}\left(\sin\theta\frac{d\Theta}{d\theta}\right) - \left[\frac{m^2}{\sin^2\theta} - l(l+1)\right]\Theta = 0$, leading to associated Legendre functions $\Theta_{lm}(\theta) \propto P_l^{|m|}(\cos\theta)$.

$$\frac{1}{r^2}\frac{d}{dr}\left(r^2\frac{dR}{dr}\right) - \left[\frac{l(l+1)}{r^2} - \frac{8\pi^2\mu}{h^2}\left(E + \frac{Ze^2}{r}\right)\right]R = 0,$$

leading to associated Laguerre polynomials: $R_{nl}(r) \propto \rho^l e^{-\rho/2} L_{n+l}^{2+1}(\rho)$, where $\rho = (8\pi^2\mu Ze^2/nh^2)r$ and $L_j{}^k(\rho) = D^k L_j(\rho)$.

8. $\lambda = \pi^2[(n/L)^2 + (m/M)^2]$, $n = 1, 2, 3, \ldots$; $m = 1, 2, 3, \ldots$; $\Psi_{n,m}(x, y) = (2/L^{1/2}M^{1/2}) \sin(n\pi x/L) \sin(m\pi x/M)$.
9. $\lambda = x_{nm}{}^2/r_0{}^2$, $\Psi_{n,m}(r, \phi) = J_m(x_{nm}r/r_0)e^{im\phi}$, not normalized.
12. $A_n \sin(\omega x/c)\cos\omega t = \frac{1}{2}A_n \sin\left[\frac{\omega}{c}(x - ct)\right] + \frac{1}{2}A_n \sin\left[\frac{\omega}{c}(x + ct)\right]$.
13. Frequency increases with increasing tension and number of nodes and with decreasing density and length; these results should be in accord with your experience.
15. (a) $c(x, t) = \sum_{n=0}^{\infty} a_n \exp(-n^2\pi^2 Dt/L^2)\cos(n\pi x/L)$.
    (b) $c(x, t) = (N/L)[1 + 2\sum_{n=1}^{\infty}(-1)^n \exp(-4n^2\pi^2 Dt/L^2)\cos(2n\pi x/L)]$.
16. The Fourier expansion of an even function has only cosine terms, since every $b_n = 0$. The Fourier expansion of an odd function has only sine terms since every $a_n = 0$.
17. (a) $\dfrac{1}{2\pi} + \dfrac{1}{\pi}\sum_{n=1}^{\infty}\cos nx$.
    (b) $\dfrac{\pi^2}{3} + 4\sum_{n=1}^{\infty}(-1)^n\dfrac{\cos nx}{n^2}$, so $\sum_{n=1}^{\infty}(-1)^n\dfrac{1}{n^2} = -\dfrac{\pi^2}{12}$.
    (c) $\dfrac{4}{\pi}\sum_{n=0}^{\infty}\dfrac{\sin[(2n+1)\pi x/L]}{2n+1}$; no. (d) $\dfrac{2}{\pi}\sum_{n=1}^{\infty}\dfrac{n^2\cos 2nx}{(n^2 - \frac{1}{4})^3}$, $\pi^2/4$.
18. (a) $\frac{2}{5}\pi^5 + \sum_{n=1}^{\infty}(-1)^n\left(\dfrac{8\pi^3}{n^2} - \dfrac{48\pi}{n^4}\right)\cos nx$. (b) $\Gamma(4)\sum\dfrac{1}{n^4} = \dfrac{\pi^4}{15}$.
19. Equations 5.13 and 5.14.
20. (a) $\dfrac{1}{2\pi}\sum_{n=-\infty}^{\infty} e^{iny}$.
    (b) Via $F\{\sum\delta(x - nd)\} = (2\pi)^{-1/2}\sum e^{-indy} = (2\pi)^{-1/2}\sum e^{+indy}$
    $= (2\pi)^{1/2}\sum\delta(dy - 2\pi m)$
22. Application of the Fourier integral theorem shows that the two integrals are equal.
23. (a) $\delta(x - x')$.
    (b) $(2\pi)^{1/2}\delta(y \mp \omega)$, $(\pi/2)^{1/2}[\delta(y - \omega) + \delta(y + \omega)]$,
    $(1/i)(\pi/2)^{1/2}[\delta(y - \omega) - \delta(y + \omega)]$.
    (c) $(2\pi)^{1/2}\sum a_n\delta(y - n\omega)$.
24. $(2\pi)^{1/2}a(\omega)$.
25. (a) $|g(\omega)| = (2/\pi)^{1/2}a\,|\sin[\frac{1}{2}(\omega - \omega_0)t_0]/(\omega - \omega_0)|$. (b) ω_0. (d) $\omega = \omega_0 \pm 2n\pi/t_0$, $n = 1, 2, \ldots$. (e) $\Delta\omega = 2\pi/t_0$. (f) 2π.
26. (a) $X(x)T(t) = \exp[-iyx - Dy^2t]$. (b) $F\{g(y)\exp(-Dy^2t)\}$.
    (c) $g(y) = F^{-1}[f(x)]$. (d) $c(x, t) = F\{F^{-1}[f(x)]\exp(-Dy^2t)\}$.
27. (a) $\exp(-Dy^2t) = F^{-1}\{(2Dt)^{-1/2}\exp(-x^2/4Dt)\}$.
    (b) $c(x, t) = (2Dt)^{-1/2}[\exp(-x^2/4Dt)*f(x)]$.
28. $c(x, t) = (N^2/4\pi Dt)^{1/2}\exp(-x^2/4Dt)$.
29. $\int\phi_x^*(\xi)\phi_{x'}(\xi)\,d\xi = \delta(x - x')$, which is zero when $x \neq x'$.
30. $L\{L_n(t)\} = \int_0^{\infty} e^{-st}L_n(t)\,dt = (s - 1)^n/s^{n+1}$. $\left(\text{Expansion of } s^{-1}\left[1 + \dfrac{1-s}{s}x\right]^{-1} \text{ is easiest.}\right)$

31. $c_n = \int_0^\infty f(x)L_n(x)e^{-x}\,dx = \int_0^\infty e^{-(a+1)x}L_n(x)\,dx = a^n/(1+a)^{n+1}$, so

$$e^{-ax} = \sum_{n=0}^{\infty} \frac{a^n}{(1+a)^{n+1}} L_n(x).$$

32. $H_{n+1}(x) - 2xH_n(x) + 2nH_{n-1}(x) = 0.$
33. $(2^n\pi^{1/2}n!)^{1/2}\delta_{mn}.$
34. $c_n = (2^n\pi^{1/2}n!)^{-1}\int_{-\infty}^{\infty} f(x)H_n(x)e^{-x^2/2}\,dx.$
35. $e^{-3/2 x^2} = \sum_{n=0}^{\infty}(-1)^n \frac{H_{2n}(x)e^{-x^2/2}}{2^{3n+1/2}n!}.$

Chapter 6

1. 1/55.
2. (a) 1/37. (b) 18/37.
3. (a) 1/16. (b) 4/16. (c) 15/16. (d) 5/16.
4. (a) $1/6^3$. (b) $1/6^3$. (c) $6/6^3$. (d) $10/6^3$.
5. (a) 1/8100. (b) 179/8100.
6. Both are equally likely; you have probably never been dealt either hand, just ones that look like the second one.
7. $1 - P(E)$.
8. 3/8.
9. $\frac{4}{52}\cdot\frac{3}{51}\cdot\frac{2}{50}\cdot\frac{1}{49}$.
10. (a) $\frac{13}{52}\cdot\frac{12}{51}\cdot\frac{11}{50}\cdot\frac{10}{49}\cdot\frac{9}{48}$. (b) $\frac{39}{52}\cdot\frac{38}{51}\cdots\frac{27}{40}$.
11. (a) $\frac{8}{52}\cdot\frac{7}{51}\cdot\frac{6}{50}\cdot\frac{5}{49}\cdot\frac{4}{48}$. (b) $\frac{12}{52}\cdot\frac{11}{51}\cdot\frac{10}{50}\cdot\frac{9}{49}\cdot\frac{8}{48}$.
    (c) $\frac{16}{52}\cdot\frac{15}{51}\cdot\frac{14}{50}\cdot\frac{13}{49}\cdot\frac{12}{48}$.
12. $\frac{52}{52}\cdot\frac{48}{51}\cdot\frac{44}{50}\cdot\frac{40}{49}\cdot\frac{36}{48}$.
13. $\frac{52}{52}\cdot\frac{39}{51}\cdot\frac{26}{50}\cdot\frac{13}{49}$.
14. (a) $\frac{6}{6}\cdot\frac{5}{6}\cdot\frac{4}{6}\cdot\frac{3}{6}\cdot\frac{2}{6}\cdot\frac{1}{6}$. (b) $\frac{52}{52}\cdot\frac{48}{51}\cdot\frac{44}{50}\cdots\frac{4}{40}$.
15. (a) $\frac{15}{27}\cdot\frac{14}{26}$. (b) $\frac{12}{27}\cdot\frac{11}{26}$. (c) $1 - \dfrac{15\cdot 14 + 12\cdot 11}{27\cdot 26}$.
16. (a) $\frac{37}{39}\cdot\frac{36}{38}\cdot\frac{35}{37}\cdots\frac{25}{27}$. (b) $\frac{37}{39}\cdot\frac{36}{38}\cdots\frac{12}{14}$.
    (c) 1 minus the sum of (a) and (b).
17. If E_1 = "individual does not show the trait," and E_2 = "individual has one *a* gene, then $P(E_2 \mid E_1) = 2p(1-p)/(1-p^2)$.
18. If E_i = "*i*th coin is gold," then $P(E_2 \mid E_1) = \dfrac{1/3}{1/2} = 2/3$.
19. Follow from Eq. 6.6.
21. $(4/52)^4$.
22. (a) $V_A/(V_A + V_B)$. (b) $[V_A/(V_A + V_B)]^N$.
    (c) $\Delta S = Nk \ln [(V_A + V_B)/V_A]$.
23. $P(A + B + C) = P(A) + P(B) + P(C) - P(AB) - P(BC) - P(AC) + P(ABC)$.
24. All equations are correct.
25. $\frac{12}{27}\cdot\frac{15}{26} + \frac{15}{27}\cdot\frac{12}{26}$.
26. (a) $N/2^N$. (b) $5^{N-1}N/6^N$.
27. $\frac{4}{9}\cdot\frac{1}{8} + \frac{3}{9}\cdot\frac{4}{8} + \frac{2}{9}\cdot\frac{3}{8}$.
28. (a) $(0.99)^{101}$. (b) $101 \times 0.01 \times (0.99)^{100}$. (c) $1 - 2.00 \times (0.99)^{100} \sim 0.267$.
29. (a) $\frac{20}{52}\cdot\frac{16}{51}\cdot\frac{12}{50}\cdot\frac{8}{49}\cdot\frac{4}{48}$. (b) 10 times (a).
30. (a) $4\cdot\frac{13}{52}\cdot\frac{12}{51}\cdot\frac{11}{50}\cdot\frac{10}{49}\cdot\frac{9}{48}$. (b) $4\cdot 5\cdot\frac{13}{52}\cdot\frac{12}{51}\cdot\frac{11}{50}\cdot\frac{10}{49}\cdot\frac{39}{48}\cdot\frac{9}{47}$ + (a).
31. (a) $\frac{32}{52}\cdot\frac{31}{51}\cdots\frac{20}{40}$. (b) $\frac{28}{52}\cdot\frac{27}{51}\cdots\frac{16}{40}$. (c) (a) minus (b).

33. (a) $\frac{4}{7} \cdot \frac{1}{2} + \frac{2}{8} \cdot \frac{1}{2}$. (b) $(\frac{4}{7} \cdot \frac{3}{6} + \frac{2}{8} \cdot \frac{1}{7})\frac{1}{2}$, $(\frac{3}{7} \cdot \frac{2}{6} + \frac{6}{8} \cdot \frac{5}{7})\frac{1}{2}$, $\frac{4}{7} \cdot \frac{3}{6} + \frac{2}{8} \cdot \frac{6}{7}$.
(c) $(\frac{4}{7} \cdot \frac{3}{6} + \frac{4}{7} \cdot \frac{2}{8} + \frac{2}{8} \cdot \frac{4}{7} + \frac{2}{8} \cdot \frac{1}{7})\frac{1}{4}$, $(\frac{3}{7} \cdot \frac{2}{6} + \frac{3}{7} \cdot \frac{6}{8} + \frac{6}{8} \cdot \frac{3}{7} + \frac{6}{8} \cdot \frac{5}{7})\frac{1}{4}$,
$2(\frac{4}{7} \cdot \frac{3}{6} + \frac{4}{7} \cdot \frac{6}{8} + \frac{3}{7} \cdot \frac{2}{8} + \frac{2}{8} \cdot \frac{6}{7})\frac{1}{4}$. (d) $\frac{6}{15} \cdot \frac{5}{14}$, $\frac{9}{15} \cdot \frac{8}{14}$, $2 \cdot \frac{6}{15} \cdot \frac{9}{14}$.
34. (a) p^2. (b) $(1 - p)^2$. (c) $2p(1 - p)$. (d) $p(1 - p)/(1 - p^2)$. (e) Square of (d).
35. (a) If E_1 = "pinochle deck chosen," E_2 = "bridge deck chosen," E_3 = "ace drawn,"
then $\dfrac{P(E_1 \mid E_3)}{P(E_2 \mid E_3)} = \dfrac{1/2}{1/2} \cdot \dfrac{8/48}{4/52}$, or $P(E_1 \mid E_3) = \frac{8}{48} \cdot \frac{52}{4}/(\frac{8}{48} \cdot \frac{52}{4} + 1) = \frac{13}{19}$.
(b) $\frac{8}{48} \cdot \frac{13}{19} + \frac{4}{52} \cdot \frac{6}{19}$.
37. $1 - (1/2)^n$.
38. $1 - \frac{48}{52} \cdot \frac{47}{51} \cdots \frac{36}{40}$.
39. (a) $1 - \left(\dfrac{n-1}{n}\right)^r$. (b) $1 - \dfrac{n}{n} \cdot \dfrac{n-1}{n} \cdots \dfrac{(n-r+1)}{n}$.
40. $1 - \frac{365}{365} \cdot \frac{364}{365} \cdot \frac{363}{365} \cdots \frac{342}{365} \sim 0.54$.
42. (a) 52^r. (b) 52^{-r}.
43. (a) 4^3. (b) 20^{124}.
44. $26^3 - 5^3 - 21^3$.
45. (a) $P_{5N}^{52} = 52!/(52 - 5N)!$. (b) $P_9^{25} = 25!/16!$.
46. $1/6!$.
49. (a) $C_2^{10} = 10 \cdot 9/1 \cdot 2$. (b) C_4^{24}.
50. (a) C_5^{52}. (b) $1/C_5^{52}$.
51. $26^3 \cdot 10^3$. (b) $C_4^{17}C_3^{72}$.
52. (a) $P_4^4/4^4 \sim 0.094$. (b) $P_6^{10}/10^6 \sim 0.015$.
53. (a) $C_1^4C_{12}^{48}/C_{13}^{52}$. (b) $(C_1^4C_{12}^{48}/C_{13}^{52})(C_1^3C_{12}^{36}/C_{13}^{39})(C_1^2C_{12}^{24}/C_{13}^{26})$.
54. (a) C_5^{13}/C_5^{52}, C_{13}^{39}/C_{13}^{52}. (b) $(C_1^{13})^4/C_4^{52}$. (c) $(C_1^4)^{13}/C_{13}^{52}$.
(d) C_{13}^{37}/C_{13}^{39}, $C_2^2C_{11}^{37}/C_{13}^{39}$, $C_1^2C_{12}^{37}/C_{13}^{39}$. (e) $(C_1^4)^5/C_5^{52}$.
(f) $1 - P_r^n/n^r$. You may like to check that these are really the same answers as those given previously.
55. (a) $C_2^{13}(C_1^{13})^3/C_5^{52}$. (b) 4 times (a). (c) $C_4^{13}(C_3^{13})^3/C_{13}^{52}$. (d) 4 times (c).
56. $C_{r-1}^{48}C_1^4/C_r^{52}$.
57. (a) $C_2^4C_{11}^{22}/C_{13}^{26}$. (b) $C_1^4C_{12}^{22}/C_{13}^{26}$. (c) $C_3^4C_{10}^{22}/C_{13}^{26}$. (d) C_{13}^{22}/C_{13}^{26}.
(e) $C_4^4C_9^{22}/C_{13}^{26}$. (f) $C_2^4C_{11}^{22}/C_{13}^{26} \sim 0.4$, $2C_1^4C_{12}^{22}/C_{13}^{26} \sim 0.5$, $2C_{13}^{22}/C_{13}^{26} \sim 0.1$.
60. (b) 1.
61. $P(0) = C_5^{48}/C_5^{52}$, $P(1) = C_1^4C_4^{48}/C_5^{52}$, $P(2) = C_2^4C_3^{48}/C_5^{52}$,
$P(3) = C_3^4C_2^{48}/C_5^{52}$, $P(4) = C_4^4C_1^{48}/C_5^{52}$.
A distribution such that $P(r) = C_r^lC_{m-r}^{n-l}/C_m^n$ is known as a HYPERGEOMETRIC DISTRIBUTION.
62. $P(0, 0) = (4/6)^2$, $P(0, 1) = 2(1/6)(4/6) = P(1, 0)$,
$P(1, 1) = 2(1/6)^2$, $P(2, 0) = (1/6)^2 = P(0, 2)$.
64. (a) $P(r, s) = C_r^4C_s^4C_{13-r-s}^{44}/C_{13}^{52}$. (b) No.
(c) No. For example, $P(4) = C_4^4C_9^{48}/C_{13}^{52}$, but $P(4 \mid 13) = 0$.
66. (a) 2^n. (b) 0.
67. 1:6:15:20:15:6:1.
68. $C_3^{10}(\frac{1}{6})^3(\frac{5}{6})^7$.
69. $C_{13}^{52}(\frac{4}{52})^{13}(\frac{48}{52})^{39}$.
70. $C_2^{50}(\frac{1}{25})^2(\frac{24}{25})^{48}C_2^{48}(\frac{1}{24})^2(\frac{23}{24})^{46}C_2^{46}(\frac{1}{23})^2(\frac{22}{23})^{44} \cdots C_2^4(\frac{1}{2})^2(\frac{1}{2})^2 = \dfrac{50!}{2^{25}25^{50}}$.
72. $P(r \mid 4) = \dfrac{C_r^4}{C_4^{10}} \cdot \dfrac{C_r^{10}(\frac{1}{2})^{10}}{(\frac{1}{2})^4} = C_{r-4}^6(\frac{1}{2})^6$.
73. $1/2e \sim 0.184$, $C_2^{52}(\frac{1}{52})^2(\frac{51}{52})^{50} \sim 0.186$.

74. $\ln P(10^6) = 10^6 \ln 10^4 - 10^4 - \ln 10^6! \sim -3.6 \times 10^6$, by Stirling's approximation, or $P(10^6) \sim 10^{-1600000}$. Therefore the U.S. population does not seem to be distributed at random.
75. (a) 2. (b) $P(31) \sim 2^{31}e^{-2}/31! \sim 3.4 \times 10^{-26}$.
77. (a) $1/\sqrt{5\pi} \sim 0.25$. (b) $1/\sqrt{50\pi} \sim 0.080$. (c) $1/\sqrt{500\pi} \sim 0.025$.
78. (a) $2\Phi(\sqrt{2/5}) \sim 0.473$. (b) $2\Phi(2) \sim 0.9546$.

(c) $2\Phi(\sqrt{40}) = \text{erf}(\sqrt{20}) \sim 1 - \dfrac{1}{\sqrt{20\pi}} e^{-20} \sim 1 - 2.6 \times 10^{-10} = .99999999974$.

79. $P(r) = C_r^{N_m}\left(\dfrac{1}{N_i}\right)^r\left(1 - \dfrac{1}{N_i}\right)^{N_m - r}$ (objects = monomers, slots = chains).

$P(r) \sim \mu^r e^{-\mu}/r!$, where $\mu = N_m/N_i$.

81. $x^4 + y^4 + z^4 + 4(x^3y + x^3z + y^3x + y^3z + z^3x + z^3y) + 6(x^2y^2 + x^2z^2 + y^2z^2) + 12(x^2yz + y^2xz + z^2xy)$.
82. (a) $32!/4!^8$. (b) $52!/10!\,11!\,31!$.
83. (a) $6!/1!^6$. (b) $6!/6^6$. (c) $12!/2!^6$. (d) $12!/2!^6 6^{12}$. (e) $50!/2!^{25}25^{50}$.
84. $\dfrac{4!}{1!^4}\dfrac{48!}{12!^4}\Big/\dfrac{52!}{13!^4}$.
85. (a) $\dfrac{4!}{2!\,1!\,1!} C_4^{13}C_4^{13}C_3^{13}C_2^{13}/C_{13}^{52}$. (b) $\dfrac{4!}{1!^4} C_5^{13}C_4^{13}C_3^{13}C_1^{13}/C_{13}^{52}$.
86. (a) $6!/1!^6 = P_6^6$. (b) $6!/4!\,2! = C_2^6$. (c) $6!/3!\,2!\,1! = M_{\{3,2,1\}}^6$.
87. (a) 9: AB|−|−, A|B|−, A|−|B, B|A|−, −|AB|−, −|A|B, B|−|A|, −|B|A, −|−|AB.

(b) 6: AA|−|−, A|A|−, A|−|A, −|AA|−, −|A|A, −|−|AA.

(c) 3: A|A|−, A|−|A, −|A|A.

88. $C_4^6 = 15$.
89. $C_3^6 = 20$.
90. $n^{-r} < 1/C_r^{n+r-1}$.
93. Via $\sum_{i,j} x_i P(x_i, y_j) = \sum_i x_i P(x_i)$.
94. From Exercise 6.92, $\overline{x + c} = \bar{x} + c$, so $(x + c) - \overline{(x + c)} = x - \bar{x}$.
95. $\bar{\xi} = 0$, $\overline{\xi^2} = \sigma^2$.
97. Via $\sum_{i,j} x_i y_j P(x_i, y_j) = \sum_i x_i P(x_i) \sum_j y_j P(y_j)$.
98. Via $\sum_{i,j} x_i y_j P(x_i, y_j) = a\overline{x^2} + b\bar{x}$.
99. (a) 45/10. (b) 285/10. (c) $\sqrt{825/100} \sim 2.9$.
100. (a) 45/10. (b) 9750/400.

(c) $\sqrt{1650/400} \sim 2.0$. Notice that this σ is just $1/\sqrt{2}$ that of the previous exercise.

101. (a) \$3.50. (b) \$$\frac{5}{13}$.
104. $\bar{n}_A + \bar{n}_K + \bar{n}_Q + \cdots + \bar{n}_2 = 5$.
105. (a) $\sum_{r=0}^{\infty} r(np)^r e^{-np}/r! = e^{-np}\sum_{r=1}^{\infty}(np)^r/(r-1)!$
$= npe^{-np}\sum_{s=0}^{\infty}(np)^s/s! = npe^{-np}e^{np} = np$.
106. (a) 100. (b) $\sqrt{500/6}$. (c) $100, 10 = \sqrt{600/6}$.

(d) $1000, \sqrt{5000/6}, 1000, \sqrt{1000}$.

107. (a) Np. (b) $\sigma = (Np)^{1/2}$.
108. (a) $[e^{h\nu/kT} - 1]^{-1}$. (b) Least cumbersome is $\frac{1}{2}h\nu \coth(h\nu/2kT)$.
110. (b) $\sum p^{r-1}(1 - p) = (1 - p)\sum p^s = (1 - p)/(1 - p) = 1$.
111. (a) $2\frac{1}{6}$. (b) (600/13)mph.
112. (b) 60%.

113. $m_k = D^k M(0)$.
114. $\exp[-np + npe^s]$.
115. $[pe^s + (1 - p)]^n$.
116. (c) $\frac{1}{2} + \frac{1}{2}e^s$, $(\frac{1}{2} + \frac{1}{2}e^s)^n$. (d) $\left[\frac{e^s}{6} \cdot \frac{1 - e^{6s}}{1 - e^s}\right]^N$
117. (a) $P(\theta)\,d\theta = \frac{1}{\pi}\,d\theta$. (b) Zero.
118. (a) Probability of a value less than x.
(b) $\int_{-\infty}^{x_{\text{med}}} P(x)\,dx = \frac{1}{2} = \int_{x_{\text{med}}}^{\infty} P(x)\,dx$.
119. $P(h)\,dh = (mg/kT)e^{-mgh/kT}\,dh$.
120. $\mu = (a + b)/2$, $\sigma = (b - a)/\sqrt{12}$.
121. Mean $= \mu$, since $\int (x - \mu)P(x)\,dx = \int yP(y + \mu)\,dy = 0$ by symmetry; standard deviation is infinite.
122. *Hint:* Eliminate x by substituting $x = t + u$.
123. (a) $\sqrt{2/L}$. (b) $L/2$. (c) $L^2\left(\frac{1}{3} - \frac{1}{2n^2\pi^2}\right)$. (d) $\left(\frac{1}{12} - \frac{1}{2n^2\pi^2}\right)^{1/2} L$.
124. (a) 0.
(b) $(n + \frac{1}{2})/\alpha^2$, via $\int x^2 \exp[-t^2 + 2\alpha xt - s^2 + 2\alpha xs - \alpha^2 x^2]\,dx$

$$= \frac{\pi^{1/2}}{\alpha^3}\left(\tfrac{1}{2} + s^2 + 2st + t^2\right)\sum_{n=0}^{\infty} \frac{2^n s^n t^n}{n!}.$$

126. (a) 0. (b) $n^2\pi^2\hbar^2/L^2$. (c) $n\pi\hbar/L$. (d) $\left(\frac{n^2\pi^2}{12} - \frac{1}{2}\right)^{1/2}\hbar$.
127. (a) $dx/\pi(A^2 - x^2)^{1/2}$. (b) 0. (c) $A^2/2$.
128. (a) $(2\pi kT/m)^{1/2}$.
(b) Probability that a molecule has its x-component of velocity between v_x and $v_x + dv_x$, its y-component between v_y and $v_y + dv_y$, and its z-component between v_z and $v_z + dv_z$.
(c) Probability that a molecule has a total speed between v and $v + dv$, that the azimuthal coordinate of its trajectory is between θ_v and $\theta_v + d\theta_v$, and that the longitudinal coordinate of its trajectory is between ϕ_v and $\phi_v + d\phi_v$.
(d) Probability that the total speed is between v and $v + dv$.
129. (a) $\int_{-\infty}^{x} P(x)\,dx$.
(b) $\int_{-\infty}^{y} P[f(x)]\,df(x)$.
131. $\langle i(t)\rangle = \frac{6}{7}i(t_{\max})$.
132. (a) $\frac{1}{2}$. (b) $\int_0^\pi \cos^2\theta \sin\theta\,d\theta / \int_0^\pi \sin\theta\,d\theta = \frac{1}{3}$.
133. $m_k = (2\pi)^{1/2} i^{-k} D^k M(0)$.
135. $m_{2n} = (2n)!/2^n n!$.
136. μ, σ, $\sigma^3\gamma$.
137. (b) ρ. (c) If and only if $\rho = 0$.
140. (a) Via $\overline{\chi^2} = 2\Gamma\left(\frac{\nu}{2} + 1\right)\Big/\Gamma\left(\frac{\nu}{2}\right)$.
(b) Column headed 0.50; for $\nu = 1$, $\chi^2_{\text{med}} = 0.455$.
141. No, no, yes (barely).
142. (a) 65. (b) $\sqrt{40} \sim 6.3$. (d) $\frac{1}{2} - \Phi(2) = 0.023$.
(e) Yes, since $88 = \bar{N} + 3.64\sigma_N$, and the probability of obtaining as many aces as 88 is $\frac{1}{2} - \Phi(3.64) = 0.0001$.

143. Yes, since $(14400 - 14173)/\sqrt{14400} \sim 1.89 > 1.645$. Notice that the critical region is the single tail corresponding to fewer counts than $14400 - 1.645 \times 120$. The probability of obtaining 14173 counts or fewer is only $\frac{1}{2} - \Phi(1.89) \sim 0.03$.
144. No, since $\chi^2 = 4.48$, and with $\nu = 3$ the critical region is $\chi^2 > 7.815$.
145. (a) 81.
(b) Apply chi-square test with $\nu = 11$; critical region is $\chi^2 > 19.675$. In 1969, the values were 35.53 and 28.32. If the two National League expansion teams are omitted, $\chi^2 = 9.90$, which is quite close to the median χ^2 for $\nu = 9$.
146. Results are too even, since with $\nu = 5$ the probability of obtaining $\chi^2 \leq 0.44$ is less than 0.01.
149. (a) 450, 75781. (b) -0.195, 0.9907.
151. $\Phi(0.6745) = 0.25$.
152. With $n = 1$, σ_x is completely indeterminate.
153. (a) -0.195, 1.017. (c) -0.195 ± 0.2075, -0.195 ± 0.415. (d) Yes.
154. (a) 450, 281. (b) 450 ± 57.4, 450 ± 115.
(c) True $\mu_x = 499.5$ *is* between 392.6 and 507.4. (d) $\sigma_x = 1000/\sqrt{12} \sim 289$.
155. $3610 \pm \sqrt{36100}/10$.
157. 358 to 512. (With $\nu = 7$, the median value of t is 0.711, which may be compared with the value 0.6745 for $\nu = \infty$.) Yes.
158. Loosely speaking, there is a 5% chance that $\sigma_x > \sqrt{6/1.145} \times 0.15\%$, and a 17.5% chance that $\sigma_x > 0.25\%$.
159. The occupation numbers are $\{10, 1, 7, 6\}$, giving $\chi^2 = 7.0$. With $\nu = 1$ and $\alpha = 0.05$, the critical region is $\chi^2 > 3.841$, so we reject the hypothesis that the 24 numbers were sampled from a normal distribution.
160. With $\nu = 5$, the critical region is $\chi^2 > 11.070$. On the basis of the observed $\chi^2 = 17.111$, we conclude that the grading curves (or the students) are different.
161. $(3.14 \pm 0.47) \times 10^6\ \text{sec}^{-1}$.
163. $\pm\sqrt{140}\% \sim \pm 12\%$.
164. Via $\Delta c/c = \Delta T/(T \ln T)$ and $\ln T_{\text{optimum}} = -1$.
165. $\pm 0.3\%$.
166. (a) 82. (b) 128. (c) $\pm\sqrt{12.2 + 4.0} \sim \pm 4.0$, $\pm\sqrt{16.8 + 4.0} \sim \pm 4.6$.
(d) $82/128 \sim 0.64$. (e) $\pm\sqrt{16.2/82^2 + 20.8/128^2} \sim 6.1\%$ (standard deviation).
167. (a) 115, 13176, 0 to 230. (b) No.
168. $\hat{a} = 7.524$, $\hat{b} = -3.607$, $r = 0.9653$.
$\Delta H^\ddagger = 13.12$ kcal/mole, $\Delta S^\ddagger = -30.62$ cal/mole deg.
169.
170.
$$\hat{a} = \frac{1}{\Delta}\begin{vmatrix} \sum x^2y & \sum x^3 & \sum x^2 \\ \sum xy & \sum x^2 & \sum x \\ \sum y & \sum x & n \end{vmatrix} \qquad \hat{b} = \frac{1}{\Delta}\begin{vmatrix} \sum x^4 & \sum x^2y & \sum x^2 \\ \sum x^3 & \sum xy & \sum x \\ \sum x^2 & \sum y & n \end{vmatrix}$$
$$\hat{c} = \frac{1}{\Delta}\begin{vmatrix} \sum x^4 & \sum x^3 & \sum x^2y \\ \sum x^3 & \sum x^2 & \sum xy \\ \sum x^2 & \sum x & \sum y \end{vmatrix} \qquad \Delta = \begin{vmatrix} \sum x^4 & \sum x^3 & \sum x^2 \\ \sum x^3 & \sum x^2 & \sum x \\ \sum x^2 & \sum x & n \end{vmatrix}.$$
172. Via $\sum d^2 = \sum y^2 - a\sum xy - b\sum y + a(a\sum x^2 + b\sum x - \sum xy)$
$+ b(a\sum x + nb - \sum y)$.
173. Two possible ways are $u = x^2$, $v = x/y$, $\beta = a$, $\alpha = b$, and $u = 1/x^2$, $v = 1/xy$, $\alpha = a$, $\beta = b$.

174. $\hat{a} = [(\sum w)(\sum wxy) - (\sum wx)(\sum wy)]/[(\sum w)(\sum wx^2) - (\sum wx)^2]$.
$\hat{b} = [(\sum wx^2)(\sum wy) - (\sum wx)(\sum wxy)]/[(\sum w)(\sum wx^2) - (\sum wx)^2]$.
175. As in Exercise 6.174, with $a = 2k$, $b = 1/A_0$, $w_i = A_i^4$, $x_i = t_i$, $y_i = 1/A_i$.
176. As in Exercise 6.174, with $a = (\alpha nF/RT)$, $b = (\alpha nF/RT)E_{1/2}$, $x = E$, $y = \ln [i/(i_d - i)]$, $w = i^2(i_d - i)^2$.

Chapter 7

2. 0, X, 0.
3. $X_1 + X_2 + X_3 + X_4 = 0$.
6. (a) 1. (b) 2. (c) Latter. (d) Former.
7. 1, N.
8. Via $\sum (x_i - x_i')E_i = 0$.
9. $(x_1, x_2, x_3) = \frac{1}{2}(-x_1 + x_2 + x_3)(0, 1, 1) + \frac{1}{2}(x_1 - x_2 + x_3)(1, 0, 1) + \frac{1}{2}(x_1 + x_2 - x_3)(1, 1, 0)$.
12. No, No.
14. The diagonals of a rhombus are perpendicular.
15. $|C|^2 = (\mathbf{A} - \mathbf{B}) \cdot (\mathbf{A} - \mathbf{B}) = |A|^2 - 2|A|\,|B| \cos\theta + |B|^2$.
18. $\dfrac{|\mathbf{A} \times \mathbf{B}|}{|A|\,|B|\,|C|} = \dfrac{|\mathbf{B} \times \mathbf{C}|}{|A|\,|B|\,|C|} = \dfrac{|\mathbf{A} \times \mathbf{C}|}{|A|\,|B|\,|C|}$, or $\dfrac{\sin\theta_{AB}}{|C|} = \dfrac{\sin\theta_{BC}}{|A|} = \dfrac{\sin\theta_{AC}}{|B|}$.
19. $\hat{\mathbf{u}} = \cos\alpha\,\hat{\mathbf{i}} + \cos\beta\,\hat{\mathbf{j}} + \cos\gamma\,\hat{\mathbf{k}}$, $\cos^2\alpha + \cos^2\beta + \cos^2\gamma = 1$.
20. (a) $2\hat{\mathbf{i}} + 3\hat{\mathbf{j}} + 4\hat{\mathbf{k}}$, $\sqrt{3}$, $\sqrt{14}$, 6, $\sqrt{6/7}$.
(b) $4\hat{\mathbf{i}} - 8\hat{\mathbf{j}} + 2\hat{\mathbf{k}}$, $\sqrt{14}$, $\sqrt{38}$, 16, $8/\sqrt{133}$.
(c) $2\hat{\mathbf{i}} + \hat{\mathbf{j}} + \hat{\mathbf{k}}$, $\sqrt{2}$, $\sqrt{2}$, 1, 1/2.
21. $\cos^{-1}(-1/3)$.
22. (a) $2\pi/3 = 120°$ (sp^2). (b) $\cos^{-1}(-1/3)$ = tetrahedral angle $\sim 109.5°$ (sp^3).
23. (a) 3.054 Å. (b) 1.716 Å. (c) 1.718 Å.
24. 1.390, 1.519, 1.247, 1.272 Å.
25. 14, −1, $\sqrt{N}$.
26. (a) $\hat{\mathbf{i}} - 2\hat{\mathbf{j}} + \hat{\mathbf{k}}$, $\sqrt{6}$; $-14\hat{\mathbf{i}} - 8\hat{\mathbf{j}} - 4\hat{\mathbf{k}}$, $2\sqrt{69}$; $\hat{\mathbf{i}} - \hat{\mathbf{j}} - \hat{\mathbf{k}}$, $\sqrt{3}$.
(b) −2.
27. $x_1y_2 - x_2y_1 = \begin{vmatrix} x_1 & y_1 \\ x_2 & y_2 \end{vmatrix}$.
28. $\hat{\mathbf{i}} - \hat{\mathbf{j}}$, $\hat{\mathbf{j}} - \hat{\mathbf{k}}$, $\hat{\mathbf{i}} - \hat{\mathbf{k}}$, $\hat{\mathbf{i}} + \hat{\mathbf{j}} - 2\hat{\mathbf{k}}$, and linear combinations of these.
31. (a) $\mathbf{A} = a_x\hat{\mathbf{i}} + a_y\hat{\mathbf{j}} + a_z\hat{\mathbf{k}}$, $\mathbf{B} = b_x\hat{\mathbf{i}}$, $\mathbf{C} = c_x\hat{\mathbf{i}} + c_y\hat{\mathbf{j}}$, and multiply out.
(b) $(\mathbf{A} \cdot \mathbf{C})(\mathbf{B} \cdot \mathbf{D}) - (\mathbf{A} \cdot \mathbf{D})(\mathbf{B} \cdot \mathbf{C})$.
33. $|A|^2 (\mathbf{B} \times \mathbf{A})$, $\mathbf{0}$.
34. (a) $\hat{\phi}_1 = 6^{-1/2}\phi_1$, $\hat{\phi}_6 = 6^{-1/2}\phi_6$. (b) $\hat{\phi}_2 = 12^{-1/2}\phi_2$, $\hat{\phi}_3 = \frac{1}{2}(0, 1, 1, 0, -1, -1)$.
(c) $\hat{\phi}_4 = 12^{-1/2}\phi_4$, $\hat{\phi}_5 = \frac{1}{2}(0, 1, -1, 0, 1, -1)$.
35. $(1/2)^{1/2}$, $(3/2)^{1/2}x$, $(5/8)^{1/2}(3x^2 - 1)$, $(7/8)^{1/2}(5x^3 - 3x)$.
36. 1, $x - 1$, $(x^2 - 4x + 2)/4$, $(x^3 - 9x^2 + 18x - 6)/36$.
37. $\mathbf{D} = a\hat{\mathbf{i}} + b\hat{\mathbf{j}} + c\hat{\mathbf{k}}$.
38. (a) $2x - y - 2z = 3$. (b) $\mathbf{N} = 4\hat{\mathbf{i}} + \hat{\mathbf{j}} - 3\hat{\mathbf{k}}$, $\mathbf{X}_0 = \hat{\mathbf{i}} + \hat{\mathbf{j}}$ or $5\hat{\mathbf{k}}/3$ or others.
39. $(\mathbf{X} - \mathbf{X}_0) \cdot (\mathbf{X}_0 - \mathbf{A}_0) = 0$.
40. $(\mathbf{X} - \mathbf{R}_1) \cdot [(\mathbf{R}_2 - \mathbf{R}_1) \times (\mathbf{R}_3 - \mathbf{R}_1)] = 0$, or others similar.
41. $\mathbf{X} = \mathbf{X}_0 + t(\mathbf{X}_1 - \mathbf{X}_0)$ or $(\mathbf{X} - \mathbf{X}_0) \times (\mathbf{X}_1 - \mathbf{X}_0) = \mathbf{0}$ or similar.

42. (a) $\mathbf{N}_0 \times \mathbf{N}_1$. (b) $\mathbf{X} = (\hat{\mathbf{i}} + \hat{\mathbf{j}}) + t(2\hat{\mathbf{i}} + 10\hat{\mathbf{j}} + 6\hat{\mathbf{k}})$ or equivalent.
43. If the lines intersect, they have a point, $\mathbf{X}_0 + t_0\mathbf{D}_0 = \mathbf{X}_1 + t_1\mathbf{D}_1$, in common. Rearranging and taking the cross product with $\mathbf{D}_0$ gives $(\mathbf{X}_1 - \mathbf{X}_0) \times \mathbf{D}_0 = (t_0\mathbf{D}_0 - t_1\mathbf{D}_1) \times \mathbf{D}_0 = t_1\mathbf{D}_0 \times \mathbf{D}_1$.
44. $(\mathbf{P}_0 - \mathbf{X}_0) \cdot \mathbf{N}/|N|$.
45. $\{(\mathbf{X}_1^0 \cdot \mathbf{n}_1)(\mathbf{n}_2 \times \mathbf{n}_3) + (\mathbf{X}_2^0 \cdot \mathbf{n}_2)(\mathbf{n}_3 \times \mathbf{n}_1) + (\mathbf{X}_3^0 \cdot \mathbf{n}_3)(\mathbf{n}_1 \times \mathbf{n}_2)\}/[n_1n_2n_3]$
$$= \sum (\mathbf{X}_i^0 \cdot \mathbf{n}_i)\boldsymbol{\nu}_i, \text{ where } \boldsymbol{\nu}_i \cdot \mathbf{n}_j = \delta_{ij}.$$
46. $\psi = \cos^{-1}(1/2)$.
47. x or $y = \pm d(1 + \cos\psi)^{1/2}/(3 + \cos\psi)^{1/2}$, $z = \pm d(1 - \cos\psi)^{1/2}/2(3 + \cos\psi)^{1/2}$. $\cos\theta = (1 - \cos\psi)/(3 + \cos\psi)$.
48. 5/3 of the C—C bond length.
49. Infinite circular cone whose apex is at the origin, whose axis is in the direction of $\hat{\mathbf{n}}$, and whose half-apical angle is $\cos^{-1} a$.
50. Infinite circular cylinder of radius r, whose axis is in the direction $\mathbf{D}$ and passes through the point $\mathbf{A}_0$.
51. (a) Function of one variable. (b) Vector field. (c) Vector field. (d) Scalar point function. (e) Vector function. (f) Vector function.
52. $G(\boldsymbol{\xi}) = (2\pi)^{-3/2} \iiint \Phi(\mathbf{X})\, e^{-i\boldsymbol{\xi}\cdot\mathbf{X}}\, dV$, with $\boldsymbol{\xi}$ expanded in the reciprocal basis.
53. (a) $\Delta d = \mathbf{X} \cdot \hat{\mathbf{v}}_0 - \mathbf{X} \cdot \hat{\mathbf{v}}$. (b) $\Delta\delta = -\mathbf{X} \cdot \boldsymbol{\xi}$.
(c) $\iiint \rho(\mathbf{X})e^{i\Delta\delta}\, dV = F\{\rho(\mathbf{X})\}$, neglecting "nuisance" factors of $(2\pi)^{1/2}$.
55. $\mathbf{X}(t) = r\cos\omega t\hat{\mathbf{i}} \pm r\sin\omega t\hat{\mathbf{j}} \mp vt\hat{\mathbf{k}}$.
57. $\mathbf{X}_1(t) = 10\cos(2\pi t/10)\hat{\mathbf{i}} + 10\sin(2\pi t/10)\hat{\mathbf{j}} + 3.4t\hat{\mathbf{k}}$.
$\mathbf{X}_2(t) = -10\cos(2\pi t/10)\hat{\mathbf{i}} - 10\sin(2\pi t/10)\hat{\mathbf{j}} + 3.4t\hat{\mathbf{k}}$.
59. $\mathbf{E}(\mathbf{R}) = -r\hat{\mathbf{e}}_r$, $\mathbf{E}(\mathbf{X}) = r\hat{\mathbf{e}}_\phi$.
60. Lines bend at the phase boundary but continue through it.
61. Expand in Cartesian coordinates.
62. Via $\dfrac{d}{dt}(\mathbf{X} \cdot \mathbf{X}) = 2\mathbf{X} \cdot \mathbf{v}$.
63. $(\mathbf{X} - \mathbf{X}_0) \times \mathbf{D} = \mathbf{0}$ or $\mathbf{X} - \mathbf{X}_0 = t\mathbf{D}$, where $\mathbf{D} = d\mathbf{X}/dt\,|_{\mathbf{X}=\mathbf{X}_0}$.
64. (a) $-\omega r\sin\omega t\hat{\mathbf{i}} + \omega r\cos\omega t\hat{\mathbf{j}} + v\hat{\mathbf{k}}$. (b) $\sin^{-1}[v/(\omega^2r^2 + v^2)^{1/2}] = \tan^{-1}(v/\omega r)$.
65. $\tan^{-1}(3.4/2\pi)$.
66. $2\pi(v^2 + \omega^2r^2)^{1/2}/\omega$.
67. $\sinh x_0$.
69. $\mathbf{X}(t_1) - \mathbf{X}(t_0)$.
70. $\rho(\mathbf{X}) \equiv dm(\mathbf{X})/dV$.
71. $\frac{3}{4}a\hat{\mathbf{k}}$.
73. $\mathbf{a}(t) = -\omega^2r\cos\omega t\,\hat{\mathbf{i}} - \omega^2r\sin\omega t\,\hat{\mathbf{j}} = -\omega^2\mathbf{X}$.
74. $\mathbf{X}(t) = -\frac{1}{2}gt^2\hat{\mathbf{k}} + t\mathbf{v}_0 + \mathbf{X}_0$.
75. (a) $\mathbf{X}(t) = x_0\cos\omega_xt\hat{\mathbf{i}} + y_0\cos\omega_yt\hat{\mathbf{j}} + z_0\cos\omega_zt\hat{\mathbf{k}}$, where $\omega_x = (k_x/m)^{1/2}$, etc.
(b) $T(t) = \frac{1}{2}(k_xx_0^2\sin^2\omega_xt + k_yy_0^2\sin^2\omega_yt + k_zz_0^2\sin^2\omega_zt)$,
$T_{max} = \frac{1}{2}(k_xx_0^2 + k_yy_0^2 + k_zz^2)$.
(c) $V(t) = \frac{1}{2}(k_xx_0^2\cos^2\omega_zt + k_yy_0^2\cos^2\omega_yt + k_zz_0^2\cos^2\omega_zt)$,
$V(\mathbf{X}) = \frac{1}{2}(k_xx^2 + k_yy^2 + k_zz^2)$.
78. (a) Southward and upward, such that the force is perpendicular to the earth's axis.
(b) Since $\boldsymbol{\omega}$ points toward the North Star, $-2m(\boldsymbol{\omega} \times \mathbf{v})$ is eastward.
(c) Westward.
79. Via $\mathbf{v} \cdot \mathbf{p} = \boldsymbol{\omega} \times \mathbf{r} \cdot \mathbf{p} = \boldsymbol{\omega} \cdot \mathbf{r} \times \mathbf{p}$.
81. $\sum m_i[(y_i^2 + z_i^2)\omega_x - x_iy_i\omega_y - x_iz_i\omega_z]$.

82. $\mathbf{T}_r = -\boldsymbol{\omega} \times \mathbf{L}_f$.
83. Via $\boldsymbol{\mu} \cdot (d\boldsymbol{\mu}/dt) = \gamma[\mu\mu H_0]$.
84. (a) $\boldsymbol{\mu}_r = \text{const.}$ (b) $\boldsymbol{\omega} = -\gamma\mathbf{H}$. (c) $|\omega| = \gamma|H|$.
86. For example, $\nabla \cdot (a\mathbf{E} + b\mathbf{F}) = a\nabla \cdot \mathbf{E} + b\nabla \cdot \mathbf{F}$.
87. (a) $\nabla\phi = yz\hat{\mathbf{i}} + xz\hat{\mathbf{j}} + xy\hat{\mathbf{k}}$, $\nabla^2\phi = 0$.
(b) $\nabla \cdot \mathbf{E} = 0$, $\nabla \times \mathbf{E} = a\hat{\mathbf{i}} + b\hat{\mathbf{j}} + c\hat{\mathbf{k}}$.
(c) $\nabla \cdot \mathbf{E} = 0$, $\nabla \times \mathbf{E} = 2y\hat{\mathbf{i}} - 2x\hat{\mathbf{j}}$. (d) $\nabla \cdot \mathbf{E} = 3z^2$, $\nabla \times \mathbf{E} = 4(x^2 + y^2)\hat{\mathbf{k}}$.
(e) $\nabla\phi = (x\hat{\mathbf{i}} + y\hat{\mathbf{j}} + z\hat{\mathbf{k}})/r = \hat{\mathbf{e}}_r$, $\nabla^2\phi = 2(x^2 + y^2 + z^2)^{-1/2} = 2/r$.
(f) $\nabla \cdot \mathbf{E} = 2(x^2 + y^2 + z^2)^{-1/2} = 2/r$, $\nabla \times \mathbf{E} = \mathbf{0}$.
89. (a) $q(x\hat{\mathbf{i}} + y\hat{\mathbf{j}} + z\hat{\mathbf{k}})/(x^2 + y^2 + z^2)^{3/2}$. (b) $(q/r^2)\hat{\mathbf{e}}_r$.
90. (a) Expand in Cartesian coordinates.
(b) Expand in spherical polar coordinates. (c) As (a).
91. $-\nabla\phi = -[(\boldsymbol{\mu} \cdot \mathbf{r})\nabla(1/r^3) + (1/r^3)\nabla(\boldsymbol{\mu} \cdot \mathbf{r})] = 3(\boldsymbol{\mu} \cdot \mathbf{r})\mathbf{r}/r^5 - \boldsymbol{\mu}/r^3$.
92. $\psi_{\mathbf{k}}(\mathbf{X}) = e^{i\mathbf{k}\cdot\mathbf{X}}$.
93. 8.
94. $(\mathbf{X} - \mathbf{X}_0) \cdot \nabla\phi(\mathbf{X}_0) = 0$.
95. $8\hat{\mathbf{i}} + 9\hat{\mathbf{j}} - 6\hat{\mathbf{k}}$ (or any multiple).
96. $\pm[D_xF(x_0, y_0)\hat{\mathbf{i}} + D_yF(x_0, y_0)\hat{\mathbf{j}} - \hat{\mathbf{k}}]$.
97. (a) $\nabla G(x, y, z)$.
(c) ∇F must be in the same direction as ∇G, or $\nabla F = t\nabla G$. For comparison with Lagrange's formula, let $t = -\lambda$ and expand in Cartesian components.
98. (a) $du/|\nabla u|$. (b) $dv/|\nabla v|$ and $dw/|\nabla w|$. (c) $du\, dv\, dw/|\nabla u|\, |\nabla v|\, |\nabla w|$.
99. $\nabla F(\{x_i\}) = \sum \left(\dfrac{\partial F}{\partial x_i}\right)\hat{\mathbf{e}}_i$, where $\hat{\mathbf{e}}_i$ is a unit vector along the ith coordinate axis in an n-dimensional space.
100. At each $\mathbf{X}$, ∇H is orthogonal to $\mathbf{X}$.
103. $F(\mathbf{r}) = \sum \dfrac{1}{n!} (\mathbf{r} \cdot \nabla)^n F(\mathbf{0})$.
105. (a) -3. (b) 0. (c) 1. (d) 0. (e) 0.
106. $\nabla \cdot (q/r^2)\hat{\mathbf{e}}_r = (1/r^2)\dfrac{dq}{dr} = 0/r^2$.
108. $\nabla^2(\phi\psi) = \nabla \cdot \nabla(\phi\psi) = \nabla \cdot [\phi\nabla\psi + \psi\nabla\phi] = \phi\nabla^2\psi + 2\nabla\phi \cdot \nabla\psi + \psi\nabla^2\phi$.
110. (a) $\mathbf{0}$. (b) $\nabla \times [f(r)\mathbf{R}] = f(r)\nabla \times \mathbf{R} + \nabla f \times \mathbf{R} = \dfrac{df}{dr}\hat{\mathbf{e}}_r \times \mathbf{R} = \mathbf{0}$.
112. If $\mathbf{B} = \nabla \times \mathbf{A}$, then $\nabla \times (\mathbf{A} + \nabla\phi)$ also equals $\mathbf{B}$, since $\nabla \times \nabla\phi = \mathbf{0}$.
113. $\nabla \cdot (\nabla\phi \times \nabla\psi) = \nabla\phi \cdot \nabla \times \nabla\psi - \nabla\psi \cdot \nabla \times \nabla\phi = \nabla\phi \cdot \mathbf{0} - \nabla\psi \cdot \mathbf{0}$.
115. $\frac{1}{3}|\mathbf{X}_1 - \mathbf{X}_0|\, \{|X_0|^2 + \mathbf{X}_0 \cdot \mathbf{X}_1 + |X_1|^2\} = \frac{1}{3}|\mathbf{X}_1 - \mathbf{X}_0|\, \{|\mathbf{X}_0 + \mathbf{X}_1|^2 - \mathbf{X}_0 \cdot \mathbf{X}_1\}$
116. $W = mg$, for whatever path you may have chosen.
117. $\displaystyle\int \mathbf{F} \cdot d\mathbf{X} = \int \frac{d\mathbf{p}}{dt} \cdot \frac{d\mathbf{X}}{dt}\, dt = \tfrac{1}{2}m \int \frac{d}{dt}(\mathbf{v} \cdot \mathbf{v})\, dt = \tfrac{1}{2}m(v^2_{\text{final}} - v^2_{\text{initial}})$.
118. 2π.
119. $\omega r^2 e^{-r^2}\sqrt{\pi}/v$.
120. (a) $R[-T_0 \ln(P_1/P_0) + (T_1 - T_0)]$ or equivalent.
(b) $R\dfrac{T_1P_0 - P_1T_0}{P_1 - P_0} \ln \dfrac{P_1}{P_0}$ or equivalent.

121. $C_P \ln (T_1/T_0) - R \ln (P_1/P_0)$, along either path.

122. $\nabla \cdot \mathbf{H} = I \oint \nabla \cdot \dfrac{\mathbf{dX} \times (\mathbf{R} - \mathbf{X})}{|\mathbf{R} - \mathbf{X}|^3} = -I \oint d\mathbf{X} \cdot \nabla \times \dfrac{\mathbf{R} - \mathbf{X}}{|\mathbf{R} - \mathbf{X}|^3}$

$= -I \oint d(\mathbf{R} - \mathbf{r}) \cdot \nabla \times \dfrac{\mathbf{r}}{|r|^3} = 0$ by Exercise 7.110b. We have also used Exercise 7.109a.

123. No, since $\oint \mathbf{F} \cdot \mathbf{dX} \propto \oint \mathbf{v} \cdot \mathbf{dX} = \oint \mathbf{v} \cdot \dfrac{\mathbf{dX}}{dt}\, dt = \oint (\mathbf{v} \cdot \mathbf{v})\, dt$, and integrand is always positive.

124. Because $\tan^{-1}$ is not a (single-valued) function.

125. If $\mu\mathbf{F} \cdot \mathbf{dX}$ is exact, then $\nabla \times (\mu\mathbf{F}) = \mathbf{0}$. By Exercise 7.109b, $\nabla \times (\mu\mathbf{F}) = \mu\nabla \times \mathbf{F} + \nabla\mu \times \mathbf{F}$. Taking the dot product with $\mathbf{F}$ gives $\mu\mathbf{F} \cdot \nabla \times \mathbf{F} + [\mathbf{F}\nabla\mu\mathbf{F}] = 0$. Since $[\mathbf{F}\nabla\mu\mathbf{F}] = 0$ and $\mu \neq 0$, we may conclude that $\mathbf{F} \cdot \nabla \times \mathbf{F} = 0$.

126. No.

127. Via $\hat{\mathbf{n}} = \left[-\dfrac{\partial\phi}{\partial x}\hat{\mathbf{\imath}} - \dfrac{\partial\phi}{\partial y}\hat{\mathbf{\jmath}} + \hat{\mathbf{k}}\right] \Big/ \left[\left(\dfrac{\partial\phi}{\partial x}\right)^2 + \left(\dfrac{\partial\phi}{\partial y}\right)^2 + 1\right]^{1/2}$ and

$\hat{\mathbf{n}}/(\hat{\mathbf{k}} \cdot \hat{\mathbf{n}}) = -\dfrac{\partial\phi}{\partial x}\hat{\mathbf{\imath}} - \dfrac{\partial\phi}{\partial y}\hat{\mathbf{\jmath}} + \hat{\mathbf{k}}.$

128. $2r \iint (r^2 - x^2 - y^2)^{-1/2}\, dx\, dy = 4\pi r^2$.

129. $2\pi(1 - \cos \frac{1}{2}\chi)$.

130. $V(z_0) = Q/z_0$ for $z_0 > a$ (same as if charge were concentrated at the origin), $V(z_0) = Q/a$ for $z_0 < a$.

131. Consider $\iiint_V \nabla \cdot (\mathbf{r}/r^3)\, dV$, where V is the volume within the arbitrary surface but outside the small sphere. By Gauss' theorem, this volume integral equals the difference of the two surface integrals. But according to Exercise 7.106, $\nabla \cdot (\mathbf{r}/r^3) = 0$ everywhere within V, so the two surface integrals are equal. And $\displaystyle\iint \frac{\mathbf{r} \cdot \mathbf{n}\, dA}{r^3}$ over the sphere equals $\displaystyle\frac{1}{r_0^2}\iint dA = 4\pi$, by Exercise 7.128.

132. The surface integral, $\oiint_\Sigma (\mathbf{r} \cdot \hat{\mathbf{n}})\, dA/r^3$ equals 4π if Σ encloses $\mathbf{0}$ and equals 0 if Σ does not enclose $\mathbf{0}$. Therefore we may write $\oiint_\Sigma (\mathbf{r} \cdot \hat{\mathbf{n}})\, dA/r^3 = 4\pi \iiint_V \delta(\mathbf{r})\, dV$, where V encloses Σ. But by Gauss' theorem, $\oiint_\Sigma (\mathbf{r} \cdot \hat{\mathbf{n}})\, dA/r^3 = \iiint_V \nabla \cdot (\mathbf{r}/r^3)\, dV$, so we may conclude that $\nabla \cdot (\mathbf{r}/r^3) = 4\pi\delta(\mathbf{r})$: According to Exercise 7.106, $\nabla \cdot (\mathbf{r}/r^3) = 0$ except at the origin. Now we see that $\nabla \cdot (\mathbf{r}/r^3)$ at the origin is so large an infinity that the integral of $\nabla \cdot (\mathbf{r}/r^3)$ over any volume containing the origin is equal to 4π.

133. Follows immediately from $dz = 0$ and $\hat{\mathbf{n}} = \hat{\mathbf{k}}$.

134. Twice the area enclosed by Γ.

135. $8 \sinh^2 \frac{1}{2}$.

136. $4\sqrt{\pi}(1 - e^{-\rho^2})$.

137. $0(r < a)$, $\pi(r = a)$, $2\pi(r > a)$. (*Hint:* Let $x = a + r\cos\phi$, $y = r \sin\phi$.) By Green's theorem, if Γ does not enclose the origin, then $\oint (-y\, dx + x\, dy)/(x^2 + y^2) = 0$, but if Γ does enclose the origin, the value of the integral is indeterminate. The earlier analysis shows that this value is 2π, and the delta function expresses these results.

Chapter 8

1. mn.
2. $\displaystyle\sum_k \sum_l a_{ik} b_{kl} c_{lj}$.

3. $\begin{pmatrix} 2 & 1 \\ 1 & 1 \end{pmatrix}, \begin{pmatrix} 1 & 1 \\ 1 & 2 \end{pmatrix}.$

4. (3 5 4), undefined.

5. $\begin{pmatrix} u_y v_z - u_z v_y \\ u_z v_x - u_x v_z \\ u_x v_y - u_y v_x \end{pmatrix} = \mathbf{u} \times \mathbf{v}.$

6. (a) $x_1 y_1 + x_2 y_2$. (b) $x_1 f_{11} y_1 + x_1 f_{12} y_2 + x_2 f_{21} y_1 + x_2 f_{22} y_2$.

 (c) $\begin{pmatrix} x_1 y_1 & x_2 y_1 \\ x_1 y_2 & x_2 y_2 \end{pmatrix}.$

7. $\begin{pmatrix} 0 & 0 & 0 \\ 0 & 0 & \sqrt{2} \\ 0 & \sqrt{2} & 0 \end{pmatrix}.$

8. $\begin{pmatrix} a_{21} & a_{22} & a_{23} \\ a_{11} & a_{12} & a_{13} \\ a_{31} & a_{32} & a_{33} \end{pmatrix}, \begin{pmatrix} a_{12} & a_{11} & a_{13} \\ a_{22} & a_{21} & a_{23} \\ a_{32} & a_{31} & a_{33} \end{pmatrix}, \begin{pmatrix} a_{22} & a_{21} & a_{23} \\ a_{12} & a_{11} & a_{13} \\ a_{32} & a_{31} & a_{33} \end{pmatrix}.$

9. Premultiplication interchanges rows i and j; postmultiplication interchanges columns i and j.

10. (a) Premultiplication by **B**, with $b_{kl} = \delta_{kl}$, except $b_{jj} = \alpha$.
 (b) Premultiplication by **C**, with $c_{kl} = \delta_{kl}$, except $c_{ij} = 1$.
 (c) Premultiplication by **D**, with $d_{kl} = \delta_{kl}$, except $d_{ij} = \alpha$.

11. $\begin{pmatrix} \cos\psi\cos\theta\cos\phi - \sin\psi\sin\phi & \cos\psi\cos\theta\sin\phi + \sin\psi\cos\phi & \cos\psi\sin\theta \\ -\sin\psi\cos\theta\cos\phi - \cos\psi\sin\phi & -\sin\psi\cos\theta\sin\phi + \cos\psi\cos\phi & -\sin\psi\sin\theta \\ -\sin\theta\cos\phi & -\sin\theta\sin\phi & \cos\theta \end{pmatrix}.$

12. (a) $(\mathbf{B})_{ij} = b_i \delta_{ij}$. (b) Postmultiplication by $\mathbf{C} = (c_i \delta_{ij})$.

13. **0**.

14. $2i\boldsymbol{\sigma}_z, 2i\boldsymbol{\sigma}_x, 2i\boldsymbol{\sigma}_y$.

15. (a) $\begin{pmatrix} 0 & 0 & 0 \\ 0 & 0 & 0 \\ 0 & 0 & 0 \end{pmatrix}, \begin{pmatrix} 0 & 0 & 0 \\ 0 & 0 & 0 \\ 1 & 2 & 0 \end{pmatrix}.$ (b) $\begin{pmatrix} 0 & 0 \\ 0 & 0 \end{pmatrix}.$
 (c) Although $ab = 0$ implies $a = 0$ or $b = 0$, it is possible that $\mathbf{AB} = \mathbf{0}$ and yet neither $\mathbf{A} = \mathbf{0}$ nor $\mathbf{B} = \mathbf{0}$.

17. No. From Exercise 8.3, $[\mathbf{A}, \mathbf{1}] = \mathbf{0} = [\mathbf{B}, \mathbf{1}]$, but $[\mathbf{A}, \mathbf{B}] = \begin{pmatrix} 1 & 0 \\ 0 & -1 \end{pmatrix}.$

19. First and last of Exercise 8.7; **i** of Exercises 8.8 and 8.9; $\boldsymbol{\sigma}_x$, $\boldsymbol{\sigma}_y$, and $\boldsymbol{\sigma}_z$ of Exercise 8.14.

21. (a) 2^N.
 (b) For example, with $N = 3$, the matrix product is $\alpha\alpha\alpha + \alpha\alpha\beta + \beta\alpha\alpha + \beta\alpha\beta + \alpha\beta\alpha + \alpha\beta\beta + \beta\beta\alpha + \beta\beta\beta$.

23. diag $(-\frac{1}{2}, -\frac{1}{2}, -\frac{1}{2})$.

24. A vector field is not necessarily linear: $\mathbf{E}(a\mathbf{X} + b\mathbf{Y}) \neq a\mathbf{E}(\mathbf{X}) + b\mathbf{E}(\mathbf{Y})$.

26. $\mathbf{s}_x = \begin{pmatrix} 0 & 1/2 \\ 1/2 & 0 \end{pmatrix}.$ $\mathbf{s}_y = \begin{pmatrix} 0 & -i/2 \\ i/2 & 0 \end{pmatrix}.$ $\mathbf{s}_z = \begin{pmatrix} 1/2 & 0 \\ 0 & -1/2 \end{pmatrix}.$

 $\mathbf{s}_+ = \begin{pmatrix} 0 & 1 \\ 0 & 0 \end{pmatrix}.$ $\mathbf{s}_- = \begin{pmatrix} 0 & 0 \\ 1 & 0 \end{pmatrix}.$ $\mathbf{s}^2 = \begin{pmatrix} 3/4 & 0 \\ 0 & 3/4 \end{pmatrix}.$

27. $\mathbf{S}^2 = \begin{pmatrix} 2 & 0 & 0 & 0 \\ 0 & 1 & 1 & 0 \\ 0 & 1 & 1 & 0 \\ 0 & 0 & 0 & 2 \end{pmatrix}$. $\mathbf{S}_z = \begin{pmatrix} 1 & 0 & 0 & 0 \\ 0 & 0 & 0 & 0 \\ 0 & 0 & 0 & 0 \\ 0 & 0 & 0 & -1 \end{pmatrix}$.

29. (a) $(\mathbf{P}_i)_{kl} = 0$ except $(\mathbf{P}_i)_{ii} = 1$ (or $(\mathbf{P}_i)_{kl} = \delta_{ik}\delta_{il}$). (c) $\mathbf{1}$.

30. (a) $(N - i)/N$. (b) i/N.
(d) $P(B_k) = \sum P(B_k \mid A_j)P(A_j) = \sum\sum P(B_k \mid A_j)P(A_j \mid E_i)P(E_i)$.
(e) If $\mathbf{p}_0$ is an $(N + 1)$-dimensional row matrix such that $(\mathbf{p}_0)_r = p_0(r)$, then $p_n(s) = (\mathbf{p}_0\mathbf{P}^n)_s$, where $\mathbf{P}$ is the $(N + 1) \times (N + 1)$ matrix of (c).

31. Clearly, $(\mathbf{P}_1\mathbf{P}_2)_{ij} = \sum (\mathbf{P}_1)_{ik}(\mathbf{P}_2)_{kj} \geqslant 0$.
Also $\sum_j (\mathbf{P}_1\mathbf{P}_2)_{ij} = \sum_j \sum_k (\mathbf{P}_1)_{ik}(\mathbf{P}_2)_{kj} = \sum_k (\mathbf{P}_1)_{ik} = 1$.

32. (a) $\begin{pmatrix} 1 & 0 & 0 & & & & \\ q & 0 & p & & & & \\ & q & 0 & p & & & \\ & & & \cdot & & & \\ & & & & \cdot & & \\ & & & & & \cdot & \\ & & & & q & 0 & p \\ & & & & & & 1 \end{pmatrix}$. (b) $\begin{pmatrix} 0 & 1 & 0 & & & & \\ q & 0 & p & & & & \\ & q & 0 & p & & & \\ & & & \cdot & & & \\ & & & & \cdot & & \\ & & & & & \cdot & \\ & & & & q & 0 & p \\ & & & & & 1 & 0 \end{pmatrix}$.

(c) $\begin{pmatrix} 1-p & p & & & & \\ 0 & 1-p & p & & & \\ & & \cdot & & & \\ & & & \cdot & & \\ & & & & \cdot & \\ & & & & 1-p & p \\ & & & & 0 & 1 \end{pmatrix}$.

(d) $\begin{pmatrix} 1 - 10^{-9}k_I I & 10^{-9}k_I I & 0 & 0 \\ 0 & 0 & 1 & 0 \\ 10^{-9}k_F & 0 & 1 - 10^{-9}(k_F + k_{\text{isc}}) & 10^{-9}k_{\text{isc}} \\ 10^{-9}k_P & 0 & 0 & 1 - 10^{-9}k_P \end{pmatrix}$.

In (d), rows and columns are in the order S_0, S_2, S_1, T_1.

33. (a)

$$\mathbf{I} = \begin{pmatrix} \sum m(y^2 + z^2) & -\sum mxy & -\sum mxz \\ -\sum mxy & \sum m(x^2 + z^2) & -\sum myz \\ -\sum mxz & -\sum myz & \sum m(x^2 + y^2) \end{pmatrix}$$

(b) With N—N bond along x-axis and F atoms in quadrants I and III of the xy-plane,

$$I_{xx} = \tfrac{3}{2}m_{\text{F}}d_{\text{NF}}^2,$$
$$I_{xy} = -\tfrac{1}{2}\sqrt{3}m_{\text{F}}d_{\text{NN}}d_{\text{NF}} - \tfrac{1}{2}\sqrt{3}m_{\text{F}}d_{\text{NF}}^2 = I_{yx},$$
$$I_{yy} = \tfrac{1}{2}m_{\text{N}}d_{\text{NN}}^2 + \tfrac{1}{2}m_{\text{F}}d_{\text{NN}}^2 + \tfrac{1}{2}m_{\text{F}}d_{\text{NF}}^2 + m_{\text{F}}d_{\text{NN}}d_{\text{NF}},$$
$$I_{zz} = \tfrac{1}{2}m_{\text{N}}d_{\text{NN}}^2 + \tfrac{1}{2}m_{\text{F}}d_{\text{NN}}^2 + 2m_{\text{F}}d_{\text{NF}}^2 + m_{\text{F}}d_{\text{NN}}d_{\text{NF}},$$
$$I_{xz} = I_{zx} = 0 = I_{yz} = I_{zy}.$$

34. (c) Since $\mathrm{E} = -\nabla\phi$, $(\nabla \mathrm{E})_{ij} = (\nabla \mathrm{E})_{ji}$. (d) diag $(2q/a^3, 2q/a^3, -4q/a^3)$.
(e) If O—C—O axis is along z-axis, then $\mathbf{Q} = \text{diag}\,(2qd^2, 2qd^2, -4qd^2)$.
(f) If N—N bond is along x-axis, and F atoms, with charge $-q$, in quadrants I and III of the xy-plane,

$$\mathbf{Q} = -\tfrac{1}{2}qd_{\mathrm{NF}}\begin{pmatrix} 4d_{\mathrm{NF}} - d_{\mathrm{NN}} & 3(d_{\mathrm{NN}} + d_{\mathrm{NF}})\sqrt{3} & 0 \\ 3(d_{\mathrm{NN}} + d_{\mathrm{NF}})\sqrt{3} & 5d_{\mathrm{NF}} - 2d_{\mathrm{NN}} & 0 \\ 0 & 0 & -2d_{\mathrm{NN}} - 4d_{\mathrm{NF}} \end{pmatrix}.$$

35. (a) $\mathrm{Tr}\,(\mathbf{A} + \mathbf{B}) = \sum (a_{ii} + b_{ii}) = \sum a_{ii} + \sum b_{ii} = \mathrm{Tr}\,\mathbf{A} + \mathrm{Tr}\,\mathbf{B}$.
(b) $\mathrm{Tr}\,(a\mathbf{A}) = \sum a(\mathbf{A})_{ii} = a \sum (\mathbf{A})_{ii} = a\,\mathrm{Tr}\,\mathbf{A}$.
(c) $\mathrm{Tr}\,(\mathbf{AB}) = \sum (\mathbf{AB})_{ii} = \sum\sum a_{ij}b_{ji} = \sum\sum b_{ji}a_{ij} = \sum (\mathbf{BA})_{jj} = \mathrm{Tr}\,(\mathbf{BA})$.

36. $$\begin{pmatrix} abc \\ bca \end{pmatrix}\begin{pmatrix} abc \\ acb \end{pmatrix} = \begin{pmatrix} abc \\ bac \end{pmatrix} \neq \begin{pmatrix} abc \\ cba \end{pmatrix} = \begin{pmatrix} abc \\ acb \end{pmatrix}\begin{pmatrix} abc \\ bca \end{pmatrix}.$$

37. E, (abc), (acb), (abd), (adb), (acd), (adc), (bcd), (bdc), $(ab)(cd)$, $(ac)(bd)$, $(ad)(bc)$, $(abcd)$, $(dcba)$, $(abdc)$, $(cdba)$, $(adbc)$, $(cbda)$, (ab), (ac), (ad), (bc), (bd), (cd). If yours differ, remember that $(abcd) = (bcda) = (cdab) = (dabc)$, and likewise for other cycles.
38. $(ab)(ab) = E$.
39. (148)(23)(5 10 97) is even.
40. (1546)(27) = (15)(14)(16)(27) or others.
41. (a) $$\begin{vmatrix} \phi_1(x_1)\alpha(x_1) & \phi_2(x_1)\beta(x_1) \\ \phi_1(x_2)\alpha(x_2) & \phi_2(x_2)\beta(x_2) \end{vmatrix} - \begin{vmatrix} \phi_1(x_1)\beta(x_1) & \phi_2(x_1)\alpha(x_1) \\ \phi_1(x_2)\beta(x_2) & \phi_2(x_2)\alpha(x_2) \end{vmatrix} = |\phi_1\alpha\ \phi_2\beta| - |\phi_1\beta\ \phi_2\alpha|.$$
(b) $2\,|\phi_1\alpha\ \phi_2\alpha\ \phi_3\beta| - |\phi_1\alpha\ \phi_2\beta\ \phi_3\alpha| - |\phi_1\beta\ \phi_2\alpha\ \phi_3\alpha|$.
43. $|\mathbf{\Pi}| = 1$.
44. $0, z^n$.
46. 37.
47. For definiteness, let $\sum c_i f_i = 0$, with $c_1 \neq 0$. Now multiply row 1 by c_1 and add c_2 times row 2, c_3 times row 3, . . . , and c_n times row n to row 1. Then $|\mathbf{W}(x)|$ equals $1/c_1$ times the determinant of a matrix whose first row is zero.
48. (a) Yes. (b) No: $-\sin^2 x - \cos^2 x + 1 = 0$.
49. (a) $D_n(x) = xD_{n-1}(x) - D_{n-2}(x)$, and $D_2(x) = x^2 - 1 = x^2 - D_0(x)$.
(b) $G(x, t) = (1 - xt + t^2)^{-1}$. (c) $D_n(2\cos\theta) = \sin[(n + 1)\theta]/\sin\theta$.
(d) If $D_n(2\cos\theta) = 0$, then $\theta = k\pi/(n + 1)$, $k = 1, 2, \ldots, n$.
And $x = 2\cos[k\pi/(n + 1)]$.
53. (a) Yes. (b) Yes. (c) No. (d) Yes.
57. $\mathbf{A} \oplus \mathbf{A}$.
60. Expanding $|(\mathbf{x}\cdot\mathbf{x})\mathbf{y} - (\mathbf{x}\cdot\mathbf{y})\mathbf{x}|^2/|\mathbf{x}|^2 \geq 0$ gives $(\mathbf{x}\cdot\mathbf{x})(\mathbf{y}\cdot\mathbf{y}) - (\mathbf{x}\cdot\mathbf{y})(\mathbf{y}\cdot\mathbf{x}) \geq 0$. Transposing and taking the square root gives the desired result.
61. $|\sum x_i^* y_i| \leqslant (\sum x_i^* x_i)^{1/2} (\sum y_i^* y_i)^{1/2}$.
62. Rewrite the Schwarz equality as $|\mathbf{y}\cdot\mathbf{x}| \leqslant |\mathbf{x}|\,|\mathbf{y}|$, add the two forms, and add $|\mathbf{x}|^2 + |\mathbf{y}|^2$ to both sides of the result, to obtain $|\mathbf{x}|^2 + |\mathbf{x}\cdot\mathbf{y}| + |\mathbf{y}\cdot\mathbf{x}| + |\mathbf{y}|^2 \leqslant |\mathbf{x}|^2 + 2\,|\mathbf{x}|\,|\mathbf{y}| + |\mathbf{y}|^2$. But from Exercise A4.1, $|\mathbf{x}|^2 + |\mathbf{x}\cdot\mathbf{y}| + |\mathbf{y}\cdot\mathbf{x}| + |\mathbf{y}|^2 \geqslant ||\mathbf{x}|^2 + \mathbf{x}\cdot\mathbf{y} + \mathbf{y}\cdot\mathbf{x} + |\mathbf{y}|^2|$. Finally, take the square root.
63. $\mathbf{x}^\dagger\mathbf{A}\mathbf{y}$, $\mathbf{y}^\dagger\mathbf{A}^\dagger\mathbf{x}$.
64. $\sum\sum x_i^*(\mathbf{e}_i\cdot\mathbf{e}_j)y_j = \mathbf{x}^\dagger\mathbf{S}\mathbf{y}$.

65. (a) $\mathbf{a}_{n+1} = \eta[\mathbf{A}_{n+1} - \sum (\mathbf{a}_i^\dagger \mathbf{A}_{n+1})\mathbf{a}_i]$.
(b) $\frac{1}{2}(1, i, 1, i)^T$ and $\frac{1}{\sqrt{12}}(1, 2-i, 2i-1, i)^T$.
66. $A^{-1}\mathbf{y} = A^{-1}A\mathbf{x} = \mathbf{x}$, so $\mathbf{A}^{-1}\mathbf{A}$ must equal $\mathbf{1}$.
67. $\mathbf{T} = \mathbf{S}^{-1}$.
69. diag $(1/d_1, 1/d_2, \ldots, 1/d_n)$.
70. $|\mathbf{A}^{-1}|\,|\mathbf{A}| = |\mathbf{A}^{-1}\mathbf{A}| = |\mathbf{1}| = 1$.
72. (b) If $\mathbf{C}^n = \mathbf{1}$, $|\mathbf{C}^n| = |\mathbf{C}|^n = \mathbf{1}$, so $|\mathbf{C}| \neq 0$. Also, multiply $\mathbf{C}^n = \mathbf{1}$ by $(\mathbf{C}^{-1})^n$ to obtain $\mathbf{1} = (\mathbf{C}^{-1})^n$.
73. Assume $|\mathbf{P}| \neq 0$. Then $\mathbf{P}^{-1}$ exists, so we may multiply $\mathbf{P}^2 = \mathbf{P}$ by $\mathbf{P}^{-1}$ to obtain $\mathbf{P} = \mathbf{1}$. That this is indeed a proof is an exercise in logic (Exercise A1.4f). It is instructive to consider two fallacious proofs: "If $\mathbf{P}^2 = \mathbf{P}$, then $\mathbf{P}^2 - \mathbf{P} = \mathbf{P}(\mathbf{P} - \mathbf{1}) = \mathbf{0}$. Therefore either $\mathbf{P} = \mathbf{0}$ or $\mathbf{P} - \mathbf{1} = \mathbf{0}$, so $|\mathbf{P}| = 0$ or $\mathbf{P} = \mathbf{1}$," and "If $\mathbf{P}^2 = \mathbf{P}$, then $|\mathbf{P}^2| = |\mathbf{P}|\,|\mathbf{P}| = |\mathbf{P}|$, which is solved by $|\mathbf{P}| = 0$ or $|\mathbf{P}| = 1$. Therefore either $|\mathbf{P}| = 0$ or $\mathbf{P} = \mathbf{1}$." The first is wrong because $\mathbf{AB} = \mathbf{0}$ does not imply $\mathbf{A} = \mathbf{0}$ or $\mathbf{B} = \mathbf{0}$ (Exercise 8.15). The second is wrong because $|\mathbf{P}| = 1$ does not imply $\mathbf{P} = \mathbf{1}$, since many matrices have determinant unity (Exercises 8.3, 8.11, 8.75, and 8.88).
74. $\frac{1}{3}\begin{pmatrix} 2 & -i \\ -i & 1 \end{pmatrix}$.
75. Its transpose.
78. $\mathbf{A} = \mathbf{H} + \mathbf{K}$, where $\mathbf{H} = \frac{1}{2}(\mathbf{A} + \mathbf{A}^\dagger)$ and $\mathbf{K} = \frac{1}{2}(\mathbf{A} - \mathbf{A}^\dagger)$.
80. (a) $(\mathbf{x}^\dagger\mathbf{H}\mathbf{y})^* = (\mathbf{x}^\dagger\mathbf{H}\mathbf{y})^\dagger = \mathbf{y}^\dagger\mathbf{H}^\dagger\mathbf{x} = \mathbf{y}^\dagger\mathbf{H}\mathbf{x}$. (b) $\mathbf{x}^\dagger\mathbf{H}\mathbf{x}$ is real.
(c) $\mathbf{x}^\dagger\mathbf{K}\mathbf{x}$ is imaginary.
81. $(\mathbf{GH} - \mathbf{HG})^\dagger = \mathbf{H}^\dagger\mathbf{G}^\dagger - \mathbf{G}^\dagger\mathbf{H}^\dagger = -[\mathbf{G}, \mathbf{H}]$.
82. If and only if they commute.
83. Since $\mathbf{G}^2$ is Hermitian, $\mathbf{x}^\dagger\mathbf{G}^2\mathbf{x}$ is real. Since $[\mathbf{F}, \mathbf{G}]$ is skew-Hermitian, $\mathbf{x}^\dagger[\mathbf{F}, \mathbf{G}]\mathbf{x}$ is imaginary. Therefore the quotient is also imaginary.
84. Yes, no.
85. $(\mathbf{UV})^{-1} = \mathbf{V}^{-1}\mathbf{U}^{-1} = \mathbf{V}^\dagger\mathbf{U}^\dagger = (\mathbf{UV})^\dagger$.
86. $(\mathbf{U}^{-1}\mathbf{HU})^\dagger = \mathbf{U}^\dagger\mathbf{H}^\dagger(\mathbf{U}^{-1})^\dagger = \mathbf{U}^{-1}\mathbf{HU}$.
87. (a) $\mathbf{x}^\dagger\mathbf{A}^\dagger\mathbf{A}\mathbf{x} = (\mathbf{A}\mathbf{x})^\dagger\mathbf{A}\mathbf{x} = |A\mathbf{x}|^2 \geqslant 0$.
(b) $\mathbf{A}^\dagger = \mathbf{F} - i\lambda\mathbf{G}$, $\mathbf{A}^\dagger\mathbf{A} = \mathbf{F}^2 + i\lambda[\mathbf{F}, \mathbf{G}] + \lambda^2\mathbf{G}^2$.
88. Equals ± 1.
92. (a) diag $(-1, -1, -1)$. (b) diag $(1, 1, -1)$.
(c) $\begin{pmatrix} \cos\phi & \sin\phi & 0 \\ -\sin\phi & \cos\phi & 0 \\ 0 & 0 & 1 \end{pmatrix}$. (d) $\begin{pmatrix} 1 & 0 & 0 \\ 0 & \cos\phi & \sin\phi \\ 0 & -\sin\phi & \cos\phi \end{pmatrix}$.
(e) $\begin{pmatrix} \cos\phi & \sin\phi & 0 \\ -\sin\phi & \cos\phi & 0 \\ 0 & 0 & -1 \end{pmatrix}$. (f) $\begin{pmatrix} -\cos\phi & -\sin\phi & 0 \\ \sin\phi & -\cos\phi & 0 \\ 0 & 0 & -1 \end{pmatrix}$.
93. (a) Reflection in the xz-plane. (b) Rotation by 90° about the z-axis.
(c) Reflection in the plane $x = y$. (d) Rotation by 120° about the axis $\hat{\mathbf{i}} + \hat{\mathbf{j}} + \hat{\mathbf{k}}$.
94. Analogous to Exercise 8.47.
95. $\{\mathbf{n}_1, \mathbf{n}_2, \mathbf{n}_3\}$ must be linearly independent. There are many equivalent geometrical statements of this property.
96. $\frac{1}{3}\begin{pmatrix} 2 & -i \\ -i & 1 \end{pmatrix}\begin{pmatrix} 3 \\ 3i \end{pmatrix} = \begin{pmatrix} 3 \\ 0 \end{pmatrix}$, or $x = 3$, $y = 0$.

97. $\mathbf{D}^{-1} - \mathbf{D}^{-1}\boldsymbol{\delta}\mathbf{D}^{-1}$.

98. $x_i = (\mathbf{A}^{-1}\mathbf{c})_i = \sum (\mathbf{A}^{-1})_{ij}c_j = \sum (-1)^{i+j}\,|A_{ji}|\,c_j = |\mathbf{A}_i|$.

99. $[Cl^-] = 0.07575M$, $[Br^-] = 0.1538M$.

100. $x_1 = -3/37$, $x_2 = 69/37$, $x_3 = -49/37$, $x_4 = -4/37$.

101. $\mathbf{A}^{-1} = \mathbf{T}^{-1}\boldsymbol{\Pi}$.

102. (a) $11Cr_2O_7^{2-} + 6Co(NO_2)_6^{3-} + 82H^+ = 22Cr^{3+} + 6Co^{2+} + 36NO_3^- + 41H_2O$.

(b) $61MnO_4^- + 5Fe(CN)_6^{4-} + 218H^+ = 61Mn^{2+}$
$+ 5Fe^{3+} + 30CO_2 + 30HNO_3 + 94H_2O$.

(c) Since any multiple of a balanced equation is still a balanced equation, one of the coefficients may be arbitrarily set equal to unity. Then when the equation is solved, the coefficients may be cleared of fractions.

104. $(3, -1, -1)^T$ or any multiple.

106. $(2, -1, 0)^T$ and $(3, 6, -5)^T$ or any orthogonal linear combinations of these.

107. $\mathbf{x}_0 + \sum c_i\mathbf{h}_i$, where $\{\mathbf{h}_i\}$ are the linearly independent solutions of $\mathbf{Ax} = \mathbf{0}$.

109. (a) $\mathbf{P}_i$ of Exercise 8.29: $(\mathbf{P}_i)_{kl} = \delta_{ik}\delta_{il}$.

(b) $\mathbf{UP}_i\mathbf{U}^{-1}$, where $(\mathbf{UP}_i\mathbf{U}^{-1})_{kl} = u_{ki}u_{li}^*$. (c) $\mathbf{UP}_i\mathbf{U}^{-1}\mathbf{UP}_j\mathbf{U}^{-1} = \mathbf{UP}_i\mathbf{P}_j\mathbf{U}^{-1}$.

(d) $\mathbf{1}$. (e) $\mathbf{U1U}^{-1} = \mathbf{1}$.

110. $\hat{\mathbf{v}} = \mathbf{x} \times \hat{\mathbf{u}}/|\mathbf{x} \times \hat{\mathbf{u}}|$, with $\mathbf{x}$ arbitrary, but for convenience, take $\mathbf{x} = \hat{\mathbf{k}}$; $\hat{\mathbf{w}} = \hat{\mathbf{u}} \times \hat{\mathbf{v}}$. In the $\{\hat{\mathbf{u}}, \hat{\mathbf{v}}, \hat{\mathbf{w}}\}$ basis, $\mathbf{O}$ is the matrix of Exercise 8.92d.

113. Multiply $\mathbf{AB} = \mathbf{BA}$ from the left by $\mathbf{1} = \mathbf{TT}^{-1}$. Then, since $\mathbf{1}$ commutes with $\mathbf{A}$ and with $\mathbf{B}$, we obtain $\mathbf{ATT}^{-1}\mathbf{B} = \mathbf{BTT}^{-1}\mathbf{A}$. Multiplying from the left by $\mathbf{T}^{-1}$ and from the right by $\mathbf{T}$, and inserting clarifying parentheses, gives $(\mathbf{T}^{-1}\mathbf{AT})(\mathbf{T}^{-1}\mathbf{BT}) = (\mathbf{T}^{-1}\mathbf{BT})(\mathbf{T}^{-1}\mathbf{AT})$, the desired result.

114. (a) $\mathbf{1}^{-1}\mathbf{A1} = \mathbf{A}$. (b) If $\mathbf{T}^{-1}\mathbf{BT} = \mathbf{A}$, then $\mathbf{S}^{-1}\mathbf{AS} = \mathbf{B}$, where $\mathbf{S} = \mathbf{T}^{-1}$.

(c) If $\mathbf{T}^{-1}\mathbf{AT} = \mathbf{B}$ and $\mathbf{S}^{-1}\mathbf{BS} = \mathbf{C}$, then $\mathbf{R}^{-1}\mathbf{AR} = \mathbf{C}$, where $\mathbf{R} = \mathbf{TS}$.

116. λ_i.

117. $\frac{3}{4}m_{\rm N}m_{\rm F}d_{\rm NN}^2d_{\rm NF}^2(\frac{1}{2}m_{\rm N}d_{\rm NN}^2 + \frac{1}{2}m_{\rm F}d_{\rm NN}^2 + 2m_{\rm F}d_{\rm NF}^2 + m_{\rm F}d_{\rm NN}d_{\rm NF})$.

118. If every $\mathbf{x}^\dagger\mathbf{Dx} \geqslant 0$, then $\mathbf{v}_i{}^\dagger\mathbf{Dv}_i = \lambda_i \geqslant 0$.

119. If $\mathbf{Ax}_i = \lambda_i\mathbf{x}_i$, then $\mathbf{A}^{-1}\mathbf{x}_i = (1/\lambda_i)\mathbf{x}_i$; $\mathbf{A}^{-1}$ has the same eigenvectors, reciprocal eigenvalues.

120. Each is an nth root of unity.

121. (a) Its determinant is equal to the determinant of a matrix whose first column contains only zeros.

(b) If $\sum_j p_{ij} = 1$, then $\sum_j (p_{ij} - \delta_{ij}) = 0$, or $\mathbf{P} - \mathbf{1}$ is singular. Therefore $|\mathbf{P} - \lambda\mathbf{1}| = 0$ for $\lambda = 1$.

122. $m = (\varepsilon - \alpha)/\beta$.

125. (a) $\lambda_1 = 2$, $\mathbf{x}_1 = (1/\sqrt{2}, -1/\sqrt{2})^T$, $\lambda_2 = 0$, $\mathbf{x}_2 = (1/\sqrt{2}, 1/\sqrt{2})^T$.

(b) $\lambda_1 = 4$, $\mathbf{x}_1 = (1/\sqrt{2}, -i/\sqrt{2})^T$, $\lambda_2 = 2$, $\mathbf{x}_2 = (1/\sqrt{2}, i/\sqrt{2})^T$.

(c) $\lambda_1 = 1 + \sqrt{2}$, $\mathbf{x}_1 = (4 + 2\sqrt{2})^{-1/2}(1 + \sqrt{2}, 1)^T$,
$\lambda_2 = 1 - \sqrt{2}$, $\mathbf{x}_2 = (4 - 2\sqrt{2})^{-1/2}(1 - \sqrt{2}, 1)^T$.

(d) $\lambda_1 = 2$, $\mathbf{x}_1 = (1/\sqrt{2}, 1/\sqrt{2})^T$, $\lambda_2 = 1$, $\mathbf{x}_2 = (1, 0)^T$.

(e) $\lambda_1 = 3$, $\mathbf{x}_1 = (1/\sqrt{5}, 2/\sqrt{5})^T$, $\lambda_2 = 2$, $\mathbf{x}_2 = (1/\sqrt{2}, 1/\sqrt{2})^T$.

(f) $\lambda_1 = 3$, $\mathbf{x}_1 = (1/3, 2/3, 2/3)^T$, $\lambda_2 = 0$, $\mathbf{x}_2 = (2/3, 1/3, -2/3)^T$,
$\lambda_3 = -3$, $\mathbf{x}_3 = (2/3, -2/3, 1/3)^T$.

126. $\lambda_1 = 2$, $\mathbf{x}_1 = (1, 0, 0, 0)^T$, $\lambda_2 = 2$, $\mathbf{x}_2 = (0, 0, 0, 1)^T$. $\lambda_3 = 2$, $\mathbf{x}_3 = (0, 1/\sqrt{2}, 1/\sqrt{2}, 0)^T$, $\lambda_4 = 0$, $\mathbf{x}_4 = (0, 1/\sqrt{2}, -1/\sqrt{2}, 0)^T$. These correspond to the eigenfunctions $\alpha\alpha$, $\beta\beta$, $(\alpha\beta + \beta\alpha)/\sqrt{2}$, and $(\alpha\beta - \beta\alpha)/\sqrt{2}$, respectively. Any linear combination of the first three is still an eigenvector of $\mathbf{S}^2$ (eigenfunction of S^2) with eigenvalue 2.

127. (a) $\lambda_1 = \sqrt{2}$, $\mathbf{x}_1 = (1/2, 1/\sqrt{2}, 1/2)^T$, $\lambda_2 = 0$, $\mathbf{x}_2 = (1/\sqrt{2}, 0, -1/\sqrt{2})^T$, $\lambda_3 = -\sqrt{2}$, $\mathbf{x}_3 = (1/2, -1/\sqrt{2}, 1/2)^T$.

(b) $\lambda_1 = 2$, $\mathbf{x}_1 = (1/\sqrt{3}, 1/\sqrt{3}, 1/\sqrt{3})^T$, $\lambda_2 = -1$, $\mathbf{x}_2 = (1/\sqrt{2}, 0, -1/\sqrt{2})^T$, $\lambda_3 = -1$, $\mathbf{x}_3 = (1/\sqrt{6}, -2/\sqrt{6}, 1/\sqrt{6})^T$, or any orthonormal linear combination of these last two vectors.

(c) $\lambda_1 = 2$, $\mathbf{x}_1 = (1/2, 1/2, 1/2, 1/2)^T$, $\lambda_2 = 0$, $x_2 = (1/\sqrt{2}, 0, -1/\sqrt{2}, 0)^T$, $\lambda_3 = 0$, $\mathbf{x}_3 = (0, 1/\sqrt{2}, 0, -1/\sqrt{2})^T$ (or any orthonormal linear combination of these last two vectors), $\lambda_4 = -2$, $\mathbf{x}_4 = (1/2, -1/2, 1/2, -1/2)^T$.

(d) $\lambda_1 = (1 + \sqrt{5})/2$, $\mathbf{x}_1 = (5 + \sqrt{5})^{-1/2}(1, \lambda_1, \lambda_1, 1)^T$,
$\lambda_2 = (-1 + \sqrt{5})/2$, $\mathbf{x}_2 = (5 - \sqrt{5})^{-1/2}(-1, -\lambda_2, \lambda_2, 1)^T$,
$\lambda_3 = (1 - \sqrt{5})/2$, $\mathbf{x}_3 = (5 - \sqrt{5})^{-1/2}(1, \lambda_3, \lambda_3, 1)^T$,
$\lambda_4 = (-1 - \sqrt{5})/2$, $\mathbf{x}_4 = (5 + \sqrt{5})^{-1/2}(-1, -\lambda_4, \lambda_4, 1)^T$.

129. $\lambda_1 = e^{i\phi}$, $\mathbf{x}_1 = (1/\sqrt{2}, i/\sqrt{2})^T$, $\lambda_2 = e^{-i\phi}$, $\mathbf{x}_2 = (i/\sqrt{2}, 1/\sqrt{2})^T$.

131. $\lambda_0 = 0$, $\mathbf{x}_0 = (u_x, u_y, u_z)^T$, $\lambda_\pm = \pm i(u_x^2 + u_y^2 + u_z^2)^{1/2}$, $\mathbf{x}_\pm = (\lambda_\pm^2 + u_x^2, u_x u_y + \lambda_\pm u_z$ $u_x u_z - \lambda_\pm u_y)^T$. (The eigenvectors are not normalized.) Notice that $\mathbf{u} \times \mathbf{x}_0 = 0\mathbf{x}_0 = \mathbf{0}$ means that $\mathbf{u} \times \mathbf{x}_0$ is orthogonal to both $\mathbf{u}$ and $\mathbf{x}_0$. Also, since $\mathbf{u}\times$ is skew-Hermitian, its eigenvectors $\mathbf{x}_+$ and $\mathbf{x}_-$ are orthogonal to $\mathbf{x}_0$ (Exercise 8.143). Therefore $\mathbf{u} \times \mathbf{x}_+ = \lambda_+\mathbf{x}_+$ and $\mathbf{u} \times \mathbf{x}_- = \lambda_-\mathbf{x}_-$ are indeed orthogonal to $\mathbf{u}$. Nevertheless, $\mathbf{u} \times \mathbf{x}_+ = \lambda_+\mathbf{x}_+$ is *not* orthogonal to $\mathbf{x}_+$, nor $\mathbf{u} \times \mathbf{x}_- = \lambda_-\mathbf{x}_-$ to $\mathbf{x}_-$, even though the cosine of the angle between $\mathbf{x}_+$ and $\lambda_+\mathbf{x}_+$, or between $\mathbf{x}_-$ and $\lambda\mathbf{x}_-$, is zero (cf. Section 8.7).

132. -2.1838, $1.5919 \pm 1.8490i$.

133. (b) 1.000101.

134. $\lambda = [d \pm (d^2 + 4x^2)^{1/2}]/2$. Both graphs are hyperbolas.

137. Multiply $\mathbf{Ax} = \lambda\mathbf{x}$ from the left by $\mathbf{T}^{-1}$, and insert $\mathbf{1} = \mathbf{TT}^{-1}$ between $\mathbf{A}$ and $\mathbf{x}$ on the left (cf. Exercise 8.113), to obtain $(\mathbf{T}^{-1}\mathbf{AT})(\mathbf{T}^{-1}\mathbf{x}) = \lambda(\mathbf{T}^{-1}\mathbf{x})$, which shows that λ is indeed an eigenvalue of $\mathbf{T}^{-1}\mathbf{AT}$.

138. (a) $(\mathbf{p})_r = C_r^N(\frac{1}{2})^N$.

139. Because the determinant equals the product of the eigenvalues.

140. (a) If $\mathbf{A}$ is not Hermitian, replace it by $\frac{1}{2}(\mathbf{A} + \mathbf{A}^\dagger)$. The quadratic form is unchanged.

(b) $x_j' = \sum_i u_{ij}^* x_i$.

(c) With $x' = (x - y)/\sqrt{2}$, $y' = (x + y + z)/\sqrt{3}$, and $z' = (x + y - 2z)/\sqrt{6}$, $a = 3$, $b = 6$, $c = 9$.

141. $U_{ii} = \cos\phi = U_{jj}$, $U_{ij} = \sin\phi = -U_{ji}$, where $\tan 2\phi = 2H_{ij}/(H_{jj} - H_{ii})$.

142. If $(\mathbf{R} + i\mathbf{I})(\mathbf{x} + i\mathbf{y}) = \lambda(\mathbf{x} + i\mathbf{y})$, with λ real, then $\mathbf{Rx} - \mathbf{Iy} = \lambda\mathbf{x}$ and $\mathbf{Ry} + \mathbf{Ix} = \lambda\mathbf{y}$. Writing these two equations in matrix form shows that $(\mathbf{x}\ \ \mathbf{y})^T$ is an eigenvector of the $2n \times 2n$ matrix given, with eigenvalue λ. Since $(\mathbf{R} + i\mathbf{I})^\dagger = \mathbf{R} - i\mathbf{I}^\dagger = \mathbf{R} + i\mathbf{I}$, $\mathbf{I}^\dagger = -\mathbf{I}$, so the $2n \times 2n$ matrix *is* Hermitian. If $(\mathbf{x}\ \ \mathbf{y})^T$ is an eigenvector belonging to the eigenvalue λ, so is $(-\mathbf{y}\ \ \mathbf{x})^T$. However, $\mathbf{x} + i\mathbf{y}$ and $-\mathbf{y} + i\mathbf{x}$ are not linearly independent, and only one is needed.

143. (a) As for $\mathbf{H}$, but add to obtain $0 = \mathbf{x}^\dagger\mathbf{Kx} - \mathbf{x}^\dagger\mathbf{Kx} = (\lambda + \lambda^*)\mathbf{x}^\dagger\mathbf{x}$, or $\lambda^* = -\lambda$.

(b) As for $\mathbf{H}$, subtraction gives $(\lambda_1^* + \lambda_2)\mathbf{x}_1^\dagger\mathbf{x}_2 = (-\lambda_1 + \lambda_2)\mathbf{x}_1^\dagger\mathbf{x}_2 = 0$.

144. (a) If $\mathbf{Ux} = \lambda\mathbf{x}$, then $\mathbf{x}^\dagger\mathbf{U}^\dagger = \lambda^*\mathbf{x}^\dagger$. Multiplying these equations gives $\mathbf{x}^\dagger\mathbf{U}^\dagger\mathbf{Ux} = \lambda^*\lambda\mathbf{x}^\dagger\mathbf{x}$, or $\mathbf{x}^\dagger\mathbf{x} = |\lambda|^2\mathbf{x}^\dagger\mathbf{x}$. Since $\mathbf{x}^\dagger\mathbf{x} \neq 0$, we may conclude that $|\lambda|^2 = 1$, or $|\lambda| = 1$.

(b) If $\mathbf{Ux}_1 = \lambda_1\mathbf{x}_1$ and $\mathbf{Ux}_2 = \lambda_2\mathbf{x}_2$, then $\mathbf{x}_2^\dagger\mathbf{U}^\dagger = \lambda_2^*\mathbf{x}_2^\dagger$ and $\lambda_2\mathbf{x}_2^\dagger\mathbf{U}^\dagger\mathbf{Ux}_1 = \lambda_2\lambda_1\lambda_2^*\mathbf{x}_2^\dagger\mathbf{x}_1$, or $\lambda_2\mathbf{x}_2^\dagger\mathbf{x}_1 = \lambda_1\mathbf{x}_2^\dagger\mathbf{x}_1$.

145. (a) If $P_n(\lambda) = 0$, then $P_n(\lambda^*) = 0^* = 0$.
(b) Complex roots λ and λ^* are paired, so at least one root must be "left over."
(c) ± 1.
(d) Rotation about $\mathbf{v}$; if the eigenvalue is -1, the rotation is followed by reflection in the plane $\mathbf{X} \cdot \mathbf{v} = 0$.

146. Let $\mathbf{X}^{-1}\mathbf{AX} = \mathbf{\Lambda}$ and $\mathbf{X}^{-1}\mathbf{BX} = \mathbf{M}$. Then since $\mathbf{\Lambda}$ and $\mathbf{M}$ are diagonal, they must commute, and so also must $\mathbf{A} = \mathbf{X\Lambda X}^{-1}$ and $\mathbf{B} = \mathbf{XMX}^{-1}$.

147. (a) $\mathbf{C}_n\mathbf{x} = (x_2, x_3, \ldots, x_n, x_1)^T$, $\mathbf{C}_n^2\mathbf{x} = (x_3, x_4, \ldots, x_2)^T$, $\mathbf{C}_n^n\mathbf{x} = \mathbf{x}$.
(b) $\lambda_j = e^{2\pi ij/n}$, $j = 0, 1, \ldots, n-1$, $(\mathbf{x}_j)_k = e^{2\pi ijk/n}/\sqrt{n}$.
(c) $(\mathbf{C}_n^{-1})_{i+1,i} = 1 = (\mathbf{C}_n^{-1})_{1n}$, other elements zero.
(d) $\lambda_j = 2\cos(2\pi j/n)$, $j = 0, 1, \ldots, n-1$. Since $\lambda_j = \lambda_{n-j}$, all eigenvalues except λ_0 (and $\lambda_{n/2}$ when n is even) are doubly degenerate.

148. (a) $\chi_1 = (\phi_1 + \phi_2 + \phi_3 + \phi_4)/2$, $\chi_2 = (\phi_5 + \phi_6)/\sqrt{2}$, $\chi_3 = (\phi_1 - \phi_2 - \phi_3 + \phi_4)/2$
$\chi_4 = (\phi_5 - \phi_6)/\sqrt{2}$, $\chi_5 = (\phi_1 + \phi_2 - \phi_3 - \phi_4)/2$, $\chi_6 = (\phi_1 - \phi_2 + \phi_3 - \phi_4)/2$.

(b) $$\boldsymbol{\beta} = \begin{pmatrix} 1 & \sqrt{2} \\ \sqrt{2} & 1 \end{pmatrix} \oplus \begin{pmatrix} -1 & \sqrt{2} \\ \sqrt{2} & -1 \end{pmatrix} \oplus (1) \oplus (-1).$$

(c) $\lambda_1 = 1 + \sqrt{2}$, $\psi_1 = (\chi_1 + \chi_2)/\sqrt{2}$, $\lambda_2 = 1 - \sqrt{2}$, $\psi_2 = (\chi_1 - \chi_2)/\sqrt{2}$, $\lambda_3 = -1 + \sqrt{2}$, $\psi_3 = (\chi_3 + \chi_4)/\sqrt{2}$, $\lambda_4 = -1 - \sqrt{2}$, $\psi_4 = (\chi_3 - \chi_4)/\sqrt{2}$, $\lambda_5 = 1$, $\psi_5 = \chi_5$, $\lambda_6 = -1$, $\psi_6 = \chi_6$.

149. (a) $\sum_{r=0}^{n} C_r^n \mathbf{A}^r$. (b) $\sum_{n=0}^{\infty} (-1)^n \mathbf{A}^n$. (c) $\sum_{n=0}^{\infty} (-1)^n \frac{(2n)!}{2^{2n} n!^2} \mathbf{A}^n$

(d) $\lim_{n\to\infty} \mathbf{A}^n = \mathbf{0}$ and $\lim_{n\to\infty} \frac{(2n)!}{2^{2n} n!^2} \mathbf{A}^n = \mathbf{0}$, respectively.

150. $\sum c_j \mathbf{D}^j = \operatorname{diag}\left(\sum c_j d_1^j, \sum c_j d_2^j, \ldots, \sum c_j d_n^j\right)$.

151. (a) $$\begin{pmatrix} 1 & 0 \\ 0 & 1 \end{pmatrix}, \begin{pmatrix} 1 & 0 \\ 0 & -1 \end{pmatrix}, \begin{pmatrix} -1 & 0 \\ 0 & 1 \end{pmatrix}, \begin{pmatrix} -1 & 0 \\ 0 & -1 \end{pmatrix}.$$

(b) $$\begin{pmatrix} \pm(1 - ab)^{1/2} & b \\ a & \mp(1 - ab)^{1/2} \end{pmatrix}.$$

152. If and only if $\mathbf{A}$ and $\mathbf{B}$ commute.

153. (a) $\mathbf{M} = \mathbf{0}$. (b) $\mathbf{M} = \mathbf{1}$. (c) $\mathbf{M}^2 = \mathbf{0}$.

154. $(e^{i\mathbf{H}})^\dagger = (\sum i^n \mathbf{H}^n/n!)^\dagger = \sum (-i)^n \mathbf{H}^n/n! = e^{-i\mathbf{H}}$ and by Exercise 8.152, $e^{i\mathbf{H}} e^{-i\mathbf{H}} = \mathbf{1}$.

155. (b) $2 \sin \mathbf{A} \cos \mathbf{A}$. (c) If and only if $\mathbf{A}$ and $\mathbf{B}$ commute.

156. $\mathbf{\Lambda} = \operatorname{diag}[\cosh(J/kT), -\sinh(J/kT)]$, $Q = 4\cosh^{N-1}(J/kT)$.

157. Expand the arbitrary vector $\mathbf{v}$ as a linear combination, $\sum c_i \mathbf{x}_i$, of eigenvectors of $\mathbf{A}$. (Since the n eigenvectors are linearly independent, they are suitable as a basis.) Then $\mathbf{P}_n(\mathbf{A})\mathbf{v} = \sum c_i \mathbf{P}_n(\mathbf{A})\mathbf{x}_i = \sum c_i P_n(\lambda_j)\mathbf{x}_j = \mathbf{0}$, since each λ_j is a root of P_n. But the only matrix that annihilates every $\mathbf{v}$ is the zero matrix, so $\mathbf{P}_n(\mathbf{A}) = \mathbf{0}$.

158. $\mathbf{A}^{-1} = (-1/c_0) \sum_{j=1}^{n} c_j \mathbf{A}^{j-1}$.

159. Expand the arbitrary vector $\mathbf{v}$ as a linear combination, $\sum c_i \mathbf{x}_i$, of eigenvectors of $\mathbf{A}$. (We are assuming that these eigenvectors are suitable as a basis.) Then $\sum f(\lambda_j)\mathbf{P}_j\mathbf{v} = \sum f(\lambda_j) c_j \mathbf{x}_j = \sum \mathbf{f}(\mathbf{A}) c_j \mathbf{x}_j = \mathbf{f}(\mathbf{A})\mathbf{v}$. Since this must hold for every $\mathbf{v}$, we may conclude that $\sum f(\lambda_j)\mathbf{P}_j = \mathbf{f}(\mathbf{A})$.

160. (a) $$\frac{d}{dt} e^{\pm it\omega} = \frac{d}{dt} \sum (\pm 1)^n i^n t^n \boldsymbol{\omega}^n/n! = \sum (\pm 1)^n i^n t^{n-1} \boldsymbol{\omega}^n/(n-1)!$$
$$= \pm i\boldsymbol{\omega} \sum (\pm 1)^m i^m t^m \boldsymbol{\omega}^m/m! = \pm i\boldsymbol{\omega} e^{\pm it\omega}.$$

(b) $-\boldsymbol{\omega}^2 e^{\pm it\omega}$, $-\boldsymbol{\omega}^2 \cos(t\boldsymbol{\omega})$, $-\boldsymbol{\omega}^2 \sin(t\boldsymbol{\omega})$.

161. (a) Since $-i\hbar \dfrac{d\boldsymbol{\psi}^\dagger}{dt} = \boldsymbol{\psi}^\dagger \mathbf{H}$,

$$\frac{d}{dt}(\boldsymbol{\psi}^\dagger\boldsymbol{\psi}) = \frac{d\boldsymbol{\psi}^\dagger}{dt}\boldsymbol{\psi} + \boldsymbol{\psi}^\dagger \frac{d\boldsymbol{\psi}}{dt} = (i/\hbar)\boldsymbol{\psi}^\dagger\mathbf{H}\boldsymbol{\psi} - (i/\hbar)\boldsymbol{\psi}^\dagger\mathbf{H}\boldsymbol{\psi} = 0.$$

(b) $\boldsymbol{\psi} = e^{-i(t/\hbar)\mathbf{H}}\boldsymbol{\psi}_0$.

162. $\mathbf{K} = \begin{pmatrix} k_1 & 0 \\ -k_1 & k_2 \end{pmatrix}$. $\quad \mathbf{X} = \begin{pmatrix} k_2 - k_1 & 0 \\ k_1 & k_1 \end{pmatrix}$,

$$\mathbf{X}^{-1} = \frac{1}{k_1(k_2 - k_1)} \begin{pmatrix} k_1 & 0 \\ -k_1 & k_2 - k_1 \end{pmatrix}, \quad \mathbf{\Lambda} = \text{diag}\,(k_1, k_2).$$

$\mathbf{C} = (A_0 e^{-k_1 t}, k_1 A_0 [e^{-k_1 t} - e^{-k_2 t}]/[k_2 - k_1])^T$.

163. (a) $\mathbf{K}_0 = \text{diag}\,(0, 0, 1/T_1)$, $\mathbf{M}_0 = (0, 0, M_0)^T$.

$$\mathbf{K} = \begin{pmatrix} 1/T_2 & -\gamma H_z & \gamma H_y \\ \gamma H_z & 1/T_2 & -\gamma H_x \\ -\gamma H_y & \gamma H_x & 1/T_1 \end{pmatrix}. \qquad \mathbf{K}' = \begin{pmatrix} 1/T_2 & \omega - \omega_0 & 0 \\ \omega_0 - \omega & 1/T_2 & -\gamma H_1 \\ 0 & \gamma H_1 & 1/T_1 \end{pmatrix}.$$

(c) $\mathbf{M}'_{ss} = (\mathbf{K}')^{-1}\mathbf{K}'_0\mathbf{M}'_0$.

164. (a) $(\mathbf{S}^\dagger)_{ij} = (\mathbf{e}_j \cdot \mathbf{e}_i)^* = (\mathbf{e}_j^\dagger \mathbf{e}_i)^* = \mathbf{e}_i^\dagger \mathbf{e}_j = (\mathbf{S})_{ij}$.

(b) $\mathbf{x}^\dagger\mathbf{S}\mathbf{x} = \mathbf{x} \cdot \mathbf{x} \geqslant 0$ for all $\mathbf{x}$.

(c) If $\mathbf{U}^\dagger\mathbf{S}\mathbf{U} = \mathbf{\Sigma} = \text{diag}\,(\sigma_1, \sigma_2, \ldots, \sigma_n)$, then since $|\mathbf{S}| \neq 0$, no $\sigma_i = 0$. Also, by Exercise 8.118, every $\sigma_i \geqslant 0$, so we may define $\mathbf{\Sigma}^{-1/2} = \text{diag}\,(\sigma_1^{-1/2}, \sigma_2^{-1/2}, \ldots, \sigma_n^{-1/2})$, which is Hermitian since every $\sigma_i^{-1/2}$ is real. And $\mathbf{S}^{-1/2} = \mathbf{U}\mathbf{\Sigma}^{-1/2}\mathbf{U}^\dagger$ is also Hermitian, by Exercise 8.86.

(d) Since both $\mathbf{S}^{-1/2}$ and $\mathbf{H}$ are Hermitian, $(\mathbf{S}^{-1/2}\mathbf{H}\mathbf{S}^{-1/2})^\dagger = \mathbf{S}^{-1/2}\mathbf{H}\mathbf{S}^{-1/2}$.

165. $$\begin{pmatrix} 1/\sqrt{2} & 1/\sqrt{2} \\ -1/\sqrt{2} & 1/\sqrt{2} \end{pmatrix} \begin{pmatrix} 1 & 1/4 \\ 1/4 & 1 \end{pmatrix} \begin{pmatrix} 1/\sqrt{2} & -1/\sqrt{2} \\ 1/\sqrt{2} & 1/\sqrt{2} \end{pmatrix} = \begin{pmatrix} 5/4 & 0 \\ 0 & 3/4 \end{pmatrix}.$$

$$\mathbf{S}^{-1/2} = \begin{pmatrix} 1/\sqrt{2} & -1/\sqrt{2} \\ 1/\sqrt{2} & 1/\sqrt{2} \end{pmatrix} \begin{pmatrix} \sqrt{4/5} & 0 \\ 0 & \sqrt{4/3} \end{pmatrix} \begin{pmatrix} 1/\sqrt{2} & 1/\sqrt{2} \\ -1/\sqrt{2} & 1/\sqrt{2} \end{pmatrix}$$

$$= \begin{pmatrix} 1/\sqrt{5} + 1/\sqrt{3} & 1/\sqrt{5} - 1/\sqrt{3} \\ 1/\sqrt{5} - 1/\sqrt{3} & 1/\sqrt{5} + 1/\sqrt{3} \end{pmatrix}.$$

166. (a) $\mathbf{A}_n = (-1/c_n) \sum_{j=0}^{n-1} c_j \mathbf{A}^j$, and higher powers likewise, so all the higher terms in the series expansion of $\mathbf{f}(\mathbf{A})$ can be reduced to $(n-1)$st-order polynomials.

(b) If $\mathbf{f}(\mathbf{A})\mathbf{x}_i = \sum b_j \mathbf{A}^{j-1}\mathbf{x}_i$, then $f(\lambda_i)\mathbf{x}_i = \sum b_j \lambda_i^{j-1}\mathbf{x}_i$. Comparing coefficients along each $\mathbf{x}_i$ leads to n simultaneous linear equations in the n unknowns $\{b_j\}$.

(c) $$\frac{1}{2S}\left(\frac{S-1}{\sqrt{1+S}} + \frac{S+1}{\sqrt{1-S}}\right)\begin{pmatrix} 1 & 0 \\ 0 & 1 \end{pmatrix} + \frac{1}{2S}\left(\frac{1}{\sqrt{1+S}} - \frac{1}{\sqrt{1-S}}\right)\begin{pmatrix} 1 & S \\ S & 1 \end{pmatrix},$$

which can be simplified.

167. $(\mathbf{GF} - \lambda\mathbf{1})\boldsymbol{\xi} = \mathbf{0}$, $(\mathbf{F} - \lambda\mathbf{G}^{-1})\boldsymbol{\xi} = \mathbf{0}$, $(\mathbf{G} - \lambda\mathbf{F}^{-1})\mathbf{F}\boldsymbol{\xi} = \mathbf{0}$, $(\mathbf{FG} - \lambda\mathbf{1})\mathbf{F}\boldsymbol{\xi} = \mathbf{0}$.

168. Center of mass is stationary.

169. If $\mathbf{C}^\dagger\mathbf{G}^{-1}\mathbf{C} = \mathbf{1}$, then $\mathbf{C}^\dagger = \mathbf{C}^{-1}\mathbf{G}$, and $\mathbf{\Lambda} = \mathbf{C}^\dagger\mathbf{F}\mathbf{C} = \mathbf{C}^{-1}\mathbf{GFC}$, so similarity transformation by $\mathbf{C}$ diagonalizes $\mathbf{GF}$ to $\mathbf{\Lambda}$.

170. $\mathbf{G}^{1/2}\mathbf{F}\mathbf{G}^{1/2} = \begin{pmatrix} k/m & k' \\ k' & k/m \end{pmatrix}$, $\mathbf{q}_0 = (A\sqrt{m}, 0)^T$,

$$\mathbf{G}^{-1/2}\mathbf{C} = \begin{pmatrix} 1/\sqrt{2} & -1/\sqrt{2} \\ 1/\sqrt{2} & 1/\sqrt{2} \end{pmatrix}, \quad \mathbf{\Lambda} = \begin{pmatrix} \omega_+^2 & 0 \\ 0 & \omega_-^2 \end{pmatrix}$$

171. (a) $\partial d_{\text{OH}_1}/\partial x_1 = 1$, $\partial d_{\text{OH}_1}/\partial x_0 = \cos\phi$, $\partial(2\phi)/\partial y_1 = y_1/d_0$, $\partial(2\phi)/\partial x_0 = -2\sin\phi$.

(b) $$\mathbf{G}^{1/2}\mathbf{F}\mathbf{G}^{1/2} = \frac{1}{16}\begin{pmatrix} 2kc^2 + 4k's^2 & 4kc & 4kc & -8k's & -8k's & 0 \\ 4kc & 16k & 0 & 0 & 0 & 4ks \\ 4kc & 0 & 16k & 0 & 0 & -4ks \\ -8k's & 0 & 0 & 16k' & 16k' & 0 \\ -8k's & 0 & 0 & 16k' & 16k' & 0 \\ 0 & 4ks & -4ks & 0 & 0 & 2ks^2 \end{pmatrix},$$

where $c = \cos\phi$, $s = \sin\phi$.

(c) $$\mathbf{U}^\dagger\mathbf{G}^{1/2}\mathbf{F}\mathbf{G}^{1/2}\mathbf{U} = \frac{1}{16}\begin{pmatrix} 2kc^2 + 4k's^2 & 4\sqrt{2}kc & -8\sqrt{2}k's \\ 4\sqrt{2}kc & 16k & 0 \\ -8\sqrt{2}k's & 0 & 32k' \end{pmatrix}$$

$$\oplus\ (0) \oplus \frac{1}{16}\begin{pmatrix} 16k & 4\sqrt{2}ks \\ 4\sqrt{2}ks & 2ks^2 \end{pmatrix}.$$

(d) The determinant of each submatrix equals zero. (e) $k(1 + \frac{1}{8}\sin^2\phi)$.

(f) $Q_6 = x_1 - x_2 + 2\sin\phi\, y_0$. If $Q_5 = 0$, then $x_1{:}x_2{:}y_0{::}2{:}-2{:}\frac{1}{4}\sin\phi$.

(g) $Q_3 = 16x_0 - (x_1 + x_2)\cos\phi + (y_1 + y_2)\sin\phi$. If $Q_1 = 0 = Q_2$, then $x_0{:}x_1{:}x_2{:}y_1{:}y_2{::}1{:}-\cos\phi{:}-\cos\phi{:}\sin\phi{:}\sin\phi$.

Chapter 9

2. $c_n = (-1)^n i\sqrt{2\pi}/n$, $c_0 = 0$. $\mathbf{x} = i\sqrt{2\pi}(\ldots, \frac{1}{3}, -\frac{1}{2}, 1, 0, -1, \frac{1}{2}, -\frac{1}{3}, \ldots)^T$.
3. $\langle\phi|(e^{i\theta})^* e^{i\theta}|\phi\rangle = \langle\phi|e^{-i\theta}e^{i\theta}|\phi\rangle = \langle\phi\mid\phi\rangle = 1$.
4. $(\mathbf{U}^\dagger\mathbf{U})_{j'j} = \sum_i (\mathbf{U}^\dagger)_{j'i}(\mathbf{U})_{ij} = \sum_i \langle i\,|\,j'\rangle^*\langle i\,|\,j\rangle = \sum_i \langle j'\mid i\rangle\langle i\mid j\rangle$ $= \langle j'\mid j\rangle = \delta_{j'j}$, or $\mathbf{U}^\dagger\mathbf{U} = \mathbf{1}$.
5. $t_{lj} = \langle l\mid j\rangle = \int \langle l\mid x\rangle\langle x\mid j\rangle\,dx = 2^{-1/2}\int_{-1}^{1} \hat{P}_l(x)e^{ij\pi x}\,dx$.
6. $\delta(x - r\cos\phi)\delta(y - r\sin\phi)$ or $\delta[r - (x^2 + y^2)^{1/2}]\delta[\phi - \tan^{-1}(y/x)]$.
7. $\langle x\mid n\rangle = (-1)^n D^n\delta(x)/n!$.
8. For all f and g $|\int f^*(x)g(x)\,dx| \leqslant (\int f^*(x)f(x)\,dx)^{1/2}(\int g^*(x)g(x)\,dx)^{1/2}$.
9. S, $*$.
10. $i\hbar\mathbf{L}$.
11. For "almost all" $|\phi\rangle$, $\sum_i \langle\phi\mid i\rangle\langle i\mid\phi\rangle < \langle\phi\mid\phi\rangle$.
12. $(|\phi\rangle\langle\phi|)^2 = |\phi\rangle\langle\phi\mid\phi\rangle\langle\phi| = |\phi\rangle\langle\phi|$.
13. $|j\rangle\langle j| + |k\rangle\langle k|$.
14. (b) Infinite (Cf. Exercise 9.1).
15. $|\psi\rangle\langle\phi|$.
16. Taking the complex conjugate gives $\int f(x)Dg^*(x)\,dx = \int g^*(x)D^\dagger f(x)\,dx$; $D^\dagger = -D$.
18. $\mathbf{T}^{-1}\boldsymbol{\Omega}\mathbf{T} = \boldsymbol{\Omega}'$, where $t_{i'j'} = \langle i'\,|\,j'\rangle$. Cf. Eq. 8.24.
19. $H_{ij} = \langle i|\,H\,|j\rangle = \langle j|\,H\,|i\rangle^* = H_{ji}^*$.
20. (a) $(\mathbf{1})_{jk} = \delta_{jk}$. (b) $(\mathbf{D})_{jk} = ik\delta_{jk} = \text{diag}(\ldots, -2i, -i, 0, i, 2i, \ldots)$.
    (c) $(\mathbf{x})_{jk} = (-1)^{k-j}/i(k-j)$, $k \neq j$, $(\mathbf{x})_{jj} = 0$.

$$\mathbf{x} = \left\|\begin{matrix} & \cdot & \cdot & \cdot & \cdot & \cdot & \\ & \cdot & \cdot & \cdot & \cdot & \cdot & \\ & \cdot & \cdot & \cdot & \cdot & \cdot & \\ \cdots & -i & 0 & i & -i/2 & i/3 & \cdots \\ \cdots & i/2 & -i & 0 & i & -i/2 & \cdots \\ \cdots & -i/3 & i/2 & -i & 0 & i & \cdots \\ & \cdot & \cdot & \cdot & \cdot & \cdot & \\ & \cdot & \cdot & \cdot & \cdot & \cdot & \\ & \cdot & \cdot & \cdot & \cdot & \cdot & \end{matrix}\right\|$$

21. $(\mathbf{F})_{jk} = (2\pi)^{-1/2}e^{-ijk}$.
22. $(\mathbf{1})_{xx'} = \langle x|\,1\,|x'\rangle = \delta(x - x')$.
23. (a) 1. (b) $(\boldsymbol{\rho})_{xx'} = \langle x\,|\Psi\rangle\langle\Psi\,|\,x'\rangle$.
(c) $\int \langle x\,|\Psi\rangle\langle\Psi\,|\,x\rangle\,dx$, which still $= 1$.
(d) $\langle\Psi|\,\Omega\,|\Psi\rangle = \sum\sum \langle\Psi\,|\,j\rangle\langle j|\,\Omega\,|i\rangle\langle i\,|\,\Psi\rangle = \sum\sum \langle i\,|\,\Psi\rangle\langle\Psi\,|\,j\rangle\langle j|\,\Omega\,|i\rangle$
$= \sum\sum (\boldsymbol{\rho})_{ij}(\boldsymbol{\Omega})_{ji} = \sum (\boldsymbol{\rho\Omega})_{ii} = \text{Tr}\,(\boldsymbol{\rho\Omega})$.
24. (a) $(\boldsymbol{\rho})_{ij} = \sum_n P_n\langle\Psi_i\,|\,\Psi_n\rangle\langle\Psi_n\,|\,\Psi_j\rangle = \sum_n P_n\delta_{in}\delta_{nj} = P_i\delta_{ij}$.
(b) $\{P_i\}$, by Exercise 8.115. By the nature of probability $0 \leqslant P_i \leqslant 1$.
(c) If and only if one $P_i = 1$ and all the rest zero.
25. $\lambda = \langle\phi\,|\,\phi\rangle$, $|\psi\rangle = |\phi\rangle/\langle\phi\,|\,\phi\rangle^{1/2}$. $\lambda = 0$, $|\psi\rangle =$ any vector orthogonal to $|\phi\rangle$.
26. As in Section 8.13, leading to $(\lambda - \lambda^*)\langle\lambda\,|\,\lambda\rangle = 0$ and $(\lambda_1 - \lambda_2)\langle\lambda_1\,|\,\lambda_2\rangle = 0$.
27. As in Section 8.13, via $\Omega_1|\lambda\rangle = \lambda|\lambda\rangle$, $\Omega_2\Omega_1|\lambda\rangle = \Omega_1\Omega_2|\lambda\rangle = \lambda\Omega_2|\lambda\rangle$ and $\Omega_2|\lambda\rangle = \mu|\lambda\rangle$. It is common to write this eigenvector of two operators as $|\lambda\mu\rangle$ to emphasize that it belongs to the two eigenvalues λ and μ.
28. $[L_z, L_\pm] = \pm\hbar L_\pm$, so $L_\pm L_z = L_z L_\pm \mp \hbar L_\pm$. Then if $L_z|l_z\rangle = l_z|l_z\rangle$, $L_\pm L_z|l_z\rangle = l_z L_\pm|l_z\rangle$, and $L_z L_\pm|l_z\rangle \mp \hbar L_\pm|l_z\rangle = l_z L_\pm|l_z\rangle$. Substituting and collecting terms gives $L_z L_\pm|l_z\rangle = (l_z \pm \hbar)L_\pm|l_z\rangle$, showing that $L_\pm|l_z\rangle$ is an eigenvector of L_z with eigenvalue $l_z \pm \hbar$.
29. (a) $(2h\nu)^{-1/2}(p - 2\pi i\nu q)$.
(b) $R^\dagger R = (h\nu)^{-1}(H + \frac{1}{2}h\nu 1)$, $RR^\dagger = (h\nu)^{-1}(H - \frac{1}{2}h\nu 1)$. (c) $h\nu R$.
(d) Operate R on $H|E\rangle = E|E\rangle$, and substitute for RH from (c), then collect terms, to obtain $HR|E\rangle = (E + h\nu)R|E\rangle$.
(e) $E + h\nu$. (f) $R^n|E_0\rangle$ belongs to eigenvalue $E_n = (n + \frac{1}{2})h\nu$.
30. (a) 1. (b) 0.
31. $\sum P(\lambda) = \sum_\lambda |\langle\lambda\,|\,\phi\rangle|^2 = \sum \langle\phi\,|\,\lambda\rangle\langle\lambda\,|\,\phi\rangle = \langle\phi\,|\,\phi\rangle = 1$.
32. If $\Omega|\omega\rangle = \omega|\omega\rangle$, then $P(\omega) = \langle\phi\,|\,\omega\rangle\langle\omega\,|\,\phi\rangle$ and
$\bar{\omega} = \sum \omega P(\omega) = \sum \omega\langle\phi\,|\,\omega\rangle\langle\omega\,|\,\phi\rangle = \sum \langle\phi|\,\Omega\,|\omega\rangle\langle\omega\,|\,\phi\rangle = \langle\phi|\,\Omega\,|\phi\rangle$.
33. $\bar{\omega} = \sum\sum \omega P_n\langle\Psi_n\,|\,\omega\rangle\langle\omega\,|\,\Psi_n\rangle = \sum\sum\sum P_n\langle\omega\,|\,\Psi_n\rangle\langle\Psi_n\,|\,\omega'\rangle\langle\omega'|\,\Omega\,|\omega\rangle$
$= \sum\sum (\boldsymbol{\rho})_{\omega\omega'}(\boldsymbol{\Omega})_{\omega'\omega} = \text{Tr}\,(\boldsymbol{\rho\Omega})$.
34. $\langle x\,|\,k\rangle = e^{ikx/\hbar}$. $P(k_0) = |\langle k_0\,|\,E\rangle|^2 = |\int \langle k_0\,|\,x\rangle\langle x\,|\,E\rangle\,dx|^2 = |\int e^{-ik_0x/\hbar}\psi(x)\,dx|^2$
$= |F\{\psi(x)\}|^2$, neglecting constants.
35. $|\phi\rangle/\langle\phi\,|\,\phi\rangle^{1/2}$ is normalized.
36. If $|\phi\rangle = |E_0\rangle + \varepsilon|\chi\rangle$, then to first order,
$W = \langle\phi|\,H\,|\phi\rangle/\langle\phi\,|\,\phi\rangle \sim \{E_0 + \varepsilon\langle\chi|\,H\,|E_0\rangle + \varepsilon\langle E_0|\,H\,|\chi\rangle\}/\{1 + \varepsilon\langle\chi\,|\,E_0\rangle + \varepsilon\langle E_0|\,\chi\rangle\}$
$= E_0\{1 + \varepsilon\langle\chi\,|\,E_0\rangle + \varepsilon\langle\chi\,|\,E_0\rangle^*\}/\{1 + \varepsilon\langle\chi\,|\,E_0\rangle + \varepsilon\langle\chi\,|\,E_0\rangle^*\} = E_0$.
Furthermore, we would obtain this result more simply if we note that we may take $\langle\chi\,|\,E_0\rangle = 0$ to first order.
37. Via $\sum_{E\neq E_0} \langle\phi\,|\,E\rangle\langle E\,|\,\phi\rangle(E - E_0) \geqslant \sum_{E\neq E_0} \langle\phi\,|\,E\rangle\langle E\,|\,\phi\rangle(E_1 - E_0)$
$= (E_1 - E_0)(1 - \langle\phi\,|\,E_0\rangle\langle E_0\,|\,\phi\rangle)$.
38. $|\langle\phi\,|\,E_0\rangle|^2 = 1 - \sum \langle\phi\,|\,E\rangle\langle E\,|\,\phi\rangle$, but $W = \sum E\langle\phi\,|\,E\rangle\langle E\,|\,\phi\rangle$ has E as a weighting factor in the sum, so eigenvectors belonging to large $|E|$ contribute more heavily to W.
39. If introduction of a new parameter were such as to increase W, it could be set equal to zero.
40. (a) $\beta = (10k)^{1/6}$ gives $W_{\min} = \frac{9}{8}(10k)^{1/6}$.
(b) Because it is an odd function and orthogonal to $|E_0\rangle$. (c) Nothing.
41. $\alpha = 8/9\pi \sim 0.283$ gives $W_{\min} = -4/3\pi \sim -0.425$.
43. Multiply the ith simultaneous equations by $\langle\phi_k\,|\,i\rangle$ and add all the equations.
44. $\dfrac{5}{2}\dfrac{\pi^2}{L^2} + \dfrac{1}{2}V_0 - \left(\dfrac{9\pi^4}{4L^4} + \dfrac{16V_0^2}{9\pi^2}\right)^{1/2}$.

45. $\frac{1}{2} + \left(\frac{1}{6} + \frac{\sqrt{3}}{8\pi}\right) V_0 - \left(\frac{1}{4} + \frac{V_0\sqrt{3}}{8\pi} + \frac{35V_0^2}{64\pi^2}\right)^{1/2}$.

46. (3) To first order, the correction to $|E_i^\circ\rangle$ is orthogonal to $|E_i^\circ\rangle$, so $|E_i'\rangle = \sum_{j\neq i} |E_j^\circ\rangle\langle E_j^\circ \mid E_i'\rangle$.

48. $E_i' = \int \psi_i^\circ{}^* H \psi_i^\circ\, dx$. $\quad \psi_i' = -\sum_{j\neq i} \frac{\int \psi_j^\circ{}^* H' \psi_i^\circ\, dx}{E_j^\circ - E_i^\circ} \psi_j^\circ$.

$$E_i'' = \int \psi_i^\circ{}^* H'' \psi_i^\circ\, dx - \sum_{j\neq i} \frac{|\int \psi_i^\circ{}^* H' \psi_j^\circ\, dx|^2}{E_j^\circ - E_i^\circ} \quad \text{(Exercise 9.47).}$$

49. The lowest eigenvalue will be little affected by going to a complete set; that this effect is small is clear from the form of the second-order correction to that eigenvalue, since the ΔE in the denominator is large. For higher eigenvalues, this ΔE is much smaller.

50. $\frac{\pi^2}{L^2} + \frac{1}{2} V_0 - \frac{16L^2V_0^2}{\pi^4} \sum_{n=1}^{\infty} \frac{n^2}{(4n^2-1)^3} = \frac{\pi^2}{L^2} + \frac{1}{2} V_0 - \frac{L^2V_0^2}{16\pi^2}$, by Exercise 5.17d.

51. $\frac{1}{6} V_0 - \frac{2}{\pi^2} V_0^2 \sum_{m=0}^{\infty} m^{-4} \sin^2 (m\pi/6)$.

52. $E_{\pm n} = n^2 + \frac{1}{6}V_0$ if $n = 3m$, $E_{\pm n} = n^2 + \frac{1}{6}V_0 \pm \frac{V_0\sqrt{3}}{4n\pi}$ if $n = 3m \pm 1$. Notice that it is necessary to choose the proper zeroth-order eigenfunctions, $\sin n\phi$ and $\cos n\phi$, and that whenever $n = 3m \pm 1$ the perturbation does split their degeneracy in first order.

53. (a) $\lambda_1^\circ - \kappa^2/(\lambda_2^\circ - \lambda_1^\circ)$ (apparently doubly degenerate), $\lambda_2^\circ - 2\kappa^2/(\lambda_1^\circ - \lambda_2^\circ)$.
(b) λ_1°, $\frac{1}{2}(\lambda_1^\circ + \lambda_2^\circ) \pm \frac{1}{2}[(\lambda_1^\circ - \lambda_2^\circ)^2 + 8\kappa^2]^{1/2}$. Notice that the degeneracy is in fact split, even in second order.
(c) It is necessary to choose the proper zeroth-order combinations of $|\lambda_{1,1}^\circ\rangle$ and $|\lambda_{1,2}^\circ\rangle$, which are degenerate in zeroth order. In Exercise 9.52 the degeneracy was split in first order by the coupling between the two unperturbed states. Here the degeneracy is split in second order by the indirect coupling of the two unperturbed states, through a third one. Notice that in both cases, energy denominators equal to zero warn of this problem.

Chapter 10

1. (a) Inverses. (b) Inverses. (c) Closure, identity. (d) Yes.
(e) Closure, identity. (f) Yes.
(g) Dot: closure, associativity, identity, inverses; cross: associativity, identity, inverses.
(h) Yes. (i) Associativity, identity (although $a - 0 = a$, $0 - a \neq a$).
(j) Inverses.
2. E, $B^{-1}A^{-1}$.
3. $n!$.
6. If X appears twice under A, say in rows B and C, then $BA = X = CA$. Multiply from the right by A^{-1}, to obtain $B = XA^{-1} = C$, or $B = C$. Likewise for X twice in a column.
8.

	A	B	C	D
A	B	C	D	E
B	C	D	E	A
C	D	E	A	B
D	E	A	B	C

Abelian.

9. $ACB = F$, $ACD = E$, $CDF = D$.
10. Isomorphic to $\mathcal{V}$ (E = rotate by $0°$).
11. If we identify (ab) with C_2 and E with E, then $\mathcal{P}_2$ is isomorphic to $\mathcal{C}_2$.
12. Identify (abc) with A, (acb) with B, (ab) with C, (ac) with D, and (bc) with F.
13. (a) $C_9^m C_9^n = C_9^{m+n}$.
    (b) Set $AB = E$, $C^3 = D^3 = CD = E$, $F = AC$, $G = AD$, $H = BC$, $J = BD$, and continue to fill in the table. This group is Abelian, is called $\mathcal{C}_3 \times \mathcal{C}_3$, and is not isomorphic to any possible point group.
14. What is $i\boldsymbol{\mu}$ or $C_2'\boldsymbol{\mu}$ or $C_n\boldsymbol{\mu}$?
16. $C_4^{-1}(x, y, z) = (y, -x, z)$; If $f(x, y, z) = x$, then $C_4 f(x, y, z) = y = f(y, -x, z)$.
17. (a) ± 1, since $\sigma_v{}^2\psi = \lambda^2\psi = \psi$. (b) nth root of unity.
    (c) nth root of unity (n even), $2n$th root of unity (n odd).
18. $C_n\sigma_h(x, y, z) = (x \cos 2\pi/n - y \sin 2\pi/n,\ x \sin 2\pi/n + y \cos 2\pi/n,\ -z)$.
19. $\begin{pmatrix} \cos 2\phi & \sin 2\phi & 0 \\ \sin 2\phi & -\cos 2\phi & 0 \\ 0 & 0 & 1 \end{pmatrix}$.
20. σ, i.
21. By closure, S_n^n is in $\mathcal{G}$. But $S_n^n = (\sigma_h C_n)^n = \sigma_n^n C_n^n = \sigma_h$ if n is odd. Then, by closure, $S_n\sigma_h = C_n$ must also be in $\mathcal{G}$.
22. $S_n^2 = C_n^2 = C_{n/2}$.
24. (a) C_2''. (b) σ_h. (c) σ_v'. (d) σ_v''. (e) S_6. (f) S_6.
25. (a) C_2' ($30°$ from first C_2'), C_2'' ($45°$ from C_2').
    (b) C_2' ($60°$ from first C_2'), C_2' ($90°$ from first C_2'). (c) n.

26. 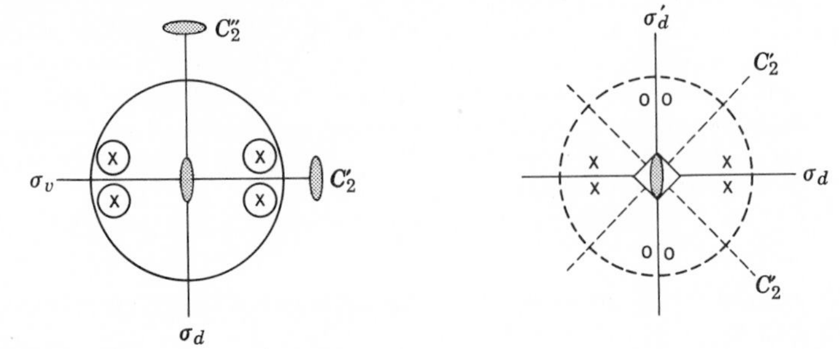

27. g.
28. (a) $\mathcal{C}_s$. (b) $\mathcal{C}_{3h}$. (c) $\mathcal{C}_s$. (d) $\mathcal{C}_2$. (e) $\mathcal{C}_{2h}$. (f) $\mathcal{C}_{2v}$.
29. $\mathcal{S}_4 \leftrightarrow \mathcal{C}_4$, $\mathcal{V} \leftrightarrow \mathcal{C}_{2h} \leftrightarrow \mathcal{C}_{2v} \leftrightarrow \mathcal{D}_2$.
30. $\mathcal{C}_{3v} \leftrightarrow \mathcal{D}_3 \leftrightarrow \mathcal{P}_3$. $\mathcal{C}_6$, $\mathcal{S}_6$, and $\mathcal{C}_{3h}$ ($= \{S_3^n\}$) are cyclic.
31. Yes. Yes. No, if $n > 2$ (since then C_r and σ_v do not commute).
32. (a) $E, C_3, C_3^2, C_2', C_2', C_2'$. (b) $E, C_4, C_2, C_4^3, C_2', C_2', C_2'', C_2''$.
33. $E, C_6, C_3, C_2, C_3^2, C_6^5, C_2', C_2', C_2', C_2'', C_2'', C_2'', \sigma_v, \sigma_v', \sigma_v'', \sigma_d, \sigma_d', \sigma_d'', \sigma_h, S_6, S_3, i, S_3^5, S_6^5$.
34. π/n, $\pi/2n$, see Exercise 10.24e.
36. C_4 through center of a face, S_6 along main diagonal, σ_h perpendicular to C_4, σ_d bisecting the angle between C_4^2s, C_2' through middle of an edge.
37. Inscribe the tetrahedron in a cube; then the C_3, S_4, and σ_d are along the cube's S_6, C_4, and σ_d, respectively. Each of the 24 operations of $\mathcal{T}_d$ is a different permutation of the four vertices.

39. 12 (four C_3s contribute 8, three C_2s contribute 3, and E contributes 1).
40. 24 (9 from $3C_4$s, 8 from $4C_3$s, 6 from $6C_2'$s, and 1 from E). 60 (24 from $6C_5$s, 20 from $10C_3$s, 15 from $15C_2$s, and 1 from E).
41. (a) $\mathcal{C}_{2v}$. (b) $\mathcal{D}_{2h}$. (c) $\mathcal{C}_s$. (d) $\mathcal{C}_{2v}$. (e) $\mathcal{D}_{3h}$. (f) $\mathcal{D}_{2h}$ (g) $\mathcal{D}_{\infty h}$. (h) $\mathcal{C}_2$. (i) $\mathcal{D}_{4d}$. (j) $\mathcal{C}_{2h}$. (k) $\mathcal{C}_{3h}$. (l) $\mathcal{C}_{3v}$. (m) $\mathcal{C}_{4v}$. (n) $\mathcal{D}_{8h}$. (o) $\mathcal{D}_{4h}$. (p) $\mathcal{D}_3$. (q) $\mathcal{C}_i$. (r) $\mathcal{D}_2$. (s) $\mathcal{D}_3$. (t) $\mathcal{D}_{2d}$. (u) $\mathcal{T}_d$. (v) $\mathcal{S}_4$.
44. From Exercise 10.42, $\{C_E\} = \{X^{-1}EX\} = \{E\}$.
45. $\{C_A\} = \{X^{-1}AX\} = \{A\}$, so each of the n elements forms a class by itself.
46. $X^{-1}AX$.
47. *Hint:* The $5\sigma_v$s are all indistinguishable; C_5 and C_5^{-1} are mirror images of each other, so $C_5^{-1}\sigma_v C_5 = \sigma_v'$.
48. $\{E\}, \{C_6, C_6^5\}, \{C_3, C_3^2\}, \{C_2\}, \{3C_2'\}, \{3C_2''\}, \{3\sigma_v\}, \{3\sigma_d\}, \{\sigma_h\}, \{S_6, S_6^5\}, \{S_3, S_3^5\}, \{i\}$.
49. If A and B are in the same class, there is an X_b such that $X_b^{-1}AX_b = B$. Also, for each X_α, $X_b^{-1}X_\alpha^{-1}AX_\alpha X_b = B$. Therefore there are n_A elements, $\{X_\beta\} = \{X_\alpha X_b\}$, such that $X_\beta^{-1}AX_\beta = B$. Similarly, for each of the c_A elements of $\{C_A\}$, there are n_A such elements, all of which are distinct. These must exhaust $\mathcal{G}$, so $n_A c_A = g$.
51. $\{E, (abc), (acb)\}, \{E, (ab)\}, \{E, (ac)\}, \{E, (bc)\}$.
53. (b) Yes.
55.

$\mathcal{C}_{4v}$	E	C_4	C_2	C_4^3	σ_v	σ_v'	σ_d	σ_d'
E	E	C_4	C_2	C_4^3	σ_v	σ_v'	σ_d	σ_d'
C_4	C_4	C_2	C_4^3	E	σ_d			
C_2	C_2	C_4^3	E	C_4	σ_v'	σ_v	σ_d'	σ_d
C_4^3	C_4^3	E	C_4	C_2	?		σ_v	
σ_v	σ_v	?	σ_v'	σ_d	E	C_2	C_4^3	?
σ_v'	σ_v'		σ_v		C_2	E		
σ_d	σ_d	σ_v	σ_d'		C_4		E	C_2
σ_d'	σ_d'		σ_d		?		C_2	E

This may be completed by invoking the Latin-square property: In the σ_v row, the ?s are C_4 and σ_d'. But C_4 cannot appear again in the C_4 column, nor σ_d' in the σ_d' column. All the other products may be assigned by a similar process of elimination.
56. (a) If $m = n + \frac{1}{2}$, then $R\Phi_m = -\Phi_m$, but $R^2\Phi_m = \Phi_m$.
(b) Twice the order of $\mathcal{G}$.
(c) From $C_2C_2' = C_2''$, multiplication by R gives $RC_2C_2' = RC_2''$, or $SC_2' = U$, etc., finally giving

$\mathcal{D}_2'$	E	C_2	C_2'	C_2''	R	S	T	U
E	E	C_2	C_2'	C_2''	R	S	T	U
C_2	C_2	R	C_2''	T	S	E	U	C_2'
C_2'	C_2'	U	R	C_2	T	C_2''	E	S
C_2''	C_2''	C_2'	S	R	U	T	C_2	E
R	R	S	T	U	E	C_2	C_2'	C_2''
S	S	E	U	C_2'	C_2	R	C_2''	T
T	T	C_2''	E	S	C_2'	U	R	C_2
U	U	T	C_2	E	C_2''	C_2'	S	R

(d) Only $\mathcal{D}_4$, $\mathcal{C}_{4v}$, and $\mathcal{D}_{2d}$ are non-Abelian groups of order 8, but in each of them there are 4 elements equal to their inverses. In $\mathcal{D}_2'$, only $E^{-1} = E$ and $R^{-1} = R$.

57. (a) 3. (b) 4. (c) 1. (d) 6. (e) 12.

58. $\mathbf{E} = \begin{pmatrix} 1 & 0 \\ 0 & 1 \end{pmatrix}, \mathbf{i} = \begin{pmatrix} 1 & 0 \\ 0 & -1 \end{pmatrix}.$

59. $\mathbf{E} = \begin{pmatrix} 1 & 0 \\ 0 & 1 \end{pmatrix}, \mathbf{C}_4 = \begin{pmatrix} 0 & -1 \\ 1 & 0 \end{pmatrix}, \mathbf{C}_2 = \begin{pmatrix} -1 & 0 \\ 0 & -1 \end{pmatrix}, \mathbf{C}_4^3 = \begin{pmatrix} 0 & 1 \\ -1 & 0 \end{pmatrix},$

$$\boldsymbol{\sigma}_v = \begin{pmatrix} 1 & 0 \\ 0 & -1 \end{pmatrix}, \boldsymbol{\sigma}_v' = \begin{pmatrix} -1 & 0 \\ 0 & 1 \end{pmatrix}, \boldsymbol{\sigma}_d = \begin{pmatrix} 0 & 1 \\ 1 & 0 \end{pmatrix}, \boldsymbol{\sigma}_d' = \begin{pmatrix} 0 & -1 \\ -1 & 0 \end{pmatrix}.$$

$\boldsymbol{\sigma}_d \boldsymbol{\sigma}_v = \mathbf{C}_4 \neq \mathbf{C}_4^3 = \boldsymbol{\sigma}_v \boldsymbol{\sigma}_d$ (or other examples).

60.
$$\mathbf{E} = \begin{pmatrix} 1 & 0 & 0 \\ 0 & 1 & 0 \\ 0 & 0 & 1 \end{pmatrix} = \mathbf{C}_2, \mathbf{C}_4 = \begin{pmatrix} 0 & 1 & 0 \\ 1 & 0 & 0 \\ 0 & 0 & -1 \end{pmatrix} = \mathbf{C}_4^3,$$

$$\boldsymbol{\sigma}_v = \begin{pmatrix} 1 & 0 & 0 \\ 0 & 1 & 0 \\ 0 & 0 & -1 \end{pmatrix} = \boldsymbol{\sigma}_v', \boldsymbol{\sigma}_d = \begin{pmatrix} 0 & 1 & 0 \\ 1 & 0 & 0 \\ 0 & 0 & 1 \end{pmatrix} = \boldsymbol{\sigma}_d'.$$

All matrices commute.

61. For all $\mathbf{A}$, $\mathbf{EA} = \mathbf{A}$. Multiplying by $\mathbf{A}^{-1}$ gives $\mathbf{E} = \mathbf{1}$.

62. The matrices of Exercise 10.59 form a faithful representation of $\mathcal{C}_{4v}$, but those of Exercise 10.60 do not.

63. If $\mathbf{RS} = \mathbf{T}$, then $\mathbf{R}^*\mathbf{S}^* = \mathbf{T}^*$.

64. In Fig. 10.5a, start numbering the xs clockwise, starting at 12:00, in the order 152634, then $\mathbf{E} = \mathbf{1}$, and

$$\mathbf{C}_3 = \begin{pmatrix} 0 & 1 & 0 & 0 & 0 & 0 \\ 0 & 0 & 1 & 0 & 0 & 0 \\ 1 & 0 & 0 & 0 & 0 & 0 \\ 0 & 0 & 0 & 0 & 1 & 0 \\ 0 & 0 & 0 & 0 & 0 & 1 \\ 0 & 0 & 0 & 1 & 0 & 0 \end{pmatrix} = (\mathbf{C})_3^{2\dagger} \qquad \boldsymbol{\sigma}_v = \begin{pmatrix} 0 & 0 & 0 & 1 & 0 & 0 \\ 0 & 0 & 0 & 0 & 0 & 1 \\ 0 & 0 & 0 & 0 & 1 & 0 \\ 1 & 0 & 0 & 0 & 0 & 0 \\ 0 & 0 & 1 & 0 & 0 & 0 \\ 0 & 1 & 0 & 0 & 0 & 0 \end{pmatrix}$$

$$\boldsymbol{\sigma}_v' = \begin{pmatrix} 0 & 0 & 0 & 0 & 1 & 0 \\ 0 & 0 & 0 & 1 & 0 & 0 \\ 0 & 0 & 0 & 0 & 0 & 1 \\ 0 & 1 & 0 & 0 & 0 & 0 \\ 1 & 0 & 0 & 0 & 0 & 0 \\ 0 & 0 & 1 & 0 & 0 & 0 \end{pmatrix} \qquad \boldsymbol{\sigma}_v'' = \begin{pmatrix} 0 & 0 & 0 & 0 & 0 & 1 \\ 0 & 0 & 0 & 0 & 1 & 0 \\ 0 & 0 & 0 & 1 & 0 & 0 \\ 0 & 0 & 1 & 0 & 0 & 0 \\ 0 & 1 & 0 & 0 & 0 & 0 \\ 1 & 0 & 0 & 0 & 0 & 0 \end{pmatrix}.$$

This same regular representation may also be obtained as follows: Construct the group multiplication table, with columns labeled E, C_3, C_3^2, σ_v, σ_v', and σ_v'', as usual, but with rows labeled E^{-1}, C_3^{-1}, C_3^{-2}, σ_v^{-1}, $\sigma_v'^{-1}$, $\sigma_v''^{-1}$. For each R, the matrix $\mathbf{R}$ is obtained by replacing R in the table by 1, and all other products by 0. This method is applicable even if no stereographic projection exists.

65. If $\mathbf{RS} = \mathbf{T}$, then $\mathbf{X}^{-1}\mathbf{RSX} = \mathbf{X}^{-1}\mathbf{TX}$, and $(\mathbf{X}^{-1}\mathbf{RX})(\mathbf{X}^{-1}\mathbf{SX}) = \mathbf{X}^{-1}\mathbf{TX}$.

67.

$$\mathbf{C}_4 = \begin{pmatrix} 0 & 0 & 0 & 1 \\ 1 & 0 & 0 & 0 \\ 0 & 1 & 0 & 0 \\ 0 & 0 & 1 & 0 \end{pmatrix} \oplus \begin{pmatrix} 0 & 0 & 0 & 1 \\ 1 & 0 & 0 & 0 \\ 0 & 1 & 0 & 0 \\ 0 & 0 & 1 & 0 \end{pmatrix} \oplus \begin{pmatrix} 0 & 0 & 0 & 1 \\ 1 & 0 & 0 & 0 \\ 0 & 1 & 0 & 0 \\ 0 & 0 & 1 & 0 \end{pmatrix}.$$

$$\boldsymbol{\sigma}_v = \begin{pmatrix} 1 & 0 & 0 & 0 \\ 0 & 0 & 0 & 1 \\ 0 & 0 & 1 & 0 \\ 0 & 1 & 0 & 0 \end{pmatrix} \oplus \begin{pmatrix} -1 & 0 & 0 & 0 \\ 0 & 0 & 0 & -1 \\ 0 & 0 & -1 & 0 \\ 0 & -1 & 0 & 0 \end{pmatrix} \oplus \begin{pmatrix} 1 & 0 & 0 & 0 \\ 0 & 0 & 0 & 1 \\ 0 & 0 & 1 & 0 \\ 0 & 1 & 0 & 0 \end{pmatrix}.$$

$$\boldsymbol{\sigma}_h = \begin{pmatrix} 1 & 0 & 0 & 0 \\ 0 & 1 & 0 & 0 \\ 0 & 0 & 1 & 0 \\ 0 & 0 & 0 & 1 \end{pmatrix} \oplus \begin{pmatrix} 1 & 0 & 0 & 0 \\ 0 & 1 & 0 & 0 \\ 0 & 0 & 1 & 0 \\ 0 & 0 & 0 & 1 \end{pmatrix} \oplus \begin{pmatrix} -1 & 0 & 0 & 0 \\ 0 & -1 & 0 & 0 \\ 0 & 0 & -1 & 0 \\ 0 & 0 & 0 & -1 \end{pmatrix}.$$

69. Γ_1.

70. Set of tangential displacements of the H atoms of NH_3, or (with the numbering system of Exercise 10.64) the linear combination $\mathsf{x}_1 + \mathsf{x}_2 + \mathsf{x}_3 - \mathsf{x}_4 - \mathsf{x}_5 - \mathsf{x}_6$ of the xs of the stereographic projection.

71. $\Gamma_1(P_i) = 1$ for all P_i. $\Gamma_2(P_i) = 1$ if P_i is even, $\Gamma_2(P_i) = -1$ if P_i is odd.

74. Let us assume that Γ is a multidimensional irreducible representation. Then, for all R and S, since $RS = SR$, $\Gamma(R)\Gamma(S) = \Gamma(S)\Gamma(R)$. Since all the matrices of Γ commute, they may all be diagonalized simultaneously, so Γ *is* reducible to a direct sum of one-dimensional representations.

75. If $\mathcal{G}$ is non-Abelian, there are R and S such that $RS \neq SR$. If Γ is faithful, $\mathbf{RS} \neq \mathbf{SR}$. Since these matrices do not commute, they cannot be diagonalized simultaneously.

77. $\mathbf{U}^\dagger\mathbf{EU} = (1) \oplus (1) \oplus (1) \oplus (1)$, $\mathbf{U}^\dagger\mathbf{C}_2\mathbf{U} = (1) \oplus (1) \oplus (-1) \oplus (-1)$, $\mathbf{U}^\dagger\mathbf{C}_2'\mathbf{U} = (1) \oplus (-1) \oplus (1) \oplus (-1)$, $\mathbf{U}^\dagger\mathbf{C}_2''\mathbf{U} = (1) \oplus (-1) \oplus (-1) \oplus (1)$, where

$$\mathbf{U} = \frac{1}{2}\begin{pmatrix} 1 & 1 & 1 & 1 \\ 1 & 1 & -1 & -1 \\ 1 & -1 & 1 & -1 \\ 1 & -1 & -1 & 1 \end{pmatrix}.$$

Rearranging columns or rows of $\mathbf{U}$ rearranges diagonal elements of the transformed matrices.

78. $\mathbf{U}^\dagger\mathbf{C}_3^2\mathbf{U} = \text{diag}\,(1, \omega^2, \omega)$.

$$\mathbf{U}^\dagger\boldsymbol{\sigma}_v\mathbf{U} = \begin{pmatrix} 1 & 0 & 0 \\ 0 & 0 & 1 \\ 0 & 1 & 0 \end{pmatrix} \qquad \mathbf{U}^\dagger\boldsymbol{\sigma}_v'\mathbf{U} = \begin{pmatrix} 1 & 0 & 0 \\ 0 & 0 & \omega \\ 0 & \omega^2 & 0 \end{pmatrix}$$

$$\mathbf{U}^\dagger\boldsymbol{\sigma}_v''\mathbf{U} = \begin{pmatrix} 1 & 0 & 0 \\ 0 & 0 & \omega^2 \\ 0 & \omega & 0 \end{pmatrix} \qquad \mathbf{U} = \frac{1}{\sqrt{3}}\begin{pmatrix} 1 & 1 & 1 \\ 1 & \omega & \omega^2 \\ 1 & \omega^2 & \omega \end{pmatrix}.$$

79. If $\mathbf{U} = \begin{pmatrix} 1/\sqrt{2} & -1/\sqrt{2} \\ 1/\sqrt{2} & 1/\sqrt{2} \end{pmatrix} \oplus (1)$, then $\mathbf{U}^\dagger\mathbf{EU} = (1) \oplus (1) \oplus (1) = \mathbf{U}^\dagger\mathbf{C}_2\mathbf{U}$,

$\mathbf{U}^\dagger\mathbf{C}_4\mathbf{U} = (1) \oplus (-1) \oplus (-1) = \mathbf{U}^\dagger\mathbf{C}_4^3\mathbf{U}$, $\mathbf{U}^\dagger\boldsymbol{\sigma}_v\mathbf{U} = (1) \oplus (1) \oplus (-1) = \mathbf{U}^\dagger\boldsymbol{\sigma}_v'\mathbf{U}$, $\mathbf{U}^\dagger\boldsymbol{\sigma}_d\mathbf{U} = (1) \oplus (-1) \oplus (1) = \mathbf{U}^\dagger\boldsymbol{\sigma}_d'\mathbf{U}$.

80. If $|\phi\rangle$ transforms as a one-dimensional representation, then $|\phi\rangle$ must be an eigenvector of R. Then apply R to $R^{-1}|\phi\rangle = R|\phi\rangle = \lambda|\phi\rangle$ to obtain $|\phi\rangle = \lambda R|\phi\rangle = \lambda^2|\phi\rangle$, or $\lambda = \pm 1$.

81. $\mathrm{x}_1 + \mathrm{x}_2 + \mathrm{o}_1 + \mathrm{o}_2$, $\mathrm{x}_1 + \mathrm{x}_2 - \mathrm{o}_1 - \mathrm{o}_2$, $\mathrm{x}_1 - \mathrm{x}_2 + \mathrm{o}_1 - \mathrm{o}_2$, $\mathrm{x}_1 - \mathrm{x}_2 - \mathrm{o}_1 + \mathrm{o}_2$.

82. $(a - b)/\sqrt{2} = -\frac{1}{2}(b - c)/\sqrt{2} + \frac{1}{2}\sqrt{3}(2a - b - c)/\sqrt{6}$.

84. Γ^*.

85. $$\begin{pmatrix} Q & J & J \\ J & Q & J \\ J & J & Q \end{pmatrix}, \begin{pmatrix} Q + 2J & 0 & 0 \\ 0 & Q - J & 0 \\ 0 & 0 & Q - J \end{pmatrix}.$$

86. In the notation of Fig. 10.9, $(s_1 + s_2)$ couples with p_z and $2s$, $(s_1 - s_2)$ with p_y.

87. C_4, S_4^3, σ_d, or one C_2'' converts ϕ_1 into ϕ_2, and S_4, C_4^3, σ_d', or the other C_2'' converts ϕ_1 into $-\phi_2$. Thus, for example, if $H\phi_1 = E\phi_1$, then $C_4 H\phi_1 = EC_4\phi_1 = E\phi_2$. And since $C_4 H = HC_4$, $C_4 H\phi_1 = HC_4\phi_1 = H\phi_2 = E\phi_2$, so ϕ_2 is also an eigenfunction of H with eigenvalue E.

88. If $H\psi = E\psi$, then $H^*\psi^* = E^*\psi^*$, but $H^* = H$, and $E^* = E$.

89. $\Gamma_j(C_n) = e^{2\pi ij/n}$, $\Gamma_j(C_n^k) = e^{2\pi ijk/n}$.

90. $$\begin{pmatrix} \cos^2(2\pi/5) - \sin^2(2\pi/5) & -2\sin(2\pi/n)\cos(2\pi/5) \\ 2\sin(2\pi/5)\cos(2\pi/5) & \cos^2(2\pi/n) - \sin^2(2\pi/5) \end{pmatrix}$$
$$= \begin{pmatrix} \cos(4\pi/5) & -\sin(4\pi/5) \\ \sin(4\pi/5) & \cos(4\pi/5) \end{pmatrix} = \mathbf{1}^{2/5}.$$

91. $+1, -1$.

92.

$\mathfrak{D}_{2h}$	E	C_2	C_2'	C_2''	i	σ_h	σ_d	σ_v
A_g	1	1	1	1	1	1	1	1
B_{1g}	1	1	−1	−1	1	1	−1	−1
B_{2g}	1	−1	1	−1	1	−1	1	−1
B_{3g}	1	−1	−1	1	1	−1	−1	1
A_u	1	1	1	1	−1	−1	−1	−1
B_{1u}	1	1	−1	−1	−1	−1	1	1
B_{2u}	1	−1	1	−1	−1	1	−1	1
B_{3u}	1	−1	−1	1	−1	1	1	−1.

$\mathfrak{D}_{3h}$	E	C_3	C_3^2	σ_v	σ_v'	σ_v''	σ_h	S_3	S_3^5	C_2'	C_2'	C
A_1'	1	1	1	1	1	1	1	1	1	1	1	1
A_2'	1	1	1	−1	−1	−1	1	1	1	−1	−1	−1
E'	$\begin{pmatrix} 1 & 0 \\ 0 & 1 \end{pmatrix}$	$\begin{pmatrix} -c & -s \\ s & -c \end{pmatrix}$	$\begin{pmatrix} -c & s \\ -s & -c \end{pmatrix}$	$\begin{pmatrix} -1 & 0 \\ 0 & 1 \end{pmatrix}$	$\begin{pmatrix} c & -s \\ -s & -c \end{pmatrix}$	$\begin{pmatrix} c & s \\ s & -c \end{pmatrix}$	$\begin{pmatrix} 1 & 0 \\ 0 & 1 \end{pmatrix}$	$\begin{pmatrix} -c & -s \\ s & -c \end{pmatrix}$	$\begin{pmatrix} -c & s \\ -s & -c \end{pmatrix}$	$\begin{pmatrix} -1 & 0 \\ 0 & 1 \end{pmatrix}$	$\begin{pmatrix} c & -s \\ -s & -c \end{pmatrix}$	$\Big(\begin{matrix} c \\ s\ - \end{matrix}$
A_1''	1	1	1	−1	−1	−1	−1	−1	−1	1	1	1
A_2''	1	1	1	1	1	1	−1	−1	−1	−1	−1	−1
E''	$\begin{pmatrix} 1 & 0 \\ 0 & 1 \end{pmatrix}$	$\begin{pmatrix} -c & -s \\ s & -c \end{pmatrix}$	$\begin{pmatrix} -c & s \\ -s & -c \end{pmatrix}$	$\begin{pmatrix} -1 & 0 \\ 0 & 1 \end{pmatrix}$	$\begin{pmatrix} c & -s \\ -s & -c \end{pmatrix}$	$\begin{pmatrix} c & s \\ s & -c \end{pmatrix}$	$\begin{pmatrix} -1 & 0 \\ 0 & -1 \end{pmatrix}$	$\begin{pmatrix} c & s \\ -s & c \end{pmatrix}$	$\begin{pmatrix} c & -s \\ s & c \end{pmatrix}$	$\begin{pmatrix} 1 & 0 \\ 0 & -1 \end{pmatrix}$	$\begin{pmatrix} -c & s \\ s & c \end{pmatrix}$	$\Big(\begin{matrix} -c\ - \\ -s \end{matrix}$

where $c = 1/2$, $s = \sqrt{3}/2$.

93.

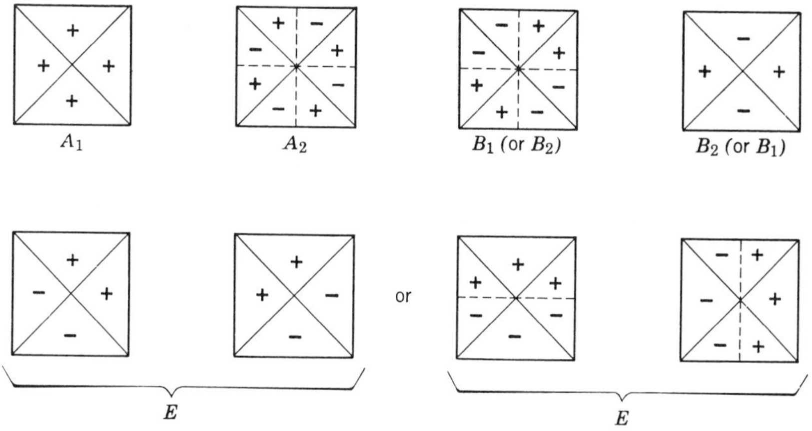

94. A_g, A_g, A_u, B_{1g}, B_{1u}, B_{3g}, B_{3u}, B_{2u}.

95. (a) T_{2g}, T_{2g}, T_{2g}, E_g, E_g, E_g.
(b) B_2 (or B_1), E, E, B_1 (or B_2), d_{z^2} as A_1.

96. Remaining is $\mathbf{E} = \mathbf{C}_4 = \mathbf{C}_2 = \mathbf{C}_4^3 = 1$, $\boldsymbol{\sigma}_v = \boldsymbol{\sigma}_v' = \boldsymbol{\sigma}_d = \boldsymbol{\sigma}_d' = -1$.

97. (a) $\{E\}, \{R\}, \{C_2, S\}, \{C_2', T\}, \{C_2'', U\}$. (b) 5. (c) None. (d) 1. (e) 4.

98.

$\mathfrak{D}_2'$	E	C_2	C_2'	C_2''	R	S	T	U
Γ_1	1	1	1	1	1	1	1	1
Γ_2	1	1	−1	−1	1	1	−1	−1
Γ_3	1	−1	1	−1	1	−1	1	−1
Γ_4	1	−1	−1	1	1	−1	−1	1
Γ_5	$\begin{pmatrix}1 & 0\\ 0 & 1\end{pmatrix}$	$\begin{pmatrix}i & 0\\ 0 & -i\end{pmatrix}$	$\begin{pmatrix}0 & i\\ i & 0\end{pmatrix}$	$\begin{pmatrix}0 & -1\\ 1 & 0\end{pmatrix}$	$\begin{pmatrix}-1 & 0\\ 0 & -1\end{pmatrix}$	$\begin{pmatrix}-i & 0\\ 0 & i\end{pmatrix}$	$\begin{pmatrix}0 & -i\\ -i & 0\end{pmatrix}$	$\begin{pmatrix}0 & 1\\ -1 & 0\end{pmatrix}$.

99. d_Γ.

100. g, 0.

101. Because the trace is invariant to similarity transformation.

102. If $X^{-1}RX = S$, then $\mathrm{Tr}\,\mathbf{R} = \mathrm{Tr}\,(\mathbf{X}^{-1}\mathbf{RX}) = \mathrm{Tr}\,\mathbf{S}$.

103.

$\mathfrak{D}_{3h}$	E	$2C_3$	$3\sigma_v$	σ_h	$2S_3$	$3C_2'$
A_1'	1	1	1	1	1	1
A_2'	1	1	−1	1	1	−1
E'	2	−1	0	2	−1	0
A_1''	1	1	−1	−1	−1	1
A_2''	1	1	1	−1	−1	−1
E''	2	−1	0	−2	1	0.

104.

$\mathcal{C}_{4v}$	E	$2C_4$	C_2	$2\sigma_v$	$2\sigma_d$
A_1	1	1	1	1	1
A_2	1	1	1	-1	-1
B_1	1	-1	1	1	-1
B_2	1	-1	1	-1	1
E	2	0	-2	0	0.

105. Setting $i = j$ and $i' = j'$ in Eq. 10.1, and summing over i and i' gives Eq. 10.2a directly.

106. Since $\chi_E^{\Gamma_\alpha} = d_\alpha$, $\chi_E^{\Gamma} = g$, and all other $\chi_R^{\Gamma} = 0$, $n_\alpha = \frac{1}{g} \cdot d_{\alpha\delta}^* g = d_\alpha$.

107. $g = \chi_E^{\Gamma} = \sum n_\alpha \chi_E^{\Gamma_\alpha} = \sum d_\alpha d_\alpha$.

108. $\chi_E^{\Gamma} = 3$, $\chi_{2C_3}^{\Gamma} = 0$, $\chi_{3\sigma_v}^{\Gamma} = 1$; $n_{A_1} = 1$, $n_{A_2} = 0$, $n_E = 1$.

109. Exercise 8.55.

110. (c)

	A	B_1	B_2	B_3
A	A	B_1	B_2	B_3
B_1	B_1	A	B_3	B_2
B_2	B_2	B_3	A	B_1
B_3	B_3	B_2	B_1	A.

111. (b) $\chi_E = 4$, $\chi_{2C_3} = 1$, $\chi_{3\sigma_v} = 0$. (c) $E \times E = A_1 + A_2 + E$.

	A_1	A_2	E
A_1	A_1	A_2	E
A_2	A_2	A_1	E
E	E	E	$A_1 + A_2 + E$.

112. Since $\chi_R^{\Gamma_\alpha \times \Gamma_1} = \chi_R^{\Gamma_\alpha} \chi_R^{\Gamma_1} = \chi_R^{\Gamma_\alpha}$, $n_\beta = \frac{1}{g} \sum (\chi_R^{\Gamma_\beta})^* \chi_R^{\Gamma_\alpha \times \Gamma_1}$

$$= \frac{1}{g} \sum (\chi_R^{\Gamma_\beta})^* \chi_R^{\Gamma_\alpha} = \delta_{\alpha\beta}, \text{ by Eq. 10.2a.}$$

113. $A_1 + A_2 + 2E + 2T_1 + 2T_2$.

114. (a) $E \times E = A_1 + A_2 + B_1 + B_2$.
(b) $x^4 + y^4$ as A_1, $x^3y - xy^3$ as A_2, $x^4 - y^4$ as B_1, $x^3y + xy^3$ as B_2.

115. $e^{\pm i\phi}$ as E_1, $e^{\pm 3i\phi}$ as E_2; $E_1 \times E_2 = E_1 + E_2$.

116. $e^{\pm i\phi}$ as E_1, $e^{3i\phi} + e^{-3i\phi}$ as B_1, $e^{3i\phi} - e^{-3i\phi}$ as B_2. $E_1 \times E_2 = B_1 + B_2 + E$.

117. $e^{6i\phi} + e^{-6i\phi}$ as A_1, $e^{6i\phi} - e^{-6i\phi}$ as A_2, $e^{\pm 4i\phi}$ as E_2. $E_1 \times E_1 = A_1 + A_2 + E_2$.

118. (b) $\chi_E^{\Gamma\cdot\Gamma} = 3$, $\chi_{2C_4}^{\Gamma\cdot\Gamma} = -1$, $\chi_{C_2}^{\Gamma\cdot\Gamma} = 3$, $\chi_{2\sigma_v}^{\Gamma\cdot\Gamma} = 1$, $\chi_{2\sigma_d}^{\Gamma\cdot\Gamma} = 1$.
(c) $E \cdot E = A_1 + B_1 + B_2 \neq A_1 + A_2 + B_1 + B_2 = E \times E$.
(d) $\chi_E^{\Gamma\cdot\Gamma\cdot\Gamma} = 4$, $\chi_{2C_4}^{\Gamma\cdot\Gamma\cdot\Gamma} = 0$, $\chi_{C_2}^{\Gamma\cdot\Gamma\cdot\Gamma} = -4$, $\chi_{2\sigma_v}^{\Gamma\cdot\Gamma\cdot\Gamma} = 0$, $\chi_{2\sigma_d}^{\Gamma\cdot\Gamma\cdot\Gamma} = 0$; $E \cdot E \cdot E = 2E$.

119. (a) Let p_z be along C_n. Then $C_2p_x = -p_x$, $C_2p_y = -p_y$, $C_2p_z = p_z$, so $\mathbf{C}_2 =$ diag $(-1, -1, 1)$, and $\chi_{C_2} = -1$. Also, $C_4p_x = p_y$, $C_4p_y = -p_x$, and $C_4p_z = p_z$, so $\chi_{C_4} = 1$.
(b) Since $C_3p_x = p_y$, $C_3p_y = p_z$, and $C_3p_z = p_x$, $\chi_{C_3} = 0$.

120. Via $C_np = \left(\cos \frac{2\pi}{n}\right) p + \left(\sin \frac{2\pi}{n}\right) C_4p$, where C_4p is orthogonal to p.

121. (a) $\chi_E = 4$, $\chi_{C_3} = 1$, $\chi_{\sigma_v} = 2$, $\chi_{\sigma_h} = -4$, $\chi_{S_3} = -1$, $\chi_{C_2'} = -2$.
(b) $\Gamma = 2A_2'' + E''$.

122. $\Gamma = 2A_1 + E$.

123. $3A_1 + 2A_2 + 2B_1 + 3B_2$.

124. $\Gamma_H = A_1' + E' + A_2'' + E''$, $\Gamma_s = A_1' + E'$, $\Gamma_p = A_1' + A_2' + 2E' + A_2'' + E''$.

125. $2A_{1g} + 2E_g + T_{1g} + 3T_{1u} + 2T_{2g} + T_{2u}$.

126. $3A_1 + E + T_1 + 4T_2$.

128. (a) $A_{1g} + A_{2g} + B_{1g} + B_{2g} + E_g + 2A_{2u} + B_{2u} + 3E_u$.
(b) $A_{1g} + B_{1g} + B_{2g} + A_{2u} + B_{2u} + 2E_u$.

129. (a) $\Gamma_r = A_{1g} + E_{2g} + B_{1u} + E_{1u}$, $\Gamma_t = A_{2g} + E_{2g} + B_{2u} + E_{1u}$,
$\Gamma_v = B_{2g} + E_{1g} + A_{2u} + E_{2u}$.
(b) $A_{1g} + B_{2g} + 2E_{2g} + B_{1u} + B_{2u} + E_{1u} + E_{2u}$.

130. $P_{\Gamma_\alpha}P_{\Gamma_\beta} = \sum\sum (\chi_R^{\Gamma\alpha})^*(\chi_S^{\Gamma\beta})^* RS = \sum\sum (\chi_R^{\Gamma\alpha})^*\chi_R^{\Gamma\beta}(\chi_R^{\Gamma\beta}\chi_S^{\Gamma\beta})^* RS$
$= \sum_R (\chi_R^{\Gamma\alpha})^*\chi_R^{\Gamma\beta} \sum_{RS} (\chi_{RS}^{\Gamma\beta})^* RS = g\delta_{\alpha\beta}P_{\Gamma_\beta}$.

131. Multiply $SP_{\Gamma_{\alpha,ij}}\phi = \sum_R (\mathbf{R}_\alpha)_{ij}^* SR\phi$ by $\delta_{li} = \sum_k (\mathbf{S}_\alpha^{-1})_{lk}^*(\mathbf{S}_\alpha)_{ki}^*$, and sum over i to obtain $SP_{\Gamma_{\alpha,ij}}\phi = \sum_{R,k} (\mathbf{S}_\alpha^{-1})_{lk}^*(\mathbf{SR})_{kj}^* SR\phi = \sum_k (\mathbf{S}_\alpha)_{kl} P_{\Gamma_{\alpha,kj}}\phi$.

132. 0, because A_2 is not contained in the representation formed by this basis.

133. They are not linearly independent.

134. (a) $1 + P_{12}$. (b) $\Phi(x_1, x_2) + \Phi(x_2, x_1)$.
(c) $1 - P_{12}$. (d) $\Phi(x_1, x_2) - \Phi(x_2, x_1)$.
(e) $\begin{vmatrix} \phi_1(x_1) & \phi_2(x_1) \\ \phi_1(x_2) & \phi_2(x_2) \end{vmatrix}$.

135. (a) $P_{\Gamma_1} = E + (123) + (132) + (12) + (13) + (23)$.
$P_{\Gamma_2} = E + (123) + (132) - (12) - (13) - (23)$.
(b) $\begin{vmatrix} \phi_1(x_1) & \phi_2(x_1) & \phi_3(x_1) \\ \phi_1(x_2) & \phi_2(x_2) & \phi_3(x_2) \\ \phi_1(x_3) & \phi_2(x_3) & \phi_3(x_3) \end{vmatrix}$.

136. $$\sum_{j=1}^{n!} (-1)^{p(P_j)} \prod_{i=1}^{n} \phi_i(x_{P_j i}) = \begin{vmatrix} \phi_1(x_1) & \phi_2(x_1) & \cdots & \phi_n(x_1) \\ \phi_1(x_2) & \phi_2(x_2) & \cdots & \phi_n(x_2) \\ \cdot & \cdot & & \cdot \\ \cdot & \cdot & & \cdot \\ \cdot & \cdot & & \cdot \\ \phi_1(x_n) & \phi_2(x_n) & \cdots & \phi_n(x_n) \end{vmatrix}.$$

137. $\Gamma = A_1 + T_2$. The linear combination $\phi_1 + \phi_2 + \phi_3 + \phi_4$ transforms as A_1. The linear combinations $3\phi_1 - \phi_2 - \phi_3 - \phi_4$, $3\phi_2 - \phi_1 - \phi_3 - \phi_4$, $3\phi_3 - \phi_1 - \phi_2 - \phi_4$, and $3\phi_4 - \phi_1 - \phi_2 - \phi_3$ transform as T_2, but only 3 are linearly independent, and they are not orthonormal. The most symmetric orthonormal set is $(\phi_1 + \phi_2 - \phi_3 - \phi_4)/\sqrt{4 - 4S}$, $(\phi_1 - \phi_2 + \phi_3 - \phi_4)/\sqrt{4 - 4S}$, and $(\phi_1 - \phi_2 - \phi_3 + \phi_4)/\sqrt{4 - 4S}$, where $S = \int \phi_1\phi_2\, dV$.

138. $\Gamma = A_{1g} + E_g + T_{1u}$. The linear combination $(\phi_1 + \phi_2 + \phi_3 + \phi_4 + \phi_5 + \phi_6)/\sqrt{6}$ transforms as A_{1g}; if we neglect overlap, it is normalized. If ϕ_1 is opposite ϕ_2, ϕ_3 opposite ϕ_4, and ϕ_5 opposite ϕ_6, then the linear combinations $2\phi_1 + 2\phi_2 - \phi_3 - \phi_4 - \phi_5 - \phi_6$, $2\phi_3 + 2\phi_4 - \phi_1 - \phi_2 - \phi_5 - \phi_6$, and $2\phi_5 + 2\phi_6 - \phi_1 - \phi_2 - \phi_3 - \phi_4$ transform as E_g, but only two are linearly independent, and they are not orthogonal. Orthonormal combinations are $(2\phi_1 + 2\phi_2 - \phi_3 - \phi_4 - \phi_5 - \phi_6)/\sqrt{12}$ and $(\phi_3 + \phi_4 - \phi_5 - \phi_6)/2$. The linear combinations $(\phi_1 - \phi_2)/\sqrt{2}$, $(\phi_3 - \phi_4)/\sqrt{2}$, and $(\phi_5 - \phi_6)/\sqrt{2}$ transform as T_{1u}.

139. (a) $\sum_{\mathbf{T}} e^{-i\mathbf{k}\cdot\mathbf{T}}\mathbf{T}$. (b) $\sum_{\mathbf{T}} e^{-i\mathbf{k}\cdot\mathbf{T}} f(\mathbf{X}-\mathbf{T})$.

140. Using $\mathcal{C}_{2v}$ symmetry, with $\sigma_v' = C_5$—C_6, gives $P_{A_1}\phi_1 \propto \chi_1$, $P_{A_1}\phi_5 \propto \chi_2$, $P_{A_2}\phi_1 \propto \chi_3$, $P_{A_2}\phi_5 \propto \chi_4$, $P_{B_1}\phi_1 \propto \chi_6$, and $P_{B_2}\phi_1 \propto \chi_5$.

141. $\mathcal{C}_{2v}$: A_1: $(\phi_1 + \phi_4 + \phi_5 + \phi_8)/2$, $(\phi_2 + \phi_3 + \phi_6 + \phi_7)/2$, $(\phi_9 + \phi_{10})/\sqrt{2}$, $\lambda = 1$, $(1 \pm \sqrt{13})/2$.
A_2: $(\phi_1 - \phi_4 + \phi_5 - \phi_8)/2$, $(\phi_2 - \phi_3 + \phi_6 - \phi_7)/2$, $\lambda = (-1 \pm \sqrt{5})/2$.
B_1: $(\phi_1 + \phi_4 - \phi_5 - \phi_8)/2$, $(\phi_2 + \phi_3 - \phi_6 - \phi_7)/2$, $\lambda = (1 \pm \sqrt{5})/2$.
B_2: $(\phi_1 - \phi_4 - \phi_5 + \phi_8)/2$, $(\phi_2 - \phi_3 - \phi_6 + \phi_7)/2$, $(\phi_9 - \phi_{10})/\sqrt{2}$, $\lambda = -1$, $(-1 \pm \sqrt{13})/2$.

142. $\mathcal{C}_{4v}$: A_1: $\chi_1 = (\phi_1 + \phi_2 + \phi_3 + \phi_4)/2$, $\chi_2 = (\phi_5 + \phi_6 + \phi_7 + \phi_8)/2$. $\lambda = 1 \pm \sqrt{2}$, $\psi_\lambda = \lambda\chi_1 + \chi_2$.
B_1: $\chi_3 = (\phi_1 - \phi_2 + \phi_3 - \phi_4)/2$, $\chi_4 = (\phi_5 - \phi_6 + \phi_7 - \phi_8)/2$. $\lambda = -1 \pm \sqrt{2}$, $\psi_\lambda = \lambda\chi_3 + \chi_4$.
E: $\chi_{5x} = (\phi_1 - \phi_3)/\sqrt{2}$, $\chi_{6x} = (\phi_5 - \phi_7)/\sqrt{2}$. $\lambda = \pm 1$, $\psi_\lambda = \lambda\chi_{5x} + \chi_{6x}$ (y exactly as x).

143. With $\mathcal{C}_{6v}$, the symmetry orbitals are the molecular orbitals.
A_1: $\chi_0 = (\phi_1 + \phi_2 + \phi_3 + \phi_4 + \phi_5 + \phi_6)/\sqrt{6}$, $\lambda = 2$
E_1: $\chi_{1x} = (2\phi_1 + \phi_2 - \phi_3 - 2\phi_4 - \phi_5 + \phi_6)/\sqrt{12}$, $\chi_{1y} = (\phi_2 + \phi_3 - \phi_5 - \phi_6)/2$, $\lambda = 1$
E_2: $\chi_{2x} = (2\phi_1 - \phi_2 - \phi_3 + 2\phi_4 - \phi_5 - \phi_6)/\sqrt{12}$, $\chi_{2y} = (\phi_2 - \phi_3 + \phi_5 - \phi_6)/2$, $\lambda = -1$
B_1 (or B_2): $\chi_3 = (\phi_1 - \phi_2 + \phi_3 - \phi_4 + \phi_5 - \phi_6)/\sqrt{6}$, $\lambda = -2$.

144. In $\mathcal{C}_{3v}$, $\Gamma = 3A_1 + 3E$. We shall give only E_x.
A_1: $\chi_1 = (\phi_1 + \phi_2 + \phi_3)/\sqrt{3}$, $\chi_2 = (\phi_4 + \phi_5 + \phi_6)/\sqrt{3}$, $\chi_3 = (\phi_7 + \phi_8 + \phi_9)/\sqrt{3}$.
$\lambda = 0$, $\psi_0 = 2\chi_1 - \chi_3$.
$\lambda = \pm\sqrt{5}$, $\psi_\lambda = \chi_1 + \lambda\chi_2 + 2\chi_3$.
E: $\chi_{4x} = (\phi_2 - \phi_3)/\sqrt{2}$, $\chi_{5x} = (\phi_5 - \phi_6)/\sqrt{2}$, $\chi_{6x} = (\phi_8 - \phi_9)/\sqrt{2}$.
$\lambda = 0$, $\psi_x = \chi_{4x} - \chi_{6x}$.
$\lambda = \pm\sqrt{2}$, $\psi_{\lambda x} = \chi_{4x} + \lambda\chi_{5x} + \chi_{6x}$.

145. In $\mathcal{C}_{6v}$, with σ_v through atoms, A_1 gives $\lambda^3 - 3\lambda^2 - \lambda + 5 = 0$, A_2 gives $\lambda + 1 = 0$, B_1 gives $\lambda^3 + 3\lambda^2 - \lambda - 5 = 0$, B_2 gives $\lambda - 1 = 0$, E_1 gives $\lambda^4 - \lambda^3 - 4\lambda^2 + 2\lambda + 2 = 0$, and E_2 gives $\lambda^4 + \lambda^3 - 4\lambda^2 - 2\lambda + 2 = 0$.

146. A_{1g}: $r_1 + r_2 + r_3 + r_4$.
A_{2g}: $t_1 + t_2 + t_3 + t_4$.
B_{1g}: $r_1 - r_2 + r_3 - r_4$.
B_{2g}: $t_1 - t_2 + t_3 - t_4$.
E_{gx}: $v_1 - v_3$, E_{gy}: $v_2 - v_4$.
A_{2u}: $v_1 + v_2 + v_3 + v_4$, v_{Pt}.
B_{2u}: $v_1 - v_2 + v_3 - v_4$.
E_{ux}: $r_1 - r_3$, $t_2 - t_4$, x_{Pt}, E_{uy}: $r_2 - r_4$, $t_1 - t_3$, y_{Pt}.

147. Let p_x point outward at each carbon, p_y clockwise, and p_z perpendicular to σ_h. Number the H atoms 1 and 2 on carbon A, etc. Then
A_1': $s_1 + s_2 + s_3 + s_4 + s_5 + s_6$, $s_A + s_B + s_C$, $x_A + x_B + x_C$.
A_2': $y_A + y_B + y_C$.
E_x': $2s_1 + 2s_2 - s_3 - s_4 - s_5 - s_6$, $2s_A - s_B - s_C$, $2x_A - x_B - x_C$, $y_B - y_C$.
E_y': $s_3 + s_4 - s_5 - s_6$, $s_B - s_C$, $x_B - x_C$, $2y_A - y_B - y_C$.
A_2'': $s_1 - s_2 + s_3 - s_4 + s_5 - s_6$, $z_A + z_B + z_C$.
E_x'': $2s_1 - 2s_2 - s_3 + s_4 - s_5 + s_6$, $2z_A - z_B - z_C$.
E_y'': $s_3 - s_4 - s_5 + s_6$, $z_B - z_C$.

149. $n_1 = \frac{1}{g} \sum (\chi_R^{\Gamma\alpha})^* \chi_R^{\Gamma\beta} = \delta_{\alpha\beta}$.

150. (a) $B_{1u} + B_{2u} + E_{1u}$. (b) E_{1u} is xy-polarized.
151. $E_{2u} \to E_{2g}$ is z-polarized, $E_{2u} \to B_{1g}$, $E_{2u} \to B_{2g}$, and $E_{2u} \to E_{1g}$ are xy-polarized.
152. A_{2u} and $2E_u$ are IR-active, A_{1g}, B_{1g}, B_{2g}, and E_g are Raman-active.
153. E_{1u} is IR-active, A_{1g} and $2E_{2g}$ are Raman-active.
154. If Q is IR-active, then $\Gamma_Q \times \Gamma_T$ must equal A_{1g}. Since Γ_T is u (eigenvalue -1 under i), Γ_Q must also be u. Therefore $\Gamma_Q \times \Gamma_\alpha$ must be some $\Gamma_u \neq A_{1g}$, and Q is Raman-forbidden.

Appendix 1

1. Yes.
2. $p \to q$.
3. A is false, but we cannot conclude that B is true.
4. (a) No. (b) Yes. (c) Yes. (d) Yes. (e) Yes. (f) Yes.
5. (a) All men are non-islands. All islands are nonmen.
   (b) Nothing in the world is good for nothing. Nothing useless exists.
   (c) If anyone must stay after 5 P.M., then someone didn't do his work carefully.
   (d) Some people cannot be fooled all of the time. There is a time when not everyone can be fooled.
   (e) Some quinine water is not a treat without the perfect martini gin.
   (a′) At least one man is an island. At least one island is a man.
   (b′) Something is good for nothing.
   (c′) Even if everyone does his work carefully, someone may have to stay after 5 P.M.
   (d′) You can fool all of the people all of the time.
   (e′) There does exist an imperfect martini gin that makes any (every) quinine water a treat.

Appendix 2

1. 29.8.
2. 2.353×10^{-2}.
3. 5.16.
4. 16.87.
5. 1.605×10^4.
6. $L \propto h$.
7. $a_0 \propto h^2/me^2$.
8. $T \propto (g/l)^{1/2}$, independent of m.
9. $\eta \propto (RTm/NA^2)^{1/2}$, independent of ρ.

Appendix 4

1. (a) Triangle rules: The sum (difference) of the lengths of two sides of a triangle is greater (less) than the length of the third side.
   (b) Multiply by 2 and add $z_1^* z_1 + z_2^* z_2$, to obtain $(z_1^* + z_2^*)(z_1 + z_2) \leq (|z_1^*| + |z_2|)^2$. Since $|z_1^*| = |z_1|$, taking the square root gives the first inequality. The second follows similarly after multiplying by -2.
2. Via trigonometric identities, $\cos(\theta_1 + \theta_2) = \cos\theta_1 \cos\theta_2 - \sin\theta_1 \sin\theta_2$ and $\sin(\theta_1 + \theta_2) = \sin\theta_1 \cos\theta_2 + \cos\theta_1 \sin\theta_2$.
3. i, $\frac{1}{2}\sqrt{3} - \frac{1}{2}i$, $-\frac{1}{2}\sqrt{3} - \frac{1}{2}i$.
4. *Hint:* Multiplying every root by one particular root leaves the set of roots unchanged. (For a proof see Eq. 1.29.)

Appendix 5

1. (a) Analytic everywhere. (b) Analytic nowhere. (c) Analytic everywhere. (d) Singularities at $z = \pm i$.
2. (a) Analytic everywhere. (b) Singularity at $z = 0$. (c) Singularity at $z = -1$.
4. From Eq. A5.1, $\partial^2 u/\partial x^2 + \partial^2 u/\partial y^2 = \partial^2 v/\partial x\partial y - \partial^2 v/\partial y \partial x = 0$, likewise for v.
5. (a) $r^2 e^{2i\theta}$. (b) $r^{1/2} e^{i\theta/2}$. (c) $re^{-i\theta}$. (d) r. (e) $r^{-1}e^{-i\theta}$.
7. Via $|\sum z_i| \leqslant \sum |z_i|$ and $|z_1 z_2| \leqslant |z_1|\,|z_2|$.
8. $dz = d(re^{i\theta}) = e^{i\theta}\,dr + ire^{i\theta}\,d\theta$.
9. 0, 0.
10. Via $2^{-1/2} = 2^{-1/2}\int_0^\infty [\cos(\tfrac{1}{2}\pi t^2) - i\sin(\tfrac{1}{2}\pi t^2)][1+i]\,dt = 2^{-1/2}[\int_0^\infty \cos(\tfrac{1}{2}\pi t^2)\,dt + \int_0^\infty \sin(\tfrac{1}{2}\pi t^2)\,dt + i\int_0^\infty \cos(\tfrac{1}{2}\pi t^2)\,dt - i\int_0^\infty \sin(\tfrac{1}{2}\pi t^2)\,dt]$.
11. $\oint dz/z = \oint d(re^{i\theta})/re^{i\theta} = i\int_0^{2\pi} e^{i\theta}\,d\theta/e^{i\theta} = 2\pi i$.
12. $-\int_0^{2\pi} \theta e^{3i\theta}\,d\theta = 2\pi i/3$.
13. (a) $f_n(x) = \dfrac{1}{2\pi i}\oint G(x, z) z^{-(n+1)}\,dz$. (b) Via $z = e^{i\theta}$, $\tfrac{1}{2}(z - z^{-1}) = i\sin\theta$.
15. C may be deformed into a small circle of radius r_0, and the integral becomes $f(z_0)\int ir_0 e^{i\theta}\,d\theta/r_0{}^n e^{in\theta} = ir_0^{1-n} f(z_0)\int_0^{2\pi} e^{(1-n)i\theta}\,d\theta = 0$.
16. (a) $\pi/4\sqrt{2}$. (b) $\pi(2/\sqrt{3} - 1/\sqrt{2})$. (c) $\pi(e^{-b}/b - e^{-a}/a)/2(a^2 - b^2)$. (d) $2\pi(a^2 - b^2 - c^2)^{-1/2}$. (e) $2\pi i$.
17. (a) e^{at}. (b) $\dfrac{1}{a}\sin at$. (c) $\cos at$. (d) $\dfrac{1}{a}\sinh at$. (e) $\cosh at$. (f) $e^{at}\cos bt$.
18. Example 4.1.
19. $\tfrac{1}{2}\cosh at + \tfrac{1}{2}\cos at$, $\cos t\cosh t$.
20. πe^{-am}.
21. (a) $\pi/2$. (b) $-\tfrac{1}{2}\pi\sin m$.
22. $\pi/2$.
23. $i\pi$.
24. (a) $2\pi i$. (b) $\pi(a^5 - 5a^3b^2 + 5a^2b^3 - b^5)/2a^3b^3(a^2 - b^2)^3$. (c) $(2n-2)!\pi/2^{2n-2}(n-1)!^2$.
25. $t\sin(at)/2a$.

Appendix 6

1.

```
   SX = 0.0
   SY= 0.0
   SX2 = 0.0
   SXY = 0.0
   DO 10  I = 1, N
   SX = SX + X(I)
   SY = SY + Y(I)
   SX2 = SX2 + X(I)*X(I)
10 SXY = SXY +X(I)*Y(I)
   A = (N*SXY -SX*SY)/(N*SX2 - SX*SX).
   B = (SX2*SX  - SX*SXY)/(N*SX2 - SX*SX).
```

2–4. These programs are sufficiently complicated that a listing will not help you determine whether your answer is correct or where you might have erred. To check your answer, run your programs on a computer and see if they will solve Exercises 8.46, 8.75, and 8.100.

5.

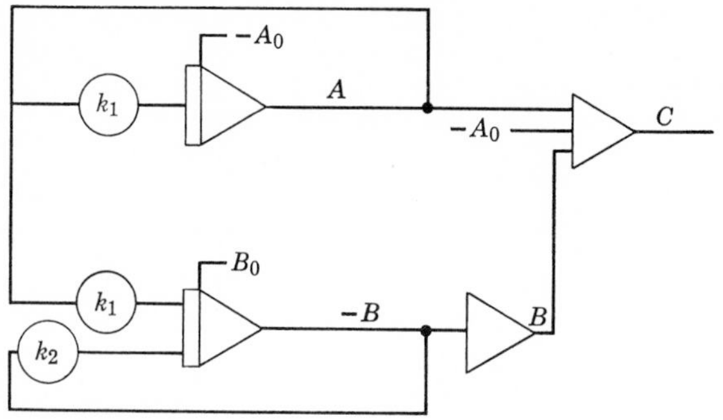

6.

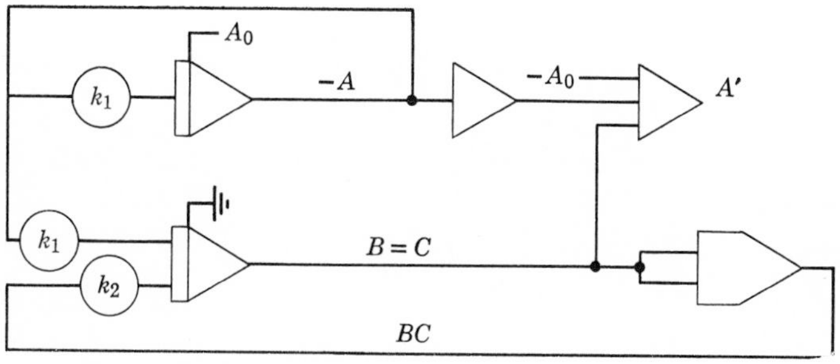

7.

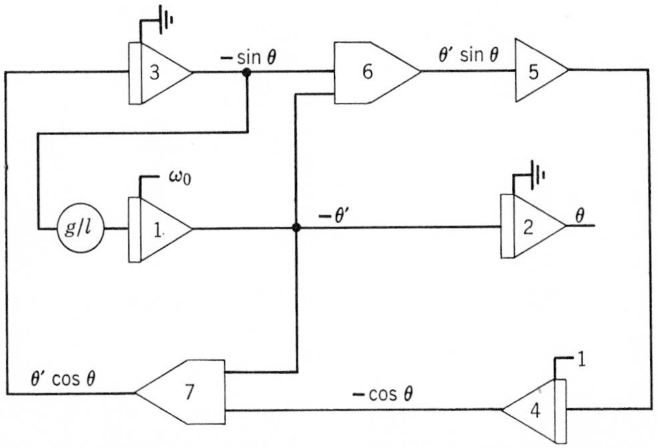

Notice how components numbered 3 through 7 generate $\sin\theta$ (and also $\cos\theta$) via

$$\frac{d}{dt}\sin\theta = \cos\theta\frac{d\theta}{dt} \text{ and } \frac{d}{dt}\cos\theta = -\sin\theta\frac{d\theta}{dt}.$$

INDEX